Methods in Enzymology

Volume 262

DNA REPLICATION

METHODS IN ENZYMOLOGY

EDITORS-IN-CHIEF

John N. Abelson Melvin I. Simon

DIVISION OF BIOLOGY
CALIFORNIA INSTITUTE OF TECHNOLOGY
PASADENA, CALIFORNIA

FOUNDING EDITORS

Sidney P. Colowick and Nathan O. Kaplan

Methods in Enzymology

Volume 262

DNA Replication

EDITED BY

Judith L. Campbell

DIVISIONS OF CHEMISTRY AND BIOLOGY
CALIFORNIA INSTITUTE OF TECHNOLOGY
PASADENA, CALIFORNIA

ACADEMIC PRESS
San Diego New York Boston London Sydney Tokyo Toronto

This book is printed on acid-free paper. ♾

Academic Press, Inc.
A Division of Harcourt Brace & Company
525 B Street, Suite 1900, San Diego, California 92101-4495

United Kingdom Edition published by
Academic Press Limited
24-28 Oval Road, London NW1 7DX

International Standard Serial Number: 0076-6879

International Standard Book Number: 0-12-182163-3

PRINTED IN THE UNITED STATES OF AMERICA
95 96 97 98 99 00 MM 9 8 7 6 5 4 3 2 1

Table of Contents

Section I. Purification and Characterization of DNA Polymerases

Section II. Structural Analysis of DNA Polymerases

Section III. Mechanisms of DNA Polymerases

Section IV. Structure–Function Studies of DNA Polymerases

Section V. Polymerase Accessory Functions, Replication Proteins, Multienzyme Replication Complexes

Section VI. *In Vitro* Replication Systems: Crude and Reconstituted

Section VII. DNA Synthesis *in Vivo*

Contributors to Volume 262

Article numbers are in parentheses following the names of contributors.
Affiliations listed are current.

EDWARD ARNOLD (15), *Center for Advanced Biotechnology and Medicine, and Chemistry Department, Rutgers University, Piscataway, New Jersey 08854-5638*

ROBERT A. BAMBARA (21), *Departments of Biochemistry, Microbiology and Immunology, and the Cancer Center, University of Rochester, Rochester, New York 14642*

MARJORIE H. BARNES (4), *Department of Pharmacology, University of Massachusetts Medical School, Worcester, Massachusetts 01655*

BLAINE BARTHOLOMEW (37), *Department of Medical Biochemistry, Southern Illinois University School of Medicine, Carbondale, Illinois 62901-6503*

DANIEL W. BEAN (29), *Department of Biology, University of North Carolina, Chapel Hill, North Carolina 27599*

WILLIAM A. BEARD (11), *Sealy Center for Molecular Science, University of Texas Medical Branch, Galveston, Texas 77555-1068*

KATARZYNA BEBENEK (18), *Laboratory of Molecular Genetics, National Institute of Environmental Health Science, Research Triangle Park, North Carolina 27709*

WILLIAM R. BEBRIN (24), *Department of Biological Chemistry and Molecular Pharmacology, Harvard Medical School, Boston, Massachusetts 02115-5747*

STEPHEN J. BENKOVIC (13, 20, 34), *Department of Chemistry, The Pennsylvania State University, University Park, Pennsylvania 16802*

ROLF BERNANDER (45), *Department of Biophysics, Institute for Cancer Research, The Norwegian Radium Hospital, 0310 Oslo, Norway*

STACY BLAIN (27), *Department of Biochemistry and Molecular Biophysics, Howard Hughes Medical Institute, Columbia University, College of Physicians and Surgeons, New York, New York 10032*

LUIS BLANCO (5, 22), *Centro de Biología Molecular "Severo Ochoa," Universidad Autónoma, Canto Blanco, 28049 Madrid, Spain*

LINDA B. BLOOM (19), *Hedco Molecular Biology Laboratories, Department of Biological Sciences, University of Southern California, Los Angeles, California 90089-1340*

ERIK BOYE (45), *Department of Biophysics, Institute for Cancer Research, The Norwegian Radium Hospital, 0310 Oslo, Norway*

BONITA J. BREWER (46), *Department of Genetics, University of Washington, Seattle, Washington 98195-7360*

NEAL C. BROWN (4, 17), *Department of Pharmacology, University of Massachusetts Medical School, Worcester, Massachusetts 01655*

GEORGE S. BRUSH (41), *Department of Molecular Biology and Genetics, The Johns Hopkins University School of Medicine, Baltimore, Maryland 21205*

MARTIN E. BUDD (12), *Department of Chemistry, California Institute of Technology, Pasadena, California 91125*

PETER M. J. BURGERS (6), *Department of Biochemistry and Molecular Biophysics, Washington University School of Medicine, St. Louis, Missouri 63110*

HONG CAI (2), *Hedco Molecular Biology Laboratories, Department of Biological Sciences, University of Southern California, Los Angeles, California 90089-1340*

CRAIG E. CAMERON (13, 20), *Department of Chemistry, The Pennsylvania State University, University Park, Pennsylvania 16802*

JUDITH L. CAMPBELL (12), *Department of Chemistry and Biology, California Institute of Technology, Pasadena, California 91125*

TODD L. CAPSON (34), *Department of Chemistry, University of Utah, Salt Lake City, Utah 84132*

CHUEN-SHEUE CHIANG (7), *Department of Biochemistry, Stanford University School of Medicine, Stanford, California 94305*

GLORIA SHEAU-JIN CHUI (10), *Department of Biochemistry, Stanford University, Stanford, California 94305-5307*

ARTHUR D. CLARK, JR. (15), *Center for Advanced Biotechnology and Medicine, and Chemistry Department, Rutgers University, Piscataway, New Jersey 08854-5638*

PATRICK CLARK (15), *SAIC-Frederick, NCI-Frederick Cancer Research and Development Center, Frederick, Maryland 21701-1013*

DONALD M. COEN (24), *Department of Biological Chemistry and Molecular Pharmacology, Harvard Medical School, Boston, Massachusetts 02115-5747*

FRANK E. J. COENJAERTS (42), *Laboratory for Physiological Chemistry, Utrecht University, 3508 TA Utrecht, The Netherlands*

NANCY COLOWICK (44), *Department of Molecular Biology, Vanderbilt University, Nashville, Tennessee 37235*

WILLIAM C. COPELAND (8, 23), *Department of Pathology, Stanford University School of Medicine, Stanford, California 94305-5324*

STEVEN CREIGHTON (19), *Hedco Molecular Biology Laboratories, Department of Biological Sciences, University of Southern California, Los Angeles, California 90089-1340*

ELLIOTT CROOKE (39), *Department of Biochemistry and Molecular Biology, Georgetown University Medical Center, Washington, DC 20007*

MILLARD G. CULL (3), *Department of Biochemistry, Biophysics, and Genetics and Program in Molecular Biology, University of Colorado Health Sciences Center, Denver, Colorado 80262*

SHIRLEY S. DAUBE (36), *Department of Biological Chemistry, The Institute of Life Sciences, The Hebrew University of Jerusalem, Givat-Ram, Jerusalem 91904, Israel*

ZEGER DEBYSER (35), *Department of Biological Chemistry and Molecular Pharmacology, Harvard Medical School, Boston, Massachusetts 02115*

MELVIN L. DEPAMPHILIS (47), *Roche Research Center, Roche Institute of Molecular Biology, Nutley, New Jersey 07110*

VICTORIA DERBYSHIRE (1, 28), *Department of Molecular Biophysics and Biochemistry, Bass Center for Molecular and Structural Biology, Yale University, New Haven, Connecticut 06520-8114*

PAUL DIGARD (24), *Department of Pathology, Division of Virology, University of Cambridge, Cambridge CB21QP, United Kingdom*

QUN DONG (8, 23), *Department of Pathology, Stanford University School of Medicine, Stanford, California 94305-5324*

KATHLEEN M. DOWNEY (9), *Department of Medicine, University of Miami School of Medicine, Miami, Florida 33101*

FRITZ ECKSTEIN (16), *Max-Planck-Institut für Experimentelle Medizin, Göttingen, Germany*

PHILIP J. FAY (21), *Departments of Medicine and Biochemistry, University of Rochester, Rochester, New York 14642*

TIM FORMOSA (31), *Department of Biochemistry, University of Utah School of Medicine, Salt Lake City, Utah 84132*

KATHERINE L. FRIEDMAN (46), *Department of Genetics, University of Washington, Seattle, Washington 98195-7360*

E. PETER GEIDUSCHEK (37), *Department of Biology, University of California, San Diego, La Jolla, California 92093-0634*

STEPHEN P. GOFF (27), *Department of Biochemistry and Molecular Biophysics, Howard Hughes Medical Institute, Columbia University, College of Physicians and Surgeons, New York, New York 10032*

MYRON F. GOODMAN (2, 19), *Hedco Molecular Biology Laboratories, Department of Biological Sciences, University of Southern California, Los Angeles, California 90089-1340*

DEBORAH M. HINTON (43), *Laboratory of Molecular and Cellular Biology, National Institute of Diabetes and Digestive and Kidney Diseases, National Institutes of Health, Bethesda, Maryland 20892-0830*

PETER H. VON HIPPEL (36), *Institute of Molecular Biology, University of Oregon, Eugene, Oregon 97403*

LISA J. HOBBS (43), *Laboratory of Molecular and Cellular Biology, National Institute of Diabetes and Digestive and Kidney Diseases, National Institutes of Health, Bethesda, Maryland 20892-0830*

STEPHEN H. HUGHES (15), *ABL-Basic Research Program, NCI-Frederick Cancer Research and Development Center, Frederick, Maryland 21701-1013*

ALFREDO JACOBO-MOLINA (15), *Center for Advanced Biotechnology and Medicine, and Chemistry Department, Rutgers University, Piscataway, New Jersey 08854-5638*

THALE C. JARVIS (36), *Ribozyme Pharmaceuticals, Inc., Boulder, Colorado 80308-7280*

CATHERINE M. JOYCE (1, 28), *Department of Molecular Biophysics and Biochemistry, Bass Center for Molecular and Structural Biology, Yale University, New Haven, Connecticut 06520-8114*

GEORGE A. KASSAVETIS (37), *Department of Biology, University of California, San Diego, La Jolla, California 92093-0634*

THOMAS J. KELLY (41), *Department of Molecular Biology and Genetics, The Johns Hopkins University School of Medicine, Baltimore, Maryland 21205*

ZVI KELMAN (32), *Cornell University Medical College, New York, New York 10021*

WILLIAM H. KONIGSBERG (26), *Department of Molecular Biophysics and Biochemistry, Yale University, New Haven, Connecticut 06510*

THOMAS A. KUNKEL (18), *Laboratory of Molecular Genetics, National Institute of Environmental Health Science, Research Triangle Park, North Carolina 27709*

JOSE M. LÁZARO (5), *Centro de Biología Molecular "Severo Ochoa," Universidad Autónoma, Canto Blanco, 28049 Madrid, Spain*

STUART F. J. LE GRICE (13), *Division of Infectious Diseases, Case Western Reserve University School of Medicine, Cleveland, Ohio 44106-4984*

I. R. LEHMAN (7), *Department of Biochemistry, Stanford University School of Medicine, Stanford, California 94305*

STUART LINN (10), *Department of Molecular and Cell Biology, University of California, Berkeley, California 94720*

LISA M. MALLABER (21), *Departments of Biochemistry, Microbiology and Immunology, and the Cancer Center, University of Rochester, Rochester, New York 14642*

KENNETH J. MARIANS (40), *Molecular Biology Program, Memorial Sloan-Kettering Cancer Center, New York, New York 10021*

STEVEN W. MATSON (29), *Department of Biology, University of North Carolina, Chapel Hill, North Carolina 27599*

KEVIN MCENTEE (2), *Department of Biological Chemistry and the Molecular Biology Institute, University of California at Los Angeles School of Medicine, Los Angeles, California 90024*

CHARLES S. MCHENRY (3), *Department of Biochemistry, Biophysics, and Genetics and Program in Molecular Biology, University of Colorado Health Sciences Center, Denver, Colorado 80262*

LYNN V. MENDELMAN (30), *Department of Biological Chemistry and Molecular Pharmacology, Harvard University Medical School, Boston, Massachusetts 02115*

PAUL G. MITSIS (7), *Department of Biochemistry, Stanford University School of Medicine, Stanford, California 94305*

ROBB E. MOSES (38), *Department of Molecular and Medical Genetics, Oregon Health Sciences University, Portland, Oregon 97201*

GISELA MOSIG (44), *Department of Molecular Biology, Vanderbilt University, Nashville, Tennessee 37235*

GREGORY P. MULLEN (14), *Department of Biochemistry, University of Connecticut Health Center, Farmington, Connecticut 06032*

VYTAUTAS NAKTINIS (32), *Institute of Biotechnology, V. Graiciuno 8, 2028 Vilnius, Lithuania*

NANCY G. NOSSAL (34, 43), *Laboratory of Molecular and Cellular Biology, National Institute of Diabetes and Digestive and Kidney Diseases, National Institutes of Health, Bethesda, Maryland 20892-0830*

MIKE O'DONNELL (32, 33), *Howard Hughes Medical Institute, Cornell University Medical College, New York, New York 10021*

JULIA K. PINSONNEAULT (28), *Department of Molecular Biophysics and Biochemistry, Bass Center for Molecular and Structural Biology, Yale University, New Haven, Connecticut 06520-8114*

MICHAEL K. REDDY (36), *Department of Chemistry, University of Wisconsin–Milwaukee, Milwaukee, Wisconsin 53201-0413*

LINDA J. REHA-KRANTZ (25), *Department of Biological Sciences, University of Alberta, Edmonton, Alberta T6G 2E9 Canada*

LARS ROGGE (8), *Department of Pathology, Stanford University School of Medicine, Stanford, California 94305-5324*

MARGARITA SALAS (5, 22), *Centro de Biología Molecular "Severo Ochoa," Universidad Autónoma, Canto Blanco, 28049 Madrid, Spain*

KIRSTEN SKARSTAD (45), *Department of Biophysics, Institute for Cancer Research, The Norwegian Radium Hospital, 0310 Oslo, Norway*

ANTERO G. SO (9), *Department of Medicine, University of Miami School of Medicine, Miami, Florida 33101*

PETER SPACCIAPOLI (43), *Laboratory of Molecular and Cellular Biology, National Institute of Diabetes and Digestive and Kidney Diseases, National Institutes of Health, Bethesda, Maryland 20892-0830*

BRUCE STILLMAN (41), *Cold Spring Harbor Laboratory, Cold Spring Harbor, New York 11724*

ALICE TELESNITSKY (27), *Department of Microbiology and Immunology, University of Michigan Medical School, Ann Arbor, Michigan 48109-0620*

JAMES B. THOMSON (16), *Max-Planck-Institut für Experimentelle Medizin, Göttingen, Germany*

RACHEL L. TINKER (37), *Department of Biology, University of California, San Diego, La Jolla, California 92093-0634*

JENNIFER TURNER (33), *Cornell University Medical College, New York, New York 10021*

PETER C. VAN DER VLIET (42), *Laboratory for Physiological Chemistry, Utrecht University, 3508 TA Utrecht, The Netherlands*

TERESA S.-F. WANG (8, 23), *Department of Pathology, Stanford University School of Medicine, Stanford, California 94305-5324*

STEPHEN E. WEITZEL (36), *Institute of Molecular Biology, University of Oregon, Eugene, Oregon 97403*

SAMUEL H. WILSON (11), *Sealy Center for Molecular Science, University of Texas Medical Branch, Galveston, Texas 77555-1068*

JACQUELINE WITTMEYER (31), *Department of Biochemistry, University of Utah School of Medicine, Salt Lake City, Utah 84132*

GEORGE E. WRIGHT (17), *Department of Pharmacology, University of Massachusetts Medical School, Worcester, Massachusetts 01655*

HONG YU (2), *Hedco Molecular Biology Laboratories, Department of Biological Sciences, University of Southern California, Los Angeles, California 90089-1340*

Preface

The increasing relevance of studies of DNA replication and DNA repair to the understanding of human genetic disease, cancer, and aging is bringing growing numbers of investigators into this field. The rich legacy of past studies of the enzymology of these processes has already had wide impact on how modern biological research is conducted in that it provided the roots for the whole field of genetic engineering. The work of the biochemist in characterizing these complex reactions is still far from done, however, since we are still short of the mark of being able to use our knowledge to prevent the devastating aberrations caused by failures of faithful copying of the genome by the self-editing DNA replication and repair apparatus.

Past study of the enzymes involved in DNA replication has given rise to a number of highly refined approaches to defining their individual enzymatic mechanisms and how they interact to carry out the process of DNA replication in the cell. These methods form the foundation on which even more detailed understanding, driven and directed by the revolutionary addition of structural information on these proteins at the atomic level, will necessarily be built. This volume contains a series of articles by the main contributors to this field which form a guide to students of nucleic acid enzymology who wish to study these types of proteins at ever increasing levels of resolution. Descriptions of functional, structural, kinetic, and genetic methods in use for analyzing DNA polymerases of all types, viral reverse transcriptases, helicases, and primases are presented. In addition, a number of chapters describe strategies for studying the interactions of these proteins during replication, in particular recycling during discontinuous synthesis and coupling of leading and lagging strands. Comprehensive descriptions of uses of both prokaryotic and eukaryotic crude *in vitro* replication systems and reconstitution of such systems from purified proteins are provided. These chapters may also be useful to investigators who are studying other multienzyme processes such as recombination, repair, and transcription, and beginning to study the coupling of these processes to DNA replication. Methods of analyzing DNA replication *in vivo* are also included.

JUDITH L. CAMPBELL

METHODS IN ENZYMOLOGY

VOLUME XVIII. Vitamins and Coenzymes (Parts A, B, and C)
Edited by DONALD B. MCCORMICK AND LEMUEL D. WRIGHT

VOLUME XIX. Proteolytic Enzymes
Edited by GERTRUDE E. PERLMANN AND LASZLO LORAND

VOLUME XX. Nucleic Acids and Protein Synthesis (Part C)
Edited by KIVIE MOLDAVE AND LAWRENCE GROSSMAN

VOLUME XXI. Nucleic Acids (Part D)
Edited by LAWRENCE GROSSMAN AND KIVIE MOLDAVE

VOLUME XXII. Enzyme Purification and Related Techniques
Edited by WILLIAM B. JAKOBY

VOLUME XXIII. Photosynthesis (Part A)
Edited by ANTHONY SAN PIETRO

VOLUME XXIV. Photosynthesis and Nitrogen Fixation (Part B)
Edited by ANTHONY SAN PIETRO

VOLUME XXV. Enzyme Structure (Part B)
Edited by C. H. W. HIRS AND SERGE N. TIMASHEFF

VOLUME XXVI. Enzyme Structure (Part C)
Edited by C. H. W. HIRS AND SERGE N. TIMASHEFF

VOLUME XXVII. Enzyme Structure (Part D)
Edited by C. H. W. HIRS AND SERGE N. TIMASHEFF

VOLUME XXVIII. Complex Carbohydrates (Part B)
Edited by VICTOR GINSBURG

VOLUME XXIX. Nucleic Acids and Protein Synthesis (Part E)
Edited by LAWRENCE GROSSMAN AND KIVIE MOLDAVE

VOLUME XXX. Nucleic Acids and Protein Synthesis (Part F)
Edited by KIVIE MOLDAVE AND LAWRENCE GROSSMAN

VOLUME XXXI. Biomembranes (Part A)
Edited by SIDNEY FLEISCHER AND LESTER PACKER

VOLUME XXXII. Biomembranes (Part B)
Edited by SIDNEY FLEISCHER AND LESTER PACKER

VOLUME XXXIII. Cumulative Subject Index Volumes I–XXX
Edited by MARTHA G. DENNIS AND EDWARD A. DENNIS

VOLUME XXXIV. Affinity Techniques (Enzyme Purification: Part B)
Edited by WILLIAM B. JAKOBY AND MEIR WILCHEK

VOLUME XXXV. Lipids (Part B)
Edited by JOHN M. LOWENSTEIN

VOLUME XXXVI. Hormone Action (Part A: Steroid Hormones)
Edited by BERT W. O'MALLEY AND JOEL G. HARDMAN

VOLUME XXXVII. Hormone Action (Part B: Peptide Hormones)
Edited by BERT W. O'MALLEY AND JOEL G. HARDMAN

VOLUME XXXVIII. Hormone Action (Part C: Cyclic Nucleotides)
Edited by JOEL G. HARDMAN AND BERT W. O'MALLEY

VOLUME XXXIX. Hormone Action (Part D: Isolated Cells, Tissues, and Organ Systems)
Edited by JOEL G. HARDMAN AND BERT W. O'MALLEY

VOLUME XL. Hormone Action (Part E: Nuclear Structure and Function)
Edited by BERT W. O'MALLEY AND JOEL G. HARDMAN

VOLUME XLI. Carbohydrate Metabolism (Part B)
Edited by W. A. WOOD

VOLUME XLII. Carbohydrate Metabolism (Part C)
Edited by W. A. WOOD

VOLUME XLIII. Antibiotics
Edited by JOHN H. HASH

VOLUME XLIV. Immobilized Enzymes
Edited by KLAUS MOSBACH

VOLUME XLV. Proteolytic Enzymes (Part B)
Edited by LASZLO LORAND

VOLUME XLVI. Affinity Labeling
Edited by WILLIAM B. JAKOBY AND MEIR WILCHEK

VOLUME XLVII. Enzyme Structure (Part E)
Edited by C. H. W. HIRS AND SERGE N. TIMASHEFF

VOLUME XLVIII. Enzyme Structure (Part F)
Edited by C. H. W. HIRS AND SERGE N. TIMASHEFF

VOLUME XLIX. Enzyme Structure (Part G)
Edited by C. H. W. HIRS AND SERGE N. TIMASHEFF

VOLUME L. Complex Carbohydrates (Part C)
Edited by VICTOR GINSBURG

VOLUME LI. Purine and Pyrimidine Nucleotide Metabolism
Edited by PATRICIA A. HOFFEE AND MARY ELLEN JONES

VOLUME LII. Biomembranes (Part C: Biological Oxidations)
Edited by SIDNEY FLEISCHER AND LESTER PACKER

VOLUME LIII. Biomembranes (Part D: Biological Oxidations)
Edited by SIDNEY FLEISCHER AND LESTER PACKER

VOLUME LIV. Biomembranes (Part E: Biological Oxidations)
Edited by SIDNEY FLEISCHER AND LESTER PACKER

VOLUME LV. Biomembranes (Part F: Bioenergetics)
Edited by SIDNEY FLEISCHER AND LESTER PACKER

VOLUME LVI. Biomembranes (Part G: Bioenergetics)
Edited by SIDNEY FLEISCHER AND LESTER PACKER

VOLUME 76. Hemoglobins
Edited by ERALDO ANTONINI, LUIGI ROSSI-BERNARDI, AND EMILIA CHIANCONE

VOLUME 77. Detoxication and Drug Metabolism
Edited by WILLIAM B. JAKOBY

VOLUME 78. Interferons (Part A)
Edited by SIDNEY PESTKA

VOLUME 79. Interferons (Part B)
Edited by SIDNEY PESTKA

VOLUME 80. Proteolytic Enzymes (Part C)
Edited by LASZLO LORAND

VOLUME 81. Biomembranes (Part H: Visual Pigments and Purple Membranes, I)
Edited by LESTER PACKER

VOLUME 82. Structural and Contractile Proteins (Part A: Extracellular Matrix)
Edited by LEON W. CUNNINGHAM AND DIXIE W. FREDERIKSEN

VOLUME 83. Complex Carbohydrates (Part D)
Edited by VICTOR GINSBURG

VOLUME 84. Immunochemical Techniques (Part D: Selected Immunoassays)
Edited by JOHN J. LANGONE AND HELEN VAN VUNAKIS

VOLUME 85. Structural and Contractile Proteins (Part B: The Contractile Apparatus and the Cytoskeleton)
Edited by DIXIE W. FREDERIKSEN AND LEON W. CUNNINGHAM

VOLUME 86. Prostaglandins and Arachidonate Metabolites
Edited by WILLIAM E. M. LANDS AND WILLIAM L. SMITH

VOLUME 87. Enzyme Kinetics and Mechanism (Part C: Intermediates, Stereochemistry, and Rate Studies)
Edited by DANIEL L. PURICH

VOLUME 88. Biomembranes (Part I: Visual Pigments and Purple Membranes, II)
Edited by LESTER PACKER

VOLUME 89. Carbohydrate Metabolism (Part D)
Edited by WILLIS A. WOOD

VOLUME 90. Carbohydrate Metabolism (Part E)
Edited by WILLIS A. WOOD

VOLUME 91. Enzyme Structure (Part I)
Edited by C. H. W. HIRS AND SERGE N. TIMASHEFF

VOLUME 92. Immunochemical Techniques (Part E: Monoclonal Antibodies and General Immunoassay Methods)
Edited by JOHN J. LANGONE AND HELEN VAN VUNAKIS

VOLUME 93. Immunochemical Techniques (Part F: Conventional Antibodies, Fc Receptors, and Cytotoxicity)
Edited by JOHN J. LANGONE AND HELEN VAN VUNAKIS

VOLUME 94. Polyamines
Edited by HERBERT TABOR AND CELIA WHITE TABOR

VOLUME 95. Cumulative Subject Index Volumes 61–74, 76–80
Edited by EDWARD A. DENNIS AND MARTHA G. DENNIS

VOLUME 96. Biomembranes [Part J: Membrane Biogenesis: Assembly and Targeting (General Methods; Eukaryotes)]
Edited by SIDNEY FLEISCHER AND BECCA FLEISCHER

VOLUME 97. Biomembranes [Part K: Membrane Biogenesis: Assembly and Targeting (Prokaryotes, Mitochondria, and Chloroplasts)]
Edited by SIDNEY FLEISCHER AND BECCA FLEISCHER

VOLUME 98. Biomembranes (Part L: Membrane Biogenesis: Processing and Recycling)
Edited by SIDNEY FLEISCHER AND BECCA FLEISCHER

VOLUME 99. Hormone Action (Part F: Protein Kinases)
Edited by JACKIE D. CORBIN AND JOEL G. HARDMAN

VOLUME 100. Recombinant DNA (Part B)
Edited by RAY WU, LAWRENCE GROSSMAN, AND KIVIE MOLDAVE

VOLUME 101. Recombinant DNA (Part C)
Edited by RAY WU, LAWRENCE GROSSMAN, AND KIVIE MOLDAVE

VOLUME 102. Hormone Action (Part G: Calmodulin and Calcium-Binding Proteins)
Edited by ANTHONY R. MEANS AND BERT W. O'MALLEY

VOLUME 103. Hormone Action (Part H: Neuroendocrine Peptides)
Edited by P. MICHAEL CONN

VOLUME 104. Enzyme Purification and Related Techniques (Part C)
Edited by WILLIAM B. JAKOBY

VOLUME 105. Oxygen Radicals in Biological Systems
Edited by LESTER PACKER

VOLUME 106. Posttranslational Modifications (Part A)
Edited by FINN WOLD AND KIVIE MOLDAVE

VOLUME 107. Posttranslational Modifications (Part B)
Edited by FINN WOLD AND KIVIE MOLDAVE

VOLUME 108. Immunochemical Techniques (Part G: Separation and Characterization of Lymphoid Cells)
Edited by GIOVANNI DI SABATO, JOHN J. LANGONE, AND HELEN VAN VUNAKIS

VOLUME 109. Hormone Action (Part I: Peptide Hormones)
Edited by LUTZ BIRNBAUMER AND BERT W. O'MALLEY

Volume 110. Steroids and Isoprenoids (Part A)
Edited by John H. Law and Hans C. Rilling

Volume 111. Steroids and Isoprenoids (Part B)
Edited by John H. Law and Hans C. Rilling

Volume 112. Drug and Enzyme Targeting (Part A)
Edited by Kenneth J. Widder and Ralph Green

Volume 113. Glutamate, Glutamine, Glutathione, and Related Compounds
Edited by Alton Meister

Volume 114. Diffraction Methods for Biological Macromolecules (Part A)
Edited by Harold W. Wyckoff, C. H. W. Hirs, and Serge N. Timasheff

Volume 115. Diffraction Methods for Biological Macromolecules (Part B)
Edited by Harold W. Wyckoff, C. H. W. Hirs, and Serge N. Timasheff

Volume 116. Immunochemical Techniques (Part H: Effectors and Mediators of Lymphoid Cell Functions)
Edited by Giovanni Di Sabato, John J. Langone, and Helen Van Vunakis

Volume 117. Enzyme Structure (Part J)
Edited by C. H. W. Hirs and Serge N. Timasheff

Volume 118. Plant Molecular Biology
Edited by Arthur Weissbach and Herbert Weissbach

Volume 119. Interferons (Part C)
Edited by Sidney Pestka

Volume 120. Cumulative Subject Index Volumes 81–94, 96–101

Volume 121. Immunochemical Techniques (Part I: Hybridoma Technology and Monoclonal Antibodies)
Edited by John J. Langone and Helen Van Vunakis

Volume 122. Vitamins and Coenzymes (Part G)
Edited by Frank Chytil and Donald B. McCormick

Volume 123. Vitamins and Coenzymes (Part H)
Edited by Frank Chytil and Donald B. McCormick

Volume 124. Hormone Action (Part J: Neuroendocrine Peptides)
Edited by P. Michael Conn

Volume 125. Biomembranes (Part M: Transport in Bacteria, Mitochondria, and Chloroplasts: General Approaches and Transport Systems)
Edited by Sidney Fleischer and Becca Fleischer

Volume 126. Biomembranes (Part N: Transport in Bacteria, Mitochondria, and Chloroplasts: Protonmotive Force)
Edited by Sidney Fleischer and Becca Fleischer

Volume 127. Biomembranes (Part O: Protons and Water: Structure and Translocation)
Edited by Lester Packer

Volume 128. Plasma Lipoproteins (Part A: Preparation, Structure, and Molecular Biology)
Edited by Jere P. Segrest and John J. Albers

Volume 129. Plasma Lipoproteins (Part B: Characterization, Cell Biology, and Metabolism)
Edited by John J. Albers and Jere P. Segrest

Volume 130. Enzyme Structure (Part K)
Edited by C. H. W. Hirs and Serge N. Timasheff

Volume 131. Enzyme Structure (Part L)
Edited by C. H. W. Hirs and Serge N. Timasheff

Volume 132. Immunochemical Techniques (Part J: Phagocytosis and Cell-Mediated Cytotoxicity)
Edited by Giovanni Di Sabato and Johannes Everse

Volume 133. Bioluminescence and Chemiluminescence (Part B)
Edited by Marlene DeLuca and William D. McElroy

Volume 134. Structural and Contractile Proteins (Part C: The Contractile Apparatus and the Cytoskeleton)
Edited by Richard B. Vallee

Volume 135. Immobilized Enzymes and Cells (Part B)
Edited by Klaus Mosbach

Volume 136. Immobilized Enzymes and Cells (Part C)
Edited by Klaus Mosbach

Volume 137. Immobilized Enzymes and Cells (Part D)
Edited by Klaus Mosbach

Volume 138. Complex Carbohydrates (Part E)
Edited by Victor Ginsburg

Volume 139. Cellular Regulators (Part A: Calcium- and Calmodulin-Binding Proteins)
Edited by Anthony R. Means and P. Michael Conn

Volume 140. Cumulative Subject Index Volumes 102–119, 121–134

Volume 141. Cellular Regulators (Part B: Calcium and Lipids)
Edited by P. Michael Conn and Anthony R. Means

Volume 142. Metabolism of Aromatic Amino Acids and Amines
Edited by Seymour Kaufman

Volume 143. Sulfur and Sulfur Amino Acids
Edited by William B. Jakoby and Owen Griffith

Volume 144. Structural and Contractile Proteins (Part D: Extracellular Matrix)
Edited by Leon W. Cunningham

Volume 145. Structural and Contractile Proteins (Part E: Extracellular Matrix)
Edited by Leon W. Cunningham

Volume 182. Guide to Protein Purification
Edited by MURRAY P. DEUTSCHER

Volume 183. Molecular Evolution: Computer Analysis of Protein and Nucleic Acid Sequences
Edited by RUSSELL F. DOOLITTLE

Volume 184. Avidin–Biotin Technology
Edited by MEIR WILCHEK AND EDWARD A. BAYER

Volume 185. Gene Expression Technology
Edited by DAVID V. GOEDDEL

Volume 186. Oxygen Radicals in Biological Systems (Part B: Oxygen Radicals and Antioxidants)
Edited by LESTER PACKER AND ALEXANDER N. GLAZER

Volume 187. Arachidonate Related Lipid Mediators
Edited by ROBERT C. MURPHY AND FRANK A. FITZPATRICK

Volume 188. Hydrocarbons and Methylotrophy
Edited by MARY E. LIDSTROM

Volume 189. Retinoids (Part A: Molecular and Metabolic Aspects)
Edited by LESTER PACKER

Volume 190. Retinoids (Part B: Cell Differentiation and Clinical Applications)
Edited by LESTER PACKER

Volume 191. Biomembranes (Part V: Cellular and Subcellular Transport: Epithelial Cells)
Edited by SIDNEY FLEISCHER AND BECCA FLEISCHER

Volume 192. Biomembranes (Part W: Cellular and Subcellular Transport: Epithelial Cells)
Edited by SIDNEY FLEISCHER AND BECCA FLEISCHER

Volume 193. Mass Spectrometry
Edited by JAMES A. MCCLOSKEY

Volume 194. Guide to Yeast Genetics and Molecular Biology
Edited by CHRISTINE GUTHRIE AND GERALD R. FINK

Volume 195. Adenylyl Cyclase, G Proteins, and Guanylyl Cyclase
Edited by ROGER A. JOHNSON AND JACKIE D. CORBIN

Volume 196. Molecular Motors and the Cytoskeleton
Edited by RICHARD B. VALLEE

Volume 197. Phospholipases
Edited by EDWARD A. DENNIS

Volume 198. Peptide Growth Factors (Part C)
Edited by DAVID BARNES, J. P. MATHER, AND GORDON H. SATO

Volume 199. Cumulative Subject Index Volumes 168–174, 176–194

VOLUME 200. Protein Phosphorylation (Part A: Protein Kinases: Assays, Purification, Antibodies, Functional Analysis, Cloning, and Expression)
Edited by TONY HUNTER AND BARTHOLOMEW M. SEFTON

Volume 201. Protein Phosphorylation (Part B: Analysis of Protein Phosphorylation, Protein Kinase Inhibitors, and Protein Phosphatases)
Edited by TONY HUNTER AND BARTHOLOMEW M. SEFTON

VOLUME 202. Molecular Design and Modeling: Concepts and Applications (Part A: Proteins, Peptides, and Enzymes)
Edited by JOHN J. LANGONE

VOLUME 203. Molecular Design and Modeling: Concepts and Applications (Part B: Antibodies and Antigens, Nucleic Acids, Polysaccharides, and Drugs)
Edited by JOHN J. LANGONE

VOLUME 204. Bacterial Genetic Systems
Edited by JEFFREY H. MILLER

VOLUME 205. Metallobiochemistry (Part B: Metallothionein and Related Molecules)
Edited by JAMES F. RIORDAN AND BERT L. VALLEE

VOLUME 206. Cytochrome P450
Edited by MICHAEL R. WATERMAN AND ERIC F. JOHNSON

VOLUME 207. Ion Channels
Edited by BERNARDO RUDY AND LINDA E. IVERSON

VOLUME 208. Protein–DNA Interactions
Edited by ROBERT T. SAUER

VOLUME 209. Phospholipid Biosynthesis
Edited by EDWARD A. DENNIS AND DENNIS E. VANCE

VOLUME 210. Numerical Computer Methods
Edited by LUDWIG BRAND AND MICHAEL L. JOHNSON

VOLUME 211. DNA Structures (Part A: Synthesis and Physical Analysis of DNA)
Edited by DAVID M. J. LILLEY AND JAMES E. DAHLBERG

VOLUME 212. DNA Structures (Part B: Chemical and Electrophoretic Analysis of DNA)
Edited by DAVID M. J. LILLEY AND JAMES E. DAHLBERG

VOLUME 213. Carotenoids (Part A: Chemistry, Separation, Quantitation, and Antioxidation)
Edited by LESTER PACKER

VOLUME 214. Carotenoids (Part B: Metabolism, Genetics, and Biosynthesis)
Edited by LESTER PACKER

VOLUME 215. Platelets: Receptors, Adhesion, Secretion (Part B)
Edited by JACEK J. HAWIGER

VOLUME 216. Recombinant DNA (Part G)
Edited by RAY WU

VOLUME 217. Recombinant DNA (Part H)
Edited by RAY WU

VOLUME 218. Recombinant DNA (Part I)
Edited by RAY WU

VOLUME 219. Reconstitution of Intracellular Transport
Edited by JAMES E. ROTHMAN

VOLUME 220. Membrane Fusion Techniques (Part A)
Edited by NEJAT DÜZGÜNEŞ

VOLUME 221. Membrane Fusion Techniques (Part B)
Edited by NEJAT DÜZGÜNEŞ

VOLUME 222. Proteolytic Enzymes in Coagulation, Fibrinolysis, and Complement Activation (Part A: Mammalian Blood Coagulation Factors and Inhibitors)
Edited by LASZLO LORAND AND KENNETH G. MANN

VOLUME 223. Proteolytic Enzymes in Coagulation, Fibrinolysis, and Complement Activation (Part B: Complement Activation, Fibrinolysis, and Nonmammalian Blood Coagulation Factors)
Edited by LASZLO LORAND AND KENNETH G. MANN

VOLUME 224. Molecular Evolution: Producing the Biochemical Data
Edited by ELIZABETH ANNE ZIMMER, THOMAS J. WHITE, REBECCA L. CANN, AND ALLAN C. WILSON

VOLUME 225. Guide to Techniques in Mouse Development
Edited by PAUL M. WASSARMAN AND MELVIN L. DEPAMPHILIS

VOLUME 226. Metallobiochemistry (Part C: Spectroscopic and Physical Methods for Probing Metal Ion Environments in Metalloenzymes and Metalloproteins)
Edited by JAMES F. RIORDAN AND BERT L. VALLEE

VOLUME 227. Metallobiochemistry (Part D: Physical and Spectroscopic Methods for Probing Metal Ion Environments in Metalloproteins)
Edited by JAMES F. RIORDAN AND BERT L. VALLEE

VOLUME 228. Aqueous Two-Phase Systems
Edited by HARRY WALTER AND GÖTE JOHANSSON

VOLUME 229. Cumulative Subject Index Volumes 195–198, 200–227

VOLUME 230. Guide to Techniques in Glycobiology
Edited by WILLIAM J. LENNARZ AND GERALD W. HART

VOLUME 231. Hemoglobins (Part B: Biochemical and Analytical Methods)
Edited by JOHANNES EVERSE, KIM D. VANDEGRIFF, AND ROBERT M. WINSLOW

VOLUME 232. Hemoglobins (Part C: Biophysical Methods)
Edited by JOHANNES EVERSE, KIM D. VANDEGRIFF, AND ROBERT M. WINSLOW

VOLUME 233. Oxygen Radicals in Biological Systems (Part C)
Edited by LESTER PACKER

VOLUME 234. Oxygen Radicals in Biological Systems (Part D)
Edited by LESTER PACKER

VOLUME 235. Bacterial Pathogenesis (Part A: Identification and Regulation of Virulence Factors)
Edited by VIRGINIA L. CLARK AND PATRIK M. BAVOIL

VOLUME 236. Bacterial Pathogenesis (Part B: Integration of Pathogenic Bacteria with Host Cells)
Edited by VIRGINIA L. CLARK AND PATRIK M. BAVOIL

VOLUME 237. Heterotrimeric G Proteins
Edited by RAVI IYENGAR

VOLUME 238. Heterotrimeric G-Protein Effectors
Edited by RAVI IYENGAR

VOLUME 239. Nuclear Magnetic Resonance (Part C)
Edited by THOMAS L. JAMES AND NORMAN J. OPPENHEIMER

VOLUME 240. Numerical Computer Methods (Part B)
Edited by MICHAEL L. JOHNSON AND LUDWIG BRAND

VOLUME 241. Retroviral Proteases
Edited by LAWRENCE C. KUO AND JULES A. SHAFER

VOLUME 242. Neoglycoconjugates (Part A)
Edited by Y. C. LEE AND REIKO T. LEE

VOLUME 243. Inorganic Microbial Sulfur Metabolism
Edited by HARRY D. PECK, JR., AND JEAN LEGALL

VOLUME 244. Proteolytic Enzymes: Serine and Cysteine Peptidases
Edited by ALAN J. BARRETT

VOLUME 245. Extracellular Matrix Components
Edited by E. RUOSLAHTI AND E. ENGVALL

VOLUME 246. Biochemical Spectroscopy
Edited by KENNETH SAUER

VOLUME 247. Neoglycoconjugates (Part B: Biomedical Applications)
Edited by Y. C. LEE AND REIKO T. LEE

VOLUME 248. Proteolytic Enzymes: Aspartic and Metallo Peptidases
Edited by ALAN J. BARRETT

VOLUME 249. Enzyme Kinetics and Mechanism (Part D: Developments in Enzyme Dynamics)
Edited by DANIEL L. PURICH

VOLUME 250. Lipid Modifications of Proteins
Edited by PATRICK J. CASEY AND JANICE E. BUSS

VOLUME 251. Biothiols (Part A: Monothiols and Dithiols, Protein Thiols, and Thiyl Radicals)
Edited by LESTER PACKER

VOLUME 252. Biothiols (Part B: Glutathione and Thioredoxin; Thiols in Signal Transduction and Gene Regulation)
Edited by LESTER PACKER

VOLUME 253. Adhesion of Microbial Pathogens
Edited by RON J. DOYLE AND ITZHAK OFEK

VOLUME 254. Oncogene Techniques
Edited by PETER K. VOGT AND INDER M. VERMA

VOLUME 255. Small GTPases and Their Regulators (Part A: Ras Family)
Edited by W. E. BALCH, CHANNING J. DER, AND ALAN HALL

VOLUME 256. Small GTPases and Their Regulators (Part B: Rho Family)
Edited by W. E. BALCH, CHANNING J. DER, AND ALAN HALL

VOLUME 257. Small GTPases and Their Regulators (Part C: Proteins Involved in Transport)
Edited by W. E. BALCH, CHANNING J. DER, AND ALAN HALL

VOLUME 258. Redox-Active Amino Acids in Biology
Edited by JUDITH P. KLINMAN

VOLUME 259. Energetics of Biological Macromolecules
Edited by MICHAEL L. JOHNSON AND GARY K. ACKERS

VOLUME 260. Mitochondrial Biogenesis and Genetics (Part A)
Edited by GIUSEPPE M. ATTARDI AND ANNE CHOMYN

VOLUME 261. Nuclear Magnetic Resonance and Nucleic Acids
Edited by THOMAS L. JAMES

VOLUME 262. DNA Replication
Edited by JUDITH L. CAMPBELL

VOLUME 263. Plasma Lipoproteins (Part C: Quantitation) (in preparation)
Edited by WILLIAM A. BRADLEY, SANDRA H. GIANTURCO, AND JERE P. SEGREST

VOLUME 264. Mitochondrial Biogenesis and Genetics (Part B) (in preparation)
Edited by GIUSEPPE M. ATTARDI AND ANNE CHOMYN

Section I

Purification and Characterization of DNA Polymerases

[1] Purification of *Escherichia coli* DNA Polymerase I and Klenow Fragment

By CATHERINE M. JOYCE and VICTORIA DERBYSHIRE

Introduction

DNA polymerase I (Pol I) of *Escherichia coli*, the first DNA polymerase to be discovered, has long served as a simple model system for studying the enzymology of DNA synthesis.[1] The original studies of Pol I relied on purification of the enzyme from *E. coli* extracts without genetic manipulation, yielding around 10 mg of purified enzyme per kilogram of cell paste.[2] Cloning of *polA*, the structural gene for Pol I, in a variety of phage λ vectors increased the level of expression about 100-fold.[3,4] Sequence analysis of the cloned *polA* gene[5] allowed construction of a plasmid-derived expression system for the Klenow fragment portion of Pol I,[6] comprising the C-terminal two-thirds of the protein and having the polymerase and $3' \rightarrow 5'$ (proofreading)-exonuclease functions of the parent molecule, but lacking the $5' \rightarrow 3'$-exonuclease that is used in nick-translation. (Earlier attempts to express whole Pol I on a plasmid vector were unsuccessful because of the lethality of wild-type *polA* in multiple copies,[3] and indicated the need for more sophisticated vectors giving tight control of the level of expression.) The ability to purify large quantities of Klenow fragment paved the way for the determination of its structure by X-ray crystallography.[7] In addition to their importance as experimental systems in their own right, both Pol I and Klenow fragment have found extensive use as biochemical reagents in a variety of cloning, sequencing, and labeling procedures. Over the years we have made improvements in the expression systems for Pol I and Klenow fragment; we describe here our most recent constructs and protocols, which typically give yields of 10 mg of pure polymerase per gram of cells.

[1] A. Kornberg and T. A. Baker, "DNA Replication," p. 113. Freeman, San Francisco (1992).

[2] T. M. Jovin, P. T. Englund, and L. L. Bertsch, *J. Biol. Chem.* **244,** 2996 (1969).

[3] W. S. Kelley, K. Chalmers, and N. E. Murray, *Proc. Natl. Acad. Sci. USA* **74,** 5632 (1977).

[4] N. E. Murray and W. S. Kelley, *Molec. Gen. Genet.* **175,** 77 (1979).

[5] C. M. Joyce, W. S. Kelley, and N. D. F. Grindley, *J. Biol. Chem.* **257,** 1958 (1982).

[6] C. M. Joyce and N. D. F. Grindley, *Proc. Natl. Acad. Sci. USA* **80,** 1830 (1983).

[7] D. L. Ollis, P. Brick, R. Hamlin, N. G. Xuong, and T. A. Steitz, *Nature* **313,** 762 (1985).

Expression Plasmids

Both whole Pol I and Klenow fragment have been substantially overexpressed using constructs derived from the pAS1 vector,[8] in which transcription is driven from the strong leftward promoter (P_L) of phage λ, and the translational start signals are derived from the λ*cII* gene. For expression of Klenow fragment, the ATG initiation codon of the expression vector replaces the codon for Val-324(GTG), the N-terminal amino acid of Klenow fragment. The construction of this plasmid has already been described.[9] It gives about a tenfold higher expression of Klenow fragment than the original expression plasmid in which the translational signals were less well optimized.[6] In the Pol I expression plasmid, whose construction is described elsewhere, the vector-derived ATG codon replaces the natural GTG start of the *polA* gene and no upstream *polA* DNA is present. This plasmid gives a much higher level of expression than the Pol I expression plasmid described previously by Minkley *et al.*[10] Not only did the earlier plasmid use the rather poor *polA* translational initiation signals, but it also retained DNA sequences derived from the *polA* promoter. Because of the lethality of a nonrepressed *polA* gene at high copy number, the latter sequences are probably responsible for the considerable problems of plasmid instability reported by Minkley *et al.*[10]

Host Strains

The highest levels of expression that we have achieved were in a strain background such as AR120,[11] in which expression is controlled by the wild-type λ repressor on a defective prophage. SOS-induction using nalidixic acid results in *recA*-mediated cleavage and inactivation of the repressor, leading to expression of the P_L-driven target gene. However, this system is not appropriate for expressing mutant derivatives of Pol I or Klenow fragment. Because the expression vector requires a wild-type chromosomal copy of *polA* for its replication, it is desirable, when expressing a mutant protein, to use a *recA*-defective host in order to minimize the possibility that exchange between plasmid and chromosomal *polA* sequences might eliminate the mutant information. Because nalidixic acid induction is ruled out in a *recA*$^-$ background, we use heat induction of a strain carrying the

[8] M. Rosenberg, Y.-S. Ho, and A. Shatzman, *Meth. Enzymol.* **101,** 123 (1983).

[9] A. H. Polesky, T. A. Steitz, N. D. F. Grindley, and C. M. Joyce, *J. Biol. Chem.* **265,** 14579 (1990).

[10] E. G. Minkley, Jr., A. T. Leney, J. B. Bodner, M. M. Panicker, and W. E. Brown, *J. Biol. Chem.* **259,** 10386 (1984).

[11] J. E. Mott, R. A. Grant, Y.-S. Ho, and T. Platt, *Proc. Natl. Acad. Sci. USA* **82,** 88 (1985).

TABLE I
OVERPRODUCER STRAINS FOR DNA POLYMERASE I AND KLENOW FRAGMENT

Protein	Plasmid	Host	Strain number	Inducing treatment
Pol I	pCJ194	AR120	CJ402	Nalidixic acid
Pol I[a]	pCJ194[a]	CJ376	—	Heat
Klenow fragment	pCJ122	AR120	CJ333	Nalidixic acid
Klenow fragment	pCJ122	CJ378	CJ379	Heat
Klenow fragment[a]	pCJ122[a]	CJ376	—	Heat

[a] Or mutant derivatives.

cI_{857} temperature-sensitive λ repressor. Our host strain, CJ376,[9] is *recA*$^-$ and carries the cI_{857} allele on a chloramphenicol-resistant plasmid, pCJ136, which is compatible with the expression vector. The CJ376 host strain is also deficient in exonuclease III, which has in the past caused concern as a possible contaminant in the purification,[12] but is now largely irrelevant with the high-resolution chromatographic methods described here. Note that the availability of the cI_{857} gene on a compatible plasmid means that virtually any strain can be converted into an expression host merely by transformation; for example, the host CJ378, obtained by transformation of BW9109,[13] is *recA*$^+$ and deficient in exonuclease III, and provides a good background for heat induction of wild-type Klenow fragment.

Induction Protocols

Typical procedures follow for the growth and induction of 1 to 2 liters of cells. The procedure can easily be scaled up, for example, for use in a fermentor. Although we routinely maintain selection pressure for the Amp^R determinant as a precaution against loss of the expression plasmid, we have not found plasmid instability to be a serious problem in this system.

Strains

The overproducer strains currently in use are listed in Table I. They are stored as glycerol cultures at $-20°$.[14] Before use they should be streaked out on plates containing carbenicillin (50 μg/ml) and, when using the CJ376 or CJ378 host, chloramphenicol (15 μg/ml). The incubation temperature is 30° for the heat-inducible strains, and 37° for the others. Strains containing

[12] P. Setlow, *Methods Enzymol.* **29,** 3 (1974).

[13] B. J. White, S. J. Hochhauser, N. M. Cintrón, and B. Weiss, *J. Bacteriol.* **126,** 1082 (1976).

[14] J. H. Miller, "Experiments in Molecular Genetics." Cold Spring Harbor Laboratories, Cold Spring Harbor (1972).

overproducer plasmids for mutant polymerase derivatives are not stored as such; to minimize the chances for exchange between wild-type and mutant information, the mutated overproducer plasmid is introduced into the CJ376 (*recA*$^-$) host only when needed.

Media

LB: 10 g tryptone, 5 g yeast extract, and 5 g NaCl per liter.[14]

MIM (maximal induction medium)[11]: 32 g tryptone and 20 g yeast extract, adjusted to pH 7.6 with 3 *M* NaOH, in a total volume of 900 ml. After autoclaving, 100 ml 10 × M9 salts, 0.1 ml 1 *M* $MgSO_4$, and 0.1 ml 0.01 *M* $FeCl_3$ are added.

10 × M9 salts[14]: 6 g Na_2HPO_4, 3 g KH_2PO_4, 5 g NaCl, and 10 g NH_4Cl dissolved in H_2O to a total volume of 100 ml, and autoclaved.

Nalidixic acid: 0.1 g nalidixic acid in 10 ml 0.3 *M* NaOH, filter-sterilized and stored at 4°.

Carbenicillin: 50 mg/ml in H_2O, filter-sterilized and stored at 4°. All media are supplemented with carbenicillin at 50 μg/ml. Ampicillin, or other related antibiotics, can be substituted.

Nalidixic Acid Induction

A 1-ml inoculum is grown from a single colony of the appropriate overproducer strain in LB/carbenicillin at 37° for approximately 8 hr. This is diluted into 40 ml MIM/carbenicillin and grown overnight. Half of this culture is inoculated into each of two 2-liter baffle flasks containing 500 ml MIM/carbenicillin. These are grown at 37° with vigorous aeration (about 250 rpm in a New Brunswick series 25 incubator shaker) to $OD_{600} \approx 1$. Nalidixic acid (2 ml per 500 ml culture) is added, giving a final concentration of 40 μg/ml. The cells (typically 5 to 6 g) are harvested by centrifugation about 8 hr later, washed with cold 50 m*M* Tris–HCl, pH 7.5, and stored frozen at −70°.

Heat Induction

A 1-ml inoculum is grown from a single colony of the appropriate overproducer strain in LB/carbenicillin at 30° for approximately 8 hr, and then diluted into 50 ml of the same medium and grown overnight. Half of this culture is inoculated into each of two 2-liter baffle flasks containing 750 ml of LB/carbenicillin. These are grown at 30° with vigorous aeration to an $OD_{600} \approx 0.6$ (approximately 4 hr). The temperature is raised by the addition to each flask of 250 ml LB, previously heated to 90°, and the flask is transferred to a shaker at 42°. After a further 2 hr, the cells (typically 3 to 5 g) are harvested as described earlier.

Monitoring Induction

For either induction method a 1-ml sample of the culture should be taken just before the inducing treatment, and when the cells are harvested. The sample is spun for 2 min in a microfuge, and the pelleted cells are resuspended in 50 μl of SDS–PAGE sample buffer and lysed by heating for 2 to 3 min at 100°. A 5- to 10-μl sample of this whole cell lysate is examined by SDS–PAGE, using a 10% gel for Klenow fragment and an 8% gel for whole Pol I. Typical results are shown in Fig. 1.

Purification Method for Klenow Fragment or DNA Polymerase I

The two methods are identical, except where noted. The procedure described here makes use of the Pharmacia fast protein liquid chromatography (FPLC) system. If this equipment is not available, published procedures[6,10] using conventional chromatography are also satisfactory.

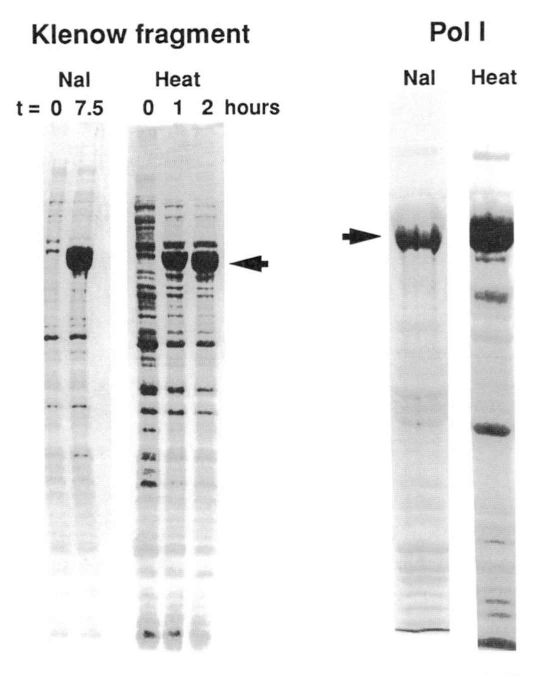

FIG. 1. Overproduction of Klenow fragment and DNA polymerase I. The Klenow fragment panel shows SDS–PAGE analysis of whole cell extracts of appropriate overproducer strains, sampled before induction ($t = 0$) and at the indicated times after the inducing treatment. The Pol I panel shows samples of the clarified crude cell lysates from cells expressing whole Pol I, after induction with nalidixic acid or with heat. The arrows indicate the positions of the respective expressed products.

General

All steps are carried out at 0 to 4°. Ammonium sulfate concentrations are expressed relative to saturation at 0°. Polymerase-containing fractions are located by SDS–PAGE, using the Laemmli formulation.[15] We have found minigels (10.3 × 8.3 × 0.1 cm) to be particularly convenient because they take only about 30 min to run.

Buffers

TED: 50 m*M* Tris–HCl, pH 7.5, 2 m*M* EDTA, and 1 m*M* dithiothreitol (DTT).
Lysis buffer: TED buffer, containing 0.02 m*M* phenylmethylsulfonyl fluoride (PMSF) and 2 mg/ml lysozyme, both added freshly before use.
Buffer A: 50 m*M* Tris–HCl, pH 7.5, 1 m*M* DTT.
Buffer B: Buffer A containing 2 *M* NaCl.
Buffer C: Buffer A containing 1.7 *M* $(NH_4)_2SO_4$.
Buffer D: 100 m*M* Tris–HCl, pH 7.5, 1 m*M* DTT.
Buffer pH values are measured at room temperature. All buffers for FPLC are prefiltered through a 0.22-μm filter.

Cell Lysis

Frozen cells are placed in a beaker on ice and allowed to soften. Lysis buffer (8 to 10 ml/g of cells) is added and a pipette or glass rod is used to break up lumps, yielding a smooth suspension, which is left on ice for at least 15 min. The cell suspension is sonicated sufficiently to break the cells and to reduce the viscosity of the initial extract. Centrifugation for 10 min at 10,000*g*, or greater, gives the clarified crude extract.

Ammonium Sulfate Fractionation: Klenow Fragment

Solid ammonium sulfate is added slowly, with stirring, to the crude extract, to 50% saturation (29.1 g per 100 ml extract). After centrifugation, the pellet is discarded, and solid ammonium sulfate is added to the supernatant to 85% saturation (an additional 23.0 g per 100 ml supernatant). For further processing, an amount of the ammonium sulfate slurry equivalent to 0.5 g cells is used, so as not to exceed the capacity of the FPLC columns described later. With larger columns the amount used can be scaled up as appropriate. The remainder of the material can be stored at 4° in 85% ammonium sulfate for many months.

[15] U. Laemmli, *Nature* **227,** 680 (1970).

Ammonium Sulfate Fractionation: Pol I

The procedure just described is followed except that the first cut is at 40% saturation (22.6 g per 100 ml extract) and the second at 60% saturation (an additional 12.0 g per 100 ml supernatant).

Mono Q Chromatography

An appropriate volume (as already discussed) of the ammonium sulfate slurry is spun down. The pellet is resuspended gently in 10 to 15 ml of Buffer A and dialyzed against 1 liter of Buffer A for 4 hr, with a buffer change after 2 hr. The dialyzed protein is filtered through a Millipore 0.22-μm filter unit; if filtration is difficult, it may be helpful to dilute further or to filter with a 0.4-μm filter before the 0.22-μm filter. The protein is then applied to a Mono Q HR 5/5 column (1-ml bed volume) equilibrated with Buffer A. The column is washed with 5 ml of Buffer A and then eluted with a 30 ml linear gradient of 0 to 0.5 *M* NaCl (i.e., from 100% Buffer A to 75% Buffer A plus 25% Buffer B). Klenow fragment elutes at 140 to 200 m*M* NaCl, and Pol I at 220 to 260 m*M* NaCl. The column is regenerated by washing with 5 ml of Buffer B.

Phenyl-Superose Chromatography

Pooled peak fractions from the Mono Q column (typically about 4 ml) are dialyzed against Buffer C (1 liter) for at least 2 hr and then loaded onto a phenyl-Superose HR 5/5 column equilibrated with Buffer C. The column is washed with 5 ml of Buffer C and then eluted with a 30 ml linear reverse ammonium sulfate gradient from 1.7 *M* (Buffer C) to zero (Buffer A). Klenow fragment elutes at 0.8 to 1.1 *M* ammonium sulfate; the pooled peak fractions (typically 3.5 ml in total) are precipitated by addition of ammonium sulfate to 85% saturation (0.46 g per ml, assuming the pool is initially at 15% saturation). Pol I elutes at 0.4 to 0.7 *M* ammonium sulfate; the pooled fractions are precipitated by addition of ammonium sulfate to 60% (0.30 g per ml, assuming the pool is initially at 10% saturation). In either case, the column is regenerated by washing with 5 ml of Buffer A.

Superose 12 Gel Filtration

The ammonium sulfate precipitate of the phenyl-Superose pool is resuspended in 150 to 200 μl of Buffer D and spun for 2 min in a microfuge at 4° to remove particulate matter. A volume not exceeding 200 μl is applied to a Superose 12 HR 10/30 column equilibrated with Buffer D. The column is developed at 0.5 ml/min with 30 ml of Buffer D, and 0.5 ml fractions are collected. Pol I elutes after 13 to 14 ml, and Klenow fragment after about 14 ml. Peak fractions (typically containing 1 to 2 mg/ml of the polymerase) are diluted with an equal volume of sterile glycerol and stored at −20°.

Rationale of the Purification Method

Figure 2 illustrates, for whole Pol I, the fractionation obtained in the various stages of the purification method just described; the purification

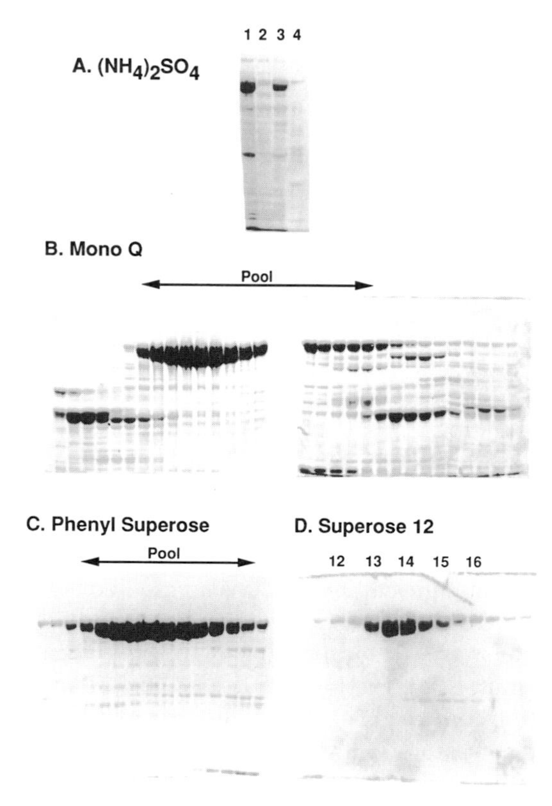

FIG. 2. Purification of DNA polymerase I. (A) Ammonium sulfate fractionation. SDS–PAGE of the clarified cell lysate (lane 1), the 0 to 40% (lane 2) and 40 to 60% (lane 3) ammonium sulfate fractions, and the material that remained soluble at 60% ammonium sulfate (lane 4). In (B–D) a 1-μl sample of each column fraction was examined on SDS–PAGE. (B) Mono Q chromatography. The gels shown correspond to the middle portion of the gradient (from about 0.15 to 0.35 *M* NaCl). The indicated fractions were pooled. (C) Phenyl-Superose chromatography. The gel corresponds to the middle one-third of the gradient. The indicated fractions were pooled. (D) Gel filtration on Superose 12. The approximate elution volumes (in ml) are noted. In each panel, Pol I is the major protein species that migrates about one-quarter of the way down the gel.

of Klenow fragment looks very similar. The initial ammonium sulfate fractionation of a crude cell extract serves primarily to remove most of the soluble lipids before the FPLC columns. Chromatography on Mono Q removes nucleic acids and gives some purification from other proteins so that the polymerase is often substantially pure (as judged by Coomassie Brilliant Blue staining) after this stage. The final two columns provide additional fractionation away from minor protein contaminants. The phenyl-Superose column is particularly useful for removal of low levels of cellular nucleases and is therefore important when studying the effects of mutations on the exonuclease activities of Pol I and Klenow fragment, but can be omitted in some other situations (for example, when studying mutations in the polymerase region). We have not investigated whether all three FPLC columns are strictly necessary when purifying Klenow fragment for use in "dideoxy" sequencing.[16] However, our experience has generally been that the most pure enzyme gives the best results. We should also stress the importance, when preparing a batch of enzyme for use in sequencing, of a careful quality control check using a template of known sequence.

Assay and Properties of Purified Enzymes

Polymerase activity is assayed by following the incorporation of labeled deoxynucleotide precursors into high molecular weight DNA.[12] Either "activated" calf thymus DNA (made by nicking with DNase I) or poly[d(AT)] can be used as the DNA substrate; poly[d(AT)], being available commercially, has the advantage of convenience. With either substrate, however, there is considerable batch-to-batch variability so that it is advisable to include a standard of known activity with each series of assays. In our hands, a specific activity for Klenow fragment of 10^4 units/mg in the poly[d(AT)] assay is typical,[6,17] where one unit catalyzes the incorporation of 10 nmol of nucleotides in 30 min at 37°. Using the same assay, we obtained a slightly lower specific activity for whole Pol I[18]; allowing for the higher molecular weight of Pol I, the turnover number is very similar to that of Klenow fragment, around 200 nucleotides added per minute. (We must stress, however, that changing the batch of poly[d(AT)] can change assay results by as much as threefold.) When activated DNA is used as the assay template, the specific activity of

[16] F. Sanger, S. Nicklen, and A. R. Coulson, *Proc. Natl. Acad. Sci. USA* **74,** 5463 (1977).
[17] V. Derbyshire, N. D. F. Grindley, and C. M. Joyce, *EMBO J.* **10,** 17 (1991).
[18] V. Derbyshire, unpublished observations (1991).

Klenow fragment is higher, and that of Pol I is lower, than when using poly[d(AT)].[2,6,19] A detailed quantitative comparison of Pol I and Klenow fragment in the two assays is complicated because the reactions respond differently to ionic strength depending on the particular combination of enzyme and assay template.[20]

The individual kinetic constants for the polymerase reaction are influenced by the nature of the DNA substrate, in particular, whether the reaction is set up so as to allow multiple rounds of processive synthesis. On a homopolymer substrate (where processive synthesis can take place), the steady-state k_{cat} is 3.8 sec^{-1} for Pol I[21] and 2.4 sec^{-1} for Klenow fragment.[9] In an experimental system where addition of only a single nucleotide per DNA molecule is possible, the steady-state k_{cat} is much lower (0.06 to 0.67 sec^{-1}), reflecting the slow rate of release of the product DNA.[22] At a more subtle level, the immediate DNA sequence context surrounding the primer terminus also exerts an influence on the kinetic parameters, so that there is variation (within a fairly narrow range) in the values obtained using different experimental systems. Typically, the K_m for dNTP utilization is in the range of 1 to 5 μM,[9,22,23] and the dissociation constant for DNA binding is 5 to 20 nM.[9,22] Even with substrates that permit extensive DNA synthesis, both Pol I and Klenow fragment have rather low processivity, adding in the range of 10 to 50 nucleotides for each enzyme–DNA encounter.[9,21,23–25]

The $3' \rightarrow 5'$-exonuclease can be assayed on a variety of single-stranded or double-stranded DNA substrates.[12,17] On single-stranded DNA, the specific activity of the $3' \rightarrow 5'$-exonuclease of Pol I is around 360 units/mg,[12] where one unit catalyzes the release of 10 nmol of nucleotides in 30 min at 37°. This corresponds to a turnover number of 12 nucleotides per minute, in good agreement with the steady-state k_{cat} of 0.09 sec^{-1} determined for Klenow fragment.[17] The degradation of duplex DNA is about 100-fold slower.[26] The $5' \rightarrow 3'$-exonuclease, assayed on a labeled DNA duplex, blocked against nuclease digestion from the 3′ terminus, gave a specific

[19] P. Setlow, D. Brutlag, and A. Kornberg, *J. Biol. Chem.* **247,** 224 (1972).

[20] H. Klenow, K. Overgaard-Hansen, and S. A. Patkar, *Eur. J. Biochem.* **22,** 371 (1971).

[21] F. R. Bryant, K. A. Johnson, and S. J. Benkovic, *Biochemistry* **22,** 3537 (1983).

[22] R. D. Kuchta, V. Mizrahi, P. A. Benkovic, K. A. Johnson, and S. J. Benkovic, *Biochemistry* **26,** 8410 (1987).

[23] W. R. McClure and T. M. Jovin, *J. Biol. Chem.* **250,** 4073 (1975).

[24] V. Mizrahi, R. N. Henrie, J. F. Marlier, K. A. Johnson, and S. J. Benkovic, *Biochemistry* **24,** 4010 (1985).

[25] C. M. Joyce, *J. Biol. Chem.* **264,** 10858 (1989).

[26] R. D. Kuchta, P. Benkovic, and S. J. Benkovic, *Biochemistry* **27,** 6716 (1988).

activity of 940 units/mg,[12] corresponding to a turnover number of 30 nucleotides per minute.

Acknowledgments

We are grateful to Xiaojun Chen Sun for excellent technical assistance and to Nigel Grindley for a critical reading of the manuscript. This work was supported by the National Institutes of Health (Grant GM-28550, to Nigel D. F. Grindley).

[2] Purification and Properties of DNA Polymerase II from *Escherichia coli*

By HONG CAI, HONG YU, KEVIN MCENTEE, and MYRON F. GOODMAN

Introduction

Three DNA polymerases have been isolated and purified from *Escherichia coli.* DNA polymerase I (Pol I) has been shown to be involved in a variety of DNA repair pathways and is responsible for removing the RNA primer portion of Okazaki fragments.[1] Pol I has $5' \rightarrow 3'$-polymerase and exonuclease activities and $3' \rightarrow 5'$ (proofreading)-exonuclease activity.[1] Pol III is required for chromosomal replication and control of spontaneous mutagenesis. It is comprised of a 10 subunit holoenzyme complex, including the α subunit containing $5' \rightarrow 3'$-polymerase activity, the ε subunit containing $3' \rightarrow 5'$ (proofreading)-exonuclease activity, and a multisubunit γ complex and β protein required for enzyme processivity.[2] *E. coli* Pol II was discovered in 1970,[3] yet its role in DNA replication and repair remains uncertain. Therefore, a brief synopsis of data relating to the biochemical properties of Pol II and the properties of cells deficient in Pol II is relevant to current efforts to determine the role of the enzyme *in vivo.*

The structural gene for Pol II is the damage-inducible *polB* gene.[4,5] Its expression is regulated by the Lex A repressor[6] as part of the SOS response

[1] A. Kornberg and T. A. Baker, *in* "DNA Replication," Chap. 4. W. H. Freeman and Company, New York, 1992.

[2] C. S. McHenry, *Ann. Rev. Biochem.* **57,** 519 (1988).

[3] R. Knippers, *Nature* **228,** 1050 (1970).

[4] C. A. Bonner, S. Hays, K. McEntee, and M. F. Goodman, *Proc. Natl. Acad. Sci. USA* **87,** 7663 (1990).

[5] H. Iwasaki, A. Nakata, G. Walker, and H. Shinagawa, *J. Bacteriol.* **172,** 6268 (1990).

[6] C. A. Bonner, S. K. Randall, C. Rayssiguier, M. Radman, R. Eritja, B. E. Kaplan, K. McEntee, and M. F. Goodman, *J. Biol. Chem.* **263,** 18946 (1988).

to DNA damage in *E. coli*,[7] and the enzyme has been classified as an α-type polymerase based on similarity in amino acid sequences to five conserved domains in eukaryotic Pol α.[4,8] Pol II has been reported to be required for bypass of abasic (apurinic/apyrimidinic) DNA template lesions in the absence of induction of heat-shock proteins Gro EL and Gro ES,[9] and we have found that strains containing a null mutant of *polB* appear to be less viable than wild type when grown in the presence of hydrogen peroxide and exhibit a threefold increase in adaptive mutation rate.[10]

Pol II exhibits several noteworthy properties *in vitro*. It incorporates nucleotides opposite abasic template sites[6] and incorporates chain terminating dideoxy- and arabinonucleotides.[11] It contains 3′-exonuclease activity, and its high exonuclease to polymerase ratio is similar in magnitude to wild-type bacteriophage T4 polymerase.[12] An unusual and potentially significant biological property of Pol II is that it interacts with Pol III accessory subunits, β and γ complex, resulting in a 150- to 600-fold increase in processivity, from about 5 nucleotides to greater than 1600 nucleotides incorporated per template-binding event.[13]

In this chapter, we describe a simple rapid procedure to obtain highly purified enzymes from wild-type *polB*$^+$ cells and from an exonuclease-deficient *polB* mutant strain (D155A/E157A), which are suitable for obtaining crystals for analysis by X-ray diffraction.[14]

Assay Method

Principle

DNA polymerase catalyzes the template-directed incorporation of deoxyribonucleotides into DNA by addition onto primer strand 3′-OH termini (5′ → 3′ synthesis) according to the reaction:

$$DNA_n + dNTP \rightarrow DNA_{n+1} + PP_i.$$

[7] G. C. Walker, *Ann. Rev. Biochem.* **54,** 425 (1985).

[8] H. Iwasaki, Y. Ishino, H. Toh, A. Nakata, and H. Shinagawa, *Mol. Gen. Genet.* **226,** 24 (1991).

[9] I. Tessman and M. A. Kennedy, *Genetics* **136,** 439 (1993).

[10] M. Escarcellar, J. Hicks, G. Gudmundsson, G. Trump, D. Touati, S. Lovett, P. L. Foster, K. McEntee, and M. F. Goodman, *J. Bacteriol.* **10,** 6221 (1994).

[11] H. Yu, Biochemical Aspects of DNA Synthesis Fidelity: DNA Polymerase and Ionized Base Mispairs (Ph.D. Thesis), University of Southern California (1993).

[12] H. Cai, H. Ya, K. McEntee, T. A. Kunkel, and M. F. Goodman, *J. Biol. Chem.* **270,** 15327 (1995).

[13] C. A. Bonner, T. Stukenberg, M. Rajagopalan, R. Eritja, M. O'Donnell, K. McEntee, H. Echols, and M. F. Goodman, *J. Biol. Chem.* **267,** 11431 (1992).

[14] W. F. Anderson, D. B. Prince, H. Yu, K. McEntee, and M. F. Goodman, *J. Mol. Biol.* **238,** 120 (1994).

Procedure

Deoxyribonucleotide Incorporation Assay

DNA polymerase activity is assayed by measuring the incorporation of [^{3}H]dTMP into acid-insoluble DNA. The reaction mixture (0.05 ml) contains 2.5 m*M* dithiothreitol (DTT), 20 m*M* Tris–HCl (pH 7.5), 7.3 m*M* $MgCl_2$, 6 m*M* spermidine hydrochloride, 1 mg/ml bovine serum albumin (BSA), 1.1 m*M* gapped primer-template DNA, 60 μM dATP, dCTP, dGTP, [^{3}H]dTTP (5×10^7 to 1×10^8 cpm/μmol), and 0.5 to 5 units of enzyme. Gapped primer-template DNA refers to salmon sperm DNA digested to about 15% acid solubility with DNase I.[15] Reactions are incubated for 15 min at 37° and are terminated by the addition of cold 0.2 *M* sodium pyrophosphate in 15% trichloroacetic acid. One Pol II polymerase unit catalyzes the incorporation of 1 pmol of [^{3}H]dTMP into acid-insoluble material in 1 min at 37°.

Exonuclease Activity Assay

Pol II has an associated $3' \rightarrow 5'$-exonuclease activity that can be assayed by measuring hydrolysis of single-stranded DNA:

$$\text{Single-stranded DNA}_n \rightarrow \text{DNA}_{n-1} + \text{dNMP}.$$

Pol II (0.1 to 1 unit) is added to 40 μl 5′-^{32}P-labeled single-stranded DNA reaction solution [180 n*M* 5′-^{32}P-labeled single-stranded synthetic DNA oligonucleotide having an arbitrary uniform length, approximately 5 μCi/pmol, 7.3 m*M* $MgCl_2$, 1 m*M* DTT, 50 m*M* Tris–HCl (pH 7.5), 40 μg/ml BSA]. Reactions are carried out at 37°, for a series of time points (e.g., approximately 10 sec to 5 min), and reactions are terminated by adding a 3-μl aliquot of the reaction mixture to 6 μl of 20 m*M* EDTA in 95% formamide. The reaction rate is determined from the slope of the linear region of a plot of percent primer degraded versus time. Procedures for 5′-end-labeling of the primers and gel electrophoresis to resolve product DNA have been described previously.[16] Integrated intensities of radiolabeled bands corresponding to primer DNA reaction products, reduced in length by the action of Pol II-associated exonuclease, can be visualized and quantified by phosphorimaging,[17] or with densitometry using X-ray film.[16] Alternatively, the exonuclease activity can be determined by measuring the release of dNMP from uniformly radiolabeled single-stranded DNA.[18] One Pol II exonuclease unit catalyzes the reduction of 1 pmol/min of single-

[15] A. E. Oleson and J. F. Koerner, *J. Biol. Chem.* **239,** 2935 (1964).

[16] M. S. Boosalis, J. Petruska, and M. F. Goodman, *J. Biol. Chem.* **262,** 14,689 (1987).

[17] H. Cai, L. B. Bloom, R. Eritja, and M. F. Goodman, *J. Biol. Chem.* **268,** 23,567 (1993).

[18] N. Muzyczka, R. L. Poland, and M. J. Bessman, *J. Biol. Chem.* **247,** 7116 (1972).

stranded DNA from n to $n-1$ nucleotides long, or equivalently, the release of 1 pmol of dNMP into acid-soluble material at 37°.

A "turnover" assay can be used to measure the action of the 3′-exonuclease coupled to DNA synthesis. This assay measures the DNA-dependent conversion of dNTP to its corresponding dNMP, as described previously.[18]

Purification of *Escherichia coli* DNA Polymerase II

Cell Growth

E. coli JM109 cells carrying the Pol II (*polB*) gene on an overproducing plasmid pHY400 (wild-type Pol II) or pHC700 (3′ → 5′-exonuclease-deficient mutant, D155A/E157A; Pol II ex 1) are grown in LB with 50 μg/ml ampicillin in a 170-liter fermenter at 37°. The overproduction of Pol II protein is induced by adding isopropyl-β-D-thiogalactoside (IPTG) to the cells at midlog phase (OD_{595} about 0.8) to a final concentration of 0.4 m*M*. The cells are grown for an additional 2 hr at 37° before harvesting. Cells are harvested and resuspended in a volume (ml) of storage buffer [sterile 50 m*M* Tris–HCl (pH 7.5), 10% (w/v) sucrose] equal to the wet weight of the cells in grams, about 600 ml to 600 g cells. A 170-liter fermenter run normally yields about 600 g of dry cells. Cells are rapidly frozen by slowly adding cell paste to liquid nitrogen and are stored at −70°.

Cell Lysis

A preparative scale purification typically starts with 300 g of dry cells and yields about 300 mg of purified Pol II (Table I). Lysis buffer [50 m*M* Tris–HCl (pH 7.5), 10% sucrose, 0.1 *M* NaCl, 15 m*M* spermidine] is added to frozen cells to achieve a final concentration of 0.2 g cells/ml. Cells are thawed at 4°. When the cells are completely thawed, the pH is adjusted to 7.7 with 2 *M* Tris base. Lysozyme is added (to the slurry of cells in lysis buffer) to achieve a final concentration of 0.2 mg/ml, and the cell slurry is incubated for 1 hr at 4°. Cells are distributed into 250-ml GSA bottles (Dupont-Sowell, Wilmington, DE) and are incubated in a water bath for an additional 4 min at 37°; the bottles are gently inverted once each minute. Centrifugation is performed in a GSA rotor at 11,800 rpm for 1 hr. The supernatant, fraction I, is saved.

Ammonium Sulfate Precipitation

Pulverized ammonium sulfate is added slowly with gentle stirring to fraction I, to a final concentration of 30% (w/v), and the suspension is allowed to sit in a cold room (4°) overnight, without stirring. The ammonium

TABLE I
PURIFICATION OF WILD-TYPE AND EXONUCLEASE-DEFICIENT (exo^-) DNA POLYMERASE II FROM *Escherichia coli*[a,d]

Polymerase II	Fraction	Volume (ml)	Protein concentration (mg/ml)[b]	Specific activity (10^3 units/mg)[c]	Recovery
Wild-type (pHY400)	I. Crude lysate	1000	12	0.50	1.0
	II. Ammonium sulfate	250	14	1.5	0.85
	III. Phosphocellulose	340	1.0	14	0.78
	IV. DEAE	1000	0.3	18	0.78
exo^- (pHC700)	I. Crude lysate	1300	8.5	0.18	1.0
	II. Ammonium sulfate	240	20	0.4	0.96
	III. Phosphocellulose	400	1	3.8	0.76
	IV. DEAE	1000	0.30	4.9	0.74

[a] Cells were induced with IPTG to overproduce Pol II.
[b] Protein concentrations were determined by the method of Bradford.[18a]
[c] One unit of enzyme catalyzes the incorporation of 1 pmol of [^{3}H]dTMP into acid-insoluble material in 1 min at 37°.
[d] Reprinted with permission from reference 12.
[18a] M. M. Bradford, *Anal. Biochem.* **72,** 248 (1976).

sulfate precipitate is collected by centrifugation in a GSA rotor at 11,800 rpm for 40 min. The supernatant is discarded. The pellet is drained while maintaining the temperature at about 4°. Buffer PC contains 50 m*M* Tris–HCl (pH 7.5), 15% glycerol, 1 m*M* EDTA, 5 m*M* DTT. A volume of buffer, PC/25, consisting of 50 m*M* Tris–HCl (pH 7.5), 15% glycerol, 1 m*M* EDTA, 5 m*M* DTT, 25 m*M* NaCl, equal to one-fifth to one-tenth of the volume of fraction I is added to the ammonium sulfate pellet to redissolve protein and create fraction II. About 6 g of fraction II protein is usually obtained when starting from 300 g of dry cells. Fraction II is dialyzed against PC/25 buffer until the conductivity reaches a value equivalent to about 40 m*M* NaCl, approximately 90 μS. After dialysis, fraction II is diluted with PC/25 buffer to a protein concentration of approximately 10 mg/ml. The conductivity should be equivalent to that of 30 to 40 m*M* NaCl, approximately 80 to 90 μS, before loading onto a phosphocellulose column.

Phosphocellulose Chromatography

Whatman cellulose phosphate ion-exchange resin P11 is used. At least a twofold excess resin is used based on the calculated capacity. P11 resin is equilibrated with buffer PC/25. The resin (800 ml) is decanted into a 5-cm i.d. × 70-cm-long Econo chromatography column (Bio-Rad, Hercules, CA) and equilibrated in buffer PC/25 at a flow rate of 2.3 ml/min. If fraction

II is turbid, it can be clarified by centrifugation in a SS-34 rotor at 16,000 rpm for 40 min before loading on the phosphocellulose column. Fraction II is loaded onto the phosphocellulose column at a flow rate of 1 ml/min or less (loading by gravity flow may be too rapid, leading to the appearance of Pol II in the column wash). The column is washed with 1 column volume of buffer PC/25 (flow rate of 2.3 ml/min). An additional column volume of buffer PC/200 (the same components as buffer PC/25 except that the NaCl concentration is 200 m*M*) is applied to the column to elute DNA polymerase III. Pol II protein is eluted with an eight-column volume gradient of 200 to 500 m*M* NaCl in buffer PC. The Pol II fractions (usually eluting at 225 to 250 m*M* NaCl) are pooled to give fraction III. Fraction III is dialyzed against buffer PK20 [20 m*M* potassium phosphate (pH 6.8), 15% glycerol, 1 m*M* EDTA, 5 m*M* DTT] until the conductivity reaches that of PK30 buffer (30 m*M* potassium phosphate), approximately 80 μS, and the pH is 6.8, before loading on the DEAE column.

A batch adsorption technique can be used as an alternative method of binding fraction II to P11. Fraction II is diluted with buffer PC/25 until the conductivity reaches that of 40 m*M* NaCl, approximately 90 μS, and is then mixed with P11 resin preequilibrated with buffer PC/25. The resin and fraction II mixture are stirred very slowly and gently for 2 hr at 4°. The resin is allowed to settle and the supernatant is discarded. An equal volume of buffer PC/25 is added to the settled resin and the mixture is poured into the column. The rest of the purification procedure is the same as described earlier except the slow loading step is omitted. This batch adsorption technique serves as a rapid way to separate most of the unbound proteins and other possible contaminants from proteins that bind to the P11 resin. Batch adsorption can also be used in the next purification step for the loading of fraction III onto DEAE cellulose.

DEAE (Diethylaminoethylcellulose) Chromatography

Whatman ion-exchange cellulose DE52 resin is used in the purification. The resin (100 ml of preswollen resin is used per 50 mg protein) is equilibrated with PK20 and decanted into a 5-cm i.d. × 70-cm-long Econo chromatography column. Fraction III is loaded onto the DEAE cellulose column at a flow rate of 1 ml/ml or less. The DEAE column is washed with 2 column volumes of PK20 followed by elution with an 8 column volume gradient of 20 to 350 m*M* potassium phosphate (PK20 to PK350). The flow rates are 2.3 ml/min. The Pol II fractions (typically eluting at 100 to 140 m*M* potassium phosphate) are pooled as fraction IV and stored at −70°.

The specific activity and recovery of Pol II following each purification step is given in Table I, and a silver-stained gel showing protein banding patterns and enrichment of Pol II during purification is shown in Fig. 1.

Purification of Pol II from Exonuclease-Deficient Mutant (D155A/E157A)[12]

The procedure used to purify the exonuclease-deficient Pol II mutant is the same used for wild-type Pol II. The specific activity and recovery of polymerase activity of the *exo*⁻ mutant of Pol II at each purification step is given in Table I; the protein bands present in each enzyme fraction are shown in Fig. 1. The specific activity of wild-type Pol II exonuclease is about 1×10^6 units/mg. When assayed at equal polymerase levels, there appears to be at least a 1000-fold reduction in exonuclease specific activity for the D155A/E157A mutant compared to wild type. Data showing degradation of a 5′-^{32}P-labeled single-stranded oligonucleotide, with increasing incubation periods, illustrates the large difference in the exonuclease activi-

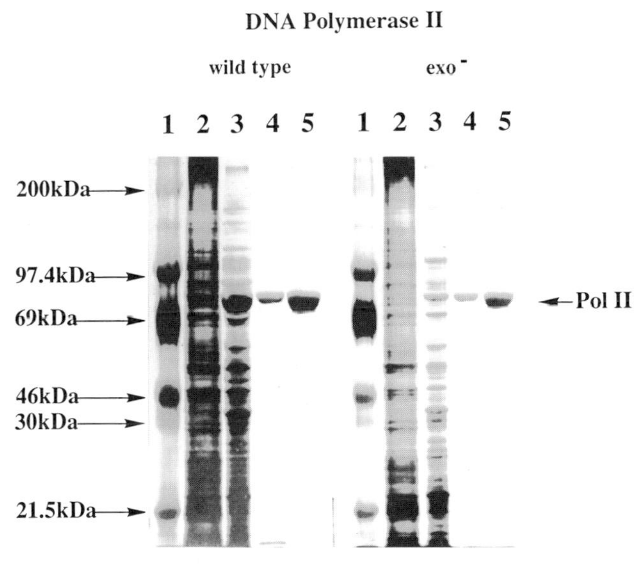

FIG. 1. Silver-stained polyacrylamide gel showing protein bands during purification of *E. coli* wild-type and exonuclease-deficient (*exo*⁻) DNA polymerase II. Lane 1, prestained molecular weight markers; lane 2, crude lysate; lane 3, ammonium sulfate fraction; lane 4, phosphocellulose fraction; lane 5, DEAE cellulose fraction. The purification procedure is described in the section on Purification of *E. coli* DNA Polymerase II. The specific activity and recovery at each stage of purification for the wild-type and exonuclease-deficient polymerases are given in Table I. [Reprinted with permission from reference 12.]

ties of wild-type and exonuclease-deficient Pol II (Fig. 2). Because the mutant protein was expressed in a background strain (JM109) containing a wild-type *polB* gene, the 1000-fold reduction represents a maximum estimate of the residual exonuclease activity contained in the mutant Pol II. We have constructed a *polB* null mutant strain that can be used to purify D155A/E157A and to obtain a more precise estimate of exonuclease activity present in the exonuclease-deficient enzyme.

Purity and Recovery of Wild-Type and Exonuclease-Deficient Pol II

The wild-type Pol II and exonuclease-deficient Pol II mutant behave similarly during purification. Starting from overproducing plasmids, the increase in specific activities are 36-fold and 27-fold for the wild-type and exonuclease-deficient enzymes, respectively, with overall recoveries of roughly 75% for both enzymes (Table I). Based on the absence of significant contaminating protein bands on silver-stained gels, the enzymes following the DEAE step are greater than 95% pure (lane 5, Fig. 1). Significantly,

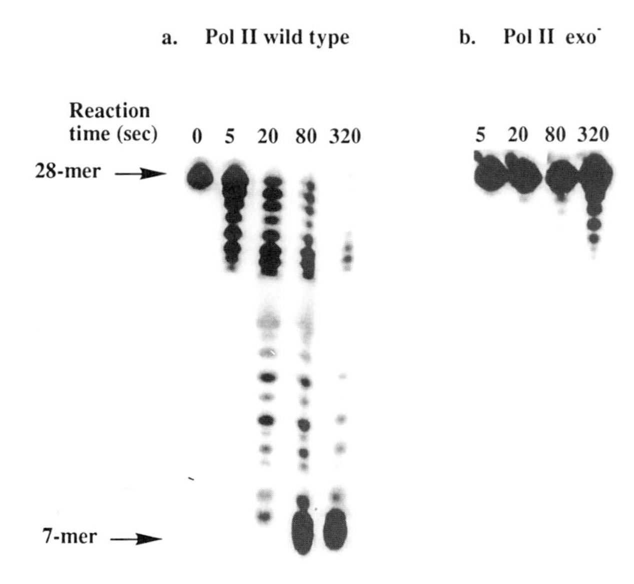

FIG. 2. Comparison of 3′ → 5′-exonuclease activities of wild-type and an exonuclease-deficient Pol II mutant (D155A/E157A). The DNA substrate is a synthetic oligonucleotide (28-mer, at a concentration of 180 n*M*) containing a ^{32}P-labeled 5′ terminus. (A) Wild-type Pol II (0.4 polymerase units) and (B) an exonuclease-deficient Pol II (4 polymerase units) are present in the reaction mix (described in the section on Exonuclease Activity Assay) for the reaction times indicated. The rate of removal of a 3′-terminal nucleotide by wild-type Pol II is about 1 to 2 nucleotides/sec. The specific activity for insertion catalyzed by wild-type Pol II is about 3.5-fold higher than Pol II exo$^-$ (Table I); the specific activity for the exonuclease is at least 1000-fold higher for the wild-type enzyme. [Reprinted with permission from reference 12.]

this degree of purification is suitable for obtaining high-quality crystals for structural analysis by X-ray diffraction.[14] A complete X-ray data set has been obtained for wild-type Pol II having a resolution of 2.8 Å, and a partial X-ray data set has also been obtained using the exonuclease-deficient mutant.

Based on active site titration measurements,[12] a minimum estimate of the fraction of active wild-type and exonuclease-deficient Pol II is 50%. There was no detectable loss in wild-type Pol II polymerase or exonuclease activities and in Pol II exopolymerase activity following storage at −70° for at least six months.

Plasmid Constructions

Construction of Pol II Overproducing Plasmid (pHY400)

A 2.4-kb DNA fragment containing the *polB* open reading frame was obtained from plasmid pSH100 by PCR (polymerase chain reaction) amplification of the *polB* coding region.[4] The PCR product was flanked by *Eco*RI and *Hin*dIII restriction sites, and the original "inefficient" GTG translation initiation codon was changed to ATG using an altered PCR primer. This 2.4-kb PCR fragment was inserted into *Eco*RI/*Hin*dIII sites of pPROK-1 vector (a 4.6-kb plasmid vector containing a Ptac promoter, from CLONTECH) to give a 7.0-kb plasmid construct, a pHY400. The sequence of *polB* was confirmed by DNA sequence analysis. The expression of *polB* is under the control of Ptac promoter, which is regulated by $LacI^q$.

Construction of Pol II 3′ → 5′-Exonuclease Mutant (D155A/E157A) *Overproducing Plasmid* (pHC700)

The *E. coli* DNA polymerase II gene containing substitutions D155A/E157A was engineered using standard oligonucleotide-directed mutagenesis procedures of the cloned *Eco*RI/*Hin*dIII fragment from pHY400.[19] Mutations in the plasmid were screened initially by restriction endonuclease mapping (the mutant oligonucleotide encoding the alanine substitution introduced a new restriction site for *AflII* endonuclease) and later by DNA sequencing of the *polB* gene. A 2.4-kb fragment containing the *polB* open reading frame with the desired mutations was inserted into the pPROK-1 vector (the same plasmid vector used for wild-type Pol II) resulting in a 7.0-kb plasmid, pHC700.

Acknowledgment

This research was supported by National Institutes of Health Grants GM42554, GM21422, and GM29558.

[19] T. A. Kunkel, K. Bebenek, and J. McClary, *Methods Enzymol.* **204,** 125 (1991).

[3] Purification of *Escherichia coli* DNA Polymerase III Holoenzyme

By MILLARD G. CULL and CHARLES S. MCHENRY

Introduction

The DNA polymerase III holoenzyme is the replicative polymerase of *Escherichia coli* and is responsible for synthesis of the majority of the chromosome.[1] The Pol III holoenzyme contains a core polymerase plus auxiliary subunits that confer its unique replicative properties, including a rapid elongation rate, high processivity, the ability to utilize a long single-stranded template coated with the single-stranded DNA binding protein, resistance to physiological levels of salt, the ability to interact with other proteins of the replicative apparatus, and the ability to coordinate the reaction through an asymmetric dimeric structure. All of these properties are critical to its unique functions. Many of these features appear to be conserved between bacterial and mammalian systems, suggesting that insight gained through studies with the Pol III holoenzyme may generalize to a variety of life forms. The replicative role of the enzyme has been established both by biochemical and genetic criteria.[2-6] Holoenzyme was biochemically defined and purified using natural chromosomal assays. Only the holoenzyme form of DNA polymerase III efficiently replicates single-stranded bacteriophages *in vitro* in the presence of other known replicative proteins,[7-9] and only the holoenzyme functions in the replication of bacteriophage λ, plasmids, and molecules containing the *E. coli* replicative origin, *oriC*.[10-13] The holo-

[1] C. S. McHenry, *Ann. Rev. Biochem.* **57,** 519 (1988).

[2] M. L. Gefter, Y. Hirota, T. Kornberg, J. A. Wechsler, and C. Barnoux, *Proc. Natl. Acad. Sci. USA* **68,** 3150 (1971).

[3] Y. Sakakibara and T. Mizukami, *Mol. Gen. Genet.* **178,** 541 (1980).

[4] H. Chu, M. M. Malone, W. G. Haldenwang, and J. R. Walker, *J. Bacteriol.* **132,** 151 (1977).

[5] W. G. Haldenwang and J. R. Walker, *J. Virol.* **22,** 23 (1977).

[6] T. Horiuchi, H. Maki, and M. Sekiguchi, *Mol. Gen. Genet.* **163,** 277 (1978).

[7] W. Wickner and A. Kornberg, *Proc. Natl. Acad. Sci. USA* **70,** 3679 (1973).

[8] J. Hurwitz and S. Wickner, *Proc. Natl. Acad. Sci. USA* **71,** 6 (1974).

[9] C. S. McHenry and A. Kornberg, *J. Biol. Chem.* **252,** 6478 (1977).

[10] K. Mensa-Wilmot, R. Seaby, C. Alfano, M. S. Wold, B. Gomes, and R. McMacken, *J. Biol. Chem.* **264,** 2853 (1989).

[11] J. S. Minden and K. J. Marians, *J. Biol. Chem.* **260,** 9316 (1985).

[12] E. Lanka, E. Scherzinger, E. Guenther, and H. Schuster, *Proc. Natl. Acad. Sci. USA* **76,** 3632 (1979).

[13] J. M. Kaguni and A. Kornberg, *Cell* **38,** 183 (1984).

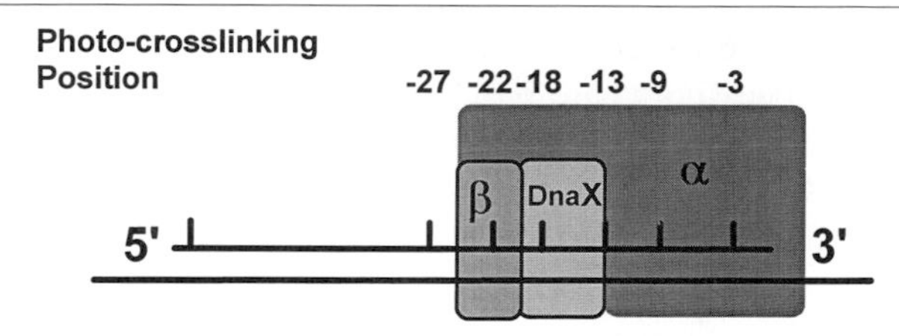

FIG. 1. The linear alignment of *E. coli* DNA polymerase III holoenzyme subunits relative to the duplex region of the primer–template at the initiation site. Schematic representation of the holoenzyme subunit–DNA contacts defined by site-specific photo-cross-linking. Within the initiation complex, α contacts roughly the first 13 nucleotides upstream of the 3′-primer terminus followed by DnaX protein at −18 and β at −22. DnaX remains part of the initiation complex connecting the α and β subunits.

enzyme appears to contain 10 subunits: α, τ, γ, β, δ, δ', ε, χ, ψ, and θ of 129, 900; 71,000; 47,400; 40,600; 38,700; 37,000; 26,900; 16,600; 15,000 and 8,800 Da, respectively.

Tripartite Structure of DNA Polymerase III Holoenzyme

The Pol III holoenzyme is composed of three subassemblies that function to create a processive enzyme. (1) The polymerase core is composed of the polymerase subunit α, the proofreading exonuclease ε, and the θ subunit of unknown function. (2) A sliding clamp, β, is required for the holoenzyme to be highly processive. X-ray crystallography[14] has revealed a bracelet-like structure for the β dimer, permitting it to slide rapidly down the DNA that it presumably encircles but preventing it from readily dissociating. Protein protein contacts between β and other components of the replicative complex tether the polymerase to the DNA, increasing its processivity. (3) A five-protein DnaX complex recognizes primer termini and closes the β bracelet around DNA. This complex remains firmly associated as part of the elongation complex between the α subunit and β. Presumably, α contacts β at a point away from the DNA[15] (Fig. 1). Key subunits of all three subassemblies contact the primer in the order α, DnaX protein, and β, starting from the primer terminus (Fig. 1).

The *dnaX* gene of *E. coli* encodes two protein products, τ and γ.[16,17] Both proteins contain a consensus ATP binding site near their amino

[14] X. P. Kong, R. Onrust, M. O'Donnell, and J. Kuriyan, *Cell* **69,** 425 (1992).
[15] J. Reems, S. Wood and C. McHenry, *J. Biol. Chem.* **270,** 5606 (1995).
[16] M. Kodaira, S. B. Biswas, and A. Kornberg, *Mol. Gen. Genet.* **192,** 80 (1983).
[17] D. A. Mullin, C. L. Woldringh, J. M. Henson, and J. R. Walker, *Mol. Gen. Genet.* **192,** 73 (1983).

terminus[18] that is used to bind and hydrolyze ATP, in concert with δ-δ'-χ-ψ, setting the β processivity clamp on the primer terminus. Protein γ arises by translational frameshifting, generating a 47,400-Da protein with sequences nearly identical to the amino-terminal two-thirds of τ.[19–22] In addition to these interactions, τ, but not γ, can bind tightly to the DNA Pol III core, causing it to dimerize.[23] It has been proposed that the two DnaX proteins might assemble asymmetrically, forming an asymmetric dimeric enzyme with distinct leading and lagging strand polymerases. The advantages of such an arrangement have been discussed.[1]

Structure of DNA Polymerase III Holoenzyme

Insight into the structure of the DNA Pol III holoenzyme has been continually evolving. Major questions relating to the placement of the γ subunit and its associated proteins within the complex and the proposed asymmetric placement of τ relative to γ remain to be resolved. However, it is clear that the τ subunit has function in addition to dimerization of the polymerase. Like γ, it serves to bind δ, δ', χ, and ψ, and in concert with these proteins, to load β onto primers to form initiation complexes that are competent for elongation.[24] Our working model for the structure of holoenzyme is shown in Fig. 2.

Methods

Cell Growth

Because it has not been possible to resolve DNA helicase II (*uvrD* gene product) chromatographically and in this way to produce DNA polymerase III holoenzyme free of this contaminant, we use an *E. coli* strain MGC1020 deleted in *uvrD* by insertion of a kanamycin-resistance cassette. *E. coli* K12 strain MGC1020 (W3110 *lexA3*, *malE*::*Tn10*, *uvrD*::*Kn*) was constructed by P1 transduction of W3110 [obtained from the American Type Culture Collection (ATCC)] to *lexA3* from phage grown on strain GW2727 (constructed in the laboratory of Graham Walker) followed by P1 transduc-

[18] K. C. Yin, A. Blinkowa, and J. R. Walker, *Nucleic Acids Res.* **14,** 6541 (1986).

[19] C. S. McHenry, M. Griep, H. Tomasiewicz, and M. Bradley, *in* "Molecular Mechanisms in DNA Replication and Recombination" (C. Richardson and I. R. Lehman, eds.), pp. 115–126, Alan R. Liss, Inc., New York, 1989.

[20] Z. Tsuchihashi and A. Kornberg, *Proc. Natl. Acad. Sci. USA* **87,** 2516 (1990).

[21] A. L. Blinkowa and J. R. Walker, *Nucleic Acids Res.* **18,** 1725 (1990).

[22] A. M. Flower and C. S. McHenry, *Proc. Natl. Acad. Sci. USA* **87,** 3713 (1990).

[23] C. S. McHenry, *J. Biol. Chem.* **257,** 2657 (1982).

[24] H. G. Dallmann and C. S. McHenry, manuscript submitted to *J. Biol. Chem.* (1995).

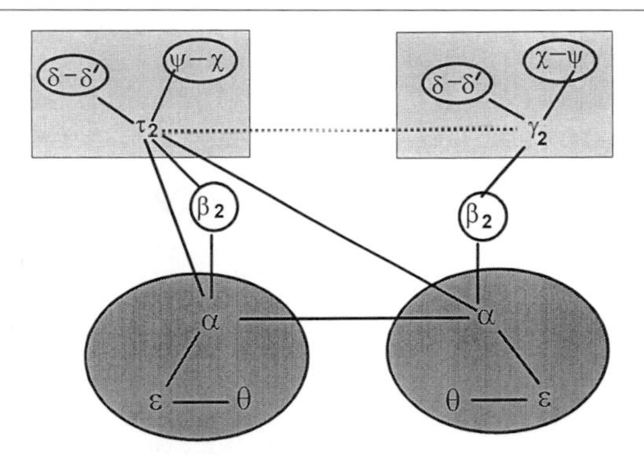

FIG. 2. Structural features of the DNA polymerase III holoenzyme. Interactions known with certainty are shown by solid lines. Lines designating direct subunit contacts extend to the specific subunit involved. Interactions between complexes extend only to the ellipse, designating an isolatable complex. Holoenzyme can be reconstituted free of γ without the loss of any detectable functions, yet native holoenzyme contains γ. Its attachment site with holoenzyme is uncertain, but it may be linked through the β subunit with which it, like τ, interacts. τ binds tightly to α, whereas γ does not. The dotted line indicates observed weak γ–τ interactions.

tion of the *uvrD* deletion from phage grown on strain SK6776.[25] MGC1020 is grown in a fermentor in F media at 37° to mid- to late-log phase (OD_{600} = 6). F medium is composed of yeast extract (14 g/liter), tryptone (8 g/liter), K_2HPO_4 (12 g/liter), and KH_2PO_4 (1.2 g/liter), pH 7.2. Glucose is added to 1% at the beginning of the fermentation, and another 1% glucose is added at an OD_{600} = 1.

Rapidly harvest cells and concomitantly chill by passing the fermentation broth through cooling coils en route to a Sharples continuous flow centrifuge. Temperature of the effluent from the flow-through centrifuge should not exceed 16°. Resuspend cells with an equal part (w/v) cold (4°) Tris–sucrose buffer (50 m*M* Tris–HCl, pH 7.5, 10% sucrose) and pour into liquid nitrogen in a stream to give a product having a "popcorn" appearance. The time from beginning harvest to popcorn should be less than 1 hr.

Assays

Stock Solutions

Enzyme dilution buffer: 50 m*M* HEPES, pH 7.5, 20% glycerol, 0.02% Nonidet P-40 (NP-40), 0.2 mg/ml bovine serum albumin (BSA), 100 m*M* potassium glutamate. Make stock without dithiothreitol (DTT) and store at −20°. Add freshly made DDT to 5 m*M* just prior to use.

[25] B. K. Washburn and S. R. Kushner, *J. Bact.* **178,** 2569 (1991).

10% Trichloroacetic acid (TCA)
250 m*M* Magnesium acetate
0.2 *M* Sodium pyrophosphate + 1 *N* HCl
0.2 *M* Sodium pyrophosphate
dNTP cocktail: 400 μM dATP, dGTP, dCTP; 150 μM [H^3] dTTP (100 cpm/pmol)
rNTP cocktail: 5 m*M* ATP, GTP, CTP, UTP
1 mg/ml Rifampicin (to inhibit RNA polymerase in assays of impure enzyme only).

The polymerase assay measures DNA synthesis at 30° from a primed M13Gori template as acid-precipitated product on GF/C filters (Millipore, Cat No. 1822 024). M13Gori DNA, single-stranded binding protein (SSB), and dnaG primase (listed below) are obtained from ENZYCO, Inc., Denver, CO. Each 25-μl reaction contains, in order of addition:

Assay component	*μl*	*Final concentration*
Enzyme dilution buffer (EDB)	14	Approximately 28 m*M* HEPES, pH 7.5, 11.2% glycerol, 0.01% NP-40, 0.1 mg/ml BSA, 56 μM potassium glutamate, 0.3 m*M* DTT
250 m*M* Magnesium acetate	1	10 m*M*
M13Gori DNA (OD_{260} = 2)	2	500 picomoles total nucleotide
SSB	2	1.6 μg
dNTP cocktail	3	48 m*M* dATP, dCTP, dGTP; 18 μM dTTP
rNTP cocktail	1	200 μM rNTPs
Rifampicin	0.2	0.2 μg
DnaG primase	0.2	55 units (100 ng)

When necessary, add EDB to bring final reaction volume to 25 μl.

Procedure. Initiate reactions by the addition of Pol III holoenzyme to the reaction mix at 0°. Transfer to a 30° bath. After 5 min, quench reactions by the addition of 2 drops 0.2 *M* sodium pyrophosphate and 0.5 ml 10% TCA acid, and filter through GF/C filters. Wash filters with 1 *M* HCl, 0.2 *M* sodium pyrophosphate. A final rinse should be made with ethanol. Dry filter and quantitate by scintillation counting. A unit is defined as 1 pmol of total deoxyribonucleotide incorporated/min.

Proteins are determined by Coomassie Plus Protein Assay Reagent (Pierce, Cat. No. 23236) according to the manufacturer's instructions.

Bovine plasma γ-globulin (Bio-Rad, Cat. No. 500-0005) is used as a standard.

Cell Lysis and Ammonium Sulfate Fractionation

Stock Solutions

Buffer T + 0.1 *M* NaCl: 50 m*M* Tris–HCl, pH 7.5, 20% glycerol, 1 m*M* EDTA, 0.1 *M* NaCl
Tris–sucrose: 50 m*M* Tris–HCl, pH 7.5, 10% sucrose (prewarm to 42°)
Lysis solution: 50 m*M* Tris–HCl, pH 7.5, 10% (w/v) sucrose, 2 *M* NaCl, 0.3 *M* spermidine hydrochloride. Adjust pH to 7.5 with 10 *N* NaOH
2.0 *M* Tris base
0.5 *M* DTT
0.20 Ammonium sulfate backwash: Add 20 g ammonium sulfate to every 100 ml Buffer T + 0.1 *M* NaCl
0.17 Ammonium sulfate backwash: Add 17 g ammonium sulfate to every 100 ml Buffer T + 0.1 *M* NaCl.

Procedure. A key feature to this procedure is the preparation of a DNA-free lysate, a requirement for the holoenzyme to bind to the Bio-Rex 70 column and for the enzyme to remain intact in a variety of manipulations. The holoenzyme is less soluble than most proteins in ammonium sulfate. The procedure we describe permits near-quantitative precipitation of holoenzyme, and removes contaminants by backwashing with decreasing concentrations of ammonium sulfate. This purification is based on 3.6 kg of cell paste (7.2 kg of "popcorn" [frozen 1:1 (w/v) suspension of cells]). The lysis step and ammonium sulfate fractionation are typically performed in four 900-g batches. Throughout the entire holoenzyme purification procedure, DTT from a 0.5 *M* stock is added to buffers just before they are needed in order to minimize oxidation. All imidazole hydrochloride and Tris–HCl buffers are prepared from 0.5 *M* stocks adjusted to the specified pH at 25°. No additional adjustments are made. Note that ammonium sulfate concentrations are reported as the amount of ammonium sulfate added *to* each ml of solution, not the amount added per each ml final volume.

1. Weigh out 1.8 kg of frozen "popcorn" into a large (~5-liter) plastic bucket.
2. Pour 2475 ml of prewarmed Tris–sucrose (42°) with stirring into the frozen popcorn. The temperature of the slurry should be monitored with a thermometer and should not be allowed to exceed 4°. Stir the slurry with an overhead stirrer; avoid foaming.
3. Add 45 ml of freshly prepared 0.5 *M* DTT.

4. Add 225 ml lysis solution.
5. Carefully adjust the pH to 8.0 with 2 *M* Tris base solution. Monitor the pH with narrow range pH indicator sticks (J. T. Baker Inc., Prod. No. 4406-01).
6. Continue moderate stirring until ice crystals have completely disappeared. Check the homogeneity of the mixture by turning off the stirrer (the ice crystals float).
7. Once a homogeneous mixture is achieved, add 0.9 g lysozyme freshly dissolved in 20 ml Tris–sucrose. Final concentration lysozyme = 0.2 mg/ml.
8. Mix thoroughly and immediately transfer to 250-ml centrifuge bottles. Leave on ice for 1 hr.
9. Swirl bottles in a 37° water bath for 4 min. Gently invert every 30 sec.
10. Return the bottles immediately to an ice bath.
11. Centrifuge at 23,000*g* for 1 hr at 4°.
12. Collect supernatant in a prechilled 4-liter cylinder; save a 2-ml sample for protein determinations, and record the volume. The supernatant is Fraction I. (Typical yield: 75 g protein in 3500 ml.)
13. Add 0.226 g ammonium sulfate to each ml of Fraction I slowly (over 30 min) while stirring with a magnetic stirbar. Record volume and remove 0.5 ml for assays.
14. Centrifuge at 23,000*g* for 30 min at 0°.
15. Using Dounce homogenizer (loose pestle), resuspend pellet in 0.125× Fraction I volume of 0.2 ammonium sulfate backwash solution + 5 m*M* DTT. Record volume and remove 2× 0.5-ml aliquots for assays.
16. Centrifuge at 23,000*g* for 45 min at 0°.
17. Using a Dounce homogenizer, resuspend pellet in 0.02× Fraction I volume of 0.17 ammonium sulfate backwash solution + 5 mm DTT. Measure volume and reserve 2× 0.5-ml aliquots to centrifuge separately and assay before dissolving the entire preparation. This will permit assessment of success of this stage of the purification and facilitate planning for subsequent steps.
18. Centrifuge at 35,000*g* for 30 min at 0°.
19. Pour off the supernatant, carefully seal the tube to prevent desiccation, and store the pellet (Fraction II) at −80°. (Typical yield for a 900 g prep: 0.5 g protein, 7.5×10^6 units.)

Cation-Exchange Chromatography on Bio-Rex 70

Stock Solutions

DMSO buffer: 50 m*M* imidazole hydrochloride, pH 6.8, 20% dimethyl sulfoxide (DMSO), 10% glycerol, 5 m*M* DTT

DMSO buffer + 0.1 *M* NaCl: 50 m*M* imidazole hydrochloride, pH 6.8, 0.1 *M* NaCl, 20% DMSO, 10% glycerol, 5 m*M* DTT

DMSO buffer + 0.2 *M* NaCl: 50 m*M* imidazole hydrochloride, pH 6.8, 0.2 *M* NaCl, 20% DMSO, 10% glycerol, 5 m*M* DTT

Bio-Rex 70 elution buffer: 50 m*M* imidazole hydrochloride, pH 6.8, 0.5 *M* NaCl, 30% glycerol, 1 m*M* EDTA, 5 m*M* DTT.

Procedure. Cation exchange provides a powerful chromatographic step for all forms of DNA Pol III. However, strong cation exchangers cause the β subunit to dissociate. The relatively weak but high-capacity cation exchanger Bio-Rex 70 permits the holoenzyme to remain intact. Its high capacity is also important since it permits the holoenzyme to be kept concentrated, minimizing losses due to its dilution sensitivity. Holoenzyme is most stable at relatively high concentrations of salt. The presence of DMSO in the loading buffers permits the holoenzyme to bind to the column at higher concentrations than is otherwise possible. The decreasing DMSO concentrations in the gradient permit elution of the enzyme.

1. Pour a Bio-Rex 70 (100 to 200 mesh) column and equilibrate it with DMSO buffer without DTT. After column equilibration run at least two more column volumes of DMSO buffer + 5 m*M* DTT through the column. The correct column size can be estimated by using 1 ml resin for every 20 mg protein in Fraction II. Bio-Rex used for the first time should be washed in both acid and base before use. Batch wash the Bio-Rex with 0.5 *M* imidazole hydrochloride, pH 6.8, before pouring the column. Used Bio-Rex can be recycled by washing with 2 *M* NaCl. Used resin gives higher yields. Bio-Rex 70 has an exceedingly high capacity and takes a long time to equilibrate. Carefully check the pH and conductivity of the column effluent.
2. Thaw Fraction II pellets on ice, centrifuge (34,800*g*, 0°, 10 min) and remove any remaining supernatant. Dissolve Fraction II pellets in ice-cold DMSO buffer + 0.1 *M* NaCl + 5 m*M* DTT. Use 1 ml of buffer for every 25 g of cells to dissolve the Fraction II pellets (144 ml for 3.6 kg preparation). Dounce-homogenize the resuspended pellets to achieve a homogeneous mixture. Save a 0.5-ml aliquot for assays and conductivity determination.
3. Centrifuge suspension (34,800*g*, 0°, 1 hr) to clarify. During centrifugation, test the conductivity of the sample from the previous step. If the sample conductivity is not in the range of the conductivity of DMSO buffer + 0.1 *M* NaCl to DMSO buffer + 0.2 *M* NaCl, dilute the clarified sample with DMSO buffer + 5 m*M* DTT to bring it within this range. Typically this dilution requires 2 volumes of DMSO buffer. (Note: Unless stated otherwise, all conductivities in this preparation are determined on 1/100 dilutions.)

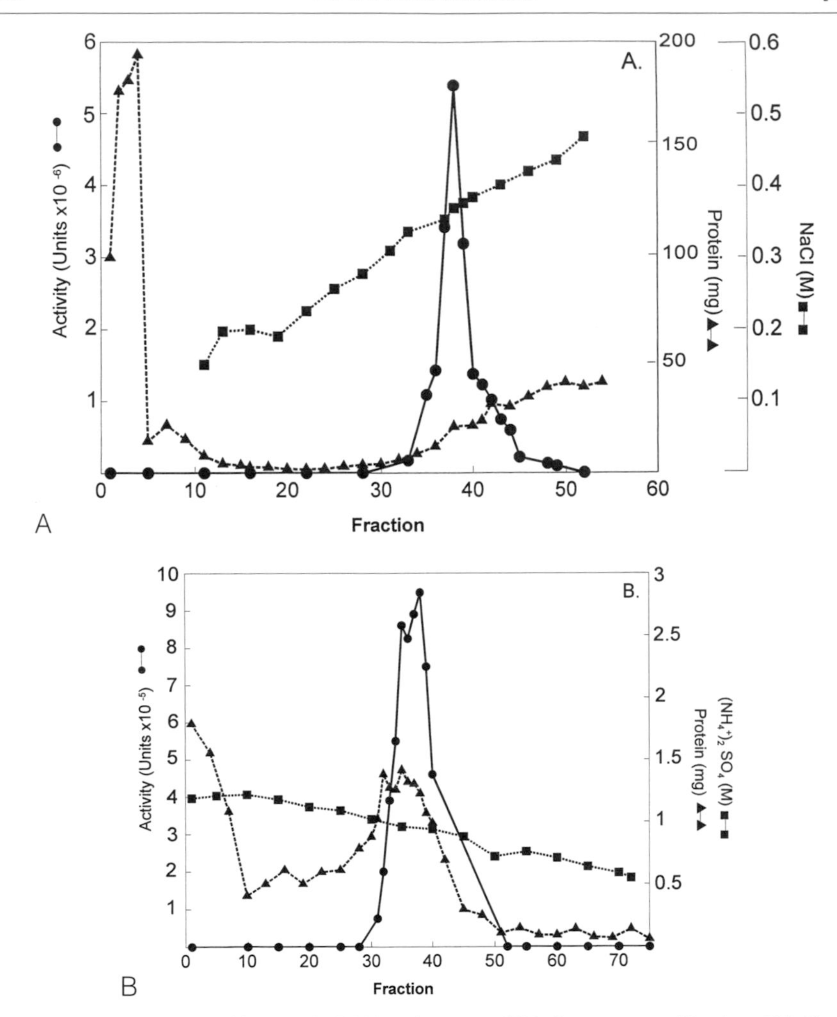

FIG. 3. Chromatographic steps in DNA polymerase III holoenzyme purification. (A) Chromatography on Bio-Rex. Flow-through fractions 1–5 were 100 ml each. Fractions 6–52 were 8 ml each. (B) Chromatography on valyl-Sepharose. Fractions were 12.5 ml each. (C) Chromatography on DEAE-Sephadex. Fractions were 1 ml each.

4. Apply the clarified Fraction II to the Bio-Rex 70 column at a rate of 2 column volumes/hr. Wash the column with 1 column volume of DMSO buffer + 0.2 *M* NaCl. The majority of the contaminating protein flows through the column (Fig. 3A).

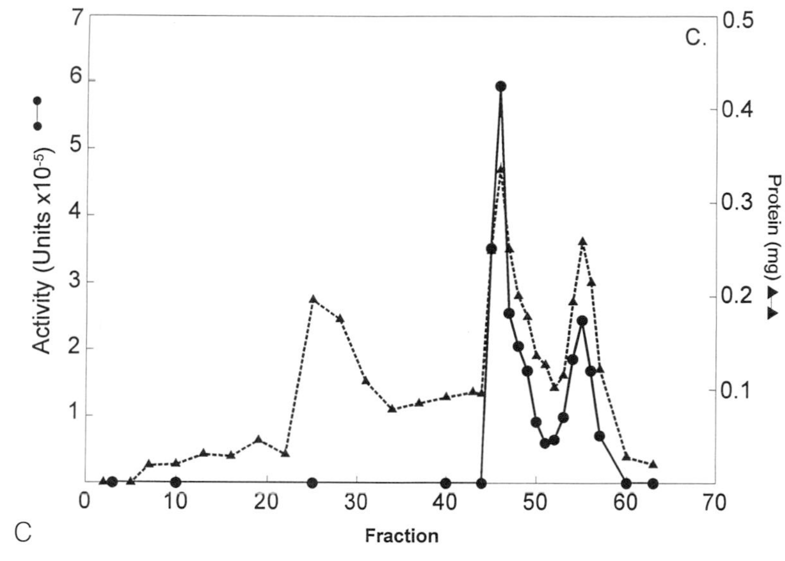

FIG. 3. (*continued*)

5. Elute the enzyme with a 4 column volume gradient at a flow rate of 2 column volumes/hr. Gradient starting buffer: DMSO buffer + 0.2 *M* NaCl. Collect ~60 fractions, save 100 μl of each fraction for assays, and immediately add an equal volume of saturated ammonium sulfate to the fractions. The enzyme is unstable during this chromatographic step, but becomes stable once ammonium sulfate is added.
6. Assay samples for activity and combine all fractions that contain at least 50% of the activity of the peak fraction. Centrifuge the ammonium sulfate precipitates repeatedly (34,800*g*, 0°, 1 hr) in two 34-ml centrifuge tubes, so that pellets can be dissolved in a minimal volume in the next step. Store pellets (Fraction III) on ice. Record volume of pooled samples, save a sample for assays, and centrifuge it separately [Typical yield: 45 mg protein, 1.3 × 10^7 units (Table I).]

Hydrophobic Interaction Chromatography on Valyl-Sepharose

Hydrophobic interaction chromatography permits purification of holoenzyme in high concentrations of stabilizing salt. This step was developed before the commercial availability of hydrophobic resins. We have not tested substitutes and continue to make our own resin. Preparation is relatively convenient, inexpensive, and does not need to be performed often since the resin can be recycled if properly handled.

TABLE I
PURIFICATION OF DNA POLYMERASE III HOLOENZYME[a]

Fraction	Units ($\times 10^{-6}$)	Protein (mg)	Volume (ml)	Specific activity
II	30	1860	390	16,100
III	13	45	85	290,000
IV	3.2	5.8	0.4	550,000
V	1.7	1.5	6.0	1,160,000

[a] 3.4 kg cells.

Stock Solutions

Buffer I + 0.1 *M* NaCl: 50 m*M* imidazole hydrochloride, pH 6.8, 0.1 *M* NaCl, 20% glycerol, 1 m*M* EDTA, 5 m*M* DTT

2.0 *M* Ammonium sulfate buffer: 50 m*M* imidazole hydrochloride, pH 6.8, 10% glycerol, 2.0 *M* ammonium sulfate, 5 m*M* DTT

1.2 *M* Ammonium sulfate buffer: 50 m*M* imidazole hydrochloride, pH 6.8, 10% glycerol, 1.2 *M* ammonium sulfate, 5 m*M* DTT

0.4 *M* Ammonium sulfate buffer: 50 m*M* imidazole hydrochloride, pH 6.8, 20% glycerol, 0.4 *M* ammonium sulfate, 5 m*M* DTT.

Resin Preparation. Valyl-Sepharose is prepared by cyanogen bromide activation of Sepharose 4B as previously described with minor modifications[26]: After activation of the beads, an equal volume of 0.2 *M* L-valine in 0.2 *M* $NaHCO_3$ is added to the beads. This slurry is allowed to incubate with shaking at 4° for 20 hr. After coupling, the beads are washed with 20 volumes each 0.1 *M* sodium acetate plus 0.5 *M* NaCl (pH to 4 with glacial acetic acid), followed by 0.1 *M* $NaHCO_3$, pH 9.5, wash, and a final wash with 0.5 *M* NaCl. The resin is stored in 0.5 *M* NaCl. The resin has about 12 μmol valine/ml as determined by acid hydrolysis, filtration, and amino acid analysis.

Procedure

1. Pour a valyl-Sepharose column and equilibrate it in 1.2 *M* ammonium sulfate buffer. The correct column size can be estimated by using 1 ml resin for every 0.78 mg protein present in Fraction III.
2. Dissolve Fraction III pellets in Buffer I + 0.1 *M* NaCl. Bring the final concentration to 1 mg/ml.
3. Separately equilibrate 10% of the volume of the valyl-Sepharose column and add the dissolved Fraction III to it (first record volume

[26] S. C. March, I. Parikh, and P. Cuatracasas, *Anal. Biochem.* **60,** 149 (1974).

and save a sample for assays). Slowly (over 15 min) add 1.5× the volume of the redissolved Fraction III of 2 *M* ammonium sulfate buffer with vortexing. This step will allow a uniform coating of protein on the beads rather than precipitation of the proteins onto the top of the column. Ammonium sulfate must be added slowly, especially the final amount, to permit uniform coating of the beads.

4. Apply the coated beads and the accompanying solution to the column. Wash the column with 1 column volume of 1.2 *M* ammonium sulfate buffer.
5. Run a 10 column volume gradient at 0.8 column volume/hr. Gradient starting buffer: 1.2 *M* ammonium sulfate buffer. Eluting buffer: 0.4 *M* ammonium sulfate buffer. Collect 60 fractions and assay the fractions for activity. Holoenzyme elutes approximately halfway through the gradient (Fig. 3B).
6. Pool fractions that have at least 50% of the activity of the peak tube and, after recording the volume and saving an aliquot for assays, add 0.262 g ammonium sulfate/ml of pooled fractions and stir overnight on ice. Centrifuge in one 34-ml tube repeatedly to permit resuspension of the resulting pellets in a small volume. This is necessary to ensure that a concentrated sample can be applied to the DEAE-Sephadex column in the next step. Store the pellets (Fraction IV) on ice. [Typical yield: 5.8 mg protein, 3.2×10^6 units (Table I).]

Ion-Filtration Chromatography on DEAE-Sephadex

Poor yields result when holoenzyme is diluted or dialyzed and then bound to standard anion-exchange columns, presumably due to its sensitivity to dilution and low salt. The ion-filtration technique[27] improves the yields obtained for holoenzyme severalfold. The method was developed by determining a salt concentration that permitted interaction and retardation of holoenzyme by the matrix without binding so tightly that it required higher salt for elution. Contaminants that do not interact with the column elute in about one-third column volume, just as in gel filtration for excluded proteins. Proteins that interact strongly elute in 1 column volume where the high salt wash elutes. Because of this delicate balance, buffers must be prepared precisely.

Stock Solutions

Buffer I + 120 m*M* NaCl: 50 m*M* imidazole hydrochloride, pH 6.8, 120 m*M* NaCl, 20% glycerol, 1 m*M* EDTA, 5 m*M* DTT

[27] L. Kirkegaard, T. Johnson, and R. M. Bock, *Anal. Biochem.* **50,** 122 (1972).

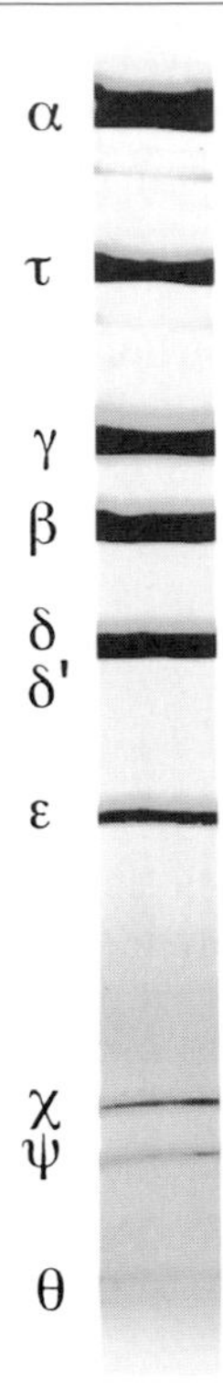

FIG. 4. Denaturing polyacrylamide gel electrophoresis (SDS–PAGE) of purified DNA polymerase III holoenzyme. The gel gradient is from 7.5 to 20% acrylamide. The identity of each subunit is indicated in the margin.

Buffer I + 210 m*M* NaCl: 50 m*M* imidazole hydrochloride, pH 6.8, 210 m*M* NaCl, 20% glycerol, 1 m*M* EDTA, 5 m*M* DTT.

Procedure

1. Pour a DEAE-Sephadex A-25, 40- to 120-μm bead size (Sigma, A-25-120) column and equilibrate it with buffer I + 120 m*M* NaCl. The correct column size can be estimated by using 1 ml resin for every 120 μg protein present in Fraction IV, typically 36 ml for 3.6 kg preparation. This step does not scale down well. As in gel filtration, this column should be long and narrow, with a height : diameter ratio of about 30 : 1.
2. Centrifuge the Fraction IV pellet at 34,800g, 0°, 10 min, and thoroughly remove the excess liquid. Redissolve the pellet in a minimal volume of Buffer I + 120 m*M* NaCl (approximately 7 × 10^6 units/

ml final concentration). This can best be accomplished by using successive small aliquots (approximately 75 μl) to resuspend the protein and transfer it to the dialysis device (step 3). The protein need not dissolve at this stage since it will dissolve during dialysis as the residual ammonium sulfate is removed.

3. In a microdialysis device (BRL), with BRL prepared dialysis membrane 12,000 to 14,000 Da exclusion limit, dialyze resuspended Fraction IV against Buffer I + 120 m*M* NaCl until the conductivity drops below that of Buffer I + 210 m*M* NaCl (typically, 6 to 10 hr if the dialysis device is continuously rocked gently). Conductivity measurements are made by adding 5 μl of sample to 1 ml of distilled water. The sample volume at the end of dialysis should be less than 1% of the column volume. Save a 10-μl aliquot of the dialyzed sample for activity and protein determinations.
4. Load the sample onto the column. After application, apply Buffer I + 210 m*M* NaCl at a flow rate of 1 column volume/18 hr. Collect about 60 fractions per column volume.
5. Assay the eluted enzyme and save fractions that have greater than 50% of the activity of the peak tube. Holoenzyme elutes at approximately 60% column volume (Fig. 3C). Occasionally, a second peak of holoenzyme activity elutes that has substoichiometric quantities of some subunits and is presumably damaged. Aliquot the individual fractions, freeze in liquid N_2, and store at $-80°$. Fractions are stable for about 1 week in an ice bucket after thawing. [Typical yield Fraction V: 1.5 mg protein, 1.7×10^6 units (Table I).] Densitometry of an SDS–polyacrylamide gel of Fraction V indicates that it is 98% pure (Fig. 4).

[4] Purification of DNA Polymerase III of Gram-Positive Bacteria

By MARJORIE H. BARNES and NEAL C. BROWN

Introduction

Bacteria possess three classes of DNA polymerase, designated I, II, and III; of these, DNA polymerase III (Pol III) is essential for replicative DNA synthesis. Pol IIIs from Gram-positive bacteria (Gr^+) constitute a unique phylogenetic class among DNA polymerases of known primary structure.[1] They show relatively little sequence homology to Gr^- Pol IIIs, and are further distinguished from them by their unique sensitivity to the "HPUra" [6-(*p*-hydroxy-phenylazo) uracil] class of inhibitory dNTP analogs.[2,3]

Gr^+ Pol IIIs also differ markedly from their Gr^- counterparts in that their editing ($3' \rightarrow 5'$)-exonuclease and polymerase functions reside in the same polypeptide chain.[4,5] It is likely that Gr^+ Pol IIIs, like Gr^- Pol IIIs, function *in vivo* as part of a multiprotein, holoenzyme structure. However, unlike the Gr^- enzymes, they display a strong tendency to purify as a single polypeptide and, accordingly, a Gr^+ Pol III holoenzyme has yet to be isolated.

Three Gr^+ Pol IIIs have been cloned and sequenced: those of *Bacillus subtilis,*[4,5] *Mycoplasma pulmonis,*[6] and *Staphylococcus aureus.*[7] All three display extensive sequence homology, particularly in their putative polymerase and exonuclease domains.[8] The following summarizes our experience and that of others in the complete purification of natural and recombinant forms of *B. subtilis* Pol III and the partial purification of enzyme from several other sources.

Native Pol IIIs

Bacillus subtilis Pol III, the class prototype, is the only Gr^+ enzyme to be purified to homogeneity.[9,10] Two methods of purification have been published and each is summarized, respectively in Tables I and II. The catalytic properties of the native enzyme are summarized in Table III. Space does not permit a detailed discussion of the procedures used in purification or enzyme analysis; therefore, the original articles should be consulted for details.

In addition to *B. subtilis* Pol III, the Pol IIIs of several other *Bacillus* strains (*B. licheniformis, B. megaterium,* and *B. pumilis*) and other Gr^+

[1] J. Ito and D. K. Braithwaite, *Nucl. Acids Res.* **19,** 4045 (1991).

[2] N. C. Brown, L. W. Dudycz, and G. W. Wright, *Drugs Exptl. Clin. Res. XII* **(6/7)**, 555 (1986).

[3] N. C. Brown and G. E. Wright, this volume [17].

[4] B. Sanjanwala and A. T. Ganesan, *Proc. Natl. Acad. Sci. USA* **86,** 4421 (1989).

[5] R. A. Hammond, M. H. Barnes, S. L. Mack, J. A. Mitchener, and N. C. Brown, *Gene* **98,** 29 (1991).

[6] M. H. Barnes, P. M. Tarantino Jr., P. Spacciapoli, N. C. Brown, H. Yu, and K. Dybvig, *Molec. Microbiol.* **13,** 843 (1994).

[7] D. Pacitti, this laboratory (1993).

[8] M. H. Barnes, R. A. Hammond, C. C. Kennedy, S. L. Mack, and N. C. Brown, *Gene* **111,** 43 (1992).

[9] R. L. Low, S. A. Rashbaum, and N. R. Cozzarelli, *J. Biol. Chem.* **251,** 1311 (1976).

[10] M. H. Barnes and N. C. Brown, *Nucl. Acids Res.* **6,** 1203 (1979).

TABLE I
PURIFICATION OF NATIVE *B. subtilis* POL III[a]

Fraction	Activity (units × 10^3)	Protein (mg)	Specific activity (units/mg)
1. High-speed supernatant	12.5	11,000	1.1
2. Streptomycin sulfate precipitation	9.7	2400	4
3. DEAE–cellulose I	7.6	810	9
4. Gel filtration	4.4	220	20
5. DEAE–cellulose II	3.9	59	66
6. Phosphocellulose	1.5	3.7	410
7. DNA–cellulose	0.32	0.06	5000
7′. Hydroxylapatite	1.05	0.52	2000

[a] Adapted from R. L. Low, S. A. Rashbaum, and N. R. Cozzarelli, *J. Biol. Chem.* **251,** 1311 (1976).

bacteria (*Enterococcus faecalis, Staphylococcus aureus, Micrococcus luteus,* and *Lactobacillus acidophilus*) have been purified as far as Fraction 5 in Table II. The behavior of each enzyme in this purification scheme was essentially equivalent to that of *B. subtilis* Pol III.[10]

Although the primary sequence of *M. pulmonis* Pol III has indicated considerable homology to the *B. subtilis* enzyme, it displays significantly different behavior during purification.[6] One of the major differences is its very low affinity for anion-exchange matrices.

TABLE II
PURIFICATION OF NATIVE *B. subtilis* POL III[a]

Fraction	Activity (units × 10^{-3})	Protein (mg)	Specific activity (units/mg)
1. Crude lysate	43	35,400	1.2
2. Streptomycin sulfate supernatant	25	19,800	1.2
3. Ammonium sulfate (55% saturated)	24	15,200	1.6
4. DEAE-cellulose (0.2 *M* ammonium sulfate)	29	14,100	2.1
5. DEAE-cellulose (low salt)	7.5	1600	4.7
6. DNA-cellulose	5	30	170
7. Bio-Rex 70	1.7	0.72	2400

[a] Adapted from M. H. Barnes and N. C. Brown, *Nucl. Acids Res.* **6,** 1203 (1979).

TABLE III
PROPERTIES OF *B. subtilis* DNA POL III[a]

Parameter	Units
Molecular weight	160–167 kDa
S value	7.6 S
Stokes radius	52 Å
pH optimum	7.4
Mg^{2+} requirement	>1 m*M*
K_m for dNTP	
dATP	0.9 μ*M*
dCTP	0.3 μ*M*
dGTP	0.6 μ*M*
dTTP	1.5 μ*M*
K_i for inhibitors	
H_2-HPUra[b]	0.4 μ*M*
TMAU[c]	0.4 μ*M*
DCBG[b]	0.5 μ*M*
DCBdGTP[b]	0.1 μ*M*

[a] Except as indicated by superscript, data are from R. L. Low, S. A. Rashbaum, and N. R. Cozzarelli, *J. Biol. Chem.* **251,** 1311 (1976).
[b] M. M. Butler, L. W. Dudycz, N. N. Khan, G. E. Wright, and N. C. Brown. *Nucl. Acids Res.* **18,** 7381 (1990).
[c] G. E. Wright and N. C. Brown, *J. Med. Chem.* **23,** 34 (1980).

Cloned Gene Products

The *polC* gene of *B. subtilis* has been cloned and engineered for overexpression in *E. coli.*[3,11] The availability of this construct has considerably simplified the purification of catalytically active Pol III. The following procedure, adapted from Ref. 12, is routinely used for its purification. Quantities given are for packed cells from a 2-liter culture. Column buffer consists of 50 m*M* potassium phosphate, pH 7.5, 1 m*M* dithiothreitol (DTT), and 20% (v/v) glycerol. All procedures are carried out at 4°.

Crude Extract. The frozen cell pellet is resuspended in 60 ml of column buffer + 1 m*M* phenylmethylsulfonyl fluoride (PMSF), and cells are ruptured in a French pressure cell at a pressure of 20,000 lb/in.^{2} The suspension is centrifuged at 27,000*g* for 45 min.

Cibacron Blue/Phenyl-Sepharose Chromatography. The cleared supernatant is applied directly to a 220-ml Cibacron Blue 3GA (Sigma) column

[11] R. A. Hammond and N. C. Brown, *Protein Expression and Purification* **3,** 65 (1992).

equilibrated in column buffer + 1 m*M* PMSF. The column is then washed with 2200 ml of the same buffer containing 0.5 *M* NaCl. The polymerase activity is then eluted with 2200 ml of column buffer + 3*M* NaCl, readjusted to pH 7.5 with KOH. Eluate is collected directly on a 24-ml Phenyl-Sepharose (Sigma) column connected in tandem to the Cibacron Blue column. After loading, the phenyl-Sepharose column is disconnected from the Cibacron Blue column and washed with (1) 50 ml of column buffer + 3 *M* NaCl, (2) 60 ml column buffer + 1 m*M* PMSF to remove salt, and (3) finally eluted with approximately 100 ml of the same buffer containing 25 m*M* taurodeoxycholate + 0.5% Triton X-100.

Mono Q FPLC Chromatography. The phenyl-Sepharose eluate is applied to an 8-ml Mono Q HR 10/10 (Pharmacia) column equilibrated in column buffer and connected to a Pharmacia FPLC system. The column is washed with 50 ml of column buffer and eluted with a 0 to 1 *M* NaCl linear gradient in column buffer. Total gradient volume is 250 ml; flow rate is 2 ml/min or less. DNA polymerase activity elutes as a sharp, symmetrical peak at 0.4 *M* NaCl. A summary of the purification scheme is shown in Table IV.

Properties of Cloned Gene Product

The catalytic properties of *B. subtilis* Pol III are essentially identical to those reported for the native enzyme in Table III. Purity is approximately 95%. Most of the peptide impurities react with a polyclonal antibody specific for native Pol III, suggesting that they are truncated versions of Pol III, formed either at the stage of transcription–translation of the cloned gene or by subsequent degradation of the complete gene product.

TABLE IV
PURIFICATION OF CLONED GENE PRODUCT[a]

Fraction	Total activity (units $\times 10^{-6}$)	Protein (mg)	Specific activity (units/mg $\times 10^{-3}$)	Yield (%)
I. Crude extract	1.2	693	1.7	100
II. Cibacron/phenyl-Sepharose	0.6	81	7.4	50
III. Mono Q FPLC	0.3	5.5	55	25

[a] Adapted from R. A. Hammond and N. C. Brown, *Protein Expression and Purification* **3,** 65 (1992).

General Considerations

In developing a purification scheme for a novel Pol III, either native or recombinant, at least three levels of purity should be considered. Accordingly, readers should decide what level will meet their needs.

The first level of purity is simply the crude cell extract freed of nucleic acids. This requires rupturing the cells in a manner that does not destroy the activity of the target enzyme. We have found disruption by use of a French pressure cell to be a simple, gentle, and reproducible method applicable to both large- and small-scale preparations. The resultant cell debris is easily removed by centrifugation, and nucleic acids can typically be removed either by precipitation with streptomycin sulfate or by passage through an anion-exchange column in high salt. Stability of the DNA polymerase activity is usually aided by storage in buffers containing salt at moderately high concentrations.

In the second level of purity, the preparation should be freed of the activity of other DNA polymerases. The presence of contaminating enzymes is conveniently indicated by activity that is insensitive to inhibition by the Pol III-specific HPUra class of inhibitors. In the purification schemes of Tables I and II, a Pol I-deficient (*polA*$^-$) strain was exploited to eliminate Pol I activity from the background, while Pol II was removed by DEAE–cellulose chromatography.

In purification of recombinant Pol III overexpressed in *E. coli,* endogenous DNA polymerases are usually a relatively minor contaminant. The background is easily estimated by comparison to an extract made from a comparable vector/host system in the absence of the cloned polymerase gene. If the primary experimental interest is, for example, comparison of catalytic properties of a group of specifically mutated DNA polymerases, a cleared cell supernatant may be an appropriate level of purity.

Third, physical analysis of the enzyme protein requires a virtually homogeneous preparation. Because Gr$^+$ DNA Pol IIIs are generally present at extremely low levels in their native hosts (we estimated a level of 0.0001% of total cellular protein for *B. subtilis*), the isolation of pure enzyme in significant amounts is a laborious and expensive process requiring fermenter-scale cell growth and liter-scale chromatography columns. If feasible, the most practical way to solve the problem of limited enzyme availability is to clone its structural gene and overexpress it as a recombinant protein.

Although the availability of an effective recombinant expression system may simplify production of pure enzyme, it is not without problems. Among the most common are the persistent presence of contaminating species, or production of catalytically inactive forms of the enzyme. The latter may result from improper folding, the presence of alternative start sites for

TABLE V
POLYMERASE AND EXONUCLEASE ASSAY MIXES

Polymerase assay mix[a]	Exonuclease assay mix[b]
18.75 m*M* Tris–HCl, pH 7.5	33.3 m*M* Tris–HCl, pH 7.5
12.5 m*M* Magnesium acetate	7.4 m*M* Magnesium chloride
31.25 μM dATP, dCTP, dGTP	3.3 m*M* DTT
12.5 μM [*methyl*-^{3}H]dTTP, 1.5 μCi/μmol	11.1% glycerol
1.25 m*M* DTT	^{3}H-labeled denatured DNA
20% glycerol	(0.05–0.2 μg/μl, ~70,000 cpm/assay)
0.5 mg/ml activated DNA	

[a] Polymerase assay mix can be stored for several months at −20° with no loss of activity.

[b] Exonuclease assay mix is made just before use, and contains ^{3}H-labeled DNA freshly denatured by boiling.

transcription, or proteolytic degradation. Folding problems are frequently ameliorated by selecting lower temperatures (i.e., less than 34°) for expression of the cloned gene product, while proteolysis can be minimized by use of either chemical inhibitors of proteolysis or protease-deficient host strains. Removal of undesirable subspecies of the recombinant gene product or unrelated proteins whose physical and chemical properties closely approximate those of the protein of interest is the most challenging barrier to the achievement of homogeneity. The solution to this problem generally will require an extensive period of trial-and-error survey of standard methods of protein purification.

Pol III Assays

For the convenience of the reader, we include below a brief description of our methods for assay of DNA polymerases. A more general discussion of polymerase assays may be found elsewhere.[12] Methods for assay of both polymerase and exonuclease activity are included.

Polymerase Assays. Five microliters of an appropriate dilution of enzyme is rapidly mixed with 20 μl of polymerase assay mix (Table V), and incubated at 30° for 10 min. Reactions are stopped by addition of 0.5 ml cold 10% trichloroacetic acid (TCA) : 10 m*M* sodium pyrophosphate. After approximately 10 min at 0°, samples are filtered on Whatman GF/A filters and washed, first with cold 1 *M* HCl : 100 m*M* sodium pyrophosphate, then with cold ethanol. Filters are dried and counted.

[12] K. B. Gass and N. R. Cozzarelli, *Meth. Enzymol.* **29,** 27 (1974).

Exonuclease Assays. Five microliters of an appropriate dilution of enzyme is quickly mixed with 45 μl of exonuclease assay mix (Table V), and incubated at 30° for 10 min. Reactions are stopped by addition of 0.5 ml 10% TCA:10 m*M* sodium pyrophosphate. Fifty microliters of a 10 mg/ml solution of bovine serum albumin is added as a coprecipitant. After approximately 10 min at 0°, samples are centrifuged at 15,000*g* for 20 min. Then 400 μl of the supernatant is removed and counted in 2 ml of an aqueous scintillant. The method given here is modified from Low *et al.*[9]

Acknowledgments

Much of the author's work reported in this article has been supported by a research grant from the USPHS–National Institutes of Health. NCB and MHB are supported by NIH grant GM45330.

[5] Purification of Bacteriophage ϕ29 DNA Polymerase

By J. M. LÁZARO, L. BLANCO, and M. SALAS

Introduction

Bacteriophage ϕ29 DNA polymerase, the product of the viral gene 2, was originally characterized as a protein involved in the initiation of ϕ29 DNA replication based on both *in vivo*[1] and *in vitro*[2,3] studies. The cloning of gene 2,[4] the overproduction and purification of its product,[5] and the development of an *in vitro* system for complete ϕ29 DNA replication[6] allowed the characterization of protein p2 as the viral DNA replicase.[7] This enzyme, with a molecular mass of only about 66 kDa, and whose enzymatic activities are summarized in Table I, catalyzes two distinguishable synthetic reactions: (1) DNA polymerization, as any other DNA-dependent DNA polymerase, the template-directed addition of dNMP units from dNTPs, occurring on a DNA or RNA primer strand, in the presence of divalent metal ions; (2) terminal protein (TP) deoxynucleotidylation, the formation of a covalent linkage (phosphoester) between the hydroxyl group of a

[1] R. P. Mellado, M. A. Peñalva, M. R. Inciarte, and M. Salas, *Virology* **104,** 84 (1980).
[2] L. Blanco, J. A. García, M. A. Peñalva, and M. Salas, *Nucleic Acids Res.* **11,** 1309 (1983).
[3] K. Matsumoto, T. Saito, and H. Hirokawa, *Mol. Gen. Genet.* **191,** 26 (1983).
[4] L. Blanco, J. A. García, and M. Salas, *Gene* **29,** 33 (1984).
[5] L. Blanco and M. Salas, *Proc. Natl. Acad. Sci. USA* **81,** 5325 (1984).
[6] L. Blanco and M. Salas, *Proc. Natl. Acad. Sci. USA* **82,** 6404 (1985).
[7] M. Salas, *Ann. Rev. Biochem.* **60,** 39 (1991).

TABLE I

ENZYMATIC ACTIVITIES OF ϕ29 DNA POLYMERASE

Activities	TP-deoxynucleotidylation	DNA polymerization	Pyrophosphorolysis	3′→5′-Exonuclease
Substrates	TP (primer) (K_d 80 nM) and dATP (K_m 5 μM)	DNA strand (primer) (K_d 10 nM) and dNTPs (K_m 1 μM)	Duplex DNA with protruding 5′-single strand (≥1 nt) and PP_i (K_m 0.1 mM)	Single-stranded DNA Mismatched primer-terminus Double-stranded DNA
Template	ϕ29 DNA-TP ϕ29 DNA ori sequences ss ϕ29 DNA (template) No template	TP-dAMP-initiated ϕ29 DNA DNA strand (hybridized to a primer)		
Activating metal[a]	Mn^{2+} >>> Fe^{2+}, Zn^{2+}, Co^{2+} >> Mg^{2+}	Mg^{2+}, Mn^{2+} > Co^{2+}	Mg^{2+}	Mg^{2+}, Mn^{2+} > Zn^{2+}, Co^{2+}
Direction		5′ → 3′	3′ → 5′	3′ → 5′
Product	TP-dAMP + PP_i	Primer-$(dNMP)_n$ + $(PP_i)_n$	DNA_{n-1} + dNTP	DNA_{n-1} + dNMP
Insertion fidelity[b]	Low (Mg^{2+} or Mn^{2+})	High (Mg^{2+}) Moderate (Mn^{2+})		
Inhibitors[c]	BuAdATP, BuPdGTP: K_i = 175 μM PLP; K_i = 1 mM NEM; K_i = 10 μM	Aphidicolin; K_i = 400 μM BuAdATP; K_i = 175 μM BuPdGTP; K_i = 200 μM PAA; K_i = 75 μM PLP; K_i = 275 μM NEM; K_i = 20 μM		Aphidicolin; K_i = 450 μM BuPdGTP; K_i = 50 μM dNMPs; K_i = 50 μM
Other relevant features[d]	OH provided by Ser-232 in the TP Directed by the second template nt Stimulated by NH_4^+ ions ϕ29 p6 reduces K_m for dATP	Highly processive (>70 kb) Able to produce strand displacement	Able to remove a mismatch Reverted by DNA polymerization	Processive (until the remaining length of the substrate is 4–5 nt)
Structural domain[e]	C terminal	C terminal	C terminal	N terminal
Physiological role	INITIATION of ϕ29 DNA replication	ELONGATION of ϕ29 DNA replication	Improvement of INSERTION FIDELITY?	PROOFREADING of DNA insertion errors

[a] Data taken from J. A. Esteban, A. Bernad, M. Salas, and L. Blanco, *Biochemistry* **31,** 350 (1992).
[b] Data taken from J. A. Esteban, M. Salas, and L. Blanco, *J. Biol. Chem.* **268,** 2719 (1993).
[c] Data taken from L. Blanco and M. Salas, *Virology* **153,** 179 (1986); A. Bernad, A. Zaballos, M. Salas, and L. Blanco, *EMBO J.* **6,** 4219 (1987).
[d] Reviewed in M. Salas, *Ann. Rev. Biochem.* **60,** 39 (1991).
[e] See [22] by L. Blanco and M. Salas, this volume.

specific serine residue (Ser-232) in ϕ29 TP and 5′-dNMP, using any of the four dNTPs as substrate, in the presence of divalent metal ions. This reaction, which requires the formation of a TP/DNA polymerase heterodimer,[8] and which can occur in the absence of DNA as template,[9] is strongly stimulated by the presence of the viral DNA replication origins. Moreover, in this case, the reaction is DNA directed, TP-dAMP being preferentially formed. By means of this reaction, in which the TP is acting as a primer, ϕ29 DNA polymerase catalyzes the initiation step of ϕ29 DNA replication.[2,10]

In addition to the synthetic activities, ϕ29 DNA polymerase has two degradative activities: (1) pyrophosphorolysis,[11] the polymerization reversal, whose physiological significance is still unclear; (2) 3′ → 5′-exonuclease,[12] or metal-dependent excision of dNMPs from a 3′-DNA end. In ϕ29 DNA polymerase, as in other replicative DNA-dependent DNA polymerases, this activity has been shown to be involved in a proofreading function.[13]

Other intrinsic properties of ϕ29 DNA polymerase are a high processivity of both DNA synthesis[14] and 3′ → 5′-exonuclease,[13] and the ability to carry out strand displacement coupled to DNA polymerization.[14]

As reviewed by Blanco and Salas elsewhere in this volume,[15] deletion and site-directed mutagenesis studies along the ϕ29 DNA polymerase sequence have demonstrated the location of its multiple enzymatic activities in two structurally separated domains: the synthetic activities (TP-primed initiation and DNA polymerization) and pyrophosphorolysis are located in the C-terminal portion, whereas the 3′ → 5′-exonuclease and strand-displacement activities reside in the N-terminal third of the polypeptide.

In this chapter we describe the protocols for the preparation of ϕ29 DNA polymerase in a highly purified form. These methods contain important variations with respect to those previously published,[4,5] and represent significant improvements at both the overproduction and the purification levels.

[8] L. Blanco, I. Prieto, J. Gutiérrez, A. Bernad, J. M. Lázaro, J. M. Hermoso, and M. Salas, *J. Virol.* **61,** 3983 (1987).

[9] L. Blanco, A. Bernad, J. A. Esteban, and M. Salas, *J. Biol. Chem.* **267,** 1225 (1992).

[10] M. A. Peñalva and M. Salas, *Proc. Natl. Acad. Sci. USA* **79,** 5522 (1982).

[11] M. A. Blasco, A. Bernad, L. Blanco, and M. Salas, *J. Biol. Chem.* **266,** 7904 (1991).

[12] L. Blanco and M. Salas, *Nucleic Acids Res.* **13,** 1239 (1985).

[13] C. Garmendia, A. Bernad, J. A. Esteban, L. Blanco, and M. Salas, *J. Biol. Chem.* **267,** 2594 (1992).

[14] L. Blanco, A. Bernad, J. M. Lázaro, G. Martín, C. Garmendia, and M. Salas, *J. Biol. Chem.* **264,** 8935 (1989).

[15] L. Blanco and M. Salas, this volume [22].

Cloning and Expression of ϕ29 DNA Polymerase

In a previous report, we described the cloning of ϕ29 DNA polymerase gene under the control of the P_L promoter of bacteriophage λ. In that case, the expression level of ϕ29 DNA polymerase in *Escherichia coli* NF2690 cells harboring the recombinant plasmid pLBw2, accounted for about 2% of the *de novo* synthesized protein.[4]

The level of expression of ϕ29 DNA polymerase was improved by modifying plasmid pLBw2 to give rise to plasmid pJLw2, in which upstream and downstream sequences neighbor to the ϕ29 DNA polymerase gene were deleted.[16] After heat induction of *E. coli* NF2690 cells harboring plasmid pJLw2, ϕ29 DNA polymerase accounted for about 0.4% of the total protein. In parallel, recombinant derivatives of plasmid pT7-4,[17] expressing ϕ29 DNA polymerase under control of the T7 RNA polymerase-specific ϕ10 promoter, were developed. Using *E. coli* BL21(DE3) or its derived pLysS or pLysE strains,[18] we could get overproduction of ϕ29 DNA polymerase at either 34° or 30°. This system, selective, highly regulated, and inducible by IPTG, was chosen for the expression of mutant derivatives of ϕ29 DNA polymerase, which could have a putative or potential thermosensitive phenotype.

Purification of ϕ29 DNA Polymerase[16]

E. coli NF2690 cells harboring plasmid pJLw2 are grown at 30° in LB medium, supplemented with 20 m*M* glucose. At a cell density of 2×10^8 cells/ml, P_L promoter is induced by shifting the temperature to 42° for 75 min. The cells are frozen at −70° for storage.

Just before purification, which is carried out at 4°, the cells (40g) are thawed and ground with twice their weight of alumina for 15 min. The slurry is resuspended in Buffer A [50 m*M* Tris–HCl, pH 7.5, 1 m*M* EDTA, 7 m*M* 2-mercaptoethanol, 5% (v/v) glycerol] supplemented with 1 *M* NaCl, by adding 6 volumes/g of cells, step by step. This mixture is centrifuged for 5 min at 1500*g* to remove alumina and intact cells. The resulting lysate is centrifuged for 15 min at 15,000*g* to separate insoluble protein (debris) from the soluble extract. After dilution to 120 OD_{260} units/ml using Buffer A, and 1 *M* NaCl, the DNA present in the soluble extract is removed by polyethyleneimine (PEI) precipitation. After a slow addition of 0.3% PEI (10% stock solution in water, pH 7.5), with stirring, the solution is stirred

[16] J. M. Lázaro, A. Bernad, A. Zaballos, L. Blanco, and M. Salas, unpublished results (1992).
[17] S. Tabor and C. C. Richardson, *Proc. Natl. Acad. Sci. USA* **82,** 1074 (1985).
[18] F. W. Studier and B. A. Moffatt, *J. Mol. Biol.* **189,** 113 (1986).

for 10 min and then centrifuged for 10 min at 15,000*g*. To remove proteins insoluble in PEI, this supernatant is diluted 2.5-fold with Buffer A and 0.04% PEI, kept for 10 min with stirring, and centrifuged for 15 min at 15,000*g*. The resulting supernatant is precipitated with ammonium sulfate to 65% saturation to obtain a PEI-free protein pellet. This pellet is resuspended in Buffer A, adjusting the conductivity of the solution to that of Buffer A and 0.2 *M* NaCl. This fraction ("ammonium sulfate" fraction in Table II), in a volume of 250 ml, is loaded onto a phosphocellulose column (15 ml of bed volume), equilibrated in Buffer A and 0.2 *M* NaCl. After washing stepwise with Buffer A containing 0.2 and 0.35 *M* NaCl, respectively, ϕ29 DNA polymerase is eluted with Buffer A and 0.4 *M* NaCl [Phosphocellulose-1 (0.4 *M*) fraction in Table II]. This eluate is diluted twofold with Buffer A containing 2 m*M* $ZnSO_4$ (Buffer B), and loaded in a second phosphocellulose (3 ml of bed volume), and equilibrated with Buffer B and 0.2 *M* NaCl.

After exhaustive washing with Buffer B and 0.35 *M* NaCl, ϕ29 DNA polymerase is eluted with Buffer B and 0.7 *M* NaCl [Phosphocellulose-2 (0.7 *M*) fraction in Table II]. This eluate, diluted 3.5-fold with Buffer B, is loaded in a Blue dextran-agarose column (50 ml bed volume), prepared essentially as previously described.[5] After washing with Buffer B and 0.25 *M* NaCl, ϕ29 DNA polymerase is eluted with the same buffer, but con-

TABLE II

PURIFICATION OF ϕ29 DNA POLYMERASE

Fraction	Total protein[a] (mg)	ϕ29 DNA polymerase[b] (mg)	Yield[c] (%)	Purity[d] (%)
Soluble extract	3480.00	15.29	100	0.44
Ammonium sulfate	1250.00	13.40	88	1.07
Phosphocellulose 1 (0.4 *M*)	94.42	13.02	85	13.80
Phosphocellulose 2 (0.7 *M*)	51.87	11.26	74	21.70
Blue dextran-agarose (0.5 *M*)	12.55	7.07	46	56.30
Heparin-Sepharose (0.8 *M*)	5.05	5.00	33	99.00

[a] Protein concentration was determined by the method of Lowry. When required, the amount of protein was determined by laser densitometry of the electrophoresed fractions, and comparison with densitometry of samples of known protein concentration.

[b] The amount of ϕ29 DNA polymerase was determined by laser densitometry of the electrophoresed fractions, using proteins of known concentration as standard.

[c] The values represent the percentage of ϕ29 DNA polymerase recovered in each purification step, considering the initial amount of enzyme as 100%.

[d] Percentage of ϕ29 DNA polymerase with respect to the total protein, in each purification step.

taining 0.5 *M* NaCl [Blue dextran-agarose (0.5 *M*) fraction in Table II]. This eluate, diluted 1.7-fold with Buffer B, is loaded in a heparin-Sepharose column (8 ml bed volume) and equilibrated with Buffer B and 0.3 *M* NaCl. After washing with Buffer B and 0.35 *M* NaCl, the fraction eluted with the same buffer, but containing 0.8 *M* NaCl [heparin-Sepharose (0.8 *M*) fraction in Table II] contains highly purified (99%) ϕ29 DNA polymerase. This fraction, adjusted to 50% (v/v) glycerol, is stored at −70°, and was stable for at least 12 months. Under these conditions, the yield is 1.44 mg of purified ϕ29 DNA polymerase/g of protein in the soluble extract. When required, this fraction can be further purified by glycerol gradient centrifugation as previously described.[12]

The specific activities of ϕ29 DNA polymerase, defined as previously described, are 9.4×10^4 U/mg (protein-primed initiation); 2.6×10^4 U/mg (DNA polymerase); and 2.6×10^4 U/mg ($3' \rightarrow 5'$-exonuclease).[19]

Role of ϕ29 DNA Polymerase in Viral DNA Replication[7]

The multiple enzymatic activities of ϕ29 DNA polymerase allow this enzyme to be the only polymerase involved in the replication of the ϕ29 genome. Furthermore, its intrinsic properties of high processivity and strand-displacement ability make unnecessary the participation of accessory proteins and DNA helicases. The primer (TP) is initially recognized by ϕ29 DNA polymerase to form a functionally active heterodimer,[9] which is presumed to be positioned at the ends of the linear DNA molecule (replication origins) by interaction with both the parental TP and the 3′ end of the template strand. The activation of the replication origins is carried out by the viral protein p6, which forms a nucleoprotein complex at both DNA ends, producing a conformational change in the DNA that probably leads to local opening of the DNA duplex.[20]

In the presence of dATP and Mg^{2+}, ϕ29 DNA polymerase catalyzes the formation of a covalent bond between dAMP and the OH group of Ser-232 of the TP acting as primer.[5,21] In this reaction, dATP is selected by base complementarity with the second 3′ nucleotide of the template strand.[22] After formation of the TP–dAMP complex, dissociation of the TP/DNA polymerase heterodimer is likely to occur to replace the protein–protein interactions required for initiation, by the protein–DNA interac-

[19] A. Bernad, L. Blanco, J. M. Lázaro, G. Martín, and M. Salas, *Cell* **59,** 219 (1989).

[20] M. Serrano, M. Salas, and J. M. Hermoso, *Science* **248,** 1012 (1990).

[21] J. M. Hermoso, E. Méndez, F. Soriano, and M. Salas, *Nucleic Acids Res.* **13,** 7715 (1985).

[22] J. Méndez, L. Blanco, J. A. Esteban, A. Bernad, and M. Salas, *Proc. Natl. Acad. Sci. USA* **89,** 9579 (1992).

tions required for the elongation of the newly created DNA primer. Concomitantly, an asymmetric translocation ("sliding-back") of only TP–dAMP, but not of the template, followed by addition of a new dAMP residue, allows the recovery of the information corresponding to the first template nucleotide.[22]

After this special translocation step, the same DNA polymerase molecule catalyzes highly processive chain elongation coupled to strand displacement,[6,14] allowing binding of the ϕ29 SSB to the parental single strand that is being displaced.[23,24] Therefore, complete replication of both strands proceeds continuously from each terminal priming event, without the need for synthesis of RNA-primed Okazaki fragments.

Fidelity of ϕ29 DNA Polymerase[25]

During DNA polymerization, the insertion discrimination of ϕ29 DNA polymerase ranged from 10^4 to 10^6, and the efficiency of mismatch elongation was 10^5- to 10^6-fold lower than that of a properly paired primer terminus. These factors, together with the fact that the $3' \rightarrow 5'$-exonuclease activity showed a strong preference for excising mismatched primer termini,[13] indicate that DNA polymerization catalyzed by ϕ29 DNA polymerase is a highly accurate process.

However, the insertion fidelity of the TP-primed initiation reaction was quite low, the discrimination factor being about 10^2. Mismatch elongation efficiency was rather high (only twofold to sixfold lower than that of a correct TP–dAMP complex). Therefore, at the insertion level, TP-primed initiation is a quite inaccurate reaction. Since the TP-dAMP initiation complex is not a substrate for the $3' \rightarrow 5'$-exonuclease, the fidelity of the TP-primed initiation reaction depends on only the insertion fidelity and the special mechanism of "sliding-back" (described before),[22] which provides an additional chance to discriminate against incorrect TP–dAMP initiation complexes.

Potential Applications of ϕ29 DNA Polymerase

In addition to those applications derived from the enzymatic properties that are common to other DNA-dependent DNA polymerases, the specific properties of ϕ29 DNA polymerase can be used for the following applications:

[23] G. Martín, J. M. Lázaro, E. Méndez, and M. Salas, *Nucleic Acids Res.* **17,** 3663 (1989).
[24] C. Gutiérrez, G. Martín, J. M. Sogo, and M. Salas, *J. Biol. Chem.* **266,** 2104 (1991).
[25] J. A. Esteban, M. Salas, and L. Blanco, *J. Biol. Chem.* **268,** 2719 (1993).

1. *Amplification of very large fragments of DNA.* The high processivity and strand-displacement ability of ϕ29 DNA polymerase make it useful in a polymerase chain reaction (PCR)-type procedure, or in a replicative-type, protein-primed, extension reaction. In the latter case, the method should be based on the use of the natural ϕ29 DNA origin sequences, flanking and located at the ends of the linear DNA molecules to be amplified, and on the use of ϕ29 TP as amplification primer.
2. *Synthesis of long (more than 70 kb) single-strand DNA probes.* These contain multiple copies of the desired sequence, obtained by strand-displacement synthesis on single-strand DNA.[14]
3. *Second-strand cDNA synthesis.* This application[26] is based on the ability to synthesize DNA by extending RNA primers, with the concomitant displacement of the RNA strand from a DNA/RNA heteroduplex.

Acknowledgments

This work is dedicated to the memory of Severo Ochoa.

This investigation has been aided by research grant 5R01 GM27242-14 from the National Institutes of Health, by grant PB90-0091 from Dirección General de Investigación Científica y Técnica, by grant BIOT CT 91-0268 from European Economic Community, and by an institutional grant from Fundación Ramón Areces.

[26] L. Blanco and M. Salas, unpublished results (1989).

[6] DNA Polymerases from *Saccharomyces cerevisiae*

By PETER M. J. BURGERS

Introduction

A combination of biochemical and genetic techniques has led to the identification of a large number of DNA-dependent DNA polymerases in the yeast *Saccharomyces cerevisiae.* Five of the six yeast DNA polymerases known to date have clear homologs in mammalian cells (Table I). The sixth DNA polymerase, predicted from the sequence of the *REV3* gene as well as genetic studies, has tentatively been identified in fractionated yeast ex-

TABLE I
PROPERTIES OF YEAST DNA POLYMERASES[a]

Property	α	β	γ	δ	ε	*REV3*
Genes (subunits)[b]	*POL1* *POL12* *PRI1* *PRI2*	*YCR14C*	*MIP1*	*POL3* 60 kDa 55 kDa	*POL2* *DPB2* *DPB3* 29 kDa	*REV3*
DNA Primase	Yes	No	No	No	No	
3′ → 5′-Exonuclease	No	Yes	Yes	Yes	Yes	
Elution from Mono Q (*M*)	0.3	<0.1[c]	0.25	0.3	0.35	<0.1[c]
Polymerase activity[d]:						
+ 50 m*M* $MgCl_2$	−	−	+++	−	−	
+ 200 m*M* NaCl	+	+++	+++	−	++	
+ Aphidicolin (10 μg/ml)	+	+++	+++	−	−	
+ BuPhdGTP (10 μM)	−	+++	+++	+++	+++	
+ ddTTP (100 μM)	+++	−	−	+++	+++	
Template primer[d]:						
Activated DNA	+++	+++	++	++	+	
poly(dA) · oligo(dT)[e]	+++	+	++	+	++	
+ PCNA	+++	+	++	++	++	
poly(dA) · oligo(dT)dC[e]	−	−	++	+	++	
Refs:						
Genes and structure	38–42	56	10	35, 57	30, 66, 67	1
Mutants and function	43–48	2, 9	10, 19	22, 35, 58–61	23, 59, 61	1
Purification	18, 49, 50	2	10, 11	16, 20	14, 25	2
Biochemical properties	40, 51–55	2, 9	10, 11, 19	12, 16, 20, 62–65	14, 25, 62, 63	

[a] For nomenclature see Ref. 4.
[b] Sizes of subunits for which no clones are available are given in kDa.
[c] These enzymes flow through the Mono Q column.
[d] General assay conditions are given in the Materials and Methods section. When inhibition by aphidicolin, BuPhdGTP, or ddTTP was measured, the concentration of dCTP was lowered to 10 μM, dGTP to 10 μM, and dTTP to 5 μM, respectively. 30 to 100% activity (+++), 10 to 30% activity (++), 3 to 10% activity (+), <3% activity (−).
[e] Nucleotide ratio, 20:1.

[38] L. M. Johnson, M. Snyder, L. M. Chang, R. W. Davis, and J. L. Campbell, *Cell* **43,** 369 (1985).
[39] A. Pizzagalli, P. Valsasnini, P. Plevani, and G. Lucchini, *Proc. Natl. Acad. Sci. USA* **85,** 3772 (1988).
[40] R. G. Brooke, R. Singhal, D. C. Hinkle, and L. B. Dumas, *J. Biol. Chem.* **266,** 3005 (1991).
[41] P. Plevani, S. Francesconi, and G. Lucchini, *Nucleic Acids Res.* **15,** 7975 (1987).
[42] M. Foiani, C. Santocanale, P. Plevani, and G. Lucchini, *Mol. Cell Biol.* **9,** 3081 (1989).
[43] M. Budd and J. L. Campbell, *Proc. Natl. Acad. Sci. USA* **84,** 2838 (1987).
[44] M. E. Budd, K. D. Wittrup, J. E. Bailey, and J. L. Campbell, *Mol. Cell Biol.* **9,** 365 (1989).
[45] G. Lucchini, M. M. Falconi, A. Pizzagalli, A. Aguilera, H. L. Klein, and P. Plevani, *Gene* **90,** 99 (1990).
[46] S. Francesconi, M. P. Longhese, A. Piseri, C. Santocanale, G. Lucchini, and P. Plevani, *Proc. Natl. Acad. Sci. USA* **88,** 3877 (1991).
[47] M. P. Longhese, L. Jovine, P. Plevani, and G. Lucchini, *Genetics* **133,** 183 (1993).
[48] M. Foiani, F. Marini, D. Gamba, G. Lucchini, and P. Plevani, *Mol. Cell. Biol.* **14,** 923 (1994).
[49] G. Badaracco, L. Capucci, P. Plevani, and L. M. Chang, *J. Biol. Chem.* **258,** 10,720 (1983).
[50] M. Brooks and L. B. Dumas, *J. Biol. Chem.* **264,** 3602 (1989).
[51] G. Badaracco, M. Bianchi, P. Valsasnini, G. Magni, and P. Plevani, *EMBO J.* **4,** 1313 (1985).
[52] G. Badaracco, P. Valsasnini, M. Foiani, R. Benfante, G. Lucchini, and P. Plevani, *Eur. J. Biochem.* **161,** 435 (1986).
[53] R. G. Brooke and L. B. Dumas, *J. Biol. Chem.* **266,** 10,093 (1991).
[54] T. A. Kunkel, J. D. Roberts, and A. Sugino, *Mutat. Res.* **250,** 175 (1991).
[55] C. Santocanale, M. Foiani, G. Lucchini, and P. Plevani, *J. Biol. Chem.* **268,** 1343 (1993).
[56] S. G. Oliver, *et al.*, *Nature* **357,** 38 (1992).
[57] A. Morrison and A. Sugino, *Nucleic Acids Res.* **20,** 375 (1992).
[58] K. C. Sitney, M. E. Budd, and J. L. Campbell, *Cell* **56,** 599 (1989).
[59] M. E. Budd and J. L. Campbell, *Mol. Cell. Biol.* **13,** 496 (1993).
[60] A. Morrison, A. L. Johnson, L. H. Johnston, and A. Sugino, *EMBO J.* **12,** 1467 (1993).
[61] A. Morrison and A. Sugino, *Mol. Gen. Genet.* **242,** 289 (1994).
[62] B. L. Yoder and P. M. J. Burgers, *J. Biol. Chem.* **266,** 22,689 (1991).
[63] P. M. J. Burgers, *J. Biol. Chem.* **266,** 22,698 (1991).
[64] K. F. Fien and B. Stillman, *Mol. Cell Biol.* **12,** 155 (1992).
[65] W. C. Brown and J. L. Campbell, *J. Biol. Chem.* **268,** 21,706 (1993).
[66] H. Araki, R. K. Hamatake, L. H. Johnston, and A. Sugino, *Proc. Natl. Acad. Sci. USA* **88,** 4601 (1991).
[67] H. Araki, R. K. Hamatake, A. Morrison, A. L. Johnson, L. H. Johnston, and A. Sugino, *Nucleic Acids Res.* **19,** 4867 (1991).

tracts.[1,2] There is no known homolog for this enzyme in mammalian cells. DNA polymerases α, δ, and ε (Pol α, Pol δ, Pol ε) are the three nuclear enzymes required for yeast cell growth. For a thorough discussion of these enzymes, their homology with the mammalian enzymes, and their function in yeast DNA metabolism, the reader is referred to some recent review articles.[3-8] An enzyme that shows sequence homology to mammalian DNA polymerase β has recently been identified as the product of the YCR14C open reading frame, resulting from the chromosome III sequencing project.[2,9] Even though it is larger in size (67 kDa) than the mammalian Pol β, which is 37 to 40 kDa, its biochemical properties clearly indicate that this enzyme belongs to the Pol β class. DNA polymerase γ (Pol γ) is the mitochondrial DNA polymerase.[10]

This chapter describes a chromatographic method that allows the separation on a small scale of Pol α, Pol δ, and Pol ε at a fairly crude stage of purification. With proper caution these partially purified enzymes can then be used in replication and repair studies. To obtain any of the DNA polymerases in pure form, the reader is referred to the original literature (see Table I for references). Anion-exchange matrices are by far the most efficient in separating these three DNA polymerases. The order of elution from anion-exchange columns is reproducibly the same, that is, Pol δ elutes first, followed by Pol α, and Pol ε elutes last. However, because proteolysis may complicate this elution pattern, it is necessary to confirm the identity of each DNA polymerase before proceeding with other studies.

In the absence of readily available specific antibodies to each of the DNA polymerases, a biochemical approach can be employed to determine their identities. The biochemical properties that allow a distinction to be made between the various DNA polymerases are listed in Table I. For

[1] A. Morrison, R. B. Christensen, J. Alley, A. K. Beck, E. G. Bernstine, J. F. Lemontt, and C. W. Lawrence, *J. Bacteriol.* **171,** 5659 (1989).

[2] K. Shimizu, C. Santocanale, P. A. Ropp, M. P. Longhese, P. Plevani, G. Lucchini, and A. Sugino, *J. Biol. Chem.* **268,** 27,148 (1993).

[3] P. M. J. Burgers, *Prog. Nucleic Acids Res. Mol. Biol.* **37,** 235 (1989).

[4] P. M. J. Burgers *et al.*, *Eur. J. Biochem.* **191,** 617 (1990).

[5] T. S. F. Wang, *Ann. Rev. Biochem.* **60,** 513 (1991).

[6] J. L. Campbell and C. S. Newlon, *in* "The Molecular and Cellular Biology of the Yeast Saccharomyces: Genome Dynamics, Protein Synthesis, and Energetics" (J. R. Broach, J. R. Pringle, and E. W. Jones, eds.), p. 41. Cold Spring Harbor Laboratory Press, New York, 1991.

[7] J. Campbell, *J. Biol. Chem.* **268,** 25,261 (1993).

[8] A. Morrison and A. Sugino, *Prog. Nucleic Acids Res. Mol. Biol.* **46,** 93 (1993).

[9] R. Prasad, S. G. Widen, R. K. Singhal, J. Watkins, L. Prakash, and S. H. Wilson, *Nucleic Acids Res.* **21,** 5301 (1993).

[10] F. Foury, *J. Biol. Chem.* **264,** 20,552 (1989).

instance, both Pol β and Pol γ are resistant to aphidicolin and sensitive to the presence of dideoxynucleotides in the assay, but only Pol γ is fully active with 50 mM $MgCl_2$ in the assay.[2,9–11] Pol α is the only enzyme that is inhibited by low levels of N^2-(*p*-*n*-butylphenyl)dGTP (BuPhdGTP).[12] Although inhibitors are useful in distinguishing Pol δ and Pol ε from the other DNA polymerases, an inhibitor-based distinction between Pol δ and Pol ε cannot be easily made. However, these enzymes can be distinguished by their preferences for certain template primers. Thus, Pol ε is most active on a poly(dA) template sparsely primed with oligo(dT),[13,14] whereas a high activity of Pol δ on this template primer depends on the presence of an accessory factor, proliferating cell nuclear antigen (PCNA), which promotes processive DNA synthesis by this polymerase.[15,16]

Additional criteria for further characterization of the DNA polymerases are not always completely reliable. DNA primase activity copurifies with Pol α, but the three nuclear RNA polymerases can also prime DNA synthesis.[17,18] Only Pol γ, Pol δ, and Pol ε, because of their associated $3' \rightarrow 5'$-exonuclease activity carry out efficient DNA synthesis on template primers with mismatched primer termini.[19–23] However, Pol α and Pol β may also be active on such substrates if a nuclease capable of removing the mismatched terminus happens to copurify with these polymerases. The latter problem can be diminished by using nuclease-deficient strains (see below).

Strain Choice and Problem of Proteolysis

Many of the problems that have plagued biochemists in their studies of the DNA polymerases can be attributed to proteolysis. Proteolysis not only leads to reduced enzyme recovery, but can also result in changes in

[11] U. Wintersberger and H. Blutsch, *Eur. J. Biochem.* **68,** 199 (1976).
[12] P. M. J. Burgers and G. A. Bauer, *J. Biol. Chem.* **263,** 925 (1988).
[13] L. M. S. Chang, *J. Biol. Chem.* **252,** 1873 (1977).
[14] R. K. Hamatake, H. Hasegawa, A. B. Clark, K. Bebenek, T. A. Kunkel, and A. Sugino, *J. Biol. Chem.* **265,** 4072 (1990).
[15] P. M. J. Burgers, *Nucleic Acids Res.* **16,** 6297 (1988).
[16] G. A. Bauer and P. M. J. Burgers, *Proc. Natl. Acad. Sci. USA* **85,** 7506 (1988).
[17] P. Plevani and L. M. Chang, *Biochemistry* **17,** 2530 (1978).
[18] P. Plevani, M. Foiani, P. Valsasnini, G. Badaracco, E. Cheriathundam, and L. M. S. Chang, *J. Biol. Chem.* **260,** 7102 (1985).
[19] F. Foury and S. Vanderstraeten, *EMBO J.* **11,** 2717 (1992).
[20] G. A. Bauer, H. M. Heller, and P. M. J. Burgers, *J. Biol. Chem.* **263,** 917 (1988).
[21] E. Wintersberger, *Eur. J. Biochem.* **84,** (1978).
[22] M. Simon, L. Giot, and G. Faye, *EMBO J.* **10,** 2165 (1991).
[23] A. Morrison, J. B. Bell, T. A. Kunkel, and A. Sugino, *Proc. Natl. Acad. Sci. USA* **88,** 9473 (1991).

chromatographic and catalytic properties of the enzyme. For example, proteolysed forms of Pol α are less sensitive to aphidicolin.[24] Multiple forms of Pol ε can be identified chromatographically as a result of proteolysis,[14,25] and Pol δ itself could only be identified when strict measures to limit proteolysis were applied.[20] Limited proteolysis of Pol δ produces a species that is unresponsive to PCNA.[26]

Proteolysis can be minimized by (1) use of protease-deficient strains whenever possible; (2) harvesting the cells when still in log phase; (3) use of protease inhibitors in all buffers; and (4) rapid purification schedule. The yeast vacuolar proteases and protease-deficient strains have been discussed previously in this series.[27] Briefly, the contribution by the main pool of cellular proteases, that is, the vacuolar proteases, can be virtually eliminated by using the vacuolar processing-deficient mutant *pep4,* preferentially in combination with a protease B mutation, *prb1.*[27] Harvesting cells before they reach stationary phase not only avoids an intracellular accumulation of proteases induced by this growth phase, but also ensures a higher DNA polymerase level, because at least some of these enzymes are only expressed at the G_1/S boundary of the cell cycle.[28–30] Even in strains deficient for vacuolar proteases, the use of additional protease inhibitors has proven to be beneficial.[20] Commonly used inhibitors of the vacuolar proteases include EDTA and EGTA (metalloproteases), phenylmethylsulfonyl fluoride (PMSF), $NaHSO_3$, benzamidine, pepstatin, chymostatin, and leupeptin (aspartic and/or serine protease inhibitors). Leupeptin also inhibits a mitochondrial protease.[31] The use of protease inhibitors becomes essential when wild-type strains rather than protease-deficient strains are used. In that case, a reasonable recovery of active enzyme may also depend on a rapid purification procedure. In fact, we have noticed that freezing wild-type cells prior to breakage may substantially reduce recoverable polymerase, presumably because of increased leakage of proteases from organelles as a result of freezing.[12] However, these latter problems are in general not evident when protease-deficient strains are used, underscoring their importance for yeast biochemistry.

[24] P. Plevani, G. Badaracco, E. Ginelli, and S. Sora, *Antimicrob. Agents Chemother.* **18,** 50 (1980).

[25] M. E. Budd, K. C. Sitney, and J. L. Campbell, *J. Biol. Chem.* **264,** 6557 (1989).

[26] P. M. J. Burgers *et al.*, unpublished results (1990).

[27] E. W. Jones, *Methods Enzymol.* **194,** 428 (1991).

[28] L. H. Johnston, J. H. White, A. L. Johnson, G. Lucchini, and P. Plevani, *Nucleic Acids Res.* **15,** 5017 (1987).

[29] G. A. Bauer and P. M. J. Burgers, *Nucleic Acids Res.* **18,** 261 (1990).

[30] A. Morrison, H. Araki, A. B. Clark, R. K. Hamatake, and A. Sugino, *Cell* **62,** 1143 (1990).

[31] A. S. Zubatov, A. E. Mikhailova, and V. N. Luzikov, *Biochim. Biophys. Acta* **787,** 188 (1984).

Materials and Methods

Yeast Strains and Growth

The protease-deficient haploid strains used in our laboratory for enzyme purification are BJ405 (*Matα, prb1-1122,prc1-407, pep4-3, trp1*) from Dr. E. W. Jones (Carnegie Mellon) and PY26 (*Mat*a, *ura3-52, trp1Δ, leu2-3,112, prb1-1122, prc1-407, pep4-3, Δnuc1 :: LEU2*) (this laboratory). The latter strain is defective for a potent mitochondrial endonuclease.[32] Cells (2 × 1.2 liter) are grown at 30° in an air shaker in 4-liter flasks in YPD (1% (w/v) yeast extract, 2% (w/v) peptone, 2% (w/v) glucose). An inoculum of 5 ml of an overnight culture per flask will grow in approximately 16 hr to an OD_{660} of 2 (about 2 to 3 × 10^7 cells/ml). At that time, another 24 g of glucose is added per flask and the cells grown further to OD_{660} of 4. The cells are then harvested for 5 min at 3000 rpm at 4° in a Sorvall GS-3 rotor. The yield of cells is 12 to 15 g, wet weight (5 to 6 g/liter). It can either be resuspended in a few milliliters of water, frozen in liquid nitrogen and stored at −70°, or immediately broken as described below.

Buffers

2× extraction buffer: 0.2 *M* Tris–HCl, pH 7.8, 10% (v/v) glycerol, 8 m*M* EDTA, 2 m*M* EGTA, 0.35 *M* ammonium sulfate, 20 m*M* $NaHSO_3$, 4 μ*M* leupeptin, 2 μ*M* pepstatin A, 10 m*M* benzamidine, 0.5 m*M* PMSF, and 6 m*M* dithiothreitol (DTT)

Buffer B: 25 m*M* potassium phosphate, pH 7.3, 10% (v/v) glycerol, 2 m*M* EDTA, 0.4 m*M* EGTA, 10 m*M* $NaHSO_3$, 2 μ*M* leupeptin, 1 μ*M* pepstatin A, 2 m*M* benzamidine, 0.5 m*M* PMSF, 3 m*M* DTT, and 50 m*M* KCl

Buffer C: 25 m*M* potassium phosphate, pH 7.3, 10% (v/v) glycerol, 1 m*M* EDTA, 5 m*M* $NaHSO_3$, 2 μ*M* leupeptin, 1 μ*M* pepstatin A, 0.02% Nonidet P-40 (NP-40), and 3 m*M* DTT.

Extraction

All steps are carried out at 0 to 4°. Buffers are precooled on ice–water. The 30-ml chamber of a bead beater (BioSpec products) is filled with glass beads (0.4 to 0.5 mm in diameter) up to the rotor level. The cells are resuspended in an equal volume of 2× extraction buffer, put in the chamber, and the chamber further filled to the top with extraction buffer, and then closed and sealed with parafilm to prevent leakage during the

[32] H. P. Zassenhaus, T. J. Hofmann, R. Uthayashanker, R. D. Vincent, and M. Zona, *Nucleic Acids Res.* **16,** 3283 (1988).

beating. The chamber is inserted in the cooling chamber filled with ice–water and turned on for 45 sec, followed by a cooling period of 2 min, for a total beating time of 5 min. The lysate is poured in a cold graduated cylinder taking care not to transfer the beads. The beads are washed with 10 ml of extraction buffer. The volume of the combined lysate is measured, and the lysate transferred to a 40-ml centrifuge tube. Forty microliters of 10% Polymin P are added per milliliter of lysate and the tube is occasionally inverted on ice–water. After 5 min, the lysate is spun for 20 min at 18,000 rpm in a SS34 rotor. The supernatant is poured into a graduated cylinder to measure the volume and then transferred to a new centrifuge tube. Then 0.28 g of solid ammonium sulfate is added per milliliter of supernatant and dissolved by occasional inversion. About 15 min after all ammonium sulfate has gone into solution, the tube is spun at 18,000 rpm for 45 min. The pellet is resuspended in 2 ml of Buffer B. At this stage the preparation can be frozen and stored at −70°. The yield is about 25 to 50 mg of protein from 12 to 15 g of cells.[33]

Desalting Column

A 50-ml Sephadex G-25 (Pharmacia) or Bio-Gel P4 (Bio-Rad) column is equilibrated at 4° in Buffer B and the excess liquid above the bed drained away. The resuspended pellet is loaded onto the column and allowed to run in by gravity. The column is washed with 3- to 5-ml portions of Buffer B until about 15 ml of eluate has been collected. At about this point the void volume containing all protein should start to elute. Fractions of 3 ml are collected by repeatedly applying 3 ml of buffer to the column and allowing it to run in. Protein concentrations in the fractions are determined using the Bradford assay[33] and salt concentrations with a conductometer. The protein containing fractions, which should be well separated from the ammonium sulfate, are further purified by phosphocellulose chromatography and by anion-exchange HPLC. Alternatively, the desalted extract can be directly injected onto the Mono Q column, although in that case care should be taken not to exceed the capacity of the column.

Batch Adsorption to Phosphocellulose (Optional)

The desalted extract is mixed for 30 min by continuous rotation or inversion with 3 ml of phosphocellulose P11 (Whatman), washed according to the manufacturers directions, and equilibrated in Buffer B. After a brief spin in a clinical centrifuge to pellet the phosphocellulose, the supernatant is poured off. The matrix is resuspended in a few milliliters of Buffer B

[33] M. M. Bradford, *Anal. Biochem.* **72,** 248 (1976).

and poured in a 10-ml column. The matrix is washed with 10 ml of Buffer B and eluted with 3-ml portions of Buffer B+0.75*M* NaCl. Fractions of 3 ml are collected. The protein is generally recovered in fractions 2–4. It is dialyzed against 200 ml of Buffer C until the conductivity is equal to that of Buffer C+50 m*M* NaCl. The yield of protein is 5 to 10 mg.

Mono Q HPLC

The desalted ammonium sulfate precipitate or the dialyzed eluate from the phosphocellulose step is spun at 18,000 rpm for 20 min to remove particulate material. A 1-ml Mono Q column (Pharmacia), assembled in a HPLC or FPLC apparatus, is equilibrated in Buffer C+50 m*M* NaCl. The protein is injected onto the column and, after all material has been loaded on, the column is washed with 2 ml of Buffer C+50 m*M* NaCl and eluted with a series of linear gradients of 5 ml 50 to 250 m*M* NaCl, 10 ml 250 to 350 m*M* NaCl, and 5 ml 350 to 600 m*M* NaCl in Buffer C. Fractions of 0.35 ml are collected, quick frozen, and stored at −70°.

DNA Substrates and Inhibitors

Salmon sperm DNA is activated as described.[34] Oligo$(dT)_{16}$ or oligo$(dT)_{16}$dC is hybridized to 5 A_{260} units/ml of poly(dA) at a 1 : 20 nucleotide ratio in 10 m*M* Tris–HCl, pH 7.5, 1 m*M* EDTA, 50 m*M* NaCl at 37° for 30 min. 3′-End-labeled single-stranded DNA is prepared from high-molecular-weight activated salmon sperm DNA (average chain length 500 to 2000 nucleotides) by reaction with DNA polymerase I, Klenow fragment, dATP, dGTP, and dCTP at 100 μM each, and dTTP (10,000 cpm/pmol) at 10 μM. The DNA is purified by phenol extraction and several ethanol precipitations, followed by size-exclusion chromatography on Bio-Gel A5m (Bio-Rad) in 10 m*M* Tris–HCl, pH 7.5, 1 m*M* EDTA, and 100 m*M* NaCl. Prior to use, the DNA is made single stranded by boiling for 3 min. All of the inhibitors used are commercially available, with the exception of BuPhdGTP, which was the generous gift of Dr. George Wright (University of Massachusetts).

DNA Polymerase Assays

The 50-μl reaction contains 20 m*M* Tris–HCl, pH 7.8, 8 m*M* $MgAc_2$, 0.2 mg/ml of bovine serum albumin (BSA), 4% glycerol, 1 m*M* DTT, 80 μM each dATP, dGTP, and dCTP, 20 μM [^{3}H]dTTP (400 cpm/pmol), 200 μg/ml activated salmon sperm DNA, 1 m*M* spermidine, and enzyme. Assays

[34] A. Spanos, S. G. Sedgwick, G. T. Yarranton, U. Hubscher, and G. R. Banks, *Nucleic Acids Res.* **9,** 1825 (1981).

are assembled on ice and incubated at 37° for 30 min. They are stopped by addition of 100 μl of 25 m*M* EDTA, 25 m*M* sodium pyrophosphate, and 50 μg/ml of salmon sperm DNA, followed by 1 ml of 10% trichloroacetic acid (TCA). After 10 min on ice, the mixture is filtered over a GF/C filter. The filter was washed with 2 × 2 ml of 1 *M* HCl, 0.05 *M* sodium pyrophosphate, rinsed with ethanol, dried, and counted in a counting fluid in a liquid scintillation counter. One unit of enzyme incorporates 1 pmol/min of nucleotide into acid-insoluble radioactivity.

When inhibition by aphidicolin, BuPhdGTP, or ddTTP is measured, the concentration of dCTP is lowered to 10 μ*M*, dGTP to 10 μ*M*, and dTTP to 5 μ*M*, respectively. When poly(dA)-oligo$(dT)_{16}$ or poly(dA)-oligo$(dT)_{16}$dC is used (at 0.2 A_{260} units/ml), spermidine is omitted. When the stimulation of Pol δ by PCNA is measured, magnesium acetate is increased to 11 m*M* in the poly(dA)-oligo$(dT)_{16}$ assay.

Exonuclease Assays

The 50-μl assay contains 20 m*M* Tris–HCl, pH 7.8, 8 m*M* magnesium acetate, 0.2 mg/ml of BSA, 4% glycerol, 1 m*M* DTT, 1.5 μg/ml of 3′-end-labeled single-stranded DNA (15,000 cpm), and enzyme. Assays are assembled on ice in microfuge tubes and incubated at 37° for 20 min. They are stopped by addition of 100 μl of 25 m*M* EDTA, 25 m*M* sodium pyrophosphate, and 50 μg/ml of salmon sperm DNA, followed by 125 μl of 10% TCA. After 10 min on ice, the tubes are spun in a microfuge for 10 min. Then 200 μl of the supernatant is added to a water-miscible counting fluid and counted in a liquid scintillation counter. One unit of enzyme releases 1 pmol/min of nucleotide into acid-soluble radioactivity.

Comments

Cell Growth

1. Care should be taken to harvest cells when in log phase growth. When cells are above OD_{660} = 4 after overnight growth, they should be diluted with an equal volume of YPD and grown for another 2 hr.
2. The glucose boost allows the cells to grow to a higher density without going into stationary phase.
3. Wild-type, that is, not protease-deficient, cells should not be frozen, but immediately processed because proteolysis problems are more severe with previously frozen cells.

Extraction

1. Protease inhibitors are made up as stock solutions in water (1 *M* $NaHSO_3$, 0.5 *M* benzamidine, 1 m*M* pepstatin and leupeptin) or 2-propanol (0.1 *M* PMSF). All are fairly stable in the buffers at 4°, except PMSF and $NaHSO_3$, which are added just prior to use.
2. When wild-type strains are used, chymostatin and aprotinin are also added (from 1 m*M* stock solutions in water) to the buffers at final concentrations of 1 and 5 μM, respectively.
3. The presence of 175 m*M* ammonium sulfate in the extraction buffer prevents coprecipitation of the DNA polymerases with the nucleic acids by Polymin P.

Desalting Column

Desalting is preferred over dialysis because it is a much more rapid procedure, thereby limiting proteolysis.

Mono Q HPLC

1. The Mono Q column was chosen here because it is a commonly used matrix in many laboratories. Other anion-exchange materials may work just as well or even better. We have previously used silica-based DEAE columns (e.g., Synchropak AX1000), which actually gave a better separation of Pol α, Pol δ, and Pol ε.[20,35] However, recovery of Pol α from this column is low due to irreversible binding. This problem can largely be overcome by including nonionic detergents in the column buffer (as in Buffer C), and reducing the salt concentration in the starting buffer of the column to C+25 m*M* NaCl. However, even under those low-salt conditions, Pol δ may occasionally fail to bind to the column, depending on the percent of detergent used in the buffers and the amount of protein loaded onto the column.
2. The loading capacity of the 1-ml Mono Q column is about 20 to 30 mg of protein; Pol δ will flow through if higher amounts of protein are loaded. If the extract is not fractionated by the phosphocellulose batch step, a larger column should be used.
3. Benzamidine is omitted from Buffer C because its absorption at 280 nm interferes with obtaining a protein profile of the Mono Q step.

[35] A. Boulet, M. Simon, G. Faye, G. A. Bauer, and P. M. J. Burgers, *EMBO J.* **8,** 1849 (1989).

Assays

1. Activated calf thymus DNA (Sigma) and activated fish DNA (USBiochemical) are both commercially available. However, caution should be used if activated calf thymus DNA is used rather than activated salmon DNA. The presence of heparin-like material in calf thymus DNA inhibits the DNA polymerases.[36] The presence of these impurities differs greatly between preparations, and they are not easily removed by extraction. This inhibition is largely overcome, either by treating the activated calf thymus DNA with heparinase or by adding ampholytes (about 0.1% w/v) into the assay.[36]
2. The 3′-end-labeled substrate used in the nuclease assays preferentially detects single-stranded DNA-dependent 3′ → 5′-exonucleases such as proofreading exonucleases. However, other nucleases can also produce a signal, albeit quantitatively lower, in this assay.[20,37]

Discussion

A Mono Q separation of an extract from strain PY26 is shown in Fig. 1. The extract was desalted and subjected to batch phosphocellulose chromatography as described. Only three DNA polymerases are indicated on the chromatogram. Pol β and the *REV3* DNA polymerase flow through the Mono Q column.[2] Most of the Pol γ is not released from the mitochondria during breakage and consequently pellets during the first spin. However, variable low levels of Pol γ may end up in the final preparation injected onto the Mono Q column. This enzyme elutes at 0.25 to 0.3 *M* NaCl, together with Pol δ and/or Pol α. Fortunately, Pol γ can easily be identified because it is resistant to aphidicolin and active in the presence of 50 m*M* $MgCl_2$ (Table I). No significant Pol γ activity was observed in the particular preparation shown in Fig. 1.

Pol δ elutes first at about 0.25 *M* NaCl. The minor peak of activity eluting at 0.1–0.15 *M* NaCl is actually a proteolysed form of Pol δ with a catalytic polypeptide of 100 kDa rather than 125 kDa, as shown by a Western analysis. Baseline separation of Pol α and Pol ε is not easily obtained, even when a shallow gradient is used. The assay with the mismatched template primer poly(dA)-oligo$(dT)_{16}$dC clearly shows a lack of activity in the Pol α region of the chromatogram, reflecting the absence of a proofreading exonuclease in Pol α. These results are most striking in extracts from strain PY26, which is deficient for a potent mitochondrially

[36] M. Goulian and C. J. Heard, *Nucleic Acids Res.* **18,** 4791 (1990).

[37] P. M. J. Burgers, G. A. Bauer, and L. Tam, *J. Biol. Chem.* **263,** 8099 (1988).

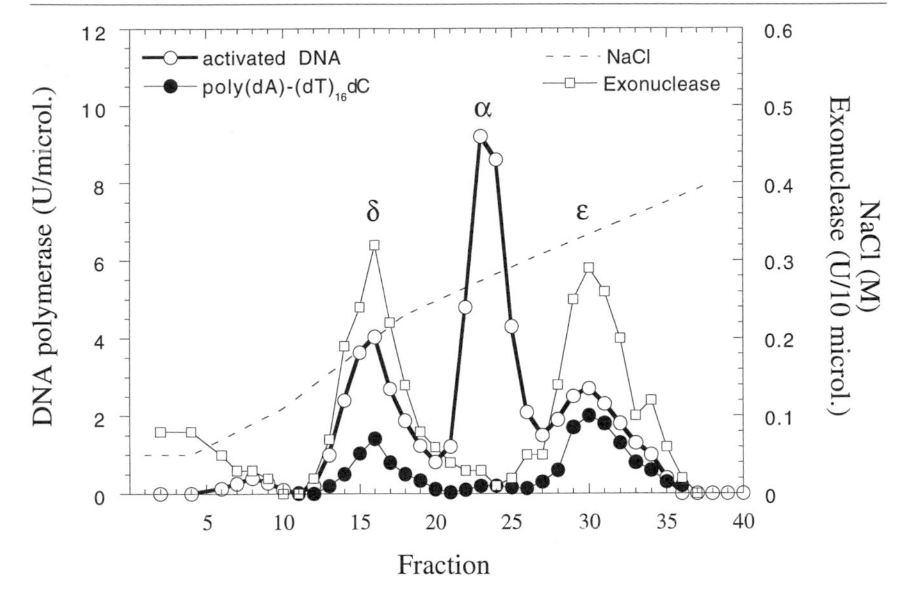

FIG. 1. Mono Q separation of three DNA polymerases. Extracts from strain PY26 were subjected to phosphocellulose chromatography and Mono Q HPLC as described in the Materials and Methods section. Aliquots (2 μl) were assayed for DNA polymerase activity on activated salmon DNA (-○-), and on the mismatched primer template poly(dA)-$(dT)_{16}$dC (-●-), and for exonuclease activity on 3′-end-labeled single-stranded calf thymus DNA (-□-). The NaCl concentration in the fractions is indicated (---).

derived nuclease, but are also evident when other strains are used. A similar separation of these three DNA polymerases is observed when the phosphocellulose step is omitted. However, there are several advantages to carrying out this step. First, because only 10 to 20% of the total cellular protein is retained on phosphocellulose, larger amounts of cells can be processed in order to obtain very active polymerase fractions. Second, because PCNA flows through phosphocellulose, it is absent from the fractions of the subsequent Mono Q step.[16] Thus, the effect of PCNA on Pol δ activity can be measured more reliably.

The activity of partially purified Pol δ obtained in this fashion was stimulated five- to tenfold by PCNA when poly(dA)-oligo(dT) (20 : 1, nucleotide ratio) was used as template primer. In contrast, the proteolysed form of Pol δ, eluting in fractions 7–9 (Fig. 1), was not stimulated by PCNA.[26] A similar degree of stimulation of Pol δ was observed with heterologous PCNAs, from calf thymus or *Drosophila melanogastor*, making yeast an excellent and easily obtainable source of Polδ for assaying PCNA preparations from heterologous organisms.[16]

A further purification of these partially purified polymerase preparations on a small scale is not easily done without incurring major losses in activity. This is particularly the case when hydrophobic or hydroxyapatite columns are used. However, cation-exchange columns, for example, a 1-ml Mono S column or a small heparin-agarose, can be used to further purify each of the polymerase peaks. To obtain homogeneous polymerase preparations, much larger quantities of yeast should be used as starting material or, at least for Pol α, immunopurification methods should be applied.

Acknowledgment

I thank Bonnie Yoder for carrying out the Mono Q HPLC separation shown in Fig. 1. This work was supported by grant GM32431 from the National Institutes of Health.

[7] Purification of DNA Polymerase–Primase (DNA Polymerase α) and DNA Polymerase δ from Embryos of *Drosophila melanogaster*

By PAUL G. MITSIS, CHUEN-SHEUE CHIANG, and I. R. LEHMAN

Introduction

Like most other eukaryotes *Drosophila melanogaster* possesses four distinct DNA polymerases: DNA polymerase–primase (Pol α),[1] DNA polymerase δ (Pol δ),[2] DNA polymerase ε (Pol ε)[2] and DNA polymerase γ, the mitochondrial DNA polymerase (Pol γ).[3] All of these enzymes have been purified to near homogeneity, with the exception of Pol ε, which has been observed only in relatively crude fractions. DNA polymerase β, which is present in many eukaryotes has not been unambiguously identified in *Drosophila*. Two of the four *Drosophila* DNA polymerases, Pol α and Pol δ, are considered here.

DNA Polymerase–Primase

Embryos of *D. melanogaster* provide an abundant and easily obtainable source of DNA polymerase–primase (Pol α). However, a major difficulty

[1] I. R. Lehman and L. S. Kaguni, *Biochem. Biophys. Acta* **950,** 87 (1988).
[2] C.-S. Chiang, P. G. Mitsis, and I. R. Lehman, *Proc. Natl. Acad. Sci. USA* **90,** 9105 (1993).
[3] C. M. Wernette and L. S. Kaguni, *J. Biol. Chem.* **261,** 14,764 (1986).

encountered during purification of this enzyme from *Drosophila* as well as from many other sources is the extensive proteolysis of the 182-kDa polymerase catalytic subunit. Initially, uncontrolled proteolysis led to uncertainty about its subunit composition.[1] However, the use of fresh embryos, protease inhibitors, and care in manipulation has largely eliminated this problem. With this in mind, it is useful to work through the initial steps of the purification as rapidly as possible. Those steps at which the procedure can be safely stopped are indicated in the purification protocol.

The purification method is essentially that of Kaguni *et al.*[1] with the final Blue A agarose column replaced by chromatography on Mono Q[4] to remove most of the proteolytic products derived from the large subunit.

Embryos

Adults of *D. melanogaster* strain Canton S are cultured en masse in population cages using standard methods,[5] and the embryos are collected on yeasted grape juice-agar plates for 16 hr. The embryos are washed and dechorionated with bleach[5,6] and processed immediately. We find that the embryos must be fresh; purification from stored or frozen embryos yields Pol α preparations that have undergone substantial degradation of the 182-kDa subunit.[1]

Chromatography Materials

Single-stranded DNA-cellulose (ssDNA-cellulose) is prepared by the method of Alberts and Herrick.[7] Because the cellulose recommended for use in this method is no longer commercially available, we found that SigmaCell Type 50 cellulose works for this purpose. Nitex screen is available from Tetko Inc. Mono Q chromatography columns are purchased from Pharmacia. P-11 phosphocellulose is purchased from Whatman. All other chemicals are of the highest grade available.

Buffers

Homogenization buffer: 15 m*M* HEPES–NaOH, pH 8.0, 5 m*M* KCl, 0.5 m*M* magnesium acetate, 0.05 m*M* EDTA, 0.35 *M* sucrose, 0.5 m*M* dithiothreitol (DTT), 1 m*M* phenylmethylsulfonyl fluoride (PMSF), 10 m*M* sodium bisulfite, 2 μg/ml leupeptin

[4] P. G. Mitsis, S. C. Kowalczykowski, and I. R. Lehman, *Biochemistry* **32,** 5257 (1993).

[5] M. Ashburner, "Drosophila, a Laboratory Handbook," Cold Spring Harbor Laboratory Press, New York, 1989.

[6] C. L. Brakel and A. B. Blumenthal, *Biochemistry* **16,** 3137 (1977).

[7] B. Alberts and G. Herrick, *Methods Enzymol.* **21,** 198 (1971).

Potassium phosphate buffers: All potassium phosphate buffers are at pH 7.6 and contain 1 m*M* 2-mercaptoethanol, 0.2 m*M* EDTA, 1 m*M* PMSF, 10 m*M* sodium bisulfite, 2 μg/ml leupeptin, and 10% (v/v) glycerol

Mono Q buffer: 20 m*M* Tris–HCl, pH 8.0, 100 m*M* KCl, 0.5 m*M* EDTA, 10% (v/v) glycerol, 0.5 m*M* DTT, 0.5 μg/ml leupeptin, 0.5 μg/ml pepstatin, 0.5 m*M* PMSF

Storage buffer: 20 m*M* HEPES–NaOH, pH 7.5, 200 m*M* sodium acetate, 10 m*M* EDTA, 0.001% (v/v) Triton X-100, 50% (v/v) glycerol, 0.5 m*M* DTT, 0.5 μg/ml leupeptin, 0.5 μg/ml pepstatin A, 0.5 m*M* PMSF.

Protein Determination and Gel Electrophoresis

Protein concentrations are determined by the method of Bradford.[8] Bovine serum albumin (BSA) is used as a concentration standard. SDS–polyacrylamide gel electrophoresis (SDS–PAGE) is performed by the method of Laemmli.[9]

DNA Polymerase Assay

Since it is less subject to contaminating activities than the primase assay (i.e., RNA polymerase), DNA polymerase activity is used to follow the purification. Reaction mixtures (0.1 ml) contain 50 m*M* Tris–HCl, pH 8.5, 5 m*M* 2-mercaptoethanol, 20 m*M* ammonium sulfate, 10 m*M* $MgCl_2$, 200 μg BSA, 25 μg of activated calf thymus DNA, 100 μM, dATP, dGTP, dCTP, and [^{3}H]dTTP (200–6000) cpm/pmol, and enzyme. Incubation is for 30 min at 37°. One unit of DNA polymerase activity is the amount that catalyzes the incorporation of 1 nmol of dNTP into acid-insoluble material in 60 min at 37°. It is important to titrate carefully the peak of activity to stay within the linear range.

DNA Primase Assay

Reaction mixtures (25 μl) contain 50 m*M* Tris–HCl, pH 8.5, 10 m*M* $MgCl_2$, 4 m*M* DTT, 5 μg BSA, 1.0 μg of poly(dT), 2 m*M* ATP, 100 μM [^{3}H]dATP (300 cpm/pmol), 0.6 unit of *Escherichia coli* DNA polymerase I, and enzyme. Incubation is at 30° for 30 min. One unit of DNA primase activity is the amount that catalyzes the incorporation of 1 nmol of dATP into acid-insoluble material in 60 min at 30°.

[8] M. Bradford, *Anal. Biochem.* **72,** 248 (1976).

[9] U. K. Laemmli, *Nature* **227,** 680 (1970).

Purification of Pol α

All operations are performed at 4° unless indicated otherwise.

Preparation of Extract

The dechorionated embryos are resuspended in homogenization buffer at 4 ml/g of embryos and homogenized in 30- to 40-ml portions by five strokes of a 50-ml stainless steel homogenizer with a Teflon pestle. The homogenate is filtered through 75-μm Nitex cloth. The retentate is homogenized in the same buffer (2 ml/g of embryos) and filtered as above. The combined fractions are centrifuged at 10,000*g* for 15 min, and the supernatant recentrifuged at 125,000*g* for 60 min. The high-speed supernatant is decanted through eight layers of sterile cheesecloth to remove residual lipid to give the S-100 fraction (Fraction I).

Fraction I is adjusted to the ionic equivalent of 80 m*M* potassium phosphate with 1 *M* potassium phosphate 7.6. This fraction can be safely frozen in liquid nitrogen and stored at −80° if desired.

Phosphocellulose Chromatography

If frozen, the S-100 fraction is quickly thawed at 30° over a period of 15 to 20 min with occasional mixing, then transferred to ice. The S-100 fraction is loaded onto a phosphocellulose column (5 ml packed phosphocellulose per gram of embryos) that had been equilibrated with 80 m*M* potassium phosphate buffer at 0.6 to 0.8 column volumes per hour. The column is washed with 100 m*M* potassium phosphate buffer at 0.9 to 1.2 column volumes per hour until protein can no longer be detected in the effluent. The column is rapidly eluted with a 2 column volume gradient of 100 to 300 m*M* potassium phosphate buffer, and washed with 500 m*M* potassium phosphate buffer. The enzyme elutes at ~200 m*M* potassium phosphate buffer but will occasionally elute at a higher salt concentration, depending on the batch of embryos. The active fractions are pooled (Fraction II).

Solid ammonium sulfate is added to 0.131 g/ml of Fraction II over a 30-min period and stirred an additional 45 min. The precipitated protein is collected by centrifugation at 32,000*g* for 40 min. The pellet can be stored at this point at 4° for 4–8 hr or frozen in liquid nitrogen and stored at −80°.

Hydroxylapatite Chromatography

The ammonium sulfate pellet is dissolved in 10 m*M* potassium phosphate buffer (Fraction III) and dialyzed against 10 m*M* potassium phosphate buffer until the ionic equivalent of 100 m*M* potassium phosphate buffer is reached. The dialyzed fraction is clarified by centrifugation at 12,000*g* for

5 min and is loaded onto a hydroxylapatite column (equilibrated with 100 m*M* potassium phosphate buffer) at 5 mg packed hydroxylapatite per milligram of protein, at a flow rate of 2 column volumes per hour. The column is washed with 100 m*M* potassium phosphate buffer and the protein is eluted with a linear gradient of 100 to 200 m*M* potassium phosphate buffer at a flow rate of 3 column volumes per hour. The fractions containing DNA polymerase activity elute at ~145 m*M* potassium phosphate buffer (Fraction IV).

Fraction IV is concentrated by ammonium sulfate precipitation as described above. The pellet can be stored overnight at 4° and resuspended as above in 20 m*M* potassium phosphate buffer (Fraction IVa).

ssDNA-Cellulose Chromatography

Fraction IVa is dialyzed to an ionic equivalent of 20 m*M* potassium phosphate buffer containing 60 m*M* NaCl and centrifuged as above. The supernatant is loaded onto a ssDNA-cellulose column (washed with 5 column volumes of 2 *M* NaCl and equilibrated with 20 m*M* potassium phosphate buffer containing 60 m*M* NaCl) at 1 to 1.5 mg protein per milliliter of ssDNA-cellulose, at a flow rate of 1 column volume per hour. The column is then washed with 1 column volume of 20 m*M* potassium phosphate buffer containing 60 m*M* NaCl at 2 column volumes per hour. The column is eluted with 3 column volumes of 20 m*M* potassium phosphate buffer containing 350 m*M* NaCl. The DNA polymerase should be present in the fractions eluting at 350 m*M* NaCl.

The active fractions (Fraction V) are pooled and mixed with a saturated solution of ammonium sulfate (pH 7.0) at 1.2 ml/ml of Fraction V. After 60 min at 0° the precipitate is collected by centrifugation at 150,000*g* for 45 min. The precipitate is resuspended in 0.5 to 1.0 ml of 20 m*M* potassium phosphate buffer (Fraction Va).

Glycerol Gradient Sedimentation

Fraction Va is layered onto two preformed 10.6 ml 10 to 30% glycerol gradients containing 50 m*M* potassium phosphate, 200 m*M* ammonium sulfate, 1 m*M* 2-mercaptoethanol, 1 m*M* EDTA, 1 m*M* PMSF, 10 m*M* sodium bisulfate, and 2 μg/ml leupeptin, prepared in polyallomar tubes suitable for use in a Beckman SW41 rotor. The capacity is about 5 mg protein per gradient, however, about 2 mg is routinely loaded. Centrifugation is at 35,000*g* for 38.5 hr, and 15-drop fractions are collected. The active fractions (Fraction VI) are pooled and stored at −80°. At this step the enzyme is 40 to 50% homogeneous; it is free of deoxyribonuclease activity and can be used for most studies. To purify Pol α to about 80% homogeneity, several preparations of Fraction VI can be combined and chromatographed on a Mono Q column. Mono Q is particularly effective for purifying Pol α preparations containing degraded forms of the 182-kDa subunit.

Mono Q Chromatography

Fraction VI (0.35 mg) is dialyzed against 500 ml of Mono Q buffer containing 100 m*M* KCl for 5 hr. The dialyzed Fraction VI is loaded onto a Mono Q column that has been equilibrated with Mono Q buffer containing 100 m*M* KCl. The column is eluted with a 20-ml linear gradient of Mono Q buffer containing 0.1 to 0.4 *M* KCl at 1 ml/min, collecting 0.5-ml fractions. The peak of DNA polymerase activity appears at approximately 300 m*M* KCl. The active fractions are dialyzed against storage buffer, and stored at −80°. The purified enzyme is stable at −80°, however, it loses activity once thawed.

Pol α can be purified to 40 to 50% homogeneity in five steps with good yield (a typical purification is shown in Table I). Following Mono Q chromatography, the enzyme consists of four subunits with molecular weights of 182,000, 73,000, 60,000, and 50,000 when examined by SDS–PAGE (Fig. 1). Quantitation by densitometry of the SDS–polyacrylamide gel after staining with Coomassie Blue G indicates that this preparation is 80% homogeneous and is substantially free of proteolytic breakdown products of the 182-kDa subunit. This subunit composition is consistent with that found for Pol α from a wide variety of sources, an indication of the high degree of conservation of the subunit structure of this enzyme.[10]

Properties of Purified Pol α

The purified Pol α requires divalent cations for activity. There is a broad optimum of DNA polymerase activity at 10 to 15 m*M* Mg^{2+} or 0.1 to 0.15 m*M* Mn^{2+}, with Mg^{2+} giving about sevenfold more activity than Mn^{2+}. It shows a broad pH optimum with a peak at pH 8.0. DNA polymerase activity is inhibited by increasing concentrations of monovalent ions, N-ethyl maleimide (NEM) (>90% inhibition at 0.1 m*M*) and aphidicolin (80% at 5 μg/ml), and is stimulated by low concentrations (40 m*M*) of ammonium sulfate. Activity is dependent on the four dNTPs and a primed template. The K_m for dTTP is 3.7 μ*M*. Of the templates tested, Pol α shows the most activity on activated DNA, followed by multiply primed homopolymer templates, and is least active on random sequence single-primed ssDNA. The level of activity with these templates is very condition dependent. For example,

[10] T. S. Wang, *Ann. Rev. Biochem.* **60,** 513 (1991).

TABLE I
PURIFICATION OF DNA POLYMERASE–PRIMASE FROM *Drosophila Melanogaster*[a]

Fraction	Volume (ml)	Protein (mg)	Polymerase		Primase		Polymerase/primase
			Activity (units × 10^{-2})	Specific activity (units/mg)	Activity (units × 10^{-2})	Specific activity (units/mg)	
S-100 (I)	360	2,581	1,792	69.4	792	30.7	2.3
Phosphocellulose (II)	493	261	704	269	266	86.5	3.1
Ammonium sulfate (III)	50	65	772	1,188	189	291	4.1
Hydroxylapatite (IV)	41	8.3	284	3,425	66.4	800	4.3
DNA-cellulose (V)	5	0.78	142	18,256	34.6	4,431	4.1
Glycerol gradient (VI)	1.5	0.19	98.3	51,737	24.6	12,947	4.0

[a] Freshly harvested embryos.

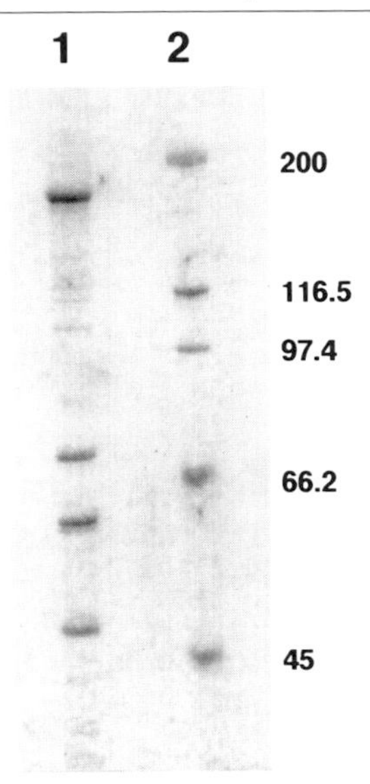

FIG. 1. SDS–PAGE of Pol α purified by Mono Q chromatography. Two micrograms of Pol α were resolved by electrophoresis on a 7.5% SDS–polyacrylamide gel and stained with Coomassie Blue G. Lane 1, Pol α; lane 2, molecular weight ($\times 10^{-3}$) markers.

addition of ammonium sulfate selectively inhibits DNA polymerase activity on poly (dA) · oligo (dT) as template by approximately 100-fold.

The primase activity of Pol α catalyzes the synthesis of oligoribonucleotides, 10 to 15 nucleotides in length, with single-stranded M13 DNA, as template. These oligoribonucleotides can then serve as primers for the DNA polymerase.[11]

Separation of Pol α Subunits

The Pol α subunits can be dissociated by treatment with either 2.5 *M* urea or 50% ethylene glycol.

Separation of DNA Polymerase from Primase[12]

The holoenzyme is treated with 3.4 urea for 4 hr at 4°, then applied to a 10 to 30% glycerol gradient (formed as in the purification protocol) containing 2.5 *M* urea. It is sedimented for 72 hr at 41,000 rpm in a Beckman SW41 rotor, and active fractions are collected.

[11] R. C. Conaway and I. R. Lehman, *Proc. Natl. Acad. Sci. USA* **79,** 2523 (1982).

[12] L. S. Kaguni, J. M. Rossignol, R. C. Conaway, G. R. Banks, and I. R. Lehman, *J. Biol. Chem.* **258,** 9937 (1983).

Starting with 20 μg of protein, two complexes consisting of the 182-kDa DNA polymerase and 73-kDa subunits, and the 60- and 50-kDa primase subunits can be recovered in approximately 5 and 10% yield of polymerase and primase activity, respectively.

Isolation of 182-kDa Subunit[13]

Pol α is applied to a 0 to 10% glycerol gradient containing 50% ethylene glycol, 20 mM potassium phosphate, pH 7.5, 2 mM DTT, and 20 mM ammonium sulfate. The gradient is sedimented in a Beckman SW56 rotor at 50,000 rpm for 64 hr at 4° and active fractions are collected.

Yields of polymerase activity from this procedure are about 10 to 15%. The ethylene glycol should be of highest quality available since even greater losses of activity can occur due to contaminants that accumulate during storage. The purified subunits are rather labile and should be stored at −80°.

Properties of Isolated Subunits

The isolated 60- and 50-kDa primase subunits show little difference from the primase activity of the intact four-subunit enzyme, with the exception that the length of primers synthesized by the isolated subunits is twice that of the holoenzyme.[14] The properties of the isolated 182-kDa polymerase subunit are also very similar to those of the intact enzyme with two exceptions: (1) It shows an enhanced processivity in the presence of the *E. coli* SSB and (2) it has a 3′ → 5′ (proofreading)-exonuclease that is undetectable in the intact four-subunit enzyme.[13,15]

Properties of 182-kDa Polymerase Subunit Overexpressed in Sf9 Cells

A full-length cDNA encoding the gene for the 182-kDa DNA polymerase subunit obtained from A. Matsukage[16] has been subcloned into a baculovirus transfer plasmid and expressed in *Spodoptera frugiperda* (Sf9, fall armyworm ovary) cells. The overexpressed protein is purified by chromatography on columns of phophocellulose, S-Sepharose, hydroxyapatite, single-strand DNA-agarose, and Q-Sepharose. The purified enzyme consists of a single, 182-kDa polypeptide as judged by SDS–PAGE followed by

[13] S. M. Cotterill, G. Chui, and I. R. Lehman, *J. Biol. Chem.* **262,** 16,105 (1987).

[14] S. M. Cotterill, G. Chui, and I. R. Lehman, *J. Biol. Chem.* **262,** 16,100 (1987).

[15] S. M. Cotterill, M. E. Reyland, L. A. Loeb, and I. R. Lehman, *Proc. Natl. Acad. Sci. USA* **84,** 5635 (1987).

[16] F. Hirose, M. Yamaguchi, Y. Nishada, M. Masutani, H. Miyazawa, F. Hanaoka, and A. Matsukage, *Nucleic Acids Res.* 4991 (1991).

staining with Coomassie blue or immunoblotting with a monoclonal antibody (MAb) directed against *Drosophila* Pol α.

Like the 182-kDa DNA polymerase subunit isolated from the four-subunit enzyme by glycerol gradient sedimentation in the presence of ethylene glycol (see above), the baculovirus expressed DNA polymerase subunit possesses an intrinsic 3′ → 5′-exonuclease activity. The DNA polymerase and 3′ → 5′-exonuclease activities of the overexpressed 182-kDa subunit bind to and coelute from a 5′ AMP-agarose column, a feature of 3′ → 5′-exonuclease containing DNA polymerases. DNA polymerase and 3′ → 5′-exonuclease activities also cosediment during glycerol gradient sedimentation of the overexpressed 182-kDa subunit.[17]

DNA Polymerase δ

Although Pol α is readily purified from extracts of *Drosophila* embryos, Pol δ has been difficult to detect. Recently, however, it has been purified to near homogeneity from extracts of 0- to 2-hr embryos. Critical to the success of the purification is the use of 0- to 2-hr rather than 0- to 16-hr embryos, and centrifugation of the extract prior to purification at low (10,000*g*) rather than high (125,000*g*) speeds. The purification method described is that of Chiang *et al.*[2]

Embryos

Drosophila melanogaster (Canton S) adults are cultured as described above, and embryos are collected on yeasted grape juice-agar plates for 2 hr.

Chromatography Materials

The ssDNA-cellulose is prepared as described above. Phosphocellulose can be obtained from Whatman. DEAE-Sephacel and Q-Sepharose Fast Flow are purchased from Pharmacia. Hydroxyapatite Bio-Gel HT is obtained from Bio-Rad. Heparin-Sepharose is prepared as described by Davison *et al.*[18]

Buffers

Buffer A: 20 m*M* HEPES–NaOH, pH 7.5, 5 m*M* magnesium acetate, 50 m*M* potassium acetate, 0.5 m*M* EGTA, 0.1% (v/v) Triton X-100, 10% (v/v) glycerol, 10 m*M* sodium bisulfite, 1 m*M* DTT; 0.5 μg leupeptin per milliliter, 0.5 μg pepstatin A per milliliter.

[17] S.-K. Lee, G. Lindberg, and I. R. Lehman, unpublished (1992).

[18] B. L. Davison, T. Leighton, and J. C. Rabinowitz, *J. Biol. Chem.* **254,** 9220 (1979).

Buffer B: 20 mM HEPES–NaOH, pH 7.5, 10% (v/v) glycerol, 10 mM sodium bisulfite, 1 mM EDTA, 0.5 mM EGTA, 1 mM DTT, 0.5 μg leupeptin per milliliter, 0.5 μg pepstatin A per milliliter, 0.5 mM PMSF.

Buffer C: 20 mM HEPES–NaOH, pH 7.6, 10% (v/v) glycerol, 5 mM $MgCl_2$, 0.5 mM EDTA, 0.5 mM EGTA, 0.5 mM DTT, 0.5 μg leupeptin per milliliter, 0.5 μg pepstatin A per milliliter, 0.5 mM PMSF.

Buffer D: Same as Buffer C but containing 0.1 M KCl and no EDTA or EGTA.

Protein Determination and Gel Electrophoresis

SDS–PAGE and protein determinations are performed as described above.

Enzyme Assays

DNA polymerase activity is measured in reaction mixtures (25 μl) containing 20 mM HEPES–NaOH, pH 7.5, 5 mM magnesium acetate, 0.1 mg per milliliter BSA per milliliter, 10% (v/v) glycerol, 0.5 mM DTT, 350 μg activated salmon sperm DNA per milliliter, 100 μM each dATP, dCTP, and dGTP, and 25 μM [α-^{32}P]dTTP (3000 cpm/pmol). After 20 min at 25°, acid-precipitable radioactivity is determined.

Exonuclease activity is measured in the same reaction buffer used for measurement of DNA polymerase activity, with 2 mM ATP replacing the deoxynucleoside triphosphates. *Hin*fI-digested, 3′-end labeled pUC 18 DNA is used as substrate. The release of acid-soluble radioactivity is measured after 12 min at 25°.

Purification of Pol δ

All operations are performed at 4° unless otherwise indicated.

Preparation of Extract

Embryos (0 to 2 hr) are collected, dechorionated with bleach, washed, and homogenized as described above, using 0.5 ml of Buffer A and 0.01 ml of 100 mM PMSF per gram of embryos. The homogenate is centrifuged at 10,000g for 15 min. The supernatant is collected and centrifuged at 10,000g for 15 min. The supernatant is collected, frozen in liquid nitrogen, and stored at −80° (Fraction I).

Phosphocellulose Chromatography

A 1-liter column (9 × 16.5 cm) of phosphocellulose is equilibrated with Buffer B containing 0.1 M KCl. The extract is thawed slowly on ice and loaded onto the column at 300 ml/hr. The column is washed with 6 liters of Buffer B containing 0.1 *M* KCl and eluted with Buffer B containing 0.5 *M* KCl at 500 ml/hr. The peak of DNA polymerase activity is pooled and stirred with 0.56 g of ammonium sulfate per milliliter of eluate for 5 hr. The precipitate is collected by centrifugation at 15,000*g* for 25 min. The pellet is resuspended in Buffer C containing 0.1 *M* KCl, dialyzed against the same buffer for 12 hr, and the solution clarified by centrifugation at 15,000*g* for 25 min (Fraction II).

Heparin-Sepharose Chromatography

One-half of Fraction II is loaded onto a 300-ml column (4.8 × 17 cm) of heparin-Sepharose CL-4B (containing 0.45 mg of heparin per milliliter of packed heparin-Sepharose) equilibrated with Buffer C containing 0.1 *M* KCl at 120 ml/hr. The column is washed with 1.2 liters of Buffer C containing 0.1 *M* KCl and then eluted at 90 ml/hr with a 1.5-liter gradient of 0.1 to 0.5 *M* KCl in Buffer C.

The DNA polymerase activity is resolved into four peaks, eluting at 150, 240, 310, and 350 m*M* KCl (Fig. 2). Immunoblot analysis of the four peaks with monoclonal antibody directed against the 182-kDa subunit of Pol α showed that peaks I and II contain epitopes recognized by the antibody. Comparison with the purified intact Pol α indicated that peak II

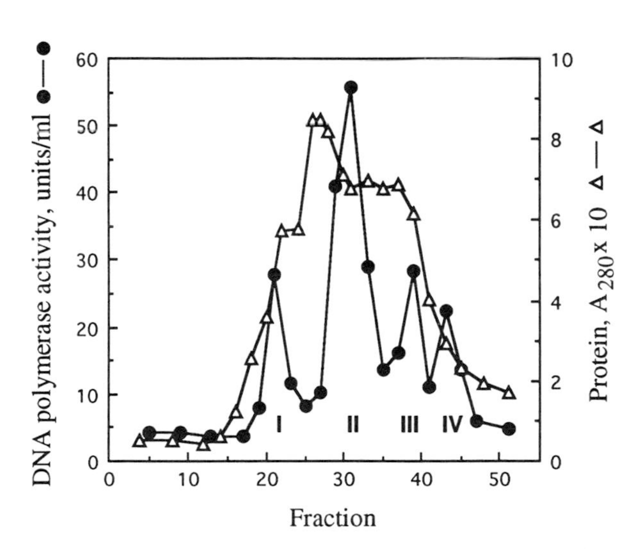

FIG. 2. Chromatography of *Drosophila* DNA polymerases on heparin-Sepharose. The identities of peaks I–IV are described in the text. Taken from Chiang *et al.*[2]

is a mixture of intact and partially degraded forms of Pol α, and peak I contains a degraded form of Pol α.

The DNA polymerase activity of peaks III and IV are not recognized by the antibody. Although both peaks contain exonuclease, they show different properties upon further characterization. The DNA polymerase activity of peak III is not retained by DEAE-Sephacel and prefers poly(dA-dT) as template. In contrast, the DNA polymerase activity of peak IV adsorbed to DEAE-Sephacel and prefers poly (dA) · oligo (dT) as template. The DNA polymerase activity of peak III shows a low degree of processivity, whereas peak II is highly processive. The properties of peaks III and IV are characteristic of Pol δ and Pol ε, respectively.[10]

Heparin-Sepharose chromatography is repeated with the remaining half of Fraction II. The active fractions from peak III are pooled, dialyzed

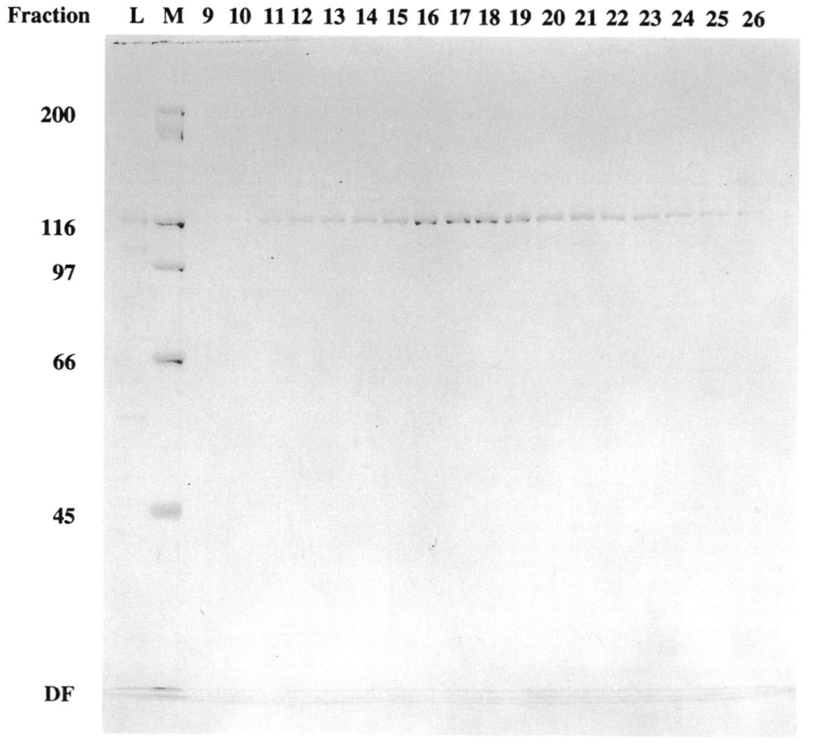

FIG. 3. SDS–PAGE of ssDNA-cellulose fractions of Pol δ. See text for details. Lane L, load (Fraction V); lane M, molecular weight ($\times 10^{-3}$) markers; DF, dye front. Taken from Chiang *et al.*[2]

against Buffer C containing 0.1 *M* KCl for 12 hr, and centrifuged at 15,000*g* for 25 min (Fraction III).

DEAE-Sephacel and Q-Sepharose Fast Flow Chromatography

Fraction III is loaded onto a 72-ml DEAE-Sephacel column (2.5 × 15.5 cm) equilibrated with Buffer C containing 0.1 *M* KCl at 72 ml/hr. The flow-through fraction is loaded directly onto a 42-ml Q-Sepharose Fast Flow column (1.6 × 21.5 cm) equilibrated with Buffer C containing 0.1 *M* KCl at 63 ml/hr. The Q-Sepharose column is washed with 210 ml of Buffer C containing 0.1 *M* KCl and eluted with a 840-ml linear gradient of 0.1 to 0.5 *M* KCl in Buffer C at 42 ml/hr. The peak of DNA polymerase activity is dialyzed against Buffer D for 12 hr and clarified by centrifugation at 15,000*g* for 25 min (Fraction IV).

Hydroxylapatite Chromatography

Fraction IV is loaded onto a 5.5-ml column of hydroxylapatite (1 × 7.5 cm) equilibrated with Buffer D at 5.5 ml/hr. The column is washed with Buffer D and eluted with a 110-ml linear gradient of 0 to 0.1 *M* sodium phosphate in Buffer D at 5.5 ml/hr; 60-drop fractions (~1.4 ml) are collected. The peak of DNA polymerase activity is dialyzed against Buffer C containing 40 m*M* KCl for 18 hr and clarified by centrifugation at 15,000*g* for 25 min (Fraction V).

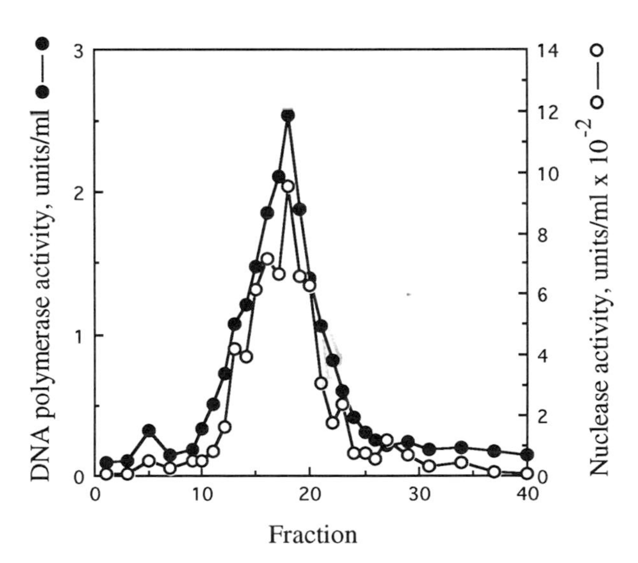

FIG. 4. Chromatography of *Drosophila* Pol δ on ssDNA-cellulose. Taken from Chiang *et al.*[2]

TABLE II
PURIFICATION OF POL δ FROM *Drosophila* EMBRYOS[a]

Fraction	Protein (mg)	Total activity (units)	Specific activity (units/mg)
I. Crude extract	9950	—	—
II. Phosphocellulose/ammonium sulfate	2219	—	—
III. Heparin-Sepharose	341.2	1215	3.6
IV. DEAE-Sephacel/Q-Sepharose Fast Flow	14.4	149.4	10.4
V. Hydroxylapatite	0.35	22.7	64.8
VI. ssDNA-cellulose/Q-Sepharose Fast Flow	0.0015	3.42	2280

[a] 0 to 2 hr.

ssDNA-Cellulose and Q-Sepharose Fast Flow Chromatography

Fraction V is loaded at 2 ml/hr onto a 2-ml column of ssDNA-cellulose (0.7 × 2.5 cm) containing 3 to 4 mg of ssDNA per gram of dry cellulose equilibrated with Buffer C containing 40 m*M* KCl. The column is washed with the same buffer and eluted with a 40-ml linear gradient of 40 to 200 m*M* KCl in Buffer C at 5 ml/hr; 15-drop fractions (~350 μl) are collected. The peak of DNA polymerase activity is concentrated by loading it onto a 100-μl Q-Sepharose Fast Flow column (0.3 × 1.2 cm) equilibrated with Buffer C containing 0.1 *M* KCl and then eluting the column with Buffer C containing 0.25 *M* KCl. The DNA polymerase peak can be frozen in liquid nitrogen and stored at −80° (Fraction VI).

The ssDNA-cellulose fraction (Fraction VI) is >95% homogeneous. It consists of a single polypeptide with a molecular mass of 120 kDa (Fig. 3), and contains both DNA polymerase and exonuclease activities (Fig. 4). It is purified >600-fold with a yield of 0.3% from the heparin-Sepharose peak III (Table II). One and one-half micrograms of purified enzyme are obtained from 99.5 g of protein (250 g of embryos). The low yield may be

TABLE III
EFFECT OF INHIBITORS ON *Drosophila* POL δ[a]

Inhibitor added	Concentration	DNA polymerase activity
None		1.00
Aphidicolin	10 μg/ml	0.16
BuPdGTP	10 μ*M*	1.10
BuAdATP	10 μ*M*	1.11
Carbonyl diphosphonate	15 μ*M*	0.44
DMSO	10%	1.40
ddTTP	500 μ*M*	1.03

[a] Taken from Chiang *et al.*[2]

attributed, in part, to difficulty in the extraction of Pol δ. As noted above, it can only be identified in low-speed supernatants of embryo extracts, but not in the high-speed supernatant commonly used for the purification of eukaryotic DNA polymerases. The low yield is also a consequence of the loss sustained during ssDNA-cellulose chromatography, a step that is required to remove the last traces of contaminating protein. However, the hydroxylapatite fraction (Fraction V) (350 μg from 250 g of embryos), which is approximately 70% pure (see Fig. 3, lane L), is suitable for most studies of the enzyme.

Properties of Pol δ

The purified *Drosophila* Pol δ prefers poly(dA-dT) as a template over activated DNA and poly(dA) · oligo(dT); it is inactive with poly(rA) · oligo(dT). It is most active in the presence of 2 m*M* $MgCl_2$ at pH 7.0, and is inhibited by 200 m*M* NaCl. The exonuclease activity is inhibited by dNTPs.

The effect of various inhibitors on the *Drosophila* Pol δ is shown in Table III. It is inhibited by aphidicolin and carbonyl diphosphonate, but is resistant to BuPdGTP, BuAdATP, and ddTTP.

Unlike Pol δ from yeast and mammalian sources, the processivity of deoxynucleotide polymerization by the *Drosophila* Pol δ is not stimulated by *Drosophila* PCNA.[10] The lack of response to PCNA may be a consequence of the absence of a 50-kDa subunit that is present in the yeast and mammalian enzymes, but is absent from the *Drosophila* Pol δ.

Acknowledgment

This work was supported by grant GM 06196 from the National Institutes of Health.

[8] Purification of Mammalian DNA Polymerases: DNA Polymerase α

By TERESA S.-F. WANG, WILLIAM C. COPELAND, LARS ROGGE, and QUN DONG

Introduction

DNA polymerase α plays a key role in the initiation and replication of the mammalian chromosome.[1] Despite intense effort by numerous laboratories, the complexity of the DNA polymerase α structure and the low abundance and lability of this enzyme have distinguished its purification

[1] T. S.-F. Wang, *Ann. Rev. Biochem.* **60,** 513 (1991).

from mammalian cells and tissues as one of the most problematic in the biochemical literature.

The production of a panel of monoclonal antibodies against human DNA polymerase α allowed the development of a three-step immunoaffinity purification protocol.[2] The monoclonal antibodies were used as the immunoligand bound to a protein A-Sepharose 4B resin to purify the DNA polymerase α from cultured human cells.[3] This protocol yielded an immunocomplex of four polypeptides and a stoichiometric amount of monoclonal antibody, which has both the DNA polymerase α and DNA primase activities.[3,4] The four polypeptides have been identified as a 180-kDa protein containing the catalytic activity of DNA polymerase α, a 70-kDa protein with no known catalytic function, and two polypeptides of 58 and 49 kDa with primase activity.[4]

Identification of the DNA polymerase α catalytic polypeptide led to the isolation of its full-length cDNA by the reverse genetic approach.[5] The cDNA was subsequently constructed into various plasmid vectors to be expressed in monkey COS 7 cells, in yeast, and into baculovirus for expression in insect cells.[6,7] The DNA polymerase α catalytic polypeptide constructed in the recombinant baculovirus is expressed in insect cells in at least 1000-fold higher levels than in normal cultured human cells.[6] Such overproduction allowed the development of a rapid one-step immunoaffinity purification protocol.[8–10] This protocol utilizes a nonneutralizing monoclonal antibody, SJK237-71, as the immunoligand. The antibody is covalently cross-linked to Sepharose 4B resin. After washing, the bound DNA polymerase α is eluted from the immunoaffinity resin by a buffer containing $MgCl_2$, similar to the procedure described by Chang *et al.*[11] This efficient single-step immunoaffinity purification protocol gives a high yield of DNA polymerase α with high specific activity and with no monoclonal antibody (MAb) present in the enzyme fraction. Like that of the three-step immunoaffinity purification protocol with immunoglobulin G (IgG)-bound protein A-Sepharose, this one-step protocol is able to purify both the recombi-

[2] S. Tanaka, S.-Z. Hu, T. S.-F. Wang, and D. Korn, *J. Biol. Chem.* **257,** 8386 (1982).
[3] T. S.-F. Wang, S.-Z. Hu, and D. Korn, *J. Biol. Chem.* **259,** 1854 (1984).
[4] S. W. Wong, L. R. Paborsky, P. A. Fisher, T. S.-F. Wang, and D. Korn, *J. Biol. Chem.* **261,** 7958 (1986).
[5] S. W. Wong, A. F. Wahl, P.-M. Yuan, N. Arai, B. E. Pearson, K.-I. Arai, D. Korn, M. W. Hunkapiller, and T. S.-F. Wang, *EMBO J.* **7,** 37 (1988).
[6] W. C. Copeland and T. S.-F. Wang, *J. Biol. Chem.* **266,** 22,739 (1991).
[7] S. Francesconi, W. C. Copeland, and T. S.-F. Wang, *Mol. Gen. Genet.* **241,** 457 (1993).
[8] L. Rogge and T. S.-F. Wang, *Chromosoma* **102,** S114 (1992).
[9] W. C. Copeland and T. S.-F. Wang, *J. Biol. Chem.* **268,** 11,028 (1993).
[10] Q. Dong, W. C. Copeland, and T. S.-F. Wang, *J. Biol. Chem.* **268,** 24,163 (1993).
[11] L. M. S. Chang, E. Rafter, C. Augl, and F. J. Bollum, *J. Biol. Chem.* **259,** 14,679 (1984).

nant single subunit DNA polymerase α catalytic polypeptide produced from baculovirus infected insect cell lysates as well as the four-subunit DNA polymerase α–primase complex from cultured human cells. The enzymatic properties of the DNA polymerase α purified by this one-step protocol are identical to those of the four-subunit DNA polymerase α–primase purified by the three-step IgG–protein A immunoaffinity protocol.[6–10]

Protocol

Preparation of Cell Extracts from Human Cell Lines

Human KB cells are grown as a 9-liter suspension culture in MEM containing 10% calf serum and are harvested at a cell density of 3.5×10^5 cells/ml by centrifugation at 800*g* for 10 min. Cell pellets are washed twice by centrifugation and resuspension of the cells in 20 ml of 50 m*M* KPO_4, pH 7.5, and 150 m*M* NaCl. The washed cell pellets are then combined into a preweighed centrifuge tube and the wet weight of the cell pellet is estimated. For cell lysis, cells are resuspended with a wide-mouth pipette in 9× wet cell weight volumes of hypotonic Dounce buffer. The hypotonic Dounce buffer is formulated by adding $MgCl_2$ to a final concentration of 2 m*M* to a buffer of 5 m*M* KPO_4, pH 8.0, 1 m*M* EDTA, 1 m*M* 2-mercaptoethanol at 4°. Immediately before suspending the cell pellet into the hypotonic Dounce buffer, the buffer is adjusted to 10 m*M* of sodium bisulfite by adding 1/100th volume of 1 *M* sodium bisulfite, and to 1 m*M* phenylmethylsulfonyl fluoride (PMSF) by adding 1/100th volume of 100 m*M* stock solution of PMSF. The cells are evenly suspended in the hypotonic Dounce buffer by aspirating with a wide-mouth pipette. The cell suspension is then allowed to swell in the hypotonic Dounce buffer for 30 to 60 min on ice, and the extent of cell volume increase is monitored by microscopic examination. Swollen cells are broken with a tight-fitting glass Dounce homogenizer and cell breakage is monitored by microscopic examination. Cell lysates are first spun at 1000*g* to remove the nuclei and then spun at 15,000*g* to remove the organelles.

Preparation of Cell Extracts from Insect Cells Infected with Recombinant Baculovirus

Insect cells (Sf98 or Sf21) are grown in ten T150 flasks in Grace medium and 10% fetal calf serum to 60 to 80% confluence. Cells are infected with recombinant baculovirus expressing the human DNA polymerase α catalytic subunit at a multiplicity of infection (MOI) of 10 pfu (plaque-forming unit) per cell. Forty-eight hours postinfection, the infected cells are dislodged from the T150 flasks by physical shaking and collected by centrifugation at $<200g$. The cell pellet is washed once with serum-free Grace medium to remove residual serum protein and resuspended in 3-wet cell weight volumes of lysis buffer containing 20% ethylene glycol, 100 m*M* Tris–HCl, pH 7.5, 100 m*M* NaCl, 0.5% Nonidet P-40 (NP-40), 1 m*M* EDTA, 1 m*M* 2-mercaptoethanol, 1 m*M* PMSF, and 1 m*M* sodium bisulfite. The PMSF and sodium bisulfite are added to the buffer immediately prior to suspending the insect cells as described for the human KB cell lysate preparation. The insect cell suspension is incubated on ice for 10 min and then sonicated for 10 sec at 4°. The extract is centrifuged for 10 min at 12,000*g* to remove the cell debris, nuclei, and organelles. The supernatant designated as "soluble crude extract" was adjusted to 100 m*M* ionic strength by dilution with 20% ethylene glycol, 1 m*M* 2-mercaptoethanol, and 1 m*M* EDTA.

One-step Immunoaffinity Purification

Cell lysates of either human cell line or recombinant baculovirus infected insect cells prepared as described above are mixed with 1 to 2 ml of an immunoaffinity Sepharose 4B resin, which is cross-linked with monoclonal antibody SJK237-71 at 3 mg of antibody per milliliter of Sepharose-4B resin.[2] The cell lysates are mixed with the immuno-Sepharose 4B resin by rotating end over end for 1 hr at 4°. The immuno-Sepharose 4B resin is collected by centrifugation at 800*g* for 5 min, and washed five times by centrifugation and resuspended with 50 ml of ice-cold 50 m*M* KPO_4, pH 7.5, 500 m*M* NaCl, and 0.5% NP-40. The Sepharose slurry is then washed once in a buffer of 50 m*M* Tris–HCl, pH 7.5, with 150 m*M* NaCl (Tris-buffered saline; TBS) to remove the KPO_4. The Sepharose 4B pellet is then resuspended again in TBS, packed into a 5-cc syringe column, and eluted with Mg buffer. The Mg buffer is prepared by adding $MgCl_2$ to a final concentration of 3.0 *M* into 50 m*M* Tris–HCl, pH 8.0, with no further pH adjustment. As the fractions are eluted from the immuno-Sepharose 4B column, the protein content of each fraction is measured immediately by Bradford assay.[12] Fractions containing protein are pooled and immediately diluted twofold with Tris–HCl, pH 8.0, and 150 m*M* NaCl, and dialyzed against a high EDTA buffer (10 m*M* EDTA, 50 m*M* Tris–HCl, pH 8.0, 1 m*M* 2-mercaptoethanol, and 20% (v/v) ethylene glycol) with three buffer changes to remove the $MgCl_2$. The enzyme fraction is further dialyzed against the same buffer without EDTA to remove the excess EDTA, and then concentrated by dialysis against a buffer containing 30% sucrose, 20% (v/v) ethylene glycol, 50 m*M* Tris–HCl, pH 8.0, 1 m*M* 2-mercaptoethanol,

[12] M. M. Bradford, *Anal. Biochem.* **72,** 248 (1976).

and 1 mM EDTA. The purified and concentrated DNA polymerase α samples are stored in small aliquots at $-80°$. The enzyme fraction could also be concentrated by dialysis in a buffer containing 50% glycerol, 50 mM Tris–HCl, pH 8.0, 1 mM 2-mercaptoethanol, 1 mM EDTA, and stored at $-20°$.

Purity, Yield, and Enzymatic Properties of DNA Polymerase α Purified by One-step Immunoaffinity Protocol

The protein structure of DNA polymerase α purified by this one-step immunoaffinity protocol, either as a four-subunit polymerase α–primase complex (from cultured human cells) or as a single-subunit polymerase α catalytic polypeptide (from recombinant baculovirus infected insect cells), is shown in Fig. 1. The specific activity of the single-subunit DNA polymerase α catalytic polypeptide purified by the one-step immunoaffinity protocol is comparable to that obtained by the three-step protein A immunoaffinity protocol (Table I). The yield of DNA polymerase α purified by the one-step protocol averages 50 to 80%, which is much higher than the average 20% yield obtained with the three-step protocol.[3,8–10]

The enzymatic properties of the single-subunit DNA polymerase α catalytic polypeptide purified by the one-step immunoaffinity protocol and by the three-step IgG–protein A protocol were compared (Table II). The

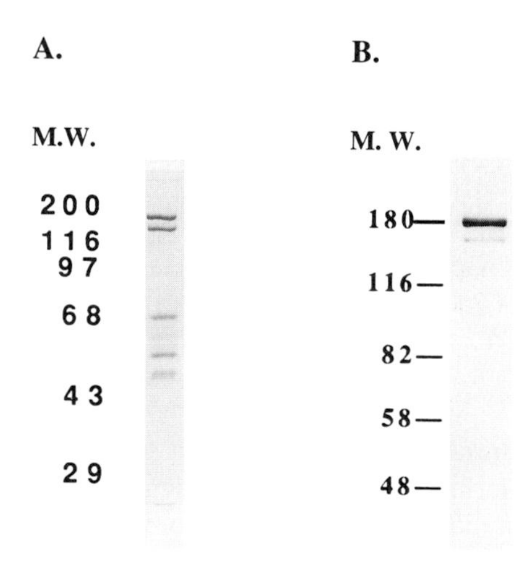

FIG. 1. DNA polymerase α purified by one-step immunopurification protocol. Coomassie blue-stained SDS gel of human DNA polymerase α purified as four-subunit DNA polymerase α–primase complex from KB cell lysates (lane A) and as recombinant single catalytic subunit from baculovirus-infected insect cell lysates (lane B). [A, Reprinted with permission from reference 9. B, Reprinted with permission from reference 10.]

TABLE I
PURIFICATION OF RECOMBINANT HUMAN DNA POLYMERASE α CATALYTIC SUBUNIT BY ONE-STEP AND THREE-STEP IMMUNOPURIFICATION PROTOCOLS[e]

Protocol	Activity[a] (units)	Protein (mg)	Specific activity (units/mg)	Yield (%)
One-step				
Crude cell lysate	156,000	370[b]	415	100
IgG-Sepharose 4B	83,000	0.61[b]	136,000	54
Three-step				
Crude cell lysate	38,400	98.4[b]	390	100
Phosphocellulose	35,710	18.7[b]	1,910	93
IgG-protein A	13,050	0.18[c]	70,000[d]	34
DNA cellulose	7680	0.038[c]	200,000[d]	20

[a] One unit of DNA polymerase α is defined as the amount of protein required to incorporate 1 nmol of labeled dNMP/hr at 37° into acid-insoluble radioactive material.

[b] Protein concentration was determined by Bradford analysis using bovine serum albumin as the standard.

[c] Estimated non-IgG protein determined by densitometric analysis of Coomassie blue-stained gels and compared with known amounts of bovine serum albumin standard.

[d] Specific activity of non-IgG protein. Specific activity of enzyme fractions estimated with IgG is 9700 and 57,000 units/mg for IgG–protein A and DNA cellulose fractions, respectively.

[e] Reprinted with permission from reference 6.

K_m values for dNTP and primer terminus, and the k_{cat} of the DNA polymerase α catalytic polypeptide purified by either method were comparable if not identical (Table II). Like that of the four-subunit DNA polymerase α–primase complex, their DNA synthetic processivity was moderate, and exhibited an average product length of 6 to 10 nucleotides per binding event (Table II). Both the one-step and three-step immunoaffinity protocol yielded purified DNA polymerase α equally sensitive to inhibitors such as aphidicolin and N^2-(*p*-*n*-butylphenyl)dGTP (Table II).

Comments

The one-step immunoaffinity purification protocol described here is a rapid and efficient method to isolate DNA polymerase α to near-homogeneity from both mammalian cell lysates and recombinant baculovirus-infected insect cell lysates. The entire operation from cell lysis to dialysis of the purified enzyme can be completed within 4 hr.

Note that the washing and elution conditions of the one-step immunopurification protocol described here apply only when the nonneutralizing monoclonal antibody, SJK237-71, is used as the immunoligand. When other

TABLE II

ENZYMATIC PROPERTIES OF RECOMBINANT DNA POLYMERASE α CATALYTIC SUBUNIT PURIFIED BY ONE-STEP AND THREE-STEP IMMUNOPURIFICATION PROTOCOLS

Property[a]	Polymerase α purified by	
	One-step protocol	Three-step protocol
Optimal pH	8.0	8.0
$K_{m(dNTP)}$, μM	1.6	1.20
$K_{m(primer\ terminus)}$, μM	0.6	0.4
k_{cat} (S^{-1})	1.8	1.6
Processivity[b]	6	7–13
50% BuPdGTP inhibition,[c] μM	0.04	0.22
50% aphidicolin inhibition,[c] μM	6–10	13–20

[a] Properties were assayed in reactions using optimally gapped salmon sperm DNA as primer–template in 20 mM Tris–HCl, pH 8.0, 2 mM 2-mercaptoethanol, 200 μg/ml bovine serum albumin, 10 mM $MgCl_2$, 50 μM dNTPs with [α-^{32}P]dATP as the label. K_m values for primer terminus were measured with oligo$(dT)_{12}$:poly$(dA)_{290}$ in 20 mM Tris–HCl, pH 8.0, 2 mM 2-mercaptoethanol, 200 μg/ml bovine serum albumin, 2 mM $MgCl_2$, and 50 μM [α-^{32}P]dTTP. All kinetic parameters were calculated from Lineweaver–Burk plots by the method of least squares.

[b] Processivity was measured with ^{32}P-labeled oligo$(dT)_{12}$ annealed to poly$(dA)_{595}$ by the DNA trap method.[9,10,13]

[c] The concentration of inhibitors required to inhibit the activity of 5 ng of DNA polymerase α in a 50-μl reaction. Inhibitor concentrations were derived from extrapolations of log graphs.

neutralizing or nonneutralizing monoclonal antibodies with different binding affinities to DNA polymerase α are used as immunoligands, washing buffers of different ionic strength with different percentages of nonionic detergent must be empirically defined.

To obtain a higher yield of the recombinant DNA polymerase α catalytic polypeptide, the insoluble insect cell pellet can be further extracted by 5-cell wet weight volumes of a high salt buffer containing 600 mM NaCl, 1 mM EDTA, 1 mM 2-mercaptoethanol, 1 mM PMSF, and 1 mM sodium bisulfite in 50 mM KPO_4, pH 7.5, with 20% ethylene glycol at 4°. The insoluble cell debris were removed by centrifugation at 12,000g and the supernatant of this high salt solubilized fraction can be combined with the

[13] A. H. Polesky, T. A. Steitz, N. D. F. Grindley, and C. M. Joyce, *J. Biol. Chem.* **265,** 14,579 (1990).

"soluble crude extract" from the original sonication as the enzyme source for the one-step immunoaffinity purification. In general, about 25% more DNA polymerase activity can be recovered by this high salt extraction.

The DNA polymerase α-bound SJK237-71 antibody-Sepharose 4B resin can also be washed under other conditions without affecting the yield and purity of the polymerase α enzyme fraction. It can be first washed once with 15 ml of 50 m*M* KPO_4, pH 7.5, 200 m*M* NaCl, 200 m*M* KCl, 20% ethylene glycol, 0.3% Triton X-100, and twice with 15 ml of 50 m*M* KPO_4, pH 7.5, 200 m*M* NaCl, 20% ethylene glycol, and 0.2% Triton X-100, and then washed twice more with 15 ml of 50 m*M* Tris–HCl, pH 8.0, 150 m*M* NaCl. The 20% ethylene glycol in the wash buffer can be also replaced by 10% glycerol with no detectable loss of activity. An alternative method is to wash the DNA polymerase α bound resin four times with 15× resin volumes of a buffer containing 60 m*M* Tris–HCl, pH 8.8, 0.4 *M* NaCl, 0.4% NP-40, and 1 m*M* EDTA. After the last wash, the Sepharose 4B resin is packed into a column with 20 m*M* Tris–HCl, pH 8.0, and 0.4 *M* NaCl and washed further with 5 volumes of 20 m*M* Tris–HCl, pH 8.0, 2 *M* NaCl.[8] DNA polymerase α bound to this washed Sepharose 4B resin is then eluted by the 3.0 *M* $MgCl_2$ buffer as described above.

Acknowledgments

We thank many members of our laboratory, past or present, who contributed to the evolution of the protocol. Methods described here were developed with the support of grant CA14835 from the National Institutes of Health, U.S. Public Health Service.

[9] Purification of Mammalian Polymerases: DNA Polymerase δ

By KATHLEEN M. DOWNEY and ANTERO G. SO

Introduction

DNA polymerase δ is one of three DNA polymerases required for replication of genomic DNA in eukaryotic cells. The enzyme has been purified from a number of mammalian tissues and cells including calf thymus,[1,2] rabbit bone marrow,[3] human placenta,[4] human HeLa cells,[5-7] human 293 cells,[8] and mouse lymphocytes.[9] DNA polymerase δ is a heterodimer with subunits of approximately 125 and 48 kDa. The enzyme is essentially nonprocessive in the absence of the accessory proteins proliferating cell nuclear antigen (PCNA) and replication factor C (RFC) but highly processive in their presence.[10-13] PCNA stimulation of the activity and processivity of DNA polymerase δ on a poly(dA) · oligo(dT) template is a hallmark of this DNA polymerase and can be used to distinguish Pol δ from the other mammalian DNA polymerases, α, β, γ, and ε.[10,11] In addition to DNA polymerase activity, DNA polymerase δ has a $3' \rightarrow 5'$-exonuclease activity, which serves a proofreading function.[14]

Purification Procedure

The following purification procedure describes the isolation of nearly homogeneous DNA polymerase δ from 1.5 kg fetal calf thymus. The DNA polymerase δ accessory protein PCNA can also be purified in the same procedure. A similar protocol has been used to obtain homogeneous DNA polymerase δ from HeLa cells.[5] Purification protocols using smaller amounts of tissue or cultured mammalian cells using fast protein liquid chromatography (FPLC, Pharmacia) usually result in highly purified but not homogeneous DNA polymerase δ.[2-4,6-9]

[1] M. Y. W. T. Lee, C.-K. Tan, K. M. Downey, and A. G. So, *Biochemistry* **23,** 1906 (1984).
[2] T. Weiser, M. Gassmann, P. Thommes, E. Ferrari, P. Hafkemeyer, and U. Huebscher, *J. Biol. Chem.* **266,** 10,420 (1991).
[3] C.-D. Lu and J. J. Byrnes, *Biochemistry* **31,** 12,403 (1992).
[4] M. Y. W. T. Lee, Y. Jiang, S. J. Zhang, and N. L. Toomey, *J. Biol. Chem.* **266,** 2423 (1991).
[5] J. Syvaoja, S. Suomensaari, C. Nishida, J. S. Goldsmith, G. S. J. Chiu, S. Jain, and S. Linn, *Proc. Natl. Acad. Sci. USA* **87,** 6664 (1990).
[6] D. H. Weinberg, K. L. Collins, P. Simancek, A. Russo, M. S. Wold, D. M. Virshup, and T. J. Kelly, *Proc. Natl. Acad. Sci. USA* **87,** 8692 (1990).
[7] S.-H. Lee, T. Eki, and J. Hurwitz, *Proc. Natl. Acad. Sci. USA* **86,** 7361 (1989).
[8] T. Melendy and B. Stillman, *J. Biol. Chem.* **266,** 1942 (1991).
[9] M. Goulian, S. M. Herrmann, J. W. Sackett, and S. L. Grimm, *J. Biol. Chem.* **265,** 16,402 (1990).
[10] C.-K. Tan, C. Castillo, A. G. So, and K. M. Downey, *J. Biol. Chem.* **261,** 12,310 (1986).
[11] G. Prelich, C.-K. Tan, M. Kostura, M. B. Mathews, A. G. So, K. M. Downey, and B. Stillman, *Nature (London)* **326,** 517 (1987).
[12] T. Tsurimoto and B. Stillman, *Proc. Natl. Acad. Sci. USA* **87,** 1023 (1990).
[13] S.-H. Lee, A. D. Kwong, Z.-Q. Pan, and J. Hurwitz, *J. Biol. Chem.* **266,** 594 (1991).
[14] D. C. Thomas, J. D. Roberts, R. D. Sabatino, T. W. Myers, C.-K. Tan, K. M. Downey, A. G. So, R. A. Bambara, and T. A. Kunkel, *Biochemistry* **30,** 11,751 (1991).

Buffers

Buffer A: 50 mM Tris–HCl, pH 8.5, 7.5% glycerol, 0.5 mM EDTA, 0.1 mM EGTA, 0.1 mg/ml bacitracin, 250 μg/ml soybean trypsin inhibitor, 10 mM benzamidine hydrochloride, 0.8 mM phenylmethylsulfonyl fluoride (PMSF), 5 μg/ml pepstatin, 5 μg/ml leupeptin, and 1 mM dithiothreitol (DTT)

Buffer B: 50 mM Tris–HCl, pH 7.8, 20% (v/v) glycerol, 0.5 mM EDTA, 0.1 mM EGTA, 0.2 mM PMSF, and 1 mM DTT

Buffer C: 20 mM potassium phosphate buffer, pH 7.0, 7.5% glycerol, 0.5 mM EDTA, 0.1 mM EGTA, 0.2 mM PMSF, and 1 mM DTT

Buffer D: 50 mM Tris–HCl, pH 8.8, 20% glycerol, 0.5 mM EDTA, 0.1 mM EGTA, 0.2 mM PMSF, and 1 mM DTT

Buffer E: 20 mM potassium phosphate buffer, pH 7.0, 20% glycerol, 0.2 mM PMSF, and 1 mM DTT

Buffer F: 25 mM Bis–Tris–HCl, pH 6.5, 20% glycerol, 0.5 mM EDTA, 0.1 mM EGTA, 0.2 mM PMSF, and 1 mM DTT.

Step 1: Preparation of Crude Extract. First, 1.5 kg fresh-frozen fetal calf thymus is broken into small pieces and homogenized in 3 liters of ice-cold Buffer A in a Waring blender for a total of 5 min (30 sec at a time with cooling between). The extract is filtered through four layers of cheesecloth, and after centrifugation for 90 min at 9000 rpm in a GSA rotor, the supernatant is filtered through glass wool.

Step 2: DEAE-Cellulose Chromatography. Step 1 enzyme (2300 ml) is batch absorbed to 2500 ml of DEAE-cellulose (DE-52) equilibrated in Buffer B, packed into a column and washed with 7 liters of Buffer B. Pol δ and PCNA are eluted from the column with an 8-liter linear gradient of 20 to 600 mM KCl in Buffer B. DNA polymerase δ is recovered in fractions eluting at 60 to 150 mM KCl and PCNA is eluted at 170 to 270 mM KCl. DNA polymerase δ (1460 ml) and PCNA (1410 ml) are pooled separately and solid ammonium sulfate is added to each to 55% saturation with stirring at 4°. The precipitates are collected by centrifugation for 30 min at 15,000 rpm in an SS34 rotor. PCNA is resuspended in Buffer C (250 ml) and stored at −70°. DNA polymerase δ is resuspended in 250 ml Buffer C and solid ammonium sulfate added to 18% saturation, centrifuged at 15,000 rpm, and the supernatant taken.

Step 3: Phenyl-Agarose Chromatography. Step 2 enzyme is batch absorbed to phenyl-agarose (425 ml) equilibrated in Buffer C with 18% saturated ammonium sulfate, and packed into a column. The column is washed with 2 column volumes of equilibration buffer and DNA polymerase δ is eluted with Buffer D. Active fractions are pooled and ammonium sulfate is added to 55% saturation. The pellet is resuspended in 80 ml Buffer B and dialyzed against 6 liters of the same buffer.

Step 4: Phosphocellulose Chromatography. Dialyzed step 3 enzyme is diluted to 20 m*M* KCl with Buffer B and loaded on a phosphocellulose (P11) column (350 ml) equilibrated with Buffer B containing 20 m*M* KCl. The column is washed with 3 column volumes of equilibration buffer and pol δ is eluted with a 2-liter gradient from 20 to 500 m*M* KCl in Buffer B. DNA polymerase δ elutes between 140 and 250 m*M* KCl. Active fractions are pooled and ammonium sufate is added to 55% saturation. The pellet is collected by centrifugation, resuspended in 30 ml Buffer E and dialyzed overnight against the same buffer.

Step 5: Hydroxylapatite Chromatography. Dialyzed step 4 enzyme is loaded on a 38-ml column of hydroxylapatite equilibrated in Buffer E and washed with 2 column volumes of equilibration buffer. DNA polymerase δ is eluted with Buffer E containing 600 m*M* KCl and dialyzed versus Buffer B.

Step 6: DEAE-Sephadex Chromatography. Dialyzed Step 5 enzyme is loaded on an 80-ml column of DEAE-Sephadex A-25 equilibrated with 20 m*M* KCl in Buffer B. After washing with 2 column volumes, DNA polymerase δ is eluted with a 600-ml linear gradient of 20 to 350 m*M* KCl. The enzyme elutes between 100 and 160 m*M* KCl.

Step 7: Heparin-Agarose Chromatography. Step 6 enzyme is adjusted to 160 m*M* KCl, loaded on a 50-ml heparin-agarose column equilibrated with 160 m*M* KCl in Buffer B and washed with 2 column volumes of equilibration buffer. The enzyme is eluted with a 250-ml linear gradient of 160 to 600 m*M* KCl in Buffer B. DNA polymerase δ elutes between 200 and 300 m*M* KCl. The purified enzyme is concentrated by stepwise elution from a small heparin-agarose column and dialyzed against Buffer F. After the addition of bacitracin (0.2 mg/ml) and soybean trypsin inhibitor (0.1 mg/ml), DNA polymerase δ is stored in small aliquots at $-70°$.

Comments on the Purification

Table I summarizes the purification of DNA polymerase δ from 1.5 kg of fetal calf thymus. The protocol is designed to produce nearly homogeneous enzyme. The tissue is disrupted in a buffer containing a cocktail of protease inhibitors and the first two steps are carried out as quickly as possible to minimize proteolysis. In spite of these precautions, preparations of DNA polymerase δ frequently contain variable amounts of a 116-kDa degradation product of the 125-kDa catalytic subunit.[15] DEAE-cellulose chromatography (step 2) removes nucleic acids and separates DNA polymerase δ from PCNA, whereas phenyl-agarose chromatography (step 3) separates the

[15] L. Ng, C.-K. Tan, K. M. Downey, and P. A. Fisher, *J. Biol. Chem.* **266,** 11,699 (1991).

TABLE I
PURIFICATION OF DNA POLYMERASE δ FROM CALF THYMUS[a]

Fraction	Protein (mg)[b]	Activity (units)[c]	Recovery (%)	Specific activity (units/mg)	Stimulation by PCNA (-fold)[d]
1. Crude extract	37,240	2,268,000	100	60.9	1.6
2. DEAE-cellulose	14,290	947,050	41.8	66.3	1.6
3. Phenyl-agarose	6,530	494,700	21.8	75.8	1.7
4. Phosphocellulose	2,080	200,910	8.9	96.6	2.2
5. Hydroxylapatite	35.1	92,820	4.1	2,644	59
6. DEAE-Sephadex	4.7	41,240	1.8	8,774	>100
7. Heparin-agarose	0.7	29,700	1.3	42,429	>100

[a] Purification is based on 1.5 kg fetal calf thymus.
[b] Protein was determined by the Bradford assay.
[c] One unit of DNA polymerase incorporates 1 nmol dNMP/hr at 37°.
[d] Ratio of activities in the presence and absence of 2 μg/ml PCNA using the poly(dA) · oligo(dT) assay.

enzyme from most of the contaminating proteases. Phosphocellulose chromatography (step 4) partially separates DNA polymerase δ from DNA polymerases α and ε and removes the remaining PCNA, which does not bind to phosphocellulose. The major separation of DNA polymerase δ from other mammalian DNA polymerases occurs on hydroxylapatite chromatography, as is evident by the large stimulation of activity on a poly (dA) · oligo(dT) template by PCNA after this step (Table I). Thus, step 5 enzyme can be used when homogeneous DNA polymerase δ is not required. DNA polymerase δ can also be eluted from hydroxylapatite with a phosphate gradient, however it is not as cleanly separated from DNA polymerases α and ε under these elution conditions. The separation of contaminating nucleases is accomplished on heparin-agarose chromatography and the proofreading 3′ → 5′-exonuclease activity of DNA polymerase δ can only be measured accurately after this step. Highly purified DNA polymerase δ is stable for months or years when stored at −70° but rapidly loses activity on repeated freezing and thawing.

Purification of PCNA

PCNA is separated from DNA polymerase δ in step 2 of the enzyme purification. It can be further purified in the following protocol.

Step 3: Phenyl-Agarose Chromatography. The ammonium sulfate precipitate of PCNA from step 2 of the enzyme purification is resuspended in 250 ml Buffer C, and solid ammonium sulfate is added to 18% saturation, centrifuged at 15,000 rpm, and the supernatant taken. PCNA is batch absorbed to phenyl-agarose (400 ml) in Buffer C with 18% saturated ammonium sulfate, packed into a column, and washed with 2 column volumes of equilibration buffer. PCNA is eluted with Buffer D and dialyzed against Buffer B.

Step 4: Phosphocellulose Chromatography. PCNA is loaded on a phosphocellulose column (300 ml) equilibrated in Buffer B containing 20 m*M* KCl and the column flow-through is collected. Because PCNA does not bind to phosphocellulose even at low ionic strength, this step separates PCNA from DNA polymerases.

Step 5: DEAE-Cellulose Chromatography. Step 4 protein is adjusted to 150 m*M* KCl, loaded on a DE52 column (160 ml) equilibrated in Buffer B containing 150 m*M* KCl and washed with 2 column volumes of equilibration buffer. The protein is eluted with a 1-liter gradient of KCl from 150 to 450 m*M* in Buffer B. PCNA elutes between 175 and 200 m*M* KCl. This step is very efficient in separating contaminating proteins and can be repeated to improve the purification.

Step 6: Hydroxylapatite Chromatography. The pooled fractions are dialyzed against Buffer E and loaded on a hydroxylapatite column (120 ml) equilibrated in Buffer E. After washing with equilibrating buffer, essentially homogeneous PCNA is eluted with a linear gradient of 20 m*M* potassium phosphate to 200 m*M* potassium phosphate in Buffer E and the pooled fractions dialyzed against Buffer B. Purified PCNA is concentrated by stepwise elution from a small DE52 column and dialyzed against Buffer F. The protein is stored in aliquots at −70°.

Assay of DNA Polymerase δ

The PCNA dependence of the activity of DNA polymerase δ on poly (dA) · oligo(dT) template–primers can be used to distinguish DNA polymerase δ from other mammalian DNA polymerases. The standard reaction mixture contains in a final volume of 50 μl: 40 m*M* Bis–Tris, pH 6.5, 6 m*M* $MgCl_2$, 40 μM [^{3}H]dTTP, 100 to 200 cpm/pmol, 40 μg/ml bovine serum albumin, 15 μg/ml poly(dA) · oligo(dT) (10 : 1 to 20 : 1 in nucleotide equivalents), 2% glycerol, 2 μg/ml PCNA, and 0.1 to 2 units of DNA polymerase δ (1 unit of enzyme catalyzes the incorporation of 1 nmol dNMP/hr at 37°). Reactions are incubated at 37° for 15 min and stopped by the addition of 2 ml 5% trichloroacetic acid containing 20 m*M* sodium pyrophosphate. Precipitates are collected on Whatman GF/C filters, washed, and counted. The same reaction mixture is used to measure PCNA activity except that PCNA is omitted and 0.5 unit of DNA polymerase δ is added.

In the absence of a source of PCNA, the activity of DNA polymerase

TABLE II
PHYSICAL PROPERTIES OF DNA POLYMERASE δ[a]

Property	Value
Stokes radius	53 Å
$s_{20,w}$	7.9 S
Calculated M_r	173,000
f/f_o	1.44
pI	5.5
Subunit M_r	125,000; 48,000

[a] Reprinted with permission from M. Y. W. T. Lee, C.-K. Tan, K. M. Downey, and A. G. So, *Biochemistry* **23,** 1906 (1984). Copyright 1984, American Chemical Society.

δ on a poly(dA–dT) template–primer can be used to follow purification. The standard reaction mixture contains in a final volume of 50 μl: 40 mM Bis–Tris, pH 6.5, 2 mM $MgCl_2$, 40 μg/ml bovine serum albumin, 40 μM dATP, 10 μM [^{3}H]dTTP (1000 cpm/pmole), 2% glycerol, 20 μg/ml activated poly(dA–dT), and 0.2 to 2 units DNA polymerase δ.

Properties of DNA Polymerase δ

Physical Properties

Table II summarizes the physical properties of homogeneous calf thymus DNA polymerase δ. The estimated molecular weight of the enzyme is 173,000 and the individual subunits are approximately 125 and 48 kDa. The cDNAs for the two subunits of bovine DNA polymerase δ have recently been cloned and sequenced.[16,17] Based on the deduced amino acid sequences the calculated molecular weights of the subunits are 123,707 and 50,908, in reasonable agreement with the values estimated from SDS–polyacrylamide gel electrophoresis (SDS–PAGE).

Enzymatic Properties

Table III summarizes the enzymatic properties of DNA polymerase δ. In the absence of PCNA, the enzyme prefers the alternating copolymer poly(dA–dT) as template and has relatively less activity on activated calf

[16] J. Zhang, D. W. Chung, C.-K. Tan, K. M. Downey, E. W. Davie, and A. G. So, *Biochemistry* **30,** 11,742 (1991).
[17] J. Zhang, C.-K. Tan, B. McMullen, K. M. Downey, and A. G. So, submitted for publication.

TABLE III
ENZYMATIC PROPERTIES OF DNA POLYMERASE δ

Template preference		K_m for substrates[a]		Inhibitor profile[b]			Ratio of Polymerase to 3'-5' exonuclease[d]	
					IC_{50}			
Template	Relative activity	Substrate	K_m (μM)	Inhibitor	40 μM dTTP	4 μM dTTP	Exonuclease substrate	Pol/Exo
poly(dA-dT)	1.00	dATP	1.8	Aphidicolin (μg/ml)	0.15	0.15	poly(dT)	5.1
Calf thymus DNA	0.40	dGTP	1.1	BuPdGTP (μM)	270	90	poly(dA)	17.8
poly(dA)·oligo(dT) (20:1)	<0.05	dCTP	0.7	BuAdATP (μM)	150	45	poly(dA-dT)	37.5
poly(dA)·oligo(dT) (20:1) with PCNA	2.9	dTTP	1.1	COMDP[c] (μM)	40	40		
				ddTTP (μM)	>800	300		

[a] Determined with activated calf thymus DNA as template.
[b] Determined with poly(dA)·oligo(dT) as template. BuPdGTP, BuAdATP, and COMDP were gifts of Dr. George E. Wright.
[c] HEPES, pH 6.5, was substituted for Bis–Tris, pH 6.5.
[d] DNA polymerase activity determined with poly(dA)·oligo(dT) as template.

thymus DNA. Although DNA polymerase δ has little activity on templates with a high template–primer ratio, for example, poly(dA) · oligo(dT) (20:1) in the absence of PCNA, it is very active in its presence, since PCNA increases the processivity of DNA polymerase δ from <10 nucleotides incorporated per enzyme binding event to >1000.[10] Kinetic studies have shown that PCNA both decreases the K_m of DNA polymerase δ for template–primer and increases the V_{max}, suggesting that PCNA increases the residence time of the enzyme on the template–primer and the rate of nucleotide incorporation.[15] The ability of PCNA to stabilize the interaction of DNA polymerase δ with template–primer has also been demonstrated directly by binding studies.[18] The K_m values for the dNTP substrates are of the order of 1 to 2 μM using a variety of template–primers and are not affected by PCNA.

The inhibitor profile of DNA polymerase δ is also shown in Table III. As is the case for DNA polymerase α and DNA polymerase ε, but unlike DNA polymerase β, DNA polymerase δ is sensitive to inhibition by aphidicolin and relatively resistant to dideoxynucleoside triphosphates (ddNTP).[2,5] DNA polymerase δ is also relatively resistant to butylphenyl-dGTP (BuPdGTP) and butylanilino-dATP (BuAdATP), as is DNA polymerase ε, and these inhibitors can be used to inhibit preferentially DNA polymerase α activity in cell extracts.[2,5] The activity of DNA polymerase δ is 50% inhibited by carbonyl diphosphonate (COMDP) at 40 μM. With the nucleotide analogs ddTTP, BuPdGTP, and BuAdATP, inhibition is dependent on the dTTP concentration, whereas with aphidicolin and COMDP, inhibition is independent of dTTP concentration.

The $3' \rightarrow 5'$-exonuclease activity of DNA polymerase δ is thought to provide a proofreading function to enhance the fidelity of DNA synthesis.[14] The activity of the $3' \rightarrow 5'$-exonuclease is dependent on substrate structure and/or composition, being most active with single-stranded DNA and least active with double-stranded DNA substrates (Table III), reflecting the preference of the $3' \rightarrow 5'$-exonuclease for a mismatched rather than a matched primer terminus.[19]

Acknowledgment

This work was supported by grant DK26206 from the National Institutes of Health.

[18] L. Ng, M. McConnell, C.-K. Tan, K. M. Downey, and P. A. Fisher, *J. Biol. Chem.* **268,** 13,571 (1993).

[19] M. Y. W. T. Lee, C.-K. Tan, K. M. Downey and A. G. So, *Prog. Nucleic Acids Res. Mol. Biol.* **26,** 83 (1981).

[10] Purification of Mammalian Polymerases: DNA Polymerase ε

By GLORIA S. J. CHUI and STUART LINN

Introduction

DNA polymerase ε (Pol ε) was first observed in yeast,[1] then it was purified from calf thymus,[2,3] and HeLa cells[4,5] as a polymerase containing a $3' \rightarrow 5'$-exonuclease activity similar to that of DNA polymerase δ (Pol δ). But, unlike Pol δ, Pol ε is highly processive in the absence of proliferating cell nuclear antigen (PCNA) and antibody raised against Pol δ does not cross react with Pol ε.

When the gene-encoding yeast Pol ε is disrupted,[6] yeast cells become nonviable, indicating that Pol ε has a role in DNA replication. The temperature-sensitive (*ts*) Pol ε mutants of the yeast are defective in the chromosomal DNA replication at the restrictive temperature, also indicating that Pol ε is involved in DNA replication.[7] No chromosomal-size DNA is formed after a shift of a yeast Pol ε *ts* mutant to the nonpermissive temperature, indicating that Pol ε is involved in elongation.[8] Pol ε from HeLa cells was first isolated as a repair factor required for permeabilized diploid human fibroblasts.[4] It has also been reported that DNA repair synthesis in permeabilized yeast nuclei excision repair is catalyzed by DNA Pol ε.[9] More studies are clearly needed to reveal whether Pol ε functions in DNA replication, DNA repair, or both.

Assay Methods

DNA Pol ε functions maximally with long single-stranded DNA as a template. The most commonly used template is poly(dA)$_{3000-5000}$ ·

[1] E. Wintersberger, *Eur. J. Biochem.* **50,** 41 (1974).
[2] J. J. Crute, A. F. Wahl, and R. A. Bambara, *Biochemistry* **25,** 26 (1986).
[3] A. F. Wahl, J. J. Crute, R. D. Sabatino, J. B. Bodner, R. L. Maraccino, L. W. Harwell, E. M. Lord, and R. A. Bambara, *Biochemistry* **25,** 7821 (1986).
[4] C. Nishida, P. Reinhard, and S. Linn, *J. Biol. Chem.* **263,** 501 (1988).
[5] J. Syvaöja and S. Linn, *J. Biol. Chem.* **264,** 2489 (1989).
[6] A. Morrison, H. Araki, A. B. Clark, R. K. Hamatake, and A. Sugino, *Cell* **62,** 1143 (1990).
[7] H. Araki, P. A. Ropp, A. L. Johnson, L. H. Johnston, A. Morrison, and A. Sugino, *EMBO J.* **11,** 733 (1992).
[8] M. E. Budd and J. L. Cambell, *Mol. Cell Biol.* **13,** 496 (1993).
[9] Z. Wang, X. Wu, and E. C. Friedberg, *Mol. Cell Biol.* **13,** 1051 (1993).

oligo (dT). To make such a template, oligo$(dT)_{12-18}$ is annealed to poly $(dA)_{3000-5000}$ at a 1:10 nucleotide ratio by incubating the two components at 65° for 30 min, after which the water bath is switched off and the template is cooled slowly to room temperature to allow annealing. Standard reaction mixtures (25 μl) contain 50 m*M* HEPES–KOH, pH 7.5, 15 m*M* $MgCl_2$, 100 m*M* potassium glutamate, pH 7.8, 10 m*M* dithiothreitol (DTT), 0.03% (v/v) Triton X-100, 20% (v/v) glycerol, 0.2 mg/ml bovine serum albumin (BSA), 50 μM ^{3}H- or ^{32}-P-labeled dTTP, and 40 μM (nucleotide-residues) poly$(dA)_{3000-5000}$ plus 4 μM oligo$(dT)_{12-18}$ to give an interprimer distance of 135 nucleotides. After incubation at 37° for 30 min, reactions are terminated by adding 1 ml of 10% (w/v) trichloroacetic acid (TCA), 0.05 *M* $Na_4P_2O_7$. After 10 min on ice, acid-precipitable radioactivity is separated by filtering the mixtures through Whatman GF/C filters and washing the filters 10 times with a cold solution of 1 *M* HCl, 0.1 *M* $Na_2P_4O_7$ then once with cold 95% ethanol. Radioactivity incorporated into the template is assessed by liquid scintillation counting of dried filters. One unit of polymerase is defined as the amount necessary to catalyze the incorporation of 1 nmol of total nucleotide into an acid-insoluble form per hour.

To achieve maximal separation of Pol α and Pol ε, DNA polymerase α is also assayed throughout the purification. Because Pol α is unique in having primase activity, primase activity can be monitored as a sensitive measure of Pol α during purification.[2] DNA primase activity is detected by using $(dT)_{2500-5000}$ as the substrate, with the addition of *Escherichia coli* DNA polymerase I Klenow fragment as a signel amplifier for the detection of DNA-dependent RNA primer synthesis.[10] The primase reaction mixture (50 μl) contains 20 m*M* Tris–HCl, pH 7.5, 0.5 m*M* DTT, 0.5 mg/ml BSA, 7.5 m*M* $MgCl_2$, 1 m*M* ATP, 40 μM poly$(dT)_{2500-5000}$, and 50 μM [^{3}H]dATP (1000 cpm/pmol). Following the addition of the Pol α sample, 0.5 unit of DNA polymerase I Klenow fragment is added. Incubation is for 30 min at 30°, and incorporated nucleotides are assayed as described above.

Pol α activity can also be assayed with activated calf thymus DNA template–primer prepared according to the method of Schlabach and coworkers.[11] Reactions (50 μl) contain 12.5 μg of the DNA, 50 m*M* Tris–HCl, pH 7.5, 7.5 m*M* $MgCl_2$, 0.5 m*M* DTT, 0.2 mg/ml BSA, and 50 μM each of dATP, dGTP, dCTP, and [α-^{3}H]dTTP (~200 cpm/pmol). After incubation at 37° for 30 min, the reactions are stopped and processed as for Pol ε activity.

[10] R. C. Conaway and I. R. Lehman, *Proc. Natl. Acad. Sci. USA* **79,** 2523 (1982).

[11] A. Schlabach, B. Fridlender, A. Bolden, and A. Weisbach, *Biochem. Biophys. Res. Commun.* **44,** 879 (1971).

TABLE I
PURIFICATION OF DNA POLYMERASE ε[a]

Fraction	Volume (ml)	Protein (mg)	Total activity (units)	Specific activity (units/mg)
I. Cell extracts after 30–50% $(NH_4)_2SO_4$	324	713	4020	5.64
II. DEAE-Sephacel	150	345	5190	15.0
III. Phosphocellulose	260	26	2130	82
IV. Hydroxyapatite	20	5.4	917	170
V. Mono S FPLC	5.5	(0.02)	175	(8800)

[a] Purification was from 60 liters of HeLa cells (3.3×10^{10} cells) as described in the text. The protein content of Fraction V is an estimate from a silver-stained gel with comparison to bovium serum albumin, because the protein concentration was too low for assay by other means.

Protein concentration are determined by the Bradford method using BSA as the standard.[12]

Purification Procedure

Crude Cell Lysate

Purification steps are carried out at 0 to 4°. A typical purification is summarized in Table I. Mid-log phase HeLa cells (5 to 8×10^5 cells/ml) are grown in suspension in Joklik's modified Eagle's medium containing 5% newborn calf serum, supplemented with penicillin, streptomycin, and L-glutamate in a humidified 5% CO_2 incubator at 37°. Cells are collected by centrifugation at 3000 rpm for 10 min, washed twice with phosphate-buffered saline, and resuspended in Buffer A [40 m*M* HEPES–KOH, pH 7.7, 300 m*M* sucrose, 1% (w/v) dextran (MW 500,000), 16 m*M* $MgCl_2$, 1 m*M* EGTA, 0.5 m*M* DTT], and 3 ml of H_2O per 8×10^8 cells. After 15 min at 0°, cells are homogenized with 35 strokes of a Dounce homogenizer with a tight-fitting pestle A. Hypertonic Buffer B [80 m*M* HEPES–KOH, pH 7.7, 600 m*M* sucrose, 2% (w/v) dextran (MW 500,000), 32 m*M* $MgCl_2$, 2 m*M* EGTA, 4 m*M* DTT, and 933 m*M* KCl] is then added (0.75 ml/ml Buffer A initially added) and the homogenate is kept on ice for about 30 min prior to centrifugation in a Type 40 Beckman rotor at 38,000 rpm for 60 min at 4°.

[12] M. M. Bradford, *Anal. Biochem.* **72,** 248 (1976).

Ammonium Sulfate Fractionation

The supernatant is made 30% saturated in $(NH_4)_2SO_4$ by adding powdered $(NH_4)_2SO_4$ with constant stirring for 1 hr. The precipitate is removed by centrifugation at 20,000*g* for 15 min at 2° and discarded, whereas the supernate is brought to 50% $(NH_4)_2SO_4$ saturation. The pellet is collected as above, and resuspended in a minimum volume of TDG buffer [50 m*M* Tris–HCl, pH 8.0, 1 m*M* DTT, 10% (v/v) glycerol], and then dialyzed twice versus 4 liters of TDG buffer. The material is centrifuged to remove precipitated material.

DEAE-Sephacel Chromatography

The dialyzed ammonium sulfate fraction is loaded onto a Pharmacia DEAE-Sephacel column preequilibrated with TDG buffer. A 150-ml column (4.4 × 10 cm) is used for 60 liters of HeLa culture. The column is washed with 300 ml of TDG buffer and subsequently eluted with a 900-ml linear gradient in TDG buffer from 0 to 500 m*M* NaCl at 100 ml/hr. The major peak of activity elutes near 210 m*M* NaCl. Pol δ elutes near 80 m*M* NaCl and is thus well separated from Pol ε in this step.

Phosphocellulose Chromatography

The DEAE-Sephacel fraction is diluted with TDG buffer to a conductivity equivalent to that of 100 m*M* NaCl and then applied to a 75-ml phosphocellulose column (Whatman P11, 2.7 × 13 cm), which has been equilibrated with TDG buffer containing 100 m*M* NaCl. The column is washed with 150 ml of the same buffer and then eluted with a 600-ml linear gradient in TDG buffer from 100 to 600 m*M* NaCl at a flow rate of 100 ml/hr. Pol α and Pol ε both eluted near 235 m*M* NaCl.

Hydroxyapatite Chromatography

The phosphocellulose fraction is diluted fourfold with 20 m*M* potassium phosphate, pH 7.5, 20% (v/v) glycerol, 5 m*M* DTT, and then loaded onto a 12-ml hydroxyapatite column (Bio-Gel HTP, 1.5 × 7 cm), which has been equilibrated with the same buffer as that used for diluting the phosphocellulose fraction. The column is washed with 24 ml of the same buffer and then eluted with a 240-ml linear gradient in 20% (v/v) glycerol and 5 m*M* DTT from 20 to 400 m*M* potassium phosphate, pH 7.5, at a flow rate of 30 ml/hr. The peak of polymerase ε activity is eluted near 105 m*M* potassium phosphate buffer, while that of Pol α is eluted near 120 m*M* potassium phosphate buffer, overlapping the Pol ε peak.

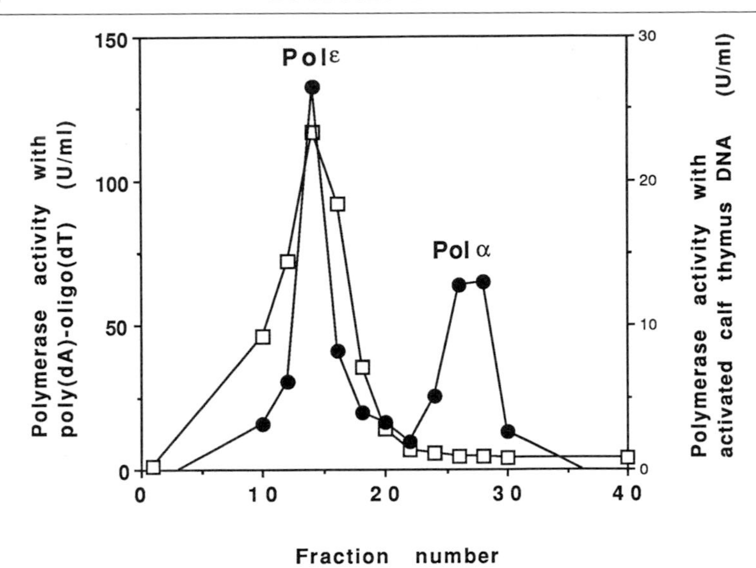

FIG. 1. Mono S FPLC chromatography of DNA polymerase ε from HeLa cells. The hydroxyapatite fraction from 60 liters of HeLa cells (5.4 mg) was loaded onto the Mono S FPLC column and eluted and assayed as described in the text. □, polymerase activity with poly(dA) · oligo(dT) template–primer; ●, polymerase activity with activated calf thymus DNA template–primer.

Mono S FPLC Chromatography

Fractions containing significant Pol ε activity are pooled, dialyzed against PDG buffer (20 m*M* potassium phosphate, pH 7.5, 1 m*M* DTT, 10% glycerol), and then loaded onto a 1-ml Mono S FPLC column (Pharmacia) equilibrated with PDG buffer. After washing with 5 ml of PDG buffer, the polymerase activities are eluted with a 24-ml linear gradient of the same buffer containing 0 to 500 m*M* NaCl at 0.5 ml/min. The two polymerases are completely separated by this chromatography procedure: Pol ε is eluted near 120 m*M* NaCl, whereas Pol α was eluted near 250 m*M* NaCl (Fig. 1).

Properties of DNA Polymerase ε

The purified HeLa Pol ε has two subunits, with M_r values of >200,000 and 55,000, as determined by SDS-gel electrophoresis. By contrast, yeast Pol ε has subunits with M_r values of 200,000, 80,000, 34,000, 30,000, and 29,000.[13] The catalytic activity resides on the large subunit, whereas the

[13] R. K. Hamatake, H. Hasegawa, A. B. Clark, K. Bebenek, T. A. Kunkel, and A. Sugino, *J. Biol. Chem.* **265,** 4072 (1990).

role of the smaller polypeptide has not been determined. Smaller forms of Pol ε have been reported from both yeast[13] and calf thymus[2] (with catalytic subunits having M_r values ranging from 120,000 to 170,000). There is a protease-sensitive region in the center of the large polypeptide of the HeLa enzyme,[14] and trypsin digestion can form active fragments with M_r values of 122,000 and 136,000. Both the polymerase and exonuclease activities are located in the 122-kDa segment of the polypeptide, which derives from the N-terminal half. The C-terminal half of the yeast *POL2* gene product seems not to be essential for cell viability.[6]

DNA Pol ε is a processive enzyme, which can elongate several thousand nucleotides without dissociating. Indeed, Pol ε alone can replicate an entire DNA-primed M13 template with 1 m*M* $MgCl_2$ present. However, it requires 15 m*M* $MgCl_2$ for maximal velocity when assayed on poly(dA) · oligo(dT) template–primer. In the absence of added salt, the pH optimum is 7.0 to 8.0 when poly(dA) · oligo(dT) is used as a template–primer. Pol ε is inhibited by NaCl, KCl, and $(NH_4)_2SO_4$; however, it is stimulated twofold to threefold by 100 m*M* potassium glutamate.

Protease inhibitors are not included in the preparations described above and Pol ε activity was stable during the purification, which was completed in five working days without freezing or thawing the fractions. The activity of purified DNA Pol ε is stable at −80° in the presence of 20% glycerol. However, since the Mono S fraction has a very low protein concentration, frequent freezing and thawing should be avoided.

[14] T. Kesti and J. E. Syvaöja, *J. Biol. Chem.* **266,** 6336 (1991).

[11] Purification and Domain-Mapping of Mammalian DNA Polymerase β

By WILLIAM A. BEARD and SAMUEL H. WILSON

Introduction

DNA polymerase β (Pol β) is a 39-kDa DNA polymerase in vertebrates lacking intrinsic accessory activities such as 3′- or 5′-exonuclease, endonuclease, dNMP turnover, or pyrophosphorolysis.[1] Higher molecular weight homologs have been reported to occur in *Saccharomyces cerevisiae*, 68

[1] K. Tanabe, E. W. Bohn, and S. H. Wilson, *Biochemistry* **18,** 3401 (1979).

kDa,[2] as well as *Drosophila melanogaster*, 110 kDa.[3] The potential for genetic analysis in these biological systems should help define the functional roles of these polymerases. Additionally, tissue-specific Pol β gene inactivation in mice has been achieved,[4] which will help identify the functional role(s) of mammalian Pol β, including the possibility that it has tissue-specific or developmental roles. Mammalian Pol β is generally believed to be involved in short-gap DNA synthesis and has been shown to play a role in short-patch excision repair.[5] In addition, a role in mismatch repair,[6] replication,[7,8] and recombination,[9] has been suggested. DNA polymerase β is present in all rodent tissues examined and is generally expressed at a low level.

DNA polymerase β is the smallest and simplest DNA polymerase identified to date and an ideal enzyme for biophysical and kinetic characterization. Characterization has been hastened by the cloning and overexpression of the rodent and human cDNAs for Pol β in *Escherichia coli* and the subsequent purification of homogeneous enzyme.[10,11] These recombinant proteins are kinetically similar to the "natural" enzyme, making them an excellent model system for detailed structure–function studies. Proteolytic mapping of Pol β has identified several structurally distinct domains with specific functional activities.[12] The crystallographic structure of Pol β[13] and the ternary complex with DNA and ddCTP[14] has defined the location of these domains relative to the tertiary structure of Pol β, in addition to increasing our understanding of the nucleotidyltransferase reaction catalyzed by RNA and DNA polymerases. This has made Pol β an important structural and functional model for other polymerases.

In this chapter, we focus on the isolation of recombinant Pol β and its

[2] R. Prasad, S. G. Widen, R. K. Singhal, J. Watkins, L. Prakash, and S. H. Wilson, *Nucleic Acids Res.* **21,** 5301 (1993).

[3] K. Sakaguchi and J. B. Boyd, *J. Biol. Chem.* **260,** 10,406 (1985).

[4] H. Gu, J. D. Marth, P. C. Orban, H. Mossmann, and K. Rajewsky, *Science* **265,** 103 (1994).

[5] R. K. Singhal, R. Prasad, and S. H. Wilson, *J. Biol. Chem.* **270,** 949 (1995).

[6] K. Wiebauer and J. Jiricny, *Proc. Natl. Acad. Sci. USA* **87,** 5842 (1990).

[7] T. M. Jenkins, J. K. Saxena, A. Kumar, S. H. Wilson, and E. J. Ackerman, *Science* **258,** 475 (1992).

[8] S. Linn, *Cell* **66,** 185 (1991).

[9] R. Nowak, M. Woszczynski, and J. A. Siedlecki, *Exp. Cell Res.* **191,** 51 (1990).

[10] T. A. Patterson, W. Little, X. Cheng, S. G. Widen, A. Kumar, W. A. Beard, and S. H. Wilson, unpublished results (1993).

[11] J. Abbotts, D. N. SenGupta, B. Zmudzka, S. G. Widen, V. Notario, and S. H. Wilson, *Biochemistry* **27,** 901 (1988).

[12] A. Kumar, S. G. Widen, K. R. Williams, P. Kedar, R. L. Karpel, and S. H. Wilson, *J. Biol. Chem.* **265,** 2124 (1990).

[13] M. R. Sawaya, H. Pelletier, A. Kumar, S. H. Wilson, and J. Kraut, *Science* **264,** 1930 (1994).

[14] H. Pelletier, M. R. Sawaya, A. Kumar, S. H. Wilson, and J. Kraut, *Science* **264,** 1891 (1994).

domains derived from proteolysis, chemical cleavage, or bacteria carrying an overexpression plasmid. Because these domains have activities that can be separated, characterization of these domains has advanced our understanding of the structure and function of the intact enzyme. Additionally, the size of the domains, in some cases, is amenable to solution structure determination by nuclear magnetic resonance (NMR) spectroscopy.

Expression and Purification of Intact Enzyme

Recombinant rat and human Pol β have been overexpressed in *E. coli*[10,11] and purified as described by Abbotts *et al.*[11] with modification after the single-stranded (ss) DNA-agarose column.[12] The cDNAs of rat and human Pol β have been subcloned in a λP_L promoter-based bacterial expression system, pRC23.[11] The rat protein is expressed at higher levels than the human enzyme due to a suppressive effect within the 3′ one-third of the human coding sequence.[12] The level of human Pol β expression was subsequently demonstrated to be dependent on *E. coli* host strain and vector sequence.[10] The recombinant protein is purified as a single polypeptide of 334 residues that starts with Ser2. The amino-terminal Met is removed during *E. coli* overexpression by an aminopeptidase.

For purification of recombinant enzyme, approximately 30 g of *E. coli* cells, induced for 2.5 hr at 42°, are resuspended in 160 ml of 50 m*M* Tris–HCl, pH 7.5, 1 m*M* EDTA, 1 m*M* phenylmethylsulfonyl fluoride (PMSF), 10 m*M* $Na_2S_2O_5$, and 1 μg/ml pepstatin A (Buffer A) supplemented with 0.5 *M* NaCl (Buffer D). This suspension is sonicated on ice, centrifuged at 20,000*g* for 20 min, and the supernatant fraction is diluted with Buffer A to bring the NaCl concentration to 0.2 *M*. This is loaded on a 200-ml DEAE-cellulose column (2.5 × 50 cm) connected in tandem with a 100-ml phosphocellulose column (2.5 × 20 cm). These columns are equilibrated and run with Buffer A including 0.2 *M* NaCl (Buffer C). Under these conditions, Pol β passes through the DEAE-cellulose column and binds to the phosphocellulose. The columns are washed with 600 ml of Buffer C, and the DEAE-cellulose column is disconnected. The phosphocellulose column is washed with an additional 400 ml of Buffer C, and Pol β is eluted with 300 ml of Buffer A with 1 *M* NaCl (Buffer E). These columns are run at 1 ml/min at 4°. Fractions with peak activity are pooled and dialyzed against Buffer A containing 0.1 *M* NaCl (Buffer B). The dialyzed fraction is loaded on a 20-ml ssDNA-cellulose column (1 × 30 cm) equilibrated with Buffer B. The column is washed with 200 ml of Buffer B, and Pol β is eluted with Buffer E. Peak fractions are pooled and dialyzed against 50 m*M* sodium acetate, pH 5.5, 75 m*M* NaCl, 0.1 m*M* DTT, and 5% glycerol at 4°. The dialyzed sample, 30 ml of approximately 0.5 mg/ml protein, is passed through a 0.22-μm filter (Millipore) and then layered on a Mono S FPLC column (HR 5/5). DNA polymerase β is eluted with a 75 to 600 m*M* NaCl gradient in the above buffer, and peak fractions ($\approx$400 m*M* NaCl) are pooled and stored at $-70°$. The protein concentration of this solution is typically 3 mg/ml. The specific activity of this preparation is approximately 600 units/mg protein, where one unit is the amount of enzyme required to catalyze the incorporation of 1 nmol of dTMP in 1 min at 20° using poly(dA)$\cdot$oligo$(dT)_{20}$ as a template–primer.

Domain Structure

Controlled proteolytic or chemical cleavage of Pol β has demonstrated that the intact enzyme is folded into discrete domains and subdomains.[12,15] This has been confirmed by X-ray crystallography.[13,14] Biochemical characterization of these purified domains has facilitated our understanding of the structural and functional relationships of these domains relative to intact enzyme. DNA polymerase β has a modular organization with an 8-kDa amino-terminal domain connected to the carboxyl-terminal domain (31 kDa) by a protease-hypersensitive hinge region (Fig. 1). Spectroscopic analysis of the amino-terminal 8-kDa domain, approximately 75 residues, indicates that it is composed primarily of α helix,[16] and this has been confirmed by NMR spectroscopy[17] and X-ray crystallography.[13] The remaining 31-kDa carboxyl-terminal domain is composed of about 250 amino acids and has a significant content of β structure as revealed from circular dichroism spectroscopy.[16]

The amino-terminal 8-kDa domain binds ssDNA nearly as tightly as the intact enzyme as revealed by chromatography on ssDNA-agarose,[12] poly(ethenoadenylate) binding,[12,16] and UV cross-linking.[18] However, the 8-kDa domain does not bind to double-stranded (ds) DNA, whereas the full-length protein possess dsDNA binding activity. The 31-kDa domain can bind dsDNA, but in contrast to the intact enzyme, does not bind ssDNA.[16] Chymotrypsin treatment of Pol β results in a 27-kDa fragment with no nucleic acid binding activity, whereas a 16-kDa peptide (residues

[15] S. H. Wilson, R. Singhal, and A. Kumar, "DNA Repair Mechanisms: Alfred Benzon Symposium 35," p. 343. Munksgaard, Copenhagen, 1992.

[16] J. R. Casas-Finet, A. Kumar, G. Morris, and S. H. Wilson, *J. Biol. Chem.* **29,** 19,618 (1991).

[17] D. Liu, E. F. DeRose, R. Prasad, S. H. Wilson, and G. P. Mullen, *Biochemistry* **33,** 9537 (1994).

[18] R. Prasad, A. Kumar, S. G. Widen, J. R. Casas-Finet, and S. H. Wilson, *J. Biol. Chem.* **268,** 22,746 (1993).

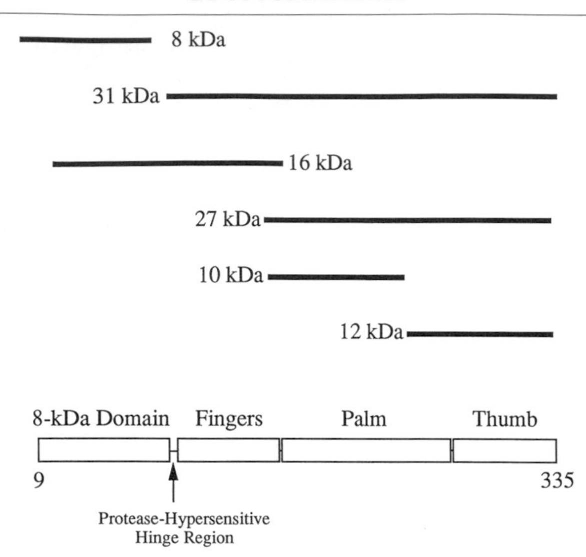

FIG. 1. Domain organization of rat and human DNA polymerase β. The relative primary sequence position of the domains derived from proteolysis or chemical cleavage of rat Pol β is indicated along with the apparent molecular weights determined by SDS–PAGE. The boundaries of these domains are summarized in Table I. The domain organization determined by X-ray crystallography is diagrammed on the bottom of the figure.[13,14] The residues prior to Glu-9 are disordered in the structure. By analogy to a right hand, the polymerase catalytic domain is composed of the fingers, palm, and thumb subdomains.[28] The protease-hypersensitive hinge region is indicated. Mild proteolysis at this site results in the formation of the 8- and 31-kDa domains.[12,15]

19–155) prepared by CNBr treatment of the intact protein has both ss- and dsDNA binding properties similar to intact enzyme.[19]

Activity gel analysis has demonstrated that the nucleotidyltransferase activity is restricted to the 31-kDa domain.[20] Further proteolysis of the 27-kDa domain with trypsin results in subdomains of approximately 10 and 12 kDa.[15] The 31-kDa catalytic domain, therefore, consists of three subdomains: 6, 10, and 12 kDa. The crystallographic structure of Pol β[13] has confirmed that the subdomains identified by proteolysis represent polymerase structural motifs: fingers, palm, and thumb (Fig. 1). Table I summarizes the activities associated with these isolated domains.

[19] J. R. Casas-Finet, A. Kumar, R. L. Karpel, and S. H. Wilson, *Biochemistry* **31,** 10,272 (1992).
[20] A. Kumar, J. Abbotts, E. M. Karawya, and S. H. Wilson, *Biochemistry* **29,** 7156 (1990).

TABLE I
CHARACTERIZATION OF RAT DNA POLYMERASE β DOMAINS GENERATED BY CHEMICAL AND PROTEOLYTIC CLEAVAGE

Domain (kDa)[a]	Residues[b]	Binding activity			Nucleotidyltransferase
		ssDNA	dsDNA	Gap[c]	
8	4–76	+	−	+	−
31	88–335	−	+	−	+
27	141–335	−	−	ND[d]	−
16	19–155	+	+	ND	−
39	2–335	+	+	+	+

[a] Apparent molecular weight determined by SDS–PAGE.

[b] Determined from amino-terminal sequencing of the purified domain. The carboxyl terminus is estimated from the apparent molecular weight and the substrate specificity of the cleavage reagent. Overexpression plasmids for the 8-, 16-, and 31-kDa domains have been constructed by S. Widen. These carry the coding sequence for residues 1–87, 1–140, and 88–335, respectively. The 10- and 12-kDa domains (Fig. 1) start with Lys-141 and Gly-231, respectively, as determined from sequencing of the purified fragments (Kumar and Wilson, unpublished result).

[c] Refers to binding to a gapped DNA substrate where there is a 5′-phosphate.

[d] Not determined. Because the 27-kDa domain does not bind nucleic acid, it is not expected to bind to a gapped substrate bearing a 5′-phosphate. In contrast, because the 8-kDa domain is part of the 16-kDa domain, the 16-kDa domain would be expected to bind.

Domain Isolation

Proteolysis and Chemical Cleavage

Controlled proteolysis of the purified recombinant enzyme with trypsin results in the generation of two domains, 8 and 31 kDa. These domains can be separated as described previously.[12] Recombinant Pol β (1 mg) is digested at 25° with trypsin (1500:1, w/w) in a final volume of 2 ml of 20 m*M* Tris–HCl, pH 7.5, 150 m*M* NaCl, 1 m*M* EDTA, and 1 m*M* PMSF. To follow the progress of the cleavage, aliquots are removed to analyze by SDS–PAGE. Nearly complete proteolysis of the intact enzyme occurs in about 100 min resulting in two fragments that are relatively resistant to further degradation. This digest is applied to a ssDNA-agarose column (1 × 5 cm) equilibrated with 25 m*M* Tris–HCl, pH 7.5, 150 m*M* NaCl, 1 m*M* DTT, 1 m*M* EDTA, and 1 m*M* PMSF. Protein is eluted with a stepwise gradient of NaCl. The 31-kDa domain does not have significant ssDNA activity and appears in the flow-through, whereas the 8-kDa domain elutes at around 400 m*M* NaCl. Further purification of these fragments is accomplished by gel filtration with a Superose 12 FPLC column. Amino-terminal sequencing indicates that the 8-kDa domain primarily begins with Arg-4 and the 31-kDa domain starts primarily with residue Ile-88.

Proteolysis of Pol β with chymotrypsin also gives rise to the 31-kDa domain. Additionally, a 27-kDa fragment is observed that is stable for several hours. Proteolysis is performed as above, but the Pol β and protease ratio is decreased to 25:1 (w/w). Very little 8-kDa domain is observed probably due to the harsh proteolytic conditions. The 31- and 27-kDa domains can be separated by phosphocellulose chromatography. The flow-through from this column contains a 16-kDa fragment whereas the 27- and 31-kDa fragments bind. These domains could be separated with a NaCl stepwise gradient. The 27-kDa fragment elutes at 500 to 600 m*M* NaCl, whereas the 31-kDa domain elutes later, that is, 800 to 1000 m*M* NaCl. Amino-terminal sequence analysis indicates that the 27-kDa domain represents the carboxyl terminus beginning with residue Lys-141.

Chemical cleavage of denatured recombinant Pol β with CNBr results in a 16-kDa domain that is resistant to further degradation consistent with the fact there are no internal methionines.[19] Recombinant Pol β (3.5 mg) in 20 m*M* Tris–HCl, pH 7.5, 500 m*M* NaCl, 1 m*M* EDTA, and 1 m*M* PMSF is made to 70% formic acid. This sample is treated at 25°, in the dark, with a freshly prepared solution of CNBr in 90% formic acid, at a protein to CNBr ratio of 3:1 for 48 hr. After digestion is complete, the sample is diluted to 30 ml with H_2O and lyophilized. This is repeated once to remove all traces of formic acid and CNBr. The lyophilized powder is resuspended in 3 ml of H_2O and dialyzed for 24 hr against 25 m*M* Tris–HCl, pH 7.5, 250 m*M* NaCl, 1 m*M* EDTA, and 1 m*M* DTT to renature the domain. The dialyzed sample is loaded on a phosphocellulose column (1×4 cm) equilibrated with resuspension buffer. The protein is eluted with a stepwise NaCl gradient. The 16-kDa protein elutes at 500 m*M* NaCl. Sequencing indicates that this fragment begins at Leu-19. From the apparent molecular weight and cleavage specificity of CNBr, the carboxyl terminus would be Met-155. The domain boundaries are summarized in Table I.

Bacterial Overexpression

At the present time, DNA polymerase β is too large to determine its solution structure by NMR spectroscopy. However, because Pol β has functionally distinct domains, the smaller domains are amenable to structural analysis by multinuclear NMR. This approach requires that the domain be cloned, overexpressed, and purified. A T7 RNA polymerase/promoter-based expression plasmid, pRSET, has been constructed with the coding sequence of each domain. This was done with the 8- (residues 1–87), 16-

(residues 1–140), and 31-kDa (residues 88–335) domains, making large amounts of recombinant protein available for biophysical and functional characterization, as well as molecular structure determination.[21] For NMR spectroscopy, the protein also needs to be isotopically labeled with ^{1}H, ^{13}C, and/or ^{15}N. This has been successfully accomplished with the ssDNA-binding 8-[17] and 31-kDa domains of Pol β. The structural determination of the 8-kDa domain is described in detail in Chapter 14 of this volume. In addition, Chapters [2]–[4] in Vol. 177 in this series describe isotopic labeling of protein samples for NMR analysis.

Recombinant rat 8-kDa domain has been overexpressed and purified for NMR structural analysis.[17] Typically, 20 to 25 mg of recombinant 8-kDa domain can be purified per liter of induced bacterial culture. For ^{15}N- and ^{13}C-correlated NMR experiments, the 8-kDa domain is overexpressed from *E. coli* grown on a minimal media. Just prior to induction, $^{15}NH_4Cl$ and/or [^{13}C]glucose is added. Uracil, thymidine, MOPS, and Tricine should be omitted from the culture media when ^{13}C-labeling is desired to eliminate alternative carbon sources. A culture of *E. coli* [BL21 (DE3) pLysS] containing the recombinant plasmid is diluted 1 : 50 in LB medium and grown at 37° until the 595-nm absorbance reaches 0.5. After a 3-hr induction with 1 m*M* isopropyl-β-D-thiogalactopyranoside, cells are pelleted by centrifugation and stored at −80° until used.

Pelleted cells are resuspended in Buffer D, sonicated, and centrifuged as described for intact enzyme. The supernatant fraction is diluted with Buffer A to bring the NaCl concentration to 75 m*M*. This is loaded on a Q-Sepharose column connected in series to a ssDNA-cellulose column preequilibrated with Buffer A including 75 m*M* NaCl. Under these conditions, the 8-kDa domain passes through the Q-Sepharose column and binds to the ssDNA-cellulose; because the net charge of the 8-kDa domain at neutral pH is +10, it does not bind to Q-Sepharose. The Q-Sepharose column is washed with equilibrating buffer and disconnected from the ssDNA-cellulose column. The ssDNA-cellulose column is washed with equilibration buffer, and bound proteins are eluted with a NaCl gradient (75 to 1000 m*M* in Buffer A). Column fractions are analyzed for 8-kDa protein by SDS–PAGE, and peak fractions (300 to 500 m*M* NaCl) are pooled and dialyzed against equilibration buffer. The dialyzed fraction is filtered (0.22-μm Millipore filter) and loaded on a Mono S FPLC column (HR 10/10). Bound proteins are eluted with a NaCl gradient as above, and 8-kDa protein containing fractions (350 to 400 m*M* NaCl) are pooled, concentrated, and stored at −70°.

Because the ssDNA binding affinity is similar for the 8- and 16-kDa

[21] S. Widen, unpublished results, 1993.

domains, the purification protocols are similar. However, the 31-kDa domain does not bind to ssDNA and must be purified by another protocol.[22] In this case, pelleted cells are resuspended in Buffer D, sonicated, and centrifuged as described above. The supernatant fraction is dialyzed against a buffer containing 10 m*M* Tris–HCl, pH 7.5, 1 m*M* EDTA, and 1 m*M* PMSF. This is loaded on a Q-Sepharose column (1 × 10 cm) equilibrated with dialysis buffer. Under these conditions, 31-kDa domain binds to the column and is eluted with a 0 to 500 m*M* NaCl gradient. Fractions containing 31-kDa are pooled (200 to 500 m*M* NaCl) and dialyzed as above. The dialyzed sample is loaded on an Affi-Gel Blue (Bio-Rad) column (3 × 10 cm) and protein is eluted with a NaCl gradient. Fractions (≈400 m*M* NaCl) containing 31-kDa protein are pooled, concentrated, and stored at −70°.

Activity Assay

The purified enzyme conducts DNA synthesis in a distributive fashion on ssDNA templates. This behavior is consistent with the weak nucleic acid binding affinity that Pol β exhibits toward such a substrate.[23] In contrast, Singhal and Wilson[24] have demonstrated processive polymerization by Pol β on short gaps (≤6 nucleotides). Processive synthesis had a strict requirement for a 5′-phosphate on the downstream polynucleotide. Prasad *et al.*[23] demonstrated that the 5′-phosphate directs binding of Pol β via its 8-kDa amino-terminal domain to the side of the gap *opposite* the free 3′-hydroxyl. The *in vitro* gap-filling reaction is consistent with a role for Pol β in "short-gap" synthesis during DNA repair and replication. The processive short gap-filling activity of Pol β also suggests that previous *in vitro* assays that followed distributive nucleotide incorporation, such as fidelity assays to determine mutation frequency[25,26] or nucleotide misincorporation,[27] need to be reexamined under conditions where Pol β processively incorporates nucleotides. Therefore, these assays need to be repeated with short gapped (≤6 nucleotides) DNA substrates where the downstream polynucleotide is 5′-phosphorylated. The increase in binding affinity of Pol β to this substrate, as compared to an ungapped or gapped substrate without a 5′-phosphate on the downstream polynucleotide, is consistent with an increase in processivity.[23] It remains to be determined if kinetic steps other than DNA binding are also influenced.

[22] R. Prasad and S. H. Wilson, unpublished results, 1993.

[23] R. Prasad, W. A. Beard, and S. H. Wilson, *J. Biol. Chem.* **289,** 18,096 (1994).

[24] R. K. Singhal and S. H. Wilson, *J. Biol. Chem.* **268,** 15,906 (1993).

[25] D. I. Feig and L. A. Loeb, *Biochemistry* **32,** 4466 (1993).

[26] T. A. Kunkel, *J. Biol. Chem.* **260,** 5787 (1985).

[27] S. Shibutani, M. Takeshita, and A. P. Grollman, *Nature* **349,** 431 (1991).

Concluding Remarks

Proteolytic domain mapping has demonstrated the modular organization of DNA polymerase β. These domains have been shown to have specific functions (Table I) by a variety of functional and biophysical assays. Most impressively, the X-ray structure determination of Pol β[13,14] has confirmed the domain organization and is consistent with a functional segregation of activities. The catalytic domains of both RNA and DNA polymerases have been described, by analogy to a right hand, as consisting of fingers, palm, and thumb subdomains.[28] The 8-kDa domain appears to be a distinct domain and does not interact with nucleic acid in the crystallographic structure, but as Pelletier *et al.*[14] suggest, this may be due to the short ssDNA overhang in their complex. A complex with a longer ssDNA overhang or with a gapped DNA substrate bearing a 5′-phosphate in the gap should confirm the role of this domain relative to intact protein. UV cross-linking suggests that it plays a critical role in the identification of a gapped DNA substrate with a 5′-phosphate.[23] In addition, NMR analysis of the 8-kDa domain complexed with ssDNA could identify protein–nucleic acid interactions, and analysis of the larger 16-kDa domain, which has both ss- and dsDNA binding activities might show how these domains coordinate their specific nucleic acid binding activities. The 16-kDa domain is composed of most of the 8-kDa domain and the fingers subdomain (Fig. 1). Finally, as demonstrated by activity gel analysis, the 31-kDa domain can catalyze the nucleotidyltransferase reaction and is structurally equivalent to the fingers, palm, and thumb subdomains, whereas the 27-kDa domain, which lacks catalytic activity, is equivalent to the palm and thumb subdomains. Proteolysis of this domain with trypsin leads to the separation of these subdomains. Therefore, the three subdomains of the catalytic domain are required for enzymatic activity and the presence of the 8-kDa domain dramatically increases catalytic efficiency. Because the 8-kDa domain and the subdomains lack catalytic activity, these protein fragments could be potentially useful in regulating Pol β functions *in vivo* and defining its cellular role(s). In this regard, the 16-kDa domain has been shown to inhibit Pol β dependent reactions in a polymerase- and domain-specific manner.[29]

Acknowledgments

We thank Drs. Amal Kumar and Rajendra Prasad for helpful discussions.

[28] L. A. Kohlstaedt, J. Wang, J. M. Friedman, P. A. Rice, and T. A. Steitz, *Science* **256,** 1783 (1992).

[29] I. Husain, B. S. Morton, W. A. Beard, R. K. Singhal, R. Prasad, S. H. Wilson, and J. M. Besterman, *Nucleic Acids Res.* **23,** 1597 (1995).

[12] Purification and Enzymatic and Functional Characterization of DNA Polymerase β-like Enzyme, *POL4*, Expressed during Yeast Meiosis

By MARTIN E. BUDD and JUDITH L. CAMPBELL

Introduction

Three nuclear DNA polymerases, designated α, δ, and ε, have been identified in *Saccharomyces cerevisiae*.[1] The corresponding genes, *POL1* (also known as *CDC17*), *POL2*, and *POL3* (also known as *CDC2*), have been cloned and sequenced and the effect of mutations in these genes has been studied.[2-5] The genes contain considerable sequence similarity, including a signature set of six highly conserved stretches in the central part of the proteins, and are often referred to as members of the α-like class of DNA polymerases. Failure to synthesize chromosomal-sized DNA at the restrictive temperature in temperature-sensitive mutants affecting DNA polymerases α, δ, and ε provides direct evidence that these three distinct polymerases are required during DNA replication.[6,7] It has been proposed, based largely on studies in the simian virus 40 (SV40) system, that during DNA replication DNA polymerase α initiates synthesis of leading and lagging strand primers with either polymerase δ or ε extending the primers. Another yeast gene, *REV3*, has significant homology to α-type DNA polymerases.[8] Strains with deletions of the *REV3* gene are viable, grow at a normal rate, show decreased induced mutation rates, and are slightly sensitive to UV, suggesting that the *REV3* protein functions in mutagenic repair but not replication. However, no DNA polymerase activity corresponding to the *REV3* gene has been identified to date.

Yet another nuclear eukaryotic DNA polymerase, DNA polymerase β,

[1] J. L. Campbell and C. S. Newlon, *Chromosomal DNA Replication*, pp. 41–146, 1991.

[2] L. M. Johnson, M. Snyder, L. M. S. Chang, R. W. Davis, and J. L. Campbell, *Cell* **43,** 369 (1985).

[3] A. Pizzagalli, P. Valsasnini, P. Plevani, and G. Lucchini, *Proc. Natl. Acad. Sci. USA* **85,** 3772 (1988).

[4] A. Boulet, M. Simon, G. Faye, G. A. Bauer, and P. M. J. Burgers, *EMBO J.* **8,** 1849 (1989).

[5] A. Morrison, H. Araki, A. B. Clark, R. K. Hamatake, and A. Sugino, *Cell* **62,** 1143 (1990).

[6] M. E. Budd, K. D. Wittrup, J. E. Bailey, and J. L. Campbell, *Mol. Cell Biol.* **9,** 365 (1989).

[7] M. E. Budd, and J. L. Campbell, *Mol. Cell Biol.* **13,** 496 (1993).

[8] A. Morrison, R. B. Christensen, J. Alley, A. K. Beck, E. G. Bernstine, J. F. Lemontt, and C. W. Lawrence, *J. Bacteriol.* **171,** 5659 (1989).

has been identified in mammalian cells but not in lower eukaryotes such as yeast.[9,10] Genes encoding DNA polymerase β have been cloned from several sources, and the sequence of the predicted proteins does not share significant similarity with the α class. DNA polymerase β is a small protein of only 40 kDa, and has received much attention from investigators interested in structure–function relationships in DNA polymerases, because it comprises the minimal functional DNA polymerase domain of any known naturally occurring polymerase. Little is known about its *in vivo* function, however. Because DNA polymerase β is expressed in nonmitotic, fully differentiated tissues such as the brain and because it is not induced in quiescent cells stimulated to undergo DNA synthesis, it is believed to function in DNA repair rather than DNA replication.[11–15] *In vitro* studies also implicate DNA polymerase β in DNA repair.[16,17] Some particularly informative experiments with purified enzyme suggest its function may be limited specifically to base excision repair.[18–20] This idea is supported by the observation that Pol β mRNA is induced by DNA damaging agents that produce lesions in single residues rather than cross-links or strand breaks.[10] A replicative function has not been entirely excluded, however, since DNA polymerase β expressed in *Escherichia coli* can complement the gap-filling deficiency of a DNA polymerase I mutant, suggesting a role in processing Okazaki fragments.[21] Finally, DNA polymerase β appears to be required to copy single-stranded DNA injected into *Xenopus* oocytes,[22] and a survey of Pol β expression reveals that the abundance of Pol β is 15–20 times higher in male germ line cells than in other tissues.[15] These

[9] D. C. Rein, A. J. Pecupero, M. P. Reed, and R. R. Meyer, "DNA Polymerase β Physiological Roles, Macromolecular Complexes, and Accessory Proteins in the Eukaryotic Nucleus," pp. 95–123, 1990.

[10] S. H. Wilson, Gene Regulation and Structure–Function Studies of Mammalian DNA Polymerase β, 200–233, 1990.

[11] L. M. S. Chang, M. Brown, and F. J. Bollum, *J. Mol. Biol.* **74,** 1 (1973).

[12] V. Bertazzoni, M. Stefanini, G. Pedrali-Noy, E. Giulotto, F. Nuzzo, A. Falaschi, and S. Spadari, *Proc. Natl. Acad. Sci. USA* **73,** 783 (1976).

[13] B. Zmudzka, A. Fornace, J. Collins, and S. H. Wilson, *Nucl. Acids Res.* **16,** 9587 (1988).

[14] F. Hirose, Y. Hotta, M. Yamaguchi, and A. Matsukage, *Exper. Cell Res.* **181,** 169 (1989).

[15] R. Nowak, J. A. Siedlecki, L. Kacmarek, B. Z. Zmudzka, and S. H. Wilson, *Biochem. Biophys. Acta* **1008,** 203 (1989).

[16] K. Wiebauer and J. Jiricny, *Proc. Natl. Acad. Sci. USA* **87,** 5842 (1990).

[17] G. Dianov, A. Price, and T. Lindahl, *Mol. Cell Biol.* **12,** 1605 (1992).

[18] D. W. Mosbaugh and S. Linn, *J. Biol. Chem.* **258,** 108 (1983).

[19] H. Randahl, G. C. Elliot, and S. Linn, *J. Biol. Chem.* **263,** 12,228 (1988).

[20] R. Singhad and S. H. Wilson, *J. Biol. Chem.* **268,** 15,906 (1993).

[21] J. B. Sweasy and L. A. Loeb, *J. Biol. Chem.* **267,** 1407 (1992).

[22] T. M. Jenkins, J. K. Saxena, A. Kumar, S. H. Wilson, and E. J. Ackerman, *Science* **258,** 475 (1992).

latter two results are consistent with a specialized function in meiosis and sexual differentiation.

To distinguish among possible physiological roles suggested by the foregoing summary, many workers have looked for DNA polymerase β in a genetically tractable organism such as yeast. A 1746 bp open reading frame (YCR14C) located on yeast chromosome III has 26% identity to rat DNA polymerase β over the 393 amino acid COOH terminus.[23,24] We present here methods to purify the protein expressed from such an open reading frame (ORF). We have tagged it with a hemagglutinin epitope, overexpressed it, and used immunoaffinity chromatography to prepare a highly purified form suitable for enzymatic characterization in a single step. Assays showing its similarity to mammalian DNA polymerase β are described. We also present methods for determining the role of this polymerase *in vivo* by deleting the ORF. The mutants show no defect in mitotic growth nor in repair of damage induced by radiation or chemical mutagens. Interestingly, the gene is highly induced during meiosis and spore viability is decreased in the *pol4*Δ/*pol4*Δ mutant. Thus, our studies emphasize the importance of the meiotic function of Pol β. Similar conclusions have been reached by others using more classical methods.[25–27]

Materials and Methods

Strains

All yeast strains are listed in Table I.

Plasmids and Oligos

The following oligonucleotides are used in the gene disruption experiment:

oligo 1 5′ GGTGTCACTGACAAGATCTCCCATGTCTCTACTGGTGGTGC
TTCTTTGGTAGCAGCACGCCATAGTGACTGACG 3′

oligo 2 GTAATAAGTAAAGGATAAACATGCGACCTGTTAGACAAATC
GCACTCATGTTTGACAGCTTATCATC

[23] P. Bork, C. Oufounis, C. Sander, M. Scharf, R. Schneider, and E. Sahnhammer, *Nature* **358,** 287 (1992).

[24] S. G. Oliver, *Nature* **357,** 38 (1992).

[25] R. Prasad, S. G. Widen, R. K. Singhal, J. Watkins, L. Prakash, and S. H. Wilson, *Nucl. Acids. Res.* **21,** 5301 (1993).

[26] K. Shimizu, C. Santocanale, P. A. Ropp, M. P. Longhese, P. Plevani, G. Lucchini, and A. Sugino, *J. Biol. Chem.* **268,** 27,148 (1993).

[27] S. H. Leem, P. A. Ropp, and A. Sugino, *Nucl. Acid. Res.* **22,** 3011 (1994).

TABLE I
YEAST STRAINS

Strain	Genotype	Source[a]
6210	*αleu2-3,112, ura3-52 his3-A200 trp1-Δ901 lys2-01 suc2-Δ9*	S. Emr
6211	*aleu2-3,112, ura3-52 his3-A200 trp1-Δ901 ade2-101 lys2*	S. Emr
6210β-2A	*aleu2-3,112, ura3-52 his3-A200 trp1-Δ901*	
6210β-4A	*aleu2-3,112, ura3-52 his3-A200 trp1-Δ901, ade2-101 suc2-Δ9*	
6210β-2C	*α(pol4Δ:URA3) ura3-52 leu2-3,112 his3-Δ200 trp1-Δ901 ade2-101 suc2-A9*	
6210β-2D	*α(pol4Δ:URA3) ura3-52 leu2-3,112 his3-Δ200 trp1-Δ901 ade2-1101*	
6210β-10C	*a(pol4Δ:URA3) ura3-52 leu2-3,112 his3-Δ200 trp1-Δ901 suc2-Δ9*	
D6212	6210 × 6211; *POL⁺/POL⁺*	
D6213	6210β-2C × 6210β-10C; *pol4Δ/pol4Δ*	
D6214	6210β-2D × 6210β-10C; *pol4Δ/pol4Δ*	
g388-1B	*a leu2 his1-1 trp2 can1 gal2*	M. Hoekstra; J. Game
g388-2D	*a hom3-10 his1-7 ade2*	M. Hoekstra; J. Game
g857	g388-1B × g388-2D	
BJ5459	*a ura3-52 trp1 lys2-801 leu2Δ1 his3 Δ200 pep4:HIS3*	Yeast Genetic Stock Center
X12-68	*a rad1-1 ade2-1 gal2*	Yeast Genetic Stock Center
M754-6D	*a rad54-3 tup7-1 his1-1 trp2-1 ade4 leu2-1*	
SK-2-1B	*a ura3-52 trp2 his1-1*	
SK-3-7B	*α ura3-52 his1-7*	
SK-2-1B4Δ	*a pol4Δ::URA3 ura3-52 trp2 his1-1*	
SK-3-7B4Δ	*α pol4Δ::URA3 ura3-52 his1-7*	

[a] Unless otherwise noted, strains were constructed in this laboratory.

oligo 3 GAATTCTATCTCCGTAATCGGTCTCGGTC
oligo 4 AATAGCTTGGCAGCAACAGGG
oligo 5 GAATTCCCATGTCTCTACTGGTGGTGGTGC
oligo 6 TTTACCCTTTAGAGAAGGACCACCCAAGCTAGCGTA
GTCTGGGACGTCGTATGGGTACATGAATTCATTGTCT
AACAGGTC

pSEY18GAL is obtained from Scott Emr and has the *GAL1* and *GAL10* promoters in a *URA3*, 2-μm plasmid. pYCR14C is obtained from Bénédicte Purnell, Universite Catholique de Louvain. pGAL18βHA contains the *polβ* gene with the influenza hemagglutinin epitope (HA) MYPYDVPDY-

ASLGGP[28] fused to the NH_2-terminal region of the gene. The fusion is accomplished by site-directed mutagenesis using oligo 6, and the technique from Ref. 29. The HA epitope–*polβ* gene fusion is confirmed by DNA sequencing and subsequently cloned into the *Eco*RI site of pSEY18GAL.

UV, X-ray, and Methylmethane Sulfonate Sensitivity

The UV source is a Sylvania G15T8 germicidal lamp. The dosimetry is determined by ultraviolet meter Model No. J-225 manufactured by Ultraviolet Products, San Gabriel, CA. The X-ray source is a Pantak H-F 160 machine operated at 70 kV and 20 mA. The cells are irradiated at 7 krad/min and 4.5 cm from the beam. The dosimetry is determined by the method of Refs. 30 and 31. Cells are harvested, washed in 10 m*M* Na_2PO_4, pH 6.8, irradiated in 10 m*M* Na_2PO_4, pH 6.8, diluted in the same buffer, plated, and counted 3 days later.

Sensitivity to methylmethane sulfonate (MMS) is determined as described by Ref. 32. Cells are grown to stationary phase for 2 days at 30°, harvested, and washed with 0.05 *M* KPO_4, pH 7.0. Cells are resuspended at 3×10^7 cells/ml in the same buffer. Cells are incubated with 0.05% MMS for 0, 20, 40, or 60 min and then diluted 100-fold into 10% cold sodium thiosulfate. Survival is determined by plating on YPD medium at the appropriate dilution.[32]

Gene Disruption

The DNA polymerase β gene, *POL4*, is disrupted by a novel PCR (polymerase chain reaction) procedure suggested to us by Dr. Shane Weber (Eastman Kodak Co., Rochester, NY) through Chris Byrd (California Institute of Technology). Oligo 1 contains 45 bp of the DNA sequence of the *polβ* gene upstream of the ATG start and 20 bp of the tetracycline gene upstream of the *Hind*III site. Oligo 2 has 45 bp of DNA sequence 280 bp downstream of the TAA stop codon of the *POL4* gene and 20 bp of the tetracycline gene 300 bp downstream from the *Hind*III site. These oligonucleotides are used as primers in a PCR reaction using YEp24 as a template. The *URA*3 gene is cloned into the tetracycline gene of pBR322 at the *Hind*III site in YEp24, and thus the PCR product is 1.5 kb and has

[28] J. Field, J.-I. Nikawa, D. Broek, B. MacDonald, L. Rodgers, I. A. Wilson, R. A. Lerner, and M. Wigler, *Mol. Cell Biol.* **8,** 2159 (1988).

[29] J. W. Taylor, J. Ott, and F. Eckstein, *Nucl. Acids Res.* **13,** 8765 (1985).

[30] H. Fricke and S. Morse, *Am. J. Roentgenology* **18,** 426 (1927).

[31] A. U. Fregene, *Radiation Res.* **31,** 256 (1967).

[32] L. Prakash and S. Prakash, *Genetics* **86,** 33 (1977).

an "insertion" of *URA3* flanked by *POL4* sequences. The PCR product is used directly to transform the *ura3-52* diploid D6212 and Ura^+ clones were selected. The oligonucleotides homologous to *POL4* direct integration at the *POL4* locus and the site of integration is marked by the *URA3* gene in the PCR product.

The 45 bp of *POL4* sequence identity at the end of the PCR product is sufficient to direct the DNA to the *POL4* locus in a small percentage of the transformants. Potential deletions are first checked in a PCR reaction using oligo 4 and oligo 3. Oligo 4 is 270 bp downstream from the *Hin*dIII site of the *URA3* gene, and oligo 3 is 120 bp upstream of the ATG in the *ypolβ*-70 gene. An expected 450 bp fragment is observed in the transformant D6212-6. A Southern blot was done with the diploid using the *URA3* gene as a probe and an expected 3.7 kb product was observed in the *Dra*I digest and an expected 3000 kb band was observed in a *Cla*I digest. The transformant is dissected and four viable segregants were observed. Southern blots were carried out on *URA3* and *ura3-52* segregants using a probe containing the *POL4* coding sequence. Expected *Dra*I bands of 1110 and 2840 bp, *Hin*dIII bands of 700 and 400 bp, and an *Xba–Eco*RI band of 2580 bp were observed in the *ura3-52* undisrupted segregant, but not in 2 *URA3* segregants. Thus the entire coding region of *POL4* was deleted from the chromosome.

Assays and Purification

All assays are carried out at 37° for 1 hr. DNA polymerase assay mixtures contained 50 m*M* Tris, pH 8.0, 10 m*M* $MgCl_2$, 50 m*M* NaCl, 0.5 mg/ml bovine serum albumin (BSA) and 100 μ*M* each dATP, dGTP, dCTP, and [^{3}H]dTTP (300 cpm/pmol). The DNA concentration is 100 μg/ml activated DNA or 100 μg/ml poly(dA) · oligo(dT) (20 : 1). $MnCl_2$ is substituted at 1 m*M* for $MgCl_2$. Assays using rA-dT as a primer–template are carried out at a concentration of 250 μg/ml. The rA-dT ratio is 20 : 1. There are approximately 16,000 pmol dA and 60 pmol of primer termini in the dA-dT (20 : 1) reaction and 40,000 pmol rA and 150 pmol of primer termini in the rA-dT reaction. Assays in the presence of 30 μg/ml aphidicolin are in the presence of 10 μ*M* dCTP. Assays measuring incorporation rUTP are carried out at a concentration of 100 μ*M* rUTP. Assays measuring sensitivity to dideoxythymidine triphosphate are carried out in the presence of 10 μ*M* dTTP. After incubation, assay mixtures are spotted on DE81 filters, washed 6 times in 0.5 *M* Na_2PO_4, washed once with H_2O, once with ethanol and counted in toluene-based scintillation fluor.

BJ5459 cells containing the pGAL18HAβ plasmid are grown in 2% raffinose synthetic media to 10^7 cells/ml. Galactose is added to 2%, and cells are harvested 6 hr later, and frozen at −70°. Cells were lysed in buffer containing 0.05 *M* Tris, pH 8.0, 0.05 *M* NaCl, 10% glycerol, 1 m*M* EDTA, 1 m*M* dithiothreitol (DTT) in liquid N_2 by grinding with a mortar and pestle. The powder is thawed, and protease inhibitors [1 m*M* phenylmethylsulfonyl fluoride (PMSF), 2 m*M* benzamidine, 2 μg/ml pepstatin A, 1 μg/ml leupeptin] are added. Cells are centrifuged at 30,000 rpm in a Ti 50 rotor for 20 min and 200 μg of protein is used in each immunoprecipitation. Yeast extract is mixed with 30 μg of 12CA5 monoclonal antibody for 2 hr at 0°. 12CA5 is a subclone of H26D08-mouse immunoglobulin G (IgG)-2b, which is raised against the influenza hemagglutinin peptide (75–110).[28] Thirty microliters of 10% protein A beads is added for 1.5 hr. Beads are washed five times with TBS + 0.1% Tween, 2 times with 2× assay buffer, resuspended in 30 μl of 2× assay buffer, and used directly for polymerase assay. Thirty microliters of [^{3}H]NTP, DNA, Mg^{2+}, or Mn^{2+} is added to start the reaction. Ten to 50 μg of peptide HA1 is used as competitor.

Meiotic Expression Experiments

RNA from cells at different stages in meiosis is purified from g357, an SK-1 derived strain that exhibits highly synchronous rapid sporulation. The g357 is grown 12 hr in YKA medium: 1% potassium acetate, 1% peptone, 0.5% yeast extract, 0.5% $(NH_4)_2SO_4$, 0.017% YNB without amino acids and $(NH_4)_2SO_4$, 10 g potassium acid phthalate, pH 5.8, as described previously.[33] Cells are resuspended in 1% potassium acetate and 5 mg/liter, histidine, and samples are taken every 2 hr and analyzed for *POL4* RNA and commitment to recombination. Cells sporulate after 10 hr and sporulation is complete after 12 hr. RNA is isolated by standard procedures. RNA formaldehyde gels are run and blotted onto Gene Screen, and washing and probing is carried out according to manufacturer's directions. The probe extends from 27 bp 5′ to the ATG of *POL4* to 200 bp 3′ to the TAA. For assessing commitment to recombination, cells are plated on histidine-deficient complete medium in a classical interrupted meiosis protocol as previously described.[33] The background frequency of His^+ recombinants at 0 time is about 10^{-5}. The numbers given in Fig. 3 (below) are the determined values normalized to the 0 time point.

Results

To investigate whether the gene encoded by ORF YCR14C is a homolog of mammalian DNA polymerase β, we first wished to demonstrate that

[33] M. E. Budd, K. Sitney, and J. L. Campbell, *J. Biol. Chem.* **264,** 6557 (1989).

the protein encoded by the ORF has DNA polymerase activity. Yeast was chosen as a host for expression of the cloned gene, because expression is efficient in yeast and allows for detection of proteins that interact with the gene product of interest. To distinguish any new polymerase from the known cellular polymerases, ORF YCR14C was tagged with an influenza HA epitope. The epitope YPYDVPDYASL plus an additional GGP were fused to the NH_2 terminus of the protein. The regions of similarity between the ORF and rat Pol β are in the COOH-terminal region, so the HA epitope is unlikely to interfere with polymerase activity. The fusion protein is referred to as HA y*pol*β. Since yeast DNA polymerase β had never been detected in extracts of vegetatively growing cells, it seemed likely that it was expressed at low levels from its native promoter. Therefore, to overproduce the putative polymerase, the 2.2 kb *Eco*RI fragment containing the HA Pol β fusion protein plus 400 bp of sequence downstream from the TAA was cloned into the *Eco*RI site of pSEY18 GAL, placing the HA y*pol*β gene under control of the inducible *GAL1,10* promoter.

An investigation of the protein and its DNA polymerase activities was then initiated. Extracts were prepared from strains carrying plasmid pSEY18GAL and pGAL18HAβ after growth in galactose for 6 hr. Proteins (20 μg) were separated on an SDS gel, transferred to nitrocellulose, and probed with the HA epitope-specific antibody, 12CA5. A 75-kDa band appeared in the lane with protein from pGAL18HAβ extracts but not in the lane with protein from the pSEY18GAL vector extracts (data not shown). The experimentally determined molecular mass of the fusion protein is reasonably close to the predicted molecular mass of 70 kDa (ORF plus epitope). Thus, the protein is expressed from the pGAL18HAβ plasmid.

DNA polymerase activity was measured in immunoprecipitates prepared from various strains using the HA epitope antibody 12CA5 as described in the Experimental Procedures section. In each set of assays, controls included immunoprecipitation in the presence of molar excess of the HA peptide, immunoprecipitation of extracts from cells transformed with the pSEY18GAL vector (no insert), and sometimes precipitations in the absence of 12CA5 antibody. Since DNase I-treated DNA ("activated" and gapped DNA) has been reported to be a preferred substrate for Pol β, initial assays were performed with this substrate. As shown in Table II, DNA polymerase activity was immunoprecipitated by 12CA5 antibody in galactose-induced cultures containing the plasmid pGALHAβ. The activity was 10-fold higher than in controls. This result demonstrates that the product of ORF YRC14C has DNA polymerase activity and is the fourth nuclear DNA polymerase in *S. cerevisiae.* We therefore designate ORF YRC14C, *POL4.*

TABLE II
DNA POLYMERASE β ACTIVITY IN 12CA5 IMMUNOPRECIPITATES

Primer–template	Immunoprecipitation assay conditions	Activity (pmol dTMP)[a]	
		Exp. 1	Exp. 2
DNA	Complete[b]	101	96
DNA	Vector[c]	13	12
DNA	HA peptide[d]	15	10
DNA	No antibody	9	—
DNA	dCTP (10 m*M*)	212	106
DNA	Aphidicolin (30 mg/ml)	162	78
DNA	NEM (10 m*M*)	8	—
dA-dT(20:1)	Complete	293	411
dA-dT(20:1)	dTTP (10 m*M*)	95	102
dA-dT(20:1)	dTTP (10 m*M*)/ddTTP (10 m*M*)	11	6
dA-dT(10:1)	Complete	105	244
dA-dT(20:1)	Vector[b]	11.8	16
dA-dT(10:1)	Vector[b]	10.5	—
dA	Complete	—	0.9

[a] The results of two separate experiments are reported. Activity represents extent of synthesis in 30 min.

[b] See the Experimental Procedures section for a complete description of the immunoprecipitation assay.

[c] Extracts were prepared from cells containing the pSEY18GAL vector alone.

[d] HA peptide was preincubated with the antibody before addition of antibody to the extracts as a specific competitor of precipitation.

The conservation of sequence between the Pol4 polymerase and mammalian Pol β and the clear sequence divergence from α-type DNA polymerases (α, δ, ε, and Rev3) or the mitochondrial polymerase establish the new DNA polymerase as a member of the β class, and we will refer to it as yeast DNA polymerase β-70, yPol β-70, where the "70" represents the molecular weight of this species. Since this is the first example of such a protein in fungi, a number of biochemical experiments were indicated to assess if the enzyme was likely to act as a polymerase *in vivo*, the degree of conservation between the yeast and mammalian activities, and possible *in vivo* roles. DNA polymerase β is distinguished from α-type DNA polymerases by resistance to the DNA polymerase-specific inhibitor aphidicolin, sensitivity to dideoxynucleoside triphosphate substrates (ddNTP), and resistance to *N*-ethylmaleimide. Data from Table II show that the new yeast DNA polymerase is resistant to aphidicolin at a concentration of 30 μg/ml, which completely inhibits DNA polymerases α, δ, and ε. The new DNA polymerase is inhibited by ddTTP, which does not inhibit DNA polymerase α, δ, and ε at the levels used here. The new polymerase is sensitive to *N*-ethylmaleimide (10 m*M*), in contrast to rat DNA polymerase β.[34,35] Because yPol β-70 shares only 26% identity with rat polβ and since the yeast enzyme carries an additional 30-kDa NH_2-terminal domain that is not present in rat Pol β, differences in activity in the presence of such nonspecific inhibitors are to be expected.

Additional characteristics that readily distinguish DNA polymerases and suggest *in vivo* roles are primer–template requirement, substrate specificity, and associated activities, such as nuclease. The yPol β-70 shows significant activity with $(dA)_{500} \cdot (dT)_{12-18}$, 20:1, and thus, like mammalian polβ, uses synthetic homopolymers as primer–templates as well as DNA (Tables II and III). Although we cannot be sure we are in template excess, the limited extent of synthesis suggests that processivity is low. In a reaction containing dA-dT as primer–template, with 16,000 pmol template and 60 pmol of primer termini, 480 pmol of dTTP is incorporated. Thus, the primer termini are utilized more than once in the reaction; but, as for mammalian Pol β, processivity is low.

Most DNA polymerases are incapable of initiating synthesis *de novo*, that is, they require a primer. During replication the primer is usually provided by a primase; during repair the primer is provided by a nick. To test for a primer requirement, synthesis on the poly(dA) template alone was measured. No synthesis was observed, indicating that the enzyme does not initiate *de novo* (Table III). The absence of synthesis also indicates that yPol β-70 is incapable of a nontemplated polymerase reaction primed by the 3′-OH terminus of the poly(dA). Thus, neither yeast, rat, nor human DNA polymerase β's have terminal transferase activity, although DNA polymerase β's are homologous to terminal transferase.[36,37]

Many DNA polymerases can, under the right conditions, copy RNA as well as DNA templates. Mammalian Pol β was, in fact, first discovered in a search for cellular reverse transcriptases. Thus, it is interesting that yPol β-70 has significant reverse transcriptase activity. The yeast activity is observed in the presence of either Mg^{2+} or Mn^{2+}, whereas the rat Pol β reverse transcriptase activity has been described only in the presence of Mn^{2+}.[35,38,39]

[34] D. H. Stalker, D. W. Mosbaugh, and R. R. Meyer, *Biochem.* **15,** 3114 (1976).

[35] S. Yoshida, M. Yamada, and S. Masaki, *J. Biochem.* **85,** 1387 (1979).

[36] R. S. Anderson, C. B. Lawrence, S. H. Wilson, and K. L. Beattie, *Gene* **60,** 163 (1987).

[37] A. Matsukage, K. Nishikawa, T. Ooi, Y. Seto, and M. Yamaguchi, *J. Biol. Chem.* **262,** 8960 (1987).

[38] T. S.-F. Wang and D. Korn, *Biochem.* **16,** 4927 (1977).

[39] K. Ono, A. Ohashi, K. Tanabe, A. Matsukage, M. Nishizawa, and T. Takahashi, *Nucl. Acids Res.* **7,** 715 (1979).

TABLE III
rNTP INCORPORATION AND REVERSE TRANSCRIPTASE ACTIVITIES OF DNA POLYMERASE β

Primer–template	[^{3}H]NTP	Mg/Mn	Activity (pmol)[a]			
			HA Pol β	+ Peptide	Vector alone	No antibody
Experiment 1[b]						
dA-dT	T	Mg	480	64	16	ND
dA-dT	T	Mn	769	ND	7.2	ND
dA-dT	U	Mg	110	11	5.1	2.2
dA-dT	U	Mn	202	23	7.0	6.2
rA-dT	T	Mg	65	8.3	7.6	6.6
rA-dT	T	Mn	191	20	3.5	0.8
Experiment 2						
dA-dT	U	Mg	187	5.4	0.8	1.3
dA-dT	U	Mn	209	23	2.7	3.8
rA-dT	T	Mg	81	4	1.6	1.0
rA-dT	T	Mn	59	5.9	5.7	2
rA-dT	U	Mg	32	2.1	10.9	0.6
rA-dT	U	Mn	55	6.3	1.8	1.6

[a] Immunoprecipitations were performed and assays carried out as described in the Experimental Procedures section. "HA Pol β" refers to the complete assay in extracts of strains carrying the tagged *POL4* gene. "+peptide" indicates that the antibody used in the immunoprecipitation was incubated with the HA peptide competitor before addition to the extract. "Vector alone" indicates that immunoprecipitation was carried out on extracts of strains lacking the tagged *POL4* gene. "No antibody" means that antibody was omitted from the immunoprecipitation.

[b] The data are reported in separate experiments because they were carried out using immunoprecipitates prepared on different days. All polymerase assays were carried out in duplicate.

Not only can yPol β-70 utilize polyribonucleotides as templates, it can also use ribonucleoside triphosphates as substrates, as evidenced by the ability to incorporate rUTP on dA-dT and rA-dT in the presence of either Mg^{2+} or Mn^{2+}. Other DNA polymerases, such as *E. coli* DNA polymerase I, also incorporate ribonucleotides. However, the reaction is only efficient in the presence of Mn^{2+}, which is known to reduce the ability of DNA polymerases to discriminate between correct and incorrect nucleotide substrates. Mammalian Pol β also shows markedly reduced ability to discriminate among substrates compared to α-type polymerases, because in copying DNA it has 20 to 50% as much activity in the presence of a single dNTP as in the presence of all four dNTPs.[9] Thus, yPol β-70 appears to exhibit a wider range of substrate specificity in the presence of Mg^{2+} than any known DNA polymerase. Despite the ability to incorporate rUTP, polymerization of dTTP with dA-dT as the primer–template is the preferred reaction of yPol β-70 suggesting that it functions as a DNA polymerase rather than as an RNA polymerase *in vivo*. (As shown by the control experiments in Table III, the polymerase activities are specific for the HA Pol β fusion protein.)

Most DNA polymerases have associated activities such as pyrophosphorolysis, pyrophosphate exchange, or exonuclease. Mammalian Pol β, however, lacks such activities. Since yeast DNA polymerase β has a 30-kDa NH_2-terminal region not present in the mammalian enzyme, it was important to assess additional activities. Yeast Pol β, like mammalian Pol β, showed no 3′-exonuclease activity on a substrate prepared by labeling poly(dT) at the 3′-end with terminal transferase (data not shown).

Deletion of *POL4* Gene

Although it cannot be denied that extensive biochemical characterization can predict *in vivo* function to some extent, biochemistry alone can never be definitive. To investigate the *in vivo* role of yPol β-70, a 1990 bp deletion of the *POL4* gene was created. A region spanning one bp upstream of the ATG to 220 bp downstream of the TAA termination codon was deleted by a novel PCR strategy. The method and verification of one deletion, D6212-6, by PCR and Southern blotting, are described in Experimental Procedures. A diploid transformant, heterozygous for the deletion, *POL4*$^+$/*pol4*Δ, was sporulated. All asci dissected gave greater than two viable spores, showing that the *POL4* gene is not essential.

Role of DNA Polymerase β in DNA Repair

Because DNA polymerase β is believed to be involved in DNA repair,[16–18] UV and X-ray survival curves were carried out with *pol4*Δ strains. As illustrated in Fig. 1, *pol4*Δ mutants are no more sensitive to UV than a related *POL4*$^+$ strain. A *rad1-1* strain, X12-6B (Yeast Genetic Stock Center) was UV irradiated as a control. The survival curve shows that yeast DNA polymerase β does not play an essential role in repair of UV-induced damage.

Because *pol4*Δ strains were not sensitive to UV, sensitivity to X rays was investigated. Diploid strains were irradiated instead of haploid strains, because they are more resistant to irradiation and their survival curve is not biphasic. Figure 2 illustrates an X-ray survival curve of four strains, a *POL4*$^+$ diploid, D6212, two *pol4*Δ diploid strains, and a *rad54-3* haploid strain. The *rad54-3* strains are temperature sensitive for double-strand break repair, and their haploid and diploid survival curves are the same. There is no significant difference in survival between *POL4*$^+$ and *pol4*Δ strains, even at the highest dose, 30 krad (300 Gy), which probably induces 20 double-strand breaks per diploid genome (Budd, unpublished). If one to

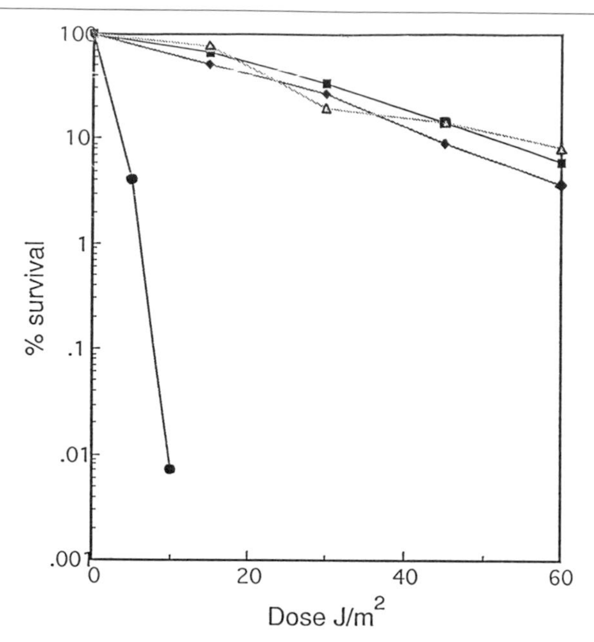

FIG. 1. UV survival curve of 6210-4A (*POL4*$^+$), 6210-2C (*pol4*Δ), 6210-2D (*pol4*Δ), and X12-6B (*rad1-1*) strains. Experiments were performed as described in Experimental Procedures. ■, *POL4*$^+$; ◆, *pol4*Δ; ▲, *pol4*Δ; ●, *rad1-1*.

two unrepaired double-strand breaks are lethal in a diploid, then the survival curve suggests that *pol4*Δ strains are completely proficient in double-strand break repair.

The genotype of the strain used in the repair studies allowed us to investigate whether yPol β-70 is required for chromosome stability after irradiation. Strains D6212, D6213, and D6214 are heterozygous for *ade2*. The appearance of red colonies shows the sum of recombination and chromosome loss. The *rad52* strains lose chromosomes at a high frequency after irradiation.[40] The frequency of red colonies in both *POL4*$^+$/*POL4*$^+$ and *pol4*Δ/*pol4*Δ strains after 6 krad of irradiation was about 1%. Thus, a high frequency of chromosome loss and recombination was not observed in *pol4*Δ diploid strains, suggesting that yPol β-70 does not play a major role in segregating chromosomes after irradiation.

Because base excision repair is best measured using an agent that introduces bulky adducts, and because mammalian Pol β is implicated in this kind of repair, we also determined sensitivity of the deletion strains to

[40] R. K. Mortimer, R. Contopoulo, and D. Schild, *Proc. Natl. Acad. Sci. USA* **78,** 5778 (1981).

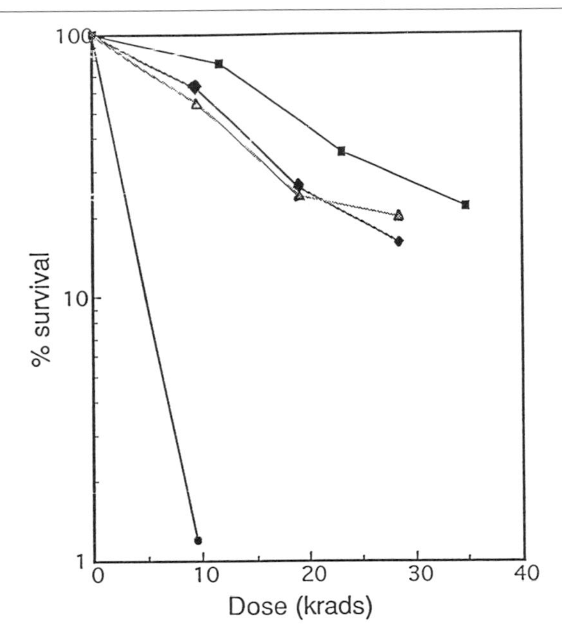

FIG. 2. X-ray survival curve of D6212 (*POL4*$^+$/*POL4*$^+$), D6213 (*pol4Δ/pol4Δ*), D6214 (*pol4Δ/pol4Δ*), and M754-3-6D strains. Experiments were performed as described in Experimental Procedures. ■, *POL4*$^+$/*POL4*$^+$; ◆, *pol4Δ/pol4Δ*; ▲, *pol4Δ/pol4Δ*; ●, *rad54-3*.

methyl methanesulfonate. Cells were treated with MMS as described in detail in Experimental Procedures and survival determined. Survival of wild type at this level of MMS was 71, 13, and 4.8% at 20, 40, or 60 min as reported previously. Again, there was no significant difference in survival between wild-type and mutant strains. In summary, Pol β does not appear to be exclusively responsible for repair of any of these types of damage.

Analysis of *POL4* mRNA Abundance in Mitosis and Meiosis

Because a comprehensive analysis of the RNA species transcribed from *S. cerevisiae* chromosome III during mitosis did not reveal an RNA at the *POL4* locus,[41] the possibility that *POL4* is induced in meiosis was tested. Strain g857 is heteroallelic at *HIS1,his1-1/his1-7*, and is derived from the SK-1 background that sporulates in 12 hr and is used for high synchrony

[41] A. Yoshikawa and K. Isono, *Yeast* **6,** 383 (1990).

sporulation experiments. As shown by the data in Fig. 3, a low abundance, 2.2 kb *POL4* message is observed in the mitotic culture and is induced 2 hr after initiating meiosis by resuspending the cells in potassium acetate medium. The message level increases to about 30-fold more than in mitosis and persists to the end of meiosis at 10 hr. After 10 hr asci can be observed in the culture but the message level remains high. To position the time of expression of DNA polymerase β mRNA with respect to other meiotic events, we used an interrupted meiosis experiment, which establishes the time of commitment to recombination in the cells during the course of meiosis.[33] At 2 hr the number of HIS^+ recombinants is the same as 0 hr. At 4 hr a 10-fold increase in HIS^+ recombinants is observed. The appearance of *POL4* message before the increase in commitment to recombination is consistent with an involvement at any stage of DNA replication or recombination, but does not constitute evidence for a role in either. All lanes also show bands at 3.2 and 1.8 kb, but since these sizes correspond to ribosomal RNAs, they may not be specific for *POL4*. The actin control shows that approximately equal amounts of RNA were loaded in each lane, except at the 10-hr time point. In summary, a 2.2 kb mRNA having the expected

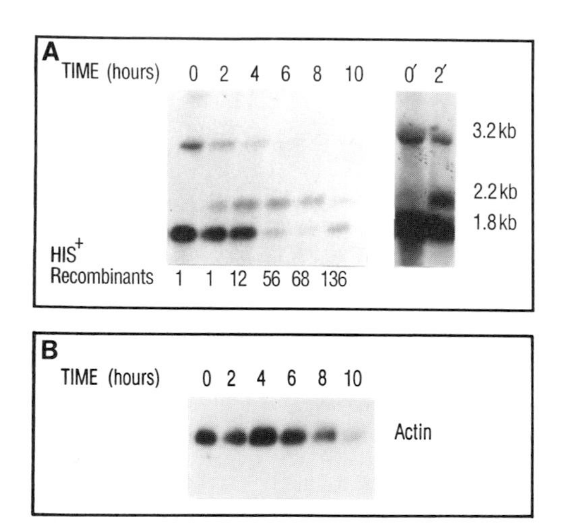

FIG. 3. Induction of DNA polymerase β message in meiosis. RNA was purified from cells at the indicated times during meiosis, electrophoresed, transferred to Gene Screen (Dupont), and probed with DNA from the *POL4* coding region. Equivalent amounts of RNA were loaded onto each lane. The numbers below the panel indicate the number of His^+ recombinants observed relative to time 0, determined as described in Experimental Procedures, and indicates commitment to meiotic recombination. Lanes labeled 0′ and 2′ are the 0- and 2-hr time points exposed long enough to reveal the small amount of *POL4* RNA in exponentially growing cells. Panel B shows the same filter hybridized to an actin probe.

size of the message of a 1749 bp ORF is expressed at low levels in vegetatively growing cells and is elevated during meiosis.

Role of yPolβ-70 in Sporulation

Because the DNA polymerase β message is induced in meiosis we further attempted to find a meiotic phenotype. The DNA polymerase β gene was disrupted in two different genetic backgrounds. Deletions in one genetic background, SEY6210 (obtained from Scott Emr), did not give rise to strains that were defective in meiosis or spore viability. The gene was then disrupted in a SK-1-derived strain background to yield a highly efficient, fast sporulating strain. The *POL4* gene was disrupted in strains SK-2-1B and SK-3-7B and they were mated. The diploid is homozygous for *pol4*Δ and is heteroallelic for *his1* (*his1-1*/*his1-7*) and can be used to study the role of DNA polymerase β in meiotic recombination. Table IV gives data from a commitment to meiotic recombination experiment. POL^+/POL^+ and *pol4*Δ/*pol4*Δ strains were incubated in potassium acetate media for the indicated times, plated onto histidine deficient media, and scored for prototrophs. As shown in Table IV, no difference in recombination frequency is observed between the *POL4* and *pol4*Δ strains. Thus, DNA polymerase β is not performing an essential role in meiotic recombination.

However, a significant percentage of the *pol4*ΔHIS^+ recombinants that appeared during the latter part of the experiment, when spores were already observable, formed smaller colonies than comparable HIS^+ recombinants from *POL4* strains. When these slow-growing *pol4*ΔHIS^+ recombinants were restreaked onto histidine-deficient media, they grew as well as the $POL4^+$ recombinants. Microscopic investigation of the slow-growing *pol4*Δ recombinants showed they exhibited haploid morphology. This suggests that a fraction of the *pol4*Δ spores may show a delay in germination.

TABLE IV
RECOMBINATION FREQUENCY IN *pol4*Δ/*pol4*Δ

Time (hr)	POL^+ HIS^+ ($\times 10^{-6}$)	*pol4*Δ HIS^+ ($\times 10^{-6}$)
0	9	35
2	16	35
4	240	307
6	4800	5200
8	5400	7500
10	1300	4000
24	6300	8000

In a second experiment, to analyze the efficiency of spore germination, about 90 tetrads from POL^{+}/POL^{+} strains and 90 tetrads from *pol4Δ/pol4Δ* strains were dissected and the viability of spores examined (Table V). There is a significant drop in spore viability in the *pol4Δ/pol4Δ* strain. The total spore viability of the *POL4/POL4* strain is 94% and the percent of 4:0's is 74%. The spore viability of the *pol4Δ/pol4Δ* strain is 62% and the percent of 4:0's is only 29%. Furthermore, a small percentage of the *pol4Δ* spores require longer than 3 days to form visible colonies, whereas all the *POL4* spores form visible colonies after 3 days. In one tetrad, no colonies appeared after 3 days, but four colonies appeared by 6 days. When the small colonies were restreaked, they grew as well as $POL4^{+}$ strains. The presence of slow-growing *pol4Δ* spores is analogous to the recombination experiment discussed above, where slow-growing $pol4\Delta HIS^{+}$ spores were observed. Thus, the slow-growth phenotype of spores after germination was probably a result of unrepaired damage after meiosis, which is repaired after the spore has grown into a colony. The drop in viability and slow growth of *pol4Δ* spores suggest that DNA polymerase β is a meiotic DNA repair enzyme. Spore viability is lower in the SK-1 background strains possibly because that sporulation takes only 12 hr to complete, whereas in the strain SEY6210 background sporulation requires 3 days. Thus, there is less time for an alternative pathway to repair damage in the SK-1 background strain than the SEY6210 background strain.

Discussion

Guided by analysis of the complete sequence of yeast chromosome III, we have identified a new DNA polymerase in *S. cerevisiae* that is similar in several respects to mammalian Pol β, most notably in primary structure and response to inhibitors. Therefore, Pol β is more highly conserved and probably more important to eukaryotic organisms than previously thought. Extensive biochemical characterization shows that the new enzyme requires

TABLE V
SPORE VIABILITY IN *pol4Δ*

Viable : inviable	*POL4/POL4*	*pol4Δ/pol4Δ*
4:0	64	26
3:1	14	26
2:2	5	11
1:3	4	19
0:4	0	8

a primer and template and is therefore probably a DNA polymerase or reverse transcriptase but not a terminal transferase *in vivo.* A strain with a deletion of the 583 amino acid ORF (*pol4*Δ) is viable, and the *pol4*Δ mutants are not sensitive to radiation or chemical mutagens. Thus, yPol β-70 is not essential for any mitotic process tested. Instead, our results redirect attention to a comparatively small body of evidence, involving analysis of the occurrence of Pol β in different mammalian tissues, that is consistent with a significant role in meiosis. We have found that the yPol β-70 message is barely detectable in mitotically growing cells, that it is induced at least 30-fold as cells enter and progress through meiosis, and that a drop in spore viability occurs in *pol4*Δ/*pol4*Δ strains.

An obvious question is whether conservation of protein sequence (26% identity) means conservation of physiological function of DNA polymerase β between yeast and mammals. yPol β-70 is 26% identical to rat Pol β over a contiguous 393 amino acid stretch, and does not contain the six conserved regions of the α-like DNA polymerases and replicases. Human and yeast DNA polymerase α's are 31% homologous; bovine and yeast pol δ's are 44% homologous; yeast and human RP-A are 26% identical. Thus, 26% identity over 393 amino acids is sufficient homology between yeast and mammalian sequences to argue for conservation of function. Furthermore, the conservation lies in particular motifs identified with DNA polymerase function. One region, [356]GSYNRGYSKCGDIDLLF[372], corresponding to [179]GSFRRGAESSGDMDVLL[195] in rat Pol β, is highly homologous to terminal transferase region [333]GGERRKKMGHDVDFLI[349] [36,37] and encompasses the motif YGDTDS found in eukaryotic and some phage replicative DNA polymerases.[42] Amino acids Arg-183, Asp-190, and Asp-192 (rat notation) are conserved from yeast to humans and are required for rat enzyme function.[43,44] Mutations in this region alter the K_m for primer termini suggesting that they participate in primer recognition. Modeling studies have suggested that the YGDTDS region is part of a site for NTP binding. In rat Pol β, Ser-55 is a site of protein kinase C phosphorylation[45] and is conserved in yPol β-70 (Ser-230), whereas another phosphorylation site, rat Ser-44, is Gln in yPol β-70 instead. Because phosphorylation of rat Pol β inactivates the enzyme, Ser-230 might regulate activity of yPol β-70. An observation for which we have not yet found an explanation is that yPol β-70 has an additional 190 amino acids at the NH_2 terminus that are not

[42] T. S.-F. Wang, S. W. Wong, and D. Korn, *FASEB J.* **3,** 14 (1989).

[43] T. Date, S. Yamamoto, K. Tanihara, Y. Nishimoto, and N. Liu, *Biochemistry* **29,** 5027 (1990).

[44] T. Date, S. Yamamoto, K. Tanihara, Y. Nishimoto, N. Liu, and A. Matsukage, *Biochemistry* **30,** 5286 (1991).

[45] T. Tokui, M. Inagaki, K. Nishizawa, R. Yatani, M. Kusagawa, K. Ajiro, Y. Nishimoto, T. Date, and A. Matsukage, *J. Biol. Chem.* **266,** 10,820 (1991).

found in mammalian polymerases. Because yPol β-70 has no 3′→5′-exonuclease, nucleolytic proofreading is not a function of this region. The predicted amino acid sequence has homology to an acidic domain of nucleolin and to several sequence motifs characteristic of transcription factors that participate in ribosomal RNA synthesis. Whether these residues allow yPol β-70 to play an additional role in yeast that it does not play in mammals, perhaps in RNA synthesis, remains to be determined.

The data in Tables II and III reveal extensive parallels between the yeast and mammalian Pol β's, especially when viewed from the perspective of how different these polymerases are from α-like replicases. One significant difference of the β polymerases from the α-like family is loose substrate specificity, which is manifested by comparatively efficient incorporation of ddNTPs and even rNTPs, and extensive noncomplementary incorporation in the presence of a single dNTP.[9] The yeast enzyme seems to be the least discriminating in that rNTPs are substrates and the activities occur in the presence of Mg^{2+} as well as Mn^{2+}. Low substrate specificity combined with another property we have documented, lack of proofreading activity, argues against a role for DNA polymerase β in any process requiring extensive and accurate synthesis. Function in a reaction requiring only limited synthesis is also indicated by the fact that the enzyme appears to synthesize only short stretches of DNA.

A second notable difference between the β and α classes is the resistance of Pol β to the drug aphidicolin, which very specifically and efficiently inhibits α-like polymerases. The aphidicolin resistance almost certainly reflects a basic structural difference in the active sites of the two types of polymerases. We suggest that an active site that is structurally more flexible as evidenced by broad-spectrum substrate acceptance can accommodate aphidicolin, whereas the drug significantly interferes with enzymes with more stringent recognition determinants. If this were true, then one would expect that mutations that relax the substrate specificity of DNA polymerase α might also decrease aphidicolin sensitivity. Mutations that increase misinsertion frequency by DNA polymerase α have been identified and, as predicted, they do increase aphidicolin resistance.[46,47]

A third significant difference of the β-type polymerases is their relatively efficient use of RNA as a template compared to DNA, that is, their reverse transcriptase activity. The availability of the *pol4Δ* mutant will allow a genetic analysis of the importance of the activity. One cellular role of reverse transcriptase is processing of telomeres. A yeast telomerase has not been definitively identified and, therefore, the possibility that yPol β-70

[46] W. C. Copeland, N. K. Lam, and T. S.-F. Wang, *J. Biol. Chem.* **268,** 11,041 (1993).
[47] W. C. Copeland and T. S.-F. Wang, *J. Biol. Chem.* **268,** 11,028 (1993).

participates in telomere synthesis deserves further investigation. The only yeast DNA polymerase directly implicated in *de novo* telomere synthesis is DNA polymerase α, since temperature-sensitive mutants synthesize extra long telomeres, perhaps due to a faulty interaction with telomerase.[1]

One feature that has been used historically to identify polymerase β is resistance to NEM. Our finding that yeast Pol β is sensitive to NEM suggests that this criterion should be reevaluated. NEM binds any -SH group and as such is very nonspecific. Aphidicolin resistance, ddNTP sensitivity, and pattern of expression offer more reliable criteria to distinguish polymerases. For instance, the ciliated protozoan *Tetrahymena* carries an aphidicolin resistant polymerase whose levels are increased in stationary phase of growth.[48,49] Although the enzyme is NEM sensitive, it may actually be the *Tetrahymena* polymerase β, since it is only 70 kDa in size, is sensitive to ddTTP, and has the monovalent cation sensitivity and pH optimum characteristic of Pol β. We suggest that Pol β is, contrary to previous opinion, conserved throughout phylogeny from yeast to man. One reason the enzyme may have gone undetected in lower eukaryotes until now could be its low abundance in exponentially growing cells (see Fig. 3). It is also possible, however, that a β-like enzyme of 70 kDa that is NEM sensitive will someday be found in mammalian cells.

DNA polymerase β is not required for movement of the replication fork under normal, vegetative growth conditions. Either yPol β-70 does not participate or another polymerase can carry out its function in its absence. Because three DNA polymerases are essential, the dispensability of Pol β is not unexpected. More surprising, because of the large body of evidence in mammalian systems that DNA polymerase β may participate in repair, is the finding that yPol β-70 mutants are not even slightly sensitive to DNA damaging agents. (Although it must be noted that it has been reported that *pol4* mutants are sensitive to high levels of MMS.[27]) Although mammalian studies suggest some role for additional polymerases in repair, some discernible defect in recovery of the yeast mutant from the wide spectrum of damage introduced by the agents we used would be anticipated, since the mammalian polymerases evidently have additive rather than compensatory functions in repair. This raises the issue of which DNA polymerase is involved in repair. The temperature-sensitive polymerase mutants have not been particularly informative on this point to date. *pol1-17* (DNA polymerase α) strains are proficient in repairing X-ray-induced single-strand breaks, suggesting that DNA polymerase α is dispensable for such repair.[6] DNA polymerase ε appears to be required for UV-induced base excision

[48] A. Sakai and Y. Watanabe, *J. Biochem.* **91,** 845 (1982a).

[49] A. Sakai and Y. Watanabe, *J. Biochem.* **91,** 855 (1982b).

repair *in vitro*, however.[50] DNA polymerase ε and DNA polymerase δ both have the potential to participate in repair *in vivo*, but one can compensate for loss of the other.[51–53]

The time of expression of the *POL4* gene in the yeast life cycle may give the best insight into its major physiological role. A 2.2 kb mRNA is present in vegetatively growing cells, though at very low levels. Strikingly, however, *POL4* mRNA is induced in meiosis and persists throughout the course of meiosis, including the period where significant spore formation is observed. Thus, although Pol β may have a mitotic role, it is likely to be more important during meiosis. The kinetics of induction demonstrates that this is an early meiotic RNA. Inspection of the *POL4* promoter reveals some elements, such as a T_4C site, found in the *IME1*-dependent UAS known to regulate genes that are repressed in mitosis and expressed exclusively in meiosis, such as *HOP1*, *RED1*, *SPO11*, or *SPO13*.[54,55] The sequences diverge considerably, however, and their spacing and organization relative to the ATG are different, making their contribution to the regulation of *POL4* somewhat questionable.

The induction of *POL4* mRNA in meiosis is likely to be functionally significant, since spores from *pol4Δ* diploid strains show a drop in viability. Unfortunately, the time course of appearance of message does not shed much light on the role of DNA polymerase β since it is consistent with a role in either replication or recombination. Premeiotic S phase, during which the bulk of meiotic DNA synthesis occurs, begins at about 4 hr in such an experiment and aspects of recombination occur over the whole time course.[56,57] DNA polymerase α is required for DNA synthesis during premeiotic S phase and both DNA polymerases α and δ are required for completion of meiosis,[6,58] so DNA polymerase β does not appear to be a replacement for a mitosis-specific DNA polymerase. Rather, Pol β is likely to perform a role in some process specific to meiosis. We have been unable to demonstrate a defect in double-strand break repair in our strain backgrounds, but Leem *et al.*[27] have reported evidence for a role in such repair.

The *pol4Δ* strains show a small drop in spore viability. Also, a small percentage of spores shows a delay in growth, which may reflect a delay

[50] Z. Wang, X. Wu, and E. C. Friedberg, *Mol. Cell Biol.* **12,** 1051 (1993).
[51] A. Blank, B. Kim, and L. Loeb, *Proc. Natl. Acad. Sci. USA* **91,** 9047 (1994).
[52] M. E. Budd and J. L. Campbell, *Mol. Cell Biol.* **15,** in press (1995).
[53] W. Suszek, H. Baranowska, J. Zuk, and W. J. Jachymczyk, *Curr. Genetics* **24,** 200 (1993).
[54] A. K. Vershon, N. M. Hollingsworth, and A. D. Johnson, *Mol. Cell Biol.* **12,** 3709 (1992).
[55] K. S. Bowdish and A. P. Mitchell, *Mol. Cell Biol.* **13,** 2172 (1993).
[56] R. E. Esposito and S. Klapholz, Meiosis and ascospore development, 211 (1981).
[57] C. Goyon and M. Lichten, *Mol. Cell Biol.* **13,** 373 (1993).
[58] D. Schild and B. Byers, *Chromosoma (Berl.)* **70,** 109 (1978).

in germination. These results are consistent with an involvement of DNA polymerase β in repairing damage, possibly breaks involved in meiosis. The drop in spore viability in strains may then be a result of chromosomal loss caused by incomplete repair of meiotic DNA damage. The reason the drop in spore viability was not observed in another genetic background in which we deleted the gene may be that when sporulation lasts 3 days, damage not repaired by DNA polymerase β is repaired by a different cellular polymerase. We have shown, for instance, that in mitosis DNA polymerase δ can compensate for loss of DNA polymerase ε in repair of UV damage, and vice versa.[52]

At least two additional periods of significant DNA synthesis have been documented in classical studies of meiosis in various *Lilium* species.[59] During the zygotene interval of meiotic prophase, semiconservative DNA synthesis, which may represent delayed replication of a specific subset of genomic sequences, occurs. During pachytene, a nonsemiconservative, repair-type DNA synthesis is observed.[59] If the same sort of phenomena occur in yeast, the properties of Pol β would be consistent with a role in either process. Repair synthesis might be neccssary during meiotic recombination, which is both quantitatively and qualitatively different from mitotic recombination and uses some gene products differentially expressed in meiosis.[60] Thus, a role in one of the several recombination pathways that have been identified may be the most likely function of Pol β.

The putative role in meiosis may be conserved in other organisms. The high level of mRNA in meiosis corresponds to the abundance of the message and protein in rat testes and in meiotic and postmeiotic male mouse germ cells, which are arrested in the pachytene stage of meiosis.[61] A special property of germ line cells that has been appreciated only recently is the *de novo* synthesis of telomeres.[62,63] Give the reverse transcriptase activity of Pol β and the reverse transcriptase activity of known telomerases, Pol β is a candidate to participate in such a process. It has been shown that a knockout of the Pol β gene in mice leads to lethality early in development.[64] This observation shows that Pol β performs an essential function and may be comparable to our observation that a fraction of *pol4*Δ spores shows a delay in germination. These results may begin to account for the otherwise puzzling conservation of the gene during evolution. The availability of the novel yeast gene and protein documented here will make

[59] Y. Hotta and H. Stern, *J. Mol. Biol.* **55,** 337 (1971).

[60] T. D. Petes, R. E. Malone, and L. S. Symington, Recombination in Yeast, pp. 407–521 (1991).

[61] P. Grippo, R. Gerenia, G. Locorotondo, and V. Monesi, *Cell Differ.* **7,** 237 (1978).

[62] F. Muller, C. Wicky, and A. Spicher, *Cell* **67,** 815 (1991).

[63] G. L. Yu and E. H. Blackburn, *Cell* **67,** 823 (1991).

[64] H. Gu, J. D. Marth, P. C. Orban, H. Mossmann, and K. Rajewsky, *Science* **265,** 103 (1994).

these and other models testable. During the preparation of this paper, three other studies employing the TRC14C ORF also appeared.[25,27,65]

Acknowledgments

We thank Bénédicte Purcell and Stephen Oliver for the plasmid YCR14C. This work was supported by grants from the USPHS (GM25508 and GM47281) and California TRDRP 3RT-0190.

[65] K. Shimizu and A. Sugino, *J. Biol. Chem.* **268,** 9578 (1993).

[13] Purification and Characterization of Human Immunodeficiency Virus Type 1 Reverse Transcriptase

By STUART F. J. LE GRICE, CRAIG E. CAMERON, and STEPHEN J. BENKOVIC

Introduction

Metal chelate chromatography is now finding widespread use in the purification of human immunodeficiency virus (HIV) reverse transcriptase (RT) directly from the high-speed supernatant of bacterial homogenates. Recombinant proteins containing a short array of histidine residues (6 His) at either their amino or carboxyl terminus[1,2] are retained at neutral pH by Ni^{2+}-nitrilotriacetic acid-Sepharose (Ni-NTA-Sepharose, Qiagen Inc. Chatsworth, CA), and eluted by application of either a descending pH or ascending imidazole gradient. In many cases, the purity of the protein exceeds 70% following this single chromatographic step and is sufficient for many enzymatic studies. In cases where highly purified enzyme is required, an additional chromatographic step can be applied (e.g., S-Sepharose). A particular feature of Ni^{2+}-NTA-Sepharose is the ability to recover selectively modified p66/p51 HIV-1 RT prepared through *in vitro* reconstitution, which allows a study of the contribution of individual subunits to functions of the parental enzyme. Examples of these strategies are presented in the following sections. In addition, an alternative strategy for preparation of an unmodified p66 subunit is presented.

[1] E. Hochuli, H. Döbeli, and A. J. Schacher, *J. Chromatogr.* **411,** 177 (1987).
[2] E. Hochuli, W. Bannwarth, H. Döbeli, R. Gentz, and D. Stüber, *Bio/technology* **6,** 1321 (1988).

Preparation of Recombinant p66/p51 HIV-1 RT in *Escherichia coli*

The biologically significant form of HIV-1 RT is a heterodimer of 66- and 51-kDa polypeptides (p66/p51), derived from the same gene, but differing in that p51 results through partial processing of p66 by the virus-coded protease (PR).[3,4] Preparation of recombinant HIV-1 RT has taken advantage of virus-mediated events by coexpressing the genes for p66 RT and HIV-1 PR, which results in accumulation of the p66/p51 heterodimer with a 1:1 subunit stoichiometry.[5,6] Typical strategies we have employed for heterodimer HIV-1 RT production are outlined in Figs. 1A–C. By genetically manipulating the polyhistidine extension onto the N terminus of p66, mature heterodimer contains the affinity label on both its p66 and p51 components (Fig. 1A). Alternatively, relocation of the affinity label to the C terminus of p66 yields a heterodimer where a single component (p66) is selectively "tagged" (Fig. 1B). Expression of recombinant p66 RT and its maturation into heterodimer is indicated in Fig. 1C.

Purification of Heterodimer HIV-1 RT

Solutions

RT Buffer A/78: 50 m*M* NaH_2PO_4, Na_2HPO_4, pH 7.8
RT Buffer A/60: 50 m*M* NaH_2PO_4, Na_2HPO_4, pH 6.0, 0.3 *M* NaCl, 10% (v/v) glycerol
RT Buffer D: 50 m*M* Tris–HCl, pH 7.0, 25 m*M* NaCl, 1 m*M* EDTA, 10% (v/v) glycerol
Storage Buffer D: As Buffer D, but containing 50% (v/v) glycerol
Phenylmethylsulfonyl fluoride (PMSF): 100 m*M* solution, prepared in 100% ethanol; stock PMSF solution can be stored at −20°
Lysozyme: 10 mg/ml in RT Buffer A/78; prepare fresh
Imidazole: 0.5 *M* solution, in RT Buffer A/60. Imidazole solutions should be stored in lightproof flasks to prevent breakdown. As the solution ages, a yellowish tinge is evident, which will interfere with UV absorption measurements.

[3] F. Di Marzo Veronese, T. D. Copeland, A. L. DeVico, R. Rahman, S. Oroszlan, R. C. Gallo, and M. G. Sarngadharan, *Science* **231,** 1289 (1986).
[4] M. M. Lightfoote, J. E. Coligan, T. M. Folks, A. S. Fauci, M. A. Martin, and S. Venkatesan, *J. Virol.* **60,** 771 (1986).
[5] V. Mizrahi, G. M. Lazarus, L. M. Miles, C. A. Meyers, and C. Debouck, *Arch. Biochem. Biophys.* **273,** 347 (1989).
[6] S. F. J. Le Grice and F. Gruninger-Leitch, *Eur. J. Biochem.* **178,** 307 (1990).

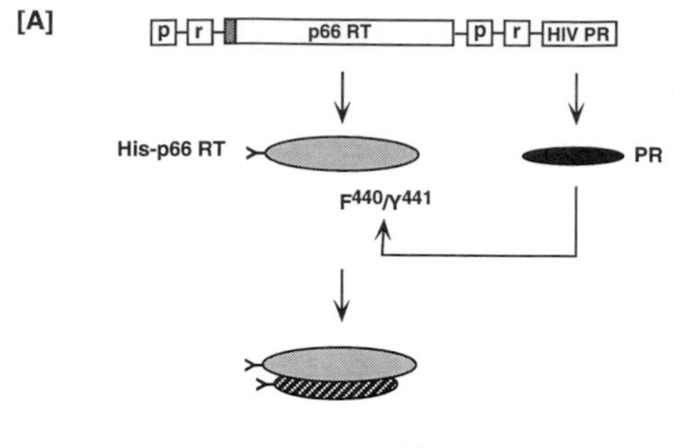
[A]
p
r
p66 RT
p
r
HIV PR
His-p66 RT
PR
F440/Y441
His-p66/His-p51 RT

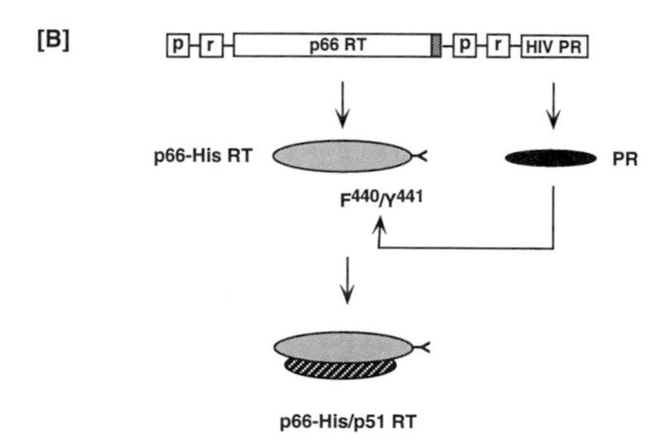
[B]
p
r
p66 RT
p
r
HIV PR
p66-His RT
PR
F440/Y441
p66-His/p51 RT

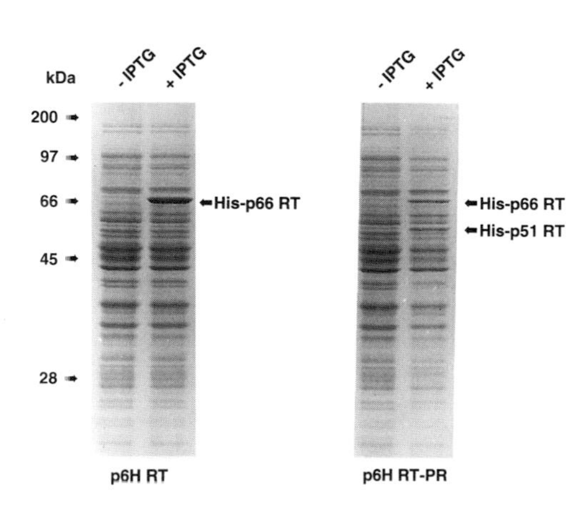
[C]
kDa
- IPTG
+ IPTG
200
97
66
45
28
His-p66 RT
p6H RT
- IPTG
+ IPTG
His-p66 RT
His-p51 RT
p6H RT-PR

Homogenization and High-Speed Centrifugation

All steps are carried out at 4°. Cells from isopropylthiogalactoside (IPTG)-induced cultures are suspended in 2 volumes of RT Buffer A/78 + 1 m*M* PMSF per gram wet weight and stirred at 4° until an even suspension is obtained. Lysozyme (in RT Buffer A/78) is added to a final concentration of 0.5 mg/ml, and stirring is continued until the suspension becomes viscous (~15 to 20 min). NaCl is added from a 5 *M* stock solution to a final concentration of 0.5 *M*, and the lysate is further disrupted by sonication over 3 × 60 sec, with 1-min cooling periods (Branson sonicator, 200 to 300 W, with pulsation). Care should be taken to ensure that the homogenate is kept cool during sonication. At this stage, the viscosity of the homogenate should be considerably reduced. Ribosomes and cell debris are removed by centrifugation at 100,000*g* for 45 min at 4°. The amber-colored high-speed supernatant is carefully decanted (avoid taking particulate matter, which will block the metal chelate column) and passed once through a narrow-gauge syringe needle to further reduce its viscosity.

Ni^{2+}-NTA-Sepharose Chromatography

The high-speed supernatant is applied to a column of Ni^{2+}-NTA-agarose (Qiagen Inc., Chatsworth, CA) equilibrated in RT Buffer A/78 containing 0.3 *M* NaCl. Due to the high selectivity of the affinity matrix, relatively small columns can be used. In general, we use 1 ml of affinity resin for the high-speed supernatant of 3 to 5 g biomass. Sample is applied at a flow rate of 1 to 2 column volumes/hr, after which the column is washed with RT Buffer A/78 + 0.3 *M* NaCl until OD_{280} equals 0.05. Weakly bound

FIG. 1. (A) Expression of polyhistidine-extended p66/p51 HIV-1 RT in *E. coli.* The recombinant plasmid p6H RT-PR contains complete expression cassettes for both the 66-kDa RT gene and HIV-1 protease (PR). *p*, repressible promoter; *r*, synthetic ribosome binding site. The shaded portion at the N terminus of the RT coding sequence represents the polyhistidine extension (6 His). Following induction of gene expression, HIV PR partially cleaves p66 RT between Phe-440 and Tyr-441, resulting in production of a heterodimer whose subunits are "tagged" at their N termini. (B) Manipulation of the polyhistidine extension for expression of selectively tagged heterodimer HIV-1 RT. The components of the expression cassette are identical to those in part (A), with the exception that the polyhistidine extension is relocated to the C terminus of p66. Following PR-catalyzed maturation, the resulting heterodimer contains the polyhistidine extension selectively at the C terminus of p66. (C) Expression of p66 HIV-1 RT and its maturation into heterodimer in the presence of the viral protease. In the absence of protease (plasmid p6H RT), a single 66-kDa protein accumulates following IPTG induction. Coexpression of p66 RT and protease (plasmid p6H RT-PR), the 66-kDa polypeptide is matured into a p66/p51 heterodimer with a 1:1 subunit stoichiometry. The migration positions of p66 and p51 RT are indicated. Recombinant HIV-1 RT was expressed in *E. coli* under inducible control of *lac* regulatory elements.

proteins are removed from the matrix by washing with ~10 column volumes RT Buffer A/60. Finally, polyhistidine extended RT is recovered by application of a 10-column volume gradient of 0 to 0.5 *M* imidazole in RT Buffer A/60. Polyhistidine-extended p66/p51 HIV-1 RT elutes at ~0.15 *M* imidazole, and can be detected enzymatically. Alternatively, we prefer to monitor the purity of the sample by immediately applying column fractions to SDS–polyacrylamide gels, followed by Coomassie blue staining. Using the Mini-PROTEAN II electrophoresis system (Bio-Rad, CA), the entire electrophoresis, staining, and destaining steps require ~90 min. If enzyme from Ni^{2+}-NTA-Sepharose is sufficiently pure for experimental purposes, it can be concentrated by overnight dialysis against 100 volumes RT Storage Buffer D (which gives an approximate threefold concentration) and stored at −20°.

Ion-exchange Chromatography

DEAE-Sepharose. Although p66/p51 HIV RT prepared by metal chelate chromatography is sufficiently pure for most purposes, trace nuclease contamination has been noted. These can be largely removed by ion-exchange chromatography through DEAE-Sepharose (Pharmacia/LKB, Piscataway, NJ). Enzyme eluted from Ni^{2+}-NTA-Sepharose is dialyzed overnight against 100 volumes RT Buffer D and applied to a column of DEAE-Sepharose, equilibrated in the same buffer, at a flow rate of ~2 column volumes/hr. Under these conditions, enzyme is recovered in the nonbinding fraction, while the majority of contaminants are retained.

S-Sepharose. Because RT recovered from DEAE-Sepharose is in a low salt buffer (25 m*M* NaCl), the column effluent can be applied immediately to a column of S-Sepharose, previously equilibrated in RT Buffer D at a flow rate of 1 to 2 column volumes/hr. In this case, enzyme is retained on the column and recovered by application of a 10-column volume gradient of 0 to 0.5 *M* NaCl in RT Buffer D. Enzyme is eluted at ~0.25 *M* NaCl, and concentrated and stored as described above.

Summary of Purification Steps

Figure 2A illustrates the three-column purification procedure we have employed for p66/p51 HIV-1 RT. The resolving power of metal chelate chromatography is clearly evident from this analysis. In addition to p66/p51 HIV-1 RT, only trace contaminants at M_r 45,000 and 30,000 are evident. These are completely removed following ion-exchange chromatography on DEAE- and S-Sepharose, yielding an enzyme with a purity that exceeds 95%. In related work, we have demonstrated that both His-p66, His-p51, and His-p66/His-p51 RT forms of equal purity can be recovered from Ni^{2+}-

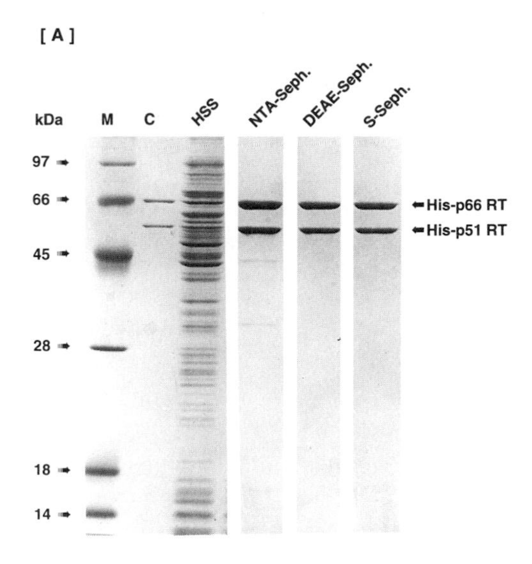

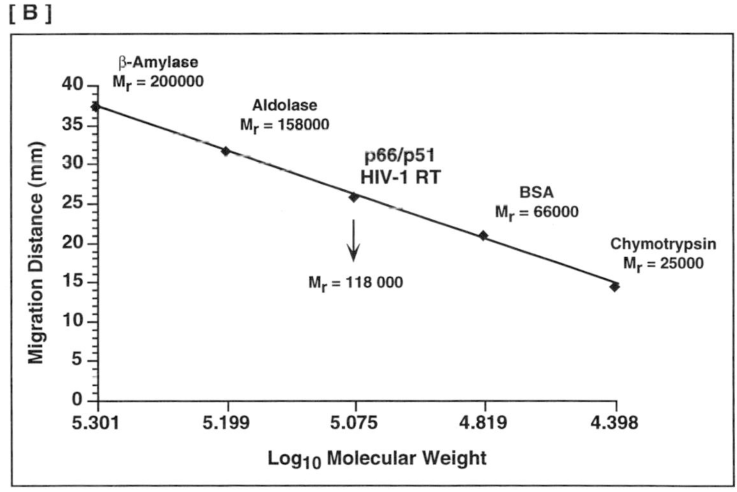

FIG. 2. (A) Summary of heterodimer RT purification. His-p66/His-p51 RT was expressed from plasmid p6H RT-PR. At each stage of purification, 15- and 7.5-μl aliquots of enzyme were analyzed for purity by SDS–polyacrylamide gel electrophoresis (SDS–PAGE) and Coomassie blue staining. Trace contaminants at M_r 45,000 and 30,000 are evident in the Ni^{2+}-NTA-Sepharose-purified enzyme. These are subsequently removed by DEAE- and S-Sepharose ion-exchange chromatography to yield a homogeneous enzyme free of DNase and RNase contamination. Lane C, control preparation of purified p66/p51 RT; Lane M, protein molecular weight standards (in kDa). (B) Determination of HIV-1 RT subunit composition by density gradient centrifugation. Purified enzyme was applied to a 4.5-ml, 15 to 35% glycerol gradient (prepared in RT Buffer D) and centrifuged at 45,000 rpm for 23 h at 4°. The gradient was fractionated and the migration position determined spectrophotometrically. A molecular weight calibration curve, in the range of 200,000 to 25,000 was obtained from parallel gradients containing purified protein M_r standards (Sigma Chemicals, St. Louis, MO).

NTA-Sepharose.[6,7] However, for studies of protein–nucleic acid complexes (e.g., RT/tRNA complexes), we recommend that the three-column strategy be followed. The molecular composition of the HIV-1 enzyme is indicated in Fig. 2B, where a sample of purified protein was subjected to density gradient centrifugation through a linear 15 to 35% (v/v) glycerol gradient. Under these conditions, enzyme was found to migrate to a position corresponding to M_r of 118,000, which would be predicted for the p66/p51 heterodimer.

A similar purification strategy has been used to purify recombinant p66 and p51 subunits of equine infectious anemia virus (EIAV) RT.[8,9] In this case, we noted that both RT subunits were tightly associated with nucleic acid following elution from Ni^{2+}-NTA-Sepharose, which influenced their chromatographic behavior on subsequent ion-exchange matrices. This could be reconciled by including 1 *M* NaCl in the homogenization buffer, since the performance of the metal chelate resin was unaffected at elevated salt concentrations (i.e., the high-speed supernatant was applied to Ni^{2+}-NTA-Sepharose in RT Buffer A/78 containing 1.0 *M* NaCl). Enzyme thus eluted was free of nucleic acid and could be further purified by DEAE- and S-Sepharose chromatography. Both EIAV RT preparations migrated as dimers through glycerol gradients (M_r = 132,000 for p66 and 102,000 for p51), indicating that ionic interactions do not contribute significantly at the dimer interface.

Purification of *in vitro* Reconstituted HIV-1 RT

As indicated earlier, heterodimer HIV-1 and HIV-2 RT are generated by a dual RT/PR expression system, which directs partial cleavage of the 66-kDa subunit by virus-coded PR. To understand the roles of these two subunits, it is necessary to prepare heterodimeric enzyme in which a single component is selectively altered. However, with the exception of the C-terminal RNase H domain, alterations in p66 RT would be duplicated in p51 when enzyme is prepared from the dual expression cassette of Fig. 1. To circumvent this problem, we have devised a protocol for *in vitro* reconstitution of selectively modified heterodimer RT,[10] the rationale for

[7] N. J. Richter-Cook, K. J. Howard, N. M. Cirino, B. M. Wöhrl, and S. F. J. Le Grice, *J. Biol. Chem.* **267,** 12,592 (1992).

[8] S. F. J. Le Grice, M. Panin, R. C. Kalayjian, N. J. Richter, G. Keith, J. L. Darlix, and S. L. Payne, *J. Virol.* **65,** 7004 (1991).

[9] B. M. Wöhrl, K. J. Howard, P. S. Jacques, and S. F. J. Le Grice, *J. Biol. Chem.* **269,** 8541 (1994).

[10] S. F. J. Le Grice, T. Naas, B. Wohlgensinger, and O. Schatz, *EMBO J.* **10,** 3905 (1991).

which is outlined in Fig. 3A. The strategy takes advantage of (1) individual bacterial strains expressing the p66 and p51 subunits and (2) inclusion of the polyhistidine extension on only one component. In the diagrammatic representation of Fig. 3A, the reconstitution partners are a wild-type p66 subunit and a mutated p51 subunit containing the polyhistidine extension. Bacterial pellets are mixed and cohomogenized, after which the high-speed supernatant is applied to the affinity matrix. Because ionic interactions do not appear to contribute to dimerization, the sample is still applied to the column in buffer containing 0.3 *M* NaCl. Metal chelate chromatography retains p66 only as a component of the reconstituted heterodimer. Because p51 is quantitatively retained by the matrix, reconstituted enzyme usually contains an excess of this subunit, which must be removed by ion-exchange chromatography. Following dialysis against RT Buffer D, the mixture of reconstituted heterodimer and His-p51 is applied to an S-Sepharose column, which is developed as indicated in the previous section. Application of a 0 to 0.5 *M* NaCl gradient has the consequence that p66/His-p51 RT is eluted at a lower salt concentration than His-p51. Gradient-eluted samples are normally analyzed by SDS–PAGE, and those containing p66 and His-p51 in a 1:1 stoichiometry are pooled and dialyzed into storage buffer.

An example of this strategy is illustrated in Fig. 3B, where two versions of "selectively deleted" HIV-1 RT were prepared. These heterodimers contain a full-length p66 component, whereas His-p51 contains C-terminal deletions of 13 (p66/p51Δ13 RT) and 19 (p66/p51Δ19 RT) residues. In the experiment of Fig. 3B, His-p51 containing a 25-residue deletion failed to reconstitute with p66, resulting in retention of solely the polyhistidine extended subunit by the affinity matrix. Such data[11] indicate the value of metal chelate chromatography as a tool for monitoring protein::protein interactions. In addition to the method presented here, the strategy of *in vitro* heterodimer reconstitution has been successful for preparation and analysis of (1) selectively mutated,[10] (2) selectively deuterated,[12] and (3) intertypic HIV-1/HIV-2 heterodimers.[13]

Purification of Unmodified p66 HIV-1 RT

Although we have not observed any alteration in the properties of polyhistidine-extended HIV and EIAV RT, there may be instances where

[11] P. S. Jacques, B. M. Wöhrl, K. J. Howard, and S. F. J. Le Grice, *J. Biol. Chem.* **269,** 1388 (1994).

[12] H. Lederer, O. Schatz, R. May, H. Crespi, J. L. Darlix, S. F. J. Le Grice, and H. Heumann, *EMBO J.* **11,** 1131 (1992).

[13] K. J. Howard, K. B. Frank, I. S. Sim, and S. F. J. Le Grice, *J. Biol. Chem.* **266,** 23,003 (1991).

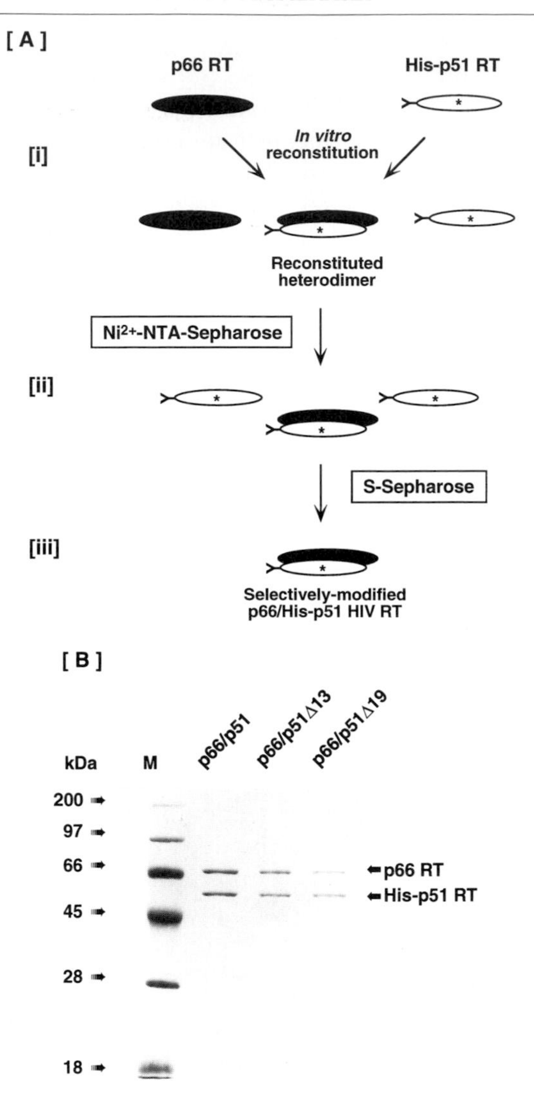

FIG. 3. (A) Strategy for preparation of selectively modified p66/p51 HIV RT. Step [i], Pellets from bacteria expressing p66 and a mutated His-p51 are mixed and co-homogenized; *in vitro* reconstitution yields heterodimer and the individual subunits. Step [ii], application of the mixed high-speed supernatant to Ni^{2+}-NTA-Sepharose results in retention of reconstituted p66/His-p51 and His-p51, while free p66 fails to bind to the matrix. Step [iii], the mixture of p66/His-p51 is separated by S-Sepharose ion-exchange chromatography. (B) Preparation of selectively deleted heterodimer HIV-1 RT. The gel indicates S-Sepharose purified p66/p51Δ13 and p66/p51Δ19 RT. Lane M, protein molecular weight markers (in kDa).

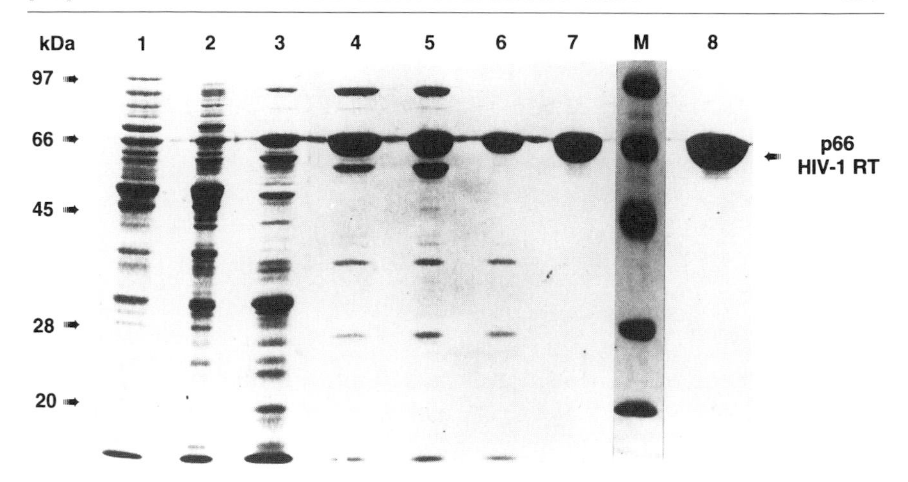

FIG. 4. Purification of unmodified p66 HIV-1 RT. Samples from each purification step were fractionated by SDS–PAGE and visualized by staining with Coomassie blue. 25 μg (lanes 1–7) or 50 μg (lanes 8–9) of protein was applied to the gel. Lane 1, post-PEI supernatant; lane 2, Pellicon concentrate; lane 3, Q-Sepharose (pH 8.0) flow-through; lane 4, CM-Trisacryl pool; lane 5, Amicon concentrate; lane 6, Q-Sepharose (pH 8.0) pool; lanes 7, 8, 25 and 50 μg, respectively, of Q-Sepharose (pH 9.3) pool.

its removal would be desirable. One means to do this is to engineer the expression cassette such that a cleavage site for one of several specific proteases (e.g., thrombin, factor Xa) is placed between the extension and mature coding sequence. RT prepared by any of the aforementioned techniques is treated with protease, then reapplied to the metal chelate matrix. Both uncleaved enzyme and the polyhistidine extension are retained, while the processed enzyme and protease are recovered in the column effluent. As an alternative, the purification procedure developed by Clark *et al.*[14] is outlined in the next section, with which the p66 subunit of HIV-1 RT was isolated. The purity of the enzyme following each chromatographic step is indicated in Fig. 4.

Buffers

Buffer A: 10 m*M* Tris–HCl, pH 8.0, 1 m*M* EDTA, 25% (w/v) sucrose
Buffer B: 50 m*M* Tris–HCl, pH 8.0, 6.25 m*M* EDTA, 0.1% (w/v) Triton X-100, 50 m*M* NaCl, 0.2 m*M* PMSF
Buffer C: As Buffer B, but containing 950 m*M* NaCl

[14] P. K. Clark, A. L. Ferris, D. A. Miller, A. Hizi, K.-W. Kim, S. M. Deringer-Boyer, M. L. Mellini, A. D. Clark Jr., G. F. Arnold, W. B. Lebhertz, E. Arnold, G. M. Muschick, and S. H. Hughes, *AIDS Res. & Human Retroviruses* **6,** 753 (1990).

Buffer D: 5 m*M* HEPES, pH 8.0, 2 m*M* DTT, 0.2 m*M* EDTA, 0.2 m*M* PMSF, 10% (v/v) glycerol
Buffer E: 20 m*M* HEPES, pH 8.0, 2 m*M* DTT, 0.2 m*M* EDTA, 5 m*M* NaCl, 0.2 m*M* PMSF, 10% (v/v) glycerol
Buffer F: As Buffer D, but pH 7.0
Buffer G: As Buffer E, but pH 7.0
Buffer H: 20 m*M* Tris–HCl, pH 8.0
Buffer I: 37 m*M* DEA, pH 9.3, 10% (v/v) glycerol.

Extract Preparation

Cells (100 g) are suspended at 4° in 220 ml of Buffer A and lysed by the addition of 45 ml 0.5 *M* EDTA and 45 ml of a 5 mg/ml solution of lysozyme. Buffer B (350 ml) is added and the lysate is incubated for 15 minutes. Next, 700 ml of Buffer C is added, followed immediately by the addition of 40 ml of a 10% polyethyleneimine (PEI) solution, pH 7.5. Insoluble material is removed by centrifugation at 5000*g* for 30 min at 4° and the supernatant filtered through glass wool. Desalting of the cell lysate to a conductivity equivalent to 40 m*M* NaCl and concentration of the supernatant is accomplished using a Millipore Pellicon ultrafiltration system equipped with a 10,000-MW cutoff cassette.

Chromatography

Q-Sepharose, pH 8.0. The extract is applied to a Q-Sepharose ion-exchange column preequilibrated with Buffer E. RT fails to bind to the matrix, while a significant proportion of the contaminants are retained. The Q-Sepharose effluent is diluted with Buffer F until its conductivity is equivalent to 20 m*M* NaCl, and the pH is adjusted to 7.0.

CM-Trisacryl. The pH 7.0 Q-Sepharose effluent is loaded onto a CM-Trisacryl column equilibrated in Buffer G. After washing the column with the same buffer, RT is eluted by application of a linear gradient of 0.005 to 0.15 *M* NaCl in Buffer G. Fractions displaying RT activity are pooled.

Q-Sepharose, pH 8.0. After desalting and concentrating with Buffer H, the sample from CM-Trisacryl is reapplied to a Q-Sepharose column equilibrated in Buffer H, and washed with the same buffer. Bound proteins are eluted with a gradient of 0 to 0.3 *M* NaCl gradient (in Buffer H). The bulk of the enzyme is recovered in the nonbinding fraction, while a small portion is found in early gradient fractions.

Q-Sepharose, pH 9.3. Pooled RT-containing samples are diluted with Buffer I until the conductivity is equivalent to that of the buffer, and the pH is adjusted to 9.3. A final round of Q-Sepharose chromatography is performed in the presence of Buffer I, the pH of which (9.3) has the

TABLE I
SUBSTRATE SPECIFICITIES OF HIV-1 REVERSE TRANSCRIPTASE[a]

Substrate	DNA polymerase activity (units/μg)[a] p66 RT[b]	p66/p51 RT[c]
poly(rA) · oligo(dT)	36.9	31.0
poly(dA) · oligo(dT)	0.014	ND
poly(rC) · oligo(dG)	3.2	8.75
poly(dC) · oligo(dG)	7.4	30.95

[a] For both p66 and p66/p51 RT, 1 unit of RT activity is defined as that amount catalyzing incorporation of 1 nmol precursor into product in 10 min at 37°.
[b] p66 RT prepared according to Clark *et al.*[14]
[c] p66/p51 RT prepared according to Le Grice and Grüninger-Leitch.[6]

consequence that the enzyme is retained by the ion-exchange matrix. Elution from Q-Sepharose is achieved by application of a gradient of 0 to 0.5 *M* NaCl in Buffer I.

Measuring RT Activities

Conventionally, the DNA polymerase activities of retroviral RT are measured with a variety of synthetic homopolymer template/primer com-

TABLE II
SELECTED INHIBITORS OF HIV-1 REVERSE TRANSCRIPTASE[a]

Inhibitor	K_i or IC_{50}	n*M*	Refs.
Nucleoside-based			
AZT[b]	K_i	2–100	15–18
ddUTP	K_i	50	19
ddTTP	K_i	7–190	15, 18, 20, 21
ddCTP	K_i	260	18
ddATP	K_i	220	18
ddGTP	K_i	30–160	18, 20, 21
Nonnucleoside-based			
Pyridinone (L-697,661)	IC_{50}	19	22
TIBO[c]	IC_{50}	700	23

[a] A comprehensive list of HIV-1 RT inhibitors can be found in de Clercq *et al.*[24]
[b] 3′-Azido-2′,3′-dideoxynucleosides.
[c] Tetrahydroimidazo[4,5,l-*jk*][1,4]benzodiazepin-2(1*H*)-one.

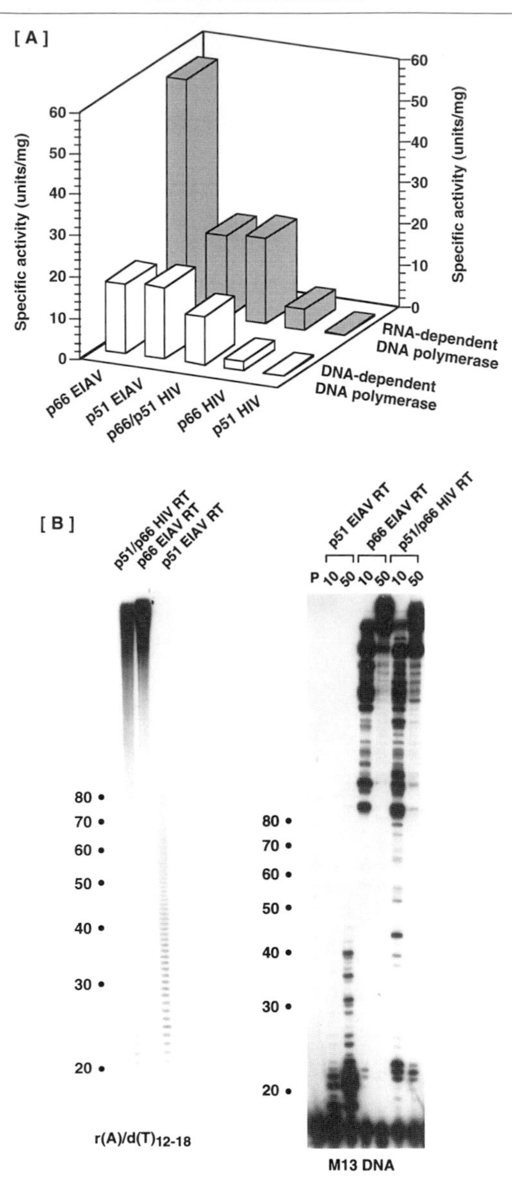

FIG. 5. Quantitative and qualitative evaluation of the DNA polymerase activities of retroviral RT. (A) Subunits of the HIV-1 and EIAV enzymes were assayed for RNA- and DNA-dependent DNA polymerase activities on the synthetic homopolymer combinations poly (rA) · oligo$(dT)_{12-18}$ and poly(dC) · oligo$(dG)_{12-18}$, respectively. Using either template–primer, 1 unit or RT activity is refined as the amount catalyzing incorporation of 1 nmol precursor into polydeoxynucleotide in 15 min at 37°. (B) Qualitative RNA- (*left*) and DNA-depen-

binations (Table I). These include poly(rA) · oligo(dT) and poly(rC) · oligo(dG) for RNA-dependent DNA polymerase function (the latter has the advantage that recombinant RT can be assayed in crude bacterial homogenates without interference from cellular polymerases) and poly(dC) · oligo(dG) for DNA-dependent DNA polymerase. Incorporation of radiolabeled precursor into polydeoxynucleotide is determined by liquid scintillation counting, from which the specific activity of the preparation can be determined. This procedure had been useful in the characterization of RT inhibitors such as those indicated in Table II.

Although they provide a quantitative evaluation of RT activity, these methods do not assess the quality of the DNA product, which in some cases can vary considerably. This is best exemplified by the data of Fig. 5, where the activities of the p66 and p51 subunits of EIAV RT are compared. Specific RNA-dependent DNA polymerase activities of the two subunits are equivalent to or greater than their HIV counterparts (Fig. 5A). However, performing the same assay in the presence of [^{32}P]TTP and fractionating the products by high-voltage electrophoresis (Fig. 5B), indicates substantial differences in their processivity. Substituting single-stranded M13 DNA

[15] Y. E. Cheng, G. E. Dutschman, K. F. Bastow, M. G. Sarngadharan, and R. Y. C. Ting, *J. Biol. Chem.* **262,** 2187 (1989).

[16] L. Vrang, H. Bazin, G. Remand, J. Chattopadhyaya, and B. Oeberg, *Antiviral Res.* **7,** 139 (1987).

[17] M. H. St. Clair, C. A. Richards, T. Spector, K. J. Weinhold, W. H. Miller, A. J. Langlios and P. A. Furman, *Antimicrob. Agents Chemother.* **31,** 1972 (1987).

[18] Z. Hao, D. A. Cooney, N. R. Hartman, C. F. Perno, A. Fridland, A. L. DeVico, M. G. Sarngadharan, S. Broder, and D. G. Johns, *Mol. Pharmacol.* **34,** 431 (1988).

[19] Z. Hao, D. A. Cooney, D. Farquar, C. F. Perno, K. Zhang, R. Masood, Y. Wilson, N. R. Hartman, J. Balzarini, and D. G. Johns, *Mol. Pharmacol.* **37,** 157 (1990).

[20] M. S. Chen and S. C. Oshana, *Biochem. Pharmacol.* **36,** 4361 (1987).

[21] W. B. Parker, E. L. White, S. C. Shaddix, L. J. Ross, R. W. Buckheit Jr., J. M. Germany, J. A. Secrist III, R. Vince, and W. M. Shannon, *J. Biol. Chem.* **266,** 1754 (1991).

[22] M. E. Goldman, J. H. Nunberg, J. A. O'Brien, J. C. Quintero, W. A. Schleif, K. F. Freund, S. L. Gaul, W. S. Saari, J. S. Wai, J. M. Hoffman, P. S. Anderson, D. J. Hupe, E. A. Emini, and A. M. Stern, *Proc. Natl. Acad. Sci. USA* **88,** 6863 (1991).

[23] Z. Debyser, R. Pauwels, K. Andries, J. Dreysmeyler, M. Kukla, P. A. J. Janssen, and E. DeClercq, *Proc. Natl. Acad. Sci. USA* **88,** 1451 (1991).

[24] E. DeClercq, *AIDS Res. & Human Retroviruses* **8,** 111 (1991).

dent DNA polymerase activities (*right*) of HIV and EIAV RT. RNA-dependent DNA polymerase activity was also measured on poly(rA) · oligo(dT)$_{12-18}$, with the exception that [^{32}P]TTP was included. Reaction products were fractionated by high-voltage electrophoresis and visualized by autoradiography. For DNA-dependent DNA polymerase analysis, poly(dC) · oligo(dG)$_{12-18}$ was substituted with single-stranded bacteriophage M13 DNA to which a 17-mer ^{32}P-labeled DNA primer was hybridized. [Reprinted with permission from reference 9.]

containing a ^{32}P-labeled primer for poly(dC)·oligo(dG) reveals similar result (Fig. 5B). Thus, while p51 EIAV RT displays considerable DNA polymerase activity, the enzyme appears to be highly distributive. The barrier to p51-catalyzed DNA synthesis arises from its inability to resolve duplex structures.[9,25] Thus, while "active" *in vitro*, it is unlikely that this RT polypeptide could complete a round of (−) strand DNA synthesis, since several structural elements are present on the viral RNA genome. When amino acid substitutions are introduced into RT that could influence its processivity, this may not be evident from specific activity measurements, and we recommend that a qualitative evaluation be undertaken.

Concluding Remarks

Metal chelate affinity chromatography offers a rapid and highly reproducible means of preparing (1) the individual p66 and p51 HIV RT subunits, (2) heterodimer p66/p51, and (3) reconstituted, selectively modified heterodimer directly from bacterial homogenates. The application of a highly selective affinity matrix as the primary purification step has the advantage that bacterial proteases are eliminated at an early stage, avoiding proteolysis of the reconstituted protein. In all cases so far studied, we have observed that inclusion of the polyhistidine extension in the purified enzyme has no deleterious effect on polymerase, RNase H, or tRNA binding properties. Finally, the ease with which HIV RT can be purified by metal chelate affinity chromatography has prompted us to develop methodologies for protein minipreparation, where four small columns can be run simultaneously and RT eluted in a batchwise fashion. This procedure has proven successful in cases where individual domains of the enzyme have been analyzed by insertional mutagenesis.

Acknowledgments

S. F. J. Le Grice is supported by grants AI 31147 and GM 46623 from the National Institutes of Health. S. J. Benkovic is supported by grant GM13306 from the National Institutes of Health. C. E. Cameron is supported by fellowship 4I 09076 from the National Institutes of Health. A portion of the work presented here was also supported by a NATO collaborative research grant (CRG 900.471) to S. F. J. Le Grice. The technical assistance of K. J. Howard in preparation of p66/p51 HIV-1 RT variants is gratefully acknowledged.

[25] B. M. Wöhrl, C. Tantillo, E. Arnold, and S. F. J. Le Grice, *Biochemistry* **34,** 5343 (1995).

Section II

Structural Analysis of DNA Polymerases

[14] Solution Structure of DNA Polymerases and DNA Polymerase–Substrate Complexes

By GREGORY P. MULLEN

Introduction

Multidimensional nuclear magnetic resonance spectroscopy (NMR),[1] nuclear Overhauser effect (NOE) studies, water proton relaxation rate (PRR) measurements, and electron paramagnetic resonance spectroscopy (EPR) have been used in the solution characterization of structure in DNA polymerases. Studies have included structure determination of a domain of DNA polymerase β (Pol β), determination of the conformations of substrates bound to DNA polymerase I, and the determination of the location and affinity of divalent metal binding sites on DNA polymerase I. NMR and EPR methods provided the first structural insight into the mechanism of *Escherichia coli* DNA polymerase I.[2,3] Studies of the cloned Klenow fragment have provided information on substrate binding sites, the interaction between substrate and metal binding sites, and substrate conformation.[4–8] Methods of 2D and 3D heteronuclear NMR are applicable to the determination of solution structures of domains of DNA polymerases with molecular weights of less than 30,000. A determination of the solution structure of DNA polymerase β, a 334 residue protein (M_r 39,000), is in the

[1] NMR, nuclear magnetic resonance; PRR, proton relaxation rate; EPR, electron paramagnetic resonance; NOE, nuclear Overhauser effect; NOESY, two-dimensional nuclear Overhauser effect spectroscopy; HMQC, two-dimensional heteronuclear multiple quantum correlation spectroscopy; HSQC, two-dimensional heteronuclear single quantum correlation spectroscopy; NOESY-HMQC, three-dimensional heteronuclear-edited nuclear Overhauser effect spectroscopy; TOCSY-HMQC, three-dimensional heteronuclear-edited total correlation spectroscopy; HNCA, three-dimensional triple resonance ^{1}H–^{15}N–^{13}Cα correlated spectroscopy; HN(CO)CA, three-dimensional triple resonance ^{1}H$_i$–^{15}N$_i$–^{13}Cα_{i-1} correlated spectroscopy; T_1, longitudinal relaxation time; dNTP, deoxynucleoside triphosphate; Pol I, DNA polymerase I; KF, Klenow fragment of Pol I; Pol β, DNA polymerase β.

[2] J. P. Slater, I. Tamir, L. A. Loeb, and A. S. Mildvan, *J. Biol. Chem.* **247,** 6784 (1972).

[3] D. L. Sloan, L. A. Loeb, A. S. Mildvan, and R. J. Feldmann, *J. Biol. Chem.* **250,** 8913 (1975).

[4] G. P. Mullen, P. Shenbagamurthi, and A. S. Mildvan, *J. Biol. Chem.* **264,** 19,637 (1989).

[5] G. P. Mullen, E. H. Serpersu, L. J. Ferrin, L. A. Loeb, and A. S. Mildvan, *J. Biol. Chem.* **264,** 14,327 (1990).

[6] L. J. Ferrin and A. S. Mildvan, *Biochemistry*, **24,** 6904 (1985).

[7] L. J. Ferrin and Mildvan, *Biochemistry*, **25,** 5131 (1986).

[8] G. P. Mullen, J. B. Vaughn Jr., P. Shenbagamurthi, and A. S. Mildvan, *Biochem. Pharmacol.* **40,** 69 (1990).

early stages. High-resolution structural analysis is presently being directed toward the single-stranded template-binding domain of the enzyme and the study of the interaction of this domain with single-stranded DNA.[9] The most recent NMR methods that are being used to study Pol β are discussed, together with of an overview of previous methods used in the study of DNA polymerase I.

Application of PRR and EPR in Studies of Metal-Binding Sites on Pol I, Klenow Fragment, and Two 3′ → 5′-Exonuclease-Deficient Mutants of Klenow Fragment

The cloning and overexpression of the Klenow fragment of Pol I and two 3′ → 5′-exonuclease-deficient mutants of the Klenow fragment allowed for a determination of the location, affinity, and substrate dependency of metal binding sites on this enzyme in solution.[5,10,11] The D355A,E357A exonuclease-deficient mutant of the Klenow fragment provides a model for metal binding in DNA polymerases lacking 3′ → 5′-exonuclease activity.[5] In characterizing metal binding, Mn^{2+} is used as a paramagnetic probe for the binding site(s) on the enzyme.[12] PRR and EPR data are used in a Scatchard plot analysis of the K_D for Mn^{2+} binding. With the PRR method, the average enhancement factors for water relaxation by the bound Mn^{2+} (ε_b) at the tight and weak sites can be evaluated. Binding at the metal binding site by Mg^{2+} is determined by competitive displacement of Mn^{2+}. The water proton relaxation enhancement, which is a result of Mn^{2+} binding, is measured in a Mg^{2+} titration of the high affinity site. Two types of Mn^{2+} binding sites were determined for Pol I and the Klenow fragment. Tight binding Mn^{2+} sites are those with a $K_D < 100\ \mu M$ and weak binding Mn^{2+} sites are those with a $K_D > 500\ \mu M$. Scatchard plots are analyzed for two types of binding sites using the cubic equation [Eq. (1)]:

$$\begin{aligned} \mathrm{Mn_f}^3 + (K_1 + C_1 + K_2 + C_2 - \mathrm{Mn_t})\,\mathrm{Mn_f}^2 + (K_1K_2 + K_1C_2 \\ + K_2C_1 - K_1\mathrm{Mn_t} - K_2\mathrm{Mn_t})\,\mathrm{Mn_f} - K_1K_2\mathrm{Mn_t} = 0, \end{aligned} \tag{1}$$

where K_1 and K_2 represent dissociation constants and C_1 and C_2 represent the concentrations of the corresponding sites. Scatchard plots are constructed using PRR and EPR data. For the PRR method, measurements

[9] D-J. Liu, E. F. DeRose, R. Prasad, S. H. Wilson, and G. P. Mullen, *Biochemistry* **33,** 9537 (1994).

[10] C. M. Joyce and N. D. F. Grindley, *Proc. Natl. Acad. Sci. USA* **80,** 1830 (1983).

[11] V. Derbyshire, P. S. Freemont, M. R. Sanderson, L. Beese, J. M. Friedman, C. M. Joyce, and T. A. Steitz, *Science*, **240,** 199 (1988).

[12] A. S. Mildvan and J. L. Engle, *Methods Enzymol.* **26C,** 654 (1972).

of the longitudinal relaxation rates ($1/T_1$) of water protons are performed using a pulsed NMR spectrometer (available from Seimco) at 24.3 MHz by employing an inversion-recovery pulse sequence (180°-τ-90°). Enzyme or enzyme–substrate complexes are formed at concentrations of Mn^{2+}, which fit into a range of 20 to 80% occupancy, and the τ delay necessary to observe a null in the FID (free induction decay) of the water signal is measured in both the presence and the absence of Mn^{2+}. The longitudinal relaxation time (T_1) is determined using the expression $\tau_{null}/\ln 2$. A similar set of measurements is performed in the absence of the enzyme or the stoichiometric enzyme–substrate complex. The paramagnetic contributions to the longitudinal relaxation (T_{1P}) have been defined previously[12] as

$$1/T_{1P} = (1/T_1) - (1/T_1)_0 \tag{2}$$
$$1/T_{1P}^* = (1/T_1)^* - (1/T_1)_0^*, \tag{3}$$

where $(1/T_1)$ and $(1/T_1)_0$ denote the relaxation rates in the presence and the absence of the paramagnetic metal and the asterisk (*) represents the presence of enzyme or enzyme–substrate complex. Concentrations of the enzyme are adjusted to ensure >90% binding of the substrate for determinations of metal binding to the enzyme–substrate complex. In studies of the Klenow fragment, TMP was used to probe metal interaction at the 3′ → 5′-exonuclease site and dGTP was used to probe metal interaction at the polymerase site.[5] Addition of TMP resulted in an increase in metal binding stoichiometry from one tight site to two tight sites. This indicated near-stoichiometric binding by TMP. Similarly, an increase in the binding stoichiometry from two tight sites in the presence of stoichiometric TMP to three tight sites in the presence of stoichiometric TMP and dGTP was observed. Independent measurements of the K_D of dGTP for the enzyme were used to determine concentrations of dGTP and enzyme required for >90% occupancy by dGTP.

An enhancement in the relaxation rate of water protons at each Mn^{2+} concentration has been defined previously[12] by

$$\varepsilon^* = (1/T_{1P}^*)/(1/T_{1P}). \tag{4}$$

Calculations of the free and bound Mn^{2+} for Scatchard analysis are determined from ε^* using

$$Mn_f/Mn_t = (\varepsilon_b - \varepsilon^*)/(\varepsilon_b - 1). \tag{5}$$

In this expression, Mn_f and Mn_t are free and total concentrations of Mn^{2+}. The ε_b values represent the *average* enhancement factors for tight or weak sites and are independently determined by measurements of $[Mn^{2+}]_{free}$ using EPR spectroscopy. The EPR measurements of $[Mn^{2+}]_{free}$ in the studies of the Klenow fragment were performed on a Varian E-4 EPR instrument. The enhancement factor for the tight sites, ε_b^{tight}, is determined using

$$\varepsilon_b^{tight} = [\varepsilon^* - Mn_f/Mn_t]/[1 - Mn_f/Mn_t] \tag{6}$$

under conditions in which the weak sites are less than 5% occupied. The assessment of the enhancement factor for the weak sites, ε_b^{weak} is determined on the basis of the mole fraction occupancy of bound Mn^{2+} at the tight and weak sites and uses Eq. (7).

$$\varepsilon_b^{total} = X_1\varepsilon_b^{tight} + X_2\varepsilon_b^{weak} \tag{7}$$

The binding constants for Mg^{2+} are determined by competitive displacement of Mn^{2+} from the high affinity site through a titration in which ε^* is measured by PRR. The titration curve can be fit by calculation of $\Delta\varepsilon^*$ using

$$\Delta\varepsilon^* = \{(\varepsilon_{init} - \varepsilon^*_{fin})[Mg^{2+}]\}/(K_D^{app} + [Mg^{2+}]) \tag{8}$$

or by a double reciprocal plot of $1/\Delta\varepsilon^*$ versus $1/[Mg^{2+}]$. The K_D for Mg^{2+} is calculated using

$$K_D^{Mg} = K_D^{app}/(1 + [Mn^{2+}]/K_D^{Mn}), \tag{9}$$

where K_D^{app} is the apparent dissociation constant, which describes the hyperbolic displacement curve. Alternatively, the titration curve can be fit using Eq. (10)

$$\varepsilon^* = (Mn_f/Mn_t)\ 1 + (Mn_b/Mn_t)\ \varepsilon_b^{tight}, \tag{10}$$

which requires calculation of the bound concentration of manganous ion (Mn_b) using a cubic equation.[4]

The binding of metals to the Klenow fragment was evaluated in terms of the X-ray structure.[5,11,13,14] Of important significance for the mechanism of this enzyme was the finding that the single high affinity metal site is a result of ligation by carboxylates of D355 and E357 at the $3' \rightarrow 5'$-exonuclease active site. This is consistent with the metal location in the X-ray structure. This high affinity site is contaminated by Zn^{2+} in some enzyme preparations.[15] The $5' \rightarrow 3'$-exonuclease domain of Pol I bound Mn^{2+} with only weak affinity. TMP and dGTP each introduce an additional metal site on the enzyme yielding a total of three tight sites in the KF–dGTP–TMP complex. The D424A mutation reduced the affinity of the tight Mn^{2+} bind-

[13] D. L. Ollis, P. Brick, R. Hamlin, N. G. Xuong, and T. A. Steitz, *Nature* **313,** 762 (1985).

[14] T. A. Steitz, L. Beese, P. S. Freemont, J. M. Friedman, and M. R. Sanderson, *Cold Spring Harbor Symp. Quant. Biol.* **52,** 465 (1987).

[15] C. F. Springate, A. S. Mildvan, R. Abramson, J. L. Engle, and L. A. Loeb, *J. Biol. Chem.* **248,** 5987 (1973).

ing site by approximately 10-fold. The interaction by D424 with the high affinity site is electrostatic in nature if coordinated similarly to the single metal in the X-ray structure.[11] Addition of TMP did not significantly alter the Mn^{2+} binding by this mutant. A single tight site was observed for the D355A,E357A mutant in the presence of dGTP, and the enhancement factor for the complex ($\varepsilon_b^{tight} = 10.8 \pm 0.9$) was greater than a binary dNTP–Mn^{2+} complex (2.2 ± 0.1) establishing formation of a KF–dGTP–Mn^{2+} complex. The affinity of Mn^{2+} for this complex is 2.9-fold higher than the K_D for dNTP under comparable conditions and is consistent with the enzyme contributing a ligand to the metal. Mg^{2+} displaced Mn^{2+} from the ternary complex with a K_D of 100 μM. At least four of eight low affinity metal sites appear to be associated with the $3' \rightarrow 5'$-exonuclease domain of the Klenow fragment. The results of metal binding to Pol I are summarized in Table I. Interesting and mechanistically important will be a determination of whether other DNA polymerases display low or high affinity sites for Mn^{2+} in the absence of dNTP substrates.

Transferred Nuclear Overhauser Effect for Determination of Substrate Conformations and Active-Site Residue Environment

One-dimensional nuclear Overhauser effect NMR studies of the conformations of enzyme-bound dNTP substrates have provided significant information with respect to the structural mechanism of dNTP binding to the Klenow fragment of Pol I.[6–8] It is likely that similar studies using multidimensional NMR methods will be highly informative for other DNA polymerases. The nuclear Overhauser effect is a result of a through-space magnetization transfer mechanism, in which the intensity of an observed resonance is altered on saturation or perturbation of the spin distribution within the energy levels of a second nucleus. The effect is short range and requires that protons be within a 5-Å proximity. For two proton spins that display a correlation time (i.e., the time constant for rotational motion) characteristic of a macromolecule, saturation of one of the proton resonances results in a decrease in intensity for the resonance of the proton in close proximity. This effect is also observed for protons within a nucleotide, when bound to a large protein (reviewed by Rosevear and Mildvan[16]) such as DNA polymerase. In these studies, the NOE for the free nucleotide is weak and positive and does not contribute significantly at short mixing times.

The one-dimensional transferred NOE method for determination of nucleotide conformation has been superseded by a quantitative 2D NOE

[16] P. R. Rosevear and A. S. Mildvan, *Methods Enzymol.* **177,** 333 (1989).

TABLE I

Mn^{2+} BINDING TO POL I, ITS LARGE FRAGMENT, AND THE $3' \rightarrow 5'$-EXONUCLEASE-DEFICIENT MUTANTS OF THE LARGE FRAGMENT[a]

Enzyme	Nucleotide	Method	Tight sites			Weak sites		
			n[b]	K_D (μM)	ε_b[c]	n[b]	K_D (μM)	ε_b[c]
DNA polymerase I	—	EPR	0.65 ± 0.15	2.5 ± 1.1	—	20 ± 10	600 ± 300	—
Large fragment	—	EPR	0.63 ± 0.15	6.8 ± 3.0	—	8.0 ± 1.0	1000 ± 500	—
Large fragment	—	PRR	0.64 ± 0.15	2.7 ± 1.2	13.2 ± 1.4	8.0 ± 1.0	1000 ± 500	5.7 ± 0.6
Large fragment	TMP	EPR	1.86 ± 0.15	10.0 ± 2.0	—	7.0 ± 1.0	1200 ± 400	—
Large fragment	TMP	PRR	1.90 ± 0.15	12.0 ± 2.0	11.0 to 7.8[d]	7.0 ± 1.0	1200 ± 400	7.8 ± 0.6
Large fragment	TMP, dGTP	EPR	2.80 ± 0.15	7.8 ± 2.0	—	7.0 ± 1.0	1800 ± 1000	—
Large fragment	TMP, dGTP	PRR	2.80 ± 0.15	6.7 ± 2.0	10.6 to 8.8[d]	7.0 ± 1.0	2600 ± 1000	8.8 ± 0.7
D424A	—	EPR	1.0 ± 0.2	67 ± 25	—	7.0 ± 1.0	3500 ± 1500	—
D424A	—	PRR	1.0 ± 0.2	69 ± 25	14.0 ± 0.7	7.0 ± 1.0	3500 ± 1500	9.7 ± 2.1
D424A	TMP	EPR	0.7 ± 0.1	72 ± 20	—	7.0 ± 1	5400 ± 2000	—
D424A	TMP	PRR	0.7 ± 0.1	61 ± 20	15.8 ± 1.5	7.0 ± 1	7000 ± 2300	10.4 ± 4.9
D355A,E357A	—	PRR	—	—	—	4.0 ± 1.0	1000 ± 400	15.7 ± 1.6
D355A,E357A	dGTP	EPR	1.0 ± 0.2	3.6 ± 1.8	—	4.0 ± 1.0	850 ± 400	—
D355A,E357A	dGTP	PRR	—	—	10.8 ± 0.9	—	—	14.2 ± 8.5

[a] Taken from G. P. Mullen, E. H. Serpersu, L. J. Ferrin, L. A. Loeb, and A. S. Mildvan, *J. Biol. Chem.* **264,** 14,327 (1990). Conditions are described therein.

[b] Stoichiometry.

[c] PRR enhancement factor due to bound Mn^{2+} in binary, ternary, or higher complexes. Error is expressed as two standard errors of the mean.

[d] A systematic decrease in the average enhancement of the tight sites was observed with increasing occupancy.

method, which utilizes a series of mixing times (the time in which antiphase magnetization is aligned along the z axis) for determination of interproton distances. In this method the initial slope of the NOE buildup curve approximates the cross-relaxation rate (σ_{AB}) between two ^1H spins. The 2D NOE method has been successfully applied in determining conformations of mononucleotides and dinucleotides bound to staphylococcal nuclease (M_r 18,000), an enzyme of lower molecular mass than DNA polymerases, but is expected to be equally applicable to DNA polymerases.[17]

In either the 1D or 2D NOE methods of evaluating nucleotide conformation, interproton distances should be determined with high accuracy (approximately ± 0.2 Å), if possible. This precision is necessary since most protons in a nucleotide are within a 5-Å proximity, and accurate distances are required for an evaluation of sugar pucker, χ_1 torsion angles, and γ torsion angles. Contributions of spin diffusion, or secondary NOE transfer pathways, between protons in an enzyme-bound nucleotide, should be carefully assessed. Spin-diffusion pathways in general become more productive as the molecular weight of the system under study increases. Spin diffusion is minimized by performing NOESY experiments at the shortest possible mixing times (i.e., <100 ms). This requires high sample concentrations and longer acquisitions since shorter mixing times result in reduced sensitivity of the NOESY spectra. A determination that spin diffusion was not substantially contributing in the 1D NOE studies of the conformation of nucleotides bound to the Klenow fragment was in part based on the finding that in the higher molecular weight KF–dNTP–template complexes, singular conformations of dGTP and ddGTP were fit to the NOE determined distances.[7,8] In the absence of template, however, the NOEs could only be explained by a set of at least two or three low-energy conformers of the nucleotides. The saturation pulse power used in 1D transferred NOE experiments is considerably lower than that required for instantaneous saturation of spins, and results in increased selectivity of the window of irradiation. As a result, the 1D NOE buildup curves display a lag time of 100 to 150 ms and are not directly comparable to NOE buildups in 2D NOESY experiments.

In the NOE determinations of enzyme-bound dNTP conformations, RNA template–primers were utilized to avoid hydrolysis by the $3' \rightarrow 5'$-exonuclease activity of the Klenow fragment. This should not be necessary in studies of DNA polymerases lacking $3' \rightarrow 5'$-exonuclease activities. However, an exonuclease-free preparation is necessary. Studies were performed for dATP-Mg^{2+} and TTP-Mg^{2+} in the absence and presence of the corresponding complementary templates, $(rU)_{54}$ and $(rA)_{50}$.[6,7] Similar

[17] D. J. Weber, G. P. Mullen, and A. S. Mildvan, *Biochemistry*, **30,** 7425 (1991).

studies were performed for dGTP in the absence and presence of noncomplementary $(rU)_{43}$ and complementary $(rC)_{37}$ templates.[6–8] In studying the effect of template and primer, the kinetically competent KF-$(rC)_{37}$-$(rI)_{14}$-dGTP-Mg^{2+} complex extended the primer at appreciable rates (2.4×10^{-2} min^{-1}) as determined by NMR experiments; therefore, the dideoxy substrate, ddGTP, was utilized to evaluate the enzyme-bound substrate conformation. The enzyme-bound conformation of ddGTP was also evaluated in the absence and presence of template.[8]

Interesting are the multiple conformations observed for enzyme-bound dGTP-Mg^{2+} and ddGTP-Mg^{2+} in the absence of templates, but the singular-type conformation that is observed in the presence of templates (Table II). For the E-$(rC)_{37}$-$(rI)_{14}$-ddG-ddGTP-Mg^{2+} complex, multiple conformations of ddGTP-Mg^{2+} are observed, that are indistinguishable from those observed for ddGTP-Mg^{2+} bound to enzyme alone. Thus, unlike other dNTP substrates, which do not display multiple conformations while bound, dGTP and ddGTP may report on the ability of one of the conformations of the enzyme to "lock" the substrate into a singular conformation, which would presumably be necessary for catalysis. Similar studies of enzyme-bound dGTP conformation for the lower fidelity DNA polymerases in the presence and absence of template–primer should provide further correlations between dNTP conformation and fidelity.

The results from transferred NOEs between the Klenow fragment and dNTP substrate, in the presence and in the absence of template–primer, indicated that binding occurs to a hydrophobic pocket consisting of an aromatic residue and possibly two long-chain aliphatic residues.[6–8] Some speculation exists on whether the *correct* dNTP binding site on the Klenow fragment is present in the absence of template–primer, since in the sequentially ordered bi-bi reaction mechanism template–primer obligatorily binds first. Similar transferred NOEs are observed both in the presence and absence of template–primer, suggesting that the dNTP binding site is equivalent in part in these complexes. The size of the Klenow fragment (M_r 68,000) made determination of sequence specific residue assignments infeasible, but the NOEs were consistent with the site of azido-ATP cross-linking (i.e., Y766).[18] These experiments were performed by applying selective irradiation within the spectrum of the protein and stepping the decoupler transmitter through the entire protein spectrum. Promise for future experiments will be to employ multidimensional isotope-edited NOE methods on proteins such as Pol β once sequence assignments have been performed.

[18] C. M. Joyce, D. L. Ollis, J. Rush, T. A. Steitz, W. H. Konigsberg, and N. D. F. Grindley, *UCLA Symp. Mol. Cell. Biol. New Ser.* **32,** 197 (1985).

TABLE II
SUBSTRATE CONFORMATIONS IN COMPLEXES OF THE LARGE FRAGMENT OF POL I[a]

Complex	Percentage contribution	Conformation	Degrees		
			χ	δ	γ
E-dATP-Mg^{2+} [b]	100	High Anti, O1′ endo, C3′ endo	50 ± 10	95 ± 10	—
E-$(rU)_{54}$-dATP-Mg^{2+} [c]	100	High Anti, O1′ endo, C3′ endo	62 ± 10	90 ± 10	—
E-TTP-Mg^{2+} [b] and E-$(rA)_{50}$-TTP-Mg^{2+} [c]	100	High Anti, O1′ endo	40 ± 10	100 ± 10	—
E-dGTP-Mg^{2+} [d]	60 ± 10	High Anti, O1′ endo, C3′ endo	45	90	—
	40 ± 10	Syn, O1′ endo, C2′ endo	212	135	—
E-$(rC)_{37}$-dGTP-Mg^{2+} [d]	100	High Anti, O1′ endo, C3′ endo	50 (35 to 60)	90 (125–85)	—
E-ddGTP-Mg^{2+} [d] and E-$(rC)_{37}$-$(rI)_{14}$-ddG-ddGTP-Mg^{2+} [d]	30 ± 10	Low Anti, C3′ endo	30	95	180[e]
	30 ± 10	High Anti, O1′ endo, C2′ endo	55	135	180[e]
	40 ± 10	Syn, O1′ endo, C2′ endo	212	135	180[e]
E-$(rC)_{37}$-ddGTP-Mg^{2+} [d]	100	High Anti, O1′ endo, C2′ endo	45 ± 10	135 ± 10	180 ± 10[e]

[a] Taken from G. P. Mullen, J. B. Vaughn Jr., P. Shenbagamurthi, and A. S. Mildvan, *Biochem. Pharmacol.* **40,** 69 (1990). Substrates are underlined. When the interproton distances yielded a single substrate conformation, errors in the conformational angles are given. When more than one conformation was required to fit the observed NOE distances, the errors are expressed in the percentage contribution of each well-defined member of the basis set.

[b] Taken from L. J. Ferrin and A. S. Mildvan, *Biochemistry* **24,** 6904 (1985).

[c] Taken from L. J. Ferrin and Mildvan, *Biochemistry* **25,** 5131 (1986).

[d] Taken from G. P. Mullen and A. S. Mildvan, *FASEB J.* **2,** A588 (1988).

[e] A γ value of 180° indicates a O5′–O1′ *gauche,* O5′–C3′ *trans* conformation about the C4′–C5′ bond.

General Introduction to Multidimensional Heteronuclear Correlated NMR Methods

Multidimensional NMR spectroscopy is the only method for analysis of the solution structure of proteins and protein–substrate complexes at atomic resolution. In addition, NMR can be used to provide information on chemical environment, chemical exchange, and macromolecular dynamics in biological systems. The determination of backbone dynamics can provide important information on those segments of a protein displaying flexibility in solution. In conjunction with appropriate titrations, NMR methods can be used to determine equilibrium binding constants. Heteronuclear correlations with ^{15}N or ^{13}C are usually necessary to characterize completely a protein or protein domain of $M_r > 10{,}000$. The first step toward structure determination of a protein is made by performing assignments of cross-peaks in several types of 2D and 3D NMR spectra. The assignments for the most part are done by visual analysis of the spectra, which are plotted as hard copies and peak-picked within NMR software programs. Both 2D and 3D NMR spectra are presented as contour plots.

The chemical shift of a nucleus describes the position of the resonance or peak along the x axis in a 1D spectrum. In 2D and 3D spectra, a resonance is described by two and three axes, respectively. Each nucleus has a characteristic chemical shift range, which increases for heavier nuclei (i.e., those with increased electron density). The chemical shift ranges for ^{1}H, ^{13}C, and ^{15}N are ~15, ~200, and ~900 parts per million (ppm), respectively. The value of the chemical shift of a proton resonance at 500.1 MHz is defined in units of ppm by Eq. (11)

$$(\delta - \delta_{\mathrm{ref}})/(500.1 \times 10^{6}), \tag{11}$$

where the values of δ represent frequency offsets from the 500.1-MHz carrier frequency for the resonance of interest and for a reference. A particular environment of chemical bonding induces characteristic chemical shift values. Certain types of shielding and deshielding are well characterized, such as the effects induced by the proximity of aromatic rings. Thus, it is known that the edge of an aromatic ring in proximity to a proton results in deshielding (higher ppm values), whereas the face of an aromatic ring in proximity to a proton will induce shielding (lower ppm values). These additional shielding or deshielding effects are additive with respect to the chemical shift for the bonding environment.

Additional structural data are derived from coupling constants, which provide information on spins separated by two or three bonds. The size of a coupling constant for protons ranges between 0 and ~11 Hz and is maximal for three-bond coupling when protons are *trans* (or *cis*) and minimal when the dihedral angle between the protons is approximately 90°. Measurement of couplings thereby provides information on conformational states. Magnetization transfers through ^{1}H–^{1}H couplings are used for obtaining correlated spectra such as DQF-COSY and TOCSY.[19,20] Much larger coupling constants (11 to 100 Hz) between ^{1}H, ^{15}N, or ^{13}C are utilized in heteronuclear correlated experiments, which are applicable to proteins with large line widths. Several types of double-resonance 2D and double- and triple-resonance 3D NMR spectra can be acquired using pulse sequences, which transfer magnetization through coupled spins (reviewed by Clore and Gronenborn[21,22]). The ^{1}H–^{15}N HMQC, ^{1}H–^{13}C HMQC, ^{1}H–^{1}H–^{15}N TOCSY-HMQC, and ^{1}H–^{1}H–^{13}C TOCSY-HMQC experiments together with the HNCA and HN(CO)CA experiment are highly useful in providing an essential set of correlation spectra for structural analysis of helical domains of up to 10 kDa. These experiments can be performed on newer and slightly modified older model (Bruker AM and General Electric GN) NMR spectrometers. Cost-efficient modification procedures for older model GN spectrometers are available on request. Other triple-resonance experiments are of greater utility for proteins displaying significant amounts of β structure. Heteronuclear correlation experiments, which transfer magnetization through ^{13}C–^{13}C couplings in side chains, are highly applicable to large molecular weight systems.[23,24] These experiments in general are best performed on recent model NMR consoles (i.e., those purchased within the last 2 years). The use of the triple-resonance experiments (i.e., those involving ^{1}H, ^{13}C, and ^{15}N) allow one to walk along the polypeptide backbone via the coupled spins.[25]

Through-space distances between protons separated by less than 5 Å are obtained using NOE in ^{15}N- and ^{13}C-edited 3D NMR experiments. These data, together with torsion angle restraints and determined hydrogen bonds from NH exchange data, are used as input into distance geometry, simulated annealing, or restrained molecular dynamics calculations of the protein structure.[26,27] NOE experiments provide a large number of proton–

[19] M. Rance, O. W. Sorensen, G. Bodenhausen, G. Wagner, R. R. Ernst, and K. Wüthrich, *Biochem. Biophys. Res. Commun.* **117,** 479 (1983).

[20] A. Bax and D. G. Davis, *J. Magn. Reson.* **65,** 355 (1985).

[21] G. M. Clore and A. M. Gronenborn, *Ann. Rev. Biophys. Chem.* **20,** 29 (1991).

[22] G. M. Clore and A. M. Gronenborn, *Prog. NMR Spectrosc.* **23,** 43 (1991).

[23] A. Bax, G. M. Clore, P. C. Driscoll, A. M. Gronenborn, M. Ikura, and L. E. Kay, *J. Magn. Reson.* **87,** 620 (1990).

[24] A. Bax, G. M. Clore, and A. M. Gronenborn, *J. Magn. Reson.* **88,** 425 (1990).

[25] L. E. Kay, M. Ikura, R. Tschudin, and A. Bax, *J. Magn. Reson.* **89,** 496 (1990).

[26] G. M. Clore and A. M. Gronenborn, *Science* **252,** 1390 (1991).

[27] A. T. Brünger, "XPLOR Manual, Version 3.0," Yale University, New Haven (1992).

proton connectivities (>1000) in proteins of M_r 10,000. Three-dimensional NMR experiments, which are highly sensitive and useful, include the ^1H–^1H–^{15}N NOESY-HMQC for measurements of backbone–backbone and backbone–side chain distance data and the ^1H–^1H–^{13}C NOESY-HMQC for measurements of side chain–side chain distance data.[28,29]

The rates of chemical exchange of amide protons determined from NMR data provide important structural information about a protein or protein domain. The chemical exchange of amide protons with bulk solvent at pH 7.0 occurs at significant rates (~10 sec^{-1}) for an extended polypeptide segment and to lesser degrees for amides that are inaccessible to solvent or for amides that are hydrogen bonded. The exchange rate can be slowed by approximately an order of magnitude for each unit of decrease in the pH (see Wüthrich[30]). Structured segments of proteins display slow NH exchange, allowing water presaturation without significant loss of amide resonance intensity near neutral pH. To observe amide protons, samples are prepared in 90% H_2O/10% D_2O. Under these conditions, good quality water resonance suppression is necessary in order to observe the 1 m*M* concentration of protein protons in the presence of the ~100 *M* concentration of water protons. Information on segments of the protein that display slow NH exchange can be determined by lyophilization, dissolution in D_2O, and spectral observation of the nonexchanged protons. In cases in which lyophilization induces denaturation, the sample can be diluted with D_2O approximately two- to threefold and concentrated to obtain a >95% D_2O solution after several such manipulations. Only very slowly exchanging amides will be observed after the latter treatment, which requires several hours. For amides that exchange more rapidly than the time required for dissolution in D_2O and acquisition of an NMR spectrum, relative rates of exchange can be determined at near neutral pH by acquiring NMR spectra with full water presaturation. Resonances that exchange rapidly with respect to their relaxation rates ($1/T_1$) will display reduced intensity. The intensities obtained in a presaturation experiment can be compared to the intensities obtained with a selective excitation sequence such as jump-and-return.[31] This method was useful in characterizing exchange in the N-terminal domain of DNA polymerase β at pH 6.7.

Isotope Labeling

Before isotopically labeling a protein or protein domain with ^{15}N and/or ^{13}C for X-nucleus-correlated NMR studies, an important first step is to

[28] M. Ikura, L. E. Kay, R. Tschudin, and A. Bax, *J. Magn. Reson.* **86,** 204 (1990).
[29] L. E. Kay, D. Marion, and A. Bax, *J. Magn. Reson.* **84,** 72 (1989).
[30] K. Wüthrich *in* "NMR of Proteins and Nucleic Acids," Wiley, New York, 1986.
[31] P. Plateau and M. Gueron, *J. Am. Chem. Soc.* **104,** 7310 (1982).

characterize the NMR behavior of a less expensive (and possibly more easily obtained) unlabeled sample. Both 1D and 2D NMR are acquired on the unlabeled protein to assess the resolution under the conditions used. The expense of isotopically labeling with ^{15}N and ^{13}C requires an overexpression system, which yields >10 mg of protein per liter of medium. $^{15}NH_4Cl$ and [$^{13}C_6$]glucose currently cost approximately $60/g and $500/g, respectively, and a liter of minimal medium requires ~1 g of $^{15}NH_4Cl$ and ~2 g of [$^{13}C_6$]glucose. Detailed procedures for labeling with $^{15}NH_4Cl$ have been published previously.[32] This procedure was developed for the highly sensitive AR120 cell line and utilizes a MOPS/Tricine medium, necessary vitamins, and both thymine and uracil (T-U) bases. This method has been applied to the BL21(DE3) cell line with good results. For $^{15}N/^{13}C$ labeling an important modification of the procedure is removal of the MOPS/Tricine and T-U, which provide carbon sources. The details for labeling of fragments of Pol β are presented by Beard and Wilson in Chapter 11 within this volume. The presence of vitamins did not decrease the ^{13}C-labeling of the N-terminal domain as judged by the HNCA experiment in which sequential ^{13}C between residues was detected by the 2-bond magnetization transfer pathway.

Sample Preparation

Protein samples for NMR are prepared at a concentration of 1 to 2 mM in a volume of 0.5 to 0.6 mL and are placed within thin-walled high-quality 5-mm NMR tubes. In sample preparation for proton NMR, care should be taken to eliminate "contaminating protons," which may result from introduction of nondeuterated buffers, glycerol (typically present on centrifugal filtration membranes), or other species such as EDTA. Samples are normally prepared in either 90% H_2O/10% D_2O for observation of exchangeable protons or 99.99% D_2O for observation of nonexchangeable protons. Deionized water and D_2O with low paramagnetic impurities should be used in sample preparation. Care should be taken to eliminate paramagnetic metals from samples. In our laboratory, we typically avoid phosphate buffer and pass all buffers used in the final stage of protein preparation through Chelex 100 resin (Bio-Rad) before use. Proteins are exchanged into a suitable NMR buffer by loading a concentrated solution of the protein (approximately 3 to 5% of the column volume) on a Sephadex G-15 or G-25 column. Sephadex G-15 is used for proteins of $M_r < 10{,}000$. The column is washed with approximately 2 column volumes of 50 mM EDTA before equilibration with the buffer. Buffers containing approximately 5 mM Tris

[32] D. J. Weber, A. P. Gittis, G. P. Mullen, C. Abeygunawardana, E. E. Lattman, and A. S. Mildvan, *Proteins, Struct. Funct. Genet.* **13,** 275 (1992).

and 50 mM or greater NaCl are used to maintain the pH at 7.5 during elution. Deuterated Tris-d_{11} buffer should be applied to the column before loading the protein. The pH of a protein sample can be gradually lowered by addition of 1- to 2-μl aliquots of 50 mM HCl under conditions of good mixing. On lowering the pH by 0.5 unit, the effect of the pH change on the NMR spectrum of the protein or protein domain is monitored. Domains are particularly susceptible to denaturation at low pH. Loss of dispersion of NH, aromatic, or aliphatic resonances is indicative of acid denaturation.

NMR Structural Characterization of Domains in DNA Polymerases

Isolated fragments of proteins that display secondary and tertiary structure similar to that of the intact protein can be defined as protein domains. These types of domains can have highly variable sizes (e.g., 5 to 30 kDa). Such a domain may display reduced stability due to loss of interactions with the intact protein [i.e., a smaller free energy difference (ΔG_d) between the folded state and the "denatured state" for this segment of the polypeptide]. As determined using NMR, the structures of domains may display more dynamic behavior. To be suitable for NMR studies, a domain in a DNA polymerase should not form aggregates in solution. Aggregation is readily determined from line broadening in an NMR spectrum. Protein domains are also assayed for aggregation or oligomerization by application of HPLC gel-filtration chromatography. Limited proteolysis to yield soluble fragments is often an initial step in identifying a domain, but not all proteolytic fragments are likely to retain structure. Further, these fragments can be susceptible to aggregation, since proteolysis can expose hydrophobic residues. Domains from DNA polymerases can display some degree of function, such as binding activity for template, template–primer, or dNTP-Mg^{2+} substrates. Large domains are found to display enzymatic activity. Not all domains of DNA polymerases will retain function, and binding activity is possible even in the absence of a significant degree of secondary and tertiary structure as was found for a peptide from DNA polymerase I.[4,33] Such binding is likely a result of ionic and/or hydrophobic interactions within extended segments, which may or may not be of significance in regard to the structure of the intact protein. Production of cloned fragments with possible function using biochemical data as a guide appears to be the best approach to the study of a domain by NMR spectroscopy. Circular dichroism can be used as an initial screen for types of secondary structure that could be present, but the possibility of aggregated secondary structure should be considered.

[33] G. P. Mullen, J. B. Vaughn Jr., and A. S. Mildvan, *Arch. Biochem. Biophys.* **301,** 174 (1993).

Peptide I from DNA Polymerase I

The first attempt at identifying domains for NMR characterization in DNA polymerases was directed toward two peptide segments in DNA polymerase I, namely, residues 728–777 (peptide I) and residues 840–888 (peptide II).[4] Peptide I displayed binding activity for both dNTP and template–primer DNA. Peptide II did not display such activities. NMR studies indicated that peptide I displayed transient structure in solution.[33] This transient structure consisted of a clustering of hydrophobicity and a nascent helix, which could also be modeled as three open turns, and an antiparallel C-terminal loop region. Such transient structure is a result of multiple conformations in equilibrium and indicates that the polypeptide is not a domain or separate folding module. The circular dichroism spectrum of peptide I under the NMR conditions was characteristic of 16% helices, 14% β structure, and 70% random coil.

Although the segment of residues from 728–777 protrude into solution in the X-ray structure of the Klenow fragment, it is apparent that this segment requires additional structural elements for folding. In the refined X-ray structure of the Klenow fragment, two additional helices (O_1 and O_2) are fit within the segment between 766 and 797.[34] The segment comprising helices M and L (710–727), helices O_1 and O_2, and the segment 728–777 are all possibly necessary for forming a stable fingers domain. As such this domain would be amenable to NMR structural studies and studies of possible template and/or template–primer binding.

Domains of DNA Polymerase β

Fragments of the mammalian DNA polymerase β retain function suggesting that this enzyme is composed of domain structure.[35,36] Limited proteolysis of Pol β results in an N-terminal 8-kDa fragment with single-stranded DNA binding activity and C-terminal 31-kDa fragment with reduced DNA polymerase activity.[35] Circular dichroism studies indicate secondary structure associated with these fragments.[37] The 1D ^{1}H spectra (Fig. 1) and 2D NOESY spectra for the cloned and overexpressed N-terminal (residues 2–87) and 31-kDa (residues 88–334) fragments indicate that these are amenable to multidimensional NMR structure determination. The NH-aliphatic region of the 2D NOESY spectrum of the N-terminal domain

[34] L. S. Beese, V. Derbyshire, and T. A. Steitz, *Science*, **260,** 352 (1993).

[35] A. Kumar, S. G. Widen, K. R. Williams, P. Kedar, R. L. Karpel, and S. H. Wilson, *J. Biol. Chem.* **265,** 2124 (1990).

[36] J. R. Casas-Finet, A. Kumar, R. L. Karpel, and S. H. Wilson, *Biochemistry*, **31,** 10,272 (1992).

[37] J. R. Casas-Finet, A. Kumar, G. Morris, S. H. Wilson, and R. L. Karpel, *J. Biol. Chem.* **266,** 19,618 (1991).

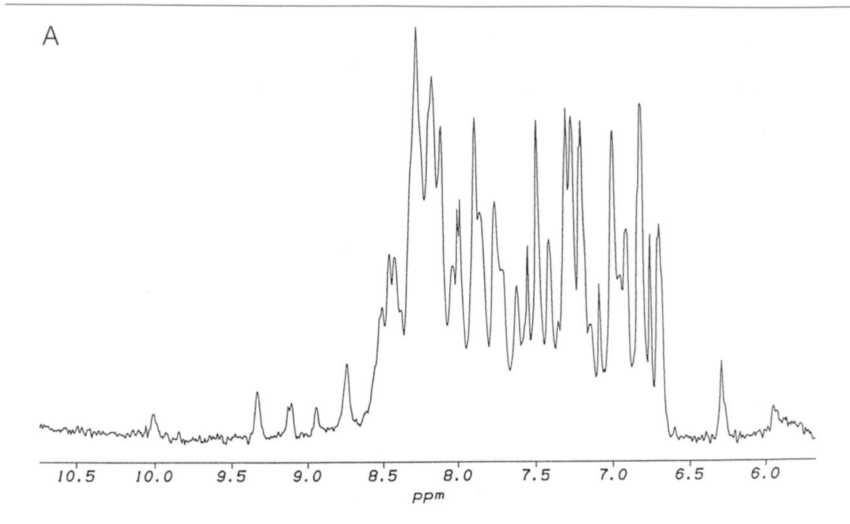

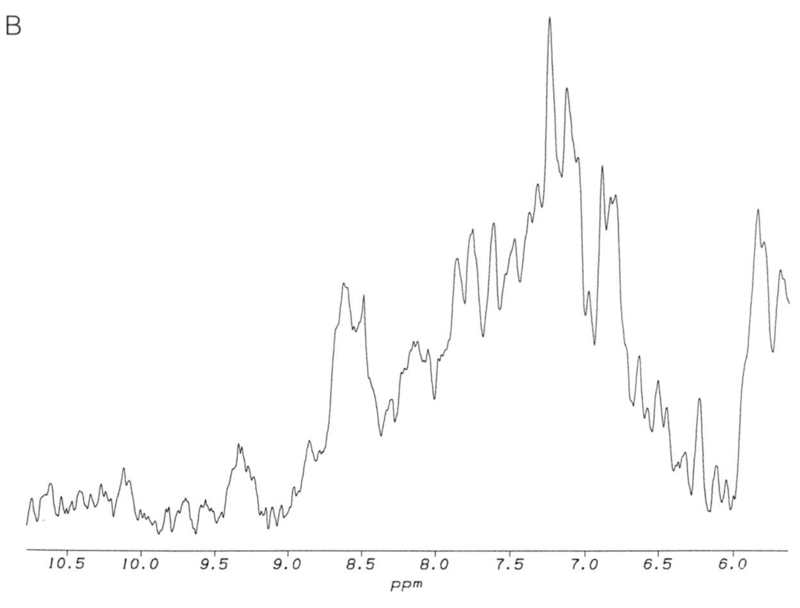

FIG. 1. The amide–aromatic regions of the 500-MHz 1D proton NMR spectra of the two domains of Pol β. (A) The N-terminal domain of DNA polymerase β (residues 2–87). (B) The catalytic domain of DNA polymerase β (residues 88–334). Spectra were acquired at 27°. The protein domains were dissolved at approximately 2 mM (A) and 1 mM (B) at pH 6.7 in 90%H_2O/10%D_2O containing 400 mM NaCl. The water resonance was suppressed with DANTE presaturation for a period of 1 sec.

(Fig. 2) illustrates the spectral crowding, which is observed for helical proteins of this size (9.5 kDa). The N-terminal domain of DNA polymerase β is composed of four helices as determined by multidimensional NMR.[9] A large degree of spectral simplification is obtained by spreading out the "contoured peaks" in a third dimension according to the ^{15}N chemical shift, which corresponds to the amide nitrogen for the attached amide proton. A 2D plane from the ^{15}N-edited 3D spectrum of the N-terminal domain at a chemical shift of 123.6 ppm for ^{15}N represents one of the most crowded planes of the 3D spectrum (Fig. 3A). By cutting vertical strips from each of the planes and laying them side by side a pseudo-2D spectrum is obtained. Each strip corresponds to a particular residue and is randomly arranged

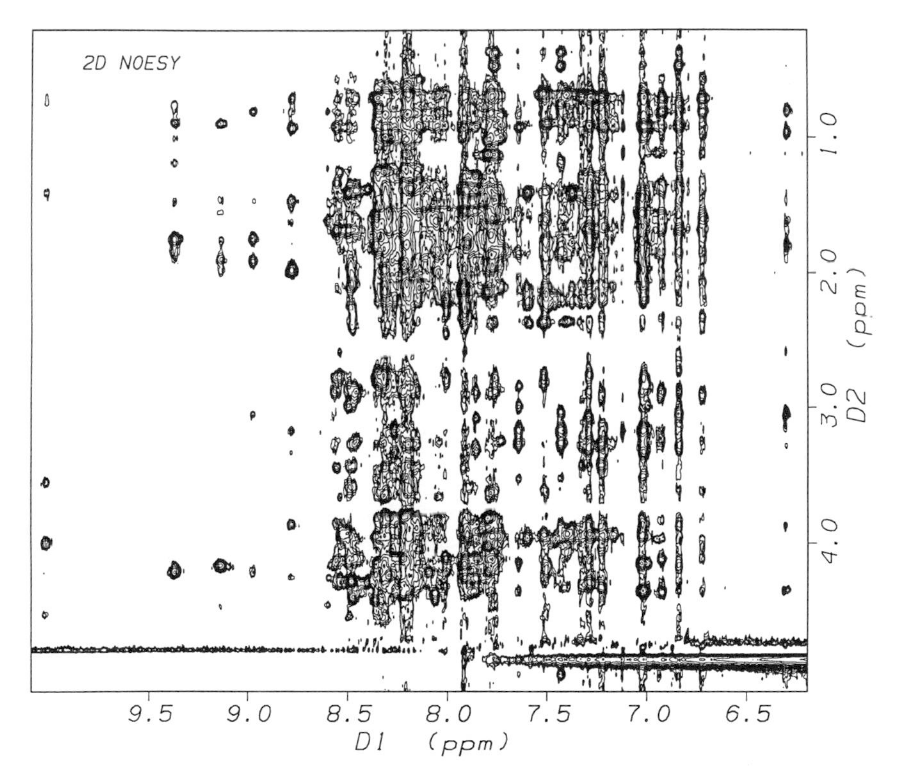

FIG. 2. The NH-aliphatic region of the 2D NOESY spectrum of the N-terminal domain of DNA polymerase β collected with an NOE mixing time of 200 ms using the conditions described in Fig. 1. The sweep widths were set to 6024 Hz centered at the water resonance at 4.8 ppm. A total of 2048 complex points were collected in t_2 and 250 complex points were collected in t_1 with quadrature detection performed using the States method. The spectrum was processed with 90° shifted squared sine bell window functions in each dimension. The acquisition time for this spectrum was approximately 1 day.

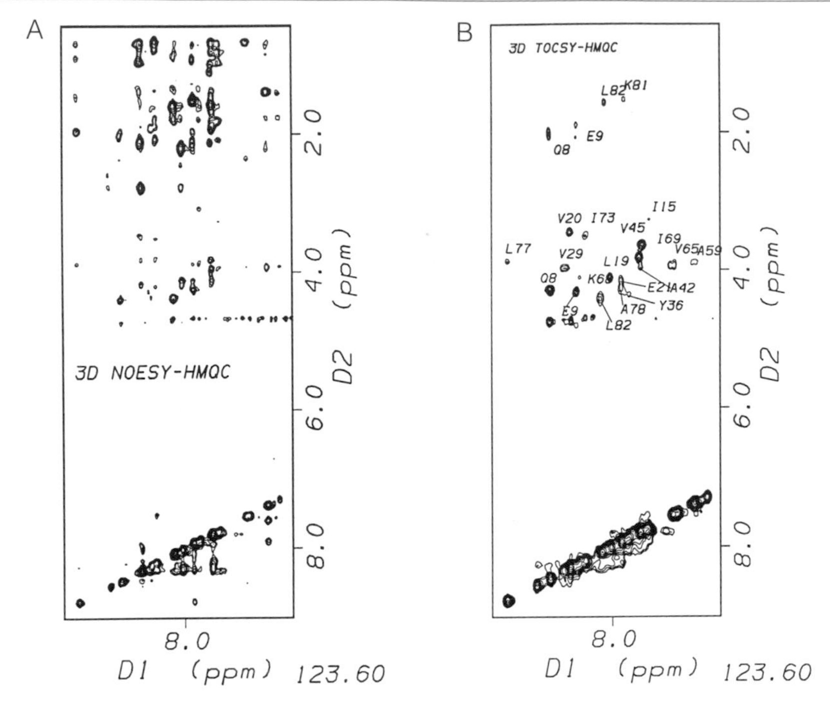

FIG. 3. The NH-aliphatic region of an ^{15}N plane taken at 123.6 ppm from the 3D ^{15}N-edited NOESY spectrum (A) and TOCSY spectrum (B). The 3D spectra were acquired with proton sweep widths as described in Fig. 2 and with an ^{15}N sweep width of 2500 Hz centered at 118.9 ppm. The ^{1}H–^{1}H–^{15}N NOESY-HMQC spectrum (Kay *et al.*[29]) was acquired with a mixing time of 200 ms. The 90° pulses were 24.5 μs for ^{1}H, 88 μs for ^{15}N, and 295 μs for ^{15}N decoupling. The 3D ^{1}H–^{1}H–^{15}N TOCSY-HMQC spectrum (Kay *et al.*[29]) was acquired with a mixing time of 30 ms. The 90° pulse lengths were 25 μs for ^{1}H, 93 μs for ^{15}N, and 305 μs for ^{15}N decoupling. A trim pulse of 1.5 ms was used before the WALTZ-16 mixing period. Delays before and after the WALTZ-16 mixing period were set to 2 ms. The carrier frequency was switched from the water resonance to 6.0 ppm for the WALTZ mixing period. For each 3D experiment, the TPPI–States phase cycling procedure was used for quadrature detection in t_1 and t_2, and 128 complex t_1, 32 complex t_2, and 512 complex t_3 data points were acquired with acquisition times of 42.5, 12.8, and 85.0 ms, respectively. GARP was used for ^{15}N decoupling in t_1 and t_3. The acquisition time for each of these spectra was approximately 3 days.

until sequence-specific assignments are made. A portion of a side-by-side strip plot is shown in Fig. 4A. A pathway of connectivities between residues is traced on the basis of NH–NH NOEs characteristic of α-helical proteins (Fig. 4B). The analogous NH–NH pathway in the 2D spectrum (Fig. 5) illustrates that sensitivity is gained in the 3D spectrum as a result of the longer overall acquisition time even though fewer scans are used. The increase in the acquisition time is a result of the use of two evolution times,

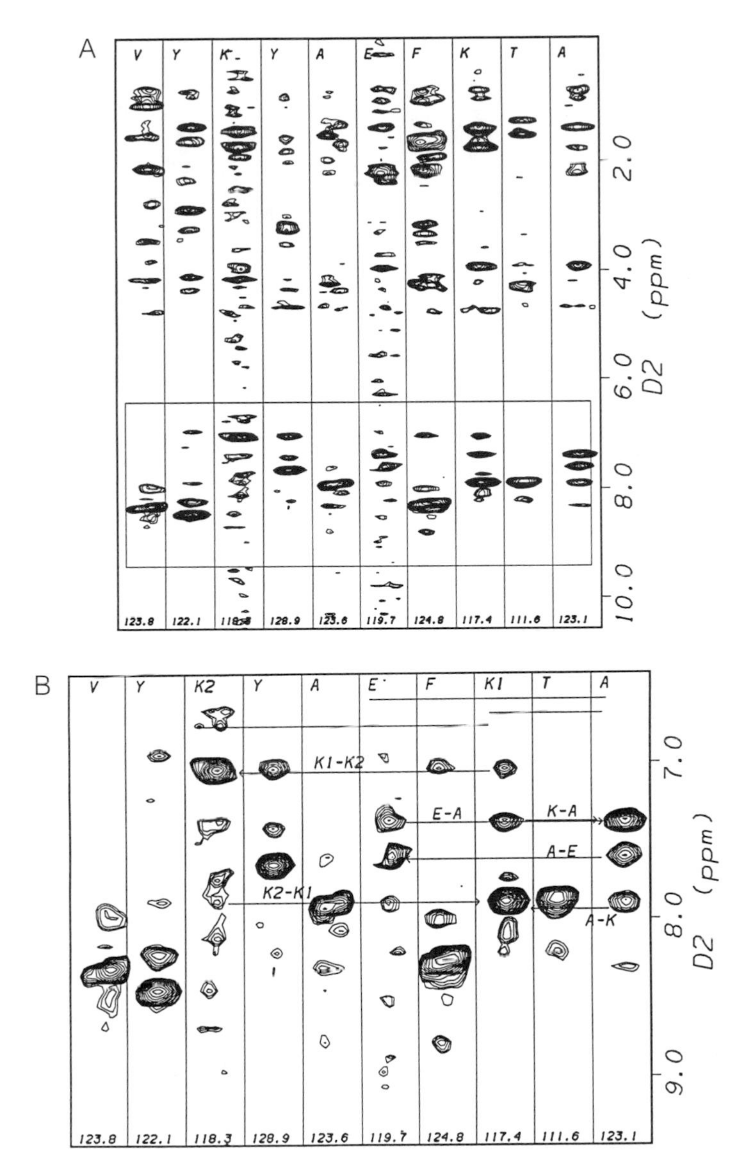

FIG. 4. A randomly arranged portion of the strip plot of the 3D ^{15}N-edited NOESY spectrum of the Pol β N-terminal domain. (A) Strips of 16 points are cut from the 3D spectrum to form a pseudo-2D spectrum. The residue type assignments are labeled for each of the strips. The ^{15}N chemical shift for the plane from which the strip was taken is indicated at the bottom. (B) Expansion of the boxed region in part (A) showing sequential NH–NH connectivities for a portion of helix-3. Lines at the top show the strips that display interconnecting cross-peaks. Lines connecting the NOE cross-peaks have their arrowhead pointing toward the diagonal peak. The labeling describes the magnetization transfer pathway from the originating proton (first label) to the destination proton (second label) for the cross-peak at the opposite end of the arrowhead.

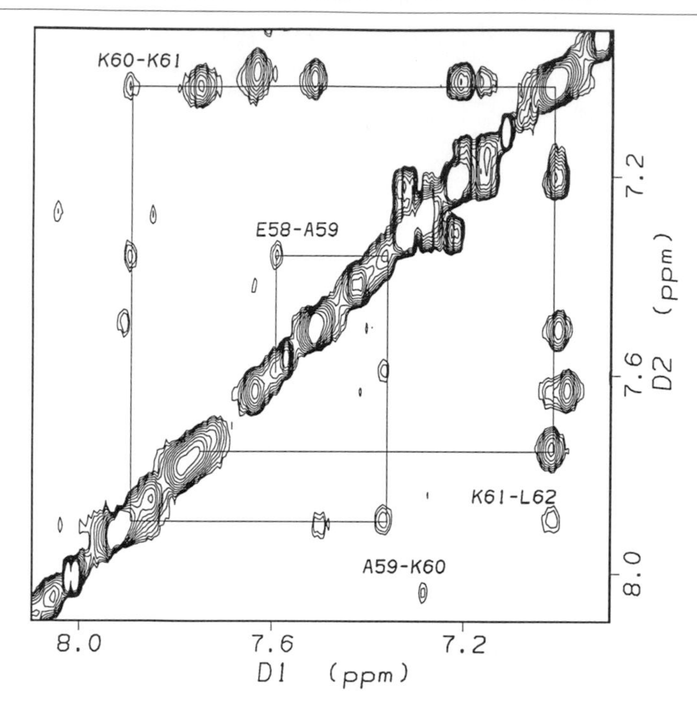

FIG. 5. The NH–NH connectivities corresponding to those illustrated in Fig. 4 in the 2D NOESY spectrum of the Pol β N-terminal domain.

which are designated t_1 for the second ^{1}H dimension and t_2 for the ^{15}N dimension. Overlaps in the 3D spectrum are resolved as is the case for K61, which displays an amide proton resonance overlap with an aromatic resonance in this example.

Residue type assignments for the 3D NOESY strips are made on the basis of through-bond connectivity patterns for each of the residues. These assignments are performed by analysis of the 3D ^{1}H–^{1}H–^{13}C TOCSY-HMQC, HNCA, and ^{1}H–^{1}H–^{15}N TOCSY-HMQC spectra.[9,25,29,38] Unambiguous side-chain proton and ^{13}C resonance assignments are made using the ^{1}H–^{1}H–^{13}C TOCSY-HMQC experiment, in which correlations are observed between protons in up to three sequential 3-bond couplings.[38] An example of correlations for the αH proton (3.84 ppm) attached to the Cα carbon (63.24 ppm) for V45 are shown in Fig. 6. The HNCA experiment is used to connect the amide ^{1}H, ^{15}N, and ^{13}Cα chemical shifts by correlating these nuclei in a single 3D spectrum. In addition, the HNCA experiment

[38] D-J. Liu, E. F. DeRose, and G. P. Mullen, unpublished results (1993).

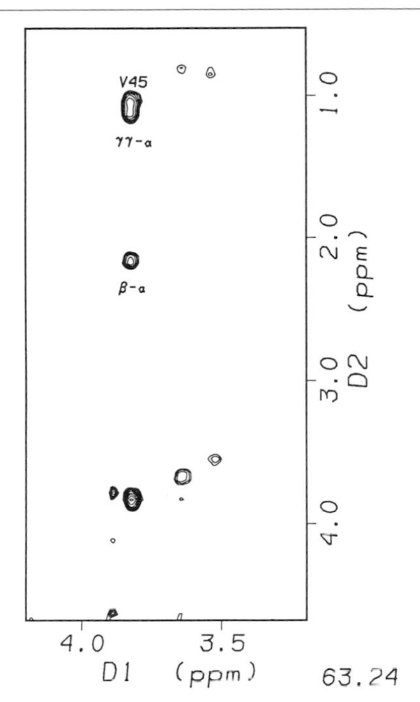

FIG. 6. A ^{13}C plane taken at 63.24 ppm from the ^{13}C-edited TOCSY spectrum. The ^{13}C-edited TOCSY experiment allows assignments of side-chain ^{1}H and ^{13}C resonances to be made on the basis of residue type. The acquisition time for the complete 3D spectrum is approximately 5 days.

provides weak sequential 2-bond connectivities of the type $C\alpha_{i-1}$–NH_i. The ^{1}H–^{1}H–^{15}N TOCSY-HMQC experiment (Fig. 3B) is used to connect the amide ^{1}H, ^{15}N, and $^{1}H\alpha$ resonances. Stronger sequential through-bond connectivities for $C\alpha_{i-1}$ and NH_i are obtained through relayed 1-bond transfers in the HN(CO)CA experiment.[39]

Proteolysis of the 31-kDa catalytic domain of Pol-β yields a 27-kDa core domain deficient in catalytic activity and truncated at the N terminus with respect to the 31-kDa polypeptide.[35] A transdomain polypeptide, which is composed of part of the 4-kDa cleavage peptide and the N-terminal domain, displays both single-stranded and double-stranded binding activities.[36] A cloned transdomain segment (i.e., residues 2–140) has NMR properties that are similar to the N-terminal domain.[40] In addition, the 2–140

[39] M. Ikura and A. Bax, *J. Biomol. Nucl. Magn. Reson.* **1,** 99 (1991).

[40] L. J. Draeger, E. F. DeRose, R. Prasad, S. H. Wilson, and G. P. Mullen, unpublished results (1993).

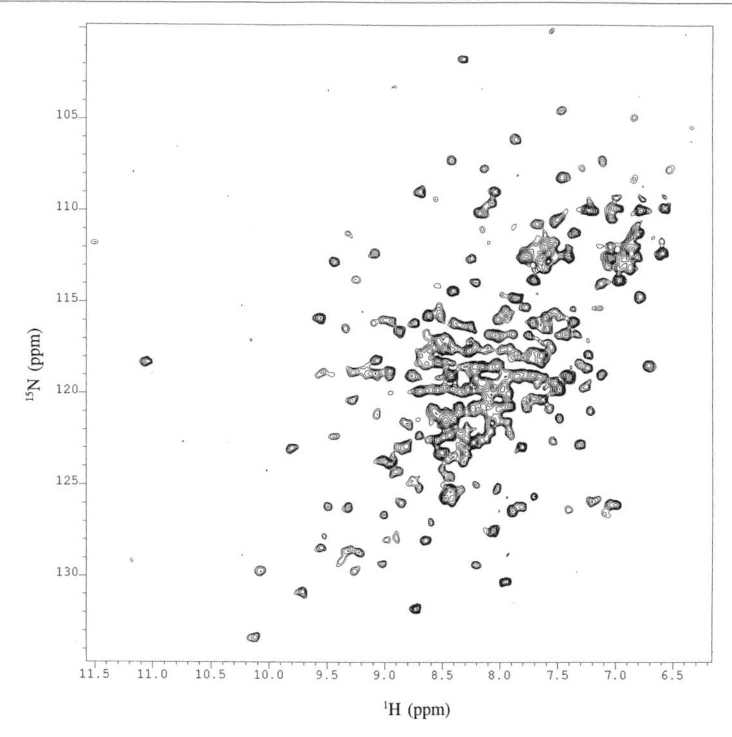

FIG. 7. The ^{1}H–^{15}N HSQC spectrum of the catalytic domain of Pol β in 90% H_2O/10% D_2O at pH 6.7 and 30° acquired on a Varian Unity-plus 600 NMR spectrometer in 19 minutes. The pulse sequence utilized z-gradients for coherence selection and water suppression (see Kay *et al.*[41]).

polypeptide exhibits an additional segment of ordered structure. HMQC comparison of the N-terminal domain and the 14-kDa polypeptide containing the transdomain segment indicates that approximately 20 additional residues display ordered structure on the basis of the slow amide exchange of these residues at pH 6.7. The finding that only 20 of the possible 53 residues are observable at pH 6.7 together with a set of resonances that displays rapid exchange and narrow line widths suggests that the transdomain segment (residues 88–140) requires additional structural support from the remainder of the enzyme for complete folding. The NMR structural results correlate to the finding that the catalytic domain requires the transdomain segment for catalytic activity.[35] More than 240 correlations are observed in the ^{1}H–^{15}N HSQC spectrum (Fig. 7) of the catalytic domain with ~220 resulting from backbone amides in secondary structure.

[41] L. E. Kay, P. Keifer, and T. Saarinen, *J. Am. Chem. Soc.* **114,** 10,663 (1992).

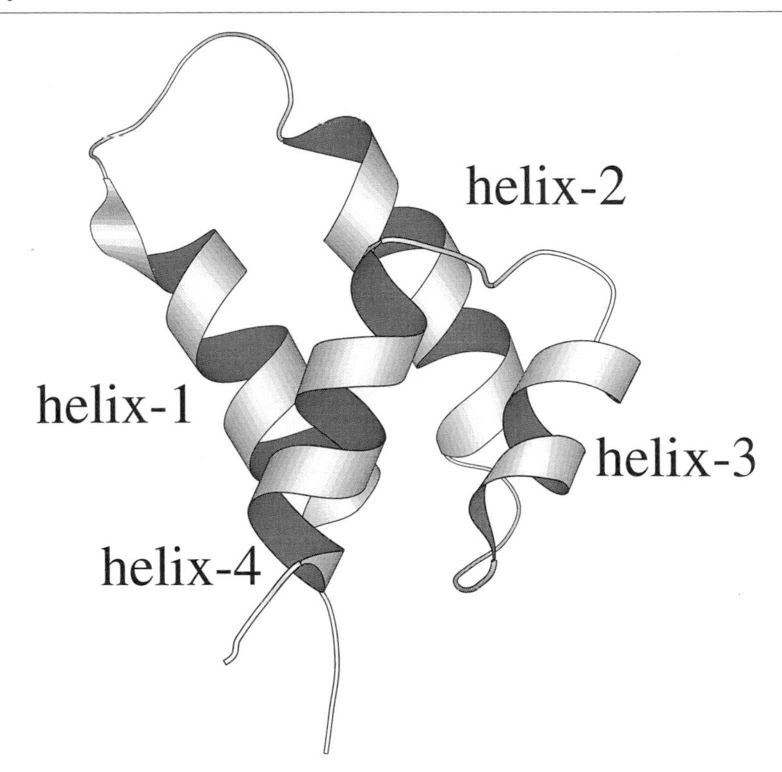

FIG. 8. A ribbon representation of the average NMR solution structure[9] of the N-terminal domain of Pol β (displayed using Molscript version 1.4[42]).

NMR Methods for Studying Substrate Binding to DNA Polymerase Domains

Whether the N-terminal domain of Pol-β constitutes a fingers- or thumb-like domain similar to that determined in the X-ray structure of the Klenow fragment remains to be assessed. The N-terminal domain contains a groove for ssDNA binding formed by helix-4 and a flexible loop connecting helices 1 and 2 (Fig. 8).[9] Three lysines (K35, K68, and K72) within the cleft participate in ssDNA binding based on chemical shift data and model building. An example of the utility of the ^1H–^{15}N HMQC experiment for assessing single-stranded DNA binding is illustrated in Fig. 9.[38] In this experiment the amide ^1H and ^{15}N chemical shifts of the free and bound states are in fast exchange. Salt titration and comparison to the spectrum in the absence of single-stranded DNA is used to determine the chemical shifts for the bound state of the domain. The advantage of the method is that several types of DNA complexes can be studied with rapid assignment of

[42] P. J. Kraulis, *J. Appl. Crystallogr.* **24,** 946 (1991).

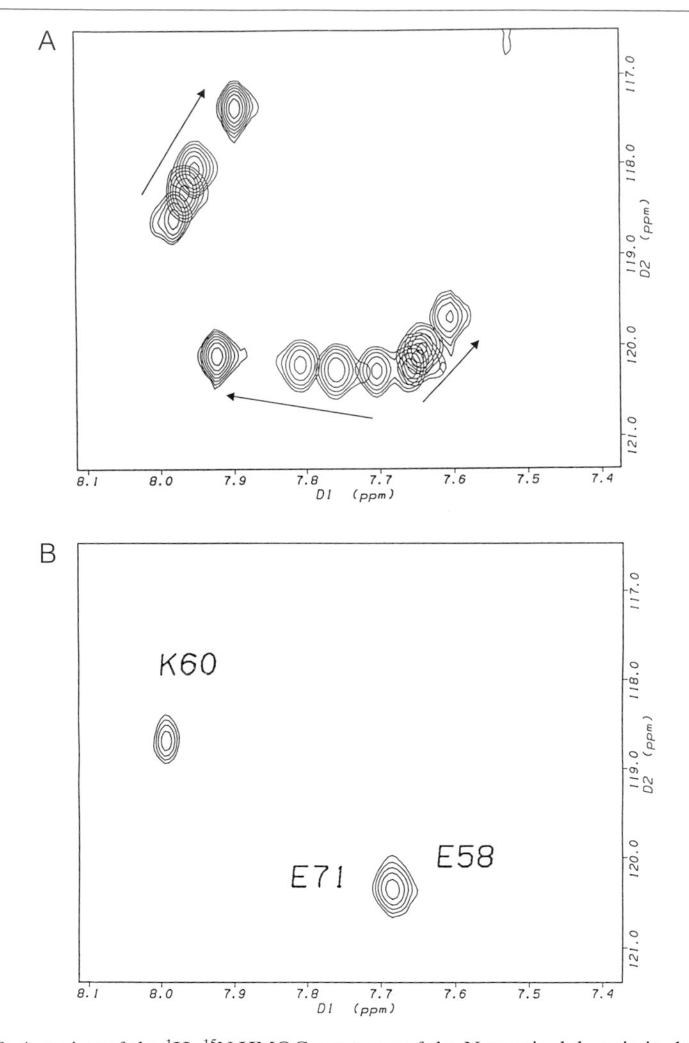

FIG. 9. A region of the $^{1}H-^{15}N$ HMQC spectrum of the N-terminal domain in the absence and the presence of $p(dT)_8$ at increasing NaCl concentrations. (A) Three of the $^{1}H-^{15}N$ correlations in this region of the spectrum exhibit chemical shift changes in the presence of DNA. The direction of the arrows indicates the following: N-terminal domain–$p(dT)_8$ at 100 m*M* NaCl after excess $p(dT)_8$ is removed to form a 1:1 complex with the N-terminal domain; the complex at 200 m*M* NaCl; the complex at 300 m*M* NaCl; and the N-terminal domain alone to 400 m*M* NaCl. No large changes were observed in the chemical shifts of the complex at 400 m*M* (not shown) as compared to 100 m*M* NaCl. (B) The three $^{1}H-^{15}N$ correlations in the presence of excess $p(dT)_8$. The cross-peaks for E71 and E58 are superimposed under these conditions.

the bound state of the domain. These chemical shifts are then used in confirming the NOE connectivity pathways in the $^{1}H-^{15}N$ NOESY-HMQC 3D spectrum without beginning assignments from scratch. On the basis of these empirical changes, binding of single-stranded DNA was mapped to helix 4 within this domain. This method of analysis, however, can be biased by conformational changes in the protein, which appear to be occurring to a small degree for helix 3 within the DNA binding to the Pol-β N-terminal domain.

Acknowledgments

Thanks are extended to those with whom I have interacted in this research. I thank Dr. Albert Mildvan for initiation of and advice in studies of the Klenow fragment of Pol I, my collaborators Dr. Samuel Wilson, Dr. Rajendra Prasad, and Dr. Steve Widen for their efforts in the production and biochemical studies of domains of Pol β, and Dr. Eugene DeRose, Dr. Dingjiang Liu, and Dr. Linda Draeger for participation in NMR structural studies of Pol β.

[15] Crystallization of Human Immunodeficiency Virus Type 1 Reverse Transcriptase with and without Nucleic Acid Substrates, Inhibitors, and an Antibody Fab Fragment

By ARTHUR D. CLARK, JR., ALFREDO JACOBO-MOLINA, PATRICK CLARK, STEPHEN H. HUGHES, and EDWARD ARNOLD

Introduction

The inherent flexibility of nucleic acid polymerases has made the crystallization of these enzymes a challenging problem. Many laboratories have conducted extensive efforts to produce crystals of the reverse transcriptase (RT) of human immunodeficiency virus type 1 (HIV-1).[1-7] The initial at-

[1] D. M. Lowe, A. Aitken, C. Bradley, G. K. Darby, B. A. Larder, K. L. Powell, D. J. M. Purifoy, M. Tisdale, and D. K. Stammers, *Biochemistry* **27,** 8884 (1988).

[2] T. Unge, H. Ahola, R. Bhikhabhai, K. Backbro, S. Lovgren, E. M. Fenyo, A. Honigman, A. Panet, J. S. Gronowitz, and B. Strandberg, *AIDS Res. Human Retroviruses* **6,** 1297 (1990).

[3] L. F. Lloyd, P. Brick, L. Mei-Zhen, N. E. Chayen, and D. M. Blow, *J. Mol. Biol.* **217,** 19 (1991).

[4] A. Jacobo-Molina, A. D. Clark Jr., R. L. Williams, R. G. Nanni, P. Clark, A. L. Ferris, S. H. Hughes, and E. Arnold, *Proc. Natl. Acad. Sci. USA* **88,** 10,895 (1991).

[5] E. Arnold, A. Jacobo-Molina, R. G. Nanni, R. L. Williams, X. Lu, J. Ding, A. D. Clark Jr., A. Zhang, A. L. Ferris, P. Clark, A. Hizi, and S. H. Hughes, *Nature* **357,** 85 (1992).

[6] L. A. Kohlstaedt, J. Wang, J. M. Friedman, P. A. Rice, and T. A. Steitz, *Science* **256,** 1783 (1992).

tempts did not produce HIV-1 RT crystals suitable for high-resolution studies. This led to attempts to "engineer" crystallizability into the protein, including cocrystallization of RT with double-stranded DNA oligomers of defined sequence, monoclonal antibody (MAb) Fab fragments, nucleotide substrates, and inhibitors. Several laboratories have succeeded in obtaining high-quality crystals of HIV-1 RT, and their structure determinations have been reported.[5,6,8–12] Most of the high-quality crystals reported to date contain the 66-kDa/51-kDa (p66/p51) heterodimeric form of the enzyme, which is the form found in virions. This chapter details the methodology used by our laboratory to produce diffraction-quality crystals of a number of RT complexes. We have included a description of the purification of the enzyme and of a noninhibitory Fab used in some of the crystallization experiments, since the reproducible preparation of high-quality crystals of HIV-1 RT is critically dependent on the protocols used to purify each of these proteins. It should be emphasized that estimating purity of the components by Coomassie or silver stained SDS–PAGE cannot be the sole criterion in determining their quality for crystallization.

Purification of HIV-1 RT Heterodimer

Preparation of HIV-1 RT 66-kDa Fractions

The 66-kDa form of HIV-1 RT is overexpressed in *Escherichia coli* recombinant strain DH-5 containing the pUC12N expression plasmid.[13,14]

---

[7] E. Y. Jones, D. I. Stuart, E. F. Garman, R. Griest, D. C. Phillips, G. L. Taylor, O. Matsumoto, G. Darby, B. Larder, D. Lowe, K. Powell, D. Purifoy, C. K. Ross, D. Somers, M. Tisdale, and D. K. Stammers, *J. Crystal Growth* **126,** 261 (1993).

[8] A. Jacobo-Molina, J. Ding, R. G. Nanni, A. D. Clark Jr., X. Lu, C. Tantillo, R. L. Williams, G. Kamer, A. L. Ferris, P. Clark, A. Hizi, S. H. Hughes, and E. Arnold, *Proc. Natl. Acad. Sci. USA* **90,** 6320 (1993).

[9] D. K. Stammers, D. O'N. Somers, C. K. Ross, I. Kirby, P. H. Ray, J. E. Wilson, M. Norman, J. S. Ren, R. M. Esnouf, E. F. Garman, E. Y. Jones, and D. I. Stuart, *J. Mol. Biol.* **242,** 586 (1994).

[10] T. Unge, S. Knight, R. Bhikhabhai, S. Lövgren, Z. Dauter, K. Wilson, and B. Strandberg, *Structure* **2,** 953 (1994).

[11] D. W. Rodgers, S. J. Gamblin, B. A. Harris, S. Ray, J. S. Culp, B. Hellmig, D. J. Woolf, C. Debouck, and S. C. Harrison, *Proc. Natl. Acad. Sci. USA* **92,** 1222 (1995).

[12] J. Ding, K. Das, C. Tantillo, W. Zhang, A. D. Clark, Jr., S. Jessen, X. Lu, Y. Hsiou, A. Jacobo-Molina, K. Andries, R. Pauwels, H. Moereels, L. Koymans, P. A. J. Janssen, R. H. Smith, Jr., M. K. Koepke, C. J. Michejda, S. H. Hughes, and E. Arnold, *Structure* **3,** 365 (1995).

[13] A. Hizi, C. McGill, and S. H. Hughes, *Proc. Natl. Acad. Sci. USA* **85,** 1218 (1988).

[14] P. K. Clark, A. L. Ferris, D. A. Miller, A. Hizi, K.-W. Kim, S. M. Deringer-Boyer, M. L. Mellini, A. D. Clark Jr., G. F. Arnold, W. B. Lebherz III, E. Arnold, G. M. Muschik, and S. H. Hughes, *AIDS Res. Human Retroviruses* **6,** 753 (1990).

E. coli cells (100 g) are treated with lysozyme and the lysate is treated with polyethyleneimine and centrifuged to remove cellular debris and nucleic acids. The resulting supernatant is concentrated and desalted using a Millipore Pellicon ultrafiltration unit equipped with a 10,000-MW cutoff polyether sulfone Centrasette (Filtron Technology Corp., Northborough, MA) and is then loaded onto a 5.0- × 15.0-cm Q-Sepharose (Pharmacia) column equilibrated with 20 m*M* HEPES, 2 m*M* dithiothreitol (DTT), 0.2 m*M* EDTA, 5 m*M* NaCl, 0.2 m*M* phenylmethylsulfonyl fluoride (PMSF), 10% glycerol, pH 8.0. The column is washed with the same buffer. Most of the RT eluted in this wash fraction. The RT solution is subsequently diluted with 5 m*M* HEPES, 2 m*M* DTT, 0.2 m*M* EDTA, 0.2 m*M* PMSF, 10% glycerol, pH 7.0, to a conductivity equal to 20 m*M* NaCl, adjusted to pH 7.0, and loaded onto a 5.0- × 15.0-cm CM-Trisacryl (Sepracor, Inc., Marlborough, MA) column. The bound RT is eluted using a linear NaCl gradient, and two peaks are collected and stored at 4°. The first peak contains undegraded p66 RT and the second peak contains a mixture of undegraded and degraded p66/p51 forms of the enzyme. The undegraded RT from peak 1 is concentrated and desalted using a Filtron Minisette ultrafiltration unit equipped with a 30,000-MW cutoff polyether sulfone membrane, and is then loaded onto a second Q-Sepharose column (2.6 × 14.0 cm) equilibrated with 20 m*M* Tris–HCl, pH 8.0. The flow-through fractions contain the 66-kDa form of RT with minimal degradation. The bound protein containing the degraded p66/p51 RT is eluted using a linear gradient from 0 to 500 m*M* in 37 m*M* diethanolamine (DEA), 10% glycerol, pH 9.3, and stored at 4°. Undegraded fractions are diluted with 4 volumes of 37 m*M* DEA, 10% glycerol, pH 9.3. Conductivity and pH are adjusted to that of the dilution buffer with NaOH or 2 *M* NaCl, and then the RT is bound to a third Q-Sepharose column (2.6 × 14.0 cm) equilibrated with the same buffer. The purified, undegraded RT is eluted with a linear gradient from 0 to 500 m*M* NaCl in the same buffer.[14]

Preparation of the HIV-1 RT p66/p51 Heterodimer

The HIV-1 RT p66/p51 heterodimer is obtained from a mixture of several fractions generated during the purification of the bacterially expressed 66-kDa form of the enzyme described above. These fractions, which contain *E. coli* protease(s) that have not been fully characterized, are responsible for the conversion of RT from p66 to p66/p51. The processing to p66/p51 is induced by combining chromatographic fractions that contained a mixture of unprocessed p66 and processed p66/p51 forms of RT, then allowing the proteolytic processing to occur over a number of days. The combined fractions consisting of (1) the CM-Trisacryl mixture of p66 and p66/p51, (2) the second Q-Sepharose column elution of p66/p51, and (3) the third Q-Sepharose column eluted peak of p66 are concentrated and diafiltered into 50 m*M* diethanolamine hydrochloride, pH 8.9, 100 m*M* NaCl buffer solution, using an Amicon (model 8050) stirred cell equipped with a YM30 low protein binding membrane and an Amicon (model RC800) reservoir. The protein solution, at a concentration of approximately 5 to 7 mg/ml, as determined by the procedure of Lowry *et al.*[15] using hen egg white lysozyme standards, is subsequently placed at 4° for 4 to 6 days and allowed to convert completely to the p66/p51 heterodimeric form. To ensure maximal yield the proteolytic processing step is monitored daily by SDS–PAGE and analytical Mono Q HR 5/5 anion-exchange chromatography (Pharmacia). The analytical column is equilibrated with 50 m*M* diethanolamine hydrochloride, pH 8.9, and eluted with a 20-min linear gradient from 0 to 240 m*M* NaCl in the same buffer at 1 ml/min. The reaction is complete when the bands corresponding to p66 and p51 appear at equal intensity on a Coomassie-stained SDS–PAGE gel, and no further increase in the area of the major peak representing the most stable p66/p51 heterodimer is seen on the analytical Mono Q column (Fig. 1). The protein solution is filtered through a 0.22-μm Millex GV filter unit and loaded onto a Mono Q HR 16/10 column previously equilibrated with 50 m*M* diethanolamine hydrochloride, pH 8.9. The column is developed with a 40-min linear gradient from 0 to 240 m*M* NaCl in the same buffer at 10 ml/min. The major peak (corresponding to the p66/p51 heterodimer) elutes at 140 m*M* NaCl. This fraction is concentrated to 10 ml using an Amicon stirred cell (model 8050) equipped with a YM30 membrane and then loaded onto two Sephadex G-75 superfine columns (2.5 × 90 cm each; Pharmacia) connected in series. The material is eluted with 20 m*M* Tris–HCl, pH 8.0, 50 m*M* NaCl at 0.25 ml/min. Lastly, the eluted peak is split into two equal fractions and loaded directly onto a Mono Q HR 10/10 column. Each fraction is rechromatographed using identical gradient conditions as the Mono Q Hr 16/10 step, at a flow rate of 4.0 ml/min. Material suitable for crystallization elutes as a single sharp peak with a reproducible retention time on an analytical Mono Q Hr 5/5 column (peak 5, retention time of 25.4 ± 0.4 min; Fig. 1). Purified p66/p51 heterodimer is subsequently concentrated in Centricon-30 microconcentrators (Amicon) concomitantly exchanging the buffer with 10 m*M* Tris–HCl, pH 8.0, 75 m*M* NaCl to a final protein concentration of 40 to 45 mg/ml. Final yields of the purified concentrates average 25 to 35 mg per 100 g of *E. coli* cells as determined by the procedure of Lowry *et al.*[15] using hen egg white lysozyme standards.

[15] O. H. Lowry, N. J. Rosebrough, A. L. Farr, and R. J. Randall, *J. Biol. Chem.* **193,** 265 (1951).

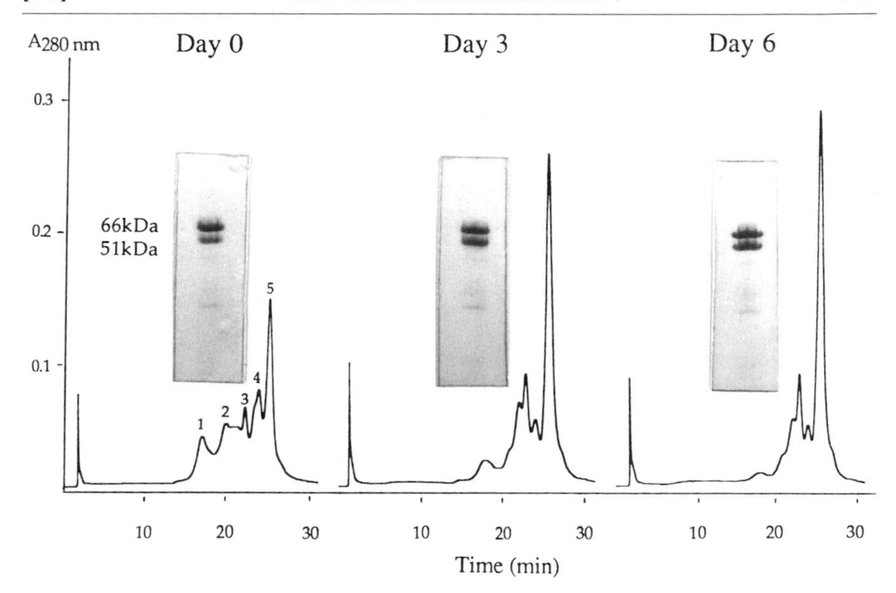

FIG. 1. SDS–PAGE and analytical Mono Q HR 5/5 (Pharmacia) anion-exchange chromatography showing conversion of the 66-kDa form HIV-1 RT to p66/p51 heterodimeric forms. Peak 1 contains the 66-kDa form, peaks 2, 3, and 4 contain minor isoelectric forms of p66/p51, and peak 5 (retention time of 25.4 ± 0.4 min) contains the major p66/p51 isotype used for crystallization. The column gradient conditions are 0 to 240 m*M* NaCl in 50 m*M* diethanolamine hydrochloride, pH 8.9, for 20 min at 1 ml/min.

The concentrate is either used immediately or divided into 2.5-mg aliquots and stored frozen at −20°.

Selection of Monoclonal Antibodies for Crystallization

Monoclonal antibody Fab fragments are used to increase the crystallizability of certain proteins.[16–18] A panel of nine MAbs directed against HIV-1 RT were considered for cocrystallization experiments. The MAbs are prioritized based on their relative affinities for the p66/p51 heterodimer, starting with the MAb with the highest affinity for the enzyme. This strategy allowed us to obtain one very versatile crystal form that is described later in this chapter.[4]

[16] W. G. Laver, *Methods* **1,** 70 (1990).

[17] A. J. Prongay, T. J. Smith, M. G. Rossmann, L. S. Ehrlich, C. A. Carter, and J. McClure, *Proc. Natl. Acad. Sci. USA* **87,** 9980 (1990).

[18] K. C. Garcia, P. M. Ronco, P. J. Verroust, A. T. Brünger, and M. Amzel, *Science* **257,** 502 (1992).

Purification of MAbs

Nine MAbs produced against the 66-kDa form of HIV-1 RT[19] are purified from ascites fluid. Ascites fluid containing the various MAbs are diluted fivefold with 1.5 *M* glycine, pH 8.9, 3 *M* NaCl and loaded at a flow rate of 1 ml/min onto a previously equilibrated 1 ml protein A-Sepharose (Pharmacia) column. After washing the column extensively with the same buffer, the bound MAb is eluted with 100 m*M* sodium citrate, pH 3.0, into a tube containing 1 *M* Tris–HCl, pH 8.0. Purified MAbs are stored at 4°.

Construction of MAb Affinity Columns

Purified MAbs are concentrated in Centricon-30 microconcentrators (Amicon) at 4° and the buffer is exchanged with a coupling buffer consisting of 0.1 *M* $NaHCO_3$, pH 8.3, 0.5 *M* NaCl, to a final volume of 2 ml. One milliliter of swelled CNBr-activated Sepharose 4B (Pharmacia) is washed with 1 m*M* HCl and placed in a 5-ml capped tube. The beads are collected by centrifugation and the supernatant removed. The concentrated MAb (in our case 0.144 to 0.780 mg, depending on the MAb) is added to the gel and mixed overnight at 4°. Essentially all the ligand present is bound to the beads as judged by measuring absorbance at 280 nm. Unreacted groups on the gel are blocked with 2 ml of 1 *M* ethanolamine, pH 8.0, for 2 hr at room temperature. The gel is then washed three times alternating with 0.1 *M* sodium acetate, pH 4.0, and coupling buffer for periods of 5 min each at room temperature. Finally, the gel is packed into a 1-cm-diameter column (Pharmacia) and stored in 20 m*M* Tris–HCl, pH 8.0, 50 m*M* NaCl at 4°. A control column without ligand is also prepared using the same protocol.

Binding Affinity of RT p66/p51 to the Panel of MAbs

Each MAb-Sepharose column is equilibrated with 10 volumes of 20 m*M* Tris–HCl, pH 8.0, 50 m*M* NaCl at 4° at a flow rate of 2.5 ml/min. After the equilibration step, the flow rate is reduced to 0.25 ml/min and the column is loaded with an excess (0.5 to 0.6 mg) of purified HIV-1 RT p66/p51 heterodimer. The breakthrough is then collected. The column is subsequently washed with equilibrating buffer at 2.5 ml/min and the first column volume is included with the breakthrough fraction. One column volume is sufficient to remove unbound HIV-1 RT completely. The total amount of RT bound is calculated as the difference between the amount of HIV-1 RT protein loaded and the amount of protein in the breakthrough.

[19] A. L. Ferris, A. Hizi, S. D. Showalter, S. Pichuantes, L. Babe, C. S. Craik, and S. H. Hughes, *Virology* **175,** 456 (1990).

The results are normalized to the amount of RT bound per milligram of MAb immobilized in the column. Nonspecific binding of RT to Sepharose is estimated using a control column that did not contain MAb. A plot showing the results of the affinity experiments is shown in Fig. 2. MAb28 binds with the HIV-1 RT p66/p51 heterodimer with the highest affinity. It is interesting to note that two of the MAbs raised against the 66-kDa form do not recognize the p66/p51 form of the enzyme.

Purification of MAb 28

Monoclonal antibody 28 was chosen because it has the highest affinity for the p66/p51 heterodimer and because it does not block the enzymatic activity of RT. It has been produced from both mouse ascites and from serum-free tissue culture medium. Typically, 10 to 50 ml of frozen ascites fluid is thawed and the lipid layer removed by Pasteur pipette and discarded.

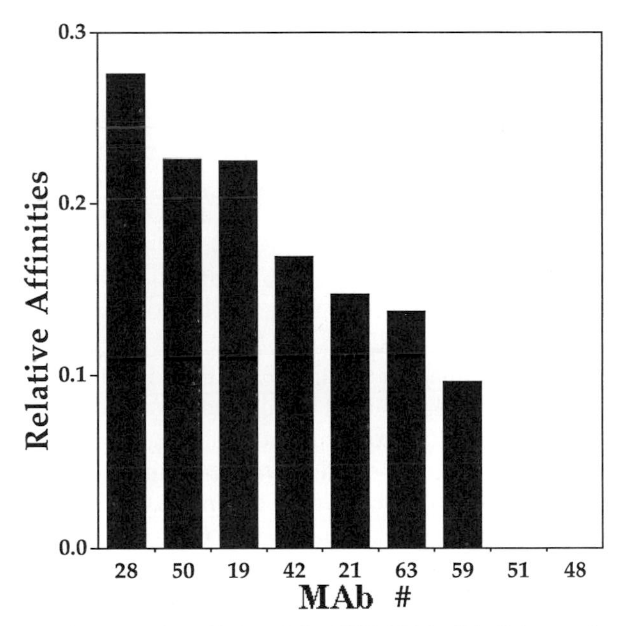

FIG. 2. Graph of the relative affinity of the HIV-1 RT p66/p51 heterodimer for a panel of nine monoclonal antibodies raised against the 66-kDa form of the enzyme. MAb affinity columns were prepared and tested for their ability to bind HIV-1 RT p66/p51 heterodimer as described in the text. The MAbs were ranked accordingly from highest affinity for RT (MAb28) to those that do not recognize this form of the enzyme (MAbs51 and 48). The first five MAbs (28, 50, 19, 42, and 21) were purified and their Fab fragments tested in cocrystallization experiments with the p66/p51 form. Only MAb28 yielded crystals suitable for structure determination.

The fluid is filtered through a 0.45-μm Millex filter unit and then diluted fivefold with 1.5 *M* glycine, pH 8.9, 3 *M* NaCl (Buffer A). Alternatively, 4 to 6 liters of serum-free medium containing MAb28 is concentrated 150-fold using a Millipore Pellicon ultrafiltration unit equipped with a 10,000-MW cutoff polysulfone cassette and then diluted twofold with Buffer A prior to chromatography. The solution containing the antibody is loaded at a flow rate of 5 ml/min onto a previously equilibrated 25-ml protein A-Sepharose (Pharmacia) column. After extensive washing, the bound antibody is eluted with 100 m*M* sodium citrate, pH 3.0, into a tube containing 1 *M* Tris–HCl, pH 8.0. Purified MAb28 is concentrated and diafiltered into 50 m*M* sodium acetate, pH 5.5, using an Amicon stirred cell equipped with a YM30 membrane to a final protein concentration of approximately 5 to 10 mg/ml, then filtered through a 0.22-μm Millex filter unit, aliquoted, and stored at −80°.

Production of Fab28

Frozen aliquots of MAb28 are thawed and diluted to 2 mg/ml with 50 m*M* sodium acetate, pH 5.5, and treated with papain cross-linked to agarose beads (Sigma) on a per weight basis of 0.3 mg insoluble papain per 1 mg antibody. A 1 *M* stock solution of L-cysteine (free base) is added to a final concentration of 50 m*M*, and a 20 m*M* stock solution of EDTA is added to a final concentration of 1 m*M*. To keep the agarose beads in suspension, the digestion mixture is placed in a shaker bath at 34° for 8 hr. The reaction is terminated by filtration through a 0.22-μm Millex filter unit. The 8-hr digestion time is optimized in a manner similar to that used to optimize the conversion of p66 form RT to p66/p51: The extent of digestion is monitored using a combination of SDS–PAGE and analytical Mono S HR 5/5 (Pharmacia) chromatography. The filtered digest is then loaded onto a Mono S HR 10/10 cation-exchange column equilibrated with 50 m*M* sodium acetate, pH 5.5. The unbound fraction containing the Fabs is concentrated and diafiltered into 20 m*M* Tris–HCl, pH 8.0, 25 m*M* NaCl to give a protein concentration of approximately 5 to 6 mg/ml, then aliquoted and frozen at −80°. Alternatively, the filtered digest could be diluted fivefold with 1.5 M glycine, pH 8.9, 3 *M* NaCl and reloaded onto a protein A-Sepharose column; the Fabs are present in the unbound fraction and the Fcs are bound to the column. Four forms of Fab28 could be detected both by isoelectric focusing and analytical Mono Q chromatography. The concentrated, diafiltered Fab28 is subsequently loaded onto a Mono Q HR 10/10 column equilibrated with 20 m*M* Tris–HCl, pH 8.0, and eluted using a linear gradient from 75 to 85 m*M* NaCl in the same buffer. The major isoelectric form is collected and analyzed by SDS–PAGE and analytical

Mono Q Hr 5/5 chromatography. The purified Fab28 solution is concentrated in Centricon-30 microconcentrators (Amicon) concomitantly exchanging the buffer with 10 m*M* Tris–HCl, pH 8.0, 75 m*M* NaCl to give a final protein concentration of 40 to 45 mg/ml. The concentrate is immediately used for mixing with freshly prepared p66/p51 RT concentrate for crystallization purposes (see below).

Oligonucleotides

Design of Oligonucleotides for Crystallization

The majority of the crystallization experiments have been performed with double-stranded DNA, although the protocol described below also has been used to cocrystallize HIV-1 RT with RNA:DNA hybrids and RNA:RNA duplexes (1994, unpublished results). The sequences of the oligomers are based on the primer binding site in the genome of HIV-1 where the complementary 18 3′-terminal nucleotides of human $tRNA^{Lys,III}$ bind the genomic RNA to prime reverse transcription *in vivo.* A variety of oligonucleotides has been successfully cocrystallized with RT. All of these oligonucleotides had duplex regions from 18 to 21 base pairs in length with single-stranded template overhang regions varying in length from 0 to 12 nucleotides. The sequence of the template–primer substrate used in the structure determination at 3.0-Å resolution[8] and further refined to 2.8 Å (Ding *et al.*, unpublished) is:

```
Template strand   5′-A TGG CGC CCG AAC AGG GAC-3′
Primer strand     3′-  ACC GCG GGC TTG TCC CTG-5′
```

All of the DNAs that have been cocrystallized with RT contained the 18-mer sequence of the primer binding site of HIV-1 (the primer strand shown). Extensions at the duplex end have varied in sequence, although GC base pairs have been used preferentially to minimize "breathing" at the end of the DNA duplex. The single-stranded template extensions have also varied in sequence, however, we have not identified specific sequences in this extension that have a direct effect on crystallization.

Purification

The oligonucleotides are synthesized by standard techniques (Applied Biosystems) using either 10- or 15-μm scale synthesis selecting the "trityl-off" option. The oligonucleotides are dissolved in 2.5 ml of distilled water, desalted using a single NAP 25 column (Pharmacia), eluted with 3.5 ml distilled water, and then purified (in multiple injections) using a Mono Q Hr 16/10 column equilibrated with 10 m*M* NaOH, pH 12, 0.5 *M* NaCl. The column is developed with a linear gradient of 0.5 to 0.9 *M* NaCl in 10 m*M* NaOH, pH 12, and the main peak is desalted, aliquoted, lyophilized, and stored at −20°.

Annealing

Approximately 1 mg of each of the complementary DNA strands is mixed in 10 m*M* Tris–HCl, pH 7.5, 100 m*M* NaCl, 0.5 m*M* EDTA to a final concentration of 2 nmol/μl (2 m*M*) and heated at 95° for 30 sec in a temperature-controlled water bath. A piece of aluminum foil is placed on top of the annealing tube to reduce condensation. The water bath is then turned off to allow the oligonucleotides to anneal for the several hours it takes for the bath to fall to room temperature. Typically, this hybridization is done overnight. After the annealing is completed, the tube is centrifuged for a few seconds at 10,000 rpm, vortexed, and spun again. The concentration of DNA is estimated by measuring absorbance at 260 nm. The annealed oligonucleotides are stored at −20°.

Crystallization of HIV-1 RT p66/p51 Heterodimer with Fab, Nucleic Acid Substrates, and Inhibitors

HIV-1 RT Complexed to Fab Fragment

The protocol we have used for the purification of Fab28 yields one of the isoelectric variants. Crystallization experiments performed with the mixture of isoelectric variants of Fab28 were not reproducible and yielded only crystals of inferior quality. Attempts to cocrystallize RT with the other isoelectric forms were unsuccessful.

Freshly purified, unfrozen concentrates of RT and Fab28 are mixed at a mass ratio of 1 RT:0.82 Fab28 and diluted with 10 m*M* Tris–HCl, pH 8.0, 75 m*M* NaCl to give a final protein concentration of 30 to 35 mg/ml. The protein complex is divided into 1.5-mg aliquots and stored at −20°. The appropriate mixing ratio used for complex formation is estimated by native polyacrylamide gel electrophoresis using the Phastsystem (Pharmacia). A similar procedure has been described to evaluate binding of Fabs to rat brain hexokinase.[20] Each protein is prepared at 1 mg/ml and a range of Fab28 is mixed with a constant amount of RT. Typically, five or six sample mixtures representing increasing amounts of Fab (1:0.5 to 1:1.5, in 0.25 increments) are tested on an 8 to 25% gradient gel. After running the samples for 550 volt hours, Fab28 migrates to the bottom of the gel

[20] J. E. Wilson and A. D. Smith, *Analyt. Biochem.* **143,** 179 (1984).

and bands corresponding to RT and RT:Fab complex migrate about halfway down the gel. A slight excess of Fab28 can be visualized at the lower end of the gel using Coomassie stain. The ratios are confirmed in crystallization experiments following the protocol described below. RT:Fab28 ratios below 0.8 and above 0.9 give nonreproducible crystals of poor quality.

Crystallization of the RT:Fab complex is carried out in hanging-drop vapor diffusion experiments[21,22] at 4° over reservoirs containing 0.5 ml of crystallization solution. Aliquots of thawed concentrate are diluted, if necessary, with 10 m*M* Tris–HCl, pH 8.0, 75 m*M* NaCl to 30 mg/ml and then spun for 2 min at 10,000 rpm in an Eppendorf centrifuge at 4° to remove aggregates. Eight microliter hanging drops are prepared by mixing equal volumes of protein concentrate and crystallization solution comprised of 100 m*M* cacodylate, pH 5.6, with varying amounts of saturated ammonium sulfate, typically 31, 32, and 33% by volume. Crystals grow as hexagonal rods reaching a maximal size, typically $0.4 \times 0.6 \times 1.2$ mm in 4 to 7 days.

HIV-1 RT Complexed with Fab Fragment and Double-Stranded DNA Substrates

Cocrystallization of RT:Fab28 complexes and double-stranded nucleic acids is carried out following the protocol described in the previous section, but including the addition of nucleic acid at a 1.5:1 molar ratio of DNA to RT. After the addition of the proper amount of annealed DNA (2 nmol/μl stock solution) to the RT:Fab mixture, the complex is diluted, if necessary, to a final protein concentration of 25 to 30 mg/ml. The sample is placed on ice for 15 min prior to centrifugation and drop preparation. Isomorphous crystals that included several different template–primers have been grown successfully in our laboratory by this method (Fig. 3a, color plate).

Crystallization of HIV-1 RT Mutants

The parent HIV-1 reverse transcriptase used in the preparation of all the crystal forms described in this chapter is a mutant RT that has serine substituted for cysteine at amino acid position 280. It is fully active in polymerization and RNase H activities and is resistant to oxidative inactivation of RNase H.[23] Crystals of RT that contain one or two amino acid substitutions have been prepared in complexes with Fab28 and dsDNA.

[21] A. McPherson, "The Preparation and Analysis of Protein Crystals." Wiley, New York, 1982.
[22] T. L. Blundell and L. N. Johnson, "Protein Crystallography," Academic Press, New York, 1976.
[23] A. Hizi, M. Shaharabany, R. Tal, and S. H. Hughes, *J. Biol. Chem.* **267,** 1293 (1992).

For example, a mutant RT that has threonine substituted for tyrosine at amino acid position 215 gives rise to AZT resistance in HIV-1[24] and this mutant has been crystallized using the protocol described for HIV-1 RT:Fab28:DNA crystallization. However, the number of nuclei formed is quite sensitive to small changes in ammonium sulfate concentration. Consequently, it is difficult to control the size and quantity of the crystals. The concentration of saturated ammonium sulfate in the crystallization solution must be carefully adjusted to 25 to 27%. Crystals of tetragonal bipyramids, 0.2 mm in thickness and width and 0.4 mm in length, grow in approximately 14 days (Fig. 3b). Substitutions of isoleucine and leucine for the tyrosines normally present at positions 181 and 188 render HIV-1 resistant to a variety of nonnucleoside inhibitors.[25,26] Crystals that contain this mutant HIV-1 RT complexed with Fab28 and dsDNA are prepared using conditions similar to those used to crystallize the tyrosine-215 to threonine mutant. The crystals grow as short, hexagonal blocks to an average size of $0.10 \times 0.15 \times 0.15$ mm in approximately 14 days (Fig. 3c).

HIV-1 RT p66/p51 Complexed with Nonnucleoside Inhibitors

The crystallization conditions described here are based on those described by Kohlstaedt *et al.*[6] However, important changes that are crucial for optimal crystal growth are included in the description. The following mixing procedure, which is designed to enhance the solubility and availability of the inhibitor upon mixing with the concentrated RT solution, has been successfully used in our laboratory to grow diffraction-quality crystals with virtually every nonnucleoside inhibitor of HIV-1 RT we have tested (22 total). The following conditions have been used with a nonnucleoside inhibitor for which the high-resolution structure of the complex with HIV-1 RT has been solved.[12] A 20 m*M* stock solution of the nonnucleoside inhibitor α-anilinophenylacetamide (α-APA, derivative R 90385; Janssen)[27] is prepared by dissolving it in 100% dimethyl sulfoxide (DMSO). In addition, a 20% (w/v) stock solution of *n*-octyl-β-D-glucopyranoside (β-OG, Sigma) is prepared in distilled water. A working solution comprised of 15 m*M* α-APA, 5% (β-OG) is prepared by adding the appropriate volume of 20%

[24] B. A. Larder and S. D. Kemp, *Science* **246,** 1155 (1989).

[25] V. V. Sardana, E. A. Emini, L. Gotlib, D. J. Graham, D. W. Lineberger, W. J. Long, A. J. Schlabach, J. A. Wolfgang, and J. H. Condra, *J. Biol. Chem.* **267,** 17,526 (1992).

[26] K. DeVreese, Z. Debyser, A.-M. Vandamme, R. Pauwels, J. Desmyter, E. DeClercq, and A. Jozef, *Virology* **188,** 900 (1992).

[27] R. Pauwels, K. Andries, Z. Debyser, P. Van Dable, D. Schols, P. Stoffels, K. De Vreese, R. Woestenborghs, A.-M. Vandamme, C. G. M. Janssen, J. Anne, G. Cauwenbergh, J. Desmyter, J. Heykants, M. A. C. Janssen, E. De Clercq, and P. A. J. Janssen, *Proc. Natl. Acad. Sci. USA* **90,** 1711 (1993).

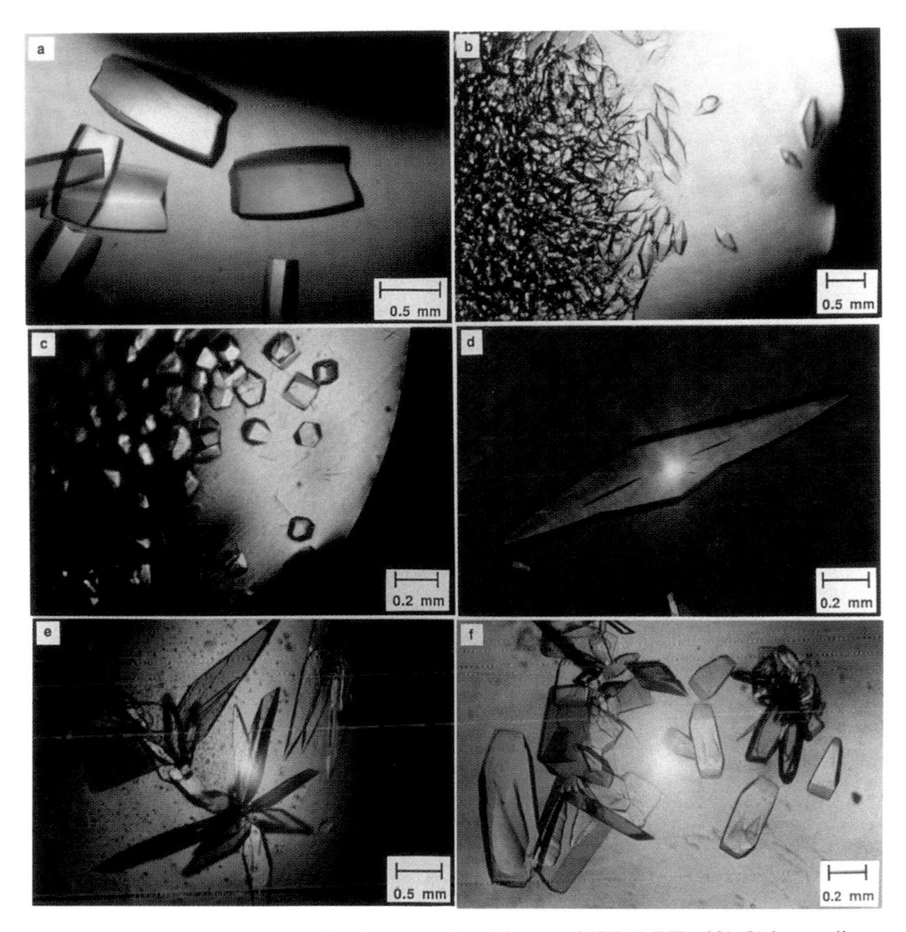

FIG. 3. Crystals of complexed and uncomplexed forms of HIV-1 RT p66/p51 heterodimer grown by hanging-drop vapor diffusion. Magnification in the photomicrographs varies. (a) Self-nucleating crystals of a ternary complex of RT, a monoclonal antibody Fab fragment, and dsDNA. Crystals grow as hexagonal rods and measure 0.4 × 0.6 × 1.2 mm. (b) Tetragonal bipyramids of an AZT-resistant mutant of RT complexed with an Fab fragment and dsDNA. In this instance, nucleation is quite sensitive to small changes in ammonium sulfate concentration and the tendency is to form many small crystals spontaneously. Average dimensions are 0.2 × 0.2 × 0.4 mm. (c) Crystals of a nonnucleoside inhibitor-resistant mutant of RT complexed with an Fab fragment and dsDNA. The crystals grow as short, hexagonal blocks to an average size of 0.1 × 0.15 × 0.15 mm in approximately 14 days. (d) A single, self-nucleating crystal of RT complexed with the nonnucleoside inhibitor α-APA, derivative R 90385 (Janssen). Dimensions are 0.2 × 0.3 × 1.5 mm. (e) Crystals of RT complexed with TIBO, derivative R 86183 (Janssen), a nonnucleoside inhibitor. The crystals, which tend to grow in clusters from a common nucleation site, are grown by streak-seeding 3-day-old equilibrated drops. Average dimensions are 0.1 × 0.4 × 1.3 mm. (f) Crystals of uncomplexed RT, originally grown by cross-seeding with microcrystals of RT: α-APA complex. Average dimensions are 0.1 × 0.2 × 0.4 mm and can be grown at 22 or 4° under identical crystallization conditions.

(w/v) β-OG stock solution to the 20 mM stock solution of the inhibitor and mixing thoroughly. Freshly prepared or thawed aliquot(s) of RT concentrate (each 2.5-mg aliquot of RT is sufficient for preparing one Linbro tray containing twelve 10-μl hanging drops) are diluted, if necessary, with 10 mM Tris–HCl, pH 8.0, 75 mM NaCl to 40 mg/ml and then quickly mixed by vortexing with the 15 mM α-APA, 5% β-OG working solution at a 2:1 molar ratio of inhibitor to RT and kept on ice for 15 min. Crystallization is performed using the hanging-drop vapor diffusion method at 4° using a crystallization solution composed of 50 mM Bis–Tris propane, pH 6.8, 100 mM $(NH_4)_2SO_4$, 10% glycerol, 12% (w/v) polyethylene glycol (PEG) 8000. The hanging drops are prepared by adding equal volumes (5 μl) of the crystallization solution with the concentrated protein solution and mixing thoroughly to give a final protein concentration of 19 to 20 mg/ml, 1.6% DMSO, and 0.1% β-OG. The initial precipitation observed in the drops upon mixing the RT with the crystallization solution dissipates after 6 to 10 hr at 4°. Crystals begin to appear 24 hr after drop preparation and attain an average size of 0.2 × 0.3 × 1.5 mm after approximately 7 days at 4° (Fig. 3d).

Microseeding techniques[28] can be beneficial in obtaining crystals of RT:inhibitor complexes in which self-nucleation occurs only occasionally under the aforementioned crystallization conditions. The 10-μl hanging drops are prepared as described above and equilibrated with the crystallization solution for 3 to 4 days at 4°. A proteinaceous "skin" surrounding the drop is broken and partially removed with a syringe needle. The drop is then streak-seeded with a cat whisker or 30-gauge syringe needle dipped into 20 μl of crystallization solution containing a few freshly crushed crystals. Crystals of RT complexed with a number of TIBO derivatives[29] were grown using this technique (Fig. 3e).

HIV-1 RT p66/p51 Alone

Crystals of uncomplexed RT have been grown at room temperature (22°) and at 4° following the procedure given for RT:inhibitor crystallization. In this case, the inhibitor is omitted and a 5% β-OG working solution is prepared by diluting a small volume of the 20% β-OG stock solution with 100% DMSO. The hanging drops require a 3- to 4-day equilibration period followed by streak-seeding in order for crystals to grow. Initially, the drops are seeded using freshly and finely crushed RT-inhibitor crystals.

[28] E. A. Stura and I. A. Wilson, *Methods* **1,** 38 (1990).

[29] R. Pauwels, K. Andries, J. Desmyter, D. Schols, M. J. Kukla, H. J. Breslin, A. Raeymaeckers, J. Van Gelder, R. Woestenborghs, J. Heykants, K. Schellekens, M. A. C. Janssen, E. De Clercq, and P. A. J. Janssen, *Nature* **343,** 470 (1990).

(After crystals of uncomplexed RT were obtained, they were used for seeding.) The inclusion of 25 m*M* magnesium acetate in the hanging drops appears to enhance crystal size while minimizing twinning. Crystals grow to an average size of 0.1 × 0.2 × 0.4 mm after 12 to 14 days (Fig. 3f). These crystals also diffracted X rays to high resolution, although they had different unit cell dimensions than crystals grown under similar conditions in the presence of inhibitor.

Potential for Cocrystallization of HIV-1 RT and Accessory Factors

Genomic replication frequently involves accessory factors in addition to polymerases.[30] Experimental evidence suggested that such factors may be required in order to complete proviral DNA synthesis *in vivo*.[31] An understanding of the possible roles of accessory factors involved in retroviral DNA synthesis is, at present, unclear. Accessory factors have been implicated in increased RT fidelity during plus strand synthesis in virus-infected cells.[32] It has been suggested that the HIV-1 nucleocapsid protein functions as a replication factor by means of increasing processivity of RT.[33] Bukrinsky *et al.*[34] have reported the association of HIV-1 integrase, matrix, and RT as protein components of a nucleoprotein (preintegration) complex in CD4$^+$ cells after acute virus infection. We describe several protocols to obtain high-quality crystals of HIV-1 RT p66/p51 heterodimer by itself, in complexes with inhibitors, with a Fab fragment, and with a Fab fragment and relevant double-stranded DNA substrates. Our laboratory has solved several of these complexes[8,12,35] and is in the process of solving others. It seems realistic to foresee the possibility of cocrystallizing this enzyme with accessory factors as new insights into their functional roles become available.

Crystallization of HIV-1 RT in Complexes Relevant for DNA Replication

Based on our experience, we are optimistic that it will be possible to obtain cocrystals of HIV-1 RT with DNAs that could mimic numerous

[30] A. Kornberg and T. A. Baker, "DNA Replication," W. H. Freeman, New York, 1991.

[31] G. J. Klarmann, C. A. Schauber, and B. D. Preston, *J. Biol. Chem.* **268,** 9793 (1993).

[32] A. Varela-Echavarria, N. Garvey, B. D. Preston, and J. P. Dougherty, *J. Biol. Chem.* **267,** 24,681 (1992).

[33] X. Ji, G. J. Klarmann, and B. D. Preston, *in* "Mtg. *Retroviruses*," Cold Spring Harbor, N.Y. (1994).

[34] M. I. Bukrinsky, N. Sharova, T. L. McDonald, T. Pushkarskaya, W. G. Tarpley, and M. Stevenson, *Proc. Natl. Acad. Sci. USA* **90,** 6125 (1993).

[35] J. Ding, K. Das, H. Moereels, L. Koymans, K. Andries, P. A. J. Janssen, S. H. Hughes, and E. Arnold, *Nat. Struct. Biol.* **2,** 407 (1995).

crucial steps during DNA replication. In particular, the RT : Fab28 complex has proven to be a very versatile system in which we have obtained high-quality crystals with a large array of DNA substrates. One explanation for this versatility could be the fact that the DNA does not appear to participate directly in crystal contacts, but is held largely by contacts with the p66 fingers, palm, thumb subdomains, together with the RNase H domain. In addition, the HIV-1 RT p66 thumb subdomain, which has conformational mobility that may be used in polymerization, does not participate in crystal contacts in the RT : Fab28 : DNA crystals. We have been successful in growing crystals of RT : Fab28 : DNA : dNTP that may ultimately yield the conformation of the enzyme just prior to dNTP incorporation (1994, unpublished results). Ultimately we are hopeful that it might be possible to visualize the time-resolved mechanism of DNA polymerization by crystallization of suitable reaction intermediates.

Acknowledgments

We are very grateful to a large number of colleagues who have contributed to development of the crystallization and purification procedures including Gail Ferstandig Arnold, Paul Boyer, Andrea Ferris, Amnon Hizi, Xiaode Lu, Raymond G. Nanni, Mukund Paidhungat, Premal Patel, Birgit M. Roy, and Chris Tantillo. We thank the other members of our laboratory for helping in the structure determination of HIV-1 RT and also NIH, CABM, Johnson & Johnson, and Janssen Research Foundation for financial support of this work. Research sponsored in part by the National Cancer Institute, DHHS, under contract NO1-CO-74101 with ABL, and the National Institute of General Medical Sciences (SHH).

Section III

Mechanisms of DNA Polymerases

[16] Phosphate Analogs for Study of DNA Polymerases

By F. ECKSTEIN and J. B. THOMSON

Introduction

Analogs of the 2′-deoxynucleoside 5′-triphosphate substrates have greatly helped in elucidating certain details of the mechanism of the reactions catalyzed by DNA polymerases. Obviously many analogs can be envisaged as the three components of the nucleotide, the base, the sugar, and the phosphate lend themselves to modification. Although many of these have been studied and have yielded interesting results, this review is restricted to substrate analogs containing modifications in the triphosphate moiety. Particular emphasis is given to the phosphorothioates, which have yielded much of the information regarding the mechanism of the reactions and at the same time have also provided interesting reaction products. This class of analogs has been reviewed in several articles.[1–5] Certain aspects, in relation to DNA polymerases are also discussed in [20] in this volume.

Nucleoside Phosphorothioates

Four properties of the 2′-deoxynucleoside 5′-(1-*O*-thiotriphosphates), commonly referred to as dNTPαS, as well as the 5′-(2-*O*-thiotriphosphates), commonly referred to as dNTPβS, have made this class of compounds particularly useful for the study of polymerases: (1) The sulfur-bearing phosphorus is chiral, which results in the existence of two diastereomers (Fig. 1). (2) Only one of the dNTPαS diastereomers is a good substrate for polymerases. (3) Sulfur and oxygen have different affinities toward metal ions, dependent on the softness of the metal ion.[1,6] (4) The product of polymerization reactions employing dNTPαS shows increased stability against the 5′ → 3′- and/or the 3′ → 5′-exonuclease activity of some polymerases.

[1] F. Eckstein, *Ann. Rev. Biochem.* **54,** 367 (1985).
[2] F. Eckstein and G. Gish, *TIBS* **14,** 97 (1989).
[3] P. A. Frey, *in* "Advances in Enzymology and Related Areas of Molecular Biology" (A. Meister, ed.), Vol. 62, p. 119. John Wiley & Sons, New York, 1989.
[4] J. A. Gerlt, *in* "The Enzymes" (D. S. Sigman, ed.), Vol. XX, p. 95. Academic Press, New York, 1992.
[5] P. A. Frey, *in* "The Enzymes" (D. S. Sigman, ed.), Vol. XX, p. 141. Academic Press, New York, 1992.
[6] V. L. Pecoraro, J. D. Hermes, and W. W. Cleland, *Biochemistry* **23,** 5262 (1984).

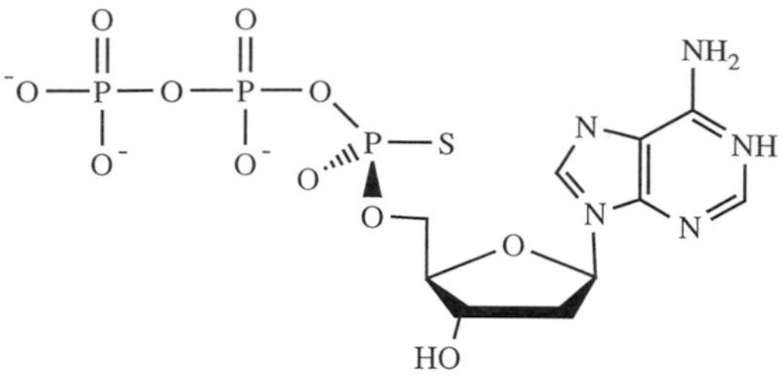

$(S_p)dATP\alpha S$

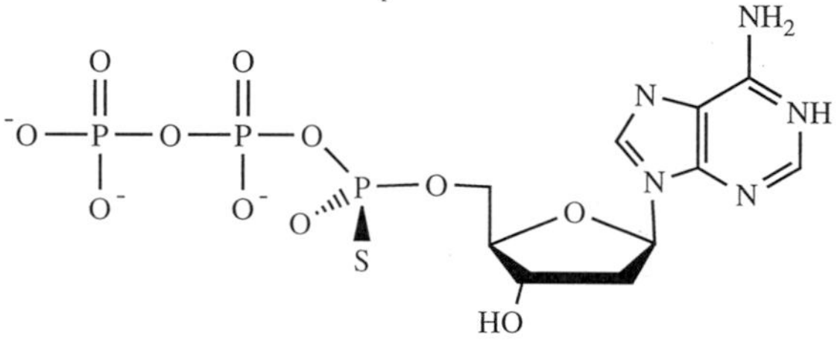

$(R_p)dATP\alpha S$

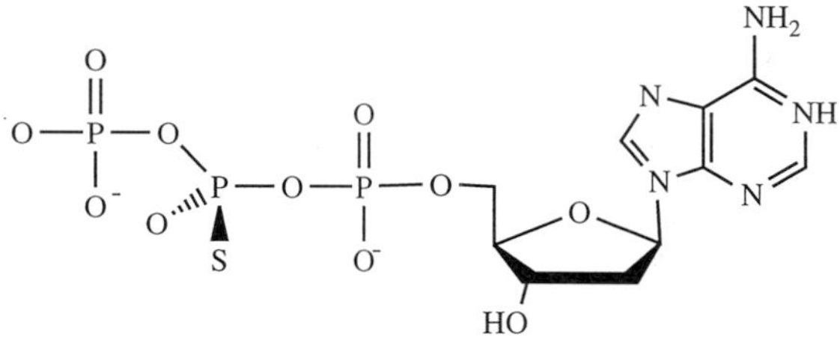

$(S_p)dATP\beta S$

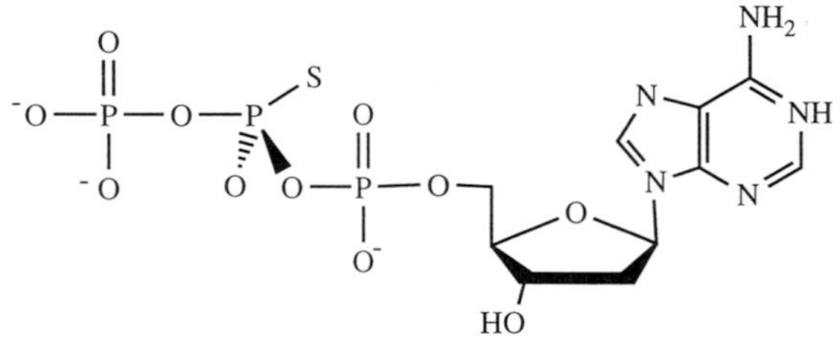

$(R_p)dATP\beta S$

FIG. 1. Structure of the diastereomers of 2′-deoxyadenosine 5′-*O*-[1-thiotriphosphate] (dATPαS) and 2′-deoxyadenosine 5′-*O*-[2-thiotriphosphate] (dATPβS).

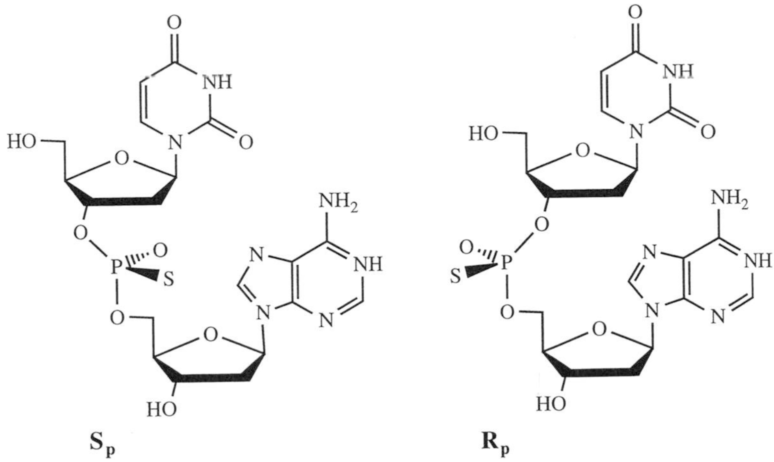

FIG. 2. Structure of the diastereomers of the phosphorothioate internucleotidic linkage.

The use of the *Sp* and *Rp* diastereomers of dNTPαS has enabled the stereochemical course of the polymerization reaction to be elucidated. The results show that all DNA polymerases and all RNA polymerases investigated so far exclusively use the *Sp* isomer as the substrate. In addition, the reactions catalyzed by all these polymerases produce an internucleotidic linkage with the *Rp* configuration (Fig. 2), and thus the reactions result in an inversion of configuration at phosphorus. The simplest interpretation of this result is that the reaction proceeds with a single inversion of configuration, although, of course, any uneven number of such reaction steps would also result in overall inversion (Fig. 3).[1,7]

The metal ion dependence of the rate of polymerization with the diastereomers of dNTPαS and dNTPβS allows a determination of the configuration of the metal–nucleoside triphosphate complex. Results obtained so far establish coordination of the metal ion to the β,γ phosphates resulting in a complex with the Δ configuration (Fig. 4).[1]

The incorporation of a phosphorothioate internucleotidic linkage of the *Rp* configuration at the 3′ end of an oligomer conveys considerable stability toward degradation by the 3′ → 5′-exonuclease of Pol I.[8] This difference in the rate of degradation between phosphate and phosphorothioate linkages has been determined to be a factor of 150.[9] Interestingly, the 3′ →

[7] J. R. Knowles, *Ann. Rev. Biochem.* **49,** 877 (1980).

[8] T. A. Kunkel, F. Eckstein, A. S. Mildvan, R. M. Koplitz, and L. A. Loeb, *Proc. Natl. Acad. Sci. USA* **78,** 6734 (1981).

[9] A. P. Gupta, P. A. Benkovic, and S. J. Benkovic, *Nucleic Acids Res.* **12,** 5897 (1984).

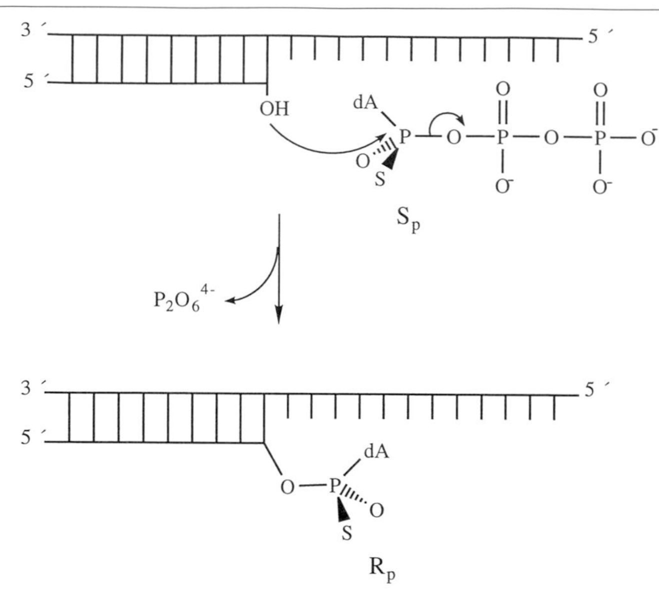

FIG. 3. Mechanism of attack of the 3′-hydroxyl group of the primer on the α-phosphorus of the incoming *Sp* dATPαS, with concomitant expulsion of inorganic phosphate from the opposite face, resulting in the *Rp* configuration of the phosphorothioate in the product.

5′-exonuclease of T4 DNA polymerase is not inhibited by a phosphorothioate.[9] The basis of this reduction in rate of hydrolysis by the 3′ → 5′-exonuclease of Pol I is discussed in another chapter of this volume, [20]. Resistance of the phosphorothioates against degradation by the 3′ → 5′-exonuclease of the *pfu* and *vent* DNA polymerases has also been described.[10] The stereochemical course of the T4 3′ → 5′-exonuclease reaction was determined by using *Sp*-2′-deoxyadenosine 5′-*O*-(1-thio[1-$^{18}O_2$]triphosphate) in an idling reaction and also proceeds with inversion of configuration.[11] The stereochemical course of the 3′ → 5′-exonuclease of Pol I was determined by using *Sp*[^{16}O,^{17}O,^{18}O]-2′dAMP, which in turn was derived from dNTPαS.[12] Again, inversion of the configuration at phosphorus was observed.

The susceptibility of the phosphorothioate internucleotidic linkage to the 5′ → 3′-exonuclease activity of DNA polymerases has not been systematically studied. However, circumstantial evidence derived from the efficiency of mutagenesis experiments with mutagenic primers elongated with

[10] A. Skerra, *Nucleic Acids Res.* **20,** 3551 (1992).
[11] A. P. Gupta, C. DeBrosse, and S. J. Benkovic, *J. Biol. Chem.* **257,** 7689 (1982).
[12] A. P. Gupta and S. J. Benkovic, *Biochemistry* **23,** 5874 (1984).

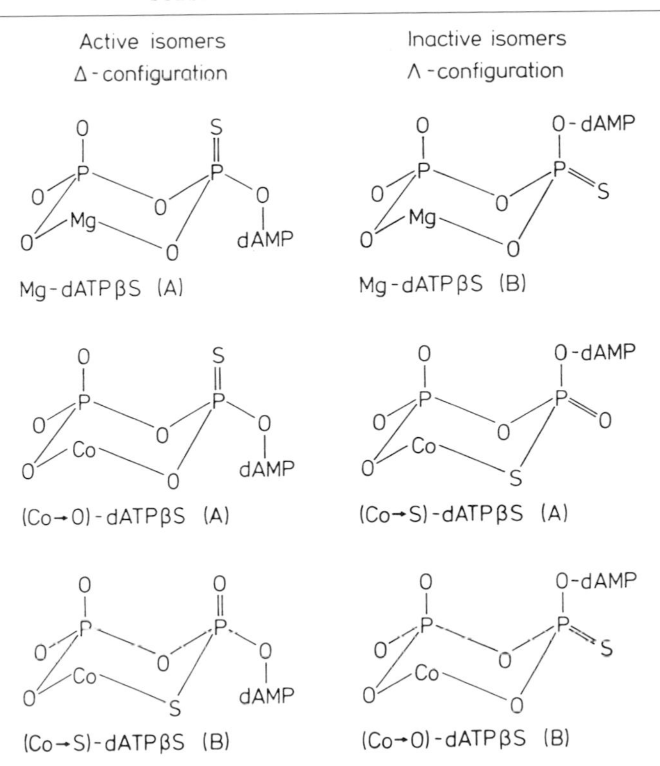

FIG. 4. Structure of metal complexes with dATPβS. From P. M. J. Burgers and F. Eckstein, *J. Biol. Chem.* **254,** 6889–6893 (1979). Copyright The American Society for Biochemistry & Molecular Biology.

either Pol I or the Klenow fragment indicates that two and three phosphorothioates at the 5′ end of the oligomer reduce degradation of the primer.[13] These experiments were performed with chemically synthesized primers, which contain a mixture of diastereomers.

All of these examples illustrate the usefulness of the phosphorothioate analogs for mechanistic studies of DNA polymerases.

Determination of Stereochemical Course of DNA Polymerase-Catalyzed Reactions

The principle of the method is the polymerization of one of the diastereomers of a dNTPαS, in all cases studied so far the *Sp* isomer, and the

[13] J. Ott and F. Eckstein, *Biochemistry* **26,** 8237 (1987).

determination of the configuration of the phosphorothioate internucleotidic linkage. This can be done in several ways. Either one follows the rate of degradation of the polymer by snake venom phosphodiesterase, which is known to degrade the *Rp* isomer much faster than the *Sp* isomer,[14] or one determines the configuration by ^{31}P NMR spectroscopy of either the polymer or, after enzymatic degradation of the polymer, the dinucleotide. A third possibility is to analyze the dinucleotide by reversed-phase HPLC.

The methods described here have also been successfully utilized for the determination of reactions catalyzed by other DNA polymerases such as DNA polymerase I from *Micrococcus luteus*[15] and T7 DNA polymerase.[16]

Materials and Methods

Reagents

dTTP, dATP, poly[d(A-T)], poly(dT), poly(A), and poly(A) · $d(pT)_{10}$, *Escherichia coli* DNA polymerase I (5000 units/mg, 0.69 mg/ml), T4 DNA polymerase (5000 units/mg; 1000 units/ml), avian myeloblastosis virus (AMV) reverse transcriptase (20,000 units/ml), bacterial alkaline phosphatase (1 mg/ml, 400 units/mg) and snake venom phosphodiesterase (*Crotalus terrificus terrenificus;* 1 mg/ml, 1.5 units/mg) from Boehringer Mannheim; the *Sp* isomers of dATPαS, [^{35}S]dATPαS, and [^{35}S]dTTPαS from Amersham International, UK. [^{14}C]poly(dA), [^{14}C]poly(A), and [^{35}S]poly(sA) were prepared as described.[17–19] The *Rp* isomer of dATPαS and the *Rp* and *Sp* isomers of dATPβS were prepared as described.[20]

Procedure

Stereochemical Course of E. coli DNA Polymerase I Reaction[14]

1. *Synthesis of [^{35}S]poly[d(sA)] with DNA polymerase I.* The reaction mixture contains, in a total volume of 5 ml, 125 m*M* HEPES, pH 7.5, 2 m*M* 2-mercaptoethanol, 0.2 m*M* poly(dT), and 4 μM $(dA)_8$, both expressed as mononucleotide concentration, 5 m*M* Mg^{2+}, 0.2 m*M* *Sp*-[^{35}S]dATPαS,

[14] P. M. J. Burgers and F. Eckstein, *J. Biol. Chem.* **254,** 6889 (1979).

[15] F. Eckstein and T. M. Jovin, *Biochemistry* **22,** 4546 (1983).

[16] R. S. Brody, S. Adler, P. Modrich, W. J. Stec, Z. J. Leznikowski, and P. A. Frey, *Biochemistry* **21,** 2570 (1982).

[17] P. M. J. Burgers and F. Eckstein, *Biochemistry* **18,** 450 (1979).

[18] F. Eckstein and H. Gindl, *Eur. J. Biochem.* **13,** 558 (1970).

[19] K. Kato, J. M. Goncales, G. E. Houts, and F. J. Bollum, *J. Biol. Chem.* **242,** 2780 (1967).

[20] F. Eckstein and R. S. Goody, *Biochemistry* **15,** 1685 (1976).

and 425 units of DNA polymerase I. The reaction is incubated at 37°. At time intervals, 20-μl aliquots are applied to strips (15 × 2 cm) of Whatman 3 MM paper, precharged at the origin with 10 μl of an 0.1 *M* EDTA solution. The strips are developed (descending chromatography) in 1 *M* ammonium acetate, pH 5.5, ethanol (1 : 1, v/v) until the solvent front has reached the edge of the strip (approximately 45 min) and dried. That part of the strip extending 1 cm to either side of the point of application is cut out and counted in a toluene-based scintillation fluid. After 3 hr, 80% of the ^{35}S label is incorporated and the reaction has reached a plateau. The reaction is then worked up by chromatography over a Sephadex G-50 column as described above. Yield 5.2 A_{260} units.

2. *Degradation of [^{35}S]poly[d(sA)] with snake venom phosphodiesterase:* The reaction mixture contains, in a total volume of 1 ml, 100 m*M* of Tris–HCl, pH 8.9, 5 m*M* of Mg^{2+}, 4 A_{260} units of polynucleotide, and 100 μg of snake venom phosphodiesterase. The reaction is incubated at 37°. Then 10- to 50-μl aliquots are taken and applied to Whatman 3MM disks. These are stirred in 5% trichloroacetic acid (TCA) as described.[21] The rate of degradation is shown in Fig. 5. The nine-fold slower rate of degradation of [^{35}S]poly[d(sA)] in comparison to that of [^{14}C]polyd(A) indicates the presence of the phosphorothioate linkage of the *Rp* configuration in the former.

Stereochemical Course of T4 DNA Polymerase Reaction[22]

1. *Synthesis of poly[d(T-sA)] with T4 DNA polymerase:* The reaction mixture contains, in a total volume of 20 ml, 100 m*M* Tris–HCl, pH 8.8, 2 m*M* 2-mercaptoethanol, 50 μ*M* poly[d(T-A)], expressed as mononucleotide concentration, 2 m*M* Mg^{2+}, 0.5 m*M* each of dTTP and *Sp*-dATPαS, and 200 units of T4 DNA polymerase. The reaction is incubated at 37°. After 18 hr, approximately 30% of the triphosphates remains, and another 100 units of enzyme is added. After a further 6 hr, the reaction is diluted with 1 ml of a 4% sodium dodecyl sulfate (SDS) solution and chilled to 0°. Protein is removed by extraction with chloroform/isoamyl alcohol (5 : 2, v/v; 2 × 2 ml). The aqueous phase is concentrated to 1 ml and then chromatographed on a Sephadex G-50 column (3 × 40 cm). The polymer product is eluted with water in the void volume. Yield 75 A_{260} units.

2. *Degradation of poly[d(T-sA)] to d[Tp(S)A)] by DNA polymerase I:* The reaction mixture contains, in a total volume of 3 ml, 100 m*M* Tris–

[21] F. J. Bollum, *in* "Proceedings of Nucleic Acids Research" (G. L. Cantoni and D. R. Davies, ed.), Vol. 1, p. 296. Harper & Row, New York, 1966.

[22] P. N. J. Romaniuk and F. Eckstein, *J. Biol. Chem.* **257,** 7684 (1982).

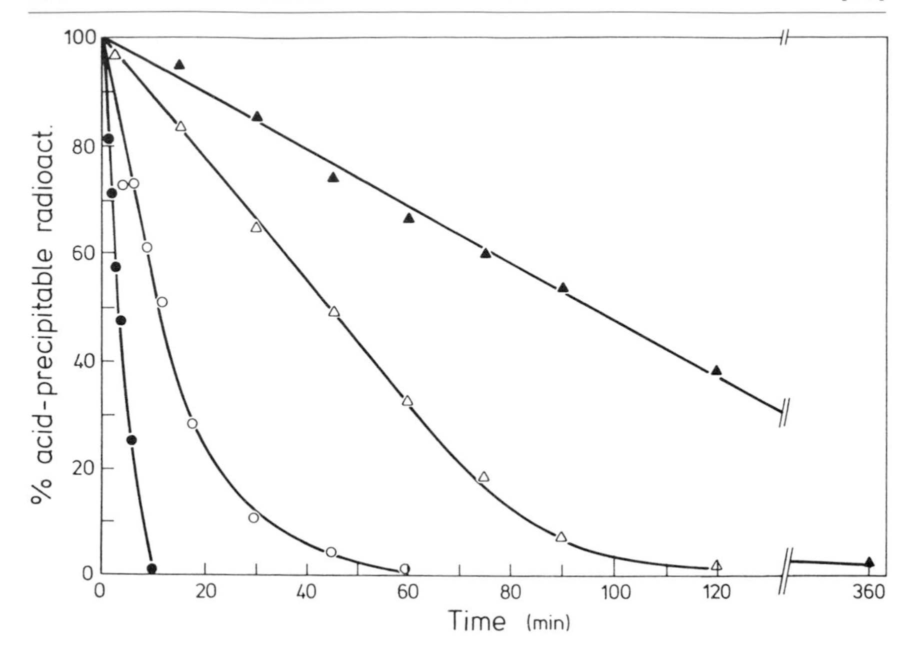

FIG. 5. Degradation of polymers by snake venom phosphodiesterase. O, [^{14}C]poly(dA); ▲, [^{35}S]poly[d(sA)]; ●, [^{14}C]poly(A); △, [^{35}S]poly(sA). From P. M. J. Burgers and F. Eckstein, *J. Biol. Chem.* **254,** 6889–6893 (1979). Copyright The American Society for Biochemistry & Molecular Biology.

HCl, pH 7.4, 10 m*M* Mg^{2+}, 1 m*M* 2-mercaptoethanol, 25 A_{260} units of poly-[d(T-sA)], 45 units of *E. coli* DNA polymerase I, and 4 units of bacterial alkaline phosphatase. After a 16-hr incubation at 37°, the reaction mixture is purified by chromatography on DEAE-Sephadex A25 (1.5 × 20 cm) eluted with a linear gradient of 0.25 liter each of 0.02 and 0.2 *M* triethylammonium bicarbonate, pH 7.5. Fractions of 20 ml are collected. Fractions 25–41 contain a UV-absorbing peak, which elutes at the same place in the gradient as authentic d[Tp(S)A)]. They are pooled and the triethylammonium bicarbonate is removed by rotary evaporation and the residue dissolved in 1 ml of water. Yield 20 A_{260} units.

The product is analyzed by reversed-phase HPLC with a buffer of 7.5% acetonitrile in 0.1 *M* triethylammonium acetate, pH 7.0, at a flow rate of 2 ml/min. Retention time of the chemically synthesized *Rp*-d[Tp(S)A)] is 10.6 min and of *Sp* d[Tp(S)A)], 16.2 min. The product isolated from the digestion of the polymer has a retention time identical to that of *Rp* isomer.

Stereochemical Course of AMV Reverse Transcriptase Reaction[23]

1. *Synthesis of poly[d(sT)] with AMV reverse transcriptase:* The reaction mixture contains, in a total volume of 5 ml, 50 m*M* Tris–HCl, pH 8.3, 40 m*M* KCl, 2 m*M* dithiothreitol (DTT), 100 μg/ml bovine serum albumin (BSA), 400 μM poly(A) · $d(pT)_{10}$, 1400 μM poly(A), all expressed as mononucleotide concentration, 5 m*M* Mg^{2+}, 1 m*M* *Sp*-[^{35}S]dTTPαS, and 1600 units of AMV reverse transcriptase. The reaction is incubated at 37° and followed by paper chromatography as described above. After 2.75 hr, an additional 1400 μM poly(A) was added. After a total of 4 hr, essentially all the dTTPαS has been polymerized. To the reaction mixture is added 0.4 ml of 100 m*M* EDTA and 1.0 ml of 1 *N* NaOH, and the solution is kept at 70° for 20 min. After cooling, the mixture is brought to pH 8 with 0.6 ml of 1 *M* acetic acid/sodium acetate, pH 4.5. The reaction is then worked up by chromatography over a Sephadex G-50 column as described above. Yield 12 A_{260} units. The ^{31}P NMR spectrum of this material showed a chemical shift of 55.87 ppm, which is identical to that of the *Rp* isomer of d[Tp(S)T)]. The chemical shift of the *Sp* isomer is 55.56 ppm (Fig. 6). This result indicates inversion of configuration at phosphorus in the polymerization reaction.

2. *Degradation of [^{35}S]poly[d(sT)] to [^{35}S]d[Tp(S)T)] by snake venom phosphodiesterase:* The reaction mixture contains, in a total volume of 360 μl, 100 m*M* Tris–HCl, pH 8.9, 5 m*M* Mg^{2+}, 350 μM of poly[d(sT)], and 0.15 μg/μl of snake venom phosphodiesterase. The reaction is incubated at 37° and is followed by applying aliquots to paper strips for chromatography, as described above. The rate of degradation is identical with that of poly[d(sT)] prepared by polymerization with *E. coli* DNA polymerase I, which is known to have the phosphorothioate linkage in the *Rp* configuration as also described above. Thus, the polymer synthesized by T4 DNA polymerase has the same configuration of the phosphorothioate as that prepared with Pol I. This confirms the result obtained by ^{31}P NMR spectroscopy that the phosphorothioate has the *Rp* configuration and thus inversion of configuration at phosphorus as a result of the reaction.

Determination of Configuration of Nucleoside Triphosphate–Metal Complex

The determination of the configuration of the nucleoside 5′-triphosphate substrate–metal complex rests on the selective coordination of Mg^{2+} to the

[23] P. A. Bartlett and F. Eckstein, *J. Biol. Chem.* **257,** 8879 (1982).

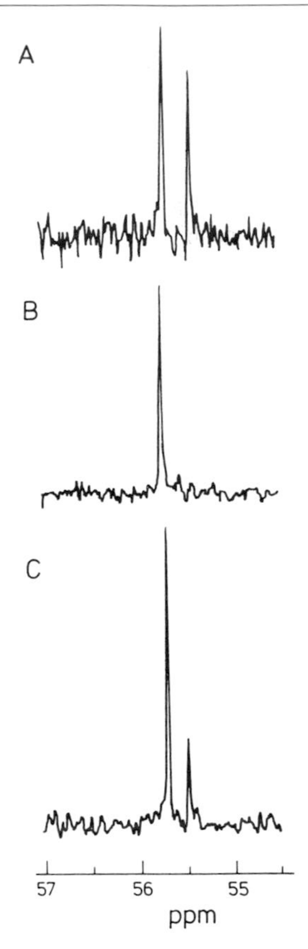

FIG. 6. ^{31}P NMR spectra of the diastereomers of d[Tp(S)T)] at 81.01 MHz. (A) Mixture of diastereomers, chemically synthesized. (B) poly[d(sT)], prepared with AMV reverse transcriptase. (C) Same as B after addition of *Sp*-d[Tp(S)T)]. From P. A. Bartlett and F. Eckstein, *J. Biol. Chem.* **257,** 8879–8884 (1982). Copyright The American Society for Biochemistry & Molecular Biology.

oxygen of the nucleoside 5′-*O*-(1-thiotriphosphates) and -(2-thiotriphosphates) and that of metal ions such as Mn^{2+}, Zn^{2+}, and Co^{2+} to both oxygen and sulfur.[1,6,24] The rational is that if only one diastereomer is a substrate in the presence of a hard ion such as Mg^{2+}, the other diastereomer should become a substrate in the presence of a soft metal ion such as Cd^{2+} or Ag^{+}, which coordinate predominantly to sulfur. However, because most enzymes do not tolerate these ions, metal ions of intermediate softness such as Mn^{2+} often have to be employed. The expectation is that both diastereomers

[24] M. Cohen, *Acc. Chem. Res.* **15,** 326 (1982).

should be active if the metal ion is coordinated to that particular phosphate. A detailed discussion of the caution that has to be exercised in this type of analysis is provided[6] and an example of such an analysis is given below.

Configuration of Substrate–Metal Complex in E. coli DNA Polymerase I-Catalyzed Reaction[14]

The rate of polymerization of phosphorothioate dATP analogs in the presence of different cations is determined. General assay conditions are as described for the polymerization reaction described above with the following exceptions: dATP or its analog and [^{14}C]dTTP (9000 cpm/nmol) are present in concentrations of 100 μM each. The enzyme concentration is 5.5 μg/ml. The concentrations of metal ions are as follows: Mg^{2+}, 7 mM; Co^{2+}, 0.25 mM; Mn^{2+}, 0.2 mM; Zn^{2+}, 0.15 mM. These optimal concentrations for each metal ion already had been determined. The reaction is incubated at 37°. The progress of the reaction is monitored by paper chromatography as described above. Initial rates are given in Table I and Michaelis–Menten kinetic parameters for the reactions, in the presence of Mg^{2+} and Co^{2+}, are given in Table II.

There is no change in specificity for the diastereomers of dATPαS indicating that there is no metal ion coordination to the α-phosphate. In contrast, the *Rp* diastereomer of dATPβS, which is not a substrate in the presence of Mg^{2+}, becomes active in the presence of Co^{2+}, Mn^{2+}, and Zn^{2+}. These data identify the Δ configuration of the metal–nucleoside triphosphate substrate complex as the active isomer (Fig. 4).

TABLE I
RATES OF POLYMERIZATION OF dATP ANALOGS BY POL I IN PRESENCE OF DIFFERENT DIVALENT CATIONS[a]

Analog	Mg^{2+}	Co^{2+}	Mn^{2+}	Zn^{2+}
dATP	100	74	13	33
dATPαS (*Sp*)	89	81	24	19
dATPαS (*Rp*)	<0.2	<0.2	0.5	<0.2
dATPβS (*Sp*)	90	59	36	31
dATPβS (*Rp*)	0.4	53	25	12

[a] General assay conditions are described in the text. All rates are relative to magnesium-dATP = 100. The absolute value for magnesium-dATP was 9 nmol of radioactive label incorporated/h/μg of protein.

TABLE II
STEADY-STATE KINETIC CONSTANTS FOR dATP AND ANALOGS IN PRESENCE OF Mg^{2+} OR Co^{2+} [a]

Analog	Mg^{2+} K_m (μM)	Mg^{2+} K_i (μM)	Mg^{2+} V_{max} (nmol/hr × μg protein)	Co^{2+} K_m (μM)	Co^{2+} V_{max} (nmol/hr × μg protein)
dATP	3.8 ± 1.2		9.0 ± 0.5	3.0 ± 1.0	0.92 ± 0.05
dATPαS(*Sp*)	4.5 ± 1.0		8.0 ± 0.5		
dATPαS(*Rp*)		30 ± 15[b]			
dATPβS(*Sp*)	7.4 ± 1.5		8.1 ± 0.5	9.0 ± 2.0	1.1 ± 0.05
dATPβS(*Rp*)		>200[b]		10.6 ± 2.0	1.1 ± 0.05

[a] General assay conditions are described in the text.
[b] Inhibition of the polymerization of dATP and dTTP. Inhibitor concentrations were 5 or 20 μM for dATPαS (*Rp*), and 10 or 40 μM for dATPβS(*Rp*).

Other Phosphate Modifications

Boranophosphates

Replacement of one of the nonbridging oxygens of the α-phosphorus by a BH_3- group has led to an interesting new class of analogs known as boranophosphates (Fig. 7A). The thymidine 5′-*O*-(1-boranotriphosphate) has been demonstrated to be a *bona fide* substrate for Sequenase, a modified form of T7 DNA polymerase, resulting in a boranophosphate diester linkage.[25] Such linkages will of course be chiral, like the phosphorothioates, but whether only one or both of the diastereomeric triphosphates can be utilized is not known.

Methylphosphonates

Unlike the boranophosphates, which retain the negatively charged phosphodiester DNA linkages, the introduction of a methyl group onto the α-phosphorus results in the formation of a neutral linkage. Thymidine 5′-1-methylphosphonyl(diphosphate) (dTTPαCH_3) (Fig. 7B) was shown to be used by terminal deoxynucleotidyltransferase to extend a 20-mer oligodeoxynucleotide initiator either by one or by two nucleotides at the 3′ terminus.[26] Subsequent degradation of the product, with nuclease P1 and alkaline phosphatase, followed by HPLC analysis indicated the presence of only two

[25] J. Tomasz, B. Ramsay-Shaw, K. Porter, B. F. Spielvogel, and A. Sood, *Angew. Chem.* **104,** 1404 (1992).
[26] H. Higuchi, T. Endo, and A. Kaji, *Biochemistry* **29,** 8747 (1990).

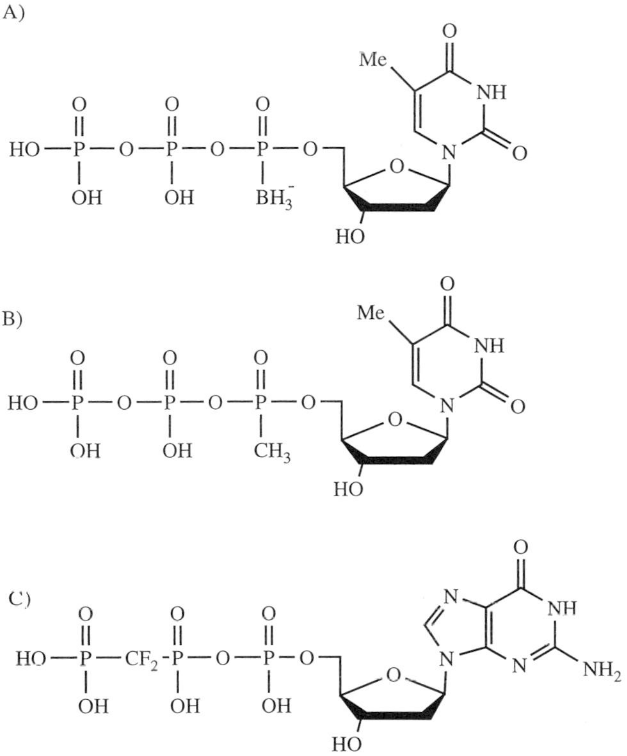

FIG. 7. Structure of phosphate-modified nucleotides. (A) Thymidine 5′-(1-boranotriphosphate). (B) Thymidine 5′-1-methylphosphonyl(diphosphate). (C) 2′-Deoxyguanosine 5′-[(βγ-difluoromethylene) triphosphate], dGMPPCF$_2$P.

fragments, d(CpCH$_3$T) and d(CpCH$_3$TpCH$_3$T), the one and two nucleotide extensions, respectively. This indicates the resistance of methyl phosphonate linkages toward nucleases. Such linkages are also resistant to 3′-exonuclease activity of snake venom phosphodiesterase. The configuration of the methyl phosphonate linkages of the di- and trinucleotide fragments have been tentatively assigned as being of the *Sp* configuration, thus indicating that nucleotide addition was accompanied with inversion of configuration at phosphorus. This result is consistent with that observed for all other polymerases studied with the phosphorothioate analogs.

βγ-Methylene Derivatives of Nucleoside Triphosphates

Replacement of the βγ-phosphate-bridging oxygen of dGTP with a difluoromethylene group, to yield 2′-deoxyguanosine 5′-[(βγ-difluoromethylene) triphosphate] ($dGMPPCF_2P$) has been reported.[27] Incorporation of the CF_2 functionality into the pyrophosphate moiety (Fig. 7C) has been postulated as being a suitable oxygen replacement since these derivatives closely mimic the metal-binding ability of the natural nucleoside triphosphates and, as such, have been termed "isosteric and isopolar" analogs.[28] $dGMPPCF_2P$ was tested as a substrate for calf thymus DNA Pol α, *E. coli* Pol I, and *Bacillus subtilis* Pol III. Although it was a suitable substrate for all three enzymes, its rate of incorporation is only efficient with *B. subtilis* Pol III, where it has a lower apparent K_m than dGTP, 3 and 10 μM, respectively. The rate of polymerization of $dGMPPCF_2P$ with *E. coli* and calf thymus polymerases were one-sixth and one-eleventh, respectively, that of the dGTP polymerization. The more hydrophobic methylene derivative, $dGMPPCH_2P$, only was a substrate for the *B. subtilis* Pol III enzyme but not for the other two enzymes, Pol I and Pol α. This demonstrates that this type of analog can be used to probe differences between polymerases with respect to the interaction with the pyrophosphate part of the substrate.

[27] L. Arabshahi, K. N. Naseema, M. Butler, T. Noonan, N. C. Brown, and G. E. Wright, *Biochemistry* **29,** 6820 (1990).

[28] G. M. Blackburn, D. E. Kent, and F. Kolkmann, *J. Chem. Soc., Perkin Trans.* **I,** 1119 (1984).

[17] Mechanisms of Inhibition of DNA Polymerases by 2′-Deoxyribonucleoside 5′-Triphosphate Analogs

By NEAL C. BROWN and GEORGE E. WRIGHT

Introduction

The purpose of this chapter is twofold: to provide the reader with an overview of the major classes of inhibitory 2′-deoxyribonucleoside 5′-triphosphate (dNTP) analogs and their mechanisms and to illustrate, with a "case study" of a complex analog, the experimental approaches applicable to the analysis of inhibitor mechanism. For a more comprehensive treatment of inhibitory dNTP analogs, the reader is referred to the review in Ref. 1.

[1] G. Wright and N. Brown, *Pharmac. Thera.* **47,** 447 (1990).

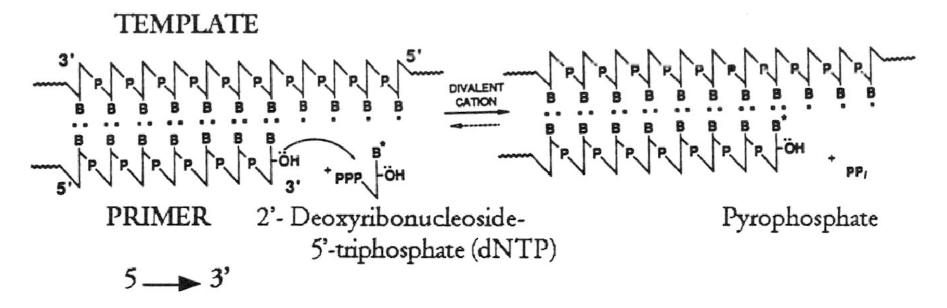

FIG. 1. Basic polymerase mechanism.

Definition of Target Enzymes and Their Basic Mechanism

The inhibitor targets considered herein are all *template*-dependent polymerases,[2] that is, enzymes that extend a primer only when its 3′-terminal end is hybridized to an antiparallel, *template* strand of polynucleotide. The template, which may be either RNA [in the case of reverse transcriptases (RTs)] or DNA, serves to guide the enzyme and dictate its choice of dNTPs as they are polymerized sequentially to the 3′ terminus of the primer. The basic reaction[2] catalyzed by this class of polymerases is shown schematically in Fig. 1.

As Fig. 1 indicates, the 3′-OH terminus of the primer acts as a nucleophile, attacking the α-phosphate of the incoming dNTP. As a consequence of the attack, inorganic pyrophosphate (PP_i) is liberated, and the primer is extended by one dNMP residue in the $5' \rightarrow 3'$ direction. As implied by the reverse arrow in Fig. 1, the polymerase reaction is reversible. Given an appropriate concentration of PP_i and a primer–template, most polymerases of this class can catalyze pyrophosphorolysis of the 3′-terminal dNMP residue to regenerate the dNTP from which it was originally derived.[2]

Polymerase Cycle

For each dNMP added to the 3′-OH terminus, the polymerase cycles through a multistep process. The major steps of this process, based on a scheme[3] proposed for *Escherichia coli* DNA polymerase I (Pol I), are summarized in Fig. 2. In step 1 (Fig. 2) the free enzyme binds to the template–primer, forming the binary complex, A, and positioning its dNTP binding site appropriately in the vicinity of the 3′-OH of the primer terminus and the first unapposed base of the template strand. In step 2 (Fig. 2) a

[2] A. Kornberg and T. Baker, "DNA Replication," 2nd ed. W. H. Freeman, New York, 1991.
[3] E. Travaglini, A. Mildvan, and L. Loeb, *J. Biol. Chem.* **250,** 8647 (1975).

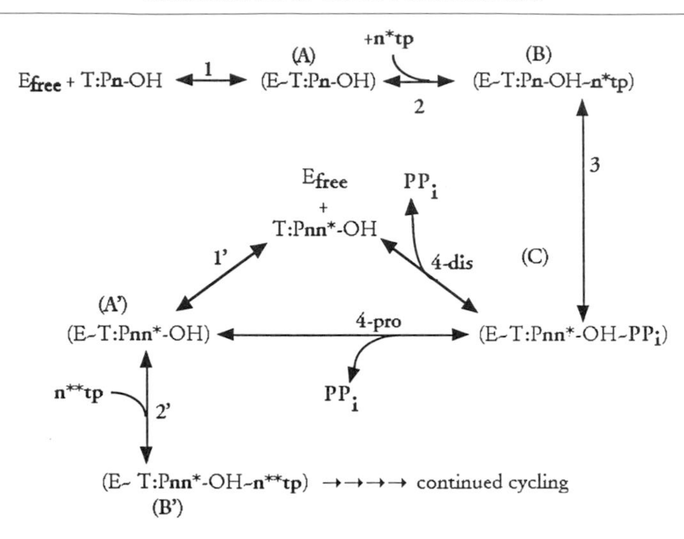

FIG. 2. Polymerization cycle. T : Pn-OH, template–primer with primer terminated by nucleotide n with a free 3′-OH group; PP_i, inorganic pyrophosphate; n*tp and n**tp, dNTPs with asterisk indicate order of addition; pro, processive enzyme; dis, distributive enzyme.

dNTP molecule (n*tp) binds to the dNTP binding site of complex A, generating the ternary complex, B. In step 3 (Fig. 2) the 3′-OH of the latter complex attacks the dNTP, extending the primer by one nucleotidyl residue (n*). In step 4 (Fig. 2) the resultant complex, C, sheds PP_i by one of two pathways, 4-pro or 4-dis. The specific path taken depends on the capacity of the enzyme to remain bound to primer–template as it repositions its active site relative to the new primer terminus. If the enzyme is *processive*, it takes pathway 4-pro, remaining bound to primer–template in step 4 as it "slides" downstream on the new terminus, creating A′, the equivalent of the first complex, A. Complex A′ then binds the next dNTP (n**tp) and reenters the cycle via step 2′. If the enzyme is purely *distributive*, it takes pathway 4-dis, dissociating from complex C, along with PP_i to yield free enzyme. The latter then reenters the cycle via step 1′.

Many polymerases either associate with or directly bear an "editing" 3′ → 5′-exonuclease (exo) activity[2] capable of hydrolyzing dNMP residues from the primer terminus. Although these editing activities prefer to remove mismatched terminal dNMP residues, they also are active against matched termini, and, therefore, their presence can introduce additional complexity into the chain elongation cycle. As shown in the case study below, 3′ → 5′-exo action also can be a consideration in the analysis of the mechanism of inhibitory dNTP analogs, in particular, those analogs that become incorporated in the primer terminus.

Bona Fide Inhibitory dNTP Analog

An agent belonging to this class of DNA polymerase inhibitor is defined as a molecule that can reduce the rate of primer extension in a manner that can be specifically and competitively antagonized by one or more dNTPs. Agents fitting this definition fall into two major categories: those that *cannot* be incorporated into the primer terminus and those that *can.* The major agents and their primary "target" enzyme(s) are summarized in Table I and discussed below; relevant structures are shown in Fig. 3.

TABLE I
MAJOR CLASSES OF INHIBITORY dNTP ANALOGS

Class	Major target polymerase
Nonincorporable agents[a]	
Carbonyl diphosphonate: (Template complementary dNTP)	Mammalian Pols δ and ε
H_2-HPUra and H_2-HPIso: (dGTP) (dATP)	Gram-positive bacterial Pol III
BuPG and BuAA: (dGTP) (dATP)	Mammalian Pol α and selected members of B polymerase family
Aphidicolin: (dCTP)	B polymerase family, e.g., T4, φ29 Pols, mammalian Pols α, δ, and ε and their equivalents in yeast and other eukaryotes; selected eukaryotic viral Pols
Incorporable agents	
Agents yielding extensible primer termini	
Triphosphate-modified dNTPs (Analogous dNTPs)	Not predictably specific (see text)
araATP and araCTP (dATP) (dCTP)	Broad spectrum
BuPdGTP and BuAdATP (dGTP) (dATP)	Same as for BuPG and BuAA
Agents yielding nonextensible termini	
acyclo-GTP (dGTP)	Polymerases encoded by Herpesviruses
2′,3′-dideoxy-NTPs (Analogous dNTP) unsubstituted 3′-azido-TTP adenosine 2′,3′-epoxide 5′-triphosphate	Broad spectrum; enzyme reactivity can depend on choice of divalent cation. Potent inhibitors of reverse transcriptase

[a] The dNTP in parentheses indicates the dNTP that is specifically competitive with the agent located just above it. The full names and structures of each agent are provided in Fig. 3.

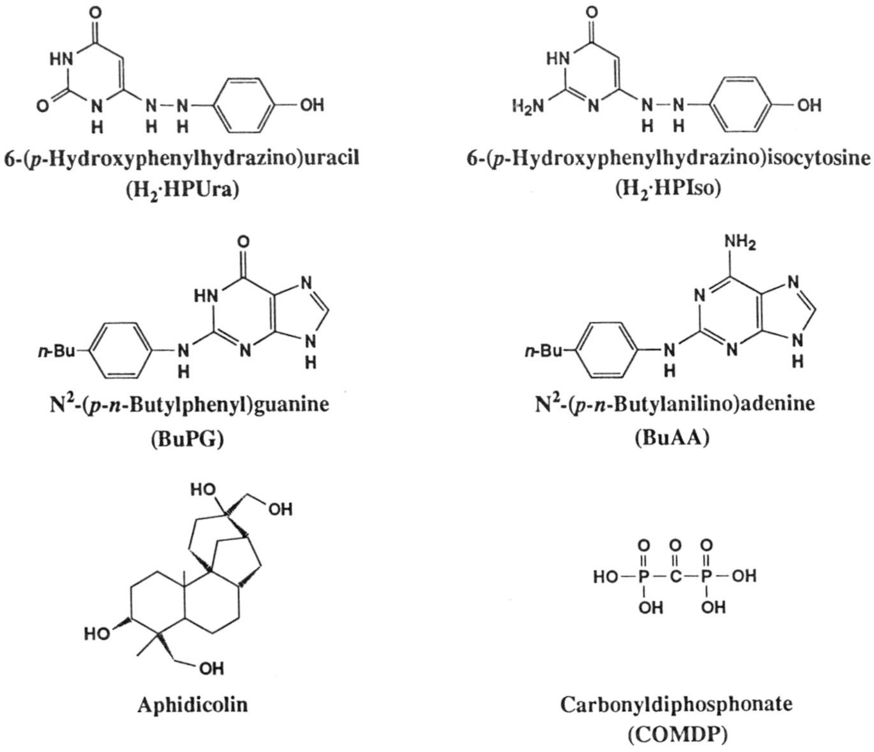

6-(*p*-Hydroxyphenylhydrazino)uracil (H_2·HPUra)

6-(*p*-Hydroxyphenylhydrazino)isocytosine (H_2·HPIso)

N^2-(*p*-*n*-Butylphenyl)guanine (BuPG)

N^2-(*p*-*n*-Butylanilino)adenine (BuAA)

Aphidicolin

Carbonyldiphosphonate (COMDP)

FIG. 3. Structures of relevant dNTP analogs.

Nonincorporable Agents

Carbonyl diphosphonate (COMDP; Fig. 3) is one of a large family of PP_i analogs that displays activity as DNA polymerase inhibitors.[4] Although the inhibitory action of most PP_i analogs is independent of dNTP concentration, that of COMDP is not.[5] Rather, its inhibitory action is competitively antagonized by dNTPs, making it the simplest of the dNTP analogs. COMDP appears to possess specific affinity for a subdomain of the dNTP binding site, which reacts with the β,γ moiety of the incoming dNTP. Its action is independent of template composition and is subject to competitive antagonism by any dNTP that is complementary to the template residue(s) being read by the polymerase.[5] For example, the inhibitory effect of

[4] B. Öberg, *Pharmac. Thera.* **40,** 213 (1989).

[5] R. Talanian, N. Brown, C. McKenna, T.-G. Ye, J. Levy, and G. Wright, *Biochemistry* **28,** 8270 (1989).

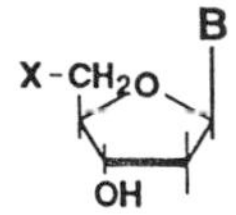

B

X-CH_2O

OH

Triphosphate Modified dNTPs

X=

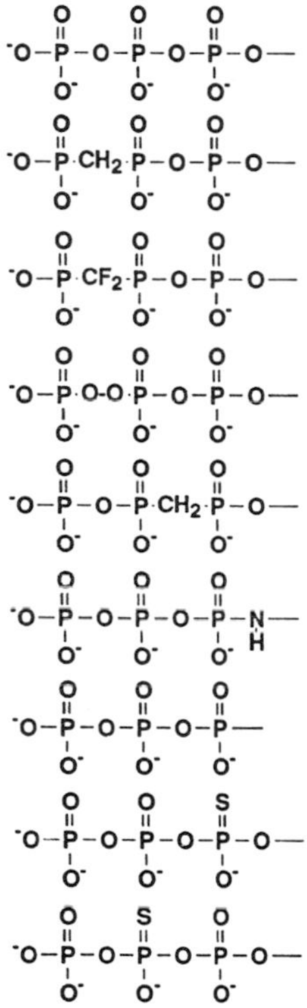

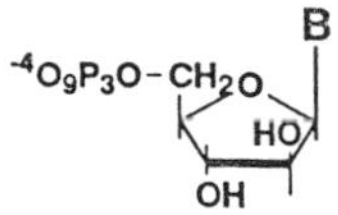

$^{-}O-P(=O)(O^{-})-O-P(=O)(O^{-})-O-P(=O)(O^{-})-O-$

$^{-}O-P(=O)(O^{-})-CH_2-P(=O)(O^{-})-O-P(=O)(O^{-})-O-$

$^{-}O-P(=O)(O^{-})-CF_2-P(=O)(O^{-})-O-P(=O)(O^{-})-O-$

$^{-}O-P(=O)(O^{-})-O-O-P(=O)(O^{-})-O-P(=O)(O^{-})-O-$

$^{-}O-P(=O)(O^{-})-O-P(=O)(O^{-})-CH_2-P(=O)(O^{-})-O-$

$^{-}O-P(=O)(O^{-})-O-P(=O)(O^{-})-O-P(=O)(O^{-})-NH-$

$^{-}O-P(=O)(O^{-})-O-P(=O)(O^{-})-O-P(=O)(O^{-})-$

$^{-}O-P(=O)(O^{-})-O-P(=O)(O^{-})-O-P(=S)(O^{-})-O-$

$^{-}O-P(=O)(O^{-})-O-P(=S)(O^{-})-O-P(=O)(O^{-})-O-$

B

$^{-4}O_9P_3O-CH_2O$

HO

OH

Arabinonucleotides

AraATP (B=adenine)

AraCTP (B=cytosine)

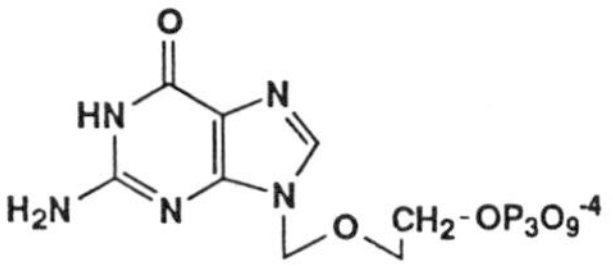

O

HN

N

H_2N

N

N

$CH_2-OP_3O_9^{-4}$

AcyclodGTP

B

$^{-4}O_9P_3O-CH_2O$

X Y

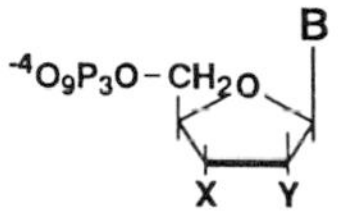

2'-3'-Dideoxy NTPs

	X=	Y=	B=
ddNTP	H	H	A,G,C,or T
AZTTP	N_3	H	T
Adenosine 2'-3'epoxide triphosphate	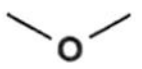 —O—		A

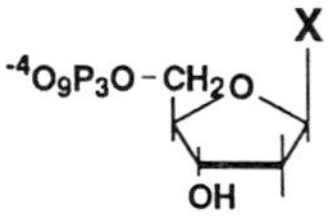

X

$^{-4}O_9P_3O-CH_2O$

OH

Butylphenyl dNTPs

X= *n*-Bu–C_6H_4–NH– (guanine, O, HN, N, N, N)

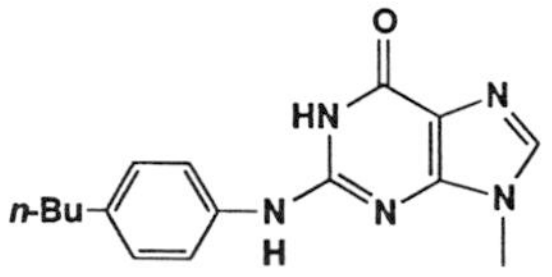

BuPdGTP

X= *n*-Bu–C_6H_4–NH– (adenine, NH_2, N, N, N, N)

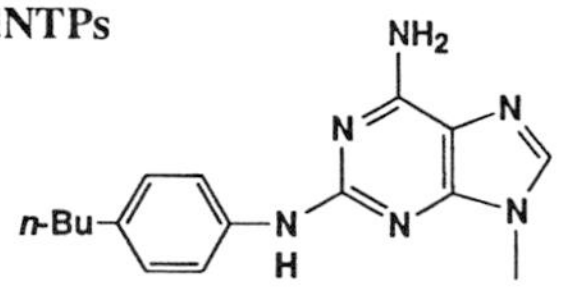

BuAdATP

FIG. 3. (*continued*)

COMDP in the presence of a poly(dA) template is specifically antagonized by dTTP; in the presence of a natural DNA template, all four dNTPs are required for full antagonism.

Although the simplicity of its structure and mechanism might suggest a broad target spectrum for COMDP, it is relatively selective for polymerases α, δ, and ε, the three mammalian members of the B polymerase family.[6] Among these enzymes, COMDP displays approximately 10-fold selectivity for Pol δ and Pol ε.[5]

The arylhydrazinopyrimidines *H_2-HPUra* and *H_2-HPIso*[1,7] are highly specific for gram-positive bacterial DNA Pol IIIs. Although formally pyrimidines, they each present a unique base-pairing domain through which they specifically mimic *purine* dNTPs—dGTP in the case of H_2-HPUra and dATP in the case of H_2-HPIso. As expected from their purine-like character, the inhibitory action of each (1) is subject to competitive antagonism by the specific purine dNTP that it mimics and (2) is absolutely dependent on the presence of the complementary pyrimidine in the template strand. Inhibition of the target polymerase occurs as a result of its sequestration[8,9] in a template–primer–inhibitor–enzyme complex equivalent to the natural "prepolymerization" complex B in Fig. 2. In this complex the inhibitor acts as a bridge between the two macromolecules. One end, the aryl ring, binds to the enzyme in the immediate vicinity of its dNTP binding site, whereas its base-pairing domain forms three H bonds with a complementary, unapposed template pyrimidine immediately downstream of the primer terminus.

BuPG and *BuAA*[7] (see Fig. 3 for structure) are, respectively, third-generation derivatives of H_2-HPUra and H_2-HPIso. In contrast to the latter agents, they formally assume the purine base structure of the dNTPs, which they mimic, and they do *not significantly inhibit Pol III.* Rather, through the installation of an N^2-(p-n-butylphenyl) substituent, they have been targeted to inhibit selectively mammalian Pol α and several other members of the so-called B family of DNA polymerases.[7] The design of BuPG and BuAA has preserved the basic sequestration mechanism of inhibition reviewed above for the arylhydrazinopyrimidines. Accordingly, against their B family targets, each of these agents acts by a mechanism identical to that found for the corresponding Pol III-specific arylhydrazinopyrimidine prototype.

Aphidicolin, a diterpene tetraol (see Fig. 3) derived from *Cephalospor-*

[6] D. Braithwaite and J. Ito, *Nucleic Acids Res.* **21,** 787 (1993).
[7] N. Brown, L. Dudycz, and G. Wright, *Drugs Exptl. Clin. Res.* **XII(6/7)** 555 (1986).
[8] K. Gass and N. Cozzarelli, *J. Biol. Chem.* **248,** 7688 (1973).
[9] J. Clements, J. D'Ambrosio, and N. Brown, *J. Biol. Chem.* **250,** 522 (1975).

ium aphidicola, is a potent and selective inhibitor of certain B family DNA polymerases.[1] It competes with each of the four dNTPs, but is most potent when competing with dCTP.[10] The molecular basis for the ability of aphidicolin to mimic dCTP is not established, although a plausible model has been proposed[11] that envisions an H-bonded complex of the inhibitor and template guanine bound to enzyme.

Aphidicolin-induced inhibition occurs by a mechanism[10] equivalent to that of the arylhydrazinopyrimidines and butylphenylpurines. The inhibitor molecule simultaneously binds the dNTP-binding site of its target enzyme and an appropriate template residue of the primer–template, creating a nonproductive ternary complex analogous to the functional complex B of Fig. 2.

Incorporable Agents

Agents of this class have the structure of a 2′-deoxyribonucleoside 5′-triphosphate or its "skeletal" equivalent. They fall into two subclasses: those whose incorporation generates *extensible* primer termini and those whose incorporation generates *nonextensible* primer termini.

Agents That Can Generate Extensible Primer Termini

Triphosphate-Modified dNTPs. This class, which is reviewed in Refs. 1 and 12, consists of dNTPs carrying a natural 2′-deoxyribonucleoside modified in the 5′-triphosphate moiety. Examples of the most common types of modification are shown in Fig. 3. They include (1) substitution/deletion of the 5′-O; (2) substitution of O with S in the α- and β-phosphates; (3) substitution of the α,β-O with the O isostere, CH_2; and (4) substitution of the β,γ-O with O–O or the isosteric CH_2 and CF_2 groups.

The nature of the response of a specific enzyme to a given triphosphate-modified dNTP is highly unpredictable. An agent simply may not bind significantly to the dNTP binding site of the enzyme and, thus, will not display significant inhibitory effect on catalysis. The agent may bind well to the dNTP binding site of another enzyme, but simply serve as a noninhibitory substrate that perfectly replaces its natural counterpart. Against a third enzyme the same agent may encounter a dNTP binding site to which it can tightly bind and from which it is poorly polymerized; on this enzyme the agent would be expected to show an inhibitory effect.

[10] R. Sheaff, D. Ilsley, and R. Kuchta, *Biochemistry* **30,** 8590 (1991).

[11] Q. Dong, W. Copeland, and T. S-F. Wang, *J. Biol. Chem.* **268,** 24,163 (1993).

[12] L. Arabshahi, N. Khan, M. Butler, T. Noonan, N. Brown, and G. Wright, *Biochemistry* **29,** 6280 (1990).

Manipulation of the triphosphate moiety of a dNTP analog can also have unpredictable effects on its potency and mechanism. Such modifications are likely to be deleterious, as illustrated by the case of the reverse transcriptase (RT) inhibitor, 3′-azido-2′,3′-dideoxy-TTP (AZTTP; see Fig. 3 and relevant section below). The replacement of the α,β and the β,γ bridge oxygens of AZTTP with CH_2 reduced its potency 200- and 600-fold, respectively.[13]

Arabinonucleoside Triphosphates (AraNTPs). At least 20 arabinofuranosyl analogs of dNTPs have been characterized with respect to target and mechanism of inhibitory action.[1] AraCTP and araATP (see Fig. 3), the respective active forms of the antitumor agent, araC,[1] and the antiviral agent, araA,[1] are the most extensively studied of this group and, thus, serve as models of araNTP action.

These model araNTPs display a broad target spectrum,[1] significantly inhibiting all eukaryotic and eukaryotic viral polymerases, bacterial Pol II and Pol III (not Pol I), and marginally inhibiting RTs. The action of each on the various enzyme targets is specifically and competitively antagonized by the analogous dNTP and requires the presence of the complementary base in the template.

With few exceptions, the potency of an araNTP for a given enzyme is governed mainly by its capacity to incorporate the araNMP moiety into the primer strand.[1] The results of Townsend and Cheng's detailed analysis of araCTP action[14] on mammalian Pol α and Pol β clearly illustrate this relationship. On incorporation of an araCMP residue, the presentation of *trans*-hydroxyls by the resultant primer terminus significantly slows the rate of polymerization of the next NTP. If the next NTP dictated by template is *not* one containing a C, the enzyme eventually polymerizes it and moves on. However, if a *tandem* C is required, and a *second* araCMP residue happens to become polymerized to the first, the rate of further primer extension approaches zero. In the absence of fresh primer–template the only way this block can be overcome is by removal of the terminal araCMP residue. For those polymerases associated with $3' \rightarrow 5'$-exonuclease activity, removal is clearly possible and, not surprisingly, such enzymes are highly likely to display araNTP resistance.[15]

Butylphenyl dNTPs. BuPdGTP and BuAdATP (Fig. 3) are the ultimate derivatives of BuPG and BuAA, the mammalian Pol α-specific purine inhibitors discussed above. The conversion of these dNTP *mimics* to their *actual dNTP forms* has created Pol α-selective inhibitors that display un-

[13] D. Hebel, K. Kirk, J. Kinjo, T. Kovács, K. Lesiak, J. Balzarini, E. DeClercq, and P. Torrence, *Biorg. Med. Chem. Lett.* **1,** 357 (1991).

[14] A. Townsend and Y.-C. Cheng, *Molec. Pharmacol.* **32,** 330 (1987).

[15] N. Cozzarelli, *Ann. Rev. Biochem.* **46,** 641 (1977).

precedented potency and faithfully retain the prototypic inhibitor mechanism,[7] including, surprisingly, immunity from polymerization.[16]

The conversion of BuAA and BuPG to their dNTP forms increased not only inhibitor potency but also target spectrum. BuPdGTP and BuAdATP inhibit several B family DNA polymerases, including yeast Pol I and the enzymes encoded by bacteriophages φ29, M2, PRD1, and T4.[1] Analysis of the mechanism of BuPdGTP action on T4 DNA Pol (see summary below in the final section), has indicated that the sensitivity of these B family enzymes is derived in great measure from their capacity to polymerize the inhibitor—a capacity that mammalian Pol α lacks.[16]

Agents That Generate Nonextensible Termini

Acyclo-GTPs. The model for this class is acycloguanosine triphosphate (9-[(2-hydroxyethoxy)methyl] guanine triphosphate; ACGTP; see Fig. 3), the active form of the antiherpesvirus agent, acycloguanosine (acyclovir).[1] The inhibitory action of ACGTP is specifically and competitively antagonized by dGTP and requires the presence of cytosine in the template.

ACGTP is 10 to 20 times more selective for herpesvirus-specific DNA polymerases than for mammalian Pol α. The degree of susceptibility of a given enzyme to ACGTP is directly related to its ability to polymerize the ACGMP moiety to the primer terminus. The ACGMP-terminated primer is the major determinant of ACGTP-induced polymerase inhibition, and it acts in two distinct ways.[17] First, as suggested from consideration of structure, its formation reduces the pool of termini available to the polymerase for extension. Second, it presents a specific structure, which strongly binds the enzyme. When presented with the next scheduled dNTP, the enzyme–DNA complex binds it to form an even stronger ternary complex. The latter has been referred to as a "dead-end" complex, and the process of its formation is described as "induced substrate inhibition."[17]

2′,3′-Dideoxy-NTPs (ddNTPs). The action of the specific agents selected for discussion below typifies that of a substantial group of inhibitory ddNTPs (see Ref. 1 for a detailed review). The inhibitory action of ddNTPs, like that of the acycloNTPs (1) is competitively antagonized by the analogous dNTP, (2) requires a complementary template base, and (3) with rare exception, depends on the incorporation of the NMP moiety into the primer terminus.

Unsubstituted derivatives. The best known of this group are the four natural ddNTPs used to create the base-specific termini in the "dideoxy"

[16] N. Khan, G. Wright, and N. Brown, *Nucleic Acids Res.* **19,** 1627 (1991).
[17] J. Reardon and T. Spector, *J. Biol. Chem.* **264,** 7405 (1989).

method of DNA sequencing.[18] These nucleotides display a broad target spectrum, which includes *E. coli* Pol I, several RTs, and, in the presence of the appropriate cation, mammalian DNA Pol α.[1,19,20] Whereas *E. coli* Pol I and the RTs are sensitive to ddNTPs in the presence of *either* Mg^{2+} or Mn^{2+} as divalent cation, mammalian Pol α displays sensitivity *only* in the presence of the latter.[19] In the case of Pol I and RT the inhibitory mechanism of ddNTP action is essentially identical to that discussed above for acyclo-GTP, that is, substrate-induced enzyme sequestration on a ddNMP-terminated primer.[17] The mechanism by which ddNTPs effect Mn^{2+}-dependent inhibition of Pol α is apparently more complex than that characterized for Mg^{2+}-dependent inhibition of Pol I and RT and is not as well understood. Evidence obtained in two different laboratories on, respectively, two different sources of Pol α (mouse[19] and human[20]) suggests the operation of at least *two* distinct mechanisms of ddNTP action: (1) competition with the analogous dNTP for occupancy of the dNTP binding site[19] *without incorporation* and (2) *incorporation* into primer termini with consequent substrate-induced sequestration like that found for Pol I and RT.[20] The basis for these conflicting observations remains to be determined.

Substituted dideoxy derivatives. The 2′,3′-ddNTP nucleus forms the structural basis for a variety of substituted dNTP analogs[1]—ranging from base-modified fluorescent derivatives used as substrates in automated DNA sequencing[21] to inhibitory nucleotides substituted in the 2′ or 3′ carbon of the deoxyribose ring. AZTTP[22] and adenosine 2′,3′-epoxide 5′-triphosphate (epoxy ATP),[23] shown in Fig. 3, exemplify the inhibitory category.

AZTTP, the active form of the anti-HIV (human immunodeficiency virus) nucleoside, AZT, is a very potent inhibitor of RTs, and it is very selective relative to Pol α, displaying at least 300- to 400-fold higher potency.[1,22,24] The inhibitory effect of AZTTP is competitively and specifically antagonized by dTTP and depends on the presence of the complementary base, adenine, in the RNA or DNA template. Although the mechanism of AZTTP inhibition of RT has not been precisely elucidated, the high potency of the agent is consistent with the production of AZTMP-terminated prim-

[18] F. Sanger, S. Nicklen, and A. Coulson, *Proc. Natl. Acad. Sci. USA* **74,** 5463 (1977).

[19] K. Ono, M. Ogasawara, Y. Iwata, and H. Nakane, *Biomed. Pharmacotherapy* **38,** 382 (1982).

[20] P. Fisher and D. Korn, *Biochemistry* **20,** 4560 (1981).

[21] J. Prober, G. Trainor, R. Dam, F. Hobbs, C. Robertson, R. Zagursky, A. Cocuzza, M. Jensen, and K. Baumeister, *Science* **238,** 336 (1987).

[22] P. Furman, J. Frye, M. St. Clair, K. Weinhold, J. Rideout, G. Freeman, S. Lehrman, D. Bolognisi, S. Broder, H. Mitsuya, and D. Barry, *Proc. Natl. Acad. Sci. USA* **83,** 8333 (1986).

[23] M. Abboud, W. Sim, L. Loeb, and A. Mildvan, *J. Biol. Chem.* **253,** 3415 (1978).

[24] B. Eriksson, C. Chu, and R. Schinazi, *Antimicrob. Agents Chemoth.* **33,** 1729 (1989).

ers and substrate-induced sequestration of enzyme described above for the unsubstituted ddNTPs and acycloGTP.[1,17,20]

Epoxy-ATP is not strictly a dideoxy-NTP. However, it is included in the ddNTP category because it acts by a similar mechanism and serves as an example with which to illustrate the experimental rigor required to distinguish such a mechanism from a confusing alternative.

Epoxy-ATP is a broad spectrum analog, displaying strong inhibitory activity against enzymes such as *E. coli* Pol I, mammalian Pol α and Pol β, and RTs.[1] The broad spectrum of epoxy-ATP and the potential chemical reactivity of the 2′,3′-epoxide suggested that its inhibitory action resulted from irreversible, "suicide" inactivation of the target enzyme effected by covalent reaction of the enzyme with the nucleotide and/or an epoxy-AMP-terminated primer. The results of the initial analysis of epoxy-ATP action, executed with *E. coli* Pol I and poly(dA-T) as primer–template, were consistent with such an irreversible mechanism.[23] A subsequent, more rigorous study of epoxy-ATP action on *E. coli* Pol I confirmed the previous observation of a strong reaction between primer–terminal epoxy-AMP and the enzyme.[25] However, this study indicated that the strength of this interaction was *not* derived from covalent bond formation, but from noncovalent binding forces that combine to create a tight but *reversible* complex of dNTP, enzyme, and template–primer equivalent to the substrate-induced, "dead-end" complexes described above for acyclo-GTP and the unsubstituted ddNTPs.

Analysis of Mechanism of BuPdGTP: Case Study

The conversion of BuPG, the Pol α-specific dGTP mimic, to its full dNTP form[26] dictated that BuPdGTP be thoroughly investigated with respect to target range and inhibitory mechanism. Because the investigation of BuPdGTP action constitutes a useful example of the practical aspects of a systematic analysis of the mechanism of a complex inhibitory dNTP analog, the investigation and the tools used in its investigation are described below.

Analytical Tools

Assays with Natural DNA. DNase I-activated calf thymus DNA is used as the primer–template in two types of assay: *complete* and *truncated.*[26,27] *Conditions for complete assay* include DNA, divalent cation, and four

[25] C. Catalano and S. Benkovic, *Biochemistry* **28,** 4374 (1989).
[26] N. Khan, G. Wright, L. Dudycz, and N. Brown, *Nucleic Acids Res.* **12,** 3695 (1984).
[27] N. Khan, L. Reha-Krantz, and G. Wright, *Nucleic Acids Res.* **22,** 232 (1994).

dNTPs, of which one is radioactively labeled in the base; DNA and dNTPs are provided at concentrations of at least $10 \times K_m$ and in large excess relative to enzyme concentration. The choice of the labeled dNTP is not trivial; it should *not* be the dNTP likely to compete selectively with the analog—particularly if the analog can be polymerized. The rationale for this choice is to avoid "pseudo" inhibition (i.e., reduction of incorporation of label), which results when an analog simply substitutes for a given dNTP without significantly inhibiting the rate of polymer formation. The inhibitory effect of dUTP on the incorporation of labeled dTTP is a classic example of pseudoinhibition.

Conditions for truncated assay are identical to those described for complete assay with one exception; the dNTP known to be competitive with a given inhibitor is *omitted.* The omission of one dNTP reduces the incorporation of the labeled dNTP by 75 to 90%, depending on the enzyme. The residual incorporation, in the absence of interference from $3' \rightarrow 5'$-exonuclease activity, reflects extension of primer termini to the point at which the template dictates the binding and incorporation of the missing dNTP. For nonpolymerizable agents like BuPG and aphidicolin, which sequester their target enzymes in nonproductive complexes with primer–template, truncated assay provides a sensitive and convenient means to monitor complex formation directly. For analogs that have the potential to serve as substrates, the appropriate truncated assay is exploited as a preliminary diagnostic probe of their incorporation.

Assay with Homopolymeric DNA. If the target enzyme can utilize homopolymeric primer–templates [for example, poly(dA) · oligo(dT) (dTTP as substrate) or poly(dC) · oligo(dG) (dGTP as substrate)], they can be used as probes to determine rigorously whether an agent is competitive or noncompetitive with a specific dNTP. Or they can be used, as described below, in the design of "trapping" experiments to detect agent-induced sequestration.

Sequestration/Trapping. If the action of the agent of interest (1) is competitively antagonized by a specific dNTP and (2) absolutely requires the presence of the complement of that dNTP in template, the capacity of that agent to sequester enzyme on a given primer–template is assessed by a simple assay method exploiting a noncomplementary template to direct agent-resistant synthesis. The approach[28] used to detect H_2-HPUra-induced sequestration of Pol III serves to illustrate the method. [^{3}H]TTP $\rightarrow$ oligo(dT) · poly(dA) is used as the primary template : primer in the presence of limiting enzyme. Given its absolute requirement for a complementary, *cytosine*-containing template, H_2-HPUra does not significantly inhibit dTTP incorporation. However, when an appropriate amount of poly(dC) · oligo

[28] K. Gass, R. Low, and N. Cozzarelli, *Proc. Natl. Acad. Sci. USA* **70,** 103 (1973).

(dG) is added to the reaction mixture, TTP incorporation is inhibited dramatically. The inhibition induced by the poly(dC) · oligo(dG) is absolutely dependent on the presence of H_2-HPUra, and it is specifically antagonized by dGTP. In sum, these and other uncited results of the experiment indicated that H_2-HPUra firmly sequesters the enzyme to the poly(dC) · oligo(dG), removing it from participation in dTTP polymerization on poly (dA) · oligo(dT).

Primer Extension. The general approach exploits sequencing gel analysis to assess residue-by-residue extension of custom-designed oligomeric 5′-^{32}P-labeled primers along appropriately designed oligomeric templates. This methodology obviates the need to synthesize radioactively labeled inhibitor forms and provides a very powerful and sensitive means to determine if a potentially polymerizable agent can, indeed, be polymerized, and if so, whether the terminus it forms can be extended.

Analysis of BuPdGTP Mechanisms

BuPdGTP was synthesized on the premise that it would be a more potent inhibitor of DNA polymerase α than the base, BuPG (see above), and that it might be a substrate for the enzyme. It was, indeed, found to be a potent, competitive, and selective (with respect to other eukaryotic and prokaryotic DNA polymerases) inhibitor of Pol α (Table II). It has

TABLE II
BuPdGTP ACTION ON THREE DNA POLYMERASES

Variable	Enzyme		
	E. coli Pol I	Mammalian Pol a	T4 Pol
Structural family[a]	A	B	B
Competitor dNTP	dGTP	dGTP	dGTP
K_i (complete synthesis)	100 μM	~1 nM	0.8 μM
IC_{50} (truncated synthesis)			
10-min assay	NA[b]	~1 nM	~1 nM
30-min assay	NA	—	~60 pM
Polymerization → primer?	Yes	No	Yes
BuPGMP primer terminus:			
extensible?	No	NA	No
exonuclease susceptible?	No	NA	Yes
sequester enzyme?	No	NA	Yes
alone?	No	NA	Yes
plus next dNTP	No	NA	Yes

[a] Classification of J. Ito and D. K. Braithwaite, *Nucleic Acids Res.* **19,** 4045 (1991),[6] based on comparative primary structure.
[b] Not applicable or measurable.

been shown, both indirectly and in primer extension assays, *not* to be a substrate for Pol α.[16] Thus, its identical K_i values obtained by classical competitor assays (complete synthesis) and in the truncated assay lacking dGTP (Table II) are in full accord with a reversible, competitive sequestration mechanism equivalent that discussed above for BuPG.

The apparent resistance of *E. coli* Pol I to BuPdGTP does not mean that the compound does not interact with this enzyme.[29] In fact, BuPdGTP is found to be readily incorporated by Pol I into a primer–template in place of dGTP, by gel analysis of primer extension assays (Table II). How could such an alternate substrate not display inhibition of the enzyme in a full assay with dGTP? Measurement of the incorporation rate for BuPdGTP in a Pol I-catalyzed primer extension assay gives one clue: The k_{cat} for BuPdGTP displayed by Pol I is at least 600 times lower than that which it displays for dGTP,[30] suggesting that BuPdGTP does not efficiently compete for incorporation in the presence of dGTP. Upon investigating the consequences of its incorporation the reason for the lack of inhibition—even in a truncated assay—becomes clear. Pol I is unable to extend a chemically synthesized 3′-BuPdG-primer–template with the next substrate; nor can its 3′ $\rightarrow$ 5′-exonuclease remove such a terminus. The suggestion that the modified primer–template no longer had affinity for the enzyme is confirmed by a sequestration assay, in which a chemically synthesized 3′-BuPdG-primer–template does not inhibit Pol I-catalyzed [^{3}H]dGTP incorporation into oligo(dG) · poly(dC), either in the absence or in the presence of the next dNTP substrate.

The effect of BuPdGTP on T4 DNA polymerase, like Pol α, a member of the B family, is even more complicated.[27] In the complete assay BuPdGTP inhibited T4 Pol competitively with dGTP and with a K_i (0.8 μM) that suggested it to be 800-fold less potent than against Pol α (Table II). However, the apparent K_i in a standard (10-min) truncated assay is 1 nM, comparable to that of Pol α under either condition, and that value decreases to 60 pM in a 30-minute assay! Primer extension reactions show that the T4 Pol, like Pol I but unlike Pol α, cannot extend the 3′-BuPdG-primer with the next nucleotide, but unlike Pol I, the 3′ $\rightarrow$ 5′-exonuclease of T4 Pol *could* remove the modified 3′-BuPdGMP residue. The latter result indicates that the primer–template resulting from incorporation of BuPdGTP must bind T4 Pol. Indeed, a sequestration experiment like that described for Pol I above results in strong inhibition of T4 Pol activity by the 3′-BuPdG-primer–template, even in the absence of the next dNTP substrate.

[29] H. Misra, N. Khan, S. Agrawal, and G. Wright, *Nucleic Acids Res.* **20,** 4547 (1992).
[30] G. E. Wright, unpublished results (1993).

Conclusions

BuPdGTP reacts with three DNA polymerases by distinct mechanisms that do not simply reflect different affinities of inhibitor and enzymes. In its sum, the complex case of BuPdGTP makes at least two strong recommendations to DNA polymerase enzymologists who would use inhibitors as enzyme probes and analyze DNA polymerase inhibitor/substrate action. First, it spells caution to those who would assume that an analog mechanism determined for a single enzyme can be generalized across the entire membership of the enzyme's structural family. Second, it illustrates the importance of taking a multifaceted approach to elucidation of mechanism: an approach that addresses the following set of questions. Does the compound inhibit DNA polymerase activity, (i.e., noncompetitor radiolabeled dNTP incorporation), and is inhibition competitive with one or more dNTPs? If so, is it selective for a single enzyme, single family, or members of different families? Does the compound inhibit activity in a truncated assay lacking the competitor, and is its IC_{50} equal to or different from that in complete assays? Is the compound a substrate for the DNA polymerase? If it is, can the enzyme extend the modified primer–templates or does the product serve as a terminator? For enzymes with $3' \rightarrow 5'$-exonuclease activity, can the 3'-modified primer–template bind the DNA polymerase, either alone or in the presence of the next dNTP specified by the template? The more of these questions that can be answered, the greater will be the potential of the agent as a probe of polymerase structure and function.

Acknowledgments

Much of the authors' work reported in this article has been supported by research grants from the USPHS-National Institutes of Health. N. C. B. is supported by NIH grant GM45330; G. E. W. is supported by NIH grant GM21747.

[18] Analyzing Fidelity of DNA Polymerases

By KATARZYNA BEBENEK and THOMAS A. KUNKEL

Introduction

Because polymerization errors are rare events relative to correct incorporations,[1] highly sensitive assays are required to measure DNA polymerase

[1] H. Echols and M. F. Goodman, *Ann. Rev. Biochem.* **60,** 477 (1991).

fidelity. Over the years several different approaches have been developed. The first fidelity assays used synthetic templates composed of only one or two nucleotides.[2] Incorporation of radioactively labeled complementary and noncomplementary nucleotides was monitored and the error rate defined as the ratio of noncomplementary to total nucleotides incorporated. The first assay that used a natural DNA template for fidelity measurements was the ΦX reversion assay.[3] The single-strand phage DNA used as templates contained *amber* codons in essential phage genes. Single-base substitution errors that reverted the *amber* codon were scored as revertant plaques on transfection of host bacterial cells with copied DNA. More recently several approaches for kinetic analysis of fidelity have been developed.[4–7] These measure kinetic constants and the degree of discrimination for each step of the polymerization reaction.

The above assays have primarily been used to describe the base substitution fidelity of DNA polymerases, usually at a small number of template nucleotides. However, mutations comprise 12 different substitution errors and several other types of mistakes occurring at many different sites within genes. To detect these, an assay is described here that monitors polymerase errors that inactivate the nonessential α-complementation activity of the *lacZ* gene in bacteriophage M13mp2. The method can be used to obtain a broad view of DNA polymerase fidelity for hundreds of different errors and the target can be manipulated to provide new sequences for examining specific hypotheses.

Experimental Outline of Fidelity Assays

The experimental approach used to measure the fidelity of DNA synthesis by purified polymerases *in vitro* is outlined in Fig. 1. A gapped M13mp2 substrate is constructed in which the single-strand gap contains the *lacZ* α-complementation target sequence. The DNA polymerase of choice is used for gap-filling synthesis. A portion of the reaction products is then analyzed by agarose gel electrophoresis to assure complete synthesis. Another aliquot of the reaction is introduced into competent *Escherichia coli* cells and these are plated onto petri dishes containing the chromogenic indicator X-Gal (5-bromo-4-chloro-3-indolyl-β-D-galactoside) and a lawn

[2] L. A. Loeb and T. A. Kunkel, *Ann. Rev. Biochem.* **52,** 429 (1982).
[3] L. A. Weymouth and L. A. Loeb, *Proc. Natl. Acad. Sci. USA* **75,** 1924 (1978).
[4] M. S. Boosalis, J. Petruska, and M. F. Goodman, *J. Biol. Chem.* **262,** 14,689 (1987).
[5] L. V. Mendelman, M. S. Boosalis, J. Petruska, and M. F. Goodman, *J. Biol. Chem.* **264,** 14,415 (1989).
[6] B. T. Eger and S. J. Benkovic, *Biochemistry* **31,** 9227 (1992).
[7] K. A. Johnson, *Ann. Rev. Biochem.* **62,** 685 (1993).

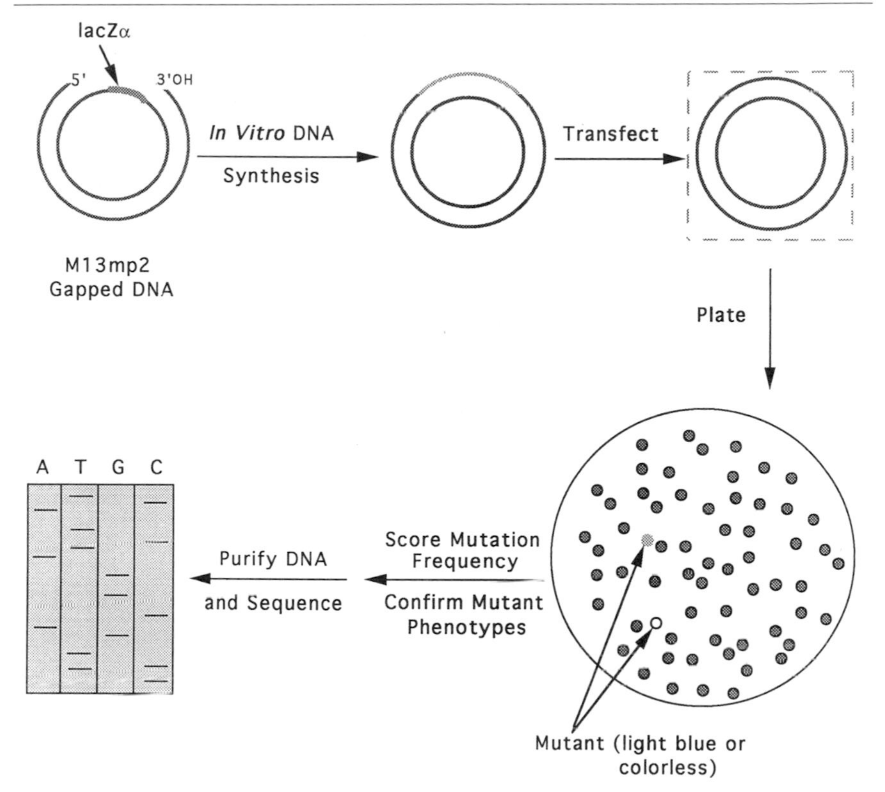

FIG. 1. Experimental outline of fidelity assay. See text for description. The 3'-OH primer terminus is nucleotide +192, where position 1 is the first transcribed nucleotide of the *lacZ* gene. The mutational target is indicated by the darker line within the gap. The square (dashed lines) represents a competent *E. coli* cell.

of *E. coli* α-complementation host cells (CSH50). If synthesis to fill in the gap in a wild-type DNA substrate is error free, then the α-peptide produced from this DNA complements the defective β-galactosidase activity of the host cell to hydrolyze the X-Gal, resulting in dark blue M13 plaques. Errors introduced during gap-filling synthesis that result in partial or complete loss of α-complementation are scored as light blue or colorless plaques. This assay for loss of function is therefore referred to as a foward mutation assay. Sequence analysis of DNA extracted from mutants can be performed to define the mutations precisely.

Because α-complementation of β-galactosidase activity can be completely inactivated without loss of phage viability, a variety of replication errors can be scored and recovered in this assay. Figure 2 illustrates the sequence of the mutational target along with the detectable single-base

```
                   GT T         AC                    AGG  A                   C    G
                   ACGCT       AGT               C    GAAACGC                 GTG    A                        G
                   CACAG  C    CTGT  G         A T    TCCCGTG      C          ACGC   CC         G     TT T    GC
GCGCA ACGCAATTAA TGTGAGTTAG CTCACTCATT AGGCACCCCA GGCTTTACAC TTTATGCTTC CGGCTCGTAT GTTGTGTGGA ATTGTGAGCG GATAACAATT
 -80                  -60                  -40                  -20                      +1                   +20

         TT         T                       A               G   G       T  AT  TC  GTT T               G     C
         CCT      G GC   C       T     C G  G   G           A   AC A   AC  GC  CTA ACC CT   T          A   A ATC
G T      AAG    AC  CGCA  GT     GC   T  T A CT  C  AT  A   TC  CCT TGT TA  TA  AGG CAA AG  TG  TA     TC  GT TGT
TCACACAGGA AACAGCTATG  ACC ATG ATT ACG AAT TCA CTG GCC GTC GTT TTA CAA CGT CGT GAC TGG GAA AAC CCT GGC GTT ACC CAA
                +40                        +60                     +80                      +100

       T   G         AC                  CT  GTT  T                    T   T
     T   C   A         GTG  G     GA  C   TCA ACC  C               C  CA   C
C  GT   A   C  TT   T  TGA  T   T  AC  AT  GAG CAA TA  T   C   T  TT  AT  TA  C             T  T
CTT AAT CGC CTT GCA GCA CAT CCC CCT TTC GCC AGC TGG CGT AAT AGC GAA GAG GCC CGC ACC GAT CGC CCT TCC CAA CAG CTG CGC
        +120                    +140                    +160                    +180
```

FIG. 2. Summary of phenotypically detectable sites in *lacZ* α target. Detectable substitutions are shown above each of the three lines of wild-type DNA sequence of the (+)-strand. Frameshifts are detectable at 153 sites of the first 51 codons and at 46 sites in the regulatory region (underlined bases). From J. D. Roberts and T. A. Kunkel, *Methods in Molecular Genetics* **2,** 295 (1993).

TABLE I
PHENOTYPICALLY DETECTABLE BASE SUBSTITUTIONS IN FORWARD MUTATION ASSAY[a]

Detectable sites				
Template base	Number	Mutation: From → to	Mispair: Template · dNTP	Number of known sites
A	33	A → G	A · dCTP	19
		A → T	A · dATP	23
		A → C	A · dGTP	17
G	29	G → A	G · dTTP	22
		G → C	G · dGTP	19
		G → T	G · dATP	25
T	33	T → C	T · dGTP	27
		T → A	T · dTTP	16
		T → G	T · dCTP	23
C	30	C → T	C · dATP	25
		C → G	C · dCTP	9
		C → A	C · dTTP	16
Total	125			241

[a] From J. D. Roberts and T. A. Kunkel, *Methods in Molecular Genetics* **2,** 295 (1993).[8]

substitution errors. Changes at 125 different sites yield mutant phenotypes. More than one substitution can be scored at many sites, yielding a total of 241 detectable substitutions (Table I, Fig. 2). The deletion or addition of a nucleotide at 199 different positions can also be scored (Table II). The forward mutation assay also scores errors involving the loss or gain of more than one base, including duplications and simple and complex deletions.

As an alternative to the forward mutation assay, the starting DNA substrate may contain a base substitution or frameshift mutation that encodes a mutant plaque phenotype. This permits the scoring of polymerase errors that restore partial or full α-complementation activity to yield a blue plaque phenotype. These reversion assays are more sensitive than the forward mutation assay, since their spontaneous background mutant frequencies are lower (10^{-6} to 10^{-5} versus 6×10^{-4}). Nonsense or missense codons in the *lacZ* α-complementation sequence that have been used to monitor substitution errors include TAA *ochre* codons at positions 57, 75, 87, 90, 108, and 147, TAG *amber* codons at positions 87 and 147, TGA *opal* codons at positions 87 and 147, and missense codons at positions 144 (AGA) and 141 (CCC).[9] Alternatively, several frameshift reversion assays have been developed to monitor nucleotide deletions or additions within

[8] J. D. Roberts and T. K. Kunkel, *Methods in Molecular Genetics* **2,** 295 (1993).
[9] L. A. Frederico, T. A. Kunkel, and B. Ramsay Shaw, *Biochemistry* **29,** 2532 (1990).

TABLE II
DETECTABLE SITES FOR ONE-NUCLEOTIDE FRAMESHIFTS[a]

Number of consecutive base pairs	A		G		T		C	
	No. of occurrences	No. of bases	No. of occurrences	No. of bases	No. of occurrences	No. of bases	No. of occurrences	No. of bases
1	27	27	26	26	16	16	28	28
2	7	14	7	14	10	20	5	10
3	0	0	1	3	3	9	5	15
4	1	4	0	0	1	4	1	4
5	0	0	0	0	0	0	1	5
Total		45		43		49		62

[a] Includes the 153 bases for the first 51 codons shown in Fig. 2 plus 46 bases (underlined in Fig. 2) of the regulatory sequence at which 1-base frameshifts have been detected.

the coding sequence.[10,11] As described in more detail below, starting mutants for reversion assays should have a colorless or faint blue plaque phenotype in order to permit quantitative scoring of blue revertants.

Materials

Stock Solutions

TE buffer: 10 m*M* Tris–HCl, pH 8.0, 0.1 m*M* EDTA.

TAE buffer (50×): Tris base, 242 g; glacial acetic acid, 57.1 ml; EDTA, 100 ml of a 500 m*M* solution, pH 8.0. Dissolve in distilled H_2O, adjust volume to 1 liter.

SSC 20X: Sodium chloride, 175.3 g; sodium citrate, 88.2 g is dissolved in 800 ml of distilled H_2O and pH adjusted to 7.0. The volume was adjusted to 1 liter and dispensed into aliquots. Sterilized by autoclaving and stored at −20°.

Sodium dodecyl sulfate dye mix: 20 m*M* Tris–HCl, pH 8.0, 5 m*M* EDTA, 5% (w/v) sodium dodecyl sulfate (SDS), 0.5% (w/v) bromphenol blue, 25% (v/v) glycerol.

X-Gal: 50 mg of 5-bromo-4-chloro-3-indolyl-β-D-galactoside is dissolved per ml in *N,N*-dimethylformamide and stored at −20°. Avoid exposure to light.

IPTG: 24 mg isopropylthiogalactoside/ml in distilled H_2O, stored at 4°.

Soft agar: 0.8% agar, 0.9% NaCl.

YT medium: Bacto-tryptone, 8 g; Bacto-yeast extract, 5 g; NaCl, 5 g. Add to 1 liter of H_2O and sterilize in an autoclave.

2× YT medium: Bacto-tryptone, 16 g; Bacto-yeast extract, 10 g; NaCl, 10 g. Add to 1 liter of H_2O, adjust pH to 7.4 with HCl and sterilize in an autoclave.

VB salts (50×): $MgSO_4 \cdot 7H_2O$, 10 g; citric acid (anhydrate), 100 g; K_2HPO_4, 500 g; $Na_2(NH_4)HPO_4 \cdot 4H_2O$, 175 g. Dissolve the above in 670 ml distilled H_2O, bring volume to 1 liter, and sterilize in an autoclave. After dilution, pH is 7.0 to 7.2.

Minimal plates: Add 16 g of Difco agar to 1 liter of distilled H_2O and sterilize in an autoclave. When the agar has cooled to 50°, add 0.3 ml of 100 m*M* IPTG, 20 ml of 50× VB salts, 20 ml of 20% glucose, and 5 ml of 1 mg thiamin hydrochloride/ml. [Each of these solutions is sterilized either by filtration (0.2-μm pore) or in an autoclave prior

[10] K. Bebenek, C. M. Joyce, M. P. Fitzgerald, and T. A. Kunkel, *J. Biol. Chem.* **265,** 13,878 (1990).

[11] K. Bebenek, J. D. Roberts, and T. A. Kunkel, *J. Biol. Chem.* **267,** 3589 (1992).

to their addition to the 50° agar.] The mixture is mixed well and dispensed into sterile, level petri dishes (30 ml/plate).

SOC medium: Add Bacto-tryptone, 20 g; Bacto-yeast extract, 5 g; NaCl, 10 ml of a 1 *M* solution; and KCl, 2.5 ml of a 1 *M* solution to 970 ml distilled H_2O. Stir to dissolve, sterilize in an autoclave, and cool to room temperature. Add 10 ml of a 2 *M* Mg^{2+} stock (1 *M* $MgCl_2 \cdot 6H_2O$ and 1 *M* $MgSO_4 \cdot 7H_2O$, filter sterilized). Add 10 ml of a 2 *M* stock of glucose (filter sterilized). Filter the complete medium through a 0.2-μm filter unit. pH should be 7.0 $\pm$ 0.1. Filter sterilizing units should be prefiltered with distilled H_2O to remove any toxic materials from the filter.

10× CTAB: Cetyltrimethylammonium bromide (CTAB), 5 g; 5 *M* NaCl, 10 ml. Add distilled H_2O to 100 ml, mix gently but thoroughly (the CTAB is slow to go into solution) and filter sterilize. Store at room temperature.

Enzymes and Reagents

Restriction enzyme *Pvu*II is obtained from New England Biolabs (Beverly, MA) and is used according to the manufacturer's instructions. Proteinase K is from Boehringer Mannheim (Indianapolis, IN). Deoxyribonucleoside triphosphates are the HPLC-purified form purchased as 100 m*M* solutions from Pharmacia LKB (Piscataway, NJ). QIAGEN columns for plasmid purification are from QIAGEN Inc. (Chatsworth, CA). Agar, tryptone, and yeast extract are from Difco Labs (Detroit, MI), PEG-8000 and CTAB are from Sigma (St. Louis, MO), IPTG is from BRL (Gaithersburg, MD) and X-Gal is from Biosynth AG (Switzerland).

Bacterial Strains. The following *E. coli* strains are used:

For phage growth to prepare gapped substrate: NR9099 [Δ(*pro-lac*)], *thi, ara, recA56/F′* (*proAB, lacI$_q$ZΔM15*). This strain is preferred because it allows less recombination between M13 DNA and the host F′, therefore reducing the number of background colorless mutants that can result from recombination (see Ref. 12).

For electroporation: MC1061 *hsdR, hsdM+, araD,* Δ(*ara, leu*), Δ(*ara, leu*), Δ(*lacIPOZY*), *galU, galK, strA.* Preferred because it yields a very high efficiency for electroporation, $\sim 10^5$ plaques/ng of DNA.

For lawn formation and ssDNA preparations for sequencing: CSH50 [Δ(*pro-lac*)], *thi, ara, strA/F′* (*proAB, lacI$_q$ZΔM15, traD36*)

Bacteriophage. M13mp2, developed by Messing and co-workers,[13] con-

[12] T. A. Kunkel, *Proc. Natl. Acad. Sci. USA* **81,** 1494 (1984).

[13] J. Messing, B. Gronenborn, B. Muller-Hill, and P. H. Hofschneider, *Proc. Natl. Acad. Sci. USA* **74,** 3642 (1977).

tains a segment of the *E. coli lac* operon within the intergenic region of M13. The bacteriophage currently used for construction of the gapped substrate for the forward assay has a single nucleotide change, T → C at position +192 (where position +1 is the first transcribed base of the *lacZ* α sequence). This substitution does not affect α-complementation but creates a new site for restriction endonuclease *Pvu*II. Double-stranded DNA with this site can then be digested with this enzyme alone to generate the fragment needed for gapped substrate construction (thus simplifying the procedure used earlier, which required use of a second restriction endonuclease[14]).

Mutant derivatives for reversion assays were either obtained from sequenced mutant collections or were prepared by oligonucleotide directed mutagenesis as described.[15]

Experimental Procedures

Construction of Gapped M13 DNA Substrate

The bacteriophage M13mp2 is plated on minimal plates (as described below), using *E. coli* NR9099 as the host strain. A single plaque is added to 1 liter of 2× YT medium containing 10 ml of an overnight culture of *E. coli* NR9099. M13mp2 infected cells are grown overnight at 37° with vigorous shaking. Cells are then harvested by centrifugation at 5000*g* for 30 min, and the RF I form of the M13mp2 bacteriophage is prepared by alkaline lysis and anion exchange chromatography on columns from QIAGEN, Inc., according to the protocol from the manufacturer. The QIAGEN purification procedure yields DNA preparations containing both single-stranded (ssDNA) and double-stranded DNA (dsDNA). The presence of the ssDNA does not interfere with restriction endonuclease digestion, and it is used during hybridization to form gapped substrate (see below).

Replicative form of M13mp2 DNA is digested to completion at 37° with restriction endonuclease *Pvu*II to produce four blunt-ended fragments of 6789, 268, 93, and 46 base pairs. These fragments are sufficiently different in size to permit separation by precipitation with polyethylene glycol (PEG).[16] The digested DNA is ethanol precipitated and resuspended in TE buffer. The solution is adjusted to 6.0% PEG-8000, 0.55 *M* NaCl, and a final DNA concentration of 0.5 mg/ml. This mixture is incubated at 0° overnight. The precipitated 6789 base pair fragment is pelleted by centrifu-

[14] T. A. Kunkel, *J. Biol. Chem.* **260,** 5787 (1985).
[15] T. A. Kunkel, K. Bebenek, and J. McClary, *Methods Enzymol.* **204,** 125 (1991).
[16] J. T. Lis, *Methods Enzymol.* **65,** 347 (1980).

gation for 10 min in a microcentrifuge and the supernatant carefully removed. After resuspending the pellet in TE buffer, an aliquot of this sample and the supernatant are analyzed by electrophoresis in a 0.8% agarose gel. The resuspended pellet fraction should not contain detectable small fragments. If necessary, the 6789 base pair fragment may be precipitated with PEG a second time to yield a preparation lacking detectable small fragment. To remove residual PEG, the large fragment is precipitated with ethanol and then resuspended in TE buffer to a final concentration of 1 mg/ml. The final yield is ~90 to 95% of the starting material, and includes most of the ssDNA originally present in the Qiagen-purified preparation.

To form the gapped template, the large fragment is hybridized to single-stranded circular, viral (plus) DNA. The DNA is diluted 10-fold with water to reduce the DNA concentration and the ionic strength, and then incubated at 70° for 5 min to denature the strands. This dilution-heating procedure was specifically developed to allow strand denaturation at relatively low temperature, in order to avoid heat-induced DNA damage to the substrate. Because denaturation at 70° only occurs at low ionic strength, it is important that the starting DNA preparation be free of residual salt from the ethanol precipitation step.

Sufficient single-strand viral DNA is added just before removal from the 70° water bath to produce a ~1:1 molar ratio of minus strand fragment to viral, circular DNA. Often sufficient ssDNA is present in the large fragment preparation such that it is not necessary to add extra ssDNA. The mixture is placed on ice for 5 min, then SSC is added to a final 2× concentration (300 m*M* NaCl, 30 m*M* sodium citrate). The mixture is incubated at 60° for 5 min and then placed on ice. The DNA is ethanol precipitated and resuspended in TE buffer, and an aliquot is analyzed by agarose gel electrophoresis (see below). Using the 1:1 fragment to viral DNA ratio, typically approximately 50% of the ssDNA is converted to double-stranded circular DNA containing a 407 base gap from position −216 to +191. The biological activity of the remaining ssDNA is very low due to the low (100-fold lower that for dsDNA) transfection efficiency. The concentration of the gapped substrate is estimated by visual inspection of the gel in comparison to a known quantity of an RF DNA standard. The above procedure can be used to construct gapped substrates for reversion assays, using a different mutant derivative of bacteriophage M13mp2.

In vitro DNA Synthesis Reactions

Reaction conditions depend on the experimental goal and the DNA polymerase used. As just one example, the reaction mixture (25 μl) contains 20 m*M* HEPES, pH 7.8, 2 m*M* dithiothreitol (DTT), 10 m*M* $MgCl_2$, 150 ng of gapped DNA, all four dNTPs, and the DNA polymerase. The amount of DNA refers to the gapped DNA itself. This is only a fraction of the total DNA in a substrate preparation, the remainder being double-stranded linear DNA and single-stranded linear and circular DNA. The amount of polymerase required for gap filling depends on the properties and quality of the DNA polymerase preparation and should be empirically determined. Typically, polymerase is required in excess over DNA substrate, because some of the enzyme binds nonproductively to the ssDNA present in the substrate preparation. Reactions are incubated at the desired temperature and terminated by adding Na_2EDTA, pH 8.0, to a final concentration 1.5 times the concentration of the $MgCl_2$.

Product Analysis of Gap-filling Reactions

Agarose gel electrophoretic analysis of 20 μl of a gap-filling synthesis reaction (by HIV-1 reverse transcriptase) is presented in Fig. 3, lane A. For comparison, lane B shows DNA standards. The upper band in lane A that migrates at the position of the RF II standard represents the product of complete gap-filling synthesis, that is, a successful reaction. Several alternative band patterns point to problems in gap-filling synthesis. Examples include (1) DNA migrating at the position of the gapped DNA standard, indicating a lack of polymerase activity or omission of a critical component (e.g., dNTPs), (2) product migrating as a discrete band above the position of gapped DNA but below the RFII DNA standard, possibly indicating

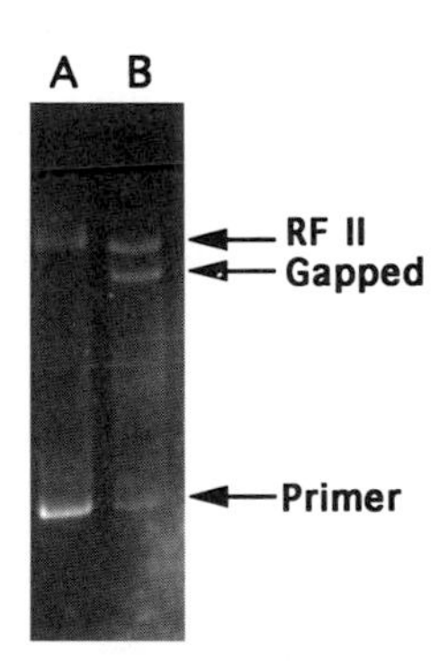

FIG. 3. Analysis of reaction products by agarose gel electrophoresis. Twenty microliters of the polymerization reaction were mixed with 5 μl of SDS dye mix and subjected to electrophoresis in an 0.8% agarose gel in TAE buffer containing 0.5 μg/ml ethidium bromide. Electrophoresis was at a constant 80 V for 16 hr. Lane A, gap-filling reaction with HIV-1 RT; lane B, standards, including an amount of uncopied gapped DNA equivalent to the copied DNA shown in lane A. Not visible is the ssDNA, which migrates well below the *Primer* DNA.

polymerase pausing at the palindromic *lac* operator sequence, (3) products migrating as a smear between the position of the uncopied DNA and the RF standard, representing products of partial gap-filling synthesis, (4) products seen as a smear, starting at the position of double-stranded RFII DNA and extending toward higher molecular weight material, indicating strand displacement synthesis, and (5) faint bands, no product, or products of lower molecular weight than the gapped DNA, indicating degradation of DNA by nuclease(s). Mutant frequencies are determined only for reactions in which the input gapped DNA migrates coincident with fully double-stranded form II DNA.

Preparation of Competent Cells

Reaction products are introduced into *E. coli* strain MC1061 by electroporation. Competent cells are prepared approximately as described.[17] A liter of 2× YT medium is inoculated with 10 ml of a fresh overnight culture. The cells are grown at 37° with vigorous shaking to an optical density at 600 nm of 0.5 to 0.8 (approximately 1.5×10^{-7} cells/ml). The flask is chilled on ice for ~30 min, the cells transferred to a cold 1-liter centrifuge bottle, and the cells pelleted by centrifugation at ~4000*g* for 30 min at 4°. The media is carefully decanted and the cells resuspended in 1 liter of ice-cold distilled H_2O, taking care to resuspend the cells completely. The cells are transferred to 250-ml conical centrifuge bottles, pelleted at ~2200*g* for 20 min at 4°, and resuspended in 500 ml of ice-cold distilled H_2O. Cells are pelleted once again (~2200*g* for 20 min at 4°), resuspended in 20 ml ice-cold 10% glycerol, transferred to a 50-ml centrifuge tube, and pelleted at 3000*g* for 15 min at 4°. This final cell pellet is resuspended in about 3 ml of ice-cold 10% glycerol and divided into 200-μl aliquots in 1.5-ml microcentrifuge tubes. Cells are then frozen rapidly in a dry ice–ethanol bath and stored at −70° until needed. Immediately prior to use, these cells are thawed and kept on ice. Cells are diluted with ice-cold 10% glycerol to a concentration of $\geq 2 \times 10^{10}$ and mixed gently. (The cell concentration is measured by plating each preparation and is usually found to be 3 to 6×10^{10} cells/ml prior to dilution.)

Electroporation

A portion of the gap-filling reaction is introduced into competent cells to score mutant frequencies. Typically, 1 μl of a 1/50 dilution in distilled H_2O is used for the forward mutation assay, while 1 μl of a 1/10 dilution is used for reversion assays. The DNA is mixed with 50 μl competent cells,

[17] W. J. Dower, J. F. Miller, and C. W. Ragsdale, *Nucleic Acids Res.* **16,** 6127 (1988).

which are then transferred to an ice-cold 0.2-cm cuvette. Electroporation is accomplished with a Bio-Rad Gene Pulser set at 2 to 2.5 kV, 400 Ω, 25 μF. Typical time constants are 9.0 to 9.3 ms. Immediately following electroporation, 2 ml of SOC medium (at room temperature) is added to the cells, which are kept at room temperature.

Plating

Cells are plated onto minimal agar plates within 20 min of electroporation, to avoid the release of progeny M13 viral particles. An aliquot (usually 5 to 20 μl for forward assays and 10 to 100 μl for reversion assays) of the electroporated cells is added to a tube at 45 to 49° containing 2.5 ml of soft agar, 500 μg IPTG and 2.5 mg of X-Gal. (Alternatively, the IPTG and X-Gal can be in the bottom agar in the plates.) Finally, 0.5 ml of a log-phase culture of CSH50 (the α-complementation strain) cells is added. The soft agar is poured onto the minimal plate and allowed to solidify. Plates are inverted and incubated overnight (~14 to 18 hr) at 37° then scored for mutant or revertant phenotypes.

An appropriate amount of the electroporated cells is used to yield 200 to 400 plaques per plate for the forward assay (Fig. 4, see color insert, plate 1) or 2000 to 10,000 M13mp2 plaques per plate for reversion assays. At densities of more than about 400 plaques per plate for the forward assay (Fig. 4, see color insert, plate 2) and 10,000 plaques per plate for the reversion assay, detection of the mutant phenotypes becomes difficult. The total number of plaques in a forward assay is determined by counting all plaques for which the color phenotype is obvious. Tiny plaques, plaques at the edge of the plate, or those on regions of the plate smeared by a drop of water are not counted. Estimates of the total number of plaques per plate in the reversion assays are calculated from 10-fold dilutions of the electroporated samples, plated separately to yield 200 to 1000 plaques per plate.

For the foward mutation assay, wild-type plaques have been arbitrarily assigned a color value of 4+ and colorless plaques a value of 0+. Plaques of intermediate color are assigned values of 1+ (light blue), 2+ (medium blue), and 3+ (medium-dark blue, i.e., almost wild-type), depending on their color (Fig. 4). For forward assays, it is helpful to examine the plates carefully within 24 hr following plating, since the 3+ phenotype is best distinguished at this time. For reversion assays, dark blue revertants are obvious within the first 24 hr, but light blue revertants may not be apparent this quickly. We typically examine reversion plates on the morning after plating and leave the plates at room temperature for an additional 48 hr. Plates are then reexamined against a white background to detect light blue (e.g., 1 and 2+) plaques. A key feature of reversion assays is that the starting mutants have a colorless (0+) or light blue (≤1+) plaque phenotype. Mutants with darker blue plaque colors (e.g., 2+) are not as useful, since it is difficult to detect reliably 3+ or 4+ revertants on plates crowded with 2+ mutants.

If needed, plates can be stored at 4° for up to several weeks prior to picking mutant plaques for further analysis. Mutant or revertant plaques can also be picked from plates and stored in a solution of 0.9% sodium chloride. When frozen, these are stable for many years.

Plaque Purification to Confirm Mutant Phenotype

A mutant plaque, containing $\sim 10^9$ plaque forming units, is placed into a 5 ml solution of 0.9% NaCl. A wild-type plaque is placed into the same tube. After setting for 20 min or more, the solution is vortexed vigorously and the phage are serially diluted 10,000-fold. Five to 20 μl is plated on minimal plates as described above, to yield about 200 mutant and 200 wild-type plaques on the same plate (Fig. 4, see color insert, plates 3 and 4). This allows a direct comparison of the colors of the putative mutant and the known wild-type phage under identical plating conditions. Plaques confirmed to have a mutant phenotype are then picked as above, being careful to avoid other plaques, and placed either into a tube of sterile 0.9% NaCl for long-term storage or into an early log culture of CSH50 for preparation of ssDNA for sequencing.

Preparation of ssDNA for Sequence Analysis

To prepare ssDNA, mutant phage are grown (6 hr to overnight) by adding a fresh plaque to ~5 ml of an early log-phase culture of CSH50 in 2× YT media at 37°. A 1.5-ml aliquot of the culture is transferred to a microcentrifuge tube and centrifuged for 10 min at room temperature. The phage are precipitated from 1 ml of the cleared supernatant by adding 0.25 ml of PEG-8000/NaCl [15% (w/v)/2.5 *M*], incubating ≥5 min at 4° and centrifuging ~2 min. Single-stranded DNA is then purified by a modification of the CTAB procedure for phage.[18] First, the phage pellet is resuspended in 690 μl of water, then 100 μl of 1 *M* Tris–HCl, pH 8, 100 μl of 0.5 *M* EDTA and 1 to 10 μl of a 10 mg proteinase K/ml solution is added and the mixture incubated at 37° for ~30 min. Next, the cationic detergent cetyltrimethylammonium bromide (CTAB) is added [100 μl, 10× CTAB, final concentration 0.5% (w/v)] and the samples incubated for ~10 min at 37°. The sample is centrifuged for 10 min and the pellet (often not visible) is resuspended in 300 μl of 1.2 *M* NaCl. DNA is then precipitated by adding

[18] G. Del Sal, G. Manfioletti, and C. Schneider, *BioTechniques* **7,** 514 (1989).

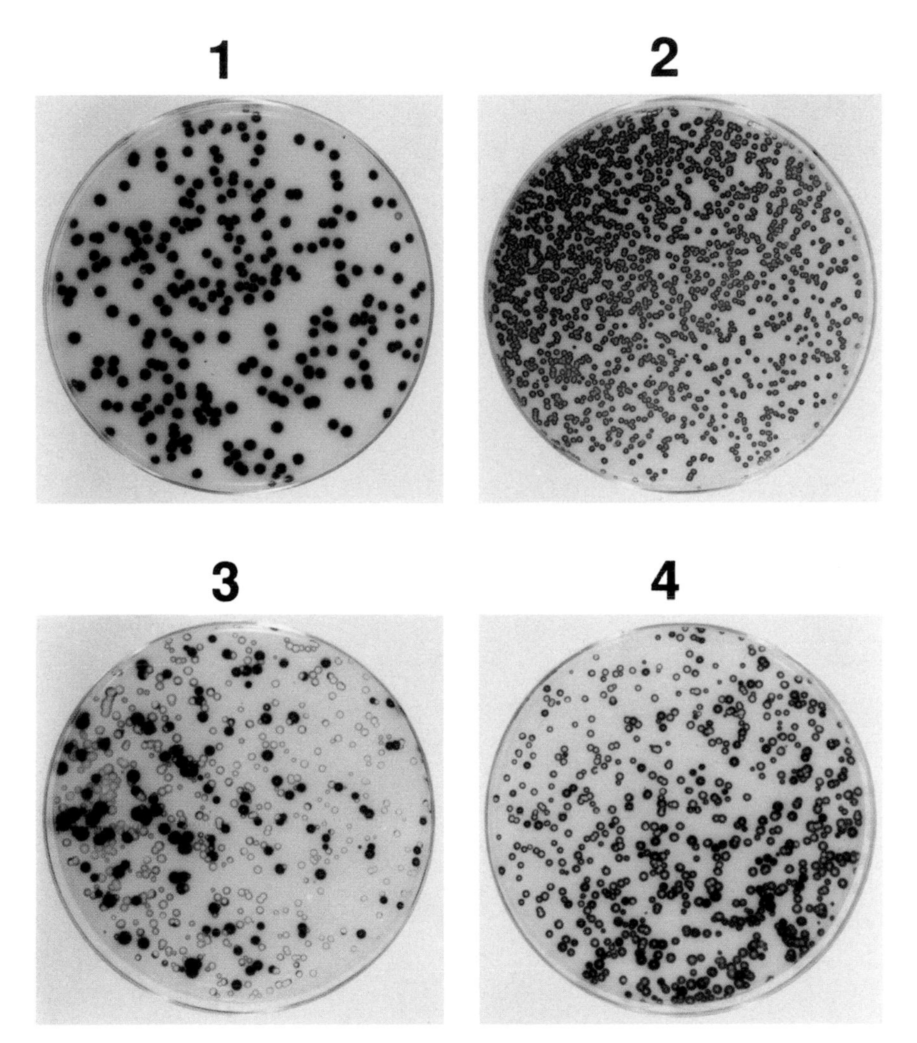

FIG. 4. M13mp2 plaques on minimal agar plates containing IPTG and X-Gal. Bacteriophage plaques on plates 1 and 2 were obtained by plating MC1061 cells transfected with the products of DNA synthesis on gapped substrate for the forward mutation assay. Plate 1 contains an appropriate number of plaques allowing easy identification of mutants. Three light blue mutant phage plaques are visible. In plate 2, the number of plaques is too high, making identification of 3+ mutant plaques difficult. Plates 3 and 4 were obtained by plating a mixture of wild-type and mutant phage.

TABLE III
FIDELITY MEASUREMENTS WITH THREE DNA POLYMERASES

Enzyme	Plaques		Mutant frequency[a]
	Total	Mutant	
Forward mutation assay			
Calf thymus Pol ε[b]	16,959	12	7×10^{-4}
Pol I Klenow fragment	16,288	85	52×10^{-4}
HIV-1 RT[c]	5,585	446	800×10^{-4}
TGA reversion assay			
Calf thymus Pol ε[d]	2,300,000	12	5×10^{-6}
Pol I Klenow fragment	1,300,000	97	77×10^{-6}
HIV-1 RT[e]	730,000	75	100×10^{-6}
Frameshift reversion assay I			
Pol I Klenow fragment	210,000	17	8×10^{-5}
HIV-1 RT	48,000	227	470×10^{-5}

[a] The background mutant frequencies for the uncopied DNA substrates were 6×10^{-4} for the forward assay, 2×10^{-6} for the opal codon reversion assay, and 1×10^{-5} for the frameshift reversion assay.

[b] From D. C. Thomas, J. D. Roberts, R. D. Sabatino, T. W. Myers, Ch-K. Tan, K. M. Downey, A. G. So, R. A. Bambara, and T. A. Kunkel, *Biochemistry* **30,** 11,751 (1991).

[c] From K. A. Eckert and T. A. Kunkel, *Nucleic Acids Res.* **21,** 5212 (1993).

[d] From T. A. Kunkel, R. D. Sabatino, and R. A. Bambara, *Proc. Natl. Acad. Sci. USA* **84,** 4865 (1987).

[e] From J. D. Roberts, K. Bebenek, and T. A. Kunkel, *Science* **242,** 1171 (1988).

0.75 ml ethanol and incubating the sample at $-20°$ for ≥ 10 min. The DNA is pelleted by centrifuging for 15 min at 4° and the pellet washed with 1 ml 70% ethanol, spun for 5 min, dried, and resuspended in 20 μl distilled H_2O. Two to 4 μl (~ 1 μg) of this ssDNA is used for dideoxy sequencing reactions[19] carried out with Sequenase version 2.0, following the protocols from US Biochemical.

Typical Results of Transfection

To illustrate typical plaque counts and mutant frequencies obtained with the forward and reversion assays, results of fidelity measurements with *E. coli* Pol I (Klenow fragment), calf thymus DNA Pol ε, and HIV-1 reverse transcriptase (RT) are presented in Table III. Accurate DNA polymerases (e.g., Pol ε) yield mutant frequency values that are close to or not significantly different from mutant frequencies of respective uncopied DNA con-

[19] F. Sanger, S. Nicklen, and A. R. Coulson, *Proc. Natl. Acad. Sci. USA* **74,** 5463 (1977).

trols. Error-prone DNA polymerases [e.g., human immunodeficiency virus type 1 (HIV-1) reverse transcriptase] yield much higher mutant frequency values.

Calculation of Error Rates

Error rates can be calculated from reversion frequencies, or from forward frequency data, followed by DNA sequence analysis of collections of independent mutants in order to define error specificity. To calculate an error rate, first subtract the background mutant frequency of uncopied DNA from the value obtained from the products of the polymerase reaction. Multiply this value by the percentage of mutants represented by the particular class of error under consideration. Divide this value by 0.6, the probability that a polymerase error in the newly synthesized minus strand will be expressed in *E. coli.* Finally, divide by the number of sites at which the error under consideration can be detected. This last step corrects for differences in target size for the various types of errors, such that all rates expressed per *detectable* nucleotide incorporated.

[19] Gel Fidelity Assay Measuring Nucleotide Misinsertion, Exonucleolytic Proofreading, and Lesion Bypass Efficiencies

By STEVEN CREIGHTON, LINDA B. BLOOM, and MYRON F. GOODMAN

Introduction

The fidelity of DNA synthesis depends on polymerase–primer–template–nucleotide interactions that can alter fidelity at different locations along a primer–template molecule. An important biological consequence is the appearance of mutational hot and cold spots. Factors known to influence fidelity include base stacking interactions, base context effects including primer–template slippage, polymerase active site constraints, protonated or ionized nucleotide substrates, polymerase accessory factors, and mutagenic metals and carcinogens.[1,2] It is desirable to have a simple fidelity assay so that an investigation of how each of these factors affects fidelity can be carried out systematically.

[1] H. Echols and M. F. Goodman, *Ann. Rev. Biochem.* **60,** 477 (1991).
[2] M. F. Goodman, S. Creighton, L. B. Bloom, and J. Petruska, *Crit. Rev. Biochem. Molec. Biol.* **28,** 83 (1993).

Template-directed synthesis of DNA involves elongation of a 3′-primer terminus by addition of deoxyribonucleotides by DNA polymerases and reverse transcriptases. In this chapter, we describe a simple general assay to measure the fidelity of polymerase reactions at arbitrary positions on DNA or RNA template strands.[3,4] Fidelity measurements are carried out by resolving single-stranded DNA differing in lengths by single nucleotides resulting from elongation of a 5-^{32}P-labeled primer by polymerase using polyacrylamide gel electrophoresis (PAGE). The only requirement is that integrated gel band intensities, corresponding to primer elongation at a template target site and at adjacent template sites, can be quantified. There are no *a priori* restrictions on the nature of the substrate and template bases; the primer–templates and dNTP substrates can consist of the four naturally occurring bases, combinations of natural and modified bases, and template abasic (apurinic/apyrimidinic) lesions.[5]

Detailed protocols are provided for assaying fidelity at specific, arbitrarily chosen template sites. Assays will be described for polymerases that possess 3′-exonuclease activity[6] and for those that do not.[3,4,6] A special application of the gel assay is described for measuring nucleotide incorporation and bypass at abasic template lesions.[5,7]

Model for Expressing Rates and Fidelity of DNA Synthesis in Terms of Integrated Gel Band Intensites

When DNA polymerase extends a ^{32}P-labeled primer by addition of either a right (R) or wrong (W) nucleotide, or when the enzyme adds a next correct nucleotide onto either a correctly paired or mispaired primer terminus, reaction products can be visualized by autoradiography or by phosphorimaging after separating primers of different lengths using PAGE. Bands corresponding to different length primers can be integrated using densitometry or by the standard phosphorimager software packages. This section describes the model used to convert the integrated band intensities into polymerase fidelities and nucleotide insertion, extension, and excision velocities. Two primer–template (P/T) constructs are discussed: running

[3] M. S. Boosalis, J. Petruska, and M. F. Goodman, *J. Biol. Chem.* **262,** 14,689 (1987).

[4] L. V. Mendelman, M. S. Boosalis, J. Petruska, and M. F. Goodman, *J. Biol. Chem.* **264,** 14,415 (1989).

[5] S. K. Randall, R. Eritja, B. E. Kaplan, J. Petruska, and M. F. Goodman, *J. Biol. Chem.* **262,** 6864 (1987).

[6] S. Creighton and M. F. Goodman, *J. Biol. Chem.* **270,** 4759 (1995).

[7] H. Cai, L. B. Bloom, R. Eritja, and M. F. Goodman, *J. Biol. Chem.* **268,** 23,567 (1993).

starts (Fig. 1A) and standing starts (Fig. 1B). Running starts are typically used to measured nucleotide insertion kinetics and fidelity. Standing starts are most commonly used to measure primer extension kinetics, comparing extension of mismatched versus correctly matched primer termini or extension at DNA sites containing "blocking" template lesions such as abasic (apurininic/apyrimidinic) sites.

Single Completed Hit Model of Polymerization

The model that we use for interpreting the intensities of bands on a gel and relating them to the kinetic parameters for DNA polymerase is referred to as the single completed hit (SCH) model,[2,6] and applies to running-start P/T molecules. SCH conditions are satisfied when the great majority of gel bands originate from addition of one or more bases in a single encounter of an individual P/T with a polymerase (Fig. 2). The single hit is from the perspective of the P/T molecule, which is acted on by a polymerase at most once; polymerase molecules will themselves encounter many P/T molecules during the course of the reaction. It is desirable that a given polymerase molecule encounter many P/T molecules during the reaction provided that the great majority of P/T molecules encounter an enzyme no more than once. Inclusion of a DNA trap in the assay, (see DNA Traps section) can also ensure that almost all hits are single completed hits in the presence of high levels of polymerase.

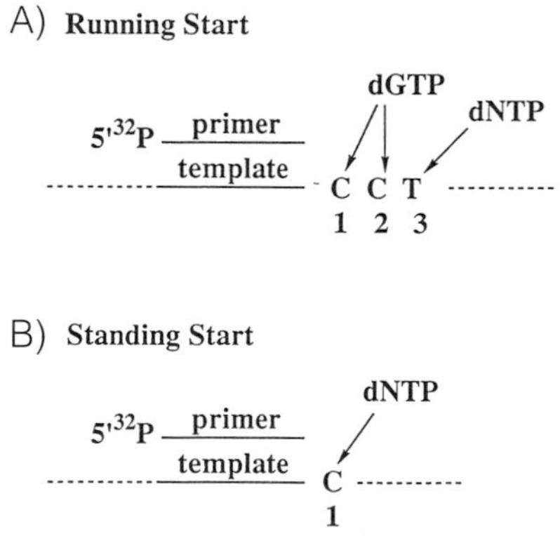

A) **Running Start**

dGTP
dNTP
$5'^{32}P$ primer
template
C C T
1 2 3

B) **Standing Start**

dNTP
$5'^{32}P$ primer
template
C
1

FIG. 1. Primer–templates used for measuring fidelity of DNA synthesis. (A) Running start used to measure reaction velocity versus dNTP concentration for insertion opposite template base T. Saturating concentration of dGTP is used for extension to reach the template T site. dNTP substrates, right or wrong, are used at variable concentrations for insertion opposite T. (B) Standing start used to measure reaction velocity versus dNTP concentration for insertion opposite template base C, using either a right or wrong dNTP substrate. Standing starts are convenient for measuring extension of matched compared to mismatched primer-3′-termini.

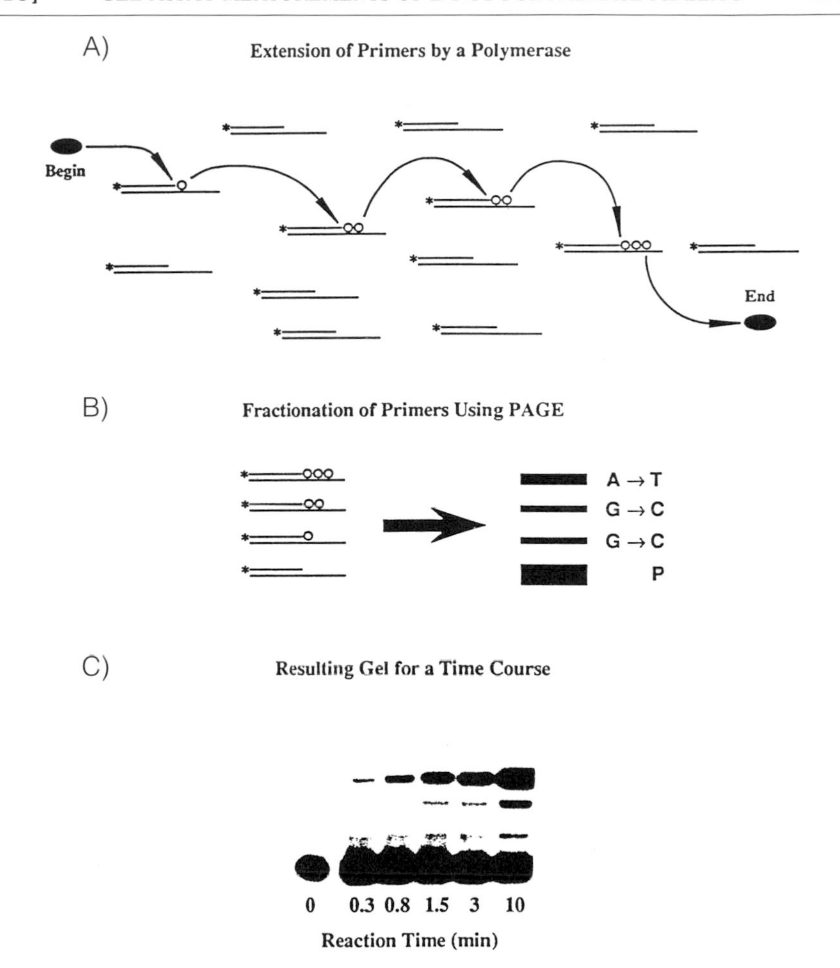

FIG. 2. Polymerization in the SCH approximation. (A) Idealized picture of polymerase action. At the beginning of the reaction, the polymerase finds a P/T and adds a variable number of bases, then dissociates. During the course of the reaction the polymerase engages a series of P/T DNAs, adding a variable number of nucleotides during each encounter. Most of the gel bands originate from single completed hits rather than from enzymes bound to P/T molecules at the end of the reaction. (B) At the conclusion of the experiment, the extended primers are loaded on a gel and fractionated according to size. The intensity of each band corresponding to the addition of $N = 1, 2, \ldots$ nucleotides, is proportional to the probability that polymerases have extended a given P/T, at most once, by incorporation of N nucleotides, in a single encounter. (C) Typical time course. *Drosophila* Pol α was present at ~1 nM, P/T DNA at 36 nM, and dGTP and dATP were present at 50 and 3 μM, respectively. Reactions were run for the indicated time at 37°. The template sequence was 3′ ... CCT ... 5′, the same as in (B).

The completed part of the SCH condition requires that most of the gel bands arise from polymerase molecules that dissociate from the P/T before the conclusion of the experiment (i.e., polymerase molecules are not interrupted by termination of the reaction, but instead have "completed" the P/T hit by dissociation). Experimentally, the completed hit condition is ensured if either the amount of time that any particular polymerase is bound to P/T is a vanishingly small fraction of the time that it is unbound, or if the concentration of polymerase molecules is far exceeded by that of the P/T.

Conditions to Achieve SCH

In practice, the conditions for single completed hits are (1) that the total amount of primer extended during the reaction must be less than 20% and (2) that the total concentration of polymerase used during the reaction must be less than 5% of the total concentration of P/T. At 20% of primer usage, there are a number of primers present that *have* been hit by more than one polymerase during the course of the reaction, but their abundance in relation to the singly hit primers is negligibly small.

Nucleotide Insertion Rates and Fidelity for Polymerases Lacking Exonuclease Activity

When SCH conditions are satisfied, integrated gel band intensities reflect *the probability that a polymerase molecule will dissociate from the P/T at that particular location* (Fig. 3A). During a single polymerase–P/T encounter, the polymerase may either dissociate prior to adding a nucleotide, or it may add one or more nucleotides to the template before dissociating. A dark gel band reflects a greater probability for the polymerase to dissociate relative to its rate of adding a nucleotide at that site; conversely, a light gel band reflects a rapid rate of polymerization compared to dissociation at that site. The behavior of the polymerase is assumed to be governed only by the relative probabilities of dissociation and polymerization, so that a polymerase reaching a given location $T - 1$ along the template (Fig. 3B) will either dissociate, probability p_{off}, or add a nucleotide to reach a target location T, probability p_{pol}. The probabilities of these events are

$$p_{\text{off}} = k_{\text{off}}/(v_{\text{pol}} + k_{\text{off}}) \tag{1}$$
$$p_{\text{pol}} = v_{\text{pol}}/(v_{\text{pol}} + k_{\text{off}}), \tag{2}$$

where k_{off} is the dissociation rate of the polymerase from the P/T at site $T - 1$ and v_{pol} is the rate of addition of a nucleotide at site T and is a function of nucleotide concentration. The intensities of the band at location $T - 1$ (I_{T-1}) and the intensities of the bands at T and beyond (I_T^*) are

A) Course of a Single Completed Hit

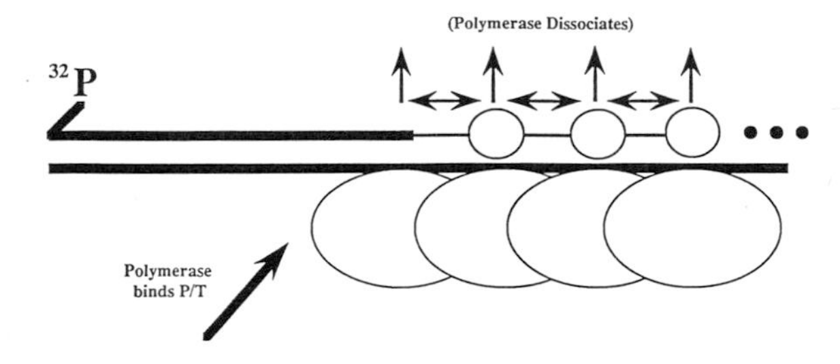

B) Standard Polymerase Model

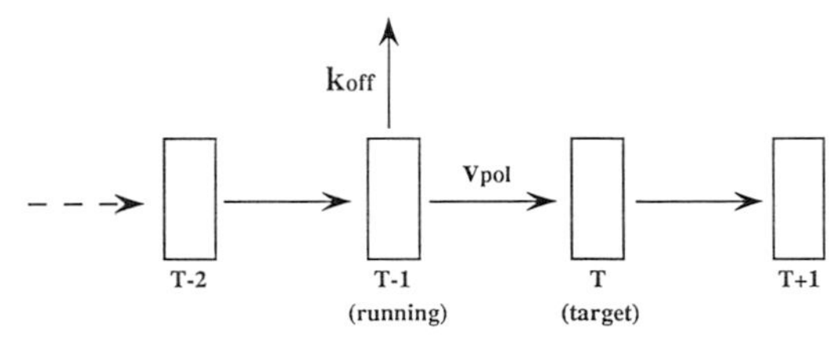

C) Exonuclease Model

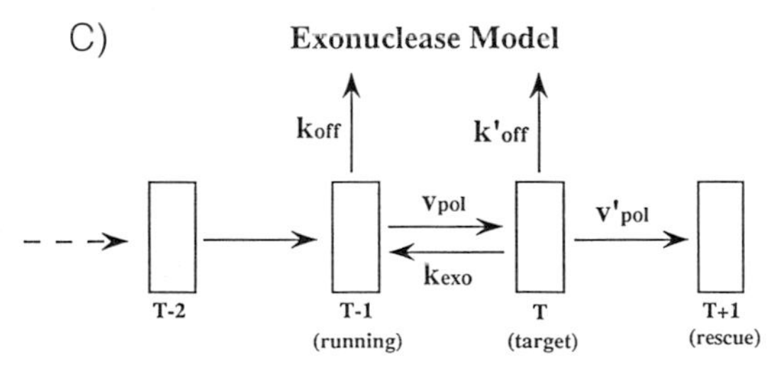

FIG. 3. Models for the events occurring during a single completed hit. (A) General picture of the action of a polymerase. After engaging the template, the polymerase adds and removes a sequence of bases until it dissociates. (B) Standard model. The polymerase adds the running start bases until it reaches location $T-1$ along the template. At this point the polymerase will either dissociate (giving rise to a band at $T-1$, denoted as I_{T-1}) or add the next base (giving rise to the bands following $T-1$, denoted as I_T^*). The rate of dissociation is assumed constant while the rate of polymerization is assumed to be a Michaelis–Menten function of the concentration of the target nucleotide in solution. (C) Exonuclease model. The standard model is used, except that after addition of the target nucleotide, the polymerase will either dissociate (giving rise to I_T), add the rescue base (giving rise to I_{T+1}^*), or cut out the newly inserted base and return to location $T-1$ (not necessarily giving rise to I_{T-1}). k_{off}, k_{off}', and k_{exo} are assumed to be constant while v_{pol} and v_{pol}' are Michaelis–Menten functions of their respective nucleotides.

related to these probabilities by $(I_T^*/I_{T-1}) = (p_{pol}/p_{off})$. I_T^* refers to the sum of the integrated intensities of bands at T, $T + 1$, ..., etc. If there are no extension bands beyond the target site T, then $I_T^* = I_T$.

Taking the ratio of the dissociation and polymerization probabilities and setting them equal to the ratio I_T^*/I_{T-1} gives:

$$v_{pol} = k_{off}\,\rho \qquad \rho = (I_T^*/I_{T-1}). \tag{3}$$

A measurement of the polymerase off-rate at template site $T - 1$ (see Measurement of Polymerase Dissociation Rates section), enables the determination of the rate-limiting nucleotide insertion step at site T, given by Eq. (3).

The rate of nucleotide insertion is a function of nucleotide concentration obeying Michaelis–Menten kinetics,

$$v_{pol} = \frac{V_{max}\,[\mathrm{dNTP}]}{K_m + [\mathrm{dNTP}]}. \tag{4}$$

The dependence of the ratio of integrated gel band intensities on substrate concentration, Eq. (5), is obtained by substituting Eq. (4) into Eq. (3),

$$\frac{I_T^*}{I_{T-1}} = \frac{\rho_{max}\,[\mathrm{dNTP}]}{K_m + [\mathrm{dNTP}]}, \tag{5}$$

where $\rho_{max} = (I_T^*/I_{T-1})_{max}$ is the ratio of the band intensities at saturating dNTP concentration. Thus, $\rho_{max} = V_{max}/k_{off}$ is analogous to the maximum velocity parameter, V_{max}.

Polymerase fidelity is determined by measuring V_{max}/K_m values for insertion of right and wrong nucleotides[8]—see Determining Insertion Fidelity section. The V_{max}/K_m ratio is equal to the slope obtained from the linear portion of a Michaelis–Menten plot of insertion velocity versus dNTP concentration. The nucleotide misinsertion efficiency (f_{ins}), defined as the ratio of wrong to right insertion efficiencies V_{max}/K_m, is expressed as

$$f_{ins} = (V_{max}/K_m)_W/(V_{max}/K_m)_R = (\rho_{max}/K_m)_W/(\rho_{max}/K_m)_R, \tag{6}$$

where ρ_{max}/K_m values are obtained from the ratio of integrated band intensities I_T^*/I_{T-1} as described in the Determining Insertion Fidelity section. Insertion fidelities are independent of k_{off};[8] since the dissociation rate at site $T - 1$ is the *same* regardless of whether a correct or incorrect nucleotide

[8] A. R. Fersht, *in* "Enzyme Structure and Mechanism," p. 112. W. H. Freeman, New York, 1985.

is inserted at site T, k_{off} cancels from Eq. (6), when dividing V_{max}/K_m ratios for wrong and right insertions.[8] The insertion fidelity $F_{ins} = 1/f_{ins}$.

Rates of Nucleotide Incorporation and Turnover for Polymerases Possessing Proofreading Exonuclease Function

To interpret gel bands in the presence of a proofreading exonuclease,[6] it is necessary to include a step that allows the enzyme to go backward on the template while excising either a mismatched or correctly matched 3′-dNMP from the primer terminus. Three template locations are required for analysis: a running-start location, $T - 1$, a target location, T, and a "rescue" location, $T + 1$ (Fig. 3C). Insertion of a next correct nucleotide at $T + 1$ acts to rescue right or wrong nucleotides, previously inserted at T, from excision by the exonuclease.

The fidelity in the presence of proofreading is the ratio of the efficiency of incorporating a single correct compared to incorrect nucleotide at a particular location (incorporation = insertion − exonucleolytic turnover). The nucleotide misincorporation efficiency (f_{inc}) is given by[9]

$$f_{inc} = f_{ins}\,[(1 - p_{exo,W})/(1 - p_{exo,R})], \qquad (7)$$

where f_{ins} is the misinsertion efficiency, Eq. (6), for the system without proofreading and p_{exo} is the probability that the exonuclease excises the newly incorporated nucleotide *in the window of time before the addition of the next nucleotide.* The incorporation fidelity $F_{inc} = 1/f_{inc}$. Incorporation fidelity depends explicitly on the concentration of rescue dNTP because the length of time prior to addition of the next nucleotide depends on the concentration of rescue base in the reaction, which in turn determines the window of time for the exonuclease to act on the newly incorporated nucleotide. This dependence on absolute dNTP pool size is commonly referred to as the "next nucleotide effect,"[9,10] and is a biologically relevant determinant of fidelity. Note that insertion fidelity in the absence of exonuclease is independent of absolute dNTP levels.[8]

Another factor affecting incorporation fidelity measured by the gel assay is the rate of polymerase dissociation from the template after addition of the target nucleotide.[2,6] While this factor has no obvious biological relevance, it can act to obscure the origin of gel bands required for determining rates and fidelities in the gel assay. In a single completed hit (Fig. 3C), the intensities of the bands $T - 1$, T, and $T + 1$ are determined by a complex series of competing processes. For example, polymerase may add a base

[9] L. K. Clayton, M. F. Goodman, E. W. Branscomb, and D. J. Galas, *J. Biol. Chem.* **254,** 1902 (1979).

[10] A. R. Fersht, *Proc. Natl. Acad. Sci. USA* **76,** 4946 (1979).

at location $T - 1$, add another base at location T, excise the base at T, and then dissociate. When the polymerase has the option of dissociation at site T it masks the competition between the polymerase and exonuclease functions at that site.

However, if there is no significant band intensity at site T, then the experiment can be analyzed exactly as described in the absence of exonuclease, where now the ratio $\rho = (I^*_{T+1}/I_{T-1})$ reflects the effect of the exonuclease on the rate of incorporation at T.[6] The nucleotide misincorporation efficiency, f_{inc}, can be defined exactly as in Eq. (6), where the ρ_{max}/K_m values refer to stable incorporation of wrong compared to right dNMPs, measured in the presence of $3' \rightarrow 5'$-exonuclease activity (see the Analyzing Single Hits Including Exonuclease section). If there is significant intensity in the T band, then the effects of polymerase dissociation at T on incorporation fidelity can either be corrected for as described previously,[2] or estimated as discussed in the Analyzing Single Hits Including Exonuclease section.

Materials and Methods

Enzymes, Templates, and Substrates

The following methods are used in all subsequent reactions. The only major changes involve optimization of reaction conditions for different polymerases. It is important that the final buffer composition in a reaction mixture be identical for all kinetics reactions with a given polymerase. Variations in buffer composition may have an effect on reaction kinetics.

Gel purified oligonucleotide primers are 5′-labeled in reactions containing 170 n*M* oligonucleotide, 320 n*M* [γ-^{32}P]ATP (4500 Ci/mmol), 56 m*M* Tris–HCl, pH 7.5, 7 m*M* $MgCl_2$, 13 m*M* dithiothreitol (DTT), and 0.2 units of T4 polynucleotide kinase. Alternatively, the ^{32}P-labeling reaction can be performed in the polymerase reaction buffer as long as T4 polynucleotide kinase is active in this buffer. Autoradiography requires the use of excess [γ-^{32}P]ATP, but if a sensitive method of detecting radioactivity, such as phosphorimaging is used, the amount of [γ-^{32}P]ATP can be decreased. Reactions are incubated for 30 min at 37° and terminated by heating at 100° for 5 min to denature the enzyme.

To 97 μl of the labeled primer solution (above) is added 133 μl of a solution containing 150 to 250 n*M* template (1.2 to 2 molar equivalents), 43 m*M* Tris–HCl, pH 8.5, at 37°, 3.5 m*M* 2-mercaptoethanol, and 0.09 mg/ml bovine serum albumin (BSA). Again, the annealing reaction can be performed in the polymerase reaction buffer. Primers are annealed to templates by placing a microcentrifuge tube of this solution in a beaker of water at 80 to 90° and allowing it to cool slowly to room temperature. An excess of template is used to ensure that greater than 95% of the primers is annealed to a template. The amount of excess template that is needed for a given P/T combination can be empirically determined from a time course of a polymerization reaction. A polymerase should be able to extend >95% of the primers if they are properly annealed. The absolute concentration of primer–template can be adjusted as needed for different experiments.

Enzymes are kept on ice and diluted to their final concentrations on ice just before the experiment begins. Enzymes are diluted in an appropriate buffer. Bovine serum albumin is typically added to concentrations of 0.1 to 1 mg/ml to stabilize dilute enzyme solutions.

FPLC-purified nucleotides are purchased from Pharmacia LKB Biotechnology, Inc., and used without further purification. Nucleotides are dissolved in 5 m*M* Tris–HCl, pH 7.5, at concentrations 10 times greater than in final kinetics reactions. Concentrations of nucleotide stocks are determined from their UV absorbances. It is important to verify that the polymerase is not inhibited at high dNTP substrate concentrations. Chelation of Mg^{2+} is likely to occur when "wrong" dNTP substrates are present at concentrations in excess of 1 to 2 m*M*, because optimal free Mg^{2+} concentrations for the polymerase reaction are typically around 1 to 10 m*M*. It may therefore be necessary to increase Mg^{2+} concentrations proportionally with [dNTP] when substrate concentrations exceed ~1 m*M*.

Determining Optimal Enzyme Concentrations and Reaction Times: Reaction Time Course

When a DNA polymerase preparation is being used for the first time, it is important to optimize reaction times and enzyme concentrations for fidelity measurements. A reaction time course is performed in which the running-start and target dNTPs are each present at saturating levels. Typically, a concentrated polymerase stock is diluted over a range of values (e.g., 1 : 100, 1 : 1000, 1 : 10,000). A separate time course is run for each enzyme dilution. For polymerases with no exonuclease activity, a primer/template/polymerase solution is made by adding polymerase to ^{32}P-labeled and annealed primer–template to make a solution that is twice as concentrated as needed in the final reaction. A nucleotide solution is prepared that contains running-start and target nucleotides as well as an appropriate reaction buffer at concentrations twice as great as in the final reaction. A 1/5 volume of a 10× solution of the polymerase reaction buffer is added.

The primer–template–polymerase solution and dNTP solution are incubated separately at the reaction temperature. Reactions are initiated by mixing equal volumes of primer–template–polymerase and dNTP solutions. A 3-μl aliquot of the reaction is removed at each time point and quenched in 6 μl of 20 m*M* EDTA in 95% formamide. For polymerases with exonuclease activity, the polymerase is included in the dNTP solution instead of the P/T solution. Figure 2C shows a gel for a typical reaction time course. From the results of these experiments, an enzyme concentration is identified that extends less than 20% of the primers in a convenient time and uses the lowest possible enzyme/DNA ratio. In addition, a reaction (at the highest polymerase concentration) is run for long enough time to demonstrate that >95% of the primers are potentially extendible by the polymerase. If a significant fraction of the primers is not annealed to templates resulting in <95% extension, then the SCH conditions may not be valid.

Insertion Kinetics

After choosing a polymerase concentration and reaction time from a time course, insertion kinetics are measured at several different concentrations of target dNTP. Kinetics reactions are performed in a manner similar to time course reactions, but using smaller volumes of primer–template–polymerase since only a single time point will be measured. A separate dNTP solution is made for each target dNTP concentration. Reactions are initiated by addition of 5 μl of dNTP solution to 5 μl of primer–template–enzyme solution and quenched with 20 to 30 μl of 20 m*M* EDTA in 95% formamide.

Gel Electrophoresis

Reaction products are separated on a polyacrylamide sequencing gel (40 cm × 30 cm × 0.4 mm). Depending on the length of the primer, 12 to 20% polyacrylamide : bisacrylamide (19 : 1) gels containing 8 *M* urea and 1× TBE buffer (89 m*M* Tris base, 89 m*M* boric acid, 2 m*M* Na_2EDTA, pH 8.3) are used.

The reaction quenching solution also serves as loading buffer. Quenched reaction mixtures are heated at 100° for 5 min and cooled on ice for 1 min before loading 4 μl into each 0.6-cm well of the gel. Markers (0.2% (w/v) bromphenol blue, 0.2% (w/v) xylene cyanole) are loaded separately. The gel is then run at 2000 V for 3 to 4 hr, until reaction products have migrated as far as possible without eluting from the bottom of the gel. The gel is then vacuum-dried onto Whatman No. 3 MM filter paper before quantitation.

Analyzing Gel Band Intensities

Gel bands can be quantified using either autoradiography and densitometry or using more advanced techniques such as phosphorimaging. In either case, a number proportional to the amount of radiation in each band is desired. Although the advanced techniques will give this number directly, autoradiography using X-ray film requires some care to get accurate quantitation,[3] because the response of X-ray film to radiation is linear over a limited range (~30-fold). In contrast, a phosphorimager (Molecular Dynamics) exhibits a linear response to radioactivity over a 10^5-fold range.

DNA Traps

A DNA trap is used to ensure that a polymerase molecule interacts with only a single labeled P/T molecule. The trap consists of either unlabeled DNA with both single- and double-stranded regions, or an analogous substance that is bound by polymerase. Preincubation of a polymerase with a labeled P/T followed by addition of dNTPs and trapping agent results in extension of only those labeled P/Ts that were initially bound to the polymerase. After dissociation of the polymerase, the probability of reassociation with a labeled P/T is small. Thus, the presence of a trap guarantees that the "single hit" condition is satisfied.

Three types of traps have been used: (1) unlabeled P/T, (2) partially digested calf thymus DNA, or (3) heparin sulfate. Unlabeled P/T used as a trap is prepared in the exact same manner as labeled P/T except that the concentration of the unlabeled P/T is at least in 100-fold molar excess in the final reaction mix. Calf thymus DNA trap is prepared as described previously.[11,12] Heparin sulfate is a good trapping agent for some polymerases for example, T4 polymerase,[13] but we have found it to be ineffective for Klenow fragment.

Choice of the appropriate trapping agent and of the appropriate concentration depends on two control experiments. First, a trap must be effective for the entire time range used in the experiment. Trap effectiveness can be determined by preincubating labeled P/T with the trapping agent before addition of the polymerase. If the trap is effective, the labeled P/T will not be extended. Second, the trapping agent should not inhibit the polymerase reaction. Thus, observed extension of labeled DNA should not decrease with increasing concentrations of trap.

Measurement of Polymerase Dissociation Rates

For running-start PTs, reactions are initiated by mixing 4 μl of a preincubated solution of polymerase (25 to 50 n*M*) and labeled P/T (100 n*M*) in

[11] C. M. Joyce, *J. Biol. Chem.* **264,** 10,858 (1989).

[12] S. Creighton, M.-M. Huang, H. Cai, N. Arnheim, and M. F. Goodman, *J. Biol. Chem.* **267,** 2633 (1992).

[13] M. K. Reddy, S. E. Weitzel, and P. H. von Hippel, *J. Biol. Chem.* **267,** 14,157 (1992).

reaction buffer with 4 μl of the running-start dNTP and "trap." After a delay time (usually between 3 and 120 s), 4 μl of a solution of a saturating concentration of the target dNTP is added. Reactions are quenched with 2 to 3 volumes of 20 m*M* EDTA in 95% formamide 15 sec after addition of the target dNTP. For a 0-sec delay time point, the target dNTP is included in the solution with the running-start dNTP and trap (Fig. 4A). Dissociation

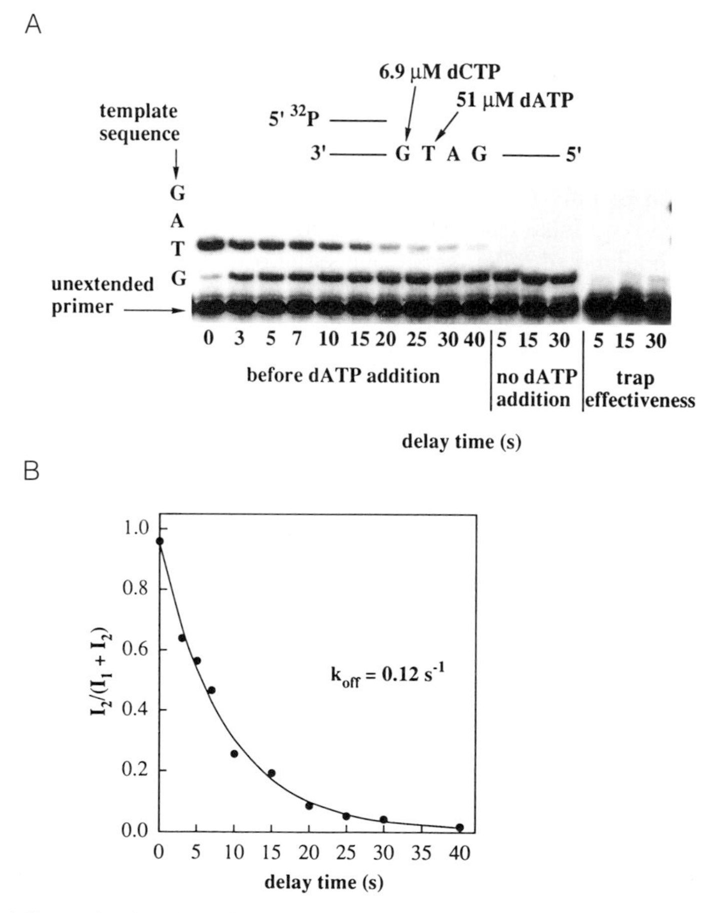

FIG. 4. Determination of the dissociation rate of Klenow exo-(D355A, E357A) at a particular location on the template in the course of polymerization. (A) Results of an experiment in which the polymerase correctly inserts dCMP opposite a running-start template G site and then is forced to wait (in the presence of DNA trap) a variable amount of time before the target nucleotide, dATP, is added. Note the decreasing ratio of $[I_2/(I_1+I_2)]$ as a function of the delay time, which reflects polymerase dissociation. (B) Plot of $[I_2/(I_1+I_2)]$ as a function of time, fit by a single exponential to a rate of 0.12 sec^{-1}.

rates for standing-start P/T can be measured in the same way except a running start dNTP does not need to be included with the trap DNA. Control reactions to determine the effectiveness of the trapping agent are performed under identical conditions except that the trapping reagent is preincubated with the labeled P/T before the addition of a solution of the polymerase and running-start dNTP (Fig. 4A). For running-start reactions, an additional control can be done by omitting the target nucleotide from a reaction to show that the running start dNTP is not misinserted opposite the template target site (Fig. 4A). Reaction products are separated by PAGE for kinetics reactions. The fraction of P/T molecules extended by addition of the target nucleotide, $I_T^*/(I_T + I_{T-1})$ for running-start P/Ts, and I_T^*/I_{total} for standing-start P/Ts is plotted against the delay time. A fit of these data to a single exponential gives the rate constant, k_{off}, for dissociation (Fig. 4B).[14]

Measuring Polymerase Insertion Fidelity without Exonuclease

The most common application of the gel methodology is the determination of polymerase misinsertion rates. There is a finite chance ($\sim 10^{-3}$ to 10^{-6}) that a wrong dNTP bound in the polymerase active site will be incorporated. The gel assay can be used to determine the relative probability that a polymerase will insert any dNTP at normal, modified (e.g., O^6-alkyl G, 8-oxo-G), or abasic DNA template sites. The objective is to provide a rapid analysis of misincorporation efficiencies in many sequence contexts.

Primer–Template Construction and Running and Standing Starts

Two types of P/T molecules have been used in the gel assay: running and standing starts (Fig. 1).[3,4] Each of these has different advantages depending on the process being studied. Running-start experiments have the advantage that for measuring insertion kinetics, the presence of a band at $T-1$ before the target band T defines the number of polymerases bound at the primer 3′-terminus that advanced to site $T-1$, and subsequently to T, allowing a simple determination of the ratio of probabilities for polymerization and dissociation at the target site [Eq. (3)].

In the example shown (Fig. 5, left-hand side), the nucleotide mix for the running-start reaction contains dGTP, complementary to the two running-start template C sites. Running-start dNTPs are present at low, but saturating, concentrations, typically 5 to 50 μM. This range of dNTP concentration is generally high enough to ensure rapid incorporation at running-

[14] L. B. Bloom, M. R. Otto, J. M. Beechem, and M. F. Goodman, *Biochemistry* **32,** 11,247 (1993).

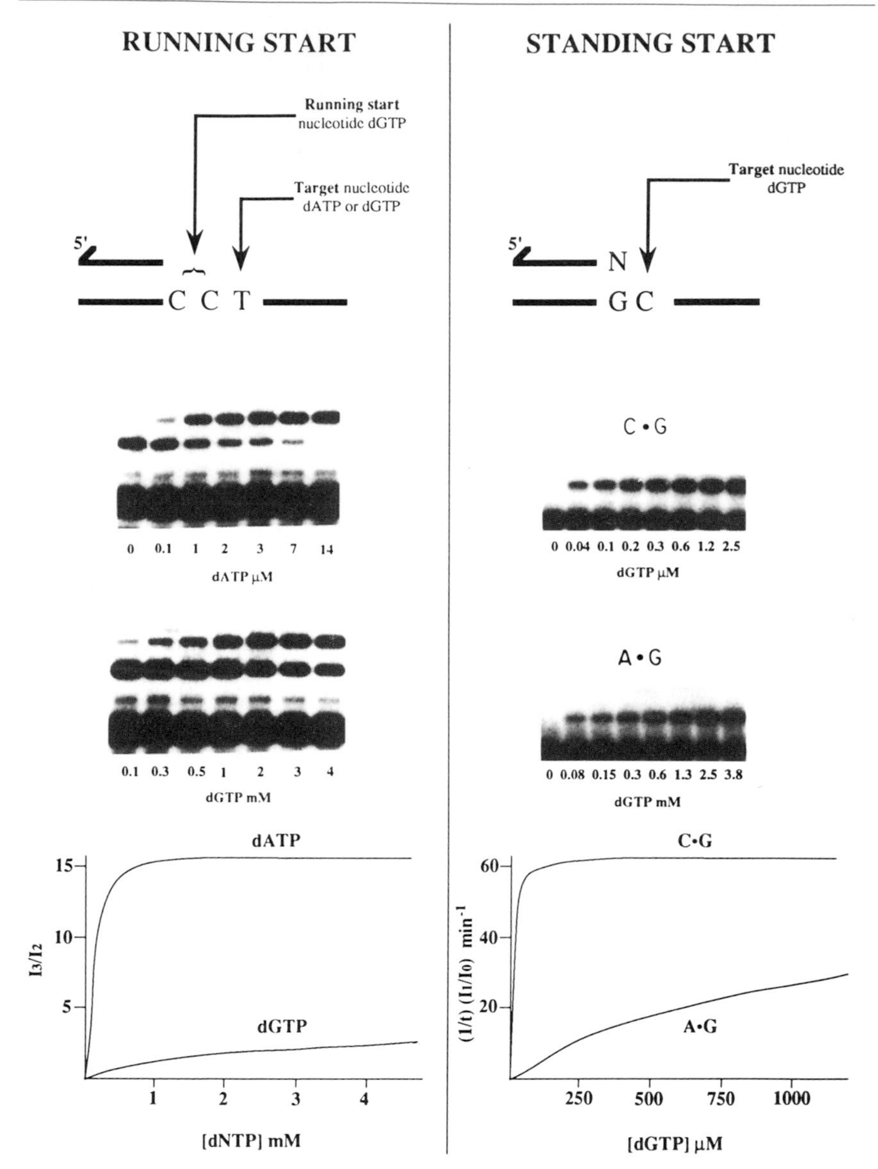

FIG. 5. Design and analysis of running- and standing-start experiments. A running-start P/T (left-hand side) contains of a number of running-start bases, usually 2, followed by the target base, shown as T. A typical kinetics experiment will produce a gel, as illustrated, in which the ratio of the target band to the previous running-start band increases as a function of the concentration of the target nucleotide, dATP or dGTP. A graph of this ratio as a function of the concentrations of complementary (dATP) or noncomplementary (dGTP) target deoxynucleotide substrates yields Michaelis–Menten curves (bottom of figure), in accor-

start template sites, but does not normally result in misincorporation of the running-start nucleotide substrate opposite the target site. Conversely, saturating levels of running-start dNTP are usually sufficient to ensure that a target dNTP substrate (present at mM levels) cannot compete for insertion at running-start template sites. Rates of polymerization at running-start sites should be maximal so that enzyme dissociations occur predominantly at the target site. The relative rates of insertion of the target nucleotide and dissociation at site $T-1$, I_T^*/I_{T-1}, are measured at a series of target dNTP concentrations, usually in a range from 10 μM to 2 mM. $\rho_{\max}$ and K_m are obtained from these data using Eq. (5), and misinsertion efficiencies are calculated using Eq. (6).

Standing-start experiments are performed on a template with no running-start sites; the target base is the first available insertion site on the template, G opposite C (Fig. 5, right-hand side). In this example, the unextended primer 3′-terminus is located opposite G, at the T-1 site, and, unlike the running-start, the $T-1$ position can no longer be used as an internal "marker" to compute the number of active polymerases involved in primer extension. However, in all other respects standing- and running-start experiments are carried out in the same manner. The analysis of this standing-start gives rise to an equation for polymerase rate[2,15]

$$v_{\text{pol}} = \alpha\, k_{\text{off}}\, \rho = \frac{1}{\text{t}} \frac{I_1^*}{I_0}, \tag{8}$$

where α is an unknown factor, but constant for each particular P/T reducing the apparent reaction rate (and also the apparent K_m value), and t is the reaction time. For the example shown, the gel bands represent extension of either a C (template)·G base pair or an A(template)·G mispair (Fig. 5, right-hand side). Roughly 1000-fold higher concentrations of next correct dNTP (dGTP) are required to obtain comparable extension of the mismatched primer terminus compared to the correctly matched terminus.

For standing-start P/T molecules, nucleotide insertion velocities can be determined as a function of dNTP concentration using Eq. (8), and

[15] J. Petruska, M. F. Goodman, M. S. Boosalis, L. C. Sowers, C. Cheong, and I. Tinoco, Jr., *Proc. Natl. Acad. Sci. USA* **85,** 6252 (1988).

dance with Eq. (5). A standing-start P/T (right-hand side) is shown having a target template base, C, adjacent to a primer 3′-terminus. The primer terminus N contains either a C, forming a correct C·G base pair, or A, forming an A·G mispair. The primer extension gel band ratios, measured as a function of dGTP concentration, are described by Michaelis–Menten curves (bottom of figure), in accordance with Eq. (8).

polymerase misinsertion efficiencies are calculated from Eq. (6). Standing-start P/Ts can also be used to compare the efficiencies of extending matched and mismatched P/T termini—see the Extending a Mismatched Primer Terminus section. Since the presence of the unextended primer–band masks the rate of enzyme dissociation at $T-1$, the absolute values of V_{max} and K_m derived from the experiment have limited mechanistic significance. In this case, association–dissociation of the polymerase from the P/T influences the interpretation of the data, and individual V_{max} and K_m values will not generally be the same for running and standing starts. However, the ratio of V_{max}/K_m for wrong compared to right insertions, Eq. (6), provides a correct value for f_{ins}.[6,8]

Determining Insertion Fidelity

Gel bands from a typical running-start experiment are shown for insertion of a right nucleotide (dAMP) and misinsertion of a wrong nucleotide (dGMP), at a target T site, as a function of dNTP concentrations (Fig. 5, left-hand side). As the concentration of each dNTP increases, the ratio of the target band to the final running-start band ($\rho = I_3/I_2$) increases. The sum of the intensities of the two bands remains approximately constant; only the partitioning changes as the concentration of the nucleotide increases. The ratio of the band intensities is plotted as a function of the nucleotide concentration (Fig. 5, bottom left-hand side), and the kinetic parameters are determined by fitting the data for $\rho = I_3^*/I_2$ to the Michaelis–Menten equation, Eq. (5), for the relative nucleotide insertion velocity ρ.[3] The misinsertion efficiency (in this case G · T relative to A · T) is given by Eq. (6).

Extending a Mismatched Primer Terminus

A mismatched primer terminus presents a strong kinetic block to continued extension by polymerase.[15–19] The mismatch can take the form of a natural base mispair (e.g., G · T), a modified base mispair (8-oxo-G · A), or possibly any primer base opposite a template lesion, X, (A · X). Allellic-specific amplification is an example of a novel PCR application[20,21] requiring knowledge of the efficiency of extending all possible natural base mis-

[16] F. W. Perrino and L. A. Loeb, *J. Biol. Chem.* **264,** 2898 (1989).
[17] L. V. Mendelman, J. Petruska, and M. F. Goodman, *J. Biol. Chem.* **265,** 2338 (1990).
[18] R. D. Kuchta, P. Benkovic, and S. J. Benkovic, *Biochemistry* **27,** 6716 (1988).
[19] I. Wong, S. S. Patel, and K. A. Johnson, *Biochemistry* **30,** 526 (1991).
[20] T. Ehlen and L. Dubeau, *Biochem. Biophys. Res. Comm.* **160,** 441 (1989).
[21] H. Li, X. Cui, and N. Arnheim, *Proc. Natl. Acad. Sci. USA* **87,** 4580 (1990).

matches by thermostable *Taq* polymerase.[22] For this and similar applications, the gel assay can be used to measure the misextension efficiency, f_{ext}, the efficiency of extension of a P/T mismatch compared to extension of a correctly paired P/T terminus.[2,17]

In contrast to insertion efficiency measurements, where right and wrong dNTP substrates compete for insertion by a polymerase *bound to the same correctly matched primer 3′-terminus*, mismatch extension efficiency measurements are performed using polymerases bound to two different primer 3′-termini, matched versus mismatched. As a consequence, f_{ext} depends explicitly on the enzyme–DNA equilibrium binding constants to matched and mismatched P/T DNA, on the concentrations of matched and mismatched P/T, and, most important, on the absolute concentration of next correct dNTP.[2,17] The relative efficiency for extending mismatches decreases dramatically with decreasing dNTP concentration, and maximum discrimination is achieved as the concentration of next correct dNTP $\rightarrow 0$.[17]

In the limit of low next correct dNTP, one can define an intrinsic mismatch extension efficiency, $f^0_{min} = [(V_{max}/K_m)_W/(V_{max}/K_m)_R]$, analogous to the nucleotide misinsertion efficiency, f_{ins}, Eq. (6). The measurement for f^0_{min} is made, using a standing-start P/T, and performed as described for f_{ins}, with one change in conditions. The extension measurement is carried out with matched or mismatched P/T in the *linear* range, $K_D >$ [P/T DNA].[12] The V_{max}/K_m values are determined by plotting the velocities of extension of matched and mismatched P/T as a function of next correct dNTP concentration and fitting to the linear portion of the Michaelis–Menten rectangular hyperbola. If the equilibrium dissociation constants are similar for the polymerase bound to matched and mismatched P/T 3′-termini, then the measurement of f^0_{min} can be carried out at any convenient P/T concentration.[12] Previous measurements using a variety of DNA polymerases and reverse transcriptases reported no significant difference in K_D values for binding matched versus mismatched P/T DNA.[12,19,22]

Determining Polymerase Dissociation Rates and Equilibrium Binding Constants

Polymerase : P/T complexes typically have equilibrium binding constants on the order of 1 to 10 n*M*. Determination of the binding constant of a particular polymerase : P/T pair can be carried out using the gel assay.[12]

Dissociation rates have been found to vary with primer–template sequence.[14,23] For many polymerases and primer/templates, k_{off} values have

[22] M.-M. Huang, N. Arnheim, and M. F. Goodman, *Nucleic Acids Res.* **20,** 4567 (1992).

[23] R. D. Kuchta, V. Mizrahi, P. A. Benkovic, K. A. Johnson, and S. J. Benkovic, *Biochemistry* **26,** 8410 (1987).

been found to be between 0.01 and 0.4 sec^{-1}.[23,24] Dissociation rates on this order and slower can be measured without the use of a rapid quench apparatus. Enzyme–DNA dissociation rates can be measured both for running- and standing-start P/Ts. The basic method involves preincubating a polymerase with a labeled DNA substrate followed by the addition of a large excess of unlabeled trap DNA. For running-start P/Ts, the running-start dNTP is included in the solution with the trap. After a variable delay time, the target nucleotide to be inserted on the labeled P/T is added before quenching the reaction. Initially, polymerase is bound to labeled P/T. During the delay time before addition of dNTP substrates, polymerase either remains bound to the labeled P/T or dissociates and interacts with the large excess of unlabeled trap DNA. When dNTP substrate is added, the fraction of polymerases remaining bound to labeled P/T can extend the labeled template. As the delay time between addition of the trap DNA and addition of the nucleotides increases, the fraction of polymerases that remains bound to the labeled substrate decreases and thus the fraction of labeled substrates that is extended decreases. Values of k_{off} can be calculated by fitting the fraction of primers extended after a given delay time to an exponential decay (Materials and Methods section, Fig. 4).

Measuring Fidelity in Presence of Proofreading Exonuclease[6]

When an exonuclease is present, the interpretation of the gel bands is more complicated, and the incorporation fidelity depends explicitly on the concentration of the next correct nucleotide (rescue dNTP).[9,10]

Analyzing Single Hits Including Exonuclease

To measure fidelity, a series of reactions with increasing amounts of the rescue nucleotide is performed. For example, if the fidelity of incorporation of G opposite T is to be measured, and the running-start nucleotide is dTTP and the rescue nucleotide is dCTP, then an experiment is performed in which [dCTP] = 0 and [dGTP] is varied over a range of concentrations. Then another experiment is performed, changing [dCTP] to a new value and varying [dGTP] as before. The experiments are repeated over a desired range of dCTP concentration. The data from all of the experiments are collected and the value of $(\rho_{max}/K_m)_{dGTP}$ at each dCTP concentration is determined. When analyzing each experiment to determine ρ_{max}/K_m, Eq. (5), we use the procedure given for the running-start experiment modified

[24] H. Yu and M. F. Goodman, *J. Biol. Chem.* **267,** 10,888 (1992).

as follows: If there is no band at the target site, T (Fig. 6A), then the ratio $\rho = I^*_{T+1}/I_{T-1}$) is plotted against [dGTP] and the ρ_{max}/K_m ratio is calculated from this graph. A plot of $(\rho_{max}/K_m)_{dGTP}$ versus [dCTP] gives the efficiency of incorporation as a function of the next correct nucleotide in the presence of an exonuclease. The misincorporation frequency, f_{inc}, is obtained from Eq. (6), where ρ_{max}/K_m values refer to nucleotide incorporation measurements.

Bacteriophage T4 exonuclease acts processively.[13] For a processive enzyme, one would expect that suppression of an intermediate band at T would occur in the presence of moderate to high concentrations of rescue dNTP. Using T4 polymerase and a rescue dNTP, we observe prominent gel bands at template locations $T+1$ and $T-1$, with no significant band intensity present at T (Fig. 6A, fourth and fifth lanes).

If there is significant intensity in the intermediate band located at site T, then either the data can be analyzed as described previously,[2] which requires a determination of k'_{off} at site T (Fig. 4), to deduce the fidelity; otherwise, the ratios $\rho_{low} = [I^*_{T+1}/(I_{T-1} + I_T)]$ and $\rho_{high} = (I^*_T/I_{T-1})$ can be plotted as a function of [dCTP]—these will give lower and upper limits for the true value of (ρ_{max}/K_m), and Eq. (6) can be used to compute a range for f_{inc}.

Determining Rate of dNTP → dNMP Turnover

Nucleotide turnover[25] occurring during a single encounter between a polymerase and a P/T is determined in experiments in which the primer is unlabeled and the dNTP substrate is α-^{32}P- or ^{33}P-labeled. The experiments are performed as running-start experiments (with a rescue nucleotide). The products of the reaction are resolved by electrophoresis on an 8% polyacrylamide gel for 1 hr at 1500 V. The gel will have bands arising from the incorporated nucleotide, the initial labeled dNTP, the hydrolyzed dNMP, and possibly other impurities in the radiolabeled dNTP (Fig. 6B). It is possible to identify the dNMP turnover spot in the presence of a much higher concentration (~100-fold) of unreacted dNTP; thus, the rate of nucleotide excision catalyzed by the proofreading exonuclease can be determined accurately for newly inserted wrong and right nucleotides.

This gel can be compared to one on which the same reaction has been run using a ^{32}P-labeled P/T. The amount of radiation, determined by the integrated intensity in the incorporated DNA bands and excised dNMP spots, is measured by densitometry using X-ray film or by phosphorimaging. Labeled P/T and dNMP markers are required to identify the location of

[25] N. Muzyczka, R. L. Poland, and M. J. Bessman, *J. Biol. Chem.* **247,** 7116 (1972).

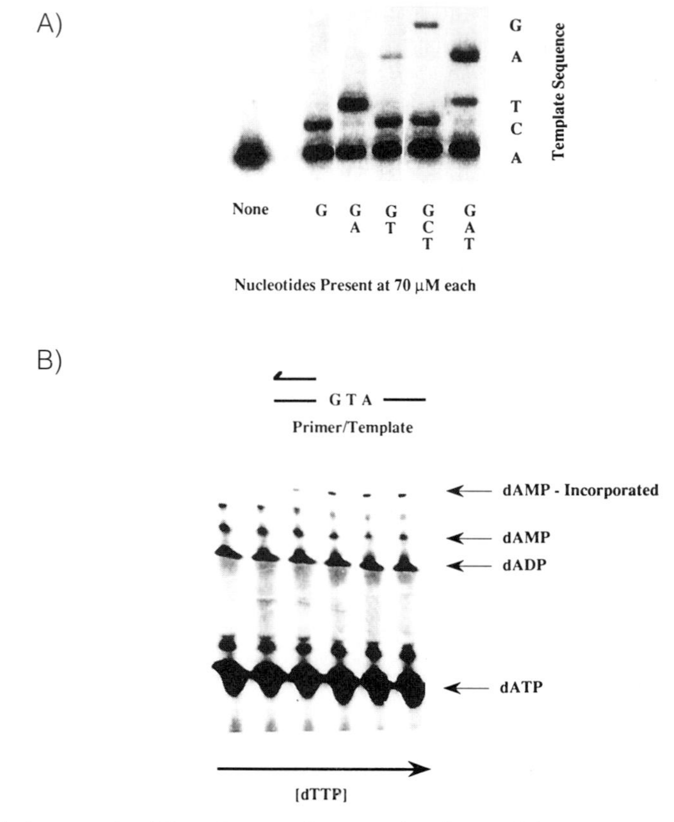

FIG. 6. An example of the gel assay using a proofreading-proficient polymerase, bacteriophage T4 DNA polymerase. (A) Demonstration of misincorporation and misextension without dissociation of the polymerase. All lanes are from a reaction containing ~0.5 n*M* T4 DNA polymerase and 30 n*M* P/T. The dNTPs indicated for each lane were present at 70 μM. Reactions were run for 1 min at 37°. The target template base is T. In the first two lanes (G and GA) the polymerase incorporates a correct nucleotide, while in the GT lane the polymerase adds G correctly, adds *T* incorrectly and follows by adding T correctly. In the GCT lane, the polymerase behaves as in the GT lane except that it adds an additional C correctly. In both cases, the polymerase makes an error but does not dissociate at the target site on the template; it either goes on to add T opposite template A or, more likely, removes the misinserted nucleotide, T. In the GAT lane, all incorporations occur with the correct dNTP; however, note the strong band at the target site, indicating that dissociation is a faster process than exonucleolytic removal at this site. (B) Demonstration of the incorporation and removal of an α-labeled nucleotide by T4 DNA polymerase. In each lane, T4 polymerase was present at ~3 n*M*, P/T at 150 n*M*, dCTP at 30 μM, α-^{33}P dATP at ~2 μM, and dTTP was present in amounts ranging from 50 n*M* to 200 μM. The reaction was run for 1 min at 37°. In a given single hit, the polymerase will first add C opposite G, ^{33}P-labeled A opposite T and then T opposite A, while possibly excising any of the newly added nucleotides. As dTTP concentration increases, the amount of excised α-^{33}P dAMP decreases, and the amount of α-^{33}P dAMP incorporated increases, demonstrating the effect of the next correct nucleotide on inhibiting the rate of exonucleolytic removal.

incorporated and excised dNMP. Nucleotide insertion rates in the presence of exonuclease are evaluated by adding the amount of nucleotide incorporated into DNA (obtained from the primer extension assay) to the amount of exonuclease catalyzed dNTP $\rightarrow$ dNMP turnover; insertion = incorporation + turnover.

Two technical points should be mentioned. First, the α-^{32}P-labeled triphosphates should be as fresh as possible to minimize the background dNMP levels as well as other radioactive decay products that may compete with dNTP in binding to polymerase. Second, it is often difficult to detect any significant amount of incorporation for single completed hit conditions using the concentration of P/T given earlier. However, by elevating P/T concentrations to high enough levels ($\sim$1 μM), it is generally possible to observe a detectable amount of incorporation under SCH conditions.

Nucleotide Incorporation and Bypass at Abasic Template Lesions

Efficiencies of nucleotide insertion opposite a template lesion and continued synthesis beyond the lesion (lesion bypass) are generally comparable in magnitude to nucleotide misinsertion and mismatch extension efficiencies involving normal dNTPs and unmodified template sites. It is convenient to use the gel assay to measure the kinetics of nucleotide insertion and bypass at site-directed coding (e.g., 8-oxo-G) and noncoding DNA template lesions (e.g., abasic sites).[5,7] A kinetic analysis of nucleotide insertion can be carried out, using a running-start P/T, exactly as described for unmodified template sites and normal dNTP substrates (Materials and Methods section).

The presence of an abasic template lesion acts as a strong impediment to continued elongation from a primer base opposite the lesion, much in the same way as normal base mispairs act as kinetic blocks to primer elongation. A kinetic analysis of lesion bypass can be carried out, using a standing-start P/T, as described for extension of primer termini containing normal mispairs.

It has generally been observed *in vitro* that insertion of A opposite abasic lesions is strongly favored over insertion of each of the other three nucleotides.[26–28] The gel assay has been used to show that primers terminating in A opposite the lesion are also strongly favored for continued synthesis past the lesion.[7] However, sequence context plays an important role in

[26] S. Boiteux and J. Laval, *Biochemistry* **21,** 6746 (1982).

[27] D. Sagher and B. S. Strauss, *Biochemistry* **22,** 4518 (1983).

[28] R. M. Schaaper, T. A. Kunkel, and L. A. Loeb, *Proc. Natl. Acad. Sci. USA* **80,** 487 (1983).

DIRECT ADDITION FAVORED FOR EXTENDING A OPPOSITE X

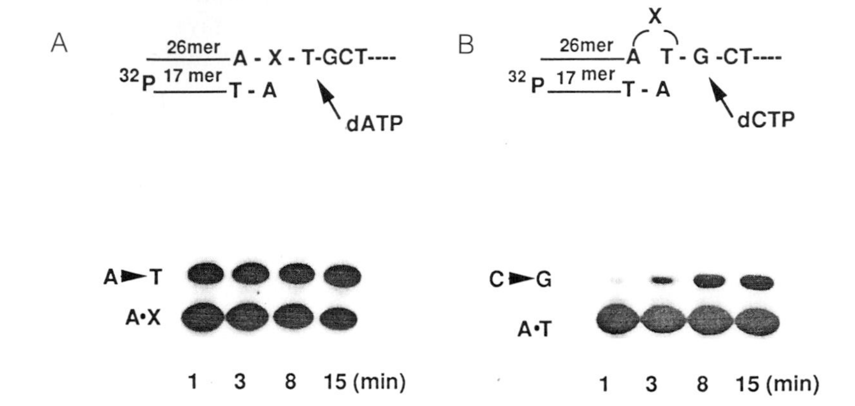

MISALIGNMENT FAVORED FOR EXTENDING C OPPOSITE X

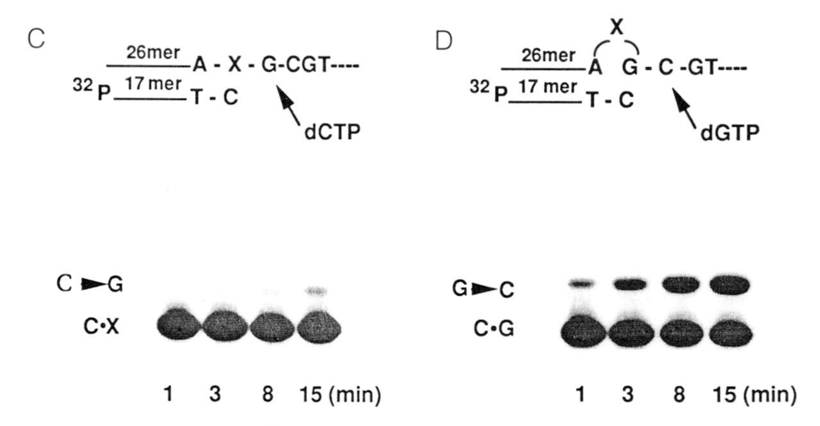

FIG. 7. Comparison of direct extension with misalignment extension at abasic template sites, X, by HIV-1 reverse transcriptase.[7] Direct extension is favored for A paired opposite X (A · X); misalignment extension is favored for C paired opposite X (C · X). For extension of A · X the concentration of next correct nucleotide is 0.5 m*m* (A) dATP and (B) dCTP, respectively. For extension of C · X the concentration is 0.5 m*m* (C) dCTP and (D) dGTP. [Adapted from reference 7.]

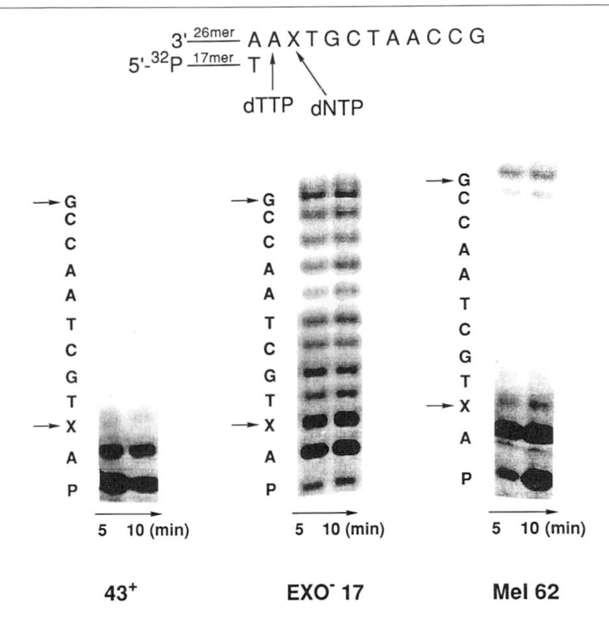

FIG. 8. Insertion and bypass at abasic site, X, by wild-type and mutant bacteriophage T4 DNA polymerases.[2] 43^+ is wild-type T4 polymerase; EXO^-17 has less than 0.1% of the exonuclease activity of 43^+, Mel62 exonuclease and polymerase activities are similar to 43^+ activity levels. [Reproduced with permission from reference 2. Copyright CRC Press, Boca Raton, Florida.]

modulating polymerase fidelity, and the gel assay is well suited for analyzing the effects of P/T perturbations on fidelity.[4,29] A time course showing lesion bypass for the enzyme HIV-1 reverse transcriptase clearly illustrates that a "reversal" in the bypass mechanism can be caused by changing the identity of a single template base located two bases downstream from the site of the lesion (Fig. 7).[7]

When A is located opposite the lesion, bypass occurs most efficiently by "direct addition," favoring incorporation of a nucleotide (dAMP) complementary to the base adjacent to the lesion (T) (Fig. 7A), rather than by "misalignment addition," where bypass occurs by addition of a complementary nucleotide (dCMP) two bases from the lesion (Fig. 7B). However, when C is opposite the lesion, a mechanism involving misalignment extension (Fig. 7D) is strongly favored over direct extension (Fig. 7C). The efficiencies of lesion bypass can be measured for each of the four P/Ts shown in Fig. 7 by choosing a convenient assay time and measuring the

[29] M. S. Boosalis, D. W. Mosbaugh, R. Hamatake, A. Sugino, T. A. Kunkel, and M. F. Goodman, *J. Biol. Chem.* **264,** 11360 (1989).

integrated intensities for the bands corresponding to A-T and C-G as function of dNTP concentration, Eq. (8), using the standing-start protocol (Materials and Methods section).

DNA synthesis is strongly inhibited in the presence of abasic template lesions.[30] Thus, one would anticipate that wild-type T4 DNA polymerase, known to contain a highly active proofreading exonuclease, would have much greater difficulty synthesizing past an abasic template lesion (Fig. 8, left-hand panel) compared to a 3′-exonuclease-deficient T4 polymerase ($EXO^{-}17$) (Fig. 8, middle panel). However, an interesting result is that another mutant T4 polymerase, Mel 62, is able to handle synthesis past the lesion despite having a "wild-type" exonuclease activity level (Fig. 8, right-hand panel). It has been postulated by Dr. L. J. Reha-Krantz, that although Mel 62 appears to have normal exonuclease activity on single-stranded DNA substrate, the enzyme is deficient in its ability to switch between synthesis and proofreading modes, and this deficiency may explain the moderately strong *mel62* mutator phenotype *in vivo*.[31]

The use of running- and standing-start protocols (Materials and Methods), coupled with a measurement of dNTP → dNMP turnover (Fig. 6B), can be used to determine the efficiencies of nucleotide insertion and net incorporation opposite the lesion and abasic bypass efficiencies for the T4 wild-type and mutant polymerases, and these studies can be extended[32] to investigate other polymerases, polymerase accessory proteins, and other proteins, for example, Rec A, Umu C, and Umu D′, which play a role in facilitating lesion bypass.[33]

Acknowledgments

This work was supported by National Institutes of Health grants GM 21422, GM42554, and AG 11398.

[30] C. W. Lawrence, A. Borden, S. K. Banerjee, and J. E. LeClerc, *Nucleic Acids Res.* **18,** 2153 (1990).

[31] L. J. Reha-Krantz, *J. Mol. Biol.* **202,** 711 (1988).

[32] M. Rajagopalan, C. Lu, R. Woodgate, M. O'Donnell, M. F. Goodman, and H. Echols, *Proc. Natl. Acad. Sci. USA* **89,** 10,777 (1992).

[33] G. C. Walker, *Microbiol. Rev.* **48,** 60 (1984).

[20] Kinetic Analysis of Nucleotide Incorporation and Misincorporation by Klenow Fragment of *Escherichia coli* DNA Polymerase I

By STEPHEN J. BENKOVIC and CRAIG E. CAMERON

Introduction

The solution of the minimal kinetic scheme for the Klenow fragment of *Escherichia coli* DNA polymerase I (KF) rests on a combination of steady-state and pre-steady-state kinetic studies on defined, short template–primers. In this chapter, we present a step-by-step analysis of how the scheme was developed for correct nucleotide incorporation as well as for single nucleotide misincorporation, and how the magnitude of each individual step was quantitated. KF retains the polymerase and $3' \rightarrow 5'$-exonuclease activities of its parent and consequently replicates DNA with high fidelity (one mispairing for every 10^6 to 10^8 bases incorporated).[1] It is ideal for study owing to its small size (68 kDa), slow polymerization rate, and lack of accessory proteins[2] and remains the only DNA polymerase for which an X-ray crystallographic structure is available.[3,4] The principal kinetic features that emerged from this solution have been shown to apply to the polymerization of DNA by the T4 and T7 polymerases.[5,6]

The overall reaction catalyzed by the polymerase is described in Eq. (1) and represents a summary of many discrete processes.[7]

$$D_N + dNTP \rightleftharpoons D_{N+1} + PP_i \tag{1}$$

The polymerase must bind at or near the 3′-hydroxyl of the substrate DNA and also bind the one correct dNTP in the presence of the other incorrect dNTPs. Formation of the chemical bond then occurs by nucleophilic attack of the 3′-hydroxyl of the growing primer strand on the α-phosphorus of

[1] V. Englisch, D. Gauss, W. Freist, S. Englisch, S. H. Sternbach, and F. vonder Haal, *Angew. Chem. Int. Ed. Engl.* **24,** 1015 (1985).

[2] S. S. Carroll and S. J. Benkovic, *Chem. Rev.* **90,** 1291 (1990).

[3] D. L. Ollis, P. Brick, R. Hamlin, N. G. Xuong, and T. A. Steitz, *Nature (London)* **313,** 762 (1985).

[4] L. S. Beese, V. Derbyshire, and T. A. Steitz, *Science* **260,** 352 (1993).

[5] T. L. Capson, J. A. Peliska, B. Fenn Kaboord, M. West Frey, C. Lively, M. Dahlberg, and S. J. Benkovic, *Biochemistry* **31,** 10,984 (1992).

[6] S. S. Patel, I. Wong, and K. A. Johnson, *Biochemistry* **30,** 511 (1991).

[7] R. D. Kuchta, V. Mizrahi, P. A. Benkovic, K. A. Johnson, and S. J. Benkovic, *Biochemistry* **26,** 8410 (1987).

the dNTP, complexed generally with Mg^{2+}. The polymerase then releases PP_i and either translocates to the next available template position or dissociates from the substrate DNA. The ordering of PP_i release and KF translocation–dissociation can be separate or a combined step. The editing by the $3' \rightarrow 5'$-exonuclease activity may happen at one or more complexes formed during the turnover. In addition, the reaction is reversed by the addition of relatively modest levels of PP_i (~50 μM). There is, however, no need to invoke enzyme-bound, covalent reaction intermediates in either the polymerase or exonuclease reactions based on observations of the stereochemical course at the α-phosphorus reaction center of these processes.[8,9]

Substrate

The DNA oligomers used varied from 9–16 bases in the primer strand and 19–20 bases in the template strand. Studies with primed M13 templates gave kinetic results indistinguishable from the shorter sequences.[10,11] The two principal oligomers used were the 9/20-mer and 13/20-mer, shown in Eq. (2):

$$\begin{aligned} \text{9/20-mer} &= \text{TCGCAGCCG-3}' \\ & \text{AGCGTCGGCAGGTTCCCAAA} \\ \text{13/20-mer} &= \text{TCGCAGCCGTCCA-3}' \\ & \text{AGCGTCGGGAGGTTCCCAAA} \end{aligned} \qquad (2)$$

Incorporation of a single dTMP or dAMP in to the 9/20-mer or 13/20-mer or multiple incorporation from a nucleotide pool was monitored by rapid-quench techniques using an EDTA or acid quench to stop the reaction progress in less than 5 ms.

Steady-State Kinetics

Binding of Substrates

The binding of DNA and dNTP can be either random, sequential, or strictly ordered. Two general methods have supported an ordered sequential binding of DNA first, followed by dNTP. In the first, inhibition of polymerization by PP_i[12] with either varying dNTP or DNA revealed a mixed

[8] P. M. J. Burgers and F. Eckstein, *J. Biol. Chem.* **254,** 6889 (1979).

[9] A. P. Gupta and S. J. Benkovic, *Biochemistry* **23,** 5874 (1984).

[10] B. T. Eger, R. D. Kuchta, S. S. Carroll, K. A. Johnson, P. A. Benkovic, M. E. Dahlberg, and S. J. Benkovic, *Biochemistry,* **30,** 1441 (1991).

[11] M. E. Dahlberg and S. J. Benkovic, *Biochemistry* **30,** 4835 (1991).

[12] W. R. McClure and T. M. Jovin, *J. Biol. Chem.* **250,** 4073 (1975).

mode of inhibition where PP_i could form a dead-end complex with the KF · DNA species or could noncompetitively inhibit the KF · DNA · dNTP complex. In the second, isotope trapping experiments in which an observable reaction can only occur if the initial substrate–enzyme complex is kinetically competent to react implicated a reactive KF · DNA but not KF · dNTP complex.[13]

Rate-Determining Steps

The fact that the steady-state rate does not correspond to one of the many possible steps involved in incorporation of a single nucleotide was determined from several studies. Positional isotopic exchange experiments using [α-^{18}O]dATP, labeled in both bridging and nonbridging positions did not reveal any observable exchange of α,β-bridging ^{18}O to β-nonbridging ^{18}O even in the presence of added PP_i, thus suggesting that the release of PP_i is rapid and not likely to be involved in limiting the steady-state rate.[14] Elemental effects—substitution of an α-thionucleotide (αS)dNTP for dNTP—on both binding of the nucleotide as well as on the rate of nucleotide incorporation into the primer have been measured. Introduction of sulfur for oxygen decreases the affinity for Mg^{2+}, lowers the electronegativity of the α-phosphorus atom, and adds to the molecular size of the α-phosphoryl moiety.[15] All such effects may influence nucleotide binding and the rate of steady-state turnover. For KF the binding of nucleotide was not changed for (αS)dNTP versus dNTP; the value of k_{cat}, however, exhibited an elemental effect ($k^{P\text{-}0}/k^{P\text{-}S} \simeq 13$) for misincorporation but was small ($k^{P\text{-}0}/k^{P\text{-}S} \simeq 2$) for correct incorporation with various DNA duplexes. Since model displacement reactions at phosphorus have elemental effects ranging from 4 to 11,[16] the rate for misincorporation is partially limited by the chemical step but correct incorporation is not; these conclusions are supported by other types of experiments.

Rapid Kinetic Methods

Single Turnover Incorporation

A typical reaction progress curve, shown in Fig. 1, in which the binary KF · DNA complex is rapidly mixed with dATP and then quenched at

[13] F. R. Bryant, K. A. Johnson, and S. J. Benkovic, *Biochemistry* **22,** 3537 (1983).

[14] V. Mizrahi, R. N. Henrie, J. Marlier, K. A. Johnson, and S. J. Benkovic, *Biochemistry* **24,** 4010 (1985).

[15] B. T. Eger and S. J. Benkovic, *Biochemistry* **31,** 9227 (1992).

[16] D. Herschlag, J. A. Piccirilli, and T. R. Cech, *Biochemistry* **20,** 4844 (1991).

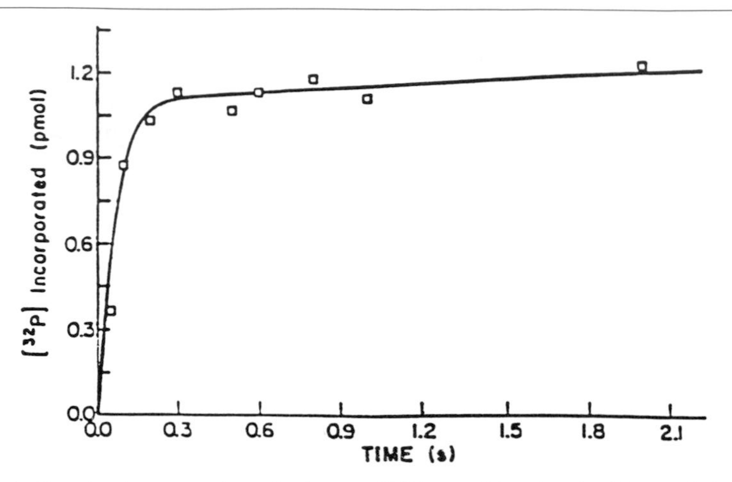

FIG. 1. Pre-steady-state polymerization on 13/20-mer. [Reprinted with permission from reference 7.]

various times demonstrates the biphasic curvature of the incorporation reaction of dATP to give 14/20-mer.[17] The amplitude of the fast phase is stoichiometric with the amount of KF present, and the rate of the slow phase is the same as the rate of steady-state incorporation. Whereas the slow phase is independent of dATP, the rapid phase is not. However, variation of the DNA sequence caused the steady-state rate to change, indicating that the dissociation of product DNA is rate limiting, an interpretation confirmed by direct measurement of the off rate for 14/20-mer dissociation from KF.

Measurement of the binding of DNA (13/20-mer) to KF took advantage of the stoichiometric amplitude of the fast phase. The amount of KF · DNA binary complex present in a solution of KF and 13/20-mer is given by the amplitude of the fast phase—the quantity of dAMP incorporated into the DNA in the first turnover—and varies as a function of 13/20-mer concentration (dATP is saturating). Scatchard analysis of the data provided a dissociation constant for the binary KF · 13/20-mer complex. Similarly, variation of dATP levels at saturating 13/20-mer yielded a set of biphasic curves for nucleotide incorporation from which the dissociation constant for dissociation of dATP from the ternary complex KF · 13/20-mer · dATP was obtained (Fig. 2). The same value was obtained by observing the partitioning of KF · 13/20-mer between dissociation of 13/20-mer and polymerization to 14/20-mer as a function of dATP levels. Combining these observations with conclusions drawn from the above steady-state measure-

[17] R. D. Kuchta, P. A. Benkovic, and S. J. Benkovic, *Biochemistry* **27,** 6716 (1988).

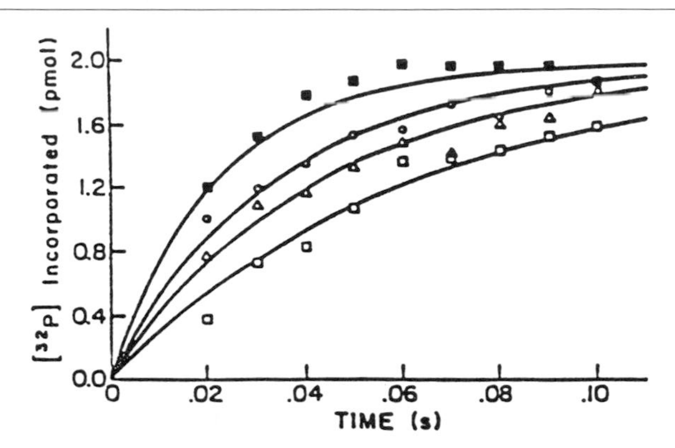

FIG. 2. Effect of varying dATP concentration on the rate of pre-steady-state polymerization. Pre-steady-state polymerization kinetics on 13/20-mer were measured with (□) 3.3, (△) 5, (○) 10, and (■) 20 μM [α-^{32}P]dATP. [Reprinted with permission from reference 7.]

ments, we can now construct the first of several kinetic schemes and attribute the fast phase of nucleotide incorporation to a nonchemical rate-limiting conformational change of the E · 13/20-mer · dATP species. For understanding we will adopt the practice of only postulating those species [Eq. (3)] required by the kinetic data discussed so far. The solid arrows indicate those rate or dissociation constants whose values are now defined.

$$\text{KF} + 13/20\text{-mer} \overset{K_D^{13/20}}{\rightleftharpoons} \text{KF} \cdot 13/20\text{-mer} \overset{K_D^{dATP}}{\rightleftharpoons} \text{KF} \cdot 13/20\text{-mer} \cdot \text{dATP} \tag{3}$$
$$\dashrightarrow \text{KF}' \cdot 13/20\text{-mer} \cdot \text{dATP} \dashrightarrow \text{KF}' \cdot 14/20\text{-mer} \overset{K_D^{14/20}}{\rightleftharpoons} \text{KF} + 14/20\text{-mer}$$

Pulse–Chase Experiments

A pictorial representation of the pulse–chase experiments is given in Fig. 3. The experimental protocol involves adding labeled dATP to the KF · 13/20-mer complex and allowing the bound intermediates to reach their steady-state equilibrium levels. These are then chased to product with an excess of unlabeled substrate, revealing those intermediates that partition favorably in the product direction (only labeled product is counted). The chase experiment is compared to an experiment that is immediately quenched without the chase, so that bound substrate species such as KF · 13/20-mer or KF · 13/20-mer · dATP are not allowed to proceed to product.

The experimental results of a pulse–chase experiment show that the burst amplitude, when the reaction is immediately quenched, is 80% of the amplitude when the reaction is chased with cold dATP (Fig. 4) indicating

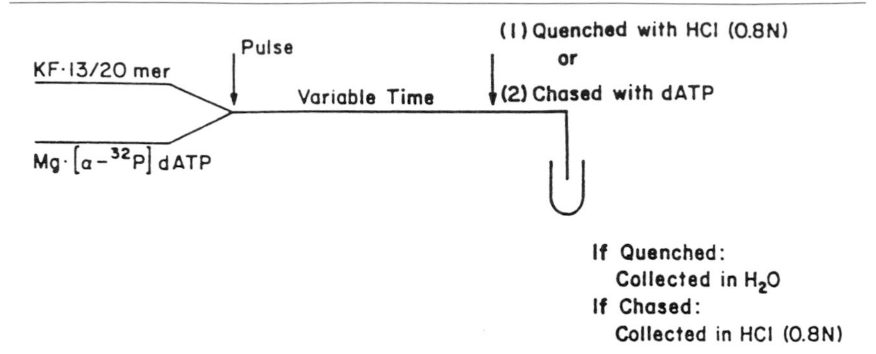

FIG. 3. Pictorial representation of the pulse–chase experiments described in the text. [Reprinted with permission from reference 11.]

that the ternary complex of KF′ · 13/20-mer · dATP exists in a form that can undergo conversion to product.[11] Other complexes, KF · 13/20-mer and KF · 13/20 · dATP, do not participate because the former would bind unlabeled dATP under the chase conditions and the latter undergoes more rapid exchange of its labeled [^{32}P]dATP with the chase pool than it converts to product.[10] Thus KF′ · 13/20-mer · dATP, which has undergone the conformational change, is the only intermediate fulfilling the requirement of

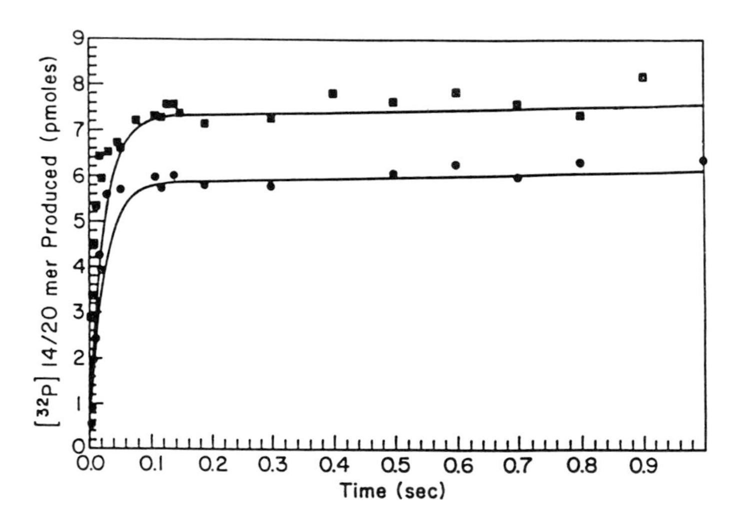

FIG. 4. Pulse–chase experimental results. The incorporation of radiolabeled dATP into 13/20-mer was measured under either pulse–chase (■) or pulse–quench (●) conditions. The curve representative of the pulse–quench data predicts the rate of formation of all DNA_{n+1} species; the simulated curve for the pulse–chase data predicts the rate of formation for all DNA_{n+1} and adds the KF′ · DNA · dNTP species. [Reprinted with permission from reference 11.]

nonreversal in order to escape dilution of the ^{32}P label. Consequently, the equilibrium across the chemical step is K_{eq} = 80%/20% = [KF′ · 14/20-mer · PP_i]/[KF′ · 13/20 mer · dATP].

For the KF′ · 14/20-mer · PP_i species to accumulate, it is necessary for a step after the chemical formation of the phosphodiester bond to prevent dissociation of PP_i from the product ternary complex. If the PP_i were to dissociate from the ternary product complex immediately after chemistry, in a step that would be both rapid and effectively irreversible, the KF′ · 13/20-mer · dATP species would never transiently build up to any significant degree. Thus the two enzyme species participating in the chemical step must be flanked by two kinetic barriers requiring that scheme 1 be extended to include a second conformational change step and a new species, KF · 14/20-mer · PP_i, imposed between KF′ · 14/20-mer · PP_i and KF + 14/20-mer.

Pyrophosphorolysis

Investigation of pyrophosphorolysis, the microscopic reverse of the polymerization reaction, permitted further comment on details of the kinetic scheme subsequent to formation of the first product complex KF′ · 14/20-mer · PP_i. Pyrophosphorolysis, under irreversible conditions (in the presence of a pool of dATP), followed the rate of formation of radiolabeled dATP from 3′-P^{32} 14/20-mer and PP_i. A single first-order exponential was observed when the concentration of KF exceeded that of the 14/20-mer. The observation of first-order kinetics is indicative of a rate-limiting step that occurs prior or during chemical cleavage. Measurement of the rate of pyrophosphorolysis at varying PP_i levels provided data for the determination of the dissociation constant for pyrophosphate from the KF · 14/20-mer · PP_i complex.

The identification of the rate-limiting conformational change step in pyrophosphorolysis with that implicated by the pulse–chase experiments was based on several lines of evidence. (1) The rate of pyrophosphorolysis displays a negligible elemental effect on substitution of a primer–template having a 5′ → 3′-thiophosphate linkage between the two terminal nucleotides. As was the case for the incorporation reaction, the absence of a large reduction in rate on substitution of α-thio for oxo indicates the chemical step for phosphoryl transfer is being masked by a slower rate. (2) The pyrophosphorolysis reaction curve is not biphasic. If there were no slow conformational change step as imposed above between KF′ · 14/20-mer · PP_i and its dissociation to unbound KF, the kinetics of pyrophosphorolysis should be biphasic since in this reverse direction a rapid chemical step would be followed by the slower first conformational change step (KF′ · 13/20-mer · dNTP → KF · 13/20-mer · dNTP). (3) The placement of the slow, nonchemical step must occur between the chemical step and the release of PP_i from the enzyme since, the rate of $[^{32}P]dATP \rightleftharpoons [^{32}P]PP_i$ exchange equals the rate of pyrophosphorolysis. Consequently, this slow step must be traversed in both exchange and pyrophorolysis, requiring this step to occur before PP_i is released, so that Model A for pyrophosphorolysis and exchange is favored over Model B [Eq. (4)].

Model A:

$$\mathrm{KF \cdot 13/20\text{-}mer} \underset{[^{32}P]dATP}{\rightleftharpoons} \mathrm{KF' \cdot 14/20\text{-}mer \cdot PP_i} \underset{k_{pyro}}{\rightleftharpoons} \mathrm{KF \cdot 14/20\text{-}mer \cdot} \overset{PP_i}{\underset{[^{32}P]PP_i}{\rightleftharpoons}} \mathrm{KF + 14/20\text{-}mer}$$

$$k_{ex} = k_{pyro}$$

Model B: (4)

$$\mathrm{KF \cdot 13/20\text{-}mer} \underset{[^{32}P]dATP}{\rightleftharpoons} \mathrm{KF' \cdot 14/20\text{-}mer \cdot PP_i} \overset{PP_i}{\underset{[^{32}P]PP_i}{\rightleftharpoons}} \mathrm{KF' \cdot 14/20\text{-}mer} \underset{k_{pyro}}{\rightleftharpoons} \mathrm{KF + 14/20\text{-}mer}$$

$$k_{ex} > k_{pyro}$$

Equation (2) is now expanded to include these results in Eq. (5):

$$\mathrm{KF + 13/20\text{-}mer} \overset{K_D^{13/20}}{\rightleftharpoons} \mathrm{KF \cdot 13/20\text{-}mer} \overset{K_D dATP}{\rightleftharpoons} \mathrm{KF \cdot 13/20\text{-}mer \cdot dATP} \rightleftharpoons \tag{5}$$

$$\mathrm{KF' \cdot 13/20\text{-}mer \cdot dATP} \overset{K_{eq}}{\rightleftharpoons} \mathrm{KF' \cdot 14/20\text{-}mer \cdot PP_i} \rightleftharpoons \mathrm{KF \cdot 14/20\text{-}mer \cdot PP_i}$$

$$\overset{K_D PP_i}{\rightleftharpoons} \mathrm{KF \cdot 14/20\text{-}mer} \overset{K_D 14/20}{\rightleftharpoons} \mathrm{KF + 14/20\text{-}mer}$$

Processivity

The rate of incorporation of successive nucleotides into the 13/20-mer was determined in this experiment. KF was incubated with 5′-end-labeled 13/20-mer and mixed with dATP and dGTP to form ultimately a 17/20-mer duplex. Time points were processed by gel electrophoresis and the levels of 13-mer, 14-mer, 15-mer, 16-mer, and 17-mer determined. The time course is shown in Fig. 5. The data show that while the first incorporation of dATP occurs at a rate commensurate with the rate of the first conformational change, addition of successive nucleotides is slowed considerably. This arises from the rate of the second slow conformational change and allows its evaluation in the forward direction.

Overall Equilibrium

To avoid the possibility that the measurement of the internal equilibrium for the polymerase reaction is compromised by action of the combined polymerase/exonuclease activities, conditions were chosen where all the DNA was enzyme bound and the reaction was constrained to a short time interval (9 sec). The equilibrium distribution of bound substrates and products ($K_{\text{bound}} = 2762 \pm 919$) is obtained in about 2 sec, a time not sufficient for an appreciable amount of exonuclease activity to occur. This value combined with the rates measured for processive synthesis and pyrophosphorolysis provided values for the rate constants associated with the conformational change steps that were still undefined.

The values of the final rate and equilibrium constants are graphed as a free-energy profile in Fig. 6. It is clear that the first nonchemical step limits the rate of a single nucleotide incorporation, whereas the second limits the rate of processive synthesis. The two are separated by a fast

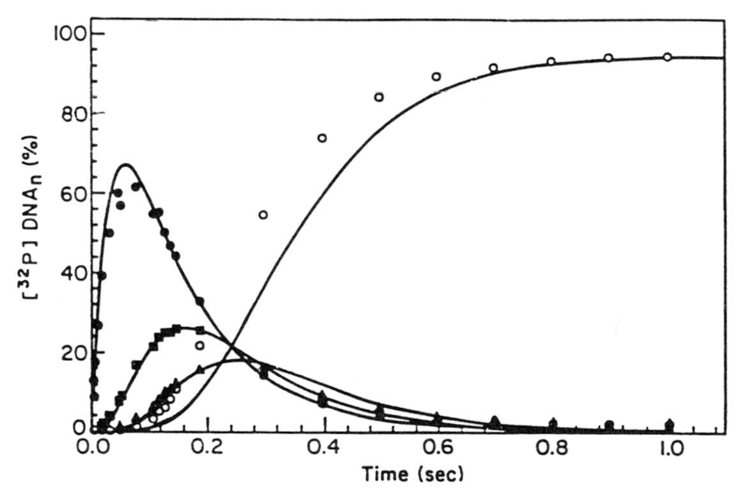

FIG. 5. Processive synthesis experiments. The rate of elongation of 5′-end-labeled 13/20-mer to form 14/20-mer (●), 15/20-mer (■), 16/20-mer (▲), and 17/20-mer (○). [Reprinted with permission from reference 11.]

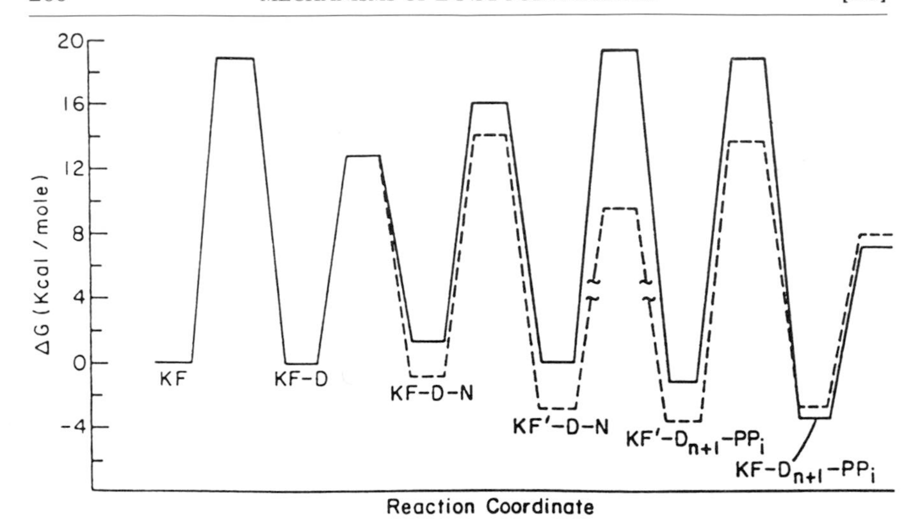

FIG. 6. Free-energy profile for correct and incorrect nucleotide incorporation by the Klenow fragment. The free energy for each reaction step was calculated from $\Delta G = RT[\ln(kT/h) - \ln(k_{obs})]$, where R is 1.99 cal K^{-1} mol^{-1} (gas constant), T is 295 K, k is 3.30×10^{-24} cal K^{-1}, Boltzmann's constant, h is 1.58×10^{-34} cal sec, Planck's constant, and k_{obs} is the first-order rate constant. The concentrations used were 5 nM 9/20-mer and 22 μM dATP. Solid lines are the free energies presented for misincorporation of dATP into 9/20-mer to form 9A/20-mer. The dashed lines are the free energies for correct nucleotide incorporation into 13/20-mer to form 14/20-mer, also calculated for 5 nM 13/20-mer and 22 μM dATP. [Reprinted with permission from reference 15.]

chemical bond formation that favors product formation although these two states are nearly equivalent in free energy. The values were used to compute the solid curves to fit the data shown in Figs. 1, 2, 4, and 5.

Misincorporation

The observation that KF can catalyze the misincorporation of nucleotides into DNA duplexes (in our present case, dATP into the 9/20-mer) led to the construction of a kinetic mechanism governing incorrect DNA synthesis.[17] The direct comparison of the kinetic stages encountered in correct and incorrect incorporation highlights the means of discrimination used by KF to increase the discrimination against incorrect nucleotide incorporation and improve fidelity of replication. Although the kinetic sequence was developed independently, in overall form it is identical to that for correct synthesis except for the inclusion of the exonuclease steps and appears as Eq. (6).

$$
\begin{array}{l}
\mathrm{KF} \underset{k_{-1}}{\overset{k_{+1},\ 9/20}{\rightleftharpoons}} \mathrm{KF\text{-}9/20} \underset{k_{-2}}{\overset{k_{+2},\ dATP}{\rightleftharpoons}} \mathrm{KF\text{-}9/20\text{-}dATP} \underset{k_{-3}}{\overset{k_{+3}}{\rightleftharpoons}} \mathrm{KF'\text{-}9/20\text{-}dATP} \underset{k_{-4}}{\overset{k_{+4}}{\rightleftharpoons}} \mathrm{KF'\text{-}9A/20\text{-}PP_i} \\
\mathrm{KF'\text{-}9A/20\text{-}PP_i} \underset{k_{-5}}{\overset{k_{+5}}{\rightleftharpoons}} \mathrm{KF\text{-}9A/20\text{-}PP_i} \underset{k_{-6},\ PP_i}{\overset{k_{+6}}{\rightleftharpoons}} \mathrm{KF\text{-}9A/20} \underset{k_{-7},\ 9A/20}{\overset{k_{+7}}{\rightleftharpoons}} \mathrm{KF} \\
\mathrm{KF\text{-}9/20} \overset{\text{Translocation}}{\rightleftharpoons} \mathrm{KF\text{-}9A/20} \\
\mathrm{KF\text{-}9A/20} \xrightarrow{k_{exo}} \mathrm{KF\text{-}9/20 + dAMP} \\
\mathrm{KF\text{-}9A/20\text{-}PP_i} \xrightarrow{k_{exo}} \mathrm{KF\text{-}9/20 + dAMP + PP_i} \\
\mathrm{KF'\text{-}9A/20\text{-}PP_i} \xrightarrow{k_{exo}} \mathrm{KF\text{-}9/20 + dAMP + PP_i}
\end{array}
\qquad (6)
$$

Their introduction is necessitated by the fact that the exonuclease rates are now comparable to the slower rates of misincorporation. Rather than again describing protocols that are similar, we select experimental approaches that differ from those mentioned so far.

There are three important differences in the quantitative rates for the kinetic sequence of misincorporation and those for correct synthesis. First, the rate of misincorporation is biphasic but the pre-steady-state rate is now the rate of the chemical step. There is an elemental effect of 13 for the misincorporation of (αS)dATP into the 9/20-mer when KF(exo$^-$) is used to remove the complication of accounting for the exonuclease activity on native KF(exo$^+$). Binding of dATP or (αS)dATP was determined in nucleotide discrimination experiments in which the binding of a single incorrect nucleotide (dATP) to the KF · 9/20-mer complex was measured by its effect on the rate of correct nucleotide (dTTP) incorporation. The observed effect is best fit by a two-step binding sequence, Eq. (7), which parallels that seen for correct incorporation, that is, nucleotide binding to the binary KF · DNA complex followed by a conformational change.

$$
\begin{array}{l}
\mathrm{KF \cdot 9/20\text{-}mer} \overset{dTTP}{\rightleftharpoons} \mathrm{KF \cdot 9/20\text{-}mer \cdot dTTP} \longrightarrow \mathrm{KF + 10/20\text{-}mer + PP_i} \\
\mathrm{KF \cdot 9/20\text{-}mer} \overset{dATP}{\rightleftharpoons} \mathrm{KF \cdot 9/20\text{-}mer \cdot dATP} \rightleftharpoons \mathrm{KF' \cdot 9/20\text{-}mer \cdot dATP}
\end{array}
\qquad (7)
$$

In general, the discrimination in the binding step is small (0 to 23-fold), and depends on the DNA sequence and the competing nucleotide.

Second, there is an increase in the rate of release of mismatched product DNA. In this case the off-rate for mismatched product, 9A′/20-mer, was measured by adding the standard duplex (13/20-mer) and its next correct nucleotide (dATP) to a solution of KF · 9A/20-mer. The lag in the production of 14/20-mer relative to its production in the absence of mismatched DNA provides the rate of release of the mismatch from KF.[15]

Third, there is a reduction in the rate of the second conformation change during misincorporation. In this case the release of PP_i during the pre-steady-state phase was monitored by converting the liberated PP_i with pyrophosphatase to P_i. The rate formation of free PP_i (measured as P_i) accompanying the misincorporation of dATP to form the 9A/20-mer (Fig. 7) occurs at a slower rate than 9A/20-mer formation. (Remember that quenching releases KF bound 9A/20-mer.) Given the high K_D for PP_i binding to KF · DNA complexes, this slower rate relative to 9A/20-mer requires that PP_i release follow a slow conformational change. This measurement provides from the amplitude and rate of release of PP_i values for

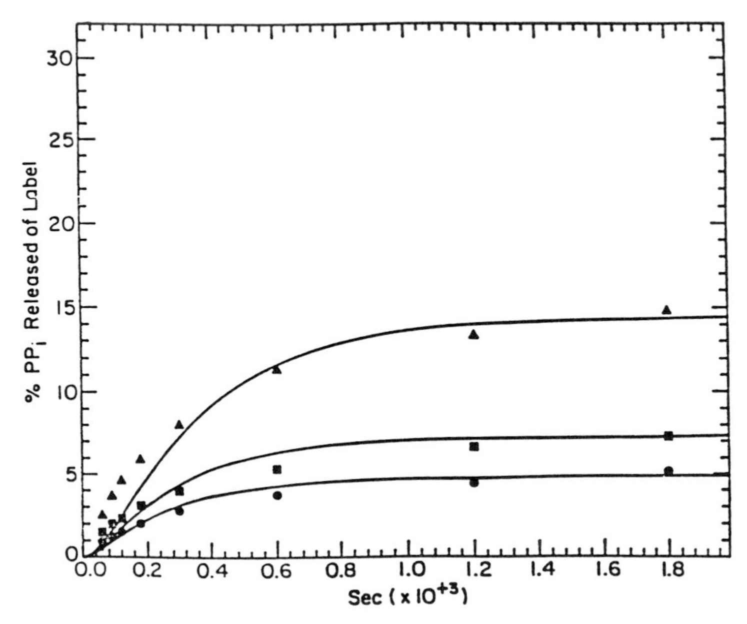

FIG. 7. Pre-steady-state burst of released pyrophosphate formed as a function of dATP and PP_i concentration. Assays were performed with excess KF(exo-) with either (▲) 10, (■) 20, of (●) 30 μM (γ^{32}P)dATP. The data are presented as the percent of PP_i released of the total radiolabel present. [Reprinted with permission from reference 15.]

the rate constants associated with the chemical step of misincorporation and the second conformational change.

Finally the rates for $3' \rightarrow 5'$-exonuclease activity were determined by direct measurement of the exonuclease activity on the 9A/20-mer in the absence or presence of saturating pyrophosphate, that is, dAMP formation from KF · 9A/20-mer and KF ·9A/20-mer · PP_i. The occurrence of the slow conformational change prior to PP_i release is also manifested in a slower dissociation of the 9A/20-mer from the KF′-9A/20-mer · PP_i complex that is produced in the chemical step. Measurement of the rate of dAMP formation at short times showed accumulation of dAMP before that predicted from editing by the KF exonuclease activity only on the latter two complexes. The data were fit by assigning an editing function to the KF′ · 9A/20-mer · PP_i complex from which the 9A/20-mer does not apparently dissociate. The value of the rate constants for exonucleolytic editing were equal for all three complexes.

The free-energy changes associated with misincorporation of dATP into the 9/20-mer are exhibited in Fig. 6 for direct comparison. Although the primer–template sequences differ, the incorporated nucleotide in both cases is dAMP and the effect of neighboring sequence is expected to be negligible. The major contributions to fidelity reside (1) in the chemical step (KF′-D_{-N} KF′-D_{N+1}-PP_i), which is minimally 6 kcal/mole; (2) the barrier to PP_i loss from the KF′-D_{N+1}-PP_i complex due to the step (KF′-D_{N+1}-PP_i KF-D_{N+1}-PP_i), which is ~7 kcal/mol more unfavorable and permits editing of the mismatch by internal pyrophosphorolysis and exonucleolytic activity at the level of the KF′-D_{N+1}-PP_i complex; and (3) not shown in the diagram the slow addition of the next correct nucleotide onto a mismatch[17,18] provides an addition factor of about 1 to 3 kcal/mol. The two steps involving conformation changes serve to lock the DNA to the central ternary complexes in order to increase the efficiency of exonucleolytic editing. In other replicative polymerases, this locking step may be achieved by the accessory proteins.

Acknowledgment

This paper was supported in part by NIH grant GM13306 (SJB). CEC is supported by NIH postdoctoral fellowship AI09076.

[18] C. M. Joyce, *J. Biol. Chem.* **264,** 10,858 (1989).

[21] Methods of Analyzing Processivity

By ROBERT A. BAMBARA, PHILIP J. FAY, and LISA M. MALLABER

Introduction

An enzyme is said to act processively if it performs the same reaction repeatedly on its substrate before the two dissociate. This situation is common for enzymes that react with biological polymers such as nucleic acids and polysaccharides. An enzyme that does not act processively is said to have a distributive mechanism. On DNA or RNA, enzymes can act processively for a variety of functions including synthesis, degradation, translocation, and strand separation. The quantitative parameter, processivity, is defined as the number of times the reaction is repeated between association and dissociation of the enzyme and substrate. Processivity is an important parameter, because it can help to distinguish between the exact biological functions of enzymes with the same catalytic activity. For example, it can suggest whether a DNA polymerase is designed to carry out extensive chromosomal replication or short patches of DNA repair. The value of processivity for an enzyme is not absolute, but must be defined for a particular substrate, under particular reaction conditions.

Procedures for measuring processivity have been based on two strategies, exemplified by studies of DNA polymerases. Both require that the primer–template be in large excess over the polymerase so that substrates virtually never react with more than one enzyme during the reaction.

In the first, synthesis is allowed to proceed, and then information is obtained on both the total number of nucleotides added in the reaction, and the number of DNA primer–templates on which the polymerase has acted. Dividing the former by the latter yields the number of nucleotides added per primer–template. We developed a procedure[1] based on this concept in which two reactions were carried out, one with all four types of dNTPs, and the other with only three. Conditions were arranged such that polymerases synthesized on one template and then moved to another at the same frequency in both reactions. In the reaction with three dNTPs, processive extension was limited to about three nucleotides. If the polymerase were normally processive for addition of 20 nucleotides, the rate of nucleotide incorporation in the reactions with four versus three dNTPs would be approximately 20 versus 3. Das and Fujimura[2] conducted primer

[1] R. A. Bambara, D. Uyemura, and T. Choi, *J. Biol. Chem.* **253,** 413 (1978).
[2] S. K. Das and R. K. Fujimura, *Nucl. Acids Res.* **8,** 657 (1980).

extension reactions in which all added nucleotides were labeled. The DNA was then digested with nucleases that made nucleoside products from the 3′-terminal residues, and nucleotide products from the internal residues. If the polymerase added 20 nucleotides processively, there would be 19 times more nucleotides than nucleosides in the digested mixture.

These methods are useful for analyzing enzymes displaying processivity of less than about 100, but become quite inaccurate at greater values. Furthermore, DNA polymerases generate a distribution of processively synthesized products. Depending on the template sequence, this distribution can be very asymmetrical. These methods provide an average value that completely lacks information about the product distribution.

In the second strategy, the polymerases are allowed to react with relatively short primers of known, fixed length. Reaction conditions are adjusted such that the length of polymer added to the primers represents the product of one round of processive synthesis by the polymerase. A round of processive synthesis consists of binding of the polymerase to the primer, elongation, and then dissociation. After extension the products are separated by gel electrophoresis, and the distribution of extended primer lengths is visualized by autoradiography. Since the earliest applications of this procedure,[3] the distribution of processively synthesized products has been used to show features of the secondary structure of the template and to distinguish different functional forms of the polymerase.

Measurement of Processivity of DNA Synthesis

Processive DNA synthesis is often measured on either homopolymeric or heteropolymeric, that is, natural, DNA templates. The advantage of using homopolymers is that a symmetrical distribution of products is obtained. The position of the peak, representing the average length of processive synthesis, is easily determined. Products made by a mixture of two polymerases or polymerase forms might readily be distinguished by the appearance of two product distributions. Use of natural templates usually causes polymerases to dissociate with high frequency at some locations, termed pause sites, while dissociating virtually not at all at other locations. The resulting product distribution consists mostly of several major products of various lengths and many minor ones.

The sizes of products in the distribution are determined by polynucleotide size standards. In the case of homopolymers, these can be purchased in some size ranges. They can also be made by a time-dependent sampling

[3] S. D. Detera and S. H. Wilson, *J. Biol. Chem.* **257,** 9770 (1982).

of the elongation of the primer–template. In this case every possible length product is made. The mobility of a particular length product can be determined by counting the bands on the gel from the primer terminus to the desired length. For natural DNA, size determinations can be made by performing Sanger dideoxynucleotide sequencing with the substrate primer–template.

Essential Criterion for Measurement

The key to accurate measurement of processivity is limiting the extension of each primer to one round of processive synthesis. This assures that the number of nucleotides added to each primer is equivalent to the processivity of the polymerase. This essential criterion is achieved in either of two ways. In the first, the quantity of polymerase is reduced relative to primer–template. Ideally, the level of polymerase should be adjusted so that only a small fraction of the primers sustains any elongation. Usually restricting primer use to less than 10% is sufficient. Procedures to verify this outcome are discussed below. This approach is adequate in most cases. However, problems could arise if the polymerase is diluted to the point that it loses activity, dissociates subunits, or changes properties.

It is generally not recommended that the reaction time be lowered to limit synthesis on the substrate. Some DNA polymerases require in the range of minutes to complete one round of processive synthesis. The reaction should be allowed to proceed into a time range in which product length is not time dependent.

Another way to achieve the criterion is the use of a trap. The trap is any molecule that can bind the DNA polymerase in a way that prevents it from binding the substrate. Typically traps are DNA primer–templates that are unlabeled and cannot support incorporation of the particular labeled nucleotide used in the reaction. In this case the polymerase and substrate are mixed in the absence of a necessary reaction component, for example, the deoxynucleoside triphosphates. Then, to start the reaction, the dNTPs are added, but along with a large excess of the trap. The polymerases already bound to the substrate carry out one round of processive synthesis and then dissociate and bind the trap. A good way to verify the effectiveness of the trap is to add it to the reaction before the polymerase. Then, when the dNTPs are added, virtually no synthesis should occur. Examples of trap molecules that have been successfully used for measurement of processivity of DNA polymerases are activated calf thymus DNA,[4] poly(dI-dC),[5] heparin,[6] and poly(A) · oligo(dT).[7]

The primary concern over use of a trap is that the trap may influence the dissociation of the DNA polymerase from the substrate. This would most likely result in a shortening of primer elongations, which would incorrectly represent the processivity of the polymerase. One can test for this effect by carrying out the reaction at two concentrations of trap, both capable of effectively sequestering the polymerase. Both experiments should yield the same value of processivity. If the higher concentration of trap yields a different processivity, one should try lowering both concentrations into a range where the measured values of processivity are the same. If the trap concentration must be lowered to a point where it is no longer effective at complete inhibition of synthesis if added before the polymerase, the trap procedure should not be used.

Verification of Appropriate Conditions for Processivity Determination

If processivity is being measured in the absence of trap, the concentration of polymerase must be sufficiently low that polymerase molecules virtually never react with the same primer twice. A simple means of verifying this criterion is measurement of the amount of nucleotide incorporation. If radiolabeled nucleotides are added, one can measure the number of nucleotides incorporated per primer terminus. The most rigorous requirement that one can apply is that the number of nucleotides incorporated be 10% or less of the number of primer termini. This assures that the number of termini utilized for synthesis is less than 10%, even if the processivity of the polymerase is one. Preliminary results indicating a processivity much greater than one would be grounds for relaxing this criterion. For many polymerases, which display processivity values from 10 to many thousands, incorporation of 1 to 10 nucleotides per primer terminus is appropriate. Care should be taken also to confirm that the level of primer termini considered 100% consists of termini that can actually sustain synthesis in polymerase excess.

Another commonly used way of verifying that primer termini are not subjected to more than one round of processive DNA synthesis is produc-

[4] C. M. Joyce, *J. Biol. Chem.* **264,** 10,858 (1989).

[5] H. E. Huber, J. M. McCoy, J. S. Seehra, and C. C. Richardson, *J. Biol. Chem.* **264,** 4669 (1989).

[6] V. Gopalakrishnan, J. A. Peliska, and S. J. Benkovic, *Proc. Natl. Acad. Sci. USA* **89,** 10,763 (1992).

[7] J. J. DeStefano, R. G. Buiser, L. M. Mallaber, T. W. Myers, R. A. Bambara, and P. J. Fay, *J. Biol. Chem.* **266,** 7423 (1991).

tion of a distribution of products under a range of ratios of enzyme to primer–template. As the ratio of primer–template concentration to enzyme concentration is increased, products will first become smaller. This is because the fraction of primer termini that have sustained more than one round of processive synthesis is being eliminated. As the ratio is increased further, there is no change in the distribution of product sizes. As explained in the examples below, this indicates that product size is determined solely by the processivity of the DNA polymerase.

Labeling of Substrate

One common approach is to label the primer with a 5′-terminal phosphate using T4 polynucleotide kinase and [γ-^{32}P]ATP. This is termed *end labeling.* The advantage is that the quantity of label in any size class of elongated primers is proportional to the number of elongated primer molecules in that size class. Therefore, the distribution of labeled products provides an accurate representation of the frequency of polymerase dissociations over the range of elongation distances. The disadvantage is that the primers are only labeled in one position, limiting the possible specific activity of labeling, and the detectability of the products.

Another approach is to incorporate labeled nucleotides using one or more [α-^{32}P]deoxynucleoside triphosphates in the synthesis reaction. This is termed *internal labeling.* An advantage of internal labeling is that the reaction is more convenient, because steps involving labeling of the 5′ terminus of the primer are not required. Also, it is possible to label the extended primers to higher specific activity, since many labeled nucleotides are usually added per primer. In this case the amount of label in any size class of product is proportional to both the number of product molecules and the number of labeled nucleotides added per molecule. The observed distribution of radioactivity then represents the number of nucleotides incorporated in each size class of product. One can readily convert the internal label distribution to an approximation of the end label distribution, using the size standard molecules. This can be done by multiplying the density of individual bands by a correction factor of one over the length of the elongated primer represented by the band.

Details of the Measurement Procedure

A typical measurement of processivity with an end-labeled primer is shown in Fig. 1. Here we have measured the processivity of T4 DNA polymerase on a homopolymeric template. The primer dT_{16}, and the template poly$(dA)_{2000-3000}$, were obtained from the Midland Certified Reagent

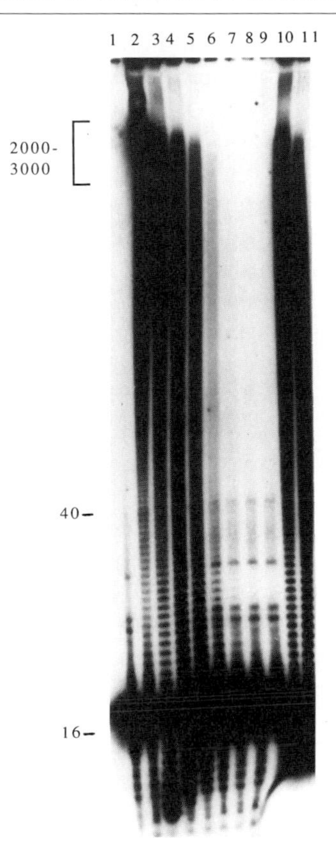

FIG. 1. End-labeled primers elongated processively by T4 DNA polymerase. Lane 1, [^{32}P](dT)$_{16}$; lane 2, primers elongated by 1 unit of Sequenase™; lanes 3–9, primers elongated by progressively less T4 DNA polymerase, 0.0625, 0.0125, 0.0025, 5×10^{-4}, 2×10^{-5}, 4×10^{-6}, and 8×10^{-7} units, respectively; lanes 10–11, primers elongated in the presence of 5 mM 5′-AMP, and 0.0625 and 0.0125 units of polymerase, respectively.

Co. The (dT)$_{16}$ was labeled with phosphate at the 5′ end using [γ-^{32}P]ATP (3000 Ci/mmol; 10 mCi/ml) from New England Nuclear, and T4 polynucleotide kinase from US Biochemical Corp. using the procedure recommended by the enzyme manufacturer. The ^{32}P-end-labeled (dT)$_{16}$ was purified away from the unreacted ATP using a NENSORB-20™ (Dupont) column according to the manufacturer's instructions. The poly(dA) and oligo(dT) were annealed by mixing at a ratio of 14 : 1 (w/w), at a poly(dA) concentration of 1 μg per μl, in 40 mM KCl, heating to 65° and then cooling over 5 min to room temperature.

Reaction mixtures (25 μl) contained 1 μg poly(dA) · oligo(dT), 1 mM

dTTP, 50 mM Tris–HCl, pH 9.0, 20 mM KCl, 1 mM $MgCl_2$, 5% glycerol, and 5 mM 2-mercaptoethanol. This mixture should be modified depending on the polymerase being tested. Reactions were started by addition of either T4 DNA polymerase or Sequenase from US Biochemical Corp. These enzymes were stored and diluted in 50 mM Tris–HCl, pH 8.5, 100 mM $(NH_4)_2SO_4$, 0.1 mM ethylenediaminetetraacetic acid (EDTA), 10 mM 2-mercaptoethanol, and 200 μg/ml bovine serum albumin. The reactions were initiated by addition of enzyme in 1 μl, and incubated at 37° for 15 min. They were stopped by addition of 10 μl of a mixture of 90% formamide, 10 mM EDTA, pH 8.0, 0.1% xylene cyanole, and 0.1% bromphenol blue. All undesignated reagents are from Sigma Chemical Corp.

The purpose of the Sequenase reaction is to perform extensive synthesis on the primer–template. This will verify that it is so long that it would not limit the processive elongation of the products made by the polymerase being tested.

Some reaction mixtures also contained 5 mM 5′-AMP, which has been shown to inhibit the 3′ → 5′-exonuclease of polymerases as diverse as *Escherichia coli* DNA polymerase I and calf DNA polymerase δ.[8]

An 8% polyacrylamide gel[9] was prepared, containing 7 M urea, 89 mM Tris–borate, pH 8.3, and 2 mM EDTA. It was 37 cm long, 30 cm wide, and 4 mm deep. The upper and lower buffers were 89 mM Tris–borate, pH 8.3, and 2 mM EDTA. The gel was subjected to electrophoresis for 30 min at 70 W. Then 8-μl reaction samples were loaded in each lane. Electrophoresis was then continued for approximately 60 min. The gel was removed and dried on a Buchler gel dryer, and then subjected to autoradiography using Kodak XAR film and a Du Pont Lightning Plus intensifying screen at −70°.

To obtain an accurate measurement of product distribution, the film must be exposed within its linear response range. The densitometer must also be used within its linear electronic response range. An alternative to the use of film is imaging equipment that scans the gel directly, producing a digital image.

In the case of T4 DNA polymerase, as the enzyme concentration was lowered, there was a substantial decrease in the sizes of primer elongation products. Clearly products made at higher enzyme concentrations resulted from visits of many enzymes to each primer terminus. At the lowest enzyme concentrations, the product size remained generally constant as the enzyme

[8] J. J. Byrnes, K. M. Downey, B. G. Que, M. Y. W. T. Lee, V. L. Black, and A. G. So, *Biochemistry* **16,** 3740 (1977).

[9] J. Sambrook, E. F. Fritsch, and T. Maniatis, "Molecular Cloning, a Laboratory Manual," 2nd ed. Cold Spring Harbor Laboratory Press, New York, 1989.

concentration was reduced. This indicates that the product distribution resulted from processive synthesis by only one enzyme at primers that have sustained synthesis. At the lowest enzyme concentrations, most of the primers were unreacted. This is also a good indication that those that have reacted have not sustained more than one round of processive synthesis.

Interpretation of Results

The autoradiogram in Fig. 1 exhibits several commonly observed phenomena. At the lowest enzyme concentrations, primer elongation products no longer appear (lanes 8–9, Fig. 1). The only labeled polymers that are longer than the unreacted primer are labeled contaminants in the primer, also visible in lane 1 (Fig. 1). These should be discounted. The distribution of products representing processive synthesis is most appropriately measured in lane 6 (Fig. 1), where products are visible, but there is a large excess of unreacted primer.

There appear to be two distinct distributions of product sizes, one very short resulting from additions of 1 to 15 nucleotides to the primers, and the other quite long, resulting from addition of hundreds of nucleotides to the primers. One possible reason for two distributions is that there are two populations of polymerase in the preparation, one exhibiting high processivity and the other low. Another possibility is that primers on the substrate have gathered into groups. Attempts at synthesis between adjacent primers might produce short products, whereas extension of primers without adjacent partners may result on long products. The issue can be resolved by further enzyme purification or use of heteropolymeric substrates.

Action of the $3' \rightarrow 5'$-exonuclease is evident from the generation of products shorter than the starting primer. As seen in lanes 10 and 11, the presence of 5′-AMP does not substantially alter the $3' \rightarrow 5'$-exonuclease activity.

Comparison of End and Internal Label

Figure 2 shows results obtained when the same analysis was performed with an internal label. The reaction mixture components were equivalent except that an unlabeled primer was used, and 25 pmol of 800 C/mmol [α-^{32}P]dTTP was added. The reaction was stopped by the addition of 2 μl of 500 m*M* EDTA, pH 8.0. After the reaction the unreacted deoxynucleoside triphosphates were separated from the primer–template by centrifuge columns.[10] These were 0.5-ml Eppendorf tubes punctured at the bottom with

[10] H. S. Penefsky, *J. Biol. Chem.* **252,** 2891 (1977).

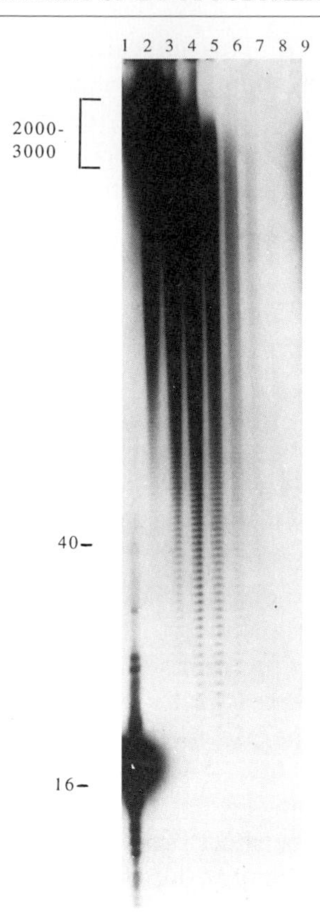

Fig. 2. Internally labeled primers elongated processively by T4 DNA polymerase. Lane 1, $[^{32}P](dT)_{16}$; lane 2, primers elongated by 1 unit of Sequenase(TM); lanes 3–9, primers elongated by progressively less T4 DNA polymerase, 0.0625, 0.0125, 0.0025, 5×10^{-4}, 2×10^{-5}, 4×10^{-6}, and 8×10^{-7} units, respectively.

a needle, and containing about 50 μl of 212- to 300-μm glass beads at the bottom and 450 μl of Sephadex G-50. The resin was swelled in 10 m*M* Tris–HCl, pH 8.0, and 1 m*M* EDTA. They were spun at 2000 rpm in a Sorvall HL-4 rotor for 4 min. Then the reaction mixture was added. They were then spun an additional 4 min. The effluent was collected in 10 μl of the formamide stop mixture, described earlier, for gel electrophoresis.

Compared to results with the end label, the region of the gel near the unreacted primer can be seen more clearly, since the unreacted primer is unlabeled. The products appear longer in general because the amount of labeling of each extended primer is proportional to the length of extension. Consequently, the longer distribution of products is more clearly visible.

This procedure has also been used to visualize products made by more processive DNA polymerases.[11]

Processivity of Nuclease Action

The same procedures that are applied to DNA polymerases can generally be modified for use with nucleases, helicases, and other enzymes potentially capable of processive functions. For example, we measured the extent of processive RNase H activity of the human immunodeficiency virus (HIV) reverse transcriptase using a trap procedure.[12] In this case, processivity is defined as the number of cleavages carried out by the nuclease from the time it binds the substrate until it dissociates. RNase H cleaves RNA hybridized to DNA. The substrate for measuring processivity consisted of a 30-mer DNA oligonucleotide annealed off center on an 83-mer RNA oligonucleotide, closer to its 3′ end. The 83-mer RNA was prepared and labeled by run-off transcription from the plasmid pBSM13+ treated with *Mse*I. This allows fixed-length transcription from a T7 promoter for 83 nucleotides. Transcription with T7 RNA polymerase was performed as in the Promega Protocols and Applications Guide. The reaction contained 18 μM [α-^{32}P]UTP at about 44 Ci/mmol and the other nucleoside triphosphates at 500 μM each. The trap was oligo$(dT)_{16}$ annealed to poly(A) at a 1:8 ratio (w/w), stored in 10 m*M* Tris–HCl, pH 8.0 and 1 m*M* EDTA. In a typical reaction, 11 ng of RT was incubated with 2 n*M* substrate for 3 min in 10 μl of 50 m*M* Tris–HCl, pH 8.0, 1 m*M* dithiothreitol, 2% glycerol (w/w), 0.1 m*M* EDTA, and 5 m*M* KCl. This allows the RT to bind to the substrate. The reaction was then initiated by addition of $MgCl_2$ to 6 m*M*, and 0.5 μg trap in 2.5 μl of reaction buffer. Samples were taken over a period of 2 to 300 sec.

The reverse transcriptase RNase H cleaved the RNA endonucleolytically, producing two sets of products, the shorter set from the original 3′-end region of the RNA, the longer from the 5′-end region. This was simply because of the offset positioning of the primer. Most of the RNA in the reaction was not cleaved at all, indicating that the substrate was in molar excess over the number of active reverse transcriptase molecules. This ensured that for nearly all substrates, cleavage was the result of the action of only one reverse transcriptase. Over the course of time, the 3′ products remained the same length, whereas the 5′ products became progressively

[11] R. D. Sabatino, T. W. Myers, R. A. Bambara, O. Kwon-Shin, R. L. Marraccino, and Paul H. Frickey, *Biochemistry* **27,** 2998 (1988).

[12] J. J. DeStefano, R. G. Buiser, L. M. Mallaber, R. A. Bambara, and P. J. Fay, *J. Biol. Chem.* **266,** 24,295 (1991).

shorter. This showed that when the reaction was initiated, the bound reverse transcriptase molecules made an initial cut, and then moved in the $3' \rightarrow 5'$ direction on the RNA to make one or more additional cuts. The conclusion was that the endonuclease action of the reverse transcriptase was directional and partially processive.

In general, processive action of any enzyme can be quantitated if (1) the distribution of its products can be measured and (2) the measurement can be made after the enzyme has bound, acted on, and then dissociated from the substrate only once. The most certain way to ensure the second condition is to have such an excess of the substrate that only a small proportion of the substrate sustains any reaction at all.

Acknowledgments

This work was supported by research grants GM24441 and GM49573, and Cancer Center Core Grant CA11198 from the National Institutes of Health.

Section IV

Structure–Function Studies of DNA Polymerases

[22] Mutational Analysis of Bacteriophage ϕ29 DNA Polymerase

By LUIS BLANCO and MARGARITA SALAS

Introduction

The fact that ϕ29 DNA polymerase is a small (66-kDa) single-subunit enzyme containing well-characterized enzymatic activities[1] makes this polymerase an appropriate system for structure–function studies. ϕ29 DNA polymerase was included in the group of eukaryotic-type DNA polymerases because of its sensitivity to the nucleotide analogs BuAdATP and BuPdGTP,[2,3] specific inhibitors of eukaryotic DNA polymerase α.[4] In addition, ϕ29 DNA polymerase contains several regions of amino acid sequence highly conserved in the C-terminal portion of eukaryotic DNA polymerases.[3,5–8] It has been proposed that these conserved regions might play an important role in the function of the enzyme, being part of the active site for synthetic activities.[5,9]

However, three conserved amino acid segments, named Exo I, Exo II, and Exo III, could be identified in the N-terminal portion of both eukaryotic-type and Pol I-type DNA polymerases.[10,11] In this particular case, the functional significance of these conserved regions could be anticipated, because they contained the critical residues forming the $3' \rightarrow 5'$-exonuclease active site of Pol I.[12,13] In this report, we compile the site-directed mutagenesis studies carried out in the most conserved regions of ϕ29 DNA polymerase (see Fig. 1).

[1] J. M. Lázaro, L. Blanco, and M. Salas, this volume [5].
[2] L. Blanco and M. Salas, *Virology* **153,** 179 (1986).
[3] A. Bernad, A. Zaballos, M. Salas, and L. Blanco, *EMBO J.* **6,** 4219 (1987).
[4] N. N. Khan, G. E. Wright, and N. C. Brown, *Nucleic Acids Res.* **19,** 1627 (1991).
[5] B. A. Larder, S. D. Kemp, and G. Darby, *EMBO J.* **6,** 169 (1987).
[6] S. W. Wong, A. F. Wahl, P.-M. Yuan, N. Arai, B. E. Pearson, K.-I. Arai, D. Korn, M. W. Hunkapiller, and T. S.-F. Wang, *EMBO J.* **7,** 37 (1988).
[7] L. Blanco, A. Bernad, M. A. Blasco, and M. Salas, *Gene (Amst)* **100,** 27 (1991).
[8] J. Ito and D. K. Braithwaite, *Nucleic Acids Res.* **19,** 4045 (1991).
[9] J. Gibbs, H. Chiou, K. Bastow, Y. Cheng, and D. M. Coen, *Proc. Natl. Acad. Sci. USA* **85,** 6672 (1988).
[10] A. Bernad, L. Blanco, J. M. Lázaro, G. Martín, and M. Salas, *Cell* **59,** 219 (1989).
[11] L. Blanco, A. Bernad, and M. Salas, *Gene* **12,** 139 (1992).
[12] V. Derbyshire, P. S. Freemont, M. R. Sanderson, L. Beese, S. M. Friedman, C. M. Joyce, and T. Steitz, *Science* **240,** 199 (1988).
[13] V. Derbyshire, N. D. F. Grindley, and C. M. Joyce, *EMBO J.* **10,** 17 (1991).

A

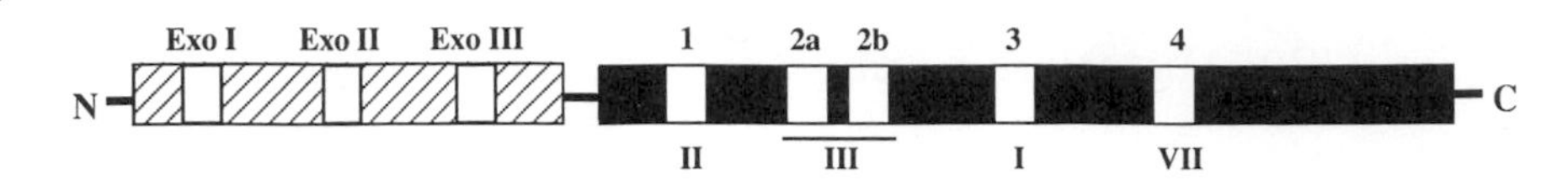

B

Structural domain	*Predicted domain*	*aa motif*	*Region*	*Original references*
N-terminal	**3′-5′ exonuclease**	**D x E**	**Exo I**	**10, 11**
		N x_{2-3}F/Y D	**Exo II**	**10, 11**
		Y x_3 D	**Exo III**	**10, 11, 19**
C-terminal	**Protein-primed initiation and DNA polymerization**	**D x_2 S L Y P**	**1 / II**	**7 / 5**
		K x_3 N S x Y G	**2a / III**	**7 / 5**
		T x_2 G/A R	**2b / III**	**7 / 5**
		Y x D T D S	**3 / I**	**7 / 5**
		K x Y	**4 / VII**	**7 / 34**

FIG. 1. (A) Relative arrangement of the most conserved regions among prokaryotic and eukaryotic DNA-dependent DNA polymerases. The amino acid sequence of ϕ29 DNA polymerase (572 aa) is represented by a bar, with the N terminus at the left. Cross-hatched and filled-in regions indicate the predicted 3′ → 5′-exonuclease and DNA polymerization domains, respectively. Boxes indicate the relative position of the most highly conserved regions (nomenclature described in B) analyzed by site-directed mutagenesis in ϕ29 DNA polymerase. (B) Description of amino acid sequence motifs. Motifs are represented in single-letter notation, where "x" indicates any amino acid. Alternative residues for a particular position are separated by a bar. The N-terminal motifs are defined taking into account the consensus for a particular region, among the two main superfamilies (Pol I-type and eukaryotic-type) of DNA-dependent DNA polymerases.[11] The C-terminal motifs are defined taking into account the three subclasses of eukaryotic-type DNA polymerases (viral, cellular, and TP-primed).[7] Alternative nomenclatures and original references describing the regions that contain the different motifs are also indicated.

Site-Directed Mutagenesis and Expression of ϕ29 DNA Polymerase Mutants

Based on amino acid sequence comparisons, ϕ29 DNA polymerase residues forming highly conserved motifs (described in Fig. 1) were selected as targets for site-directed mutagenesis. Single changes were designed taking into account secondary structure predictions[14,15] and the proposals for conservative changes made by Argos.[16] In addition, the set of aligned sequences

[14] P. Y. Chou and G. D. Fasman, *Adv. Enzymol.* **47,** 45 (1978).
[15] J. Garnier, D. J. Osguthorpe, and B. Robson, *J. Mol. Biol.* **120,** 97 (1978).
[16] P. Argos, *J. Mol. Biol.* **197,** 331 (1987).

provided complementary criteria either for permitted or prohibited substitutions.

The wild-type φ29 DNA polymerase gene, cloned into M13 derivatives, was used for site-directed mutagenesis, carried out essentially as described.[17] DNA fragments carrying the different mutations were recloned by substituting the corresponding wild-type fragment in recombinant T7 RNA polymerase-based vectors containing the φ29 DNA polymerase gene.[1] Overproduction and purification of the mutant proteins were carried out as described.[1]

Identification of 3′ → 5′-Exonuclease Active Site

Based on amino acid sequence comparisons[10,11] and crystallographic studies of Klenow complexed with TMP,[18] at least five amino acid residues were proposed to form an evolutionarily conserved 3′ → 5′-exonuclease active site in most prokaryotic and eukaryotic DNA polymerases.[10] To test this hypothesis, we carried out single substitutions in each putative active site residue of φ29 DNA polymerase: D12 and E14 (Exo I motif), D66 (Exo II motif), and Y165 and D169 (Exo III motif), together with double substitutions involving motifs Exo I and Exo II (see Fig. 1). The mutants obtained, together with their functional analysis, are shown in Table I. Whereas the protein-primed initiation and DNA polymerization were not impaired, the exonuclease activity was in all cases strongly inhibited.[10,19] The analysis of the steady-state kinetic parameters of the 3′ → 5′-exonuclease activity of the wild-type and mutant derivatives[20] allows us to conclude that the acidic residues D12, E14, D66, and D169 (proposed to be homologous to the essential Pol I residues D355, E357, D424, and D501) would be the ones most directly involved in catalysis of the exonucleolytic reaction. However, φ29 DNA polymerase residue Y165 seems to play a secondary role in the exonucleolytic reaction, as was also proposed for its homologous Pol I residue Y497.[13]

Therefore, our results are in good agreement with the catalytic role of the corresponding residues in Pol I,[13] and support the idea that the geometry of the Pol I 3′ → 5′-exonuclease active site, and the two metal ion mechanism of 3′ → 5′-exonucleolysis proposed for this enzyme,[21] can be extrapolated to φ29 DNA polymerase and the rest of proofreading-DNA polymerases.

[17] T. A. Kunkel, J. D. Roberts, and R. A. Zakour, *Methods Enzymol.* **154,** 367 (1987).

[18] D. L. Ollis, R. Brick, R. Hamlin, N. G. Xuong, and T. A. Steitz, *Nature* **313,** 762 (1985).

[19] M. S. Soengas, J. A. Esteban, J. M. Lázaro, A. Bernad, M. A. Blasco, M. Salas, and L. Blanco, *EMBO J.* **11,** 4227 (1992).

[20] J. A. Esteban, M. S. Soengas, M. Salas, and L. Blanco, unpublished results (1993).

[21] L. Beese and T. A. Steitz, *EMBO J.* **10,** 25 (1991).

TABLE I
MUTATIONAL ANALYSIS OF MOST CONSERVED N-TERMINAL AMINO ACID MOTIFS OF ϕ29 DNA POLYMERASE AND IDENTIFICATION OF $3' \rightarrow 5'$-EXONUCLEASE ACTIVE SITE RESIDUES

Amino acid motif	Enzyme[a]	Initiation[b] (%)	Polymerase[c] (%)	$3' \rightarrow 5'$-exonuclease[d] K_{cat} (sec^{-1})	TP–DNA replication[e] (%)	Strand displacement[f] (%)
—	WT	100	100	>80	100	100
"DxE" (Exo I)	D12A	69	176	1.5×10^{-3}	1.1	2
	E14A	76	247	2.4×10^{-3}	0.4	6
"$Nx_{2-3}F/YD$" (Exo II)	D66A	87	294	$<1 \times 10^{-3}$	0.2	4
(Exo I/Exo II)	D12A/D66A	103	282	8.6×10^{-4}	0.5	nd[g]
	E14A/D66A	90	200	7.8×10^{-4}	0.1	nd[g]
"Yx_3D" (Exo III)	Y165F	50	135	3.2	1.2	7
	Y165C	60	88	0.5	0.5	8
	D169A	47	135	1.7×10^{-3}	0.5	5

[a] WT stands for wild-type; mutant proteins are designated by the original amino acid (in single-letter notation), its position, and the replacing amino acid; i.e., D12A = Asp-12 to Ala.
[b] Use of a protein as primer, measured as TP-ϕ29 DNA-directed formation of TP-dAMP. Data taken from A. Bernad, L. Blanco, J. M. Lázaro, G. Martín, and M. Salas, *Cell* **59,** 219 (1989), and from M. S. Soengas, J. A. Esteban, J. M. Lázaro, A. Bernad, M. A. Blasco, M. Salas, and L. Blanco, *EMBO J.* **11,** 4227 (1992).
[c] Filling-in of 3′ recessive DNA ends. Data taken from A. Bernad, L. Blanco, J. M. Lázaro, G. Martín, and M. Salas, *Cell* **59,** 219 (1989), and from M. S. Soengas, J. A. Esteban, J. M. Lázaro, A. Bernad, M. A. Blasco, M. Salas, and L. Blanco, *EMBO J.* **11,** 4227 (1992).
[d] Assayed on a 15-mer oligonucleotide as ssDNA substrate. Assay conditions described in A. Bernad, L. Blanco, J. M. Lázaro, G. Martín, and M. Salas, *Cell* **59,** 219 (1989).
[e] This process requires TP-primed initiation and highly processive elongation coupled to strand displacement. Data taken from A. Bernad, L. Blanco, J. M. Lázaro, and M. Salas, unpublished results (1989), and from M. S. Soengas, J. A. Esteban, J. M. Lázaro, A. Bernad, M. A. Blasco, M. Salas, and L. Blanco, *EMBO J.* **11,** 4227 (1992).
[f] Rate of DNA synthesis coupled to strand displacement. Assay conditions described in M. S. Soengas, J. A. Esteban, J. M. Lázaro, A. Bernad, M. A. Blasco, M. Salas, and L. Blanco, *EMBO J.* **11,** 4227 (1992).
[g] Not determined.

Structural and Functional Integrity of $3' \rightarrow 5'$-Exonuclease Domain[22]

As described above, site-directed mutagenesis studies in ϕ29 DNA polymerase demonstrated that the $3' \rightarrow 5'$-exonuclease active site is located in the N-terminal third of the polypeptide, being formed by the evolutionarily conserved Exo I, Exo II, and Exo III amino acid motifs.[10,19] However, an important question related to the hypothesis of a modular organization

[22] L. Blanco, L. Villar, J. M. Lázaro, A. Zaballos, and M. Salas, unpublished results (1993).

of enzymatic activities in DNA polymerases[10] was one of whether the N-terminal third of ϕ29 DNA polymerase, when expressed independently, would have structural and functional integrity, and would therefore constitute a separable $3' \rightarrow 5'$-exonuclease domain. Thus, based on site-directed mutagenesis studies and refined amino acid sequence comparisons,[11] we constructed a C-terminal deletion derivative that contained the first 188 N-terminal amino acid residues (and therefore includes the Exo I, Exo II, and Exo III motifs). This truncated polypeptide was overproduced in *Escherichia coli* cells, yielding large amounts of soluble protein, and purified to homogeneity.

As expected from our mutational and sequence comparison data, the putative exonuclease domain did not catalyze protein-primed initiation and DNA polymerization; but, in agreement with its predicted structural integrity, it retained more than 10% of the $3' \rightarrow 5'$-exonuclease activity. This activity was shown to be intrinsic to the N-terminal fragment of ϕ29 DNA polymerase by *in situ* exonuclease activity assay and by glycerol gradient cosedimentation. Interestingly, the mode of digestion displayed by the exo domain was distributive in comparison with the strong processivity exhibited by the intact enzyme. This finding suggests that the optimal stabilization and/or translocation along the ssDNA substrate involves additional contacts, which would be located in the C-terminal domain.

Mapping Functional Residues of Polymerization Domain

Here, we compile site-directed mutagenesis studies in the most conserved C-terminal regions of ϕ29 DNA polymerase (Fig. 1). These studies indicate that the C-terminal domain contains the synthetic activities (protein-primed initiation and DNA polymerization) of ϕ29 DNA polymerase. DNA pyrophosphorolysis, the polymerization reversal, has also been mapped in the polymerization domain of ϕ29 DNA polymerase.[23] Furthermore, a detailed study of each mutant protein allowed us to demonstrate particular roles of several residues that form evolutionarily conserved amino acid sequence motifs.

Motif "Dx$_2$SLYP"

Considering the amino acid sequence alignment of this motif in 53 sequences of eukaryotic-type DNA polymerases belonging to the three subgroups of viral, cellular, and protein-primed enzymes, the Asp and Tyr residues are invariant in all the sequences compared, the Ser, Leu, and Pro

[23] M. A. Blasco, A. Bernad, L. Blanco, and M. Salas, *J. Biol. Chem.* **266,** 7904 (1991).

residues being conserved in 48, 41, and 47 sequences, respectively. Table II summarizes the data obtained with highly purified mutants in motif "Dx$_2$SLYP" of ϕ29 DNA polymerase.[7,24,25] As expected from the N-terminal location of the 3′ → 5′-exonuclease domain of ϕ29 DNA polymerase, none of the mutations in this conserved motif significantly affected exonuclease activity, but altered behaviors in synthetic activities were observed. The conservative change Asp249 to Glu (D249E) drastically reduced synthetic activities, although this mutant enzyme was able to interact normally with the ϕ29 terminal protein (TP), and it was only slightly affected in template–primer DNA binding. Our results suggest that the Asp residue of the "Dx$_2$SLYP" motif, invariant in all eukaryotic-like DNA polymerases described to date, could have a direct role in catalysis, perhaps acting as a metal ligand. Two site-directed mutants were obtained in ϕ29 DNA polymerase residue Ser252, S252R and S252G. Both mutants were greatly affected in binding template–primer DNA molecules, indicating that residue Ser252 could have a direct role in DNA binding. Mutant S252R was inactive in all the synthetic activities, including TP-deoxynucleotidylation in the absence of template. These results, together with the fact that this mutant was able to interact normally with the TP, suggest that the change, S252 to Arg, has an additional effect on catalysis. Mutant Y254F was strongly affected in the TP-primed initiation step of ϕ29 DNA replication, showing a decreased affinity for dATP, the initiating nucleotide.[24] The increased dNMP turnover coupled to DNA synthesis observed for mutant Y254F[7] could explain its reduced activity in synthesizing long tracts of DNA (primed-M13 DNA replication), and suggests that this mutation may be affecting enzyme-DNA translocation. Phenotypes associated to other mutants in motif "Dx$_2$SLYP" are a strong reduction of the TP-primed initiation reaction, metal ion-dependent phenotypes in synthetic reactions, and altered sensitivity to the PP_i analog PAA.

Motif "Kx$_3$NSxYG"

The Lys, Asn, and Gly residues of this motif are invariant in all the sequences (53) compared, and the Tyr and Ser residues are conserved in 51 and 38 sequences, respectively. Table II summarizes the data obtained with highly purified mutants in motif "Kx$_3$NSxYG" of ϕ29 DNA polymerase, and also in a neighbor residue (Phe393), specially conserved in DNA polymerases that use a protein as primer.[7,24,26] The 3′ → 5′-exonuclease

[24] M. A. Blasco, J. M. Lázaro, A. Bernad, L. Blanco, and M. Salas, *J. Biol. Chem* **267,** 19,427 (1992).
[25] M. A. Blasco, J. M. Lázaro, L. Blanco, and M. Salas, *J. Biol. Chem.* **268,** 24,106 (1993).
[26] M. A. Blasco, J. M. Lázaro, L. Blanco, and M. Salas, *J. Biol. Chem.* **268,** 16,763 (1993).

activity was not greatly affected by the mutations introduced, as expected. Mutant N387Y was affected both in initiation and polymerization reactions, showing a 3-fold higher K_m value for dATP and more than 11-fold lower V_{max} value than the wild-type enzyme in the initiation reaction; moreover, it was affected in enzyme–DNA translocation. Mutant S388G retained initiation and polymerization activities; interestingly, this mutation significantly increased the efficiency of dNTP incorporation in nontemplated reactions. Two mutants were obtained at residue Y390 of this motif, Y390F and Y390S. Mutant Y390F was greatly affected in replication assays, whereas mutant Y390S was partially active. In both cases, an abnormally high turnover of dNTPs to dNMPs coupled to polymerization was observed, probably reflecting a slow rate of translocation of these mutant polymerases along the DNA template. Interestingly, mutants Y390F and Y390S were hypersensitive to the dNTP analogs BuAdATP and BuPdGTP, specific inhibitors of eukaryotic DNA polymerases. This hypersensitivity was clearly related to template complementarity, in contrast with the wild-type ϕ29 DNA polymerase, and paralleling the mode of action of these drugs in most eukaryotic-type DNA polymerases. Mutant G391D was unable to bind template–primer DNA molecules, being drastically affected in template-dependent dNTP incorporation both in initiation and polymerization reactions. However, the efficiency of the nontemplated deoxynucleotidylation of the TP was even higher than that of the wild-type protein, ruling out a defect of this mutant enzyme in direct dNTP binding and catalysis. Mutation Phe393 to Tyr (F393Y) severely decreased initial binding to template–primer molecules, resulting in a reduced activity in DNA primer-dependent polymerization reactions, but not in TP-dependent ones. The results obtained from this mutational analysis indicate that the conserved amino acid motif "Kx$_3$NSxYG" is involved in template-primer DNA binding and dNTP selection.

Motif "Tx$_2$GR"

Out of a total of 53 sequences of eukaryotic-type DNA polymerases, the Thr residue is conserved in 40 sequences, the Gly residue in 35 sequences (being substituted for Ala in 10 sequences belonging to the subgroup of protein-primed DNA polymerases), and the Arg residue, the most conserved one, is invariant in 47 sequences.[27] Interestingly, this amino acid motif is also present in the sequence of Pol I-like DNA polymerases, Thr664, Gly667, and Arg668 being the corresponding residues in Pol I.[7] By site-directed mutagenesis studies in Pol I, Arg668 has been shown to be im-

[27] J. Méndez, L. Blanco, J. M. Lázaro, and M. Salas, *J. Biol. Chem.* **269,** 30,030 (1994).

TABLE II
MUTATIONAL ANALYSIS OF MOST CONSERVED C-TERMINAL AMINO ACID SEQUENCE MOTIFS OF ϕ29 DNA POLYMERASE

Amino acid motif	Enzyme[a]	$3' \rightarrow 5'$-exonuclease[b] activity (%)	DNA[c] binding	Initiation[d] TP–DNA template (%)	Initiation[d] No template (%)	Polymerization[e] Filling-in (%)	Polymerization[e] Primed-M13 DNA replication (%)	Polymerization[e] TP–DNA replication (%)	Other phenotypes
—	WT	100	+++	100	100	100	100	100	
"Dx_2SLYP"[f]	D249E	177	++	<1	<1	<1	<0.5	<0.5	
	S252G	218	+	28	20	60	90	8	PAA[hs g]
	S252R	254	—	<1	<1	<1	<0.5	<0.5	
	L253V	272	+++	9	50	70	120	1	Altered metal activation
	Y254F	200	+++	3	14	97	13	3	Increased turnover
	P255S	125	++	240	126	70	122	54	PAA[hs g]
"Kx_3NSxYG"[h]	N387Y	98	++	6	12	16	nd[i]	<1	
	S388G	126	+++	44	590	161	nd[i]	9	
	Y390S	87	+++	66	17	87	18	13	BuAdATP[hs j]; increased turnover
	Y390F	82	+++	40	55	48	<1	<1	BuAdATP[hs j]; increased turnover
	G391D	61	—	<1	470	<1	nd[i]	<1	
	F393Y	21	—	87	150	10	nd[i]	50	
"Tx_2GR"[k]	T434N	100	+	4	5	2	2.5	0.2	Weak interaction with TP
	A437G	100	+++	80	130	80	120	120	
	R438K	*[l]	++	26	7	70	33	30	
	R438I	*[l]	—	<1	3	4	7	0.1	Weak interaction with TP

"YxDTDS"[m]	Y454F	149	nd[i]	107	94	60	<2	8	
	C455G	255	nd[i]	45	30	290	110	56	Altered metal activation
	D456G	130	nd[i]	10	61	30	<2	<2	
	T457P	131	nd[i]	0.5	0.2	1	<1	<2	
	D458G	198	nd[i]	<0.5	<0.1	<1	<1	<2	
"KxY"[n]	K498R	340	+++	620	940	8	0.1	<0.5	Increased turnover
	K498T	430	+++	2	8	1	<0.1	<0.5	Increased turnover
	Y500S	270	+++	60	145	4	<0.1	<0.5	Increased turnover

[a] WT stands for wild-type; mutant proteins are designated by the original amino acid (in single-letter notation), its position, and the replacing amino acid; i.e., D12A = Asp-12 to Ala.

[b] Assayed on single-stranded DNA.

[c] Analyzed by retardation assays on low-ionic-strength polyacrylamide gels.

[d] DNA-dependent (TP-DNA template) or DNA-independent (no template) formation of TP-dAMP by ϕ29 DNA polymerase.

[e] Assayed either on a short (filling-in) or on a long (primed-M13 DNA replication) DNA template; TP-DNA replication requires TP-primed initiation and highly processive elongation coupled to strand displacement.

[f] Mutational analysis of this motif as reported by M. A. Blasco, J. M. Lázaro, L. Blanco, and M. Salas, *J. Biol. Chem.* **268,** 24,106 (1993).

[g] Hypersensitive to PAA, phosphonoacetic acid.

[h] Mutational analysis of this motif as reported by L. Blanco, A. Bernad, M. A. Blasco, and M. Salas, *Gene* (*Amst*) **100,** 27 (1991); M. A. Blasco, J. M. Lázaro, A. Bernad, L. Blanco, and M. Salas, *J. Biol. Chem.* **267,** 19,427 (1992); and M. A. Blasco, J. M. Lázaro, L. Blanco, and M. Salas, *J. Biol. Chem.* **268,** 16,763 (1993).

[i] Not determined.

[j] Hypersensitive to BuAdATP, 2-(*p*-*n*-butylanilino)dATP.

[k] Mutational analysis of this motif taken from J. Méndez, L. Blanco, J. M. Lázaro, and M. Salas, *J. Biol. Chem.* **269,** 30,030 (1994).

[l] Glycerol gradient fractionation of mutants R438K and R438I demonstrated their normal 3′→5′-exonuclease activity, although a quantitative value could not be estimated due to the presence of a contaminating nuclease activity.

[m] Mutational analysis of this motif as reported by A. Bernad, J. M. Lázaro, M. Salas, and L. Blanco, *Proc. Natl. Acad. Sci. USA* **87,** 4610 (1990).

[n] Mutational analysis of this motif as reported by M. A. Blasco, J. A. Esteban, J. Méndez, L. Blanco, and M. Salas, *Chromosoma* **102,** S32 (1992); M. A. Blasco, J. Méndez, J. M. Lázaro, L. Blanco, and M. Salas, *J. Biol. Chem.* **270,** 2735 (1995).

portant both in DNA binding and catalysis of the polymerization reaction.[28,29]

Table II summarizes the data obtained[27] with highly purified mutants in motif "Tx_2GR" of ϕ29 DNA polymerase. In agreement with other mutational analysis carried out in the C-terminal domain, all the mutants had normal 3′ → 5′-exonuclease activity. Mutants T434N and R438I presented a decreased affinity for DNA template–primer structures and were severely affected in all the synthetic activities (protein-primed initiation and DNA polymerization). The ability of these mutant polymerases to interact with TP was also found to be lower than that of the wild-type ϕ29 DNA polymerase. Conversely, mutant A437G retained all its enzymatic activities, emphasizing the equivalence of Gly (mutation) and Ala (wild-type) at this position of ϕ29 DNA polymerase. Our results indicate that the Thr and Arg residues of motif "Tx_2GR" are important for template–primer binding (involving either DNA or TP) and could also have a role in catalysis.

Motif "YxDTDS"

The Thr and second Asp residue of this motif are invariant in all the sequences (53) compared, and the Tyr, first Asp, and Ser residues are conserved in 46, 50, and 45 sequences, respectively. The "x" residue of the motif is a Gly in 32 sequences, being unconserved in the subgroup of protein-primed DNA polymerases. As summarized in Table II, site-directed point mutants in the YCDTD sequence of ϕ29 DNA polymerase[30] did not impair the 3′ → 5′-exonuclease activity of the enzyme. On the other hand, the processive elongation activity of ϕ29 DNA polymerase was completely inhibited by all mutations with the exception of C455G. Interestingly, as described above, a Gly residue is frequently present at that position in other DNA polymerases. However, when a short elongation reaction (filling-in) was analyzed, only the T457P and D458G mutations appeared to be critical for a minimal DNA polymerase activity. The template-dependent initiation activity was also essentially abolished in mutants T457P and D458G, and severely affected by the D456G mutation. None of the mutations affected the interaction of the DNA polymerase with the TP. Also interesting is the fact that mutant C455G showed differences in the optimum activating metal ion concentration with respect to the wild-type ϕ29 DNA polymerase. These results indicated that the "YxDTDS" motif plays a catalytic role

[28] A. H. Polesky, T. A. Steitz, N. D. F. Grindley, and C. M. Joyce, *J. Biol. Chem.* **265,** 14,579 (1990).

[29] A. H. Polesky, M. E. Dahlberg, S. J. Benkovic, N. D. F. Grindley, and C. M. Joyce, *J. Biol. Chem.* **267,** 8417 (1992).

[30] A. Bernad, J. M. Lázaro, M. Salas, and L. Blanco, *Proc. Natl. Acad. Sci. USA* **87,** 4610 (1990).

in the processes of synthesis (initiation and/or elongation) of ϕ29 DNA polymerases, the two Asp residues probably being involved in metal binding at the dNTP site. These results constituted the first evidence of the existence, in protein-primed DNA polymerases, of a common active site for the initiation and elongation activities.

Motif "KxY"

This strict motif has been defined taking into account the amino acid sequence alignment among 53 sequences of eukaryotic-type DNA polymerases belonging to the three subgroups of viral, cellular, and protein-primed enzymes. The Lys residue is invariant in all the sequences compared, and the Tyr residue is conserved in 52 sequences. If the group of protein-primed DNA polymerases is not considered, the motif has the more extensive sequence "KKK/RY." Table II shows the results obtained with highly purified ϕ29 DNA polymerase mutants K498R, K498T, and Y500S.[31] Measurements of the activity of these mutant proteins indicated that the invariant Lys and Tyr residues play a critical role in DNA polymerization. In the three cases, an abnormally high turnover of dNTPs to dNMPs coupled to polymerization was observed, which could be the consequence of a decreased stability of the primer terminus at the polymerization active site. This idea is supported by the increased $3' \rightarrow 5'$-exonuclease activity of these mutants, not only on ssDNA substrates, but also on template-primer DNA structures. Interestingly, substitution of the invariant Lys either by Arg or Thr produced enzymes with an increased or a greatly reduced, respectively, ability to use a protein as primer.[31,32] These results outline the catalytic differences between protein-primed initiation and DNA-primed polymerization, and suggest that the Lys and Tyr residues of motif "KxY" could play a role in primer binding.

Structural Location of Strand-Displacement Activity

Strand displacement coupled to DNA synthesis is one of the most peculiar functions of ϕ29 DNA polymerase, and allows this enzyme to carry out the complete replication of the linear double-stranded ϕ29 DNA genome (19,285bp) in the absence of any accessory proteins or helicases.[33]

[31] M. A. Blasco, J. Méndez, J. M. Lázaro, L. Blanco, and M. Salas, *J. Biol. Chem.* **270,** 2735 (1995).

[32] M. A. Blasco, J. A. Esteban, J. Méndez, L. Blanco, and M. Salas, *Chromosoma* **102,** S32 (1992).

[33] L. Blanco, A. Bernad, J. M. Lázaro, G. Martín, C. Garmendia, and M. Salas, *J. Biol. Chem.* **264,** 8935 (1989).

[34] P. A. J. Leegwater, M. Strating, N. B. Murphy, R. F. Kooy, P. C. van der Vliet, and J. P. Overdulve, *Nucleic Acids Res.* **19,** 6441 (1991).

Unexpectedly, all the mutants in the Exo I, Exo II, and Exo III motifs were almost inactive when assayed for TP-ϕ29 DNA replication (Table I). This defect was shown to be mainly due to a 10- to 50-fold decrease in the rate of DNA synthesis coupled to strand displacement. Therefore, taking into account these results and the fact that none of the mutations in the C-terminal portion of ϕ29 DNA polymerase specifically affected strand-displacement synthesis, we propose that this activity resides in the N-terminal domain, probably overlapping with the $3' \rightarrow 5'$-exonuclease active site.[19] Our model makes use of the putative ssDNA binding cleft of the $3' \rightarrow 5'$-exonuclease domain to bind the displaced strand during ϕ29 DNA replication.

Acknowledgments

This work is dedicated to the memory of Severo Ochoa. We are grateful to A. Bernad, M. A. Blasco, J. A. Esteban, J. M. Lázaro, J. Méndez, and M. S. Soengas for their contribution to the work presented in this chapter. This investigation has been aided by research grant 5R01 GM27242-14 from the National Institutes of Health, by grant PB90-0091 from Dirección General de Investigación Científica y Técnica, by grant BIOT CT 91-0268 from the European Economic Community, and by an institutional grant from Fundación Ramón Areces.

[23] Rationale for Mutagenesis of DNA Polymerase Active Sites: DNA Polymerase α

By WILLIAM C. COPELAND, QUN DONG, and TERESA S.-F. WANG

Introduction

In a living cell, DNA replication is one of the most fundamental and tightly regulated enzymatic reactions. Error-free replication of chromosomal DNA by the replicative DNA polymerases plays an essential role in transmission and maintenance of genetic information of an organism from one generation to the next.[1] During the incorporation of deoxynucleotides into DNA by a DNA polymerase, the amino acid residues in the active site of the DNA polymerase must interact in a precise and accurate manner with the incoming dNTP to position it for correct Watson–Crick base pairing with the template nucleotide. These residues in the active site must also be able to distinguish a correctly base-paired primer terminus

[1] A. Kornberg and T. A. Baker, in "DNA Replication," 2nd ed., Chap. 15. W. H. Freeman and Company, New York, 1991.

from a mispaired primer terminus for incorporation of the nucleotide.[2] Thus, it is fundamentally important to have a structural and biochemical understanding of the active site of a DNA polymerase.

Although the three-dimensional structures of two DNA polymerases, *Escherichia coli* polymerase I large fragment (Klenow fragment) and human immunodeficiency virus 1 reverse transcriptase (HIV-1 RT)[3-5] as well as that of the RNA polymerase of T7 bacteriophage[6] have been determined, it is still not clear precisely how the side chains of the amino acid residues in a DNA polymerase active site interact at the molecular level with the incoming dNTP or the primer terminus. Knowledge of the functions of the essential amino acid residues in the active site combined with high-resolution structures of the polymerase active sites will advance the understanding of the basic mechanisms of DNA synthesis, as well as provide invaluable information for designing anticancer and antiviral therapeutic drugs.

To obtain informative structure–function knowledge of the active site of a DNA polymerase, the rationale to identify the essential amino acid residues in the active site and the criteria utilized to verify the functions of these residues are (1) identification of potentially essential residues in the active site by amino acid alignments; (2) introduction of rational mutations of these potentially essential residues; and (3) definition of the function of the mutagenized amino acid residues by in-depth comparative kinetic analyses of the mutants and wild-type DNA polymerases. By these criteria, the studies of structure–function relationship of human DNA polymerase α active site is used in here as an example for the rationale of mutational studies of the active site of a DNA polymerase.[7-10]

Identification of Potentially Essential Residues in Active Site by Sequence Alignments

Potentially essential amino acids within the active site of a polymerase can be identified by both sequence alignment and structural similarity to

[2] H. Echols and M. F. Goodman, *Ann. Rev. Biochem.* **60,** 477 (1991).
[3] D. L. Ollis, R. Brick, R. Hamlin, N. G. Xuong, and T. A. Steitz, *Nature* **313,** 762 (1985).
[4] L. A. Kohlstaedt, J. Wang, J. M. Friedman, P. A. Rice, and T. A. Steitz, *Science* **256,** 1783 (1992).
[5] A. Jacobo-Molina, J. Ding, R. G. Nanni, A. D. Clark Jr., X. Lu, C. Tantillo, R. L. Williams, G. Kamer, A. L. Ferris, P. Clark, A. Hizi, S. H. Hughes, and E. Arnold, *Proc. Natl. Acad. Sci. USA* **90,** 6320 (1993).
[6] R. Sousa, Y. J. Chung, J. P. Rose, and B. C. Wang, *Nature* **364,** 593 (1993).
[7] W. C. Copeland and T. S.-F. Wang, *J. Biol. Chem.* **268,** 11,028 (1993).
[8] W. C. Copeland, N. K. Lam, and T. S.-F. Wang, *J. Biol. Chem.* **268,** 11,041 (1993).
[9] Q. Dong, W. C. Copeland, and T. S.-F. Wang, *J. Biol. Chem.* **268,** 24,163 (1993).
[10] Q. Dong, W. C. Copeland, and T. S.-F. Wang, *J. Biol. Chem.* **268,** 24,175 (1993).

the polymerases with known structure. Extensive compilations of all known DNA polymerase sequences from bacterial, viral, and eukaryotic origin have been assembled and described.[11,12] These compilations have identified several amino acid residues that are highly conserved among DNA polymerases of species over a vast phylogenetic spectrum. Physical structure data of *E. coli* polymerase I, HIV-1 RT, and T7 RNA polymerase show that the active sites of these three enzymes share the same tertiary folds. Their structures resemble a right hand with subdomains delineating as fingers, palm, and thumb.[3–6]

Human DNA polymerase α is the prototypic DNA polymerase of the so-called α-like DNA polymerases.[11–15] The deduced primary amino acid sequence of human DNA polymerase α has revealed six regions that are highly conserved in DNA polymerases from both prokaryotic and eukaryotic organisms, as well as bacteriophage and animal DNA viruses.[11–15] These six regions are designated regions I through VI according to their extent of similarity. Region I is the most conserved and region VI is the least conserved.[13] The three most conserved regions, regions I, II, and III[13–15] (or also known as motif C, A, and B,[12] respectively) are the basic core regions that are conserved in DNA polymerases from human to bacteriophage.[11–15] The conservation of these three regions in DNA polymerases over such a broad spectrum of phylogenetic species strongly suggests that these regions are conserved for maintaining the basic catalytic function of the α-like DNA polymerases. Primary sequence alignment of regions I, II, and III of human DNA polymerase α with that of the active sites of two other DNA polymerases (*E. coli* polymerase I and HIV-1 RT) and an RNA polymerase (T7 RNA polymerase) that have known crystal structures identified several conserved amino acids (12) (Fig. 1). The two aspartate residues Asp-1002 and Asp-1004 in region I (or motif C) of human DNA polymerase α are aligned with the Asp-882 and Glu-883 in the active site of *E. coli* polymerase I, with Asp-812 and Ser-813 in the active site of T7 RNA polymerase, and with Asp-185 and Asp-186 in the active site of HIV-1 RT (Fig. 1). In these polymerases, this region forms a tight turn between two antiparallel β sheets in the "palm" domain of the active site.[3–6]

In region II of human DNA polymerase α, Asp-860 is aligned with the Asp-705 of *E. coli* polymerase I, Asp-537 of T7 RNA polymerase, and Asp-110 of HIV-1 RT, whereas Tyr-865 of human DNA polymerase α is

[11] D. K. Braithwaite and J. Ito, *Nucleic Acids Res.* **21,** 787 (1993).
[12] M. Delarue, O. Poch, N. Tordo, N. D. Moras, and P. Argos, *Protein Eng.* **3,** 461 (1990).
[13] S. W. Wong, A. F. Wahl, P.-M. Yuan, N. Arai, B. E. Pearson, K.-I. Arai, D. Korn, M. W. Hunkapillar, and T. S.-F. Wang, *EMBO J.* **7,** 37 (1988).
[14] T. S.-F. Wang, S. W. Wong, and D. Korn, *FASEB. J.* **3,** 14 (1989).
[15] T. S.-F. Wang, *Ann. Rev. Biochem.* **60,** 513 (1991).

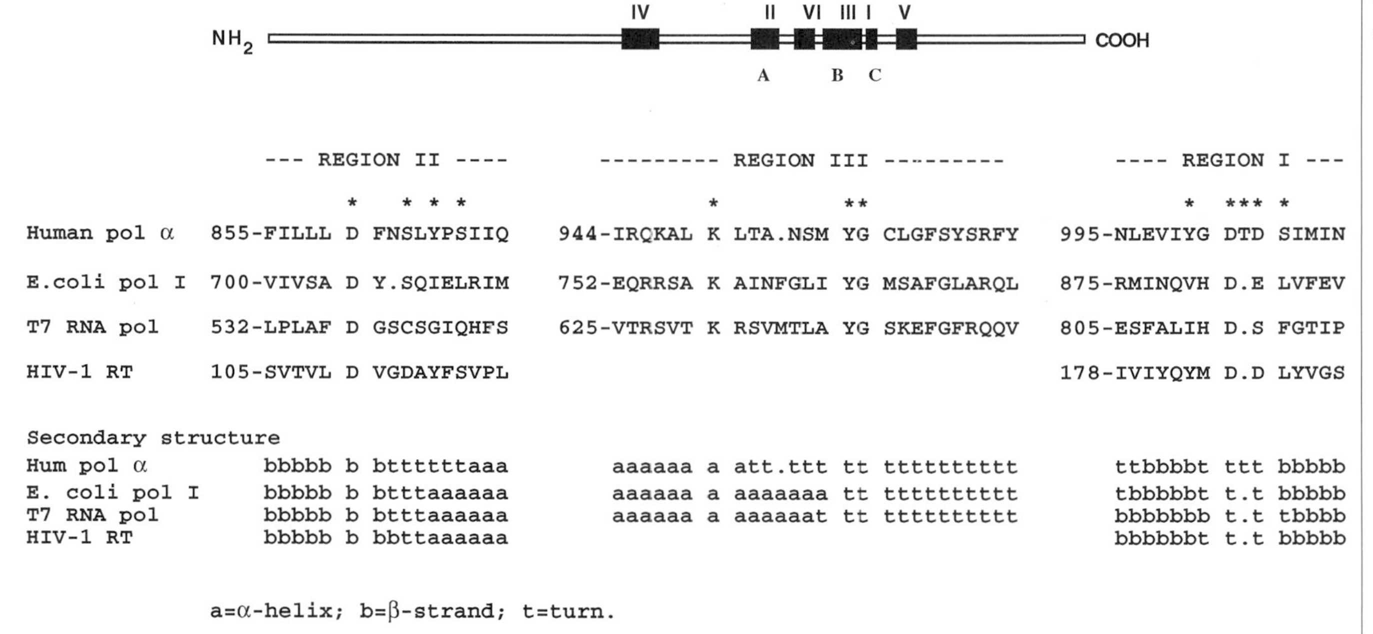

IV
II VI III I V
NH2
COOH
A B C
--- REGION II ----
--------- REGION III ---------
---- REGION I ---
Human pol α 855-FILLL D FNSLYPSIIQ 944-IRQKAL K LTA.NSM YG CLGFSYSRFY 995-NLEVIYG DTD SIMIN
E.coli pol I 700-VIVSA D Y.SQIELRIM 752-EQRRSA K AINFGLI YG MSAFGLARQL 875-RMINQVH D.E LVFEV
T7 RNA pol 532-LPLAF D GSCSGIQHFS 625-VTRSVT K RSVMTLA YG SKEFGFRQQV 805-ESFALIH D.S FGTIP
HIV-1 RT 105-SVTVL D VGDAYFSVPL 178-IVIYQYM D.D LYVGS
Secondary structure
Hum pol α bbbbb b btttttttaaa aaaaaa a att.ttt tt tttttttttt ttbbbbt ttt bbbbb
E. coli pol I bbbbb b btttaaaaaa aaaaaa a aaaaaaa tt tttttttttt tbbbbbt t.t bbbbb
T7 RNA pol bbbbb b btttaaaaaa aaaaaa a aaaaaat tt tttttttttt bbbbbbb t.t tbbbb
HIV-1 RT bbbbb b bbttaaaaaa bbbbbbt t.t bbbbb
a=α-helix; b=β-strand; t=turn.

FIG. 1. Sequence alignment and secondary structure comparison of the three most conserved regions of human DNA polymerase α with the active sites of *E. coli* polymerase I, T7 RNA polymerase, and HIV-1 RT. A schematic diagram depicting the spatial relationship of the six conserved regions of human DNA polymerase α[13] is shown on the top. Amino acid residues in the three most conserved regions, I, II, and III [or A,B, and C motifs in α-like DNA polymerase[12]] of human DNA polymerase α are aligned with the amino acid residues in the active sites of *E. coli* polymerase I, T7 RNA polymerase, and HIV-1 RT. Amino acid residues of human DNA polymerase α marked with * are the residues targeted for mutagenesis (Refs. 7–10; Q. Dong and T. S.-F. Wang, *J. Biol. Chem.* (submitted, 1995). Predicted secondary structure of human DNA polymerase α regions I, II, and III are compared with the secondary structures of the active sites of *E. coli* polymerase I, T7 RNA polymerase, and HIV-1 RT.

aligned with Tyr-115 of HIV-1 RT. These region II amino acids form a β sheet followed by an α helix also within the "palm" domain of the active site.[3–6]

In region III of human DNA polymerase α, there are three residues, Lys-950, Tyr-957, and Gly-958, that are aligned with *E. coli* polymerase I residues Lys-758, Tyr-766, and Gly-767, and T7 RNA polymerase residues Lys-631, Tyr-639, and Gly-640 (Fig. 1). Region III in the *E. coli* polymerase I and T7 RNA polymerase forms an α helix in the "fingers" domain of the active site.[3,6] The alignment of these amino acid residues of the human DNA polymerase α with residues in the active sites of *E. coli* polymerase I, HIV-1 RT, and T7 RNA polymerase strongly suggests that these residues are essential for basic catalytic function. Moreover, the predicted secondary structure of human DNA polymerase α in the regions containing these highly conserved residues identified by sequence alignment can be superimposed on the active site subdomains of *E. coli* polymerase I,[3] HIV-1 RT,[4,5] and T7 RNA polymerase[6] (Fig. 1). Thus, these conserved amino acid residues were selected as targets for mutational studies.

Rationale of Mutations

Site-directed mutations were introduced to each of the highly conserved amino acid residues in the active site of human DNA polymerase α to test their functions. The objective is to analyze the function of the side chain of each amino acid by introducing mutations that yield a mutant enzyme with altered catalytic activity but not devoid of catalytic activity. Thus, similar or altered size changes by removing part of or all of the functional group of the side chain are preferred, and extreme and drastic changes in the charge and size of the side chains are avoided. It is also important to introduce more than one mutation to the target amino acid residue in order to address the precise function of the side chain. For example, to address the function of the side chain of Tyr-865 in region II of human DNA polymerase α, Tyr-865 was changed to phenylalanine for testing the function of the hydroxyl group of the phenol side chain, or to serine for testing the function of the aromatic ring of the phenyl side chain.[9]

Site-directed mutations are introduced by oligonucleotide directed mutagenesis[16] using uracil containing DNA template.[17] Mutation is verified by DNA sequence analysis. The presence of introduced mutations is monitored at each stage of the subcloning steps by designing the oligonucleotides used for generating mutations containing silent mutations, which either add

[16] M. J. Zoller and M. Smith, *Nucleic Acids Res.* **10,** 6487 (1982).

[17] T. A. Kunkel, J. D. Roberts, and R. A. Zakour, *Meth. Enzymol.* **154,** 367 (1987).

or abolish a restriction site.[9] The rationale of the mutations introduced into the conserved amino acid residues within the active site of human DNA polymerase α is illustrated in Table I.

Defining Catalytic Function of Mutated Amino Acid Residues in Active Site

To test whether the mutated residue participates in catalysis or in substrate binding, steady-state kinetic analyses of K_m for substrates, dNTPs, and primer–template, k_{cat} values, and processivity of each of the mutant enzymes should be analyzed and compared to that of the wild-type DNA polymerase. Mutant DNA polymerases that show altered k_{cat} values but not K_m values for either dNTPs or primer–template suggest that this particular

TABLE I
RATIONALE OF MUTATIONS INTRODUCED INTO CONSERVED AMINO ACID RESIDUES IN ACTIVE SITE OF HUMAN DNA POLYMERASE α[a]

Residue	Mutated to	Function of side chain altered
Region I: -**Y**-G-**D**-**T**-**D**-**S**-[b]		
Y^{1000}	F	Loss of $-OH$
D^{1002}	N	Change of charge $-COOH$ to $-C(=O)NH_2$
T^{1003}	S	Loss of γ-CH_3
D^{1004}	N	Change of charge $-COOH$ to $C(=O)NH_2$
S^{1005}	T	Addition of γ-CH_3
Region II: -**G**---**D** F N **S** L **Y** P **S** I I-[a]		
G^{841}	A	Addition of β-CH_3 side chain
D^{860}	S	Replace $-COOH$ by $-OH$
D^{860}	N	Loss of charge side chain $-COOH$ to $-C(=O)NH_2$
D^{860}	A	Abolish acidic charge side chain $-COOH$
S^{863}	T	Addition of γ-CH_3 in side chain
S^{863}	A	Loss of $-OH$ side chain
Y^{865}	F	Loss of $-OH$ side chain
Y^{865}	S	Loss of phenyl ring side chain
S^{867}	T	Addition of β-CH_3 in side chain
S^{867}	A	Loss of $-OH$ side chain

[a] From Q. Dong, W. C. Copeland, and T. S.-F. Wang, *J. Biol. Chem.* **268,** 24,163 (1993). Copyright The American Society for Biochemistry & Molecular Biology.

[b] Amino acid residues mutated.

TABLE II

FIDELITY OF WILD-TYPE AND MUTANT DNA POLYMERASE α^{ab}

DNA polymerase	Correct dTTP		Incorrect dATP		
	K_m (μM)	V_{max} (%/min)	K_m (μM)	V_{max} (%/min)	f_{ins} [d]
(A) Misinsertion fidelity[c]					
Wild type	0.68 ± 0.16	0.18 ± 0.09	1609 ± 454	0.21 ± 0.02	4.9×10^{-4}
Y865S	1.1 ± 0.25	0.19 ± 0.01	69 ± 17	0.09 ± 0.04	7.6×10^{-3}
Y865F	7.8 ± 3.1	1.5 ± 1.3	1669 ± 278	0.25 ± 0.16	7.8×10^{-4}

DNA polymerase	Matched primer–template[e]			Mismatched primer–template[f]			
	K_m (dTTP) (μM)	k_{cat} (sec^{-1})	k_{cat}/K_m (sec^{-1} μM^{-1})	K_m (dTTP) (μM)	k_{cat} (sec^{-1})	k_{cat}/K_m (sec^{-1} μM^{-1})	f_{ext} [g] (mismatched)
(B) Mispaired primer extension fidelity							
Wild type	1.5 ± 0.08	18 (±2.9) × 10^{-2}	0.12	183 ± 11	7.7 (±1.5) × 10^{-2}	4.2×10^{-4}	1.0
S867A	0.25 ± 0.01	24 (±0.55) × 10^{-2}	0.97	21 ± 4.5	8.5 (±0.62) × 10^{-2}	4.1×10^{-3}	9.6
S867T	0.35 ± 0.09	14 (±1.6) × 10^{-2}	0.41	130 ± 25	12 (±0.81) × 10^{-2}	9.3×10^{-4}	2.2

[a] From Q. Dong, W. C. Copeland, and T. S.-F. Wang, *J. Biol. Chem.* **268,** 24175 (1993). Copyright The American Society for Biochemistry & Molecular Biology.

[b] The fidelity experiments were performed with Mg^{2+} as metal activator.

[c] The primer template used for misinsertion fidelity assay was:

5′-TGA CCA TGT AAC AGA GAG-3′
3′-ACT GGT ACA TTG TCT CTC **A**TT CTC TCT CTC TCT CTC-5′

The boldface nucleotide on the template indicates the position of insertion of dTTP or dATP.

[d] $f_{ins} = \frac{V_{max}(\text{dATP})/K_m(\text{dATP})}{V_{max}(\text{dTTP})/K_m(\text{dTTP})}$.

[e] The matched primer-template used for mispaired primer extension assay was:

5′-CGC CCA CGC GGC AGA GA**A**-3′
3′-GCG GGT GCG CCG TCT CT**T** **A**CC TCT TCT CTC TTC TCT-5′

[f] The mismatched primer-template used for mispaired primer extension assay was:

5′-CGC CCA CGC GGC AGA GA**G**-3′
3′-GCG GGT GCG CCG TCT CT**T** **A**CC TCT TCT CTC TTC TCT-5′

The boldface and underlined nucleotides indicate the position of extension with dTMP.

[g] $f_{ext}(\text{mismatched}) = \frac{k_{cat}(\text{mutant})/K_m(\text{mutant})}{k_{cat}(\text{wild type})/K_m(\text{wild type})}$.

mutated residue is involved in catalysis but not in substrate binding. For example, mutations of amino acid residues in region I of human DNA polymerase α yielded mutant enzymes with decreased k_{cat} but no changes in K_m for either dNTPs or primer–template.[7] Mutations of residues in region II of human DNA polymerase α yielded mutants with increased K_m values for dNTP or increased K_m and K_D values for primer terminus suggest that these mutated amino acid residues are involved in dNTP binding or primer terminus interaction, respectively.[9,10] Thus, residues in region I are mainly involved in catalysis, whereas several residues in region II are involved in binding the incoming dNTP or primer terminus. The structural feature of dNTP recognized by the mutated amino acid residue can be identified by comparative analyses of K_m values of dNTP analogs having modifications of the nucleotide base or ribose ring and compared with the wild-type DNA polymerase. An example is shown in the Tyr-865 mutations of human DNA polymerase α.[9] Mutants with altered K_m values for the primer terminus can be further investigated to identify the structural feature of the primer recognized by the mutated amino acid residue. Measuring the K_D values of the mutant for binding primer terminus containing nucleotide analogs at the 3′ terminus and comparing it to the K_D value of the wild-type DNA polymerase will indicate which structural feature of the primer terminus is recognized by the amino acid residue in the active site.[10] A typical example is demonstrated in the Ser-867 mutation of human DNA polymerase α.[10]

Mutations of amino acid residues involved in dNTP binding may affect the fidelity of nucleotide misinsertion.[9] Mutations of amino acid residues that interact with the primer terminus may affect the mispaired primer extension capacity of the mutant enzymes.[10] Thus, these mutants should be further investigated for their misinsertion and mispaired primer extension fidelity. Gel kinetic fidelity assays[18] or forward mutation analysis[19] can be used to analyze the effects of these mutations on misinsertion or mispaired primer extension fidelity and comparing to the wild-type DNA polymerase. These fidelity studies provide insight into the function of these amino acid residues in the active site of a DNA polymerase. An example of the fidelity studies of active site mutants of human DNA polymerase α is depicted in Table II.

By the rationale and approaches presented here, key residues and regions of human DNA polymerase α involved in either catalysis or substrate binding have been identified.[7–10] Mutational analysis of the most conserved region (region I) has shown that the residues in this region are involved

[18] M. S. Boosalis, J. Petruska, and M. F. Goodman, *J. Biol. Chem.* **262,** 14,689 (1987).

[19] K. Bebenek and T. A. Kunkel, [18], this volume.

in binding the activator metal, which is absolutely essential for catalysis.[7,8] Mutations of residues in region II have shown that the phenyl ring of Tyr-865 interacts directly with the incoming nucleotide,[9] whereas Ser-867 binds the 3′-OH terminus of the primer.[10] Thus, despite the lack of structural information for the human DNA polymerase α, the function of several residues in the active site has been identified for substrate binding and/or catalysis. The rationale and approach described here should be generally applicable to any of the cloned DNA polymerases. Knowledge gained by these kinds of rational mutagenesis studies in conjunction with the emergence of more DNA or RNA polymerase structural data will improve the understanding of basic DNA or RNA synthesis mechanisms and lay the foundation for rational design of anticancer and antiviral therapeutic inhibitors.

Acknowledgments

This work was supported by research grant CA14835 from the National Institutes of Health, U.S. Public Health Service.

[24] Mutational Analysis of DNA Polymerase Substrate Recognition and Subunit Interactions Using Herpes Simplex Virus as Prototype

By PAUL DIGARD, WILLIAM R. BEBRIN, and DONALD M. COEN

Introduction

Catalytic subunits of DNA polymerases recognize substrates and interact with proteins, particularly other polymerase subunits. The DNA polymerase encoded by herpes simplex virus (HSV) belongs to the α-like class of DNA polymerases (also known as class B), which includes the eukaryotic replicative polymerases.[1,2] In part, because of the genetic and pharmacological tools available in the HSV system, the HSV DNA polymerase has served as a prototype of the α-like polymerases, especially in terms of identifying amino acid residues involved in substrate recognition and subunit interactions. The analysis of substrate recognition has been due largely

[1] S. W. Wong, A. F. Wahl, P.-M. Yuan, N. Arai, B. E. Pearson, K.-I. Arai, D. Korn, M. W. Hunkapiller, and T. S.-F. Wang, *EMBO J.* **7,** 37 (1988).
[2] J. Ito and D. K. Brathwaite, *Nucleic Acids Res.* **19,** 4045 (1991).

to the availability of HSV mutants that were selected for resistance to drugs that mimic the natural deoxynucleoside triphosphate (dNTP) or pyrophosphate substrates of Pol and/or inhibit competitively with these substrates. Roughly 20 such mutant *pol* genes have now been mapped and sequenced.[3–5] Nearly all of the mutations lie within regions of homology shared by members of the α-like class of DNA polymerases (Fig. 1). This indicates that these conserved regions are involved directly or indirectly in the recognition of the natural dNTP and pyrophosphate substrates that the drugs mimic.

In contrast, the analysis of subunit interactions has begun from *in vitro* assays of engineered mutations in either the catalytic subunit (Pol) or the accessory subunit, UL42, which increases processivity.[6,7] These studies have identified the extreme C terminus of Pol, which is not conserved among different polymerases, as being specifically critical for subunit interactions.[8–12] As yet, no corresponding small region of UL42 has been mapped, although one mutation that specifically disrupts subunit interactions has been identified.[13]

This chapter discusses the methods used to isolate and to analyze drug-resistant mutants that identify important amino acid residues in polymerases. It then discusses methods used to investigate subunit interactions in polymerases. Although some of the methods employed are peculiar to the HSV system, the approaches could be generally applicable to any polymerase.

Drug-Resistant Mutants as Probes of Polymerase Active Sites

The underlying rationale for using drug resistance to probe the active sites of DNA polymerase can be summarized as follows: Polymerase (*pol*) mutants with altered drug sensitivities specify polymerases that are func-

[3] D. M. Coen, *Antiviral Res.* **15,** 287 (1991).

[4] P. Collins, B. A. Larder, N. M. Oliver, S. Kemp, I. W. Smith, and G. Darby, *J. Gen. Virol.* **70,** 375 (1989).

[5] C. B. C. Hwang, K. L. Ruffner, and D. M. Coen, *J. Virol.* **66,** 1774 (1992).

[6] T. R. Hernandez and I. R. Lehman, *J. Biol. Chem.* **265,** 11,227 (1990).

[7] J. Gottlieb, A. I. Marcy, D. M. Coen, and M. D. Challberg, *J. Virol.* **64,** 5976 (1990).

[8] P. Digard and D. M. Coen, *J. Biol. Chem.* **265,** 17,393 (1990).

[9] P. Digard, W. Bebrin, K. Weisshart, and D. M. Coen, *J. Virol.* **67,** 398 (1993).

[10] N. Stow, *Nucleic Acids Res.* **21,** 87 (1993).

[11] D. J. Tenney, P. A. Micheletti, J. T. Stevens, R. K. Hamatake, J. T. Matthews, A. R. Sanchez, W. W. Hurlburt, M. Bifano, and M. G. Cordingley, *J. Virol.* **67,** 1959 (1993).

[12] H. Marsden, M. Murphy, G. McVey, K. MacEachran, A. Owsianka, and N. Stow, *J. Gen. Virol.* **75,** 3127 (1994).

[13] P. Digard, C. Chow, L. Pirrit, and D. M. Coen, *J. Virol.* **67,** 1159 (1993).

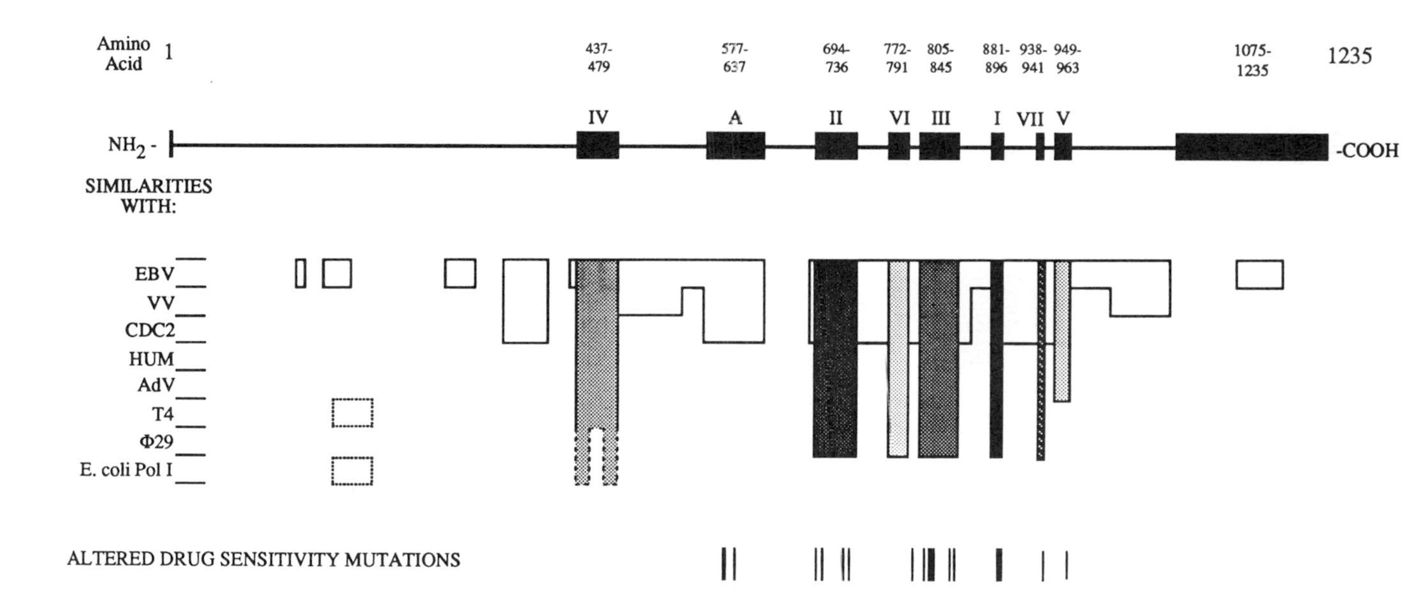

FIG. 1. Location of altered polymerase residues in drug-resistant HSV mutants relative to regions conserved among various polymerases. The top line is a schematic of the HSV Pol polypeptide with the locations of conserved regions I–VII and A and the UL42 binding domain shown as filled boxes. The numbers above the line refer to amino acid residues. Below the line are boxed areas indicating homologies with various polymerases (abbreviations provided below), with the degree of shading indicating the degree of homology, with region I having the most, region VI, the least, while region VII comprises a very short, but highly conserved segment. The vertical lines below the boxes refer to the locations of mutations conferring altered drug sensitivity phenotypes from HSV mutants isolated for drug resistance. EBV, Epstein-Barr virus; VV, vaccinia virus; CDC2, yeast DNA polymerase δ; HUM, human DNA polymerase α; AdV, adenovirus; T4, bacteriophage T4; ϕ29, bacteriophage ϕ29; *E. coli* Pol I, *E. coli* DNA polymerase I. Adapted from D. M. Coen, *Semin. Virol.* **3,** 3 (1992), with permission.

tional, yet exhibit altered specificity. When the drug used mimics and/or competes for binding with the natural substrate, mutations that confer resistance or hypersensitivity to that drug can be expected to alter the binding site for the substrate. In the case of herpes viruses, the drugs used have included nucleoside analogs such as acyclovir (ACV), ganciclovir, and vidarabine (araA), which are converted to triphosphates and inhibit competitively with dNTPs; pyrophosphate (PP_i) analogs such as phosphonacetic acid (PAA) and foscarnet (phosphonoformic acid, PFA); and the tetracyclic diterpenoid, aphidicolin, which inhibits competitively with dNTPs. Thus, mutations with altered sensitivities to these drugs can be expected to identify amino acids that are involved directly or indirectly in dNTP and PP_i recognition. In the HSV system, this approach identified amino acids in six regions that are conserved among α-like polymerases.[3,5] In principle, this approach should be applicable to any polymerase.

Isolation of Drug-Resistant Polymerase Mutants

Several steps are involved in isolating drug-resistant polymerase mutants. The first is to choose an appropriate drug. The second is to choose whether or not to mutagenize the organism. The third is to choose a selection strategy—single step versus passage. The fourth is to clone the resistant mutant biologically. Many of the considerations involved in isolating drug-resistant HSV polymerase mutants are similar to those in other systems and have been discussed nicely by Thompson and Baker in their paper describing methods to isolate mammalian cell mutants.[14] Rather than detail specific virological methods (e.g., plaque assays, plaque purification, and marker rescue), the focus will be on more generally applicable aspects.

Choice of Drug. To isolate a drug-resistant polymerase mutant, one can either use a drug that interacts with polymerase directly or following conversion to an active metabolite. Examples of the former class of drugs are aphidicolin and PFA. Examples of the latter class of drugs are the nucleoside analogs araA and ACV. Both classes of drugs have been used to select drug-resistant polymerase mutants of HSV. A possible advantage of the former class of drugs for mutant isolation is that one may be less likely to obtain unwanted mutants that are resistant, for example, simply because they fail to activate the drug. For example, most HSV mutants isolated for resistance to the nucleoside analog, acyclovir, are deficient in the virus-encoded thymidine kinase, which activates acyclovir to its mono-

[14] L. H. Thompson and R. M. Baker, *in* "Methods in Cell Biology" (D. M. Prescott, ed.), p. 209. Academic Press, New York, 1973.

phosphate.[15,16] However, all HSV mutants isolated for araA resistance that have been examined to date are polymerase mutants[17–19] because this drug is activated by cellular rather than viral enzymes.[20] Even so, one could imagine resistance to any drug that inhibits competitively with dNTPs arising from mutations that increased the activity of HSV ribonucleotide reductase. For example, cellular mutants that are resistant to aphidicolin can specify an altered ribonucleotide reductase.[21] Moreover, in attempting to isolate cellular polymerase mutants, remember that resistance to nucleoside analogs can result from mutations affecting cellular kinases and other activating enzymes.

A second set of considerations in drug choice involves potency and selectivity. A drug has to be reasonably potent against the organism of choice so that one does not have to use huge, possibly expensive quantities of the compound. The isolation and analysis of resistant mutants requires even more drug than is required to select against a sensitive wild-type parent. A drug has to be reasonably selective so that one can actually obtain mutants and to ensure a reasonable probability that the mutations will be in the gene of interest. For antiviral drugs, selectivity basically means that the drug inhibits viral replication directly rather than making the host cell so incapacitated that it cannot support viral replication. Indeed, the best test of selectivity is the isolation of a virus mutant that is resistant to the drug. For cellular systems, selectivity means that the drug will inhibit the organism due to effects on the polymerase of interest rather than on other cellular processes. For example, certain nucleoside analogs can have effects as potent against RNA synthesis as against DNA polymerase. Sometimes drugs that might be reasonably selective against polymerases are difficult to use because they are not very potent against the organism of choice. For example, wild-type *Saccharomyces cerevisiae* strains are usually not susceptible to many polymerase inhibitors due to poor uptake. In these cases, it might be possible to find mutations that facilitate uptake, thereby permitting susceptibility and selection of desired mutations.

[15] D. M. Coen and P. A. Schaffer, *Proc. Natl. Acad. Sci. USA* **77,** 2265 (1980).

[16] J. A. Fyfe, P. M. Keller, P. A. Furman, R. L. Miller, and G. B. Elion, *J. Biol. Chem.* **253,** 8721 (1978).

[17] J. S. Gibbs, H. C. Chiou, J. D. Hall, D. W. Mount, M. J. Retondo, S. K. Weller, and D. M. Coen, *Proc. Natl. Acad. Sci. USA* **82,** 7969 (1985).

[18] J. S. Gibbs, H. C. Chiou, K. F. Bastow, Y.-C. Cheng, and D. M. Coen, *Proc. Natl. Acad. Sci. USA* **85,** 6672 (1988).

[19] H. C. Chiou, Ph.D. thesis, Harvard University (1988).

[20] V. Verhoef, J. Sarup, and A. Fridland, *Cancer Res.* **41,** 4478 (1981).

[21] B. Ullman, L. J. Gudas, I. N. Caras, S. Eriksson, M. A. Wormsted, and D. W. Martin, *J. Biol. Chem.* **256,** 10,189 (1981).

Mutagenesis. HSV exhibits a high spontaneous frequency of drug-resistance mutations such that most wild-type stocks contain about 1 in 10^3 to 10^4 polymerase mutants resistant to PFA, ACV, and/or araA.[22] Therefore, chemical mutagenesis would most likely add little except unwanted additional mutations. However, in many systems, even the closely related human cytomegalovirus (HCMV),[23] the spontaneous frequency appears to be much lower. In these cases, it may be desirable to mutagenize stocks prior to selection.

Even in the HSV system, isolation and sequencing of a large number of spontaneous drug-resistant polymerase mutants failed to identify any mutations in the most highly conserved region of sequence homology, region I. To address why this was so, a portion of the *pol* gene was cloned into an M13 vector and mutagenized by use of a randomized oligonucleotide corresponding to region I. The mutant genes were introduced back into HSV by marker rescue techniques and the resulting viruses tested for viability and drug resistance. By this approach, three new drug-resistant mutants with lesions in region I were identified.[24] Similar directed mutagenesis and gene replacement methods can be employed in many systems.

Selection Procedures. Various procedures are available to select drug-resistant mutants. For HSV, it has been most advantageous to employ a single-step selection procedure.[15,25] Single-step selection is simple, incorporates a biological cloning step, and does not subject the organism to numerous selection events that could lead to multiple mutations. Figure 2 illustrates the procedure. One performs a dose–response experiment assaying, in this case, plaque formation versus acyclovir concentration, shown here with plaque formation on a log scale. This permits evaluation of heterogeneity throughout the dose range. For drug concentrations above a certain threshold dose, there is a rapid decline in survival over a relatively narrow dose interval.[14] Where the curve tails off, the presence of discrete survivors usually indicates a resistant subpopulation. In Fig. 2, this dose is seen to be ~20 μM, with a resistant subpopulation of about 0.3%. Most of the mutants shown in Fig. 2 were isolated by picking survivors in this dose range and biological cloning.

However, if one picks survivors at lower doses (in the steep range of the dose–response curve), these will not usually represent resistant mutants. Rather, they reflect the statistical nature of growth under partially selective conditions.[14] With some drugs, however, one is not so fortunate as to see

[22] D. M. Coen, *J. Antimicrob. Chemother.* **18** (Suppl. B), 1 (1986).

[23] V. Sullivan and D. M. Coen, *J. Infect. Dis.* **164,** 781 (1991).

[24] A. I. Marcy, C. B. C. Hwang, K. L. Ruffner, and D. M. Coen, *J. Virol.* **64,** 5883 (1990).

[25] D. M. Coen, P. A. Furman, P. T. Gelep, and P. A. Schaffer, *J. Virol.* **41,** 909 (1982).

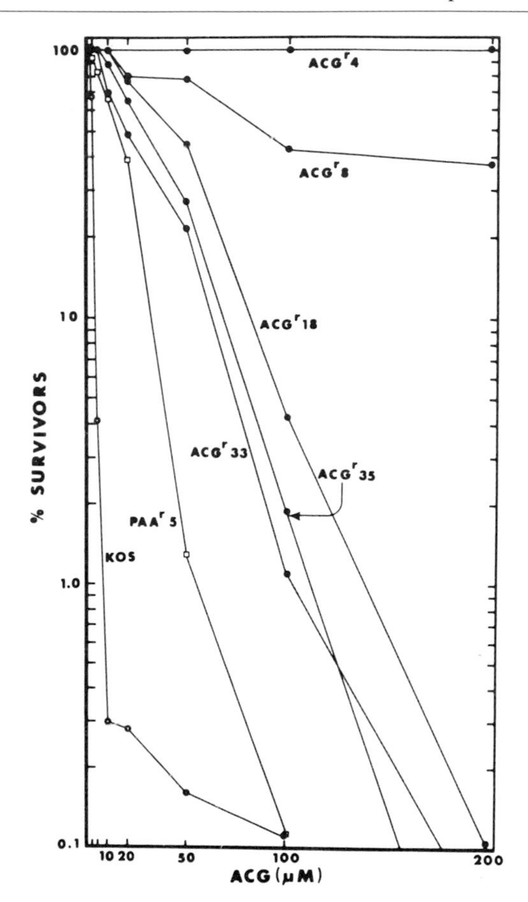

FIG. 2. Isolation of acyclovir-resistant mutants from wild-type HSV strain KOS. Viruses were assayed for the efficiency of plaque formation (log scale on the ordinate) versus varying concentrations of acyclovir (ACG). Wild-type virus, curve KOS; various viruses isolated for acyclovir-resistance (curve ACG^{r}4-35); and virus originally isolated for phosphonoacetic acid resistance (curve PAA^{r}5). Adapted from D. M. Coen and P. A. Schaffer, *Proc. Natl. Acad. Sci. USA* **77,** 2265 (1980).

such a clear "plateau" as in Fig. 2; in such cases, one simply picks survivors at the highest possible dose and hopes for the best.

When single-step selection methods fail, it is often possible to obtain mutants by passaging the organisms in drug over multiple generations. One strategy is to begin passaging at a drug concentration that permits ~50% survival and then to escalate the dose at each subsequent passage, for example, by twofold increments. Such a strategy has been successful in isolating HCMV mutants resistant to foscarnet (PFA) and ganciclovir

(DHPG).[23,26] In the latter case, the resulting mutant proved to be a double mutant containing both a kinase and a *pol* mutation, illustrating a disadvantage of passage protocols.[27,28]

Biological Cloning. Biological cloning of drug-resistant mutants is as crucial in viral systems as in any other organism to permit proper biochemical and genetic characterization. As with most viruses, plaque purification or limiting dilution has been the method used to clone HSV and HCMV drug-resistant mutants. Single rounds of biological cloning are often not sufficient. For example, at least three different phenotypes were detected in the progeny of a single plaque of HSV isolated for resistance to araA.[29] In our laboratory, we routinely clone at least three times before performing detailed analyses.

Analyses of Drug-Resistant Mutants

Once drug-resistant mutants have been isolated they can be analyzed genetically, phenotypically, and biochemically.

Genetic Analysis. An important question with a drug-resistant mutant is "What mutation confers resistance?" Depending on the organism, this question can be easy or difficult to answer and the methods used can vary widely. The initial step is usually to determine what gene contains the mutation. For herpes viruses, one preferred method is marker transfer, illustrated in Fig. 3. Briefly, individual fragments of viral DNA from the resistant mutant are each mixed with intact, infectious DNA from a drug-sensitive, usually wild-type virus, and each mixture is cotransfected into cells. If the fragment contains the resistance mutation, homologous recombination occurs between the fragment and the infectious DNA, resulting in an increased frequency of drug-resistant virus compared to the progeny of control transfections. If so, then a resistance mutation maps to that DNA fragment. The process can then be repeated with smaller fragments to map the mutation more finely to a particular gene or region of a gene. The drug-resistant virus can be isolated and its degree of resistance assayed to ascertain if the mutation in the fragment accounts for all of the resistance in the original mutant. Other phenotypes (see below) can be similarly

[26] K. K. Biron, J. A. Fyfe, S. C. Stanat, L. K. Leslie, J. B. Sorrell, C. U. Lambe, and D. M. Coen, *Proc. Natl. Acad. Sci. USA* **83,** 8769 (1986).

[27] V. Sullivan, C. L. Talarico, S. C. Stanat, M. Davis, and D. M. Coen, *Nature* **358,** 162 (1992).

[28] V. Sullivan, K. K. Biron, C. Talarico, S. C. Stanat, M. Davis, L. M. Pozzi, and D. M. Coen, *Antimicrob. Agents Chemother.* **37,** 19 (1993).

[29] D. M. Coen, H. C. Chiou, H. E. Fleming, Jr., L. K. Leslie, and M. J. Retondo, *in* "Herpesvirus" (F. Rapp, ed.), UCLA Symposia on Molecular and Cellular Biology, New Series, Vol. 21, p. 373. Alan R. Liss, New York, 1984.

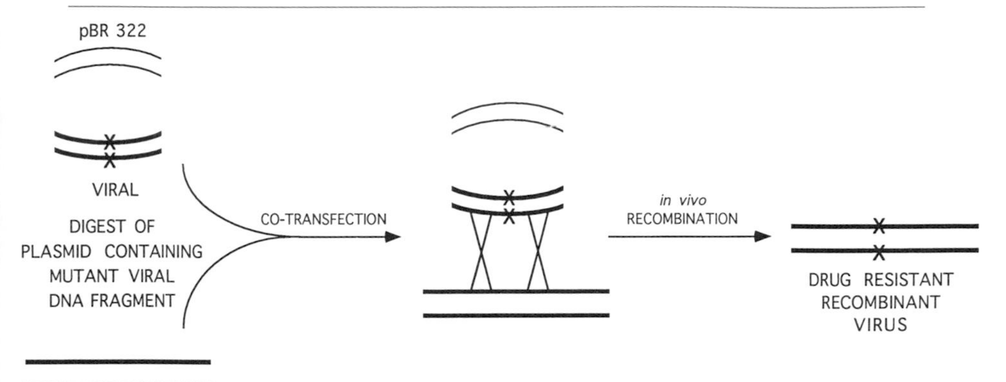

FIG. 3. Mapping of drug-resistance mutations in the HSV system using marker transfer. Plasmid DNA contains mutant viral sequences (-X-), which should confer a drug-resistant phenotype, when incorporated into the viral genome. The restriction enzyme digested plasmid is mixed with intact, infectious wild-type viral DNA and the mixture is transfected into cells. Homologous recombination between the cloned mutant sequences and the wild-type genome occurs in the transfected cells. Drug-resistant viruses can be selected and the percentage of drug-resistant progeny quantified. When the percentage of drug-resistant progeny arising from the transfection containing the mutant plasmid is substantially higher than that arising from control transfections, then the mutation is said to map to the viral sequences within the plasmid. From D. M. Coen and R. F. Ramig, *in* "Fields' Virology," 3rd ed. (B. N. Fields, D. M. Knipe, P. M. Howley, R. M. Chanock, J. L. Melnick, T. P. Monath, B. Roizman, and S. E. Straus, eds.), Raven Press, New York, 1995. Used with permission.

assessed. The region containing the mutation can then be sequenced to determine the base and amino acid changes responsible for the drug resistance.

This general approach has succeeded in mapping many different drug resistance mutations to the HSV *pol* gene, where the overwhelming majority affect regions of the catalytic subunit that are conserved among α-like polymerases (Fig. 1). These results argue that these residues are involved directly or indirectly in recognition of the drugs and, by extension, the substrates the drugs mimic. They also suggest that the region between the most distant mutations is included within a domain of the polymerase that contains polymerase activity. These most distant mutations exhibit very similar phenotypes, suggesting that protein folding brings the conserved regions together near the polymerase active site. Further genetic, biochemical, and biophysical studies are under way to test these hypotheses, which spring from the genetic and phenotypic analyses of the mutants.

Phenotypic Analysis. While, or even before, genetic analysis is under way, much information can be obtained by a phenotypic analysis of a drug-resistant organism. Two phenotypes that are especially germane to polymerase mutants are altered sensitivities to other drugs and mutation rates. Cross-resistance or hypersensitivity to other drugs can help distinguish different mutants from each other, can suggest whether or not drug resistance is due to mutation in genes other than those encoding polymerases, and can help in understanding the molecular manifestations of the mutation. Drug-hypersensitivity mutations can be especially valuable in genetic analyses because they allow one to select for the wild-type phenotype and against the mutant, while resistance permits selection only in the other direction. In the HSV polymerase system, drug hypersensitivity has also been exploited to identify suppressor mutations, some of which confer altered drug sensitivities of their own. Interestingly, these map outside the conserved regions identified by the resistance mutations, revealing additional layers of complexity in the polymerase.[30]

In many systems, certain polymerase mutants exhibit altered mutation rates. In the HSV system, drug-resistant mutants are frequently antimutators, evidently due to more stringent base selection.[31]

Biochemical Analysis. To understand fully the molecular mechanisms of drug resistance and other phenotypes exhibited by drug-resistant mutants, it is necessary to characterize the mutant polymerase biochemically. Methods for purification, assay, and characterization of DNA polymerases are presented elsewhere in this volume. Important parameters in assessing drug resistance include whether the drug inhibits competitively or not, the K_i for the drug, K_m for competing substrates, and k_{cat}, if the drug is incorporated into primer–template. HSV polymerases from drug-resistant mutants are frequently affected for K_m as well as K_i, reinforcing the idea that they are affected in substrate recognition. Much work remains, however, in correlating amino acid changes in conserved residues with changes in polymerase function.

Mutational Analysis of Subunit Recognition

What follows is a description of the techniques our laboratory has found useful in analyzing protein–protein interactions between the two subunits of the herpes simplex virus DNA polymerase.[8,9,13] Although applied here to a specific problem, similar approaches have been used to investigate a multitude of macromolecular interactions. We therefore hope that a discus-

[30] Y. Wang, S. Woodward, and J. D. Hall, *J. Virol.* **66,** 1814 (1992).

[31] J. D. Hall, P. A. Furman, M. H. St. Clair, and C. W. Knopf, *Proc. Natl. Acad. Sci. USA* **82,** 3889 (1985).

sion of the techniques used and problems encountered will be of general interest.

Our objective was to investigate the interaction between the catalytic subunit (Pol) of the herpes simplex virus DNA polymerase, and its processivity factor, UL42. The two proteins associate to form a heterodimeric polymerase capable of synthesizing long stretches of DNA.[6,7] Specifically, we wished to identify the regions of each protein responsible for recognizing the other. The experimental approach was to introduce mutations into the Pol and UL42 genes, express the altered polypeptides, and assay their ability to bind to a wild-type partner in a coimmunoprecipitation assay. Each stage is discussed below.

Expression of HSV Pol and UL42 in Vitro

Since we want to examine the phenotype of a large number of Pol and UL42 mutants, we use the technique of *in vitro* transcription followed by translation in rabbit reticulocyte lysate as a convenient and rapid means of expressing small amounts of radiolabeled protein. Standard protocols as detailed elsewhere[32,33] are used, with the exception that *in vitro* translation reactions are incubated at 37° instead of the commonly recommended 30°. This is found to increase the yield of larger polypeptides such as Pol (135 kDa). Systems for coupled *in vitro* transcription and translation (i.e., the Promega T.N.T. system) also produce satisfactory results.

For some experiments, both polypeptides are cotranslated in reticulocyte lysate,[8,9] while in others, the mutant polypeptide is translated alone, and the wild-type partner polypeptide supplied by the posttranslational addition of purified protein.[13] The latter approach has the advantage, in practical terms, that it makes it easier to control the relative amounts of the two polypeptides, because, achieving equal levels of *in vitro* translation from two transcripts that may translate with quite different efficiencies can sometimes be difficult. Also in practical terms, it can provide cleaner autoradiographic data if the proteins under examination migrate similarly on SDS–PAGE, or contain widely different numbers of methionine residues (or other radiolabeled moiety). It also has the advantage that, as discussed later, the sensitivity of the assay can be increased by adding larger amounts of the directly immunoprecipitated protein. However, as well as the obvious need for a source of large amounts of preferably purified protein, this approach relies on the polypeptides in question being able to associate

[32] D. A. Melton, P. A. Krieg, M. R. Rebagliati, T. Maniatis, K. Zinn, and M. R. Green, *Nucleic Acids Res.* **12,** 7035 (1984).

[33] M. J. Clemens, *in* "Transcription and Translation: A Practical Approach" (B. D. Hames and S. J. Higgins, eds.), p. 231. IRL Press, Oxford, 1984.

posttranslationally, which is not always the case.[34] In our studies, Pol and UL42 were purified from insect cells infected with the appropriate recombinant baculoviruses,[7,35] although any expression system from which active protein can be produced will suffice. If the protein of interest is amenable to expression in *Escherichia coli,* it is worth considering expressing it as a fusion with glutathione *S*-transferase[36] (plasmid vectors available from Pharmacia) or maltose binding protein[37] (plasmid vectors available from New England Biolabs). As well as facilitating purification, the interaction of the expressed polypeptides with the appropriate solid-phase-linked ligand can be used as a nonantibody dependent means of precipitation.[38]

Other convenient expression methods that can be used include the two hybrid vaccinia virus system, or microinjected *Xenopus* oocytes. In the former technique, tissue culture cells are first infected with a recombinant vaccinia virus that expresses the desired bacteriophage RNA polymerase,[39–41] and then transfected with plasmids containing the appropriate gene and promoter.[39] In the latter, *in vitro* transcribed transcripts are microinjected into isolated oocytes from the frog, *Xenopus laevis.*[42] In either case, the expressed polypeptides can be metabolically labeled and analyzed as described below. Expression *in vivo* may perhaps have the advantage that the target polypeptides are more likely to fold correctly than *in vitro.* Conversely, more problems with endogenously labeled polypeptides that can confound the analysis and with degradation of unstable mutant polypeptides might be expected.

Detection of Pol–UL42 Interaction by Immunoprecipitation

The principle of detecting interactions between polypeptides by immunoprecipitation is straightforward; given an antibody directed against one protein, precipitation of other polypeptides can be indicative of a protein–protein interaction. The difficulty arises in testing whether the observed indirect precipitation really reflects a specific biologically relevant interac-

[34] P. S. Masters and A. K. Banerjee, *J. Virol.* **62,** 2651 (1988).

[35] A. I. Marcy, P. D. Olivo, M. D. Challberg, and D. M. Coen, *Nucleic Acids Res.* **18,** 1207 (1990).

[36] D. B. Smith and K. S. Johnson, *Gene* **67,** 31 (1988).

[37] C. Guan, P. Li, P. D. Riggs, and H. Inouye, *Gene* **67,** 21 (1987).

[38] W. G. Kaelin, D. C. Pallas, J. A. DeCaprio, F. J. Kaye, and D. M. Livingston, *Cell* **64,** 521 (1991).

[39] T. R. Fuerst, E. G. Niles, F. W. Studier, and B. Moss, *Proc. Natl. Acad. Sci. USA* **83,** 8122 (1986).

[40] D. Rodriguez, Y. Zhou, J.-R. Rodriguez, R. K. Durbin, V. Jimenez, W. T. McAllister, and M. Esteban, *J. Virol.* **64,** 4851 (1990).

[41] T. B. Usdin, M. J. Brownstein, B. Moss, and S. N. Isaacs, *Biotechniques* **14,** 222 (1993).

[42] J. B. Gurdon and M. P. Wickens, this series, Vol. 101, p. 370.

tion. Therefore this section describes the methods we found useful in analyzing the interaction between HSV Pol and UL42; a subsequent section discusses control experiments that can be performed to assess the validity of a putative interaction.

The basic immunoprecipitation (IP) protocol is as follows: protein samples are prepared by *in vitro* translation as described above, and then microcentrifuged for 10 min to remove particulates. In experiments where Pol is supplied posttranslationally, 5-μl aliquots of reticulocyte lysate containing UL42 are first incubated with 140 ng of Pol for 1 hr at room temperature to allow the polypeptides to interact. The reticulocyte lysate (2 to 10 μl, depending on the efficiency of the translation reaction) is then diluted with 100 μl of IP buffer [100 m*M* KCl, 50 m*M* Tris–Cl, pH 7.6, 5 m*M* $MgCl_2$, 0.1% Nonidet P-40 (NP-40), 1 m*M* phenylmethylsulfonyl fluoride (PMSF), 0.02% sodium azide], in siliconized microcentrifuge tubes (National Scientific Supply Co.), followed by the addition of 1 to 2 μl of antisera and 50 μl of a 10% (w/v) slurry of protein A-Sepharose (Sigma) in IP buffer. The reactions are then incubated for 2 hr at room temperature, or overnight at 4° with gentle rotation. Next, the Sepharose beads are collected by microcentrifugation, and washed twice with 750 μl of low sodium dodecyl sulfate (SDS) IP buffer (containing 0.1% SDS, 1% sodium deoxycholate, 1% Trition X-100 in place of the 0.1% NP-40), and once with IP buffer, after which bound proteins are eluted by the addition of 40 μl of SDS–PAGE sample buffer. Labeled proteins are then separated by gel electrophoresis and detected by fluorography. The final wash step with IP buffer prevents the deoxycholate present in low SDS IP buffer from interfering with the subsequent SDS gel electrophoresis where it can produce blurred bands.

The use of siliconized microcentrifuge tubes is found to reduce the background of nonspecific precipitation, especially with certain mutant polypeptides. The stringency of the assay can be varied by altering the composition of the IP buffer. The mildest IP buffer we used contained 0.1% NP-40, with the increased detergent concentrations of the low SDS IP buffer representing an increase in stringency. Further increases in stringency that are still compatible with the primary antibody–antigen interaction (for most polyclonal rabbit sera at least) can be obtained by raising the SDS concentration to 0.5%, or by the inclusion of up to 1 *M* NaCl. To avoid the formation of detergent micelles when preparing the higher stringency IP buffers, the Triton X-100 should be added first to the diluted salts and buffer solution, followed by SDS, and finally sodium deoxycholate [from a freshly prepared 10% (w/v) solution in water].

If affinity tagged bacterial fusion proteins are to be used as the directly precipitated protein, antibodies are omitted from the reaction, and the

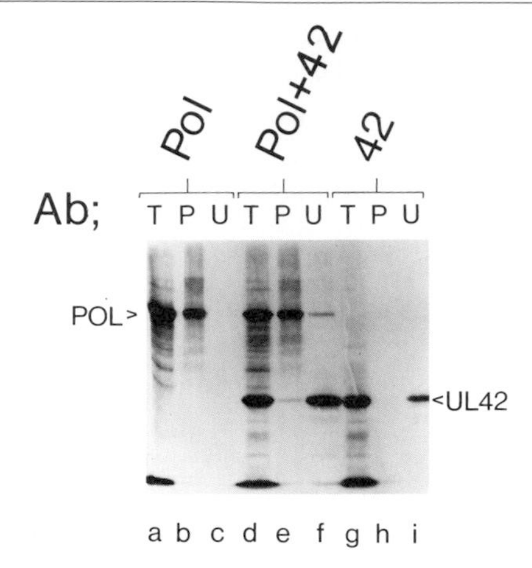

FIG. 4. Recreation of the HSV Pol-UL42 complex *in vitro*. Transcripts encoding Pol or UL42 were translated in rabbit reticulocyte lysates in the indicated mixtures and the lysates analyzed by SDS–PAGE before (T) or after immunoprecipitation with antisera specific for Pol (P) or UL42 (U). Also indicated are the polypeptide bands corresponding to the full-length proteins. From P. Digard and D. M. Coen, *J. Biol. Chem.* **265,** 17,393 (1990). Copyright The American Society for Biochemistry & Molecular Biology.

appropriate affinity resin substituted for protein A-Sepharose. This approach has been widely used with glutathione *S*-transferase (GST) fusion proteins, precipitated with glutathione-agarose, while we have successfully used maltose-binding protein fusion proteins and amylose resin. However, when a similar approach was tried with histidine-tagged fusion proteins and Ni^{2+}-charged Sepharose,[43] we found that many nontagged proteins, including reticulocyte globin, also bound to the solid phase.

Assessing Specificity of Binding. At the start of our analysis, we set out to look for an interaction *in vitro* between two proteins (Pol and UL42) that were already strongly suspected to associate *in vivo* from the results of biochemical fractionation experiments.[44] Nevertheless, several control experiments were necessary to confirm that the coprecipitation we observed between *in vitro* translated Pol and UL42 was likely to be a valid recreation of the *in vivo* Pol–UL42 complex.

The primary control for specificity was to show that although Pol and UL42 could be coprecipitated by an antibody to either protein when translated together, they were not precipitated by heterologous antisera when translated alone (Fig. 4). This controlled for cross-reactivity among the

[43] R. Janknecht, G. de Martynoff, J. Lou, R. A. Hipskind, A. Nordheim, and H. G. Stunnenberg, *Proc. Natl. Acad. Sci. USA* **88,** 8972 (1991).

[44] M. D. Challberg and T. J. Kelly, *Ann. Rev. Biochem.* **58,** 671 (1989).

antisera and nonspecific "stickiness" on the part of the polypeptides. For experiments performed by mixing *in vitro* translated UL42 with purified Pol (Fig. 5), the primary control is to mix another aliquot of UL42 with β-galactosidase. Because the samples are then immunoprecipitated with an antibody raised against a Pol β–galactosidase fusion protein, this also controlled against nonspecific binding to antigen–antibody complexes. A related control is that other polypeptides including the varicella zoster and cytomegalovirus DNA polymerases, endogenous reticulocyte polypeptides, and mutant polymerase subunits do not coprecipitate in the presence of Pol or UL42. Further reasons for believing that their coprecipitation represented a stable interaction are that the phenomenon is not dependent on a particular antibody or directly precipitated protein, since a variety of sera directed against either polypeptide could precipitate both subunits.

Although the above experiments led us to conclude that the coprecipitation of Pol and UL42 is the result of a specific interaction, they did not address the question of whether the proteins are binding directly to each

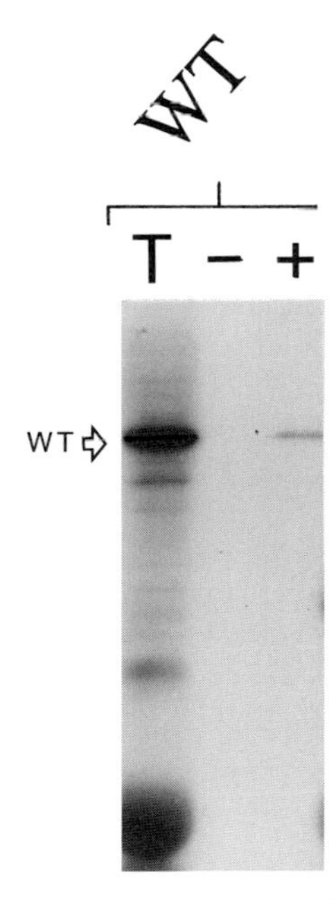

FIG. 5. Coprecipitation of *in vitro* translated UL42 and purified Pol. Aliquots of radiolabeled UL42 synthesized in reticulocyte lysates were analyzed by SDS–PAGE before (T) or after immunoprecipitation with antisera reactive against HSV Pol and *E. coli* β-galactosidase in the presence of β-galactosidase (−) or Pol (+). The arrowhead indicates the migration of UL42. Adapted from P. Digard, C. Chow, L. Pirrit, and D. M. Coen, *J. Virol.* **67,** 1159 (1993).

other, or through an intermediate molecule—a possibility that is often neglected in the literature. In our case, a direct interaction was both the simplest explanation and consistent with sedimentation data, suggesting that Pol and UL42 purified from virus-infected cells formed a heterodimer.[7,45] The finding that Pol and UL42 separately purified to apparent homogeneity coimmunoprecipitated after mixing further supported a direct interaction[46] as did "Far-Western" blotting experiments.[47] Nevertheless, the presence of an intermediate molecule that mediates binding can be hard to exclude formally. In the case of Pol and UL42, because both are DNA binding proteins, one intermediate that is important to exclude is DNA. Arguing most forcefully against DNA as an intermediate are experiments in which mutant forms of Pol and UL42 that retained DNA binding failed to interact,[13,47] whereas a mutant UL42 that fails to bind DNA is still able to interact with Pol.[48] In addition, DNase treatment did not abrogate the interaction, whereas heterologous DNA binding proteins, such as other polymerases, did not substitute for HSV Pol.

Depending on which, if any of the proteins of interest are available in excess, it can be informative to perform dose response or competition experiments. The amount of the indirectly precipitated protein should vary proportionally with the amount of directly precipitated protein, and inversely with the amount of unlabeled competitor, assuming unsaturated and saturated starting conditions respectively. Manipulating the amount of directly precipitated protein can also be used as a means of adjusting the sensitivity of the assay, and might also shed light on the affinity of the interaction between the two polypeptides. For example, the binding affinity of Pol and UL42 has been estimated to be between 1×10^{-8} and 5×10^{-9} *M*.[49,50] This is consistent with our detecting an interaction between the polypeptides when synthesized in reticulocyte lysate, which given typical *in vitro* translation yields of 0.1 to 1 ng/μl of lysate implies concentrations of Pol and UL42 in the n*M* range. However, if the interaction between the two proteins had been substantially weaker, this approach might have failed because the protein concentrations were too low to favor complex formation. In such a case, posttranslational addition of larger amounts of the directly precipitated protein would then be expected to facilitate detec-

[45] J. J. Crute and I. R. Lehman, *J. Biol. Chem.* **264,** 19,266 (1989).

[46] P. Digard, K. Weisshart, and D. M. Coen, unpublished results (1992).

[47] K. Weisshart, A. A. Kuo, C. B. C. Hwang, K. Kumura, and D. M. Coen, *J. Biol. Chem.* **269,** 22,788 (1994).

[48] C. Chow and D. M. Coen, unpublished results (1994).

[49] R. K. Hamatake, M. Bifano, D. J. Tenney, W. W. Hurlburt, and M. G. Cordingley, *J. Gen. Virol.* **74,** 2181 (1993).

[50] J. Gottlieb and M. Challberg, *J. Virol.* **68,** 4937 (1994).

tion of lower affinity complexes. We found that the addition of 140 ng of purified Pol to 5-μl aliquots of *in vitro* translated UL42 provide a sensitive and specific assay for analyzing UL42 mutants; others have used up to microgram quantities of GST fusion proteins for a variety of systems. Immobilizing one protein on a solid matrix for column chromatography may provide an even more sensitive method for detecting protein–protein interactions. Unfortunately, to a certain extent, specificity and sensitivity are mutually incompatible; an interaction only detected under mild conditions using large amounts of the directly precipitated protein is perhaps more likely to be spurious than one that survives harsher buffers and lower concentrations of the reactants. Overly mild conditions may also obscure the phenotypes of mutant proteins that are partially defective for the interaction.

Mapping Sites of Interaction between Pol and UL42

To map the sites of interaction between Pol and UL42, mutant forms of each gene are constructed and expressed *in vitro* and subjected to the assays for protein–protein interaction described above. The mutations constructed are primarily deletions and linker insertions. The next section describes the construction of the mutants and the following section describes their analysis.

Mutagenesis of Pol and UL42 Genes. All manipulations are performed on plasmid-borne copies of the Pol and UL42 genes under the control of either the bacteriophage SP6 or T7 RNA polymerase promoter, thus permitting immediate expression of the altered gene as described above. The genes are modified by both deletion and insertion mutagenesis. Deletion mutagenesis is accomplished by removing the appropriate coding regions by standard DNA cloning procedures. C-terminally truncated proteins are generated by the insertion of a synthetic oligonucleotide containing stop codons in all three translational reading frames into any convenient restriction site, or simply by linearizing the plasmid within the coding region before *in vitro* transcription. Although the latter route avoids the delay of a cloning step, we found that preferential transcription of residual supercoiled template could lead to unacceptably high background expression of the wild-type protein. Although the problem is to a certain extent dependent on the restriction enzyme used, introducing a stop codon at the desired point in the gene avoids this problem, and also allows expression of the mutant in systems that require supercoiled plasmid (see previous section).

To construct N-terminally deleted polypeptides it is necessary to supply a new initiating ATG codon. This can be supplied either by the plasmid vector or by a fortuitous in-frame ATG codon in the gene of interest, as long as there are no intervening start codons upstream, and the start codon is in a reasonable initiation context.[51] There are several suitable commercially available plasmids containing T7 RNA polymerase promoters upstream of ATG codons and multiple cloning sites, such as Invitrogen's pRSET series, or Novagen's pET plasmids. The pRSET plasmids are particularly convenient because variants are available with the polylinker in all three reading frames with respect to the ATG codon, facilitating the in-frame insertion of foreign DNA. Although these plasmids are designed for bacterial expression, we have found that they work well for *in vitro* transcription and translation in rabbit reticulocyte lysate. However, they direct expression of the inserted sequence as a fusion protein with a polyhistidine tag (for the purpose of affinity purification),[43] which could potentially interfere with the function of the region of interest.

To generate linker insertion mutations, plasmids containing full-length copies of the Pol or UL42 genes are pseudorandomly linearized by digestion with the 4-bp recognition sequence restriction enzymes *Hae*III, *Tha*I, or *Acc*II. *Acc*II was later used in preference to its isoschizomer *Tha*I because it was found to exhibit less site preference. The digests are done in the presence of an empirically determined concentration of ethidium bromide sufficient to inhibit further digestion of linear plasmid molecules generated by the first endonuclease cut.[52] The digested plasmid is then linker tailed by ligation in the presence of an excess of the unphosphorylated 12-bp palindromic oligonucleotide TGCATCGATGCA, generating predominantly linear DNA molecules containing a copy of the linker at both ends. To facilitate mapping of the insertion site, the oligonucleotide is designed to contain a *Cla*I restriction endonuclease site, a sequence not found in the Pol or UL42 genes. The linker is also designed so as to not contain either termination or proline codons in any of the three possible reading frames. High-molecular-weight DNA is separated from noncovalently bound oligonucleotide by centrifugation through a Centricon-100 filter (Amicon) in the presence of 10% dimethyl sulfoxide.[53] Linear unit length plasmid is gel purified from a 0.7% low-melting-point agarose gel, allowed to reanneal, and transformed into *Escherichia coli.* Colonies are screened for intact plasmids containing a single insertion within the gene of interest by standard minipreparation and restriction enzyme digest procedures. Time spent at the gel purification stage achieving good resolution of full-length linear DNA from residual supercoiled plasmid and multiply cut DNA is repaid here by reducing the percentage of clones found to contain either no

[51] M. Kozak, *Cell* **44,** 283 (1986).

[52] R. C. Parker, this series, Vol. 65, p. 415.

[53] R. Lathe, M. P. Kieny, S. Skory, and J. P. Lecocq, *DNA* **3,** 173 (1984).

insertion or additional deletions. The precise locations of insertion mutants that could not be unambiguously located by restriction mapping are determined by dideoxynucleotide sequencing, as are those of insertions that altered the polypeptide phenotype.

Analysis of Mutants. In the case of Pol, initial deletion mapping experiments are sufficient to identify a relatively small segment (≈20%) of the protein sufficient for binding UL42.[8] However, a similar approach was less successful when applied to UL42, with several studies failing to narrow down the Pol-binding site to less than the amino-terminal two-thirds of the protein.[13,49,54,55] This probably reflects the difference in domain (in the strict sense of an independently folding polypeptide unit[56]) structure of the two polypeptides. A large multidomain protein like Pol[47] may be more amenable to deletion analysis since proportionally more of the polypeptide can be removed without destroying the ability of the remainder to fold correctly into a functional unit. In the case of UL42, although the C-terminal third of the protein was determined to be dispensable for Pol binding,[13,55] deletion analysis of the N-terminal portion was less informative. Several deletion mutations were found in this region that abrogated the protein–protein interaction,[13,54] but they could not be interpreted as defining sequences directly involved in Pol binding since their phenotype could also have resulted from an indirect effect on the folding of the binding site. The fact that in almost every case examined, these mutants are also defective for other biochemical functions supports the latter interpretation.[13] This illustrates the major caveat of drawing conclusions from the phenotypes of deletion mutants; in the absence of supporting data, only mutants that retain the function in question are likely to be informative.

To delineate further sequences within the C-terminal 20% of Pol involved in binding UL42, we use a combination of linker insertion and deletion mutagenesis. Our strategy for the former is designed to pseudorandomly insert four amino acids (avoiding proline residues) with possibly one substitution depending on the reading frame of the insertion. It is hoped that this would provide a relatively subtle form of mutagenesis that was less likely to have gross effects on the folding of the polypeptide domain but would still result in a mutant phenotype if placed in or close to a region directly involved in binding UL42. On examination of 11 insertion mutants spaced throughout the C-terminal 200 amino acids of Pol, eight were found

[54] S. J. Monahan, T. F. Barlam, C. S. Crumpacker, and D. S. Parris, *J. Virol.* **67,** 5922 (1993).

[55] D. J. Tenney, W. W. Hurlburt, M. Bifano, J. T. Stevens, P. A. Micheletti, R. K. Hamatake, and M. G. Cordingley, *J. Virol.* **67,** 1959 (1993).

[56] C. Branden and J. Tooze, "Introduction to Protein Structure." Garland Publishing, Inc., New York and London, 1991.

to bind UL42 indistinguishably from wild-type Pol, while three either did not detectably bind or did so at substantially reduced levels. The latter three insertions were clustered at the C terminus of Pol, at amino acids 1203, 1208, and 1216, raising the possibility that this region of the protein was directly involved in binding UL42, although an indirect effect could not formally be excluded. Fortuitously though, all but one of the insertions were in the same reading frame, and could therefore be used to create (via the unique *Cla*I restriction enzyme site in the linker) a set of small in-frame deletions spanning the C terminus of Pol. Such deletions would help delineate amino acid residues not necessary for UL42 binding, and therefore potentially indicate whether the disruptive effect of the three C-terminal insertions was direct or indirect. When these deletion mutants were examined, we found that individual small deletions of all but the C-terminal 35 amino acids of Pol (the region encompassing the three disruptive insertion mutants) could be tolerated without totally abolishing complex formation, thus strongly implicating the C terminus as containing the actual site of interaction.

Again, by way of contrast, insertional mutagenesis was less successful when applied to mapping the Pol-binding site of UL42. We generated 17 linker insertions throughout the entire 488 amino acids of UL42 (a density of one insertion per 29 residues, compared to one per 21 for the C terminus of Pol), and of these, only one (at position 160) affected Pol binding. This I-160 mutant polypeptide bound DNA normally, suggesting that the insertion did not affect the overall folding of the polypeptide, but in the absence of other disruptive insertion mutations, it was not possible to conclude whether the mutation exerted its effect on Pol-binding directly or indirectly. Moreover, as mentioned above, further deletion analysis of UL42 was not particularly informative, because even relatively small deletions created using adjacent linker insertions mostly destroyed the ability of the protein to bind DNA as well as Pol,[13] favoring the cautious interpretation that the mutations were acting indirectly through a global effect on protein folding. Therefore it has not so far been possible to identify a discrete sequence or region of UL42 that is necessary and sufficient for Pol binding.

Acknowledgments

The authors thank C. Chow and J. Brown for help with manuscript and figure preparation and laboratory colleagues for helpful discussions. Grant support from the NIH (RO1 AI19838 and UO1 AI26077) is gratefully acknowledged.

[25] Use of Genetic Analyses to Probe Structure, Function, and Dynamics of Bacteriophage T4 DNA Polymerase

By LINDA J. REHA-KRANTZ

Introduction

Genetics provides a powerful tool to study protein structure and function. With respect to DNA polymerases, genetic selection techniques and screens have been applied successfully to bacteriophage T4,[1-8] herpes simplex virus,[9] vaccinia virus,[10] retroviruses,[11] *Escherichia coli* DNA polymerases I (*polA*)[12-13] and III (*polC*),[14-20] yeast DNA polymerases,[21-23] and phage ϕ29.[24] Although many of the initial genetic studies were useful in

[1] R. H. Epstein, A. Bolle, C. M. Steinberg, E. Kellenberger, E. Boy de la Tour, R. Chevalley, R. S. Edgar, M. Susman, G. H. Denhardt, and A. Lielausis, *Cold Spring Harbor Symp. Quant. Biol.* **28,** 375 (1963).
[2] J. Chao, M. Leach, and J. Karam, *J. Virol.* **24,** 557 (1977).
[3] L. J. Reha-Krantz, *J. Mol. Biol.* **145,** 677 (1981).
[4] L. J. Reha-Krantz, E. M. Liesner, S. Parmaksizoglu, and S. Stocki, *J. Mol. Biol.* **189,** 261 (1986).
[5] L. J. Reha-Krantz, *J. Mol. Biol.* **202,** 711 (1988).
[6] L. J. Reha-Krantz, S. Stocki, R. L. Nonay, E. Dimayuga, L. D. Goodrich, W. H. Konigsberg, and E. K. Spicer, *Proc. Natl. Acad. Sci. USA* **88,** 2417 (1991).
[7] L. J. Reha-Krantz and R. L. Nonay, *J. Biol. Chem.* **269,** 5635 (1994).
[8] S. A. Stoki, R. L. Nonany, and L. J. Reha-Krantz, submitted.
[9] P. Digard and D. M. Coen, this volume [24].
[10] J. A. Taddie and P. Traktman, *J. Virol.* **65,** 869 (1991).
[11] B. A. Larder, S. D. Kemp, and G. Darby, *EMBO J.* **61,** 169 (1987).
[12] M. B. Clements and M. Syvanen, *Cold Spring Harbor Symp. Quant. Biol.* **45,** 201 (1981).
[13] E. B. Konrad and I. R. Lehman, *Proc. Natl. Acad. Sci. USA* **71,** 2018 (1974).
[14] F. Bonhoeffer and H. Schaller, *Biochem. Biophys. Res. Commun.* **20,** 93 (1965).
[15] J. A. Wechsler and J. D. Gross, *Molec. Gen. Genet.* **113,** 273 (1971).
[16] C. G. Sevastopoulos, C. T. Wehr, and D. A. Glaser, *Proc. Natl. Acad. Sci. USA* **74,** 3485 (1977).
[17] E. B. Konrad, *J. Bacteriol.* **133,** 1197 (1978).
[18] T. Horiuchi, H. Maki, and M. Sekiguchi, *Molec. Gen. Genet.* **163,** 277 (1978).
[19] H. Maki, J. Mo, and M. Sekiguchi, *J. Biol. Chem.* **266,** 5055 (1991).
[20] I. Fijalkowska and R. M. Schaaper, *Genetics* **134,** 1023 (1993).
[21] L. H. Hartwell, R. K. Mortimer, J. Culotti, and M. Culotti, *Genetics* **74,** 267 (1973).
[22] K. C. Sitney, M. E. Budd, and J. L. Campbell, *Cell* **56,** 599 (1989).
[23] A. Boulet, M. Simon, G. Faye, G. A. Bauer, and P. M. J. Burgers, *EMBO J.* **8,** 1849 (1989).
[24] K. Matsumoto, C. I. Kim, H. Kobayashi, H. Kanehiro, and H. Hirokawa, *Virology* **178,** 337 (1990).

determining whether a particular gene was essential for viability of the organism, later studies were targeted specifically to DNA polymerase genes. These studies provided information on the locations of residues that function in exonucleolytic proofreading and residues that participate in 5′ → 3′ DNA replication, such as binding sites for dNTPs, PP_i, and the primer terminus. Bacteriophage T4 DNA polymerase is one of the most genetically characterized DNA polymerases with more than 100 mutants, which have been selected by several methods.[25] Because techniques for the selection of informative T4 DNA polymerase mutants may be adapted for the identification of mutant DNA polymerases in other organisms, methods developed for T4 are described here. In addition, T4 DNA polymerase shares protein sequence homologies with a family of sequence-related DNA polymerases called family B[26] or α-like[27] DNA polymerases, which includes the human α and δ DNA polymerases, yeast α, δ, and ε DNA polymerases, several viral DNA polymerases (herpes, vaccinia, adeno), *E.coli* DNA Pol II, and other phage DNA polymerases (ϕ29). Thus, information learned from T4 DNA polymerase studies may be applicable to these DNA polymerases.

Temperature-Sensitive and Nonsense Mutants

Phage T4 conditional lethal mutants, most frequently temperature-sensitive (*ts*) and nonsense mutants (*amber*), were isolated in screens for mutants that could not produce progeny under the nonpermissive conditions. These mutants were used to identify essential genes and to construct a genetic map of the T4 genome.[1] One complementation group, *g43,* was later shown to be the structural gene for the T4 DNA polymerase. There are now more than 50 *ts* and *amber* mutants in the T4 DNA polymerase collection.[25] Mutations have been identified that span the gene, but there is a gap without mutations between amino acid residues 511 to 587. The most C-terminal *ts* mutation encodes an amino acid substitution at codon 881 (E881K) and the most C-terminal *amber* mutation is at codon 882. Thus, full-length (898 residues) or nearly full-length T4 DNA polymerase is required for DNA polymerase function *in vivo.*[25]

The collection of *amber* mutants was useful in learning about the structural organization of the protein. The order of the *amber* mutants was determined by recombination and this order along with the size of the

[25] L. J. Reha-Krantz, *in* "Molecular Biology of Bacteriophage T4" (J. Karam, ed.). American Society for Microbiology, Washington, D.C., 1994, pp. 307–312.

[26] D. K. Braithwaite and J. Ito, *Nuc. Acids Res.* **21,** 787 (1993).

[27] W. C. Copeland, Q. Dong, and T. S.-F. Wang, this volume [23].

Strain	DNA sequence and protein sequence beginning at codon 652	Results of mutations
Wild type	5′ GGT - AAG - AAA - AAG - ATG G K K K M	
tsP26	"+" 5′ GGT - AAG - AAA - AA[G - AT]G - ATG G K K K M M	Addition of one Met residue.
sup	"−" "+" 5′ GG[T - AA]G - AAA - AA[G - AT]G - ATG G K K M M	Loss of one Lys residue.

FIG. 1. An example of suppressor analysis. Insertion of three nucleotides in the T4 DNA polymerase gene at the site indicated produces the *ts* phenotype (*tsP26*). The *ts* phenotype is suppressed by a second mutation that removes another three nucleotides upstream of the first mutation (*sup*). The mutant DNA polymerase in the *sup* strain has the same number of amino acid residues as the wild-type DNA polymerase, but the *sup* mutant has a methionine residue replacing one lysine residue. Nevertheless, the mutant DNA polymerase appears to function as wild type *in vivo*. Brackets indicate nucleotides added (+) or deleted (−).

peptides produced by the *amber* mutants was used to determine the orientation and the approximate location of the 5′ end of the gene.[28] None of the *amber* peptides retained DNA replication activity, but the *amB22* allele produced a 730 residue protein that had 3′ → 5′-exonuclease activity.[29] This result indicates that residues required for exonuclease activity are located in the N-terminal region of the protein.

Temperature-sensitive (*ts*) and cold-sensitive (*cs*) mutants can be used to study protein folding, structure, and function, and to identify proteins that may interact with each other. The experimental approach is to isolate second-site mutations that suppress the phenotype of the parental mutation. (Note examples of this genetic method by Jarvik and Botstein[30] and King.[31]) The second-site mutations may be located within the same gene as the original mutation (intragenic suppressors) or in another gene (intergenic suppressors). One example of this technique applied to T4 DNA polymerase uses the *tsP26* allele, which has an insertion of three nucleotides that produces a polymerase with one additional methionine residue (Fig. 1). DNA replication in the mutant strain is blocked at high temperature, but a new mutant phage strain was derived from the parent that could grow

[28] W. M. Huang and I. R. Lehman, *J. Biol. Chem.* **247,** 7663 (1972).

[29] N. G. Nossal, *J. Biol. Chem.* **244,** 218 (1969).

[30] J. Jarvik and D. Botstein, *Proc. Natl. Acad. Sci. USA* **72,** 2738 (1975).

[31] J. King, *BIO/TECHNOLOGY* **4,** 297 (1986).

at the formerly nonpermissive temperature. This mutant strain had a suppressor mutation in addition to the original mutation. The suppressor mutation removed three nucleotides upstream of the initial mutation, which restored the wild-type number of amino acids, but with a methionine residue in place of a lysine residue (Fig. 1). Thus, the genetic data suggest that this region of the protein requires a precise number of residues. Other examples of suppressor analysis are given in later sections.

Although selection of the initial collection of T4 DNA polymerase conditional lethal mutants was not designed to isolate mutants with particular alterations in function, some of the *ts* mutants at the permissive temperature had increased spontaneous mutation rates (mutator phenotype),[32] while other mutant DNA polymerases made fewer of some types of DNA replication errors (antimutator phenotype).[33] These phenotypes have been used to select additional mutant T4 DNA polymerases.

Isolation of Mutant T4 DNA Polymerases Based on Mutator Phenotype

Because of the central role of DNA polymerases in maintaining the fidelity of DNA replication, strains with increased or reduced spontaneous mutation rates may indicate the presence of a mutant DNA polymerase. The goal of genetic selections is to design conditions so that only the desired mutant strains survive; hence, the key for the selection of mutator DNA polymerases is to provide conditions so that survival is contingent on the mutator phenotype. We have used *rII* alleles as selectable markers for the isolation of mutator T4 DNA polymerases.[3–6] The *rII* genes are dispensable for phage T4 growth except if the bacterial host is a lambda lysogen. Thus, for cultures of phage carrying an *rII* mutation, there will be only a few rII^+ revertant phage in each lysate that will be able to infect permissively a lambda lysogenic host. If the phage also carries a mutation that confers the mutator phenotype, however, there will be more rII^+ revertants.

The experimental conditions used are the following. Phage carrying either the *rII P7oc* mutation,[4–6] which has a revision frequency of about 1×10^{-9}, or phage carrying two *rII ochre* mutations, *rUV199oc* and *rUV183oc* with a combined reversion frequency of 1×10^{-12},[3] are used to infect a permissive *E. coli* host. After 20 min, 10^8 infective centers are mixed with 3×10^9 lambda lysogenic bacteria, soft agar is added, and the

[32] J. F. Speyer, J. D. Karam, and A. B. Lenny, *Cold Spring Harbor Symp. Quant. Biol.* **31,** 693 (1966).

[33] J. W. Drake, E. F. Allen, S. A. Forsberg, R. M . Preparata, and E. O. Greening, *Nature* **221,** 1128 (1969).

mixture is poured over hard agar in a petri dish. Phage plaques result only from infective centers in which one or more *rII*$^+$ revertant phage are produced in the first cycles of replication in the permissive host. Because reversion of the selected *rII* markers is rare in a wild-type DNA polymerase background, there will be few *rII*$^+$ revertants and correspondingly few plaques. DNA replication by mutator DNA polymerases increases the probability of reverting the target *rII* mutations. Thus, any phage plaques that do appear are likely to have been formed by mutator phage strains.

This selection strategy was used to identify several mutations in the T4 DNA polymerase gene that confer the mutator phenotype. Most of the mutations encode amino acid changes in two regions of the protein: (1) between residues 82 to 131 and (2) between residues 255 to 363 (Fig. 2).[5] This collection contains mutations that encode the D112N and the D324G amino acid substitutions, which remove $3' \rightarrow 5'$-exonuclease active site residues.[34] Though the locations of mutations that confer the mutator phenotype were useful in identifying residues that are required for $3' \rightarrow 5'$-exonuclease activity, another essential residue, D219, was not identified by selection for mutator mutants. Instead, this residue was identified by protein sequence comparisons to *E. coli* DNA pol I.[35] Possible reasons for the absence of mutations affecting this residue in the collection of mutator mutants are discussed by Reha-Krantz and Nonay.[34]

In addition, some of the mutations identified on the basis of the mutator phenotype do not appear to affect exonuclease activity directly. For example, the G82D DNA polymerase has wild-type levels of $3' \rightarrow 5'$-exonuclease activity, but this mutant DNA polymerase can replicate past abasic template lesions.[36] Bypass replication is not observed for the wild-type T4 DNA polymerase. This new activity observed for the G82D DNA polymerase suggests that another aspect of proofreading may be affected. Thus, further biochemical studies of the G82D DNA polymerase and other mutator DNA polymerases will likely reveal additional information about the mechanism of DNA polymerase proofreading.

Isolation of Second-Site Mutations That Suppress Need for High Concentrations of dNTPs

T4 DNA polymerase antimutator strains cannot replicate DNA if the host bacteria carries the *optA1* mutation.[37] The dGTP pool is reduced in

[34] L. J. Reha-Krantz and R. L. Nonay, *J. Biol. Chem.* **268,** 27,100 (1993).

[35] L. Blanco, A. Bernad, M. A. Blasco, and M. Salas, *Gene* **100,** 27 (1991).

[36] M. F. Goodman, S. Creighton, L. B. Bloom, and J. Petruska, *Critical Reviews in Biochemistry and Molecular Biology* **28**(2), 83 (1993).

[37] P. Gauss, D. H. Doherty, and L. Gold, *Proc. Natl. Acad. Sci. USA* **80,** 1669 (1983).

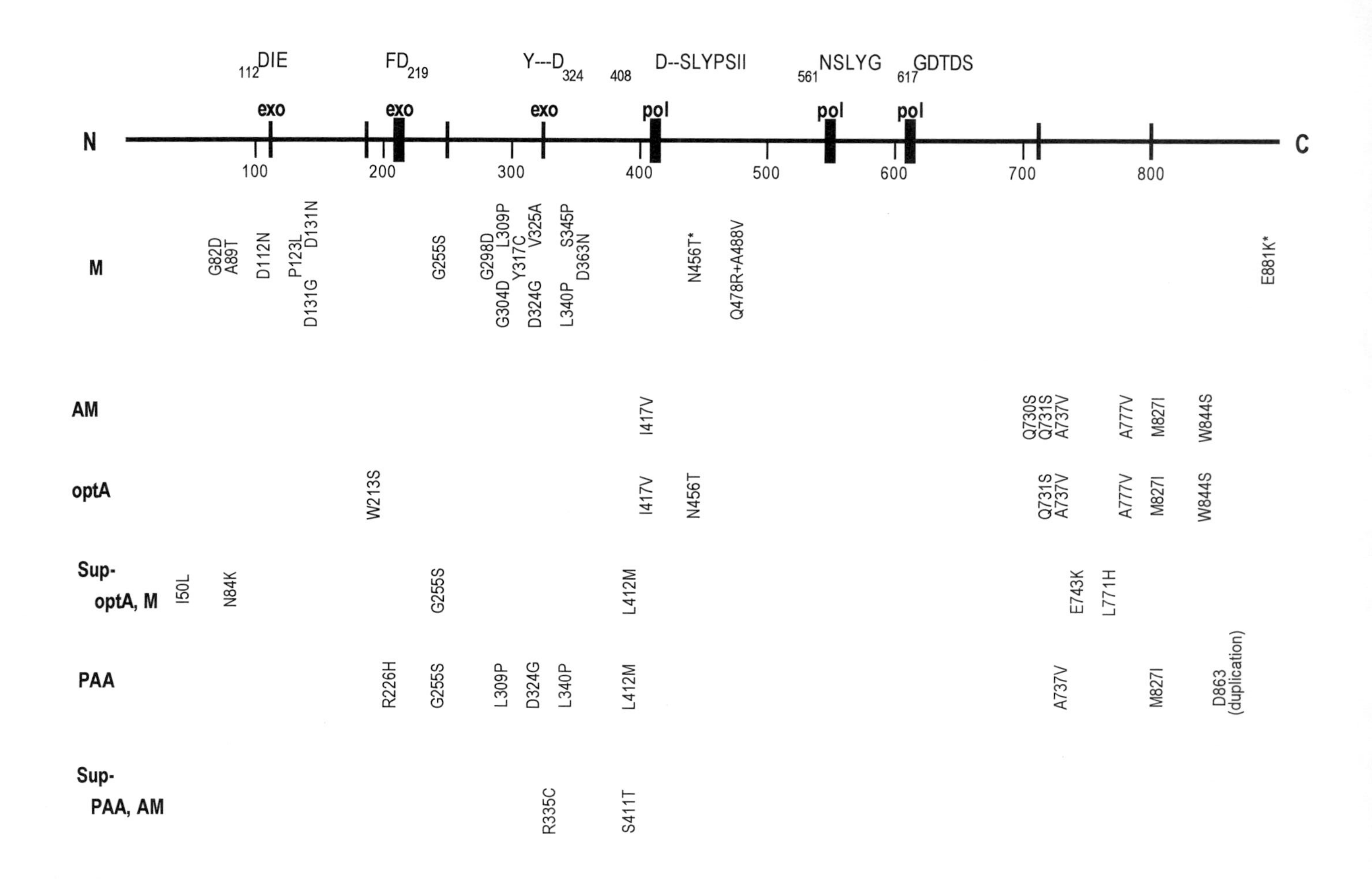

N
C
$_{112}$DIE
FD$_{219}$
Y---D$_{324}$
$_{408}$D--SLYPSII
$_{561}$NSLYG
$_{617}$GDTDS
exo
exo
exo
pol
pol
pol
100
200
300
400
500
600
700
800
M
G82D
A89T
D112N
P123L
D131G
D131N
G255S
G298D
G304D
L309P
Y317C
D324G
V325A
L340P
S345P
D363N
N456T*
Q478R+A488V
E881K*
AM
I417V
Q730S
Q731S
A737V
A777V
M827I
W844S
optA
W213S
I417V
N456T
Q731S
A737V
A777V
M827I
W844S
Sup-optA, M
I50L
N84K
G255S
L412M
E743K
L771H
PAA
R226H
G255S
L309P
D324G
L340P
L412M
A737V
M827I
D863 (duplication)
Sup-PAA, AM
R335C
S411T

optA1 bacteria due to increased expression of a novel dGTPase.[38,39] The high demand for dNTPs by antimutator DNA polymerases, which have high levels of exonucleolytic proofreading activity, combined with reduction of the dGTP pool by elevated host dGTPase activity may reduce the nucleotide pool below a critical level required for phage DNA synthesis. *OptA1* sensitivity can be suppressed by second-site mutations that "correct" the antimutator phenotype. For two T4 antimutator DNA polymerases that have been studied biochemically, the antimutator phenotype appears to be due to increased movement of the primer terminus from the polymerase to the $3' \rightarrow 5'$-exonuclease active center during DNA replication.[7,40] Intragenic suppressor mutations decrease this movement. The suppressor mutations in many cases have a mutator phenotype, as expected, if these mutant DNA polymerases restrict the opportunity for proofreading. Four regions of the T4 DNA polymerase were identified that contain suppressor mutations, two regions in the proposed polymerase domain and two regions in the proposed $3' \rightarrow 5'$-exonuclease domain (Fig. 2).[8] Both protein domains are expected to participate in shuttling the DNA between the two active centers if the regions are functionally and physically interconnected by DNA. Further biochemical studies of these informative mutant DNA polymerases are predicted to provide information on the dynamics of this aspect of DNA polymerase function.

[38] D. Seto, S. K. Bhatnager, and M. J. Bessman, *J. Biol. Chem.* **263,** 1494 (1988).
[39] S. S. Wurgler and C. C. Richardson, *Proc. Natl. Acad. Sci. USA* **87,** 2740 (1990).
[40] P. Spacciapoli and N. G. Nossal, *J. Biol. Chem.* **269,** 438 (1994).

FIG. 2. Amino acid substitutions encoded by some T4 DNA polymerase mutations. The T4 DNA polymerase primary structure is shown with intervals of 100 amino acid residues indicated. Vertical bars indicate the approximate locations of highly conserved sequences in family B DNA polymerases.[26,27] Exonuclease motifs are indicated by "exo" and the conserved residues are indicated above each motif.[26,35] Dashes indicate nonconserved residues. Polymerase motifs are indicated by "pol" and the conserved sequences are indicated for the three most highly conserved regions.[26,27] Amino acid substitutions are given for mutations that were selected on the basis of the mutator phenotype **(M),** mutations that confer the antimutator phenotype **(AM),** mutations that confer *optA* sensitivity **(optA),** mutations that were selected as suppressors of *optA* sensitivity and confer the mutator phenotype **(Sup-optA, M),** mutations that confer sensitivity to phosphonoacetic acid **(PAA),** and mutations that were isolated as suppressors of PAA sensitivity and confer the antimutator phenotype **(Sup-PAA, AM).** The asterisk (*) for the N456T and the E881K amino acid substitutions indicates that both substitutions are required to produce the mutator phenotype.

Amino acid substitutions are identified by the single letter code for the wild-type residue, followed by the codon number, and then the single letter code for the amino acid in the mutant DNA polymerase. For example, the G82D-DNA polymerase is a mutator DNA polymerase with an Asp substitution for the Gly residue at codon 82.

Mutations in the T4 DNA polymerase gene were also isolated as suppressors of mutations in gene *42* that reduce dCMP-hydroxymethylase activity.[2] T4 uses 5-hydroxymethyl-dCTP instead of dCTP, but T4 DNA polymerase readily incorporates both nucleotides. Low concentrations of 5-hydroxymethyl-dCTP produced in gene *42* mutants may lead to incorporation of dCMP, which would then make this DNA susceptible to attack by phage-induced cytosine-specific DNases. Two of the suppressor mutations encode *opal* (*UGA*) nonsense mutations in the DNA polymerase gene. Mistranslation of the *UGA* codon produces a low amount of DNA polymerase. The low level of 5-hydroxymethyl-dCTP may be sufficient for the reduced DNA synthesis under these conditions. Alternatively, the dCMP-hydroxymethylase may interact with the DNA polymerase. These suggestions and others are discussed by Chao *et al.*[2]

Isolation of Second-Site Mutations That Suppress Sensitivity to PP_i and Nucleotide Analogs; Isolation of Antimutator DNA Polymerases

One suppressor of *optA1* sensitivity of the I417V DNA polymerase (Fig. 2) was found to confer a new phenotype not previously observed for T4 DNA polymerase, sensitivity to the PP_i analog, phosphonoacetic acid (PAA).[7] This suppressor mutation encoded the L412M amino acid substitution. A screen of the T4 DNA polymerase mutant collection revealed eight additional mutants with PAA sensitivity.[41] The L412M DNA polymerase has been purified and characterized and found to be sensitive to ddNTPs in addition to PAA.[7,41] PAA sensitivity was used to isolate another collection of second-site suppressor mutations. Some of these mutants were found to be antimutators (Fig. 2).[7,42] Thus, antimutator DNA polymerases can be selected as suppressors of the PAA-sensitive phenotype. The availability of antimutator DNA polymerases provides the means to determine mechanisms used by DNA polymerases to achieve high-fidelity DNA replication.[43]

The isolation of mutant DNA polymerases based on resistance to nucleotide analogs has been a particularly successful approach for the study of reverse transcriptases[11] and herpes simplex virus DNA polymerase.[9] Additional information on this method is presented in [24] of this volume.

[41] L. J. Reha-Krantz, R. L. Nonay, and S. Stocki, *J. Virol.* **67,** 60 (1993).

[42] L. J. Reha-Krantz and C. Wong, *Mutation Research* (in press).

[43] L. J. Reha-Krantz, *Trends in Biochemical Chemistry* **20,** 136 (1995).

Summary

Functionally distinct mutant DNA polymerases have been isolated by the genetic selection strategies described here. These methods can be supplemented by the use of targeted mutagenesis procedures to enhance mutagenesis of DNA polymerase genes and to direct mutagenesis to specific sites in cloned DNA polymerases (see [22–24, 28], this volume). The power of genetic selection is in the ability to identify amino acid residues that are critical for protein structure and function that may not be obvious from studies of structural data alone. For the study of DNA polymerases, it is essential to identify residues involved in the movement of the DNA polymerase along the DNA template and in shuttling the DNA between the polymerase and exonuclease active centers. Ongoing studies are directed toward these goals.[8,44]

Acknowledgments

I thank S. Stocki for comments on the manuscript. I also thank F. Liesner, P. Svendsen, S. Stocki, R. Nonay, and C. Wong for contributions to Fig. 2. Research was supported by grants from the National Sciences and Research Council and from the National Cancer Institute of Canada with funds from the Canadian Cancer Society. L. J. R. K. is a Scientist of the Alberta Heritage Foundation for Medical Research.

[44] L. B. Bloom, M. R. Otto, R. Eritja, L. J. Reha-Krantz, M. F. Goodman, and J. M. Beechem, *Biochemistry* **33,** 7576 (1994).

[26] Limited Proteolysis of DNA Polymerases as Probe of Functional Domains

By WILLIAM H. KONIGSBERG

Introduction

DNA polymerases catalyze the formation of phosphodiester bonds in a template-directed, sequential fashion. They are the essential enzymatic components of DNA replication complexes, which usually consist of multiple proteins with specialized functions that assemble into functional units (holoenzymes) and carry out the process of duplicating the genetic information in every cell of an organism. To understand the role of each component and the mechanism of this reaction, *in vitro* systems have been developed and these, together with the results of genetic complementation, have yielded much valuable information. DNA replication systems that have been well characterized include those from *Escherichia coli,* T4 bacteriophage, which encodes all of the proteins needed for its own replication, and more recently those from yeast and mammalian cells (for reviews, see Refs. 1–7). With several of these systems the *in vitro* reactions have been reconstructed in such a way as to reveal the biochemical functions of the individual components.[8] Of the three phases that characterize DNA replication, initiation, elongation, and termination, elongation has been subjected to the most intense scrutiny. Since DNA chain growth, involving phosphodiester bond formation, always proceeds $5' \rightarrow 3'$ and, given the requirement that both DNA strands have to be duplicated at the same overall rate, nature has devised a clever way to accomplish this by using different priming mechanisms for each strand. The elongation phase, which consists of leading and lagging strand synthesis, can be studied separately *in vitro.* Leading strand synthesis, in its simplest form, requires only the DNA polymerase to extend a primer that is complementary to a template with a 5′ overhang; however, rates of elongation in a reaction containing only a primer–template and a DNA polymerase are extremely slow compared to estimated rates *in vivo.* As other "accessory" protein components are added, such as single-strand DNA binding protein, "sliding clamp" protein, and primer–template loading proteins, the synthetic rates increase dramatically. In the T4 DNA replication system, for example, 44P/62P, a tight complex of two different subunits, helps to "load" the DNA polymerase and the gene *45* protein (45P) "sliding clamp" onto the 3′ end of the primer–template junction. The addition of the T4 single-strand DNA binding protein, gene *32* protein (32P), orients the bases in the template strand so that they will be properly positioned to base pair with the incoming dNTPs, which are bound to the T4 DNA polymerase (43P). In this five-protein component system, leading strand synthesis can approach the rates estimated for *in vivo* DNA replication. In the T4 system, lagging strand synthesis requires

[1] A. Kornberg and T. A. Baker, *in* "DNA Replication," 2nd ed., pp. 113–196. Freeman, New York, 1992.

[2] M. D. Challberg and T. J. Kelly, *Ann. Rev. Biochem.* **58,** 671 (1989).

[3] B. Stillman, *Ann. Rev. Cell Biol.* **5,** 197 (1989).

[4] M. L. DePamphilis, *Ann. Rev. Biochem.* **62,** 29 (1993).

[5] M. C. Young, M. K. Reddy, and P. H. von Hippel, *Biochemistry* **31,** 8675 (1992).

[6] N. G. Nossal and B. M. Alberts, *In* "Bacteriophage T4," pp. 71–81. ASM, Washington, DC (1983).

[7] B. M. Alberts, J. Barry, P. Bedinger, T. Formosa, C. V. Jongeneel, and K. N. Kreuzer, *CSHSQB* **47,** 655 (1983).

[8] C. S. McHenry, *Ann. Rev. Biochem.* **57,** 519 (1988).

all of the components needed for the leading strand reaction but, in addition, a primase (61P) and a helicase (41P) are required.[6] In the seven-protein system, leading and lagging strand synthesis occurs in a coordinated fashion, implying an interaction between the replication protein complex that carries out leading strand synthesis with the complex that is responsible for lagging strand synthesis. For the most part, the so-called accessory proteins (44P/62P, 45P, 32P in the T4 DNA replication system) have only one function each, whereas the T4 DNA polymerase has two activities, a synthetic $5' \rightarrow 3'$ polymerizing activity and a $3' \rightarrow 5'$-exonuclease activity located in a single polypeptide chain of 898 residues. Other DNA polymerases such as pol I from *E. coli* have three enzymatic activities in a single polypeptide, a synthetic Pol activity, a $3' \rightarrow 5'$-exonuclease, and a $5' \rightarrow 3'$-exonuclease activity. Many other DNA polymerases from various sources also have several distinct enzymatic activities that reside in a single protein.

It has been well established that proteins with multiple activities are often organized so that each activity is located in a substructure of the protein termed a domain. Domains can be distinct compactly folded entities, which are connected by relatively unstructured "linker" regions that are exposed to solvent. In situations like this, it is possible to probe for the existence of domains using limited proteolysis.

This section deals with the use of partial proteolysis to detect and, in some cases, to obtain these substructures on a preparative scale from the parent protein. Most often, the results of limited proteolysis are used to identify and characterize fragments that can serve as a basis for cloning and expressing the relevant portion of the DNA polymerase cDNA encoding the fragment or domain of interest. I describe strategy, approaches, procedures for limited proteolysis then discuss methods for separation and characterization of the products. Finally I comment on interpretation and exploitation of the results so that they can be used for structural and biological studies. I present a specific example of the limited proteolysis of T4 DNA polymerase because it illustrates many of the principles of the approach.

General Considerations

Following Course of Proteolytic Digestion

If the DNA polymerase has more than one activity (exonuclease as well as polymerase, for example), follow the change in both of the activities as a function of digestion time. This should be done along with monitoring the course of digestion, which can best be accomplished by SDS–PAGE.

Amount and Purity of Polymerase Required

The limiting factor in determining the amount of DNA polymerase required is the sensitivity of the methods used for detection of the fragments. SDS–PAGE followed by staining with Coomassie Brilliant Blue is the technique most commonly used. With standard small size (10 × 7 cm) gels (Hoefer Scientific Co.), 0.05 to 0.1 μg of protein can be visualized. This is within a range that will permit testing of a number of different proteases and conditions requiring only a modest amount of the DNA polymerase. The DNA polymerase should be reasonably pure (80% or more) but the size distribution of contaminants is important. Obviously a single impurity amounting to 20% of the total could lead to problems in interpretation, whereas smaller amounts of contaminants with widely differing molecular weights would pose much less of a problem.

Selection of Protease

Accessibility and susceptibility to protease cleavage are the fundamental requirements for the success of this approach. Since there is no way of predicting which region of the protein will be solvent accessible without knowledge of the three-dimensional structure, the most practical way of dealing with this situation is to select proteases with the broadest specificity for preliminary experiments. In our experiments with many different systems, subtilisin Carlsberg has proven to be useful for the first screening experiments. Other proteases that recognize amino acid residues with charged side chains would be the next choice because they have a better chance of being solvent accessible, whereas hydrophobic residues have a greater probability of being hidden away in a tightly packed, folded structure. A survey should also include trypsin, chymotrypsin, and staphylococcal V8 protease because of the chance that the target polymerase might have residues that would be compatible with the cleavage specificity of these proteases. A list of commercially available proteolytic enzymes that can be obtained in pure form can be found in Table I along with the pH optima, conditions for storage, inhibitors, and the type of peptide bond that is cleaved.

Monitoring Rate and Extent of Cleavage

To follow the rate and extent of protease cleavage, the reactions should be set up so that the total reaction volume is slightly larger than the sum of the number of aliquots multiplied by the volume removed at each time point. To get a rough estimate of the course and the extent of the reaction, take (1) a zero time point without enzyme; (2) a 1-min time point with

TABLE I

USEFUL PROTEOLYTIC ENZYMES FOR LIMITED CLEAVAGE

Protease	Specificity[a]	pH optima	Storage conditions	Inhibitors[b]	Source (supplier)
Chymotrypsin	P_1 = W, Y, F P_1' = nonspecific	7.5–8.5 Ca^{2+}-activated	4° below pH 5 or −20°	DFP, PMSF, TPCK	Bovine pancreas (Sigma and others)
Elastase	P_1 = A, V, I, L, G, S, T P_1' = nonspecific	7.5–8.5	4° below pH 5	DFP, PMSF	Porcine pancreas (Calbiochem and others)
Endoproteinase Asp-N	P_1 = D P_1' = nonspecific	6.0–8.0	−70° do not freeze–thaw	EDTA 1,10-phenanthroline	*Pseudomonas fragi* (Boehringer)
Endoproteinase Glu-C	P_1 = E or D P_1' = nonspecific	7.8	4°	DFP, PMSF	*Staphylococcus aureus* V8 (Sigma and others)
Endoproteinase Lys-C	P_1 = Lys P_1' = nonspecific	8.5	4°	TFP, TLCK	*Lysobacter enzymogenes* (Fluka and others)
Pepsin	P_1 = nonspecific P_1' = nonspecific but cannot be V, A, G	2.0–4.0	4°	Pepstatin pH > 6	Porcine gastric mucosa (Sigma and others)
Proteinase K	P_1 = nonspecific P_1' = nonspecific	7.5–12	4°, pH 8 in 1 m*M* $NaCl_2$	PMSF, DFP	*Tritirachium album* (Sigma and others)
Subtilisin Carlsberg	P_1 = nonspecific P_1' = nonspecific	7.0–8.0	4° dry or −70°	PMSF, DFP	*Bacillus subtilis* (Sigma and others)
Thermolysin	P_1 = nonspecific P_1' = L, F, I, V, M, A	7.0–9.0	4° dry or −70°	EDTA	*Bacillus thermoproteolyticus* (Sigma and others)
Trypsin	P_1 = K, R P_1' = nonspecific cannot be Pro	8.5 20 m*M* Ca^{2+}	4° dry or −70°	DFP, TLCK, PMSF	Bovine pancreas (Sigma and others)

[a] P_1 ——↓—— P_1' (cleavage sites)

[b] DFP, diisopropyl fluorophosphate; EDTA, ethylenediaminetetraacetic acid; PMSF, phenylmethylsulfonyl fluoride; TLCK, tosyllysine chloromethyl ketone; TPCK, tosylamido-2-phenylethyl chloromethyl ketone.

enzyme; (3) a 5-min time point; (4) a 15-min time point; and (5) a 1-hr time point. Aliquots removed at each time point (5 to 10 μl) should be transferred to a small Eppendorf tube containing 5 μl of the appropriate inhibitor solution, which will immediately stop the digestion and inhibit the enzyme irreversibly. When all the aliquots have been removed, SDS–PAGE sample loading buffer (5 μl) is added to each tube, mixed thoroughly, boiled for 1 min, and loaded onto the SDS–PAGE gel. The small gels (Hoeffer) require less than 1 hr of running time. Staining with a fresh solution of Coomassie Brilliant Blue should take no longer than 15 min followed by destaining, according to standard procedures, with gentle agitation, which can be done in about 15 to 30 min. Destaining is accelerated if a Kimwipe and a small amount of anion-exchange resin is added to the destaining solution. Results from pilot experiments like these can be obtained in a few hours, enabling readjustment and refinement of conditions. For example, if a set of peptides appears quickly and then disappears as the digestion time is extended, lowering the concentration of the protease would be advisable before deciding against using the proteolytic enzyme used for the experiment. If no cleavage occurs after 1 hr at an enzyme-to-polymerase ratio of 1:100 then drop the proteolytic enzyme from the screen. If the amount of the DNA polymerase decreases with increasing digestion time but only small peptides appear, abandon the particular protease since the results indicate that digestion is proceeding in a "zipper" or "all or none" fashion without producing stable intermediates. If this happens with several proteases of widely different specificity, the chances are that it reflects the properties of the DNA polymerase substrate rather than the protease that was used. Sometimes proteins are so delicately poised that cleavage of a single peptide bond can destabilize the structure so that no stable domains can be observed. The addition of metal ions or glycerol can have a profound effect on the nature and extent of cleavage. All DNA polymerases that have been studied have metal ion requirements. These include Zn^{2+}, Mg^{2+}, and Mn^{2+}, which can substitute for both Zn^{2+} and Mg^{2+}. These metal ions have been shown to be necessary for both polymerization and exonuclease activity and their location in the three-dimensional structure of the Klenow fragment,[9,10] β polymerase (from rat),[11–13] and the

[9] L. S. Beese and T. A. Steitz, *EMBO J.* **10,** 25 (1991).

[10] V. Derbyshire, P. S. Freemont, and M. R. Sanderson, *Science* **240,** 199 (1988).

[11] H. Pelletier, M. R. Sawaya, A. Kumar, S. H. Wilson, and J. Kraut, *Science* **264,** 1891 (1994).

[12] M. R. Sawaya, H. Pelletier, A. Kumar, S. H. Wilson, and J. Kraut, *Science* **264,** 1930 (1994).

[13] J. F. Davies II, R. J. Almassy, Z. Hostomska, R. A. Ferri, and Z. Hostomsky, *Cell* **76,** 1123 (1994).

N388 fragment from T4 DNA polymerase[14] are known. Amino acid residues whose side chains act as donors are mainly Asp and Glu. The concentration of these metal ions can be drastically reduced by the addition of EDTA and the removal of metals may influence the outcome of the cleavage reaction. In any case, the inclusion or exclusion of these metal ions is another variable that can be adjusted if necessary. Glycerol will stabilize proteins in solution and it is almost always included in DNA polymerase storage buffers. In general, it tends to slow the rate of proteolytic digestion and, in the case of T4 DNA polymerase, it affects the yield of the 27-kDa fragment, as is discussed later. The addition of one of the substrates for the polymerase reaction, either dNTPs or primer–template tend to stabilize the DNA polymerase in a particular conformation and could therefore influence the course and outcome of limited proteolytic digestion. As with the addition of metal ions, or glycerol, the presence of substrates generally causes a reduction in cleavage rates. Ideally, cleavage should proceed in a stepwise sequential fashion to give a relatively simple pattern of fragment bands that can be visualized after SDS–PAGE and that remain stable after prolonged exposure to proteolytic enzymes under conditions that provide optimum stability for the polymerase. The sum of the molecular weights of the fragment bands remaining at the end of the digestion should add up to give the molecular weight of the undegraded polymerase. In practice this rarely happens because part of the protein is almost invariably destabilized and, as a result, is extensively cleaved producing small peptides that are not useful for the purpose of locating and isolating domains. As an example, in the course of chymotryptic digestion of T4 DNA polymerase, which is shown in Fig. 1, it can be seen that rapid cleavage of this polymerase leads first to an 80-kDa product, which then is split into 45- and 35-kDa fragments. The 45- and 35-kDa fragments appear to be present in equimolar amounts, a supposition that was later confirmed by quantitative amino acid analysis. Nearly all of the intact polymerase has disappeared by the end of the digestion period as has the 80-kDa intermediate (Fig. 1). We were able to show by Edman degradation that the 80-kDa fragment and the 45-kDa subfragment were derived from the NH_2-terminal region. The COOH-terminal region, from residues 698–898, never appeared as a discrete entity so we presume that it must have been relatively unstructured and therefore digested by chymotrypsin into small peptides that could not be detected on SDS–PAGE. Another minor band of 27 kDa appeared toward the end of the digestion. It was derived from an internal region of the 45-kDa

[14] G. Karam, T-C. Lin, and W. H. Konigsberg, *J. Biol. Chem.,* **269,** 19,286–19,294 (1994).

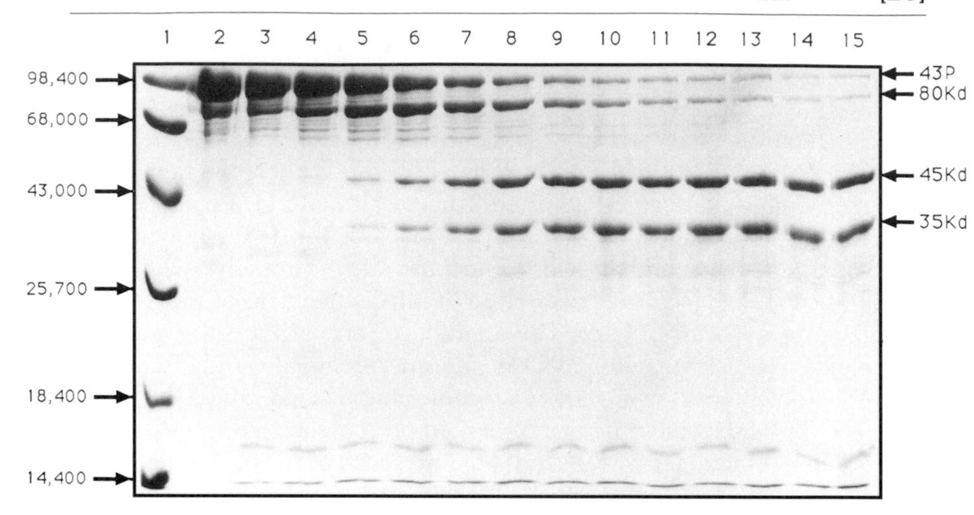

FIG. 1. SDS–PAGE of a chymotryptic digest of T4 DNA polymerase (43P) as a function of time. Samples of 43P were incubated with chymotrypsin at 37°. The 43P concentration was 0.5 mg/ml. The chymotrypsin to 43P ratio was 1:12 (w/w). The reaction was carried out in 50 m*M* Tris–HCl buffer, pH 8.0. Reactions were terminated by addition of PMSF and 2-mercaptoethanol at a final concentration of 2 m*M* each. This was followed by immediate immersion in dry ice–acetone. Aliquots of digested 43P were then subjected to SDS–PAGE in 12.5% gels. Lane 1, high molecular weight protein markers; lanes 2–15, aliquots of the digest after 0, 1, 5, 15, 30, 60, 90, 120, 150, 180, 210, 240, 270, and 300 min, respectively.

fragment and contained the $3' \rightarrow 5'$-exonuclease active site as is discussed later.

Separation of Proteolytic Fragments

As already mentioned, the products of limited proteolysis can be separated by SDS–PAGE if they differ sufficiently in molecular weight, but this procedure uses denaturing conditions. The information obtained from SDS–PAGE is useful for analytical rather than for preparative purposes. Furthermore, because there is no assurance that the fragments can be renatured after recovery from SDS–PAGE, checking for enzymatic activity using any of the fragments is problematical.

Once the conditions for the limited proteolysis have been established, the reaction can be scaled up without difficulty. The next problem is to isolate the fragments, preferably without denaturation, so that they can be tested for function. Conventional ion-exchange chromatography should be tried first since there is a high probability that the fragments have different iso-electric points. The choice of ion-exchange supports will depend on what separation systems are being used in the laboratory. With conventional open columns, use the Pharmacia Q-Sepharose (anion exchanger) or S-Sepharose (cation exchanger). If an FPLC system (Pharmacia) is available then Mono Q or Mono S columns should be tried. For laboratories that have access to HPLC, the QH and SH columns (Perspective Biosystems), which have high resolving power, are a good choice. For directions on how to run the various columns, it is best to follow the operating manuals from the supplier. In general, the pH is kept constant and the ionic strength is increased during the elution process. If partial separation is achieved, resolution can usually be improved by changing the slope and conditions of the gradient used for eluting the column. If the fragments fail to separate with conventional ion-exchange chromatography, hydrophobic chromatography can be attempted with a column of phenyl-Sepharose using a gradient of decreasing ionic strength. In this procedure, the sample, containing the fragments, can be precipitated with ammonium sulfate, then the precipitate redissolved, using an ammonium sulfate solution at a concentration just below the concentration that gave the initial precipitation, and the solution loaded on the column. A gradient of decreasing ammonium sulfate is used to elute the sample. If these methods fail to resolve the mixture of fragments, urea, usually at a concentration of 6 *M*, can be used together with the elution buffers and salt gradients in an attempt to separate the components. Sometimes all of these methods will fail if the fragments retain their conformation after limited proteolysis and, in addition, have high affinity for one another because of a large complementary surface area at their interface. This latter situation was what we found with the 80-, 45-, and 35-kDa fragments obtained by limited chymotryptic digestion of T4 DNA polymerase.[14] None of the ion-exchange or hydrophobic chromatography methods provided any separation despite the charge and size differences among the fragments. When this situation occurs, an alternative approach is required to isolate and characterize these domains, as is described below.

Characterization of Fragments, Cloning, and Expression of cDNA Encoding Domains

To obtain these fragments for further studies, we had to clone and express the cDNA encoding each of the fragments. To accomplish this, we had to know their location in the polypeptide chain. This information was obtained from the fragments that were separated by SDS–PAGE. The NH_2-terminal sequence was obtained by automated Edman degradation after transferring the fragments to polyvinylidene difluoride (PVDF) membranes. The COOH-terminal residue or residues can sometimes be obtained by carboxypeptidase digestion. Carboxypeptidase Y has the broadest specificity but sometimes carboxypeptidase A and B can also be used. Unfortunately, carboxypeptidase requires a substantial amount of material. Alternatively the fragment, after elution from SDS–PAGE and after removal of SDS, can be subjected to mass spectrometry, which gives a very accurate measure of the mass. These results, together with the NH_2-terminal sequence of the fragment and knowledge of the primary structures can be used to prepare the requisite oligonucleotide primers so that the corresponding cDNA region, from the parental cDNA containing vector, can be obtained by PCR and conventional cloning techniques.

In the case of the T4 DNA polymerase, the information obtained from sequencing and mass spectrometry allowed us to reconstruct the proteolytic cleavage pattern of the intact enzyme as shown in Fig. 2. Attempts to express the cloned cDNA corresponding to the 80- and 45-kDa fragments

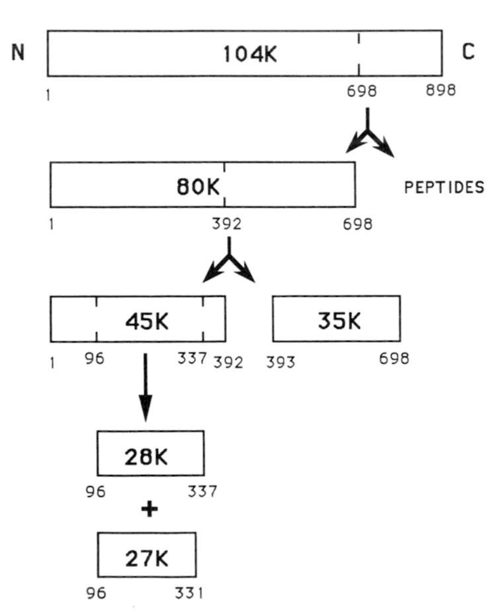

FIG. 2. Cleavage sites and product distribution resulting from limited proteolysis of 43P by chymotrypsin. The molecular weights of the products were based on comparison with known markers on SDS–PAGE. The orientation of the fragments and the assignment of the NH_2-terminal residues were based on NH_2-terminal sequencing of the fragments. Assignment of COOH-terminal residues of the fragments was based on the positions of known chymotrypsin-sensitive residues and by careful comparison of mobilities on SDS–PAGE.

were successful and provided high yields of these regions of the protein, which were then subjected to further study. Among the results, obtained from experiments performed with these cloned regions, was the surprising difference in their stability under the assay conditions used to detect polymerase or exonuclease activities.[14] As can be seen in Fig. 3, the exo activity of the intact polymerase decayed more rapidly than that of the cloned 80-kDa fragment (protein 698) or the cloned 45-kDa fragment (protein N388), which did not lose any detectable activity during the incubation period. This observation prompted us to try crystallizing the N388 protein. The crystallization attempts were successful and led to the structural determination of the N388 protein at a level of 2.4 Å (J. Wang *et al.,* unpublished results, 1994). The finding that protein subdomains are more stable and yield useful crystals more easily than the parent protein is a rather general observation and provides another impetus for exploring limited proteolysis as a way of obtaining more information about structure (T. Steitz, personal communication).

Attempts to express the 35-kDa fragment and the 27-kDa fragment from T4 DNA polymerase failed. Both of these fragments are located internally in T4 DNA polymerase, whereas the 80- and the 45-kDa fragments are derived from the NH_2 terminus of the native protein. Our current working hypothesis is that the NH_2-terminal regions of T4 DNA polymerase are necessary for the proper folding of regions of the protein that are located distal to the NH_2 terminus.

With regard to obtaining preparative amounts of the 27-kDa fragment, we found that higher yields of this fragment could be obtained if we started with the cloned 45-kDa fragment (protein N388) rather than the intact polymerase. This is probably due to protection of the scissile bond in the 45-kDa fragment by the 35-kDa fragment, which forms a tight complex with it in the chymotryptic digest of the native protein. The inclusion of 10% glycerol in the chymotryptic digest of the cloned 45-kDa protein improved the yield of the 27-kDa fragment.

Once the cloned cDNA encoding the individual fragments is expressed, there is still the problem of determining whether the protein produced by recombinant DNA techniques has the same conformation as the fragment released from the native protein by partial proteolysis. Several methods can be used to approach this problem. One strategy makes use of the method used to identify the domains in the first place. With the larger cloned fragments, such as the one corresponding to the 80-kDa fragment from T4 DNA polymerase, chymotrypsin was employed to demonstrate that the cloned 698 residue protein would give rise to the 45- and 35-kDa subfragments under the same conditions and at the same rate as observed when the intact polymerase was subjected to chymotryptic digestion. This

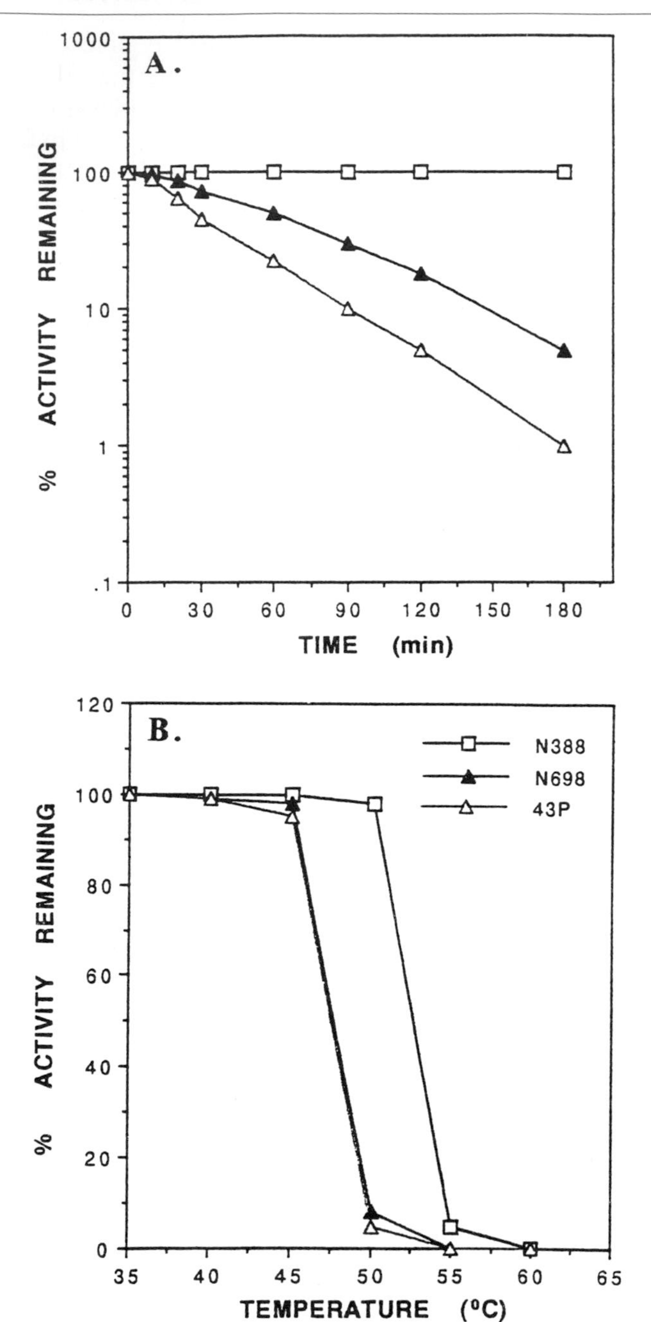
A.
1000
100
10
1
.1
% ACTIVITY REMAINING
0
30
60
90
120
150
180
TIME (min)
B.
N388
N698
43P
120
100
80
60
40
20
0
% ACTIVITY REMAINING
35
40
45
50
55
60
65
TEMPERATURE (°C)

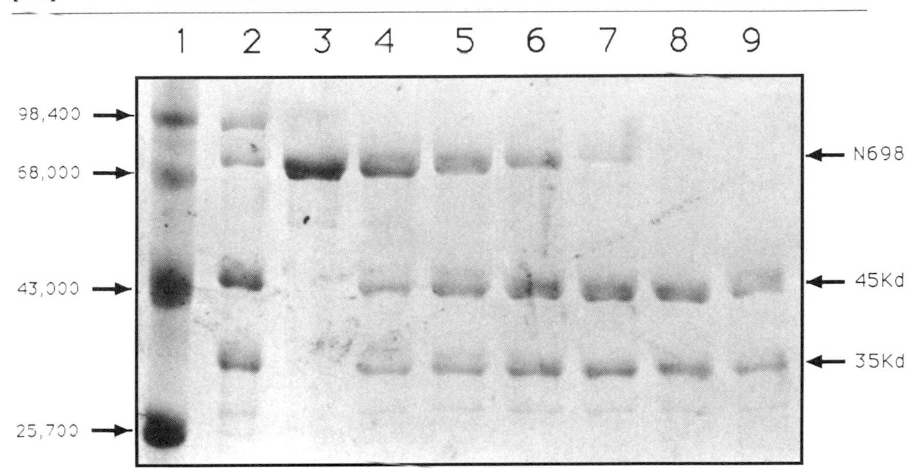

FIG. 4. SDS–PAGE of chymotryptic digest of protein N698 as a function of time. Protein N698 (the cloned 80-kDa fragment) was treated with chymotrypsin under the same conditions used for 43P. Samples were subjected to SDS–PAGE in 12.5% gels. Lane 1, high molecular weight protein markers; lane 2, limited digest of 43P; lanes 3–9, aliquots of protein N698 taken at 0, 15, 30, 60, 120, 180, and 240 min.

was indeed the case, as shown in Fig. 4. If the cloned 698 residue region has been misfolded, it might have ended up in inclusion bodies or been completely chewed up by the endogenous proteases present in *E. coli.* Even if the cloned 698 region survived expression in *E. coli,* it might still be less stable than the 80-kDa fragment generated by limited cleavage of the native polymerase. If this situation obtained, then one might expect the cloned fragment to be susceptible to complete cleavage by proteolytic enzymes and that the rate of cleavage would be greater than that of the native enzyme. This can be tested as a function of temperature; the less stable protein will start to unfold and be degraded more rapidly as the temperature is raised. Other physicochemical methods such as circular dichroism (CD) and fluorescence quenching can be used to compare the relative stabilities of the cloned and proteolytically derived fragments as a function of temperature. Finally loss of enzymatic activity as a function of time and temperature

---

FIG. 3. Relative stabilities of T4 DNA polymerase (43P, △), protein N698 (▲) and protein N388 (□). Thermal stabilities of 43P and its truncated derivatives were determined as follows: (A) The percent of 3′→5′-exonuclease activity retained by each protein, after being incubated at 30° in assay buffer but without substrate, was plotted against incubation time. (B) The percent of the 3′→5′-exonuclease activity retained by each protein, after being held for 5 min at a given temperature, was plotted against temperature.

can also be used to evaluate the folding and stability of the cloned versus protease-derived fragments as illustrated in Fig. 3.

Analysis of the functional properties of the isolated fragments should be carried out with several different substrates. This reason for this can best be explained by the results we obtained with T4 DNA polymerase. We knew that limited proteolysis completely destroyed the polymerase activity and that there was some $3' \rightarrow 5'$-exonuclease activity remaining at the end of the digest. The test substrate was $p(dT)_{16}$. The cloned 45-kDa fragment (protein N388) had only about 0.1% of the specific exo activity of the intact DNA polymerase when $p(dT)_{16}$ was the substrate. We then examined the $3' \rightarrow 5'$-exonuclease activity as a function of nucleotide length and found that the specific $3' \rightarrow 5'$-exonuclease activities of the intact polymerase, the cloned 80-kDa fragment (protein N698), and the cloned 45-kDa fragment converged when the substrate was $p(dT)_3$. The effect of substrate length on the rate of excision of the $3'$-terminal nucleotide is shown in Fig. 5 for these three proteins with the Klenow fragment included as a control. If $p(dT)_{16}$ had been the only substrate used to assay the exo activity of the fragments, we would have had difficulty interpreting the results. With the knowledge that the intact enzyme and the truncated forms of T4 DNA polymerase, including the 27-kDa fragment, had the same specific exo activity with $p(dT)_3$, we were able to assign confidently the region required for catalysis of phosphodiester bound hydrolysis to residues 96–388.[14] This example serves to illustrate the necessity of using carefully constructed assays before drawing conclusions about the relative activities of fragments obtained by limited proteolysis.

Other Examples Where Limited Proteolysis of DNA Polymerases Produces Fragmentation Products with Separate Activities

(1) Perhaps the first and most widely known example is that of Pol I from *E. coli,* which was split by subtilisin into two fragments, the NH_2-terminal region, which had $5' \rightarrow 3'$-exonuclease activity, and the COOH-terminal region, known as the Klenow fragment, which displayed $3' \rightarrow 5'$-exonuclease and polymerase activity.[15] As discussed in this section, chymotryptic digestion of T4 DNA polymerase yields an NH_2-terminal domain with $3' \rightarrow 5'$-exonuclease activity but the polymerase activity in the COOH-terminal region is lost.[14] (3) The DNA polymerase β from rat liver can be split into an 8- and a 31-kDa fragment with the 31-kDa region retaining the polymerase activity.[16] (4) The 581 residue *E. coli* primase (product of

[15] H. Klenow, K. Overgaard-Hansen, and S. A. Patkar, *Eur. J. Biochem.* **22,** 371 (1971).

[16] A. Kumar, S. G. Widen, K. R. Williams, P. Kedar, R. L. Karpel, and S. H. Wilson, *J. Biol. Chem.* **265,** 2124 (1990).

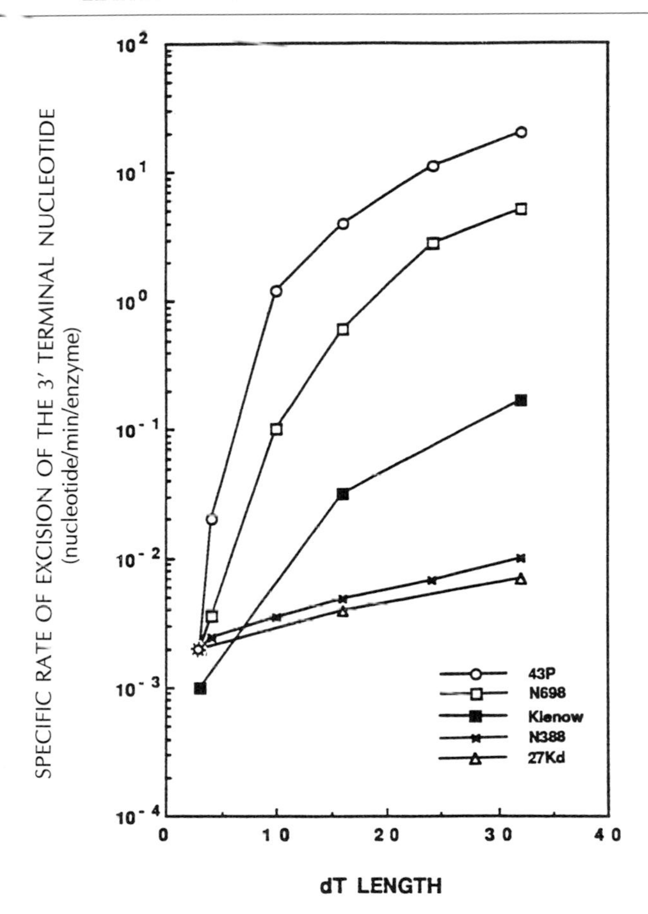

FIG. 5. Semilogarithmic plot of the specific initial rates of excision of the 3′-terminal nucleotide by T4 DNA polymerase, various 43P derivatives, and the Klenow fragment from oligo(dT) substrates as a function of substrate length: 3, 4, 8, 16, 24, and 32 nucleotides. The * represents convergence of rates of excision by 43P and its derivatives at $p(dT)_3$.

the *dnaG* gene), which is part of the DNA replication apparatus, has been cleaved by endoproteinase Asp-N into large fragments.[17,18] One of these, a 47-kDa fragment from the NH_2 terminus can synthesize pRNA in an *in vitro* system.[17]

Sometimes cleavage by intracellular proteases produces protein fragments with separate activities from DNA polymerases that were synthesized

[17] W. Sun, J. Tormo, T. A. Steitz, and G. N. Godson, *Proc. Natl. Acad. Sci. USA,* **91,** 11,462–11,466 (1994).

[18] K. Tougu, H. Peng, and K. J. Marians, *J. Biol. Chem.* **269,** 4675 (1994).

as single polypeptide chains. Reverse transcriptase from human immunodeficiency virus type 1 (HIV-I) is a well-studied case where intracellular cleavage trims one of the 66-kDa subunits to 15 kDa giving a 66/51-kDa heterodimer, which is the form that is found in the virus.[19] Another example, recently reported, is the α-like DNA polymerase from a thermoacidophilic archaeon *Sulfolobus solfataricus.*[20] This enzyme, which is synthesized as a 100-kDa polypeptide, has a site that is sensitive to endogenous proteases. During isolation of the polymerase, two proteolytic fragments of 50 and 50-kDa are observed in addition to the intact 100-kDa enzyme. The 50- and 40-kDa fragments are derived, respectively, from the COOH- and NH_2-terminal regions of the 100-kDa protein. The 50-kDa fragment has the DNA polymerizing activity whereas the 40-kDa fragment has the $3' \rightarrow 5'$-exonuclease activity.[20] This arrangement, where the $3' \rightarrow 5'$-exonuclease activity is in the NH_2-terminal half of the molecule and the polymerase activity is located in the COOH-terminal half, appears to be a feature common to all DNA polymerases that have both these activities in a single polypeptide.

Summary

Various proteolytic enzymes have been used to probe for domains in DNA polymerases. Results with several DNA polymerases that have been subjected to partial proteolysis demonstrated that there is a modular organization with different activities located in separate domains. In the case of the Klenow fragment, these domains appear to be independent of each other.[21] With other DNA polymerases, the question of modular independence is not settled. Limited proteolysis for probing structure has been used with many other proteins in addition to DNA polymerases and the information obtained has been helpful in interpreting function–structure relationships.[22] It is a general approach and can be applied in situations where the existence of domains is suspected. The simplicity of the method and the ease of monitoring the outcome is probably the main reason for its widespread and increasing use in enzymology.

[19] B. Muller, T. Restle, H. Kuhnel, and R. S. Goody, *J. Biol. Chem.* **266,** 14,709 (1991).

[20] R. Pisani and M. Rossi, *J. Biol. Chem.* **269,** 7887 (1994).

[21] C. M. Joyce and T. A. Steitz, *Ann. Rev. Biochem.* **63,** 777 (1994).

[22] N. C. Price and C. M. Johnson, *in* "Proteolytic Enzymes, A Practical Approach" (R. J. Beynon and J. S. Bonds, eds.), pp. 163–180. IRL Press, New York, 1989.

[27] Assays for Retroviral Reverse Transcriptase

By ALICE TELESNITSKY, STACY BLAIN, and STEPHEN P. GOFF

Introduction

The enzyme responsible for retroviral DNA synthesis, reverse transcriptase (RT), is both a DNA polymerase and a nuclease (for a comprehensive review, see Ref. 1). The DNA polymerase activity of RT is unique in its ability to catalyze DNA synthesis on both RNA and DNA templates; the ability to copy an RNA template to form DNA is the source of the enzyme's name, and is in large part responsible for the unusual life cycle of the retroviruses. The associated nuclease activity, termed ribonuclease II (RNase H), degrades the RNA strand of RNA : DNA hybrid duplexes. Although the RNA-dependent DNA polymerase activity of retroviral RTs has attracted the most attention, both polymerase activities and the RNase H activity are important in the generation of retroviral DNA.

Retroviral RTs are standard DNA polymerases in many respects: They use deoxynucleoside triphosphates to extend a growing 3′-OH end of a primer, which must be annealed to a template. The enzymes can use either DNA or RNA primers. There are some qualitative differences, however. Compared with host DNA polymerases, RTs show poor fidelity, misincorporating bases at fairly high frequency; and they do not show proofreading activities. In addition, the polymerase activities of RTs show generally low processivity. The RNase H activity of RT degrades RNA either in RNA : DNA hybrid or RNA : RNA duplex form, and can act either as an endonuclease or an exonuclease (Refs. 3, 4; for review, see Refs. 1–2). The products of extended digestion are short oligonucleotides with 5′-PO_4 and 3′-OH groups.

It has become important to a growing array of molecular biologists, virologists, and clinicians to perform assays for the various activities of retroviral RTs. In this chapter, we summarize some of the popular assays for RT and their use in quantifying levels of viral particles, viral RNA templates, and mutant RTs with altered polymerase and nuclease activities. In the first section, we describe a simple homopolymer-based assay for RT

[1] A. M. Skalka and S. P. Goff, eds., *Reverse Transcriptase.* Cold Spring Harbor Laboratory Press, New York, 1993.

[2] A. Jacobo-Molina and E. Arnold, *Biochemistry* **30,** 6354 (1991).

[3] H. Ben-Artzi, E. Zeelon, M. Gorecki, and A. Panet, *Proc. Natl. Acad. Sci. USA* **89,** 937 (1992).

[4] S. W. Blain and S. P. Goff, *JBC* **268,** 23,585 (1993).

DNA polymerase activity that can be used to detect retroviruses in the culture media of infected cells, or alternatively to detect or quantify RT during its purification. In the second section we describe the so-called "endogenous assay," a method to analyze the products of reverse transcription in purified virions. In the third section, we present a method for screening libraries of bacterially expressed RT mutants as a means of isolating particular variant enzymes displaying selected properties. In the final section, we present an *in situ* method that allows RT RNase H activity to be assayed separately from contaminating cellular RNase H activities.

Assay I. DNA Synthesis on Homopolymers: "Exogenous" DNA Polymerase Assay

We describe here a rapid assay for RT DNA polymerase activity, essentially unchanged from the original description of the method for detection of the Moloney murine leukemia virus enzyme.[5] This assay uses synthetic homopolymeric polyriboadenylic acid [poly(rA)] as a template, and oligodeoxythymidylic acid [oligo(dT)] as primer. Samples are incubated with this primer–template and α-radiolabeled dTTP; the resulting dTMP incorporation is monitored by spotting reaction aliquots onto DEAE paper and washing away unincorporated dTTP. When performed in a 96-well tray as described here, the end product of the assay is a piece of DEAE paper with a grid of radioactive dots that can be visualized by autoradiography. RT levels can be estimated by visual inspection of the dots' relative intensities on autoradiograms, and/or quantified by scintillation counting or beta scanning of the DEAE paper.

Assaying tissue culture media for RT activity is a sensitive way to detect retroviruses. We use this assay for such purposes as screening large numbers of candidate virus-producer cell lines, monitoring viral spread, and roughly quantifying levels of virus production. Tissue culture medium is added directly to the reaction cocktail; detergent in the cocktail lyses the virions and allows the viral RT to use exogenously added template. We routinely prepare a cocktail containing reaction components at 1.2 times their final concentrations: Adding 10 μl culture media to 50 μl of 1.2$\times$ reaction cocktail yields a final 1$\times$ concentration. The presence of serum and other components of the culture medium does not generally inhibit the reaction at this concentration (although rare batches of serum have been found to cause problems). Culture supernatants can be assayed immediately upon harvesting or else frozen at $-20°$ for later use. Limited thawing and refreezing does not significantly affect RT activity in these media samples. An

[5] S. Goff, P. Traktman, and D. Baltimore, *J. Virol.* **38,** 239 (1981).

example of the use of the method to follow the time course of virus replication is shown in Fig. 1.

Method

1. Prepare enough 1.2× reaction cocktail to provide 50 μl for each sample. The volumes below are sufficient for 20 samples.

	1.2 × cocktail
1 *M* Tris–HCl, pH 8.3	60 μl
0.1 *M* $MnCl_2$	7
5 *M* NaCl	15
10% Nonidet P-40 (NP-40)	6
1 mg/ml oligo(dT)	6
1 mg/ml poly(rA)	12
2 m*M* dTTP	6
500 m*M* DTT (dithiothreitol)	48
[α-^{32}P] dTTP (10 mCi/ml; 400 to 800 Ci/mmol)	1
Distilled H_2O	840 μl
	1 ml

2. For each sample to be assayed, place 50 μl cocktail into the well of a 96-well plate. Add 10 μl of sample to be assayed (for example, tissue culture medium from infected cells).
3. Cover plate to minimize evaporation and incubate at 37° for 1 hr.

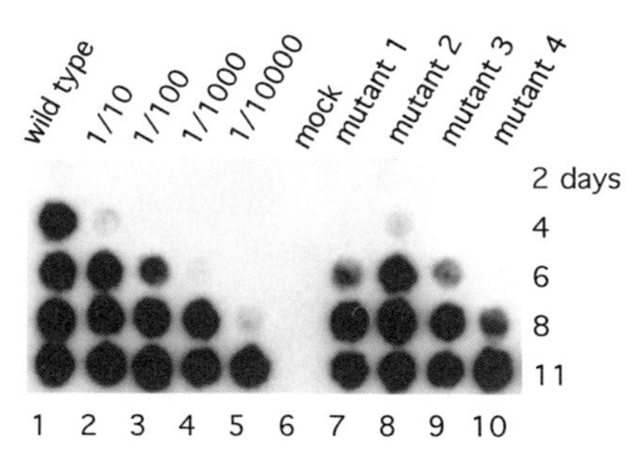

FIG. 1. RT homopolymer assay to detect rate of viral spread by wild-type and mutant Moloney murine leukemia viruses. NIH 3T3 cells were infected with wild-type virus (lane 1), the indicated dilutions of wild-type virus (lanes 2–5), mock infected (lane 6), or infected with various mutant viruses at the same titer as wild type in lane 1 (lanes 7–10). At the postinfection time points indicated at right, media samples were collected and the RT homopolymer assay was performed in a microtiter plate as described. Reproduced from Ref. 28, with permission of Oxford University Press.

4. Cut a piece of DEAE paper and place it on a flat Styrofoam block covered with plastic wrap. A piece of paper 14 × 20 cm is large enough to accommodate two replicas of a 96-well plate.
5. Spot aliquots of reaction mixtures (usually 1 to 10 μl) onto dry DEAE paper using a replica plater or a multichannel pipette. See practical note below. Spot a duplicate set of aliquots onto the same piece of paper.
6. Wash the paper three times in 2× SSC (300 m*M* NaCl, 30 m*M* sodium citrate, pH 7) by placing the dry paper in a shallow dish with 2× SSC approximately 1 cm deep. Shake gently on a rotary shaker for 5 min at room temperature, then change the 2× SSC twice. The final two washes are 15 min each.
7. Rinse the DEAE paper briefly with 95% ethanol twice. Blot the paper and dry under a heat lamp or with a hair dryer.
8. Cover paper with plastic wrap and expose film at −70° with an intensifying screen 4 to 16 hr.

Sources of Materials

96-well plates: round-bottom wells, nonsterile microtest plates, Sarstedt
DEAE paper: DE-81, Whatman, Inc.
Poly(rA): Sigma
Oligo(dT): pd$(T)_{12-18}$, Pharmacia, approximately 20 A_{260} units/mg
Replica plater: Sigma-Aldrich Techware; see alternatives described below.

Practical Notes and Points to Consider

The method presented here readily detects murine retroviruses present in tissue culture medium at levels as low as 1% or less of wild-type virus levels. Lower virus titers may be detectable by concentrating culture supernatants by ultracentrifugation or polyethylene glycol precipitation prior to assay. Considerations in the detection of virus are the host cell line and the condition of the cells. For example, culture medium of uninfected Rat2 cells generates a significant background signal, equal to about 1% of the level of wild-type virus, whereas the background level of NIH 3T3 cells is much lower. It is preferable to assay the media of healthy, confluent cells, since background levels increase and virus production decreases in postconfluent cells. Maximal levels of virus in the medium from most cells are achieved about 12 hr after the last addition of fresh medium.

The RTs of various retroviruses have distinctive preferences for templates, salts, pH, and divalent cation. Thus, whereas 1.2× concentrations for the reaction cocktail given above are appropriate for RTs such as

Moloney murine leukemia virua (M-MuLV), the reaction conditions should be changed for other viral RTs. A suitable 1.2× reaction cocktail for the human immunodeficiency virus type 1 (HIV-1) RT would include 60 mM Tris–Cl, pH 8.0, 12 mM $MgCl_2$, 72 mM NaCl, 0.06% NP-40, 24 mM DTT, 6 μg/ml oligo(dT), 12 μg/ml poly(rA), and 12 μM [α-^{32}P] dTTP (specific activity: 1 Ci/mmol). The HIV-1 enzyme can use a oligo(dG) · poly(rC) substrate essentially as well as the oligo(dT) · poly(rA) substrate.

This assay can be adapted for purposes other than virus detection. Suitably diluted fractions can be assayed to monitor RT during purification, or to quantify the enzyme.[6] In these assays, one unit of RT is typically defined as the amount of enzyme that will incorporate 1 nmol of [^{32}P]dTMP into a form retained on DE81 paper in 15 min at 37°. Note that some other polymerases, such as *Escherichia coli* DNA polymerase I (Pol I), can also incorporate dTMP using a poly(rA) · oligo(dT) primer–template, albeit at low efficiency. When expressing murine RT in *E. coli* with our wild-type M-MuLV RT producer plasmid, pRT30-2,[7] about 5% of the apparent RT activity in crude extracts is due to Pol I. Pol I does not use poly(rC) · oligo(dG), and hence this primer–template can be used in assays for HIV RT to give a more specific readout of the retroviral enzyme. The activity of M-MuLV RT is much greater on poly(rA) · oligo(dT) than on poly(rC) · oligo(dG), and hence the mouse enzyme is best assayed using the former primer–template.

A convenient way to estimate RT expression in bacterial crude extracts is to use a very simple method for the extraction of soluble RT from *E. coli.*[8] Briefly, this procedure involves washing cell pellets from 1 ml of late log to stationary cultures with ice cold buffer (100 mM NaCl, 20 mM Tris–Cl, pH 7.4, 1 mM EDTA). Bacterial pellets are then suspended in 250 μl of lysis buffer (200 mM NaCl, 25 mM Tris–Cl, pH 8, 2 mM DTT, 1% Triton X-100, 1 mM EDTA, 20% glycerol) and incubated on ice for 15 min. Insoluble material is removed by centrifugation in a microfuge and the supernatant is either assayed directly or stored at −20°. We typically perform 10- or 15-min assays with several fivefold serial dilutions of lysates from experimental and control cultures in parallel, in order to assess relative levels of expression and to determine the appropriate dilutions needed to ensure that the assay is in the linear range.

Stock mixtures of stable RT reaction components can be prepared, and stocks of more labile components can be stored at −20°. For example, buffer, salts, and detergent can be combined and stored at room tempera-

[6] M. J. Roth, N. Tanese, and S. P. Goff, *J. Biol. Chem.* **260,** 9326 (1985).
[7] A. Telesnitsky, S. W. Blain, and S. P. Goff, *J. Virol.* **66,** 615 (1992).
[8] A. Hizi, C. McGill, and S. H. Hughes, *Proc. Natl. Acad. Sci. USA* **85,** 1218 (1988).

ture indefinitely. Store the following stock solutions at −20°: 1 mg/ml oligo(dT), 1 mg/ml poly(rA), 2 m*M* dTTP, and 500 m*M* DTT. Alternatively, reaction cocktail complete except for [^{32}P]dTTP can be aliquoted and stored for a few months at −20°.

To spot samples onto DEAE paper, place the prongs of the replicator into the 96-well plate and then press the replicator down on the DE81 paper for about 10 sec. Return the replicator to the 96-well plate and make a second imprint on the same sheet of DEAE paper. Creating uniform assay dots may take a bit of practice but is not difficult. Traces of product remaining on metal replica platers from a previous assay can be degraded by flaming. First, clear the area of flammable materials. Dip the replica plater into 95% ethanol, move or cover the ethanol, and then ignite the replica plater. Cool the replica plater half an hour before using.

A significant drawback of using the commercially available replica platers is that a very small volume of fluid (about 1 μl) is transferred, and hence the signal will be lower than if a larger volume were applied to the paper. In our laboratory, we use a hand-made replicator that has flat-ended stainless steel screws in place of the smooth prongs of the replica plater described here. Screws allow the transfer of a larger volume (about 10 μl); however, to our knowledge, such a replicator must be custom-made. An alternative way of making larger dots is to use a multichannel pipette set at 5 or 10 μl to transfer the samples to DEAE paper.

Assay II. DNA Synthesis in Virion: "Endogenous Reaction"

In many settings, it is important to monitor the ability of RT to carry out the various steps of reverse transcription on the normal substrate for the enzyme, the viral RNA genome (for a detailed summary, see Ref. 9). Two strands of DNA must be formed in this reaction: the first (minus) DNA strand copied from the RNA, and the second (plus) DNA strand, copied from the minus strand. The steps required for the minus strand DNA include the initiation of DNA synthesis at the tRNA primer; the formation of a short intermediate termed "minus strand strong stop DNA" by elongation to the 5′ end of the RNA template; and further elongation only after a translocation or "jump" of the minus strand strong stop DNA to the 3′ end of the RNA template. The steps required for the plus strand synthesis include the generation of the RNA primer (a polypurine), by RNase H action on the genomic RNA; the formation of an intermediate termed "plus strand strong stop DNA" by elongation of the primer to the 5′ end of the minus strand; and finally, the completion of the plus strand

[9] E. Gilboa, S. W. Mitra, S. P. Goff, and D. Baltimore, *Cell* **18,** 93 (1979).

after its "jump" to the 3′ end of the minus strand. All of these steps are mediated by RT action on the viral genome.

Reverse transcriptase normally initiates this complex program of DNA synthesis when the core of a viral particle enters the cytoplasm of an infected cell. However, in the so-called "endogenous reaction," retroviral DNA synthesis can be artificially initiated in purified virions by permeabilizing them with detergent and providing nucleotide substrates.[9–11] Unlike the homopolymer assay, which uses exogenously added template, the products of reverse transcription in this assay are templated by the encapsidated viral RNA. Completed duplex DNA is only very inefficiently produced, but such intermediate products as minus strand strong stop DNA, and to a lesser extent, plus strand strong stop DNA, are readily detected. The products are normally analyzed by providing labeled triphosphates during the synthesis, and displaying the DNAs by gel electrophoresis and autoradiography.

The protocol presented here is optimized for murine retrovirions.[9] Endogenous reaction conditions for HIV-I have also been developed.[12]

Method

Preparation of Virus. Virus is concentrated by centrifugation from culture media collected from producer cell lines. To maximize recovery of virus, the cells should be subconfluent or just barely confluent. Repeated harvests can be collected from a given cell line: 40 to 50% of the normal amount of culture media can be used, starting from when the cells are approximately 75% confluent and harvesting the media at several successive 12-hr time points. The medium is cleared by filtration (0.45-μm filters) or low-speed centrifugation, and the virus is collected by ultracentrifugation (20,000 rpm, 1 to 4 hr, 4° in an SW28 rotor) and resuspended in TNE buffer (100 m*M* Tris–HCl, pH 8.3, 100 m*M* NaCl, 1 m*M* EDTA). The virus can be further purified by centrifugation through a sucrose cushion (25% sucrose in TNE buffer). Because some serum proteins are also concentrated in this procedure, we find it useful to use a low-serum synthetic media supplement when harvesting virus. The use of 10% Nuserum (Collaborative Research) in place of calf serum decreases the viscosity of the final virus preparation. Virions are stored in TNE buffer at −70°. Although infectivity is significantly reduced, the RT levels are quite stable to storage and repeated freeze–thaw cycles.

[10] E. Rothenberg and D. Baltimore, *J. Virol.* **17,** 168 (1976).
[11] E. Rothenberg and D. Baltimore, *J. Virol.* **21,** 168 (1977).
[12] K. Borroto-Esoda and L. R. Boone, *J. Virol.* **65,** 1952 (1991).

Endogenous Reaction Conditions

1. Typically 20 μl thawed virus is mixed gently with 50 μl reaction buffer (50 m*M* Tris–HCl, pH 8.3, 50 m*M* NaCl, 6 m*M* $MgCl_2$, 0.01% Nonidet P-40, 1 m*M* DTT, dATP, dCTP, and dGTP each at 2 m*M*, 1 m*M* [α-^{32}P] dTTP at 1 Ci/mmol), and incubated at 40° for 2 to 12 hr.
2. Reactions are stopped by the addition of 2 μl 500 m*M* EDTA and 10 μl 10% SDS, 0.5 μg carrier tRNA and 4 μg proteinase K are added, and the proteins are digested by incubation at 37° for 30 min.
3. The reaction mixtures are phenol extracted, ammonium acetate is added to a final concentration of 1 *M*, and the nucleic acids are precipitated with 2.5 volumes ethanol.
4. The nucleic acid pellets are suspended in 20 μl 0.33 *N* NaOH and incubated at 55° for 20 min to degrade residual primer RNA. Additional salt and 2.5 volumes ethanol are added to reprecipitate the DNA. Pellets should be washed with 70% ethanol and care should be taken in removing all traces of fluid if additional enzymatic steps are to be performed.
5. Samples are dried and suspended in TE, electrophoresis sample buffer is added, and samples are heated to 90° for 10 min prior to analysis on 7.5% denaturing polyacrylamide gels. After electrophoresis, the gels are dried and exposed by autoradiography. Typical gels reveal the formation of minus strand strong stop DNA and longer minus strand products (Fig. 2). The products of extended reactions include plus strand strong stop DNA and full-length molecules.

Protocol Variations

If DNA synthesis is to be limited to RNA-dependent DNA synthesis, actinomycin D should be included during DNA synthesis to a final concentration of 100 μg/ml.

The concentration of nucleotides must be high for efficient synthesis of longer products in the endogenous reaction. When reactions are carried out to detect minus strand strong stop DNA alone, the concentration of nucleotides can be reduced: dATP, dCTP, and dGTP can be present at 1 m*M* each, and [α-^{32}P] dTTP (400 Ci/mmol) can be present at 2.5 μ*M*. Incubation time can also be reduced to 10 to 30 min.

The most critical components of endogenous reactions are the concentrations of the magnesium ion[10,11] and the detergent (Nonidet P-40). For long DNA synthesis, and especially for long plus strand synthesis, the NP-40 concentration must be close to 0.01%. There is some batch-to-batch variation, however, and individual lots of the detergent may have to be titered to determine the optimal concentrations. When only minus strand

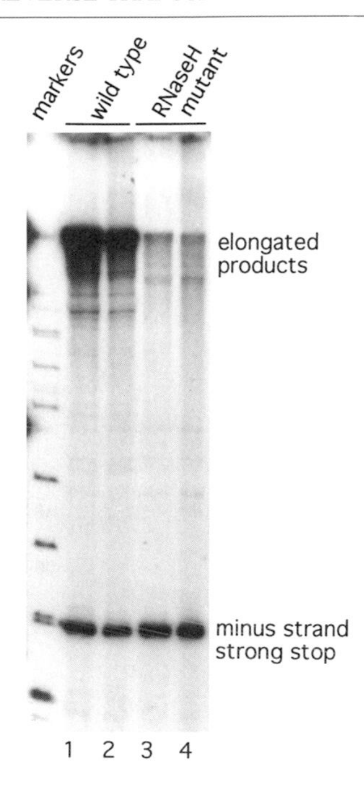

FIG. 2. Endogenous assay of products generated in wild-type and mutant M-MuLV virions. Reactions were performed as described in the text. Lanes 1 and 2, Wild-type virions; lanes 3 and 4, RNase H mutant virions; The migration positions of minus strand strong stop DNA and of elongated products are indicated. Note that fewer elongated products are present in the mutant virions' lanes, reflecting the requirement for RNase H for the generation of long reverse transcription products.

strong stop is to be analyzed, the NP-40 concentration is not as critical, and indeed higher concentrations (up to 0.05%) work well.

As we present it here, this assay allows the examination of some of the products of reverse transcription generated on endogenous viral templates. These reverse transcription products can be further manipulated to monitor the efficiency of specific substeps of the process of reverse transcription, such as the removal of the primer by RNase H, or the extent of the minus strand transfer reaction. Altering gel types—for example, using alkaline agarose gels in place of acrylamide gels—may permit a more accurate assessment of longer products. Because premature chain termination and other "weak stop" products may be prominent among endogenous reaction products, it is essential to run DNA size standards in parallel.

"tRNA primer tagging" is another variation of the endogenous reaction. In this assay only one or two deoxynucleotides are included, specifically those that are incorporated as the very first bases added to the tRNA primer for minus strand DNA synthesis. (For M-MuLV, these bases are A and T.) The result is the labeling of the 3′ end of the tRNA. The product of the reaction can be analyzed by electrophoresis on acrylamide gels, and readily quantified by autoradiography. The assay provides a quick readout of whether genomic RNA was properly packaged into virions, whether the tRNA primer was properly annealed to the genome, and whether RT was properly positioned to initiate synthesis. It is useful when a mutant RT has low processivity and cannot make complete strong stop DNA; it is also useful to help identify the tRNA serving as the primer in a given virus.

Assay III: DNA Polymerase Assay in Bacterial Colonies *in Situ*

In many settings, it is helpful to have a very high-throughput screen for the DNA polymerase activity of recombinant RTs expressed in bacterial colonies. Our group developed an *in situ* colony screening method to detect RT mutants in such a screen. The initial application of this assay was to identify a bacterial colony expressing an active revertant of a defective mutant of the HIV-1 RT[13]; it has subsequently been used to identify a dideoxyguanosine triphosphate-resistant mutant of HIV RT from a library of mutagenized RT expression plasmids,[14] and should be applicable to identifying mutants of various RTs with other selected properties. The general experimental scheme is as follows: Mutagenize an RT gene in a bacterial expression plasmid, grow a library of bacterial colonies containing the mutagenized gene, replicate the colonies to nitrocellulose filters, lyse the colonies and immobilize proteins *in situ,* incubate the filters in RT reaction cocktail containing radiolabeled substrate (see "homopolymer assay" section above), wash filters, and expose for autoradiography. The composition of the RT cocktail can be varied according to the enzyme property to be assayed. For example, to screen for the dideoxyguanosine triphosphate-resistant RT mutant, ddGTP was added to the reaction cocktail, and the relative incorporation signals of various colonies were compared on autoradiograms.

Method

1. Establish a library of HIV-1 RT mutants in bacteria. Our laboratory has had good results using both the chemical mutagen, EMS, and a *mutD5* mutator strain of *Escherichia coli* to generate mutations. Most of our experience has been with the nonmutator strain of *E. coli,* HB101, as host in the colony screen itself.

[13] V. R. Prasad and S. P. Goff, *J. Biol. Chem.* **264,** 16,689 (1989).

[14] V. R. Prasad, I. Lowy, T. de los Santos, L. Chiang, and S. P. Goff, *Proc. Natl. Acad. Sci. USA* **88,** 11,363 (1991).

2. Grow bacteria at a density of approximately 10^3 colonies/10 cm agar plate until the colonies are visible but still <1.0 mm in diameter.
3. Place nitrocellulose filter circles on colonies. After 10 min, peel filters off these reference plates and place filters, colony side up, on fresh agar plates containing any drugs required to induce RT expression (typically IPTG or other inducers, depending on the promoter). Incubate both the nitrocellulose-containing and the reference plates an additional 6 to 8 hr. Store reference plates at 4°.
4. Place filters, colony side up, on 0.4-ml drops of lysozyme (20 μg/ml) in TEND buffer (50 m*M* Tris–Cl, pH 8, 0.5 m*M* EDTA, 300 m*M* NaCl, 1 m*M* DTT) on plastic wrap for 30 min at 4°.
5. Place filters on 1-ml drops of 0.4% NP-40 in TEND buffer for 15 min at room temperature. For these two steps, care should be taken to prevent the fluid from flowing over the filters and dislodging the colonies.
6. Place filters on Whatman 3MM paper and expose to long-wave length UV light (such as from a hand-held Mineralight lamp model UVGL-25) for 10 min. Of several methods of fixing proteins that were tried, brief exposure to UV was found to yield the highest signal with the lowest background.
7. Soak filters at room temperature for 3 hr in 3 changes of 0.08% bovine serum albumin (BSA) in 30 m*M* HEPES, pH 7.5, to saturate nonspecific binding sites. The filters will be coated with gelatinous material from the lysate bacteria, and unbound bacterial products will begin to slough off.
8. Remove bacterial debris by washing filters at 4° overnight in 50 m*M* Tris–Cl, pH 7.5, 1 m*M* EDTA, 200 m*M* NaCl, 2 m*M* DTT, 10% glycerol, and 0.1% NP-40.
9. Remove filters from wash slowly, allowing the excess glycerol to drain. Place the filters in RT reaction cocktail without nucleotides. It is convenient to do this by placing the filters in petri plates containing about 0.5 ml cocktail per filter. Shake at room temperature 30 min.
10. Transfer filters into a similar volume RT cocktail with nucleotides. Incubate 30 min at 37° with shaking.
11. Fix the filters with three 10-min washes in cold 10% trichloroacetic acid (TCA). Air dry and expose to film for 3 to 4 hr.

Practical Notes

The most important issues for the successful use of the assay seem to be the linking of the protein to the filter after lysis, and blocking nonspecific binding of the label. We have tried several methods to fix the proteins to the filter and have had the most success with UV cross-linking; organic solvents were not as useful. Repeated soaking in albumin seemed effective in blocking background. During the reaction, much of the DNA product is lost from the filter and is found in solution; we have not found any easy way to prevent this loss of signal.

A number of expression constructs have been successfully used with this assay, but the level of expression probably needs to be very high. Most of our experience has been with *trpE*–HIV-1 RT fusions, and we obtain better signals with this parental construct than with nonfused RT proteins. The *trpE* portion of the fusion protein may be helpful in keeping the protein trapped in the lysed colony before cross-linking. For some enzymes, we have found that exposing the filter to 8 *M* guanidinium hydrochloride in TEND buffer for 1 hr, followed by an incubation in TEND buffer for 24 hr, yielded higher signals.

Assay IV: *In Situ* Gel Assay for RNase H Activity of RT

RNases H can be assayed most simply by their ability to convert the labeled RNA strand of an RNA:DNA hybrid into acid-soluble form. Unfortunately, it is difficult to assay retroviral RTs for their RNase H activity in this way: Very low levels of contaminating host RNases H—either from mammalian hosts or from bacterial cells used to express recombinant enzymes–can interfere with these assays. Tests of various preparations of retroviral RTs in many laboratories have suggested that even the most highly purified preparations often still contained residual *E. coli* nuclease activities.[3,4,7,15,16] To avoid this problem, we assay for RNase H activity *in vitro* using an *in situ* gel assay, which utilizes a radioactive RNA:DNA or RNA:RNA substrate cast into a standard SDS–PAGE gel. Following electrophoresis, the gel is soaked in a renaturation buffer, allowing the proteins to refold and degrade the substrate in the vicinity of their migration. This assay has been shown to be semiquantitative and linear with respect to the amount of extract or purified protein run on the gel.[4,17,18] The major advantage is that the activity of interest can be tested after separation of RT from contaminating activities based on its molecular weight. This is particularly beneficial when assaying RTs from crude bacterial extracts, because *E. coli* has numerous nuclease activities that complicate analysis

[15] H. Ben-Artzi, E. Zeelon, S. F. J. Le-Grice, M. Gorecki, and A. Panet, *Nucleic Acids Res.* **20,** 5115 (1992).

[16] Z. Hostomsky, G. O. Hudson, S. Rahmati, and Z. Hostomska, *NAR* **20,** 5819 (1992).

[17] N. Tanese and S. P. Goff, *Proc. Natl. Acad. Sci. USA* **85,** 1777 (1988).

[18] A. Hizi, S. H. Hughes, and M. Shaharabany, *Virology* **175,** 575 (1990).

in solution. RNases HI and II,[19,20] DNA Pol I,[21,22] exonuclease III,[21,23] and T7 phage gene 6 exonuclease[24] can all degrade RNA : DNA hybrids; RNase III degrades RNA : RNA duplex substrates.[4,16,25]

Method

Preparation of Labeled RNA : DNA Substrate. We first prepare single-stranded M13 phage by standard techniques, precipitating phage from bacterial culture medium by addition of 0.25 volume of 20% (w/v) polyethylene glycol (PEG) 8000, 2.5 *M* NaCl. The phage are resuspended TE and the DNA is extracted repeatedly with phenol and collected by precipitation with ethanol.[26] One milliliter of a phage stock yields roughly 1 to 10 μg of ssDNA. The RNA : DNA hybrid is prepared by using this M13 single-stranded circular DNA as a template and *E. coli* RNA polymerase holoenzyme in the presence of ribonucleotides, including [α-^{32}P]CTP. The product is an RNA : DNA hybrid in which the RNA strand is radioactively labeled. The quality of the single-stranded DNA is important: Old single-stranded DNA can serve as an effective template for RNA synthesis, but the product is not degraded as efficiently by RNase H as that made from new DNA. We have used *E. coli* RNA polymerase with and without sigma factor, from both Promega and Boeringher Mannheim, with comparable results.

The RNA synthesis reactions are carried out in a total volume of 250 μl of buffer consisting of 40 m*M* Tris–HCl, pH 8.0, 8 m*M* $MgCl_2$, 100 m*M* KCl, 2 m*M* DTT, 1 m*M* each of ATP, GTP, and UTP, and 150 μ*M* [α-^{32}P]CTP at 3 to 5 Ci/mmol. About 3 to 5 μg ss M13 DNA is added as template. The reaction is started by the addition of RNA polymerase (typically 10 to 20 units total) and allowed to proceed at 37° for 1 hr. The reaction is stopped by the addition of 10 μl 0.5 *M* EDTA, and the products are extracted twice with phenol and once with chloroform. The nucleic acid is recovered by addition of ammonium acetate to 2 *M* final concentration and precipitation with 2.5 volumes of ethanol. The product is resuspended in 75 μl of TE buffer.

To remove single-stranded radiolabeled RNA contaminating the

[19] S. Kanaya and R. J. Crouch, *Proc. Natl. Acad. Sci. USA* **81,** 3447 (1984).
[20] M. Itaya, *Proc. Natl. Acad. Sci. USA* **87,** 8587 (1990).
[21] W. Keller and R. Crouch, *Proc. Natl. Acad. Sci. USA* **69,** 3360 (1972).
[22] I. Berkower, J. Leis, and J. Hurwitz, *J. Biol. Chem.* **248,** 5914 (1973).
[23] B. Weiss, S. L. Rogers, and A. F. Taylor, *in* "DNA Repair Mechanisms" (P. C. Hanawalt, E. C. Friedberg, and C. F. Fox, eds.), p. 191. Academic Press, New York, 1978.
[24] K. Shinozaki and T. Okazaki, *Nucleic Acids Res.* **5,** 4245 (1978).
[25] R. J. Crouch, *J. Biol. Chem.* **249,** 1314 (1970).
[26] J. Sambrook, E. F. Fritsch, and T. Maniatis, "Molecular Cloning: A Laboratory Manual," 2nd ed. Cold Spring Harbor Laboratory Press, New York, 1989.

RNA:DNA hybrid, the samples are treated with RNase A in high salt. The hybrid is digested in a total volume of 200 μl of buffer (50 m*M* sodium acetate, pH 5.0, 0.3 *M* NaCl) using 20 μg/ml RNase A. The reaction is incubated for 30 min at room temperature. Following this step, carrier tRNA is added (1 μg), and the sample is extracted two to three times with 0.3 *M* NaCl-saturated phenol, extracted one to two times with chloroform, and finally precipitated with ethanol. The RNA:DNA hybrid is resuspended in 100 μl TE. Typically, we recover 0.5 to 5 μg of labeled RNA (between 1×10^6 to 1×10^7 cpm total). The clean hybrid should be stored at 4° and can be used for 2 to 3 weeks.

Preparation of RNA : RNA Duplex. This substrate is prepared by annealing two complementary RNAs templated by Bluescript KS plasmid DNA (Stratagene), which contains T3 and T7 phage promoters directing RNA synthesis in opposite orientations. The two RNAs are prepared by incubating the DNA with [α-^{32}P]CTP, three unlabeled triphosphates, and either T3 RNA polymerase or T7 RNA polymerase in separate reactions. Because the templates are circular DNAs, the products are long heterogeneous RNAs of opposite polarities. The sense and antisense RNAs are annealed and treated with RNase to remove residual single-stranded RNA, resulting in radiolabeled duplex RNA.

RNA synthesis reactions are carried out in 25 μl total volume containing 40 m*M* Tris–HCl, pH 7.5, 6 m*M* $MgCl_2$, 2 m*M* spermidine, 10 m*M* NaCl, 10 m*M* DTT, 500 μ*M* each of ATP, GTP, and UTP, 12 μ*M* [α-^{32}P]CTP at 100 Ci/mmol, 1 unit/ml RNase inhibitor, 0.5 to 1 μg circular Bluescript DNA, and either T3 and T7 RNA polymerase (20 units). After 1 hr at 37°, the reaction products are treated with DNase I (about 1 unit; Promega) for 15 min at 37°, phenol extracted twice, chloroform extracted once, and precipitated with the addition of 0.1 volume of 3 *M* sodium acetate and 2.5 volumes of ethanol. Each RNA preparation is resuspended in 20 μl of hybridization buffer (40 m*M* PIPES, pH 6.4, 1 m*M* EDTA, 400 m*M* NaCl, 80% (v/v) formamide), and heated to 85° for 5 min to denature any secondary structure. Portions of the two RNA preparations (15 μl of each) are mixed and annealed at 45° for 1 hr. The remaining 5 μl of each RNA is reserved as a control for the following RNase treatment. To the duplex, and separately to each of the reserved single RNA samples, we add 300 μl of an RNase solution (10 m*M* Tris–HCl, pH 7.5, 300 m*M* NaCl, 5 m*M* EDTA, RNase A at 40 μg/ml) and incubate the reaction at 30° for 1 hr to remove any unannealed RNA species. These reactions are stopped by the addition of 20 μl of 10% SDS and 5 μl of 20 μg/ml proteinase K for 30 min at 37°, followed by phenol extraction, chloroform extraction, and ethanol precipitation in the presence of 1 μg carrier tRNA. The duplex RNA is suspended in 100 μl of TE containing 0.1% SDS. If the annealing is success-

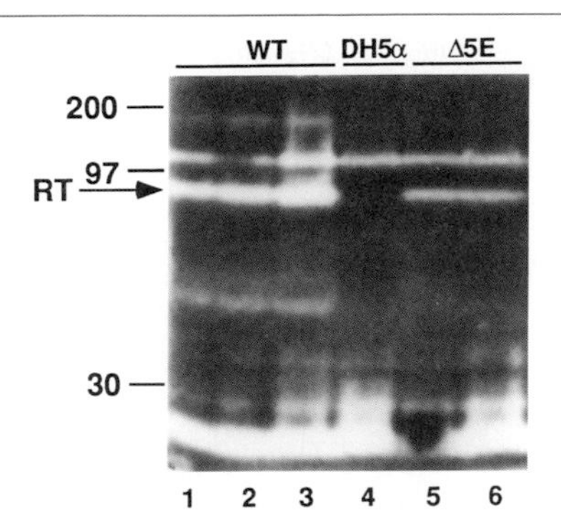

FIG. 3. Analysis of RNase H activity of wild-type and mutant RTs using the *in situ* gel assay. Crude protein extracts were prepared from bacterial cultures expressing the wild-type M-MuLV RT (lanes 1–3), no RT (lane 4), or Δ5E, a mutant RT (lanes 5–6), and applied to an SDS gel containing a uniform distribution of labeled RNA : DNA hybrid. After separation, the proteins were allowed to renature and digest the labeled substrate. Lanes 1–3 show increasing twofold dilutions of the extract. The position of migration of marker proteins is indicated at the left. The wild-type RT is indicated by the arrow. Several bacterial enzymes are detected in this assay, including activities migrating at about 100 and 20 kDa.

ful, the total radioactivity recovered for the duplex RNA will be about 100 times that recovered from digestion of the single-stranded T3 or T7 RNAs alone. Typically, we recover between 1×10^7 to 1×10^8 cpm total. This substrate should be stored at 4° and can be used for 2 to 3 weeks.

Preparation of Uniformly Labeled Gel. SDS–polyacrylamide gels for measurement of activity on either RNA : DNA or RNA : RNA substrates are prepared and run according to standard recipes[26,27] except that radiolabeled RNA : DNA hybrid or RNA : RNA duplex is added to the resolving gel solution immediately before casting the gel. This results in a uniformly radioactive gel. We typically add sufficient RNA : DNA or RNA duplex to give between 1×10^5 and 1×10^6 cpm per 3-ml resolving gel (roughly 0.75 mm × 5 cm × 8 cm); the stacking gel is prepared and cast without substrate. In our hands, the 71-kDa M-MuLV RT is separated well from all host enzymes on either a 7.5 or 10% polyacrylamide resolving gel. Regular protein sample buffer is used, but the samples should not be boiled prior to loading. When we assay protein from crude bacterial lysates, the DNA present in the lysates must be sheared by forcing the lysate through a 1-ml tuberculin syringe repeatedly prior to adding sample buffer. The gels are run at room temperature until the dye front has reached the bottom of the gel.

[27] U. K. Laemmli, *Nature* (*London*) **227,** 680 (1970).

[28] A. Telesnitsky and S. P. Goff, *EMBO J.* **12,** 4433 (1993).

Development of Activity Gel and Visualization of Areas of Clearing

When the gel has finished running, the stack is removed and the gel put into 100 to 200 ml of renaturation/reaction buffer, typically in a small deep dish. Renaturation/reaction buffer for M-MuLV RT consists of 50 m*M* Tris–HCl, pH 7.5, 2 m*M* DTT, 50 m*M* NaCl, and 2 m*M* $MnCl_2$. The buffer for HIV-1 RT is the same as above with the substitution of 10 m*M* $MgCl_2$ for the $MnCl_2$. The DTT should be as fresh as possible and added to the buffer last in order to prevent oxidation of the $MnCl_2$. If the buffer turns brown, the $MnCl_2$ is being oxidized and no activity will be detected. We have found that lowering the pH of the Tris–HCl from 8.0 to 7.5–7.2 helps prevent this problem. The gel should shake for about 48 hr at either room temperature or 37°, changing the buffer approximately every 12 hr. Multiple changes of buffer are important, helping remove residual SDS and the products of digestion from the gel. After 48 hr, the wet gel is exposed to film at room temperature to see if enough renaturation and degradation has occurred. The gel is placed in a sealable plastic bag, which prevents leakage better than plastic wrap. A wet gel typically only needs to be on film overnight in order to get an adequate exposure. If not enough degradation has occurred (i.e., if no zones of clearing are detected), we let the gel continue renaturing for one or two more days. Finally the gel is dried under vacuum and exposed to film by standard methods. An example of an autoradiogram is shown in Fig. 3.

Practical Notes

We have tried using shorter RNA:DNA and RNA:RNA substrates, on the order of 500 bp long. However, we observed that these substrates tended to wash out of the gel during the electrophoresis or the renaturation process.

As mentioned previously, several RNA:DNA and RNA:RNA nuclease activities are found in *E. coli,* and it may be necessary to experiment with different gel sizes and percentages of acrylamide to resolve the protein of interest from contaminating RNases H. *Escherichia coli* RNase H1 (18 kD) and RNase III (30 kD) migrate far from RT, and serve as positive controls. RNase H1 and RNase III-deficient strains of *E. coli* can be used to eliminate these activities if necessary.[4,16,19]

Acknowledgment

This article was written with the support of Public Health Service and CA30488 to SPG and ACS fellowship PF 3342 to AT. SPG is an investigator of the Howard Hughes Medical Institute.

[28] Structure–Function Analysis of 3′ → 5′-Exonuclease of DNA Polymerases

By VICTORIA DERBYSHIRE, JULIA K. PINSONNEAULT, and CATHERINE M. JOYCE

Introduction

The 3′ → 5′-exonuclease activity of DNA polymerases acts in opposition to the polymerase activity and serves as a proofreader, by removing polymerase errors.[1] This activity is present in the majority of DNA-dependent DNA polymerases but absent in the reverse transcriptase (RT) family. The 3′ → 5′-exonuclease is usually part of the same polypeptide chain as the DNA polymerase; an exception is the multisubunit DNA polymerase III of *Escherichia coli,* where the editing function is present on a separate subunit (ε) within the core polymerase.[2] In the structure of the Klenow fragment of DNA polymerase I, the 3′ → 5′-exonuclease is located on a discrete structural domain,[3] and it seems likely that other DNA polymerases are arranged in a similar modular fashion. The preferred substrate for the exonuclease is single-stranded DNA, and a variety of data are consistent with the idea that the primer terminus of a duplex DNA substrate is bound as a "frayed" or single-stranded end at the exonuclease active site.[4] Of all the reactions catalyzed by DNA polymerases, the 3′ → 5′-exonuclease is probably the best understood at a mechanistic level, thanks to the crystallographic data obtained with the Klenow fragment of *E. coli* DNA polymerase I. By studying cocrystals containing either the substrate DNA or the product (dNMP) at the 3′ → 5′-exonuclease site, it has been possible to identify the side chains that interact with substrate or product and form the active site[3,5,6] (Fig. 1). The structural data provided the basis for a detailed mutagenesis study of the roles of these side chains in the exonuclease reaction.[7,8] The mutational study contributed to the crystallographic characterization

[1] A. Kornberg and T. A. Baker, "DNA Replication" p. 113. Freeman, San Francisco, 1992.
[2] R. H. Scheuermann and H. Echols, *Proc. Natl. Acad. Sci. USA* **81,** 7747 (1984).
[3] D. L. Ollis, P. Brick, R. Hamlin, N. G. Xuong, and T. A. Steitz, *Nature* **313,** 762 (1985).
[4] L. S. Beese, V. Derbyshire, and T. A. Steitz, *Science* **260,** 352 (1993).
[5] P. S. Freemont, J. M. Friedman, L. S. Beese, M. R. Sanderson, and T. A. Steitz, *Proc. Natl. Acad. Sci. USA* **85,** 8924 (1988).
[6] L. Beese and T. A. Steitz, *EMBO J.* **10,** 25 (1991).
[7] V. Derbyshire, P. S. Freemont, M. R. Sanderson, L. Beese, J. M. Friedman, C. M. Joyce, and T. A. Steitz, *Science* **240,** 199 (1988).
[8] V. Derbyshire, N. D. F. Grindley, and C. M. Joyce, *EMBO J.* **10,** 17 (1991).

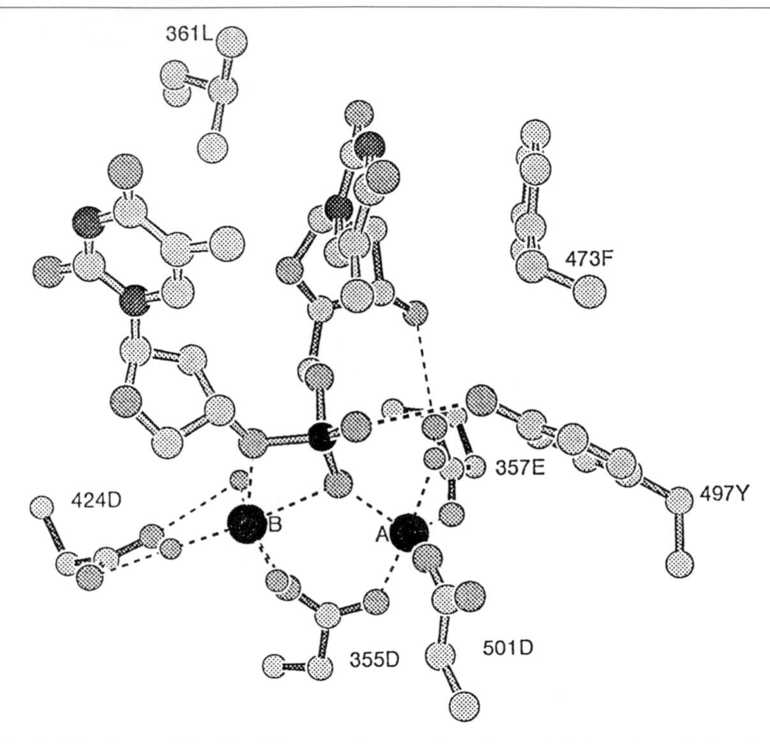

FIG. 1. Structure of the 3′ →5 ′-exonuclease active site containing a bound dinucleotide. The catalytically essential metal ions, A and B, are shown as large black balls. Other atoms are represented as smaller balls, with phosphorus black, and carbon, oxygen, and nitrogen represented by increasingly darker shades of gray. Water molecules are shown as smaller gray spheres. Reproduced, with permission from Oxford University Press, from L. Beese and T. A. Steitz, *EMBO J.* **10,** 25 (1991).

of two divalent metal ion sites at the 3′ → 5′-exonuclease site and showed the importance of the metal ions in the reaction.[7,8] Based on the structural data, a reaction mechanism was proposed involving catalysis by the two metal ions.[5,6]

It has become clear from protein sequence alignments that all polymerases with an editing function possess the same group of crucial active site residues identified in Klenow fragment, although the surrounding protein sequence may be very dissimilar.[9–12] Moreover, preliminary crystallographic

[9] A. Bernad, L. Blanco, J. M. Lázaro, G. Martin, and M. Salas, *Cell* **59,** 219 (1989).

[10] D. K. Braithwaite and J. Ito, *Nucleic Acids Res.* **21,** 787 (1993).

[11] A. Morrison, J. B. Bell, T. A. Kunkel, and A. Sugino, *Proc. Natl. Acad. Sci. USA* **88,** 947 (1991).

[12] L. Blanco, A. Bernad, and M. Salas, *Gene* **112,** 139 (1992).

data for the N-terminal 45-kDa domain of T4 DNA polymerase confirm the prediction of a similar active site geometry.[13] The sequence alignments provided the rationale for mutagenesis of the 3′ → 5′-exonuclease in a variety of DNA polymerases, which served both to confirm the hypothesis of a conserved 3′ → 5′-exonuclease active site and to provide exonuclease-deficient enzymes for a variety of applications.

In this chapter, we describe first the mutational studies that were conducted on the Klenow fragment 3′ → 5′-exonuclease, and then summarize similar studies on other DNA polymerases. These structure–function studies, and the resulting mechanistic deductions, have been reviewed elsewhere.[14]

Mutagenesis of 3′ → 5′-Exonuclease of Klenow Fragment

Choice of Mutations

Residues were chosen for mutagenesis based on high-resolution crystallographic data for the complexes of Klenow fragment with substrate or product at the exonuclease active site[3,5,6] (Fig. 1). Active-site residues can be divided into three fairly distinct groups: those that serve as ligands anchoring the two divalent ions, those that contact the substrate around the terminal phosphodiester bond that is to be cleaved, and those that contact the upstream portion of the DNA chain, more remote from the site of catalysis (Table I). In every case a mutation of the target residue to alanine was studied; this is likely to be the best approximation to a simple removal of the side chain, since alanine is less likely to be structurally disruptive than the alternative, glycine. In many cases, other more conservative changes were made; most of these were substitutions giving side chains of a similar size to the wild-type residue but with altered hydrogen-bonding properties. The rationale for such changes should be apparent as we describe the experimental results below. In addition, changes from Asp to Glu were made to test the tolerance for altered geometry at positions 424 and 501, and the L361M mutation was made to increase the similarity to the sequence of the ε subunit of DNA polymerase III.

Construction of Mutations

Site-directed mutagenesis was carried out following established procedures.[15] Synthetic oligonucleotides encoding the desired mutations were used as primers on uracil-containing M13 templates containing appropriate

[13] J. Wang, P. Yu, W. H. Konigsberg, and T. A. Steitz, unpublished observations (1993).
[14] C. M. Joyce and T. A. Steitz, *Ann. Rev. Biochem.* **63,** 777 (1994).
[15] T. A. Kunkel, J. D. Roberts, and R. A. Zakour, *Methods Enzymol.* **154,** 367 (1987).

TABLE I
AMINO ACID RESIDUES AT 3′ → 5′-EXONUCLEASE ACTIVE SITE OF KLENOW FRAGMENT

Residue	Observed contact[a]	Mutations[b]
Ligands to the metal ions		
Asp-355	Shared ligand to metal A and metal B	D355A, D355N
Glu-357	Ligand to metal A (also see below)	E357A, E357Q
Asp-424	Ligand to metal B (via bridging H_2O molecules)	D424A, D424E, D424N
Asp-501	Ligand to metal A	D501A, D501E, D501N
Substrate contacts at 3′ terminus		
Glu-357	Hydrogen-bonded to 3′-OH and may also orient attacking H_2O	E357A, E357Q
Leu-361	Inserted between terminal two bases	L361A, L361M
Phe-473	Stacks with 3′-terminal base	F473A
Tyr-497	Hydrogen-bonded to terminal phosphodiester bond and may also orient attacking H_2O	Y497A, Y497F
Upstream contacts with the substrate sugar–phosphate backbone[c]		
Gln-419	Interacts with penultimate phosphodiester bond	Q419A, Q419E
Arg-455	Ion-pair interaction with third phosphodiester bond from terminus	R455A

[a] From L. Beese and T. A. Steitz, *EMBO J.* **10,** 25 (1991).

[b] Mutations are abbreviated using the following convention: The residue number from the DNA polymerase I sequence [C. M. Joyce, W. S. Kelley, and N. D. F. Grindley, *J. Biol. Chem.* **257,** 1958 (1982)] is preceded by the symbol (in the one-letter code) for the wild-type amino acid and followed by the symbol for the mutant amino acid. Thus D355A denotes a mutation from Asp to Ala at position 355.

[c] Additional contacts with the nucleotide bases and deoxyribose positions are described by Beese and Steitz (see footnote *a*).

regions of the *polA* gene, chosen so as to facilitate subsequent cloning of the mutations into a Klenow fragment expression plasmid.[8] M13 isolates carrying the desired mutation were obtained typically at frequencies of 10 to 50% and were identified by direct sequencing of randomly chosen clones. Before subcloning into the expression plasmid, the region to be subcloned (300 to 500 bp, see below) was sequenced in its entirety to check that no additional mutations were present.

In a large-scale mutagenesis study such as the present one, it is extremely helpful to have a series of unique restriction sites that divides the target gene into modules or cassettes of a convenient size for cloning and sequencing. The 3′ → 5′-exonuclease region of Klenow fragment can be divided into two modules, defined by three unique restriction sites: the *Bst*XI site in λ control sequences 70 bp upstream of the Klenow fragment translational start (codon 324 of the *polA* structural gene[16]), the *Xho*I site at codons 402–403, and the *Sac*I site at codons 558–559. Mutations upstream of residue 402 were therefore retrieved from the M13 clone on a 302 bp *Bst*XI–*Xho*I fragment, and those beyond residue 403 on a 470 bp *Xho*I–*Sac*I fragment. The relevant fragment was inserted into the pCJ122 expression plasmid, in which Klenow fragment is under the control of the strong leftward promoter (P_L) of phage λ.[17,18] The cloning procedure was facilitated by the use of deletion derivatives of pCJ122 in which the appropriate pair of unique sites (*Bst*XI and *Xho*I, or *Xho*I and *Sac*I) was joined by a short (~10 bp) adaptor containing a *Bam*HI site. The primary advantage of this cloning strategy was that there was no danger of recovering wild-type information instead of the desired mutation because the corresponding region was absent from the recipient plasmid. Additional advantages were that the cloning efficiency could be increased by digestion of the ligation mixture with *Bam*HI to reduce the background of starting plasmid, and that plasmids containing the desired mutations could easily be differentiated from the starting plasmid by the increase in size of an appropriate restriction fragment. The overproducer plasmids for the mutant proteins were obtained and characterized in strain CJ388, a wild-type λ lysogen of the *recA*$^-$ host DH1.[19] We chose a *recA*$^-$ host to minimize the chances for genetic exchange between the mutant *polA* information on the plasmid and the wild-type chromosomal *polA* copy, needed for maintenance of many of the commonly used cloning vectors, including those used in this study.

Because this study required the cloning of many mutations, the investment of time in constructing appropriate "recipient plasmids" for the mutated fragments was clearly worthwhile; if only a small number of mutations were to be studied, the benefit of such an approach would be questionable. In the Klenow fragment system, the presence of the unique *Xho*I restriction site within the region under study was also beneficial in that it allowed the easy construction of some double mutations (e.g., E357A,D501N). We have likewise found the "cassette" approach to be valuable in our studies of the polymerase region of Klenow fragment, although in this instance it was also necessary to create, by mutation, the unique sites that defined the modules within the coding sequence.[17]

[16] C. M. Joyce, W. S. Kelley, and N. D. F. Grindley, *J. Biol. Chem.* **257,** 1958 (1982).

[17] A. H. Polesky, T. A. Steitz, N. D. F. Grindley, and C. M. Joyce, *J. Biol. Chem.* **265,** 14,579 (1990).

[18] C. M. Joyce and V. Derbyshire, this volume, [1].

[19] D. Hanahan, *J. Mol. Biol.* **166,** 557 (1983).

Overproduction and Purification of Mutant Proteins

Detailed procedures are given elsewhere in this volume.[18] Klenow fragment derivatives with mutations at the exonuclease site were overproduced by heat induction of strains carrying a temperature-sensitive λ repressor. Because the overproduction vector, carrying the mutated copy of the gene, requires host DNA polymerase I functions for its replication, two potential problems exist: (1) Contamination of a mutant Klenow fragment with wild type (derived by endogenous proteolysis of the host *polA* gene product) could give the appearance of exonuclease activity in a mutant protein that was, in reality, exonuclease deficient. (2) Recombination between mutant *polA* information on the expression plasmid and the wild-type chromosomal copy could convert the overproducer plasmid to wild type in a subpopulation of the culture. We circumvented both problems completely in the overproduction of the D355A,E357A double mutant protein by constructing an overproduction strain, CJ375, in which all the *polA* information is derived from the same exonuclease-deficient allele (this was possible because the D355A,E357A mutation does not affect plasmid replication).[20] Because it was impractical to construct a host strain of this type for every mutant protein being studied, other mutant proteins were expressed in a *recA* host strain, CJ376,[18] to reduce the likelihood of genetic exchange between mutant and wild-type information. As a further precaution against genetic exchange, overproducer strains were not stored as such; instead, cells for induction were grown from a fresh transformant in every case. The use of the CJ376 host strain addresses the problem of recombination, but does not deal with the possibility of low-level contamination with wild-type Klenow fragment.[21] Since the D355A,E357A overproducer system provides a more satisfactory solution to both problems, it is the D355A,E357A mutant protein that is available commercially as an exonuclease-deficient Klenow fragment, even though there are several other mutant proteins having similarly low levels of exonuclease activity (see Table II).

The mutant proteins were purified using the Pharmacia FPLC system.[18] We have found the superior fractionation obtained by FPLC to be particularly important for removing trace contaminants of cellular nucleases that might otherwise confuse the analysis of the mutant proteins. We avoid

[20] C. M. Joyce, unpublished observations (1987).

[21] The quantitative effect of both problems seems likely to be small, as shown by comparing the exonuclease activities of D355A,E357A (prepared from a homogenotized background), D424A (prepared from a *polA(D355A,E357A) recA*$^+$ background),[7] and D424N (prepared from a *polA*$^+$ *recA*$^-$ background).[8] The values (from Table II) are, respectively, 1.4×10^{-5}, 1.3×10^{-5}, and 2.5×10^{-5} of wild type.

TABLE II
ENZYMATIC ACTIVITY OF WILD-TYPE AND MUTANT DERIVATIVES OF KLENOW FRAGMENT[a]

Protein	Polymerase[b] activity	Ratio of exonuclease to polymerase activity[c]	
		Double-stranded DNA	Single-stranded DNA
Wild type	1	100	100
Mutations affecting metal ligands			
D355A	1.4	0.0083[d]	
D355N	0.7	≤0.01[e]	
E357A	1.3	0.18	
E357Q	1.2	0.03[d]	
D424A	1.0	0.0013[d]	
D424E	1.2	4.0	8.3
D424N	1.3	0.0025[d]	
D501A	1.3	0.0075[d]	
D501E	1.1	0.56	
D501N	1.1	50	
D355A, E357A	1.0	0.0014[d]	
E357A, D501N	0.8	≤0.002[f]	
Mutations affecting contacts to terminal nucleotide			
E357A[g]	1.3	0.18	
E357Q[g]	1.2	0.03[d]	
L361A	1.0	4.0	37
L361M	1.2	8.3	
F473A	1.1	0.03[d]	
Y497A	1.0	5.6	2.9[d]
Y497F	1.2	4.3	1.6
Q419A	1.1	23	20
Q419E	1.5	0.1[d]	
R455A	0.8	36	84

[a] All values are the average of several determinations; standard deviations are given in the original reports.[7,8]

[b] Assayed on poly[d(AT)] template.[23] Values are given relative to wild type (defined as 1.0).

[c] Relative to wild type (defined as 100 in both assays); note, however, that the reaction rates for wild-type Klenow fragment in the two assays are not the same. The substrates and assay methods were those described in the text. A value of 0.001 represents the lower limit of the assay using double-stranded DNA. A value of 0.4 is the lower limit of the single-stranded DNA assay so that it was not possible to assay those proteins having very low activity.

[d] These assays were not necessarily under V_{max} conditions, as judged by comparing the rates at different substrate concentrations.

[e] This value was calculated from the rate of degradation of a 5′-labeled duplex DNA. As indicated in the text, this assay is less sensitive than our standard duplex DNA assay. For comparison, the D355A protein, assayed at the same time, gave a very similar exonuclease rate, implying that the lower limit of this assay is around 0.01 (relative to wild type as 100).

[f] No reaction detectable after a 300-min incubation.

[g] Glu-357 is placed in both groups since the carboxylate side chain makes contacts both as a metal ligand and as a substrate-binding residue.

using phosphate buffers in the purification since we have found that some component used in our earlier procedure[22] (presumably the phosphate buffer) serves as source of pyrophosphate, resulting in a low level of Klenow fragment-catalyzed pyrophosphorysis of duplex DNA assay substrates, which can interfere with the assay of mutant proteins having very low $3' \rightarrow 5'$-exonuclease activity.[7]

Characterization of Mutant Derivatives of Klenow Fragment

Measurement of Specific Activity of Polymerase

Polymerase activity was measured by the standard poly[d(A-T)] assay.[23] Protein concentrations were determined by the Bradford colorimetric assay,[24] using the reagent supplied by Bio-Rad. Either homogeneous Klenow fragment or bovine serum albumin (BSA) (of accurately determined concentration) has been used as the standard, with identical results. A polymerase-specific activity close to that of wild-type Klenow fragment (typically $\sim 10^4$ units/mg in this assay) was taken as evidence that the mutant proteins were not grossly misfolded. Because of the variability of the assay, it is less important that the polymerase-specific activity of the mutant protein have a particular numerical value than that it be similar to that of a wild-type standard assayed at the same time with the same reagents.

Exonuclease Assay on Double-Stranded DNA

1. Preparation of Assay Substrate.[25] The assay substrate was a heterogeneous mixture of restriction fragments carrying a single ^{32}P label at the 3′-terminal phosphodiester bond. The labeled substrate was prepared from *E. coli* chromosomal DNA digested to completion with *Sau*3AI. Digested DNA (24 μg, approximately 290 pmol of ends) was 3′-end-labeled using 1 unit of Klenow fragment in a 50-μl reaction containing 10 nmol unlabeled dGTP and 30 pmol [α-^{32}P]dATP (3000 Ci/mmol) for 10 min at room temperature. Excess unlabeled dATP (1 nmol) was added and incubation was continued for a further 1 min to ensure that all the 3′ ends were extended to the same extent (leaving a 2-nucleotide 5′ extension). The reaction was terminated by addition of EDTA to 20 m*M*, and the Klenow fragment was inactivated by heating at 70° for 15 min. The labeled DNA was phenol

[22] C. M. Joyce and N. D. F. Grindley, *Proc. Natl. Acad. Sci. USA* **80,** 1830 (1983).
[23] P. Setlow, *Methods. Enzymol.* **29,** 3 (1974).
[24] M. M. Bradford, *Anal. Biochem.* **72,** 248 (1976).
[25] P. S. Freemont, D. L. Ollis, T. A. Steitz, and C. M. Joyce, *Proteins* **1,** 66 (1986).

extracted, passed through a 1-ml column of Sephadex G-50 to remove unincorporated nucleotides, and recovered by ethanol precipitation.

2. *Assay Method.*[8,25] The standard reaction (20 μl) contained $\sim 3 \times 10^{-7}$ *M* DNA 3′ termini in 6 m*M* Tris–HCl, pH 7.5, 6 m*M* $MgCl_2$, 6 m*M* 2-mercaptoethanol, and 50 m*M* NaCl. Enzyme was added and the mixture was incubated at 37°. Samples were removed at intervals into 0.5 ml of a solution containing 1 mg/ml BSA (as a precipitation carrier) and 10 m*M* EDTA. The DNA was precipitated by addition of an equal volume of 10% (w/v) trichloroacetic acid (TCA). After 5 to 10 min on ice, the DNA was pelleted by spinning for 2 min in a microfuge and the supernatant was transferred to a fresh tube. The radioactivity in both supernatant and pellet was determined by Cerenkov counting. To correct for the additional quenching in the supernatant fraction, 10 μl of a ^{32}P-containing solution was added to both supernatant and pellet from a "blank" precipitation (prepared by mixing 0.5 ml each of the BSA and TCA solutions). The ratio of the counts observed in the two tubes gave the correction factor. The fraction of the substrate ^{32}P that was solubilized was plotted versus time, giving the rate of the exonuclease reaction, which was normalized to the number of polymerase units in the assay. The exonuclease activity (per polymerase unit) of each mutant protein was then expressed relative to that of wild type, which was arbitrarily set at 100 (see Table II). For most of the proteins in this study we were able to show that this value reflected V_{max} because the same exonuclease rate was observed with a threefold higher substrate concentration.

Exonuclease Assay on Single-Stranded DNA[8]

1. *Preparation of Assay Substrate.* The substrate was a ^{32}P-labeled DNA homopolymer, which was synthesized enzymatically using terminal deoxynucleotidyltransferase. An octanucleotide primer, $p(dA)_8$ (typically 80 nmol) was incubated with 4 μmol [α-^{32}P]dATP (5 to 10 μCi/μmol) in a 100-μl reaction containing 100 m*M* Tris–HCl, pH 7.5, 100 μg/ml BSA, 1 m*M* dithiothreitol (DTT), 10 m*M* $MgCl_2$ and 80 units of terminal deoxynucleotidyltransferase for 16 hr at 37°. Unincorporated dATP was removed by gel filtration on a 1-ml Biogel P4 column. A 2-μl sample of the reaction mix was taken before and after the P4 column and was applied to a DE81 filter and washed as described below. Comparison of the counts on the two filters gave the yield of labeled DNA from the P4 column, from which could be calculated the molarity of the assay substrate and the radioactivity per mole. The average chain length of the substrate (typically around 30 nucleotides) was determined by fractionation of a sample on a 10% polyacrylamide-urea sequencing gel followed by densitometric scanning

and integration of all peaks visible on the resulting autoradiograph. A similar substrate was made by extension of $p(dT)_8$ with dTTP.

2. Assay Method. The standard reaction (20 μl) contained $\sim 1 \times 10^{-4}$ *M* DNA 3′ termini in 50 m*M* Tris–HCl, pH 7.5, 8 m*M* $MgCl_2$. Reactions were initiated by addition of enzyme and were incubated at 37°. At intervals, 2-μl samples were removed and quenched in 53 μl of 30 m*M* EDTA. The radioactivity remaining in single-stranded DNA was determined by applying 50 μl of each quenched solution to a 2.5-cm-diameter DE81 filter (Whatman). The released [^{32}P]dAMP was removed by washing the filters three times by gentle agitation for 5 min in 0.3 *M* ammonium formate, pH 8.0, followed by two washes in 95% (v/v) ethanol and one wash in ether.[26] After air-drying, the radioactivity present on each filter was determined by scintillation counting in 5 ml of Optifluor (Packard), and was plotted as a function of time to give the exonuclease rate for each protein. As in the previous method, this was normalized to the number of polymerase units in the assay and expressed relative to wild type, which was arbitrarily set at 100. For most of the proteins assayed, we could show that the exonuclease rate corresponded to V_{max} since the same rate was observed with a threefold higher substrate concentration. This assay method has also been used in our investigation of the pH dependence of the 3′→5′-exonuclease reaction.[8]

3. Steady-State Kinetics. Measurement of k_{cat} and K_m for wild-type Klenow fragment was carried out using poly(dT), made as described above for poly(dA), except that the specific activity of the labeled nucleotide was approximately 100-fold higher because lower concentrations of the poly(dT) substrate were to be used. The reaction mix (100 μl) contained 7.4×10^{-9} *M* Klenow fragment and poly(dT) (1×10^{-7} to 2×10^{-6} *M* of 3′ ends) in 6 m*M* Tris–HCl, pH 7.5, 6 m*M* $MgCl_2$, 6 m*M* 2-mercaptoethanol, 50 m*M* NaCl, and 100 μg/ml BSA. Samples (5 μl) were removed at intervals during incubation at 37° and processed as described above. Initial rates, determined by a least-squares analysis, were used to generate a Lineweaver–Burk double-reciprocal plot, from which k_{cat} and K_m were calculated.

Assessment of Exonuclease Assay Methods

Of the two exonuclease assays described above, the method using the duplex substrate is the more sensitive for assaying mutant proteins having very low exonuclease activity (see footnotes to Table II). This greater sensitivity results from two features of the duplex DNA assay. One is the

[26] F. R. Bryant, K. A. Johnson, and S. J. Benkovic, *Biochemistry* **22,** 3537 (1983).

quantitation of both the dNMP product and the remaining substrate. (In the single-stranded DNA assay, only the substrate is quantitated and, therefore, at low extents of conversion, the amount of product formed corresponds to the difference between two very large numbers.) The second important difference is that only the 3′-terminal residue of the duplex substrate is labeled, whereas the single-stranded substrate contains a sizable tract of labeled residues. As a result, a low level of exonuclease activity will give a greater proportionate release of radioactivity from the duplex substrate (though this greater sensitivity comes at the expense of a less linear time course; see below). Although we ourselves have not made these modifications, the sensitivity of the single-stranded DNA assay could be improved by having a shorter tract of labeled residues and by using a separation method (such as thin-layer chromatography) that allows quantitation of the released dNMP as well as the remaining substrate. Assay methods used by other workers follow the same general principles as the assays we have described, but may differ in the precise details of the substrates used or the methods for separation and quantitation of substrate and product. Examples can be found in the references cited in Table III.

The exonuclease assay methods that we have used were chosen as being the most appropriate, given the technology available to us at the time this work was carried out. With subsequent improvements in technology, however, some methods that we had found unsatisfactory have now become more feasible. In particular, we had rejected methods that required the separation and quantitation of products on sequencing gels because the quantitation of the DNA bands by densitometry of autoradiographs was insufficiently accurate for a detailed kinetic study. More recently, the ability to quantitate gels of this type accurately and easily using phosphorimage technology has meant that measurement of a $3' \rightarrow 5'$-exonuclease rate by following the degradation of a 5′-labeled oligonucleotide has become an attractive alternative, and we are starting to use this approach increasingly in our studies of the exonuclease reaction.

Although we have not used the gel assay extensively for the characterization of mutant proteins, our experience to date has allowed us to assess its strengths and weaknesses compared with the assays described above. Gel analysis of the degradation of a 5′-labeled substrate is less sensitive for measuring low levels of $3' \rightarrow 5'$-exonuclease activity (see footnotes to Table II) since it is difficult to quantitate a small amount of product in the presence of a large excess of starting material. The gel assay is also more time consuming and therefore less well suited for the routine screening of large numbers of mutant proteins. Advantages of the gel assay are the use of well-defined substrates, the potential for using a variety of different substrates (single- or double-stranded DNAs, oligonucleotides of different

TABLE III
MUTATIONS THAT HAVE BEEN STUDIED IN THE CONSERVED "EXO" SEQUENCE MOTIFS OF DNA POLYMERASES

Enzyme	Mutation	Effect of mutation[a]	
		Exonuclease activity	Other
Exo I (Asp-355) motif[b]			
φ29 DNA polymerase	D12A[c]	~10^3-fold decrease	Defective in strand displacement
	E14A[c]	~300-fold decrease	Defective in strand displacement
T4 DNA polymerase	D112A, E114A[d]	~10^4-fold decrease	Mutator *in vivo*
	D112N[e]	Not measured	Mutator *in vivo*
T5 DNA polymerase	D164A, E166A[f]	Not detectable	
T7 DNA polymerase	D5A, E7A[g]	~10^5-fold decrease	
E. coli DNA polymerase II	D155A[h]	10^4-fold decrease	
	E157A[h]	10^3-fold decrease	
	D155A, E157A[h]	2×10^3-fold decrease	
B. subtilis DNA polymerase III	E427A[i]	Not detectable	Large decrease in polymerase activity
	E427Q[i]	~10^3-fold decrease	10-fold decrease in polymerase activity
	G430E[i]	10-fold decrease	2-fold decrease in polymerase activity
T. litoralis ("Vent") DNA polymerase	D141A, E143A[j]	Not detectable	
Yeast DNA polymerase II	D290A, E292A[k]	≥100-fold decrease	Mutator, *in vivo*
Yeast DNA polymerase III	D321A[l]	Not measured	Mutator, *in vivo*
	D321V[l]	Not measured	Mutator, *in vivo*
	E323A[l]	Not measured	Mutator, *in vivo*
Yeast mitochondrial DNA polymerase (MIP1)	D171G[m]	≥100-fold decrease	Mutator, *in vivo;* polymerase less processive
Exo II (Asp-424) motif			
φ29 DNA polymerase	D66A[c]	~400-fold decrease	Defective in strand displacement
T4 DNA polymerase	D219A[n]	>10^7-fold decrease	Mutator *in vivo*
E. coli DNA polymerase II	D228A[h]	10^3-fold decrease	

Yeast DNA polymerase III	D405A[l]	Not detectable in crude extract	Mutator *in vivo*
Yeast mitochondrial DNA polymerase (MIP1)	D230A[m]	≥100-fold decrease	Mutator *in vivo;* polymerase less processive
Exo III (Asp-501) motif			
ϕ29 DNA polymerase	D169A[p]	10^3-fold decrease	Defective in strand displacement
	Y165C[p]	24-fold decrease	Defective in strand displacement
	Y165F[p]	13-fold decrease	Defective in strand displacement
T4 DNA polymerase	D324A[d]	~10^4-fold decrease	Mutator *in vivo*
	D324G[q]	100-fold decrease[r]	Mutator *in vivo*
E. coli DNA polymerase II	Y330F[h]	60-fold decrease	
	D334A[h]	50-fold decrease	
Yeast mitochondrial DNA polymerase (MIP1)	D347A[m]	~500-fold decrease	Mutator *in vivo;* polymerase less processive
	C344G[m]	~3-fold decrease	Weak mutator *in vivo*

[a] Relative to wild type. Except where noted, the polymerase activity was essentially the same as wild type.

[b] To facilitate comparison with the Klenow fragment results, the important carboxylate metal ligand present in each motif of the Klenow fragment sequence is noted.

[c] A. Bernad, L. Blanco, J. M. Lázaro, G. Martin, and M. Salas, *Cell* **59,** 219 (1989).

[d] L. J. Reha-Krantz and R. L. Nonay, *J. Biol. Chem.* **268,** 27,100 (1993).

[e] L. J. Reha-Krantz, *J. Mol. Biol.* **202,** 711 (1988).

[f] D. K. Chatterjee and A. J. Hughes, unpublished work (1991).

[g] S. S. Patel, I. Wong, and K. A. Johnson, *Biochemistry* **30,** 511 (1991).

[h] Y. Ishino, H. Iwasaki, I. Kato, and H. Shinagawa, *J. Biol. Chem.* **269,** 14,655 (1994).

[i] M. H. Barnes, R. A. Hammond, C. C. Kennedy, S. L. Mack, and N. C. Brown, *Gene* **111,** 43 (1992).

[j] H. Kong, R. B. Kucera, and W. E. Jack, *J. Biol. Chem.* **268,** 1965 (1993).

[k] A. Morrison, J. B. Bell, T. A. Kunkel, and A. Sugino, *Proc. Natl. Acad. Sci. USA* **88,** 9473 (1991).

[l] M. Simon, L. Giot, and G. Faye, *EMBO J.* **10,** 2165 (1991).

[m] F. Foury and S. Vanderstraeten, *EMBO J.* **11,** 2717 (1992).

[n] M. W. Frey, N. G. Nossal, T. L. Capson, and S. J. Benkovic, *Proc. Natl. Acad. Sci. USA* **90,** 2579 (1993).

[p] M. S. Soengas, J. A. Esteban, J. M. Lázaro, A. Bernad, M. A. Blasco, M. Salas, and L. Blanco, *EMBO J.* **11,** 4227 (1992).

[q] L. J. Reha-Krantz, S. Stocki, R. L. Nonay, E. Dimayuga, L. D. Goodrich, W. H. Konigsberg, and E. K. Spicer, *Proc. Natl. Acad. Sci. USA* **88,** 2417 (1991).

[r] Double mutation with E191A, which alone has little effect on exonuclease activity.

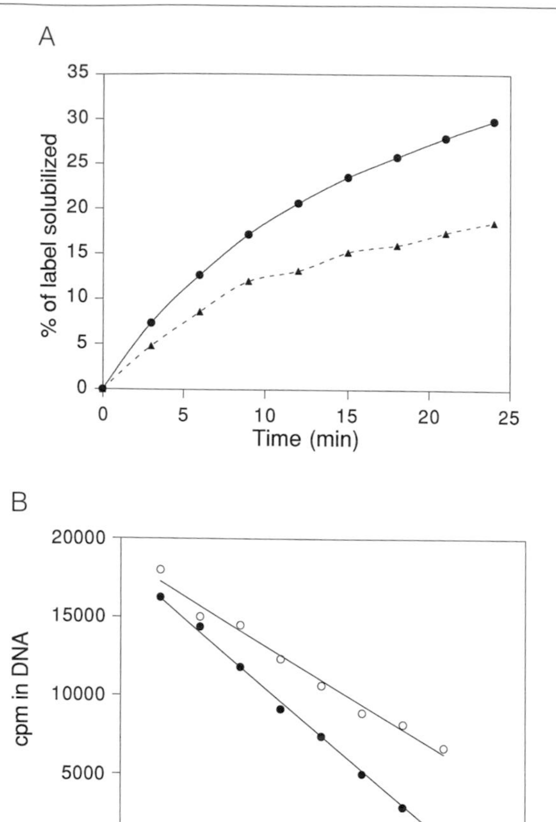

FIG. 2. Determination of 3′→5′-exonuclease activity. (A) Degradation of a 3′-end-labeled duplex DNA by wild-type Klenow fragment (●) assayed at 1.3 nM, and the R455A mutant protein (▲) assayed at 3.6 nM. (B) Degradation of uniformly labeled single-stranded poly(dA), of average length 34, by wild-type Klenow fragment (●) assayed at 0.67 μM, and the L361A mutant protein (○) assayed at 0.87 μM. (C) Gel electrophoretic analysis of 3′→5′-exonuclease activity, exemplified by the degradation of 5′-end-labeled p(dT)$_{14}$ by wild-type Klenow fragment. The reaction contained 6 μM oligonucleotide and 1.5 μM enzyme. Samples were removed at 15-sec intervals, as indicated. (D) Quantitation of the experiment shown in part (C). The extent of reaction (in μM) is presented either as the amount of 14-mer hydrolyzed (○), or as the number of phosphodiester bonds hydrolyzed (●) (calculated as described in the text). Early in the reaction, the two quantities are the same but, as the reaction proceeds, the second calculation method, which takes account of all species that can serve as substrates, is more satisfactory.

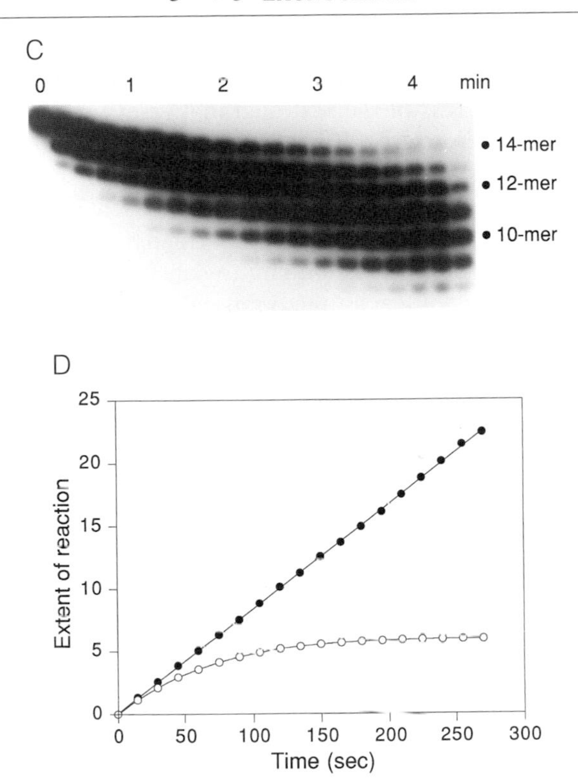

FIG. 2. (*continued*)

lengths, even extremely short oligonucleotides), and the ability to visualize the spectrum of reaction products as the reaction proceeds. As a result, the gel assay seems better suited for addressing more detailed mechanistic questions following the initial characterization of a mutant protein.

When using a particular assay method it is also important to be aware of how the reaction rate will change as the reaction proceeds and a substantial portion of the substrate is converted to product. With a substrate in which only the terminal residue is labeled (such as our duplex DNA substrate) a single exonuclease event will consume labeled substrate and produce an unlabeled competing substrate. Consequently, the reaction is only linear below 10 to 20% conversion of the substrate (Fig. 2A). By contrast, the single-stranded substrate described above contains more than 20 labeled residues so that the product from one round of exonuclease action can serve as a substrate in subsequent rounds, with the result that the reaction rate remains constant for quite a large extent of reaction (Fig.

2B). This same issue is nicely illustrated by the gel assay of the degradation of a 5′-labeled single-stranded DNA oligonucleotide (Fig. 2C). If one merely focuses on the rate of the first degradative event by measuring the rate of loss of the full-length substrate, then the time course will show substantial curvature as the reaction proceeds, due to the production of competing substrates. A more satisfactory approach is to consider the shorter species as potential substrates, as described by Cheng and Kuchta.[27] For this calculation, bands corresponding to substrate and all the reaction products are quantitated at each time point, and then the mole fraction of each species is multiplied by the number of exonuclease events required to generate that species, giving the amount of substrate degraded at each time point. Thus, for the degradation of 5′-^{32}P-labeled $(dT)_{14}$, the molar quantity of substrate degraded is given by {(fraction 14-mer)0 + (fraction 13-mer)1 + (fraction 12-mer)2 + (fraction 11-mer)3 + ...} × (moles of DNA in assay). Provided that the substrate is sufficiently long so that all the species under consideration are degraded at comparable rates, the reaction rate measured in this way remains linear for a substantial time (Fig. 2D).

Results and Interpretation of Mutational Studies of 3′ → 5′-Exonuclease of Klenow Fragment

Table II summarizes the assay results previously reported for Klenow fragment derivatives having mutations at the exonuclease active site,[7,8] together with some previously unreported data for mutations not included in our earlier study. As in any structure–function study involving data from mutant proteins, any meaningful interpretation of the data relies on the assumption that changes in protein structure due to the mutations are confined to the position of the altered side chain. Crystallographic studies of the single mutants D424A[7] and D355A[4] and of the D355A,E357A double mutant[7] validated the assumption for these proteins, and moreover suggested that the more conservative amide substitutions at these carboxylate positions would also have the same structure. For the other proteins in the study, we were able to draw on circumstantial evidence (wild-type levels of polymerase activity, similar overproduction yields and chromatographic behavior) that argued against gross structural perturbations in any of the mutant proteins. However, in the absence of further crystallographic data, we cannot rule out the possibility that some of the mutations may cause

[27] C.-H. Cheng and R. D. Kuchta, *Biochemistry* **32,** 8568 (1993).

subtle rearrangements within the active-site region, and this is an important caveat in a study of this type.

Reaction Mechanism

The structural and mutational data together have led to a proposed mechanism for the chemical step of the exonuclease reaction[5,6] (Fig. 3). In this mechanism, the pair of divalent metal ions (A and B), 4 Å apart, plays a pivotal role in the bond making and breaking processes. Metal A coordinates and polarizes the attacking water molecule. As the nucleophilic displacement takes place, the metal ions stabilize the developing negative charge on the pentacovalent phosphorus center, positioned between the two metal sites. Metal B is also available to stabilize nega-

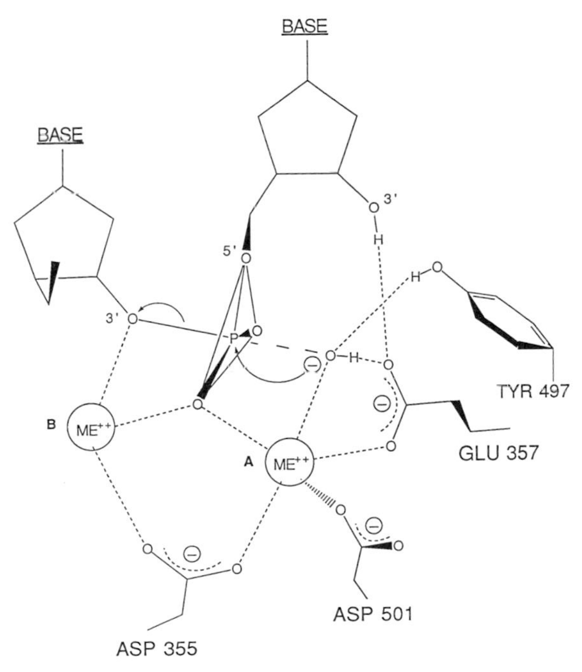

FIG. 3. The proposed transition state for the $3' \rightarrow 5'$-exonuclease reaction. The mechanism is thought to involve catalysis mediated by the two bound divalent metal ions (ME^{++}). Metal ion A facilitates the formation of the attacking hydroxide ion, whose lone pair electrons are orientated toward the phosphorus by interactions with metal A, Tyr-497, and Glu-357. Metal ion B stabilizes the geometry and charge of the pentacovalent transition state and facilitates the departure of the 3′ hydroxyl group. Reproduced, with permission from Oxford University Press, from L. Beese and T. A. Steitz, *EMBO J.* **10,** 25 (1991).

tive charge on the leaving group and thus facilitate its departure. It has become increasingly apparent that two-metal-ion catalysis of this type may be a recurrent theme in phosphoryl transfer reactions,[28] so that the studies described here may have more generality than was at first supposed.

Although the mechanism of the chemical step of catalysis seems clear, we do not have a comparable understanding of the other steps that make up the kinetic pathway of the exonuclease reaction, nor do we know with certainty which step is rate limiting. It is clear from the steady-state kinetic parameters that there are substantial differences between the hydrolysis of single-stranded DNA and that of a duplex substrate. The exonucleolytic degradation of a duplex terminus is extremely slow ($k_{cat} \sim 10^{-3}$ sec^{-1}), but substrate binding appears to be very tight (K_m estimated to be in the nanomolar range).[29] For this and other reasons, it has been suggested that a duplex DNA substrate binds first to the polymerase site of Klenow fragment and is then transferred to the exonuclease site for hydrolysis, with the transfer step being rate limiting for the wild-type enzyme.[30] The degradation of single-stranded DNA is likely to be simpler kinetically since no transfer step is required, and the faster reaction ($k_{cat} \sim 0.1$ sec^{-1}) and higher K_m (5.6×10^{-7} M) are consistent with this expectation.[8] In either reaction, one cannot assume that the same step will be rate limiting when comparing a mutant enzyme with wild type, so that the measured change in exonuclease rate will not necessarily reflect the decrease in rate of a single step of the reaction. Bearing this caveat in mind, it is still possible to provide a structural rationale for the observed properties of the mutant proteins (Table II). It is probably reasonable to infer that mutations that weaken the binding of the active-site metal ions have caused the chemical step to become rate limiting; in these cases the extremely low exonuclease rates would reflect a failure of chemical catalysis. (Moreover, since the chemical step was not rate limiting for wild type, the decrease in the rate of this step as a result of the mutation must be in excess of the 10^4- to 10^5-fold decrease measured for the overall reaction rate.) Other mutations clearly have an effect because they remove an important substrate contact, even though one cannot say at present whether the effect of this loss is manifested kinetically in substrate binding (in the hydrolysis of single-stranded DNA), or transfer from the polymerase site (in the reaction with duplex DNA), or in the

[28] T. A. Steitz and J. A. Steitz, *Proc. Natl. Acad. Sci. USA* **90,** 6498 (1993).

[29] R. D. Kuchta, P. Benkovic, and S. J. Benkovic, *Biochemistry* **27,** 6716 (1988).

[30] C. E. Catalano, D. J. Allen, and S. J. Benkovic, *Biochemistry* **29,** 3612 (1990).

chemical step. Since some mutations have quantitatively different effects depending on whether the substrate is single or double stranded, it seems that the processes involved in bringing the substrate to the exonuclease site are not entirely analogous in the two cases, perhaps because of the additional requirement for melting of a duplex terminus.

Mutations in Carboxylate Ligands to the Metal Ions

Figure 1 shows the details of the coordination of the two metal ions at the exonuclease active site. Metal A is bound in distorted tetrahedral coordination by one phosphate oxygen and the carboxylate groups of Asp-355, Glu-357, and Asp-501, with a water molecule (the proposed attacking nucleophile) as a fifth ligand. Metal B has octahedral coordination to two phosphate oxygens, Asp-355 (shared with metal A) and Asp-424, the latter acting, via bridging water molecules, as a bidendate ligand. Crystallographic studies of three different mutant proteins have shown that loss of a metal ligand results in failure to bind one or more of the metal ions.[4,7] From a mechanistic standpoint, two crystallographic results are particularly significant. The D424A mutant protein binds metal A and substrate (or product) in an apparently normal manner, but fails to bind metal B.[5–7] Conversely, the D355A mutant protein binds substrate and metal B but not metal A.[4] Because both the D355A and D424A mutant proteins have extremely low exonuclease activity (Table II), these data support the proposed involvement of both metal ions in catalysis.

The dramatic effect on exonuclease activity of mutations in the active-site carboxylates is consistent with the important role proposed for the metal ions. Additional inferences can be made by examining the data in detail. Mutations at Glu-357 are less severe than those at the other three carboxylates, suggesting that Glu-357 is the least important metal ligand. The more important function of Glu-357 may involve other interactions made by this residue that contribute to positioning the terminal nucleotide and the attacking nucleophile (see below). The results of asparagine substitutions at Asp-355, Asp-424, and Asp-501 are entirely consistent with the detailed coordination described for these residues. Both Asp-355 and Asp-424 use both of the carboxylate oxygens in metal coordination, and therefore the Asn substitution is not tolerated at these positions. By contrast, Asp-501 uses only one oxygen in metal binding and the D501N mutation has very little effect on the exonuclease activity. Interestingly, the E357A,D501N mutation causes a much greater loss in exonuclease activity than would be expected from the combination of the effects of the two single mutations. This might be the consequence of losing two negative charges in the metal-binding region or, perhaps, Asp-501 can assist in binding the attacking nucleophile when Glu-357 is absent.

Mutations in Residues That Contact the Substrate

The effect on the exonuclease reaction of mutations in these residues is variable but quantitatively smaller than the effect of mutations in the metal ligands (Table II). We have therefore concluded that these residues play an important but less pivotal role than the metal ions and their ligands. The structural data suggest a probable role for these side chains in presenting the DNA substrate and the attacking nucleophile in the correct orientation for efficient catalysis. One of the most important residues in this category is likely to be Glu-357, whose carboxylate side chain is involved in a complex network of interactions.[6] One carboxylate oxygen interacts with metal A, while the other serves as a hydrogen-bond acceptor, both to the 3′ hydroxyl of the substrate and to the attacking water molecule. Given the involvement of both oxygens in these interactions, the severe effect of the glutamine substitution (E357Q) is as expected. Intriguingly, the E357A mutation seems better tolerated than E357Q, perhaps because the smaller alanine side chain allows access of a water molecule. The results of mutations at Tyr-497 confirm the importance of the observed hydrogen-bonding interaction between the phenolic hydroxyl and one of the oxygens of the terminal phosphodiester bond, since removal of the hydroxyl group alone (Y497F) has a similar effect to removal of the entire side chain (Y497A). The properties of the remaining mutations presumably reflect the varying degrees of importance of the interactions between the protein and the single-stranded terminal region of the DNA substrate. The stacking interaction between the terminal base and Phe-473 is clearly of primary importance. The Leu-361 residue is particularly interesting in that the L361A mutation has a much greater effect on the hydrolysis of duplex DNA than on hydrolysis of single-stranded DNA, implying that the intercalation of Leu-361 between the nucleotide bases at the 3′ terminus may be particularly important in the fraying that must accompany movement of a duplex substrate into the exonuclease site. The observed interactions between Gln-419 and Arg-455 and the phosphodiester backbone upstream of the point of hydrolysis (see Table I) appear to make very little contribution to the overall reaction, at least with the assay methods currently used. Clearly, however, the introduction of a negative charge in this region (Q419E) has severe consequences, presumably by interfering with DNA binding.

3′ → 5′-Exonuclease Active Site of Other DNA Polymerases

It is now clear that all DNA polymerases that have an editing function possess three small sequence motifs, named Exo I, II, and III by Bernad *et al.*[9] Conversely, in polymerases that do not have this function, the motifs are either completely absent (as in *Taq* DNA polymerase) or lack critical catalytic residues (as in the eukaryotic α DNA polymerases).[11,12] The original alignment of the three Exo motifs[9] has been modified in subsequent work as it became apparent that a few polymerase sequences had been aligned incorrectly.[11,12] (This incorrect alignment had significant repercussions in the studies of T4 DNA polymerase, since mutagenesis of residues that had been mistakenly assigned to the Exo I region yielded exonuclease-proficient proteins,[31] and led to a lively debate as to the universality of exonuclease active site structures.[32]) The three Exo sequence motifs parallel almost exactly the active site residues noted in the Klenow fragment structure (Fig. 1; Table I), leading to the obvious inference that the sequence conservation reflects a similarity in the active sites of these enzymes. The Exo I motif contains the core sequence DXE, where the two acidic residues correspond to Asp-355 and Glu-357 of Klenow fragment. Exo II has the sequence $NX_{2\text{-}3}$(F/Y)D; in Klenow fragment the conserved residues are Asp-424 and Asn-420, the latter interacting with the substrate just upstream of the 3′ terminus.[6] The Exo III motif has the sequence YX_3D, containing the active site residues Tyr-497 and Asp-501 in Klenow fragment.

Table III summarizes the results of mutations in the conserved exonuclease motifs of a number of DNA polymerases. (For simplicity, mutations outside of the highly conserved residues have been omitted from consideration since, in the absence of structural data, their significance cannot be assessed.) The data of Table III provide strong support for the proposal of a common active-site architecture for all proofreading polymerases. Many of the exonuclease-deficient derivatives have wild-type levels of polymerase activity, consistent with the idea of polymerase and exonuclease active sites that are structurally independent of one another. Moreover, when detailed studies have been carried out, individual kinetic constants for the polymerase reaction have been found to be unaffected by mutations at the exonuclease site.[33–35] In a few enzymes (herpes simplex

[31] L. J. Reha-Krantz, S. Stocki, R. L. Nonay, E. Dimayuga, L. D. Goodrich, W. H. Konigsberg, and E. K. Spicer, *Proc. Natl. Acad. Sci. USA* **88,** 2417 (1991).

[32] L. J. Reha-Krantz, *Gene* **112,** 133 (1992).

[33] B. T. Eger, R. D. Kuchta, S. S. Carroll, P. A. Benkovic, M. E. Dahlberg, C. M. Joyce, and S. J. Benkovic, *Biochemistry* **30,** 1441 (1991).

[34] S. S. Patel, I. Wong, and K. A. Johnson, *Biochemistry* **30,** 511 (1991).

[35] M. W. Frey, N. G. Nossal, T. L. Capson, and S. J. Benkovic, *Proc. Natl. Acad. Sci. USA* **90,** 2579 (1993).

virus DNA polymerase,[36] ϕ29 DNA polymerase,[37] *Bacillus subtilis* DNA polymerase III,[38] and yeast mitochondrial DNA polymerase[39]), mutations in the exonuclease region have been found to influence some aspects of the polymerase reaction. The reasons for this are unclear at present.

Uses of Exonuclease-Deficient DNA Polymerases

Many of the mutations listed in Table III were made solely to test the hypothesis of a conserved exonuclease active site. Others were constructed, using the sequence alignments as a guide, to facilitate particular experiments or to provide a research tool with wider applications. For the T4 and T7 DNA polymerases, both of which have very active $3' \rightarrow 5'$-exonucleases, removal of the exonuclease was a necessary prerequisite for detailed kinetic studies of the polymerase function.[34,35] Exonuclease-deficient derivatives of Klenow fragment have proved invaluable in studying reactions involving mispaired bases at the polymerase site.[40] Inactivation of the exonuclease is also necessary for many biophysical experiments investigating the interaction between a polymerase and its DNA substrate in the presence of catalytically important metal-ion cofactors.

The particular attributes of an exonuclease-deficient polymerase that may make it useful as a research tool in biochemical manipulations are well illustrated by the properties of exonuclease-deficient derivatives of T7 DNA polymerase, which have found widespread use in DNA sequencing.[41,42] Because exonuclease-deficient T7 DNA polymerase cannot carry out the "idling" reaction (the turnover of dNTPs to dNMPs resulting from repeated incorporation and exonucleolytic excision by a stalled polymerase), it is better able to carry out strand-displacement synthesis or to synthesize through regions of secondary structure. Moreover, analogs such as dideoxynucleotides or α-thionucleotides are stably incorporated, instead of being rapidly removed by the exonuclease (as is the case with wild-type T7 DNA polymerase). The differences in the behavior of wild-type and exonuclease-deficient enzymes is much more pronounced for polymerases

[36] J. S. Gibbs, K. Weisshart, P. Digard, A. de Bruynkops, and D. M. Coen, *Mol. Cell. Biol.* **11,** 4786 (1991).

[37] M. S. Soengas, J. A. Esteban, J. M. Lázaro, A. Bernad, M. A. Blasco, M. Salas, and L. Blanco, *EMBO J.* **11,** 4227 (1992).

[38] M. H. Barnes, R. A. Hammond, C. C. Kennedy, S. L. Mack, and N. C. Brown, *Gene* **111,** 43 (1992).

[39] F. Foury and S. Vanderstraeten, *EMBO J.* **11,** 2717 (1992).

[40] C. M. Joyce, X. C. Sun, and N. D. F. Grindley, *J. Biol. Chem.* **267,** 24,485 (1992).

[41] S. Tabor and C. C. Richardson, *Proc. Natl. Acad. Sci. USA* **84,** 4767 (1987).

[42] S. Tabor and C. C. Richardson, *J. Biol. Chem.* **264,** 6447 (1989).

such as the T7 enzyme, which have a very active exonuclease, than for Klenow fragment with a slower exonuclease. Thus, wild-type Klenow fragment can itself carry out strand-displacement synthesis, is unable to degrade dideoxy- or α-thionucleotide termini, and gives the same pattern on sequencing gels as an exonuclease-deficient derivative.[20] Even when using Klenow fragment, however, an exonuclease-deficient enzyme may be preferable in circumstances where it is desirable to eliminate wasteful turnover of nucleotides, for example, when trying to incorporate a particular nucleotide analog or isotopically labeled nucleotide that is available only in small quantities. An important caveat when using an exonuclease-deficient enzyme in a "filling-in" reaction is that removal of the exonuclease activity may result in accumulation of products that are one nucleotide longer than expected.[43]

Other applications of exonuclease-deficient DNA polymerases take advantage of the inability of these enzymes to excise a mismatched DNA primer terminus. Although not widely used, procedures have been developed for mutagenesis by forced misincorporation and mismatch extension using an exonuclease-deficient polymerase, a strategy that is particularly well-suited for random mutagenesis over a defined region.[44] The detection of particular genetic traits by allele-specific amplification is based on the inability of the exonuclease-deficient polymerase either to remove a mismatched terminus or, under the chosen reaction conditions, to extend the mismatch.[45,46] Our understanding of the structure and mechanism of the $3' \rightarrow 5'$-exonuclease active site of DNA polymerases (as summarized in this chapter) is key to the development of exonuclease-deficient polymerases as biotechnology tools. Because of this knowledge, it should be a simple matter to design mutations to make any polymerase exonuclease deficient, and therefore there are no restrictions on choosing the polymerase with the most appropriate characteristics for the desired application.

Acknowledgments

During this work we have benefitted immensely from the insights into the $3' \rightarrow 5'$-exonuclease structure provided by Tom Steitz and colleagues. We are also grateful to Xiaojun Chen Sun for excellent technical assistance and to Nigel Grindley for a critical reading of the manuscript. This work was supported by the National Institutes of Health (grant GM-28550 to Nigel D. F. Grindley).

[43] J. M. Clark, C. M. Joyce, and G. P. Beardsley, *J. Mol. Biol.* **198,** 123 (1987).

[44] X. Liao and J. A. Wise, *Gene* **88,** 107 (1990).

[45] C. R. Newton, A. Graham, I. E. Heptinstall, S. J. Powell, C. Summers, and N. Kalsheker, *Nucleic Acids Res.* **17,** 2503 (1989).

[46] G. Sarkar, J. Cassady, C. D. K. Bottema, and S. S. Sommer, *Anal. Biochem.* **186,** 64 (1990).

Section V

Polymerase Accessory Functions, Replication Proteins, Multienzyme Replication Complexes

[29] Purification and Biochemical Characterization of Enzymes with DNA Helicase Activity

By STEVEN W. MATSON and DANIEL W. BEAN

Introduction

The term *DNA helicase* refers to a burgeoning group of enzymes able to catalyze the biochemical reaction depicted in Fig. 1. Helicases disrupt the hydrogen bonds that hold the two strands of duplex DNA together, in an energy-requiring reaction referred to as an *unwinding reaction*.[1-5] The covalent bonds in the backbone of the DNA molecule are left intact during the course of a helicase reaction, clearly distinguishing this class of proteins from the topoisomerases. The unwinding reaction catalyzed by a helicase exhibits a specific polarity defined with respect to the strand of DNA on which the helicase is presumed to be bound. Thus the reaction depicted in Fig. 1 is a $3' \rightarrow 5'$ unwinding reaction. If the helicase were bound on the opposite strand the reaction would be classified as a $5' \rightarrow 3'$ unwinding reaction. The energy requirement is satisfied by the hydrolysis of a nucleoside 5′-triphosphate (NTP), typically ATP, to yield NDP and P_i. Precisely how ATP binding/hydrolysis and the disruption of hydrogen bonds are coupled remains unknown. ATP binding and hydrolysis may be required to fuel unidirectional translocation of the enzyme along the DNA substrate through a series of conformational changes accompanied by changes in the affinity of the enzyme for single-stranded DNA (ssDNA) versus double-stranded DNA (dsDNA). Alternatively ATP hydrolysis may be directly coupled to hydrogen bond disruption. Note, however, that these possibilities are not mutually exclusive.

Tremendous progress has been made, particularly in the prokaryote *Escherichia coli,* in terms of isolating enzymes with helicase activity and defining roles for these proteins in the cell. It is now clear from both genetic and biochemical studies that the cell contains multiple helicases that catalyze the same basic biochemical reaction.[3,4] Each of these enzymes has a specific role in the cell as a participant in a particular biochemical pathway (Table I). There appears to be remarkably little redundancy of

[1] T. M. Lohman, *J. Biol. Chem.* **268,** 2269 (1993).
[2] S. W. Matson and K. A. Kaiser-Rogers, *Ann. Rev. Biochem.* **59,** 289 (1990).
[3] S. W. Matson, *Prog. Nucleic Acid Res. Mol. Biol.* **40,** 289 (1991).
[4] P. Thommes and U. Hubscher, *Chromosoma* **101,** 467 (1992).
[5] T. M. Lohman, *Mol. Microbiol.* **6,** 5 (1992).

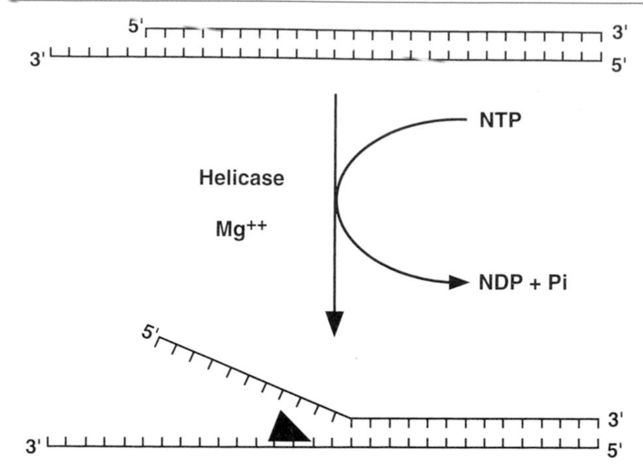

FIG. 1. Schematic representation of the unwinding reaction catalyzed by a helicase. See text for details.

TABLE I
Escherichia coli DNA HELICASES

Helicase	Molecular weight	Gene	Polarity	Physiological role
DnaB protein	52,265	*dnaB*	5′ → 3′	DNA replication
PriA protein	81,700	*priA*	3′ → 5′	DNA replication
Rep protein	76,400	*rep*	3′ → 5′	?
UvrAB complex	103,874	*uvrA*	5′ → 3′	Excision repair
	76,118	*uvrB*		
Helicase II	82,116	*uvrD*	3′ → 5′	Excision repair Mismatch repair
Helicase IV	78,033	*helD*	3′ → 5′	Recombination
RecQ protein	80,000	*recQ*	3′ → 5′	Recombination
RecBCD complex	133,973	*recB*	*a*	Recombination
	129,000	*recC*		
	66,973	*recD*		
RuvAB complex	22,000	*ruvA*	5′ → 3′	Recombination
	37,000	*ruvB*		
Helicase I	192,068	*traI* (F plasmid)	5′ → 3′	Conjugation

[a] The RecBCD helicase has a strong preference for unwinding blunt-ended duplex DNA substrates and does not appear to exhibit a polarity as defined for the other helicases.

function. Thus the *E. coli* DnaB protein serves as the helicase that unwinds duplex DNA ahead of the advancing replication fork and no other helicase in the cell is able to substitute in this role. Similarly, *E. coli* helicase II is involved in the methyl-directed mismatch repair pathway (and other repair pathways) and the closely related Rep protein cannot substitute in this role. Although many helicases have now been isolated from eukaryotic cells, little is known regarding their functions in the cell. Undoubtedly the roles played by helicases in the eukaryotic cell will be as rich and varied, perhaps even more so, than those already discovered in prokaryotic cells.

A first step in understanding the physiological roles played by this important class of proteins in eukaryotic cells is isolation and characterization of proteins with helicase activity. This chapter focuses on methods and techniques developed to study helicases in *E. coli* that also have utility in the study of helicases from eukaryotic cells. Indeed, many of these techniques are already being successfully applied to the isolation and characterization of helicases from sources ranging from yeast to human cells. In addition, many of these approaches can be modified for the study of helicases that unwind DNA–RNA hybrid molecules or regions of duplex RNA.

Biochemical Characterization of Purified Helicases

Biochemical characterization of the reaction(s) catalyzed by a helicase is accomplished by analyzing both the DNA-dependent nucleoside 5′-triphosphatase (NTPase) reaction and the unwinding reaction catalyzed by the purified protein. Specific biochemical and kinetic parameters are measured to provide a description of the enzymatic properties of each protein. This is useful in distinguishing one helicase from another in the absence of genetic data and provides insight into the cellular role played by the enzyme.

Characterization as DNA-Dependent NTPase

The DNA-dependent NTPase reaction can be analyzed using one of several assays available to measure the hydrolysis of an NTP to NDP and P_i. The assays most commonly used include (1) a thin-layer chromatography assay, (2) a charcoal absorption assay, and (3) a spectrophotometric assay. The thin-layer chromatography assay[6] separates NTP from NDP on a polyethyleneimine plate by ascending chromatography in a formic acid/LiCl solvent system. The helicase-catalyzed NTPase reaction is quenched using EDTA and a high concentration of unlabeled NTP and NDP. This allows the spots corresponding to NDP and NTP to be visualized by UV light,

[6] S. W. Matson and C. C. Richardson, *J. Biol. Chem.* **258,** 14,009 (1983).

cut out, and directly counted in a liquid scintillation counter to measure the conversion of NTP to NDP. The assay is relatively fast, simple to perform, and [^{3}H]NTP can be utilized as the substrate. The charcoal absorption assay[7] is also easy to perform using [γ-^{32}P]NTP as the substrate. In this case the charcoal binds NDP and NTP leaving the $^{32}PO_4$ in solution. After centrifugation to pellet the charcoal, the supernatant can be counted in a liquid scintillation counter to measure the release of $^{32}PO_4$. Finally, a spectrophotometric assay[8] can be used that monitors the conversion of NAD to NADH in the presence of phosphoenolpyruvate, pyruvate kinase, and lactate dehydrogenase. This assay is useful for kinetic studies since the hydrolysis of NTP to NDP can be continuously monitored using a spectrophotometer.

All helicases isolated to date are either DNA-stimulated or DNA-dependent NTPases. Determination of the preferred nucleic acid effector in the NTPase reaction is accomplished by titration of nucleic acids of known structure (and sequence) into reactions containing an NTP at saturating concentration, Mg^{2+}, and the helicase. Generally ssDNA in the form of M13 ssDNA (or some other phage chromosome) and poly(dT) are tested as effectors of the NTPase reaction. In some cases one ssDNA will be preferred over another due to sequence or structural considerations. For example, *E. coli* DNA helicase I utilizes poly(dT) very poorly as an effector of its ATPase reaction due to the linear structure of poly(dT). The circular M13 ssDNA molecule, on the other hand, is the preferred nucleic acid cofactor in the ATPase reaction catalyzed by helicase I. This is due to the processivity of the enzyme and the fact that helicase I, which translocates $5' \rightarrow 3'$ on ssDNA, does not readily dissociate from the $3'$ end of a linear DNA molecule.[9] Double-stranded DNA in the form of a linear dsDNA molecule and a supercoiled DNA molecule can be tested in a similar manner. Often supercoiled DNA proves to be more stimulatory in the NTPase reaction than linear dsDNA, presumably due to the ssDNA character present in superhelical DNA. RNA should also be tested as a nucleic acid effector to help distinguish between proteins that interact with and unwind DNA and those that interact with and unwind RNA or RNA and DNA. T antigen, for example, utilizes both DNA and RNA as effectors of the NTPase reaction.[10] The biochemical characteristics of each reaction are slightly different.

[7] G. Siegal, J. J. Turchi, C. B. Jessee, T. W. Myers, and R. A. Bambara, *J. Biol. Chem.* **267,** 13,629 (1992).

[8] T. M. Lohman, K. Chao, J. M. Green, S. Sage, and G. T. Runyon, *J. Biol. Chem.* **264,** 10,139 (1989).

[9] E. E. Lahue and S. W. Matson, *J. Biol. Chem.* **263,** 3208 (1988).

[10] M. Scheffner, R. Knippers, and H. Stahl, *Cell* **57,** 955 (1989).

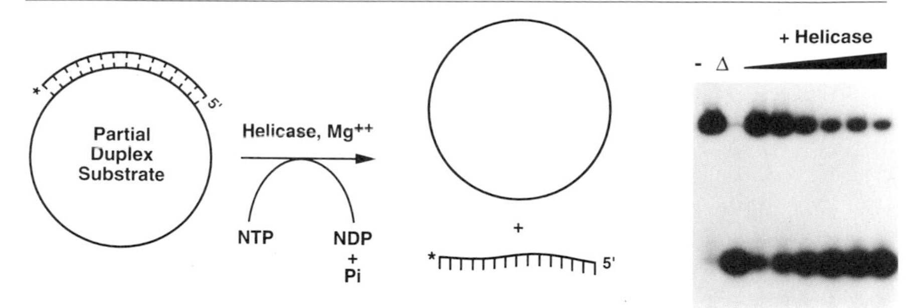

FIG. 2. Assay for measuring helicase activity *in vitro*. See text for details.

It is also useful to know which NTP is the preferred substrate for the protein as an NTPase. Determination of K_m, V_{max}, and k_{cat} parameters is accomplished at saturating concentrations of the appropriate nucleic acid effector. Many helicases utilize either ATP or dATP as evidenced by equivalent K_m and V_{max} values obtained using either substrate. In some cases these are the only (d)NTPs that can be efficiently hydrolyzed by the protein. In other cases all eight (d)NTPs can be utilized with nearly equal efficiency. Knowledge of the preferred NTP, and the kinetic parameters of the NTPase reaction, are useful when designing helicase activity assays and when comparing two helicases with similar physical properties. In most cases it is necessary to measure kinetic parameters with only one or a few (d)NTPs since the remaining (d)NTPs fail to support the unwinding reaction catalyzed by the protein.

Characterization as DNA Helicase

The assay most commonly used to measure helicase activity is shown in Fig. 2. The original DNA substrates used in this assay were prepared by annealing the complementary strand of an appropriate restriction fragment onto M13 or ϕX174 ssDNA.[11,12] In the presence of an NTP and Mg^{2+} the helicase unwinds the duplex region present in this partial duplex substrate, causing displacement of the DNA fragment from the larger circular ssDNA. The DNA fragment to be displaced can be labeled at either the 5′ end using polynucleotide kinase or at the 3′ end using the large fragment of DNA polymerase I. We routinely employ one of the restriction fragments generated by *Hae*III digestion of M13 replicative form I DNA to construct partial duplex DNA substrates that are conveniently labeled at the 3′

[11] S. W. Matson, S. Tabor, and C. C. Richardson, *J. Biol. Chem.* **258,** 14,017 (1983).
[12] M. Venkatesan, L. L. Silver, and N. G. Nossal, *J. Biol. Chem.* **257,** 12,426 (1982).

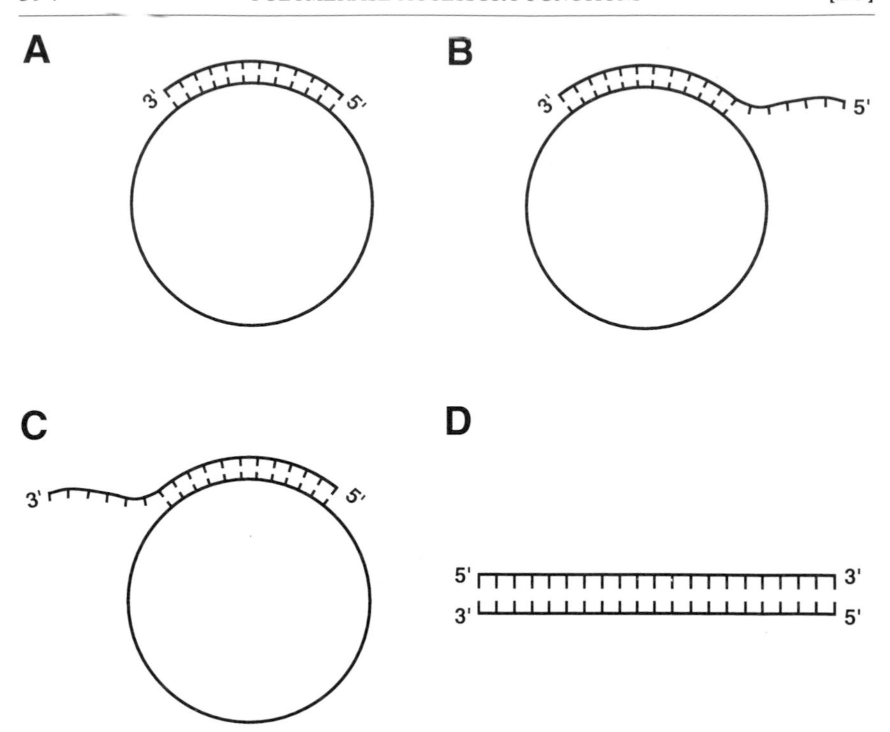

FIG. 3. DNA substrates commonly used in helicase assays. (A) Partial duplex DNA substrate. (B) Partial duplex DNA substrate with a single-stranded 5′ tail on the fragment to be displaced. (C) Partial duplex DNA substrate with a single-stranded 3′ tail on the fragment to be displaced. (D) Fully duplex DNA substrate.

end by the incorporation of two [α-^{32}P]dCMP residues (Fig. 3A). After chromatography on Bio-Gel A5M or Sepharose, to remove both the unincorporated [α-^{32}P]dCTP and the noncomplementary strand of the restriction fragment, the substrate can be used directly in unwinding reactions. More recently, oligonucleotides[13] or the product of asymmetric PCR (polymerase chain reaction)[14] have been used to provide the DNA strand that is annealed onto the M13 ssDNA molecule. When preparing partial duplex substrates for use in helicase activity assays it is important to note that some oligonucleotides may be too short to reliably measure bona fide helicase activity. Marians and colleagues[15] have shown that PriA protein unwinds very short duplex regions (i.e., <25 bp) with different kinetic

[13] J. J. Turchi, R. S. Murante, and R. A. Bambara, *Nucleic Acids Res.* **20,** 6075 (1992).
[14] H. Naegeli, L. Bardwell, and E. C. Friedberg, *J. Biol. Chem.* **267,** 392 (1992).
[15] M. S. Lee and K. J. Marians, *J. Biol. Chem.* **265,** 17,078 (1990).

parameters than longer duplex regions, perhaps because the enzyme simply "knocks" the short oligonucleotide off the DNA as opposed to unwinding the longer duplex region actively. We have obtained similar results with *E. coli* helicase II and short (50 bp) duplex regions (unpublished results). Thus duplex regions in excess of 50 bp are more appropriate than very short duplex regions. The circular partial duplex molecule offers advantages over a linear substrate in that a helicase with either polarity can catalyze unwinding of the duplex region. In addition, this substrate may be somewhat more stable in the presence of nucleases than a linear substrate.

In the presence of a helicase, ATP, and Mg^{2+} the duplex region of the substrate is disrupted and the [^{32}P]DNA fragment is displaced from the circular ssDNA molecule. This event is conveniently assayed by resolving the products of the reaction on a polyacrylamide gel (Fig. 2). We routinely assay helicase activity using partial duplex molecules ranging from 71 to 851 bp in length by this method. Alternatively, an agarose gel can be used when substrates containing longer duplex regions are employed. In each case the gel is run under nondenaturing conditions to determine the fraction of the substrate that has been unwound (i.e., rendered single stranded) by the helicase. Denaturation of the substrate in a control reaction using either heat (3 to 5 min at 95°) or NaOH (0.2 *M*) provides an appropriate size marker to distinguish between nuclease activity and unwinding of the substrate. Polyacrylamide gel electrophoresis can also be used to resolve the products of an unwinding reaction involving a fully duplex DNA fragment (see below). The ssDNA products of the unwinding reaction will migrate to a different position in the gel than the duplex DNA substrate. In fact, the two ssDNA products of this reaction are often resolved on polyacrylamide gels of the appropriate percentage and run under the correct conditions. These conditions depend on the sequence and length of the DNA substrate and must be determined empirically.

It is necessary to establish the Mg^{2+} and NTP hydrolysis dependence of the reaction being measured. The Mg^{2+} dependence is established by removing Mg^{2+} from the reaction and including a modest concentration (<1 m*M*) of EDTA to chelate any Mg^{2+} that may be present in buffers or the DNA. Different helicases display different Mg^{2+} concentration dependence and it is useful to determine the Mg^{2+} concentration required for maximal helicase activity. It is also important to establish the NTP hydrolysis dependence of the unwinding reaction. To a first approximation, this can be accomplished by removing the NTP from the reaction mixture. In the absence of added NTP there should be little or no unwinding of the DNA substrate. To establish a requirement for NTP hydrolysis, the hydrolyzable NTP can be replaced in the helicase reaction with a nonhydrolyzable NTP analog such as β,γ-methylene-NTP or a poorly hydrolyzed analog such as

NTP(γ)S; both are commercially available. More rigorous demonstration of the hydrolysis dependence of the reaction would include the direct demonstration that the β,γ-methylene-NTP or the NTP(γ)S is a competitive inhibitor of the NTP hydrolysis reaction. This serves to ensure that the NTP analog binds at the NTP binding site on the protein.

The results of the helicase assay are easily quantified by one of three methods: (1) slicing the gel and directly counting gel slices in a liquid scintillation counter, (2) densitometry of the autoradiograph obtained by exposing the gel to X-ray film, or (3) phosphorimage analysis of the polyacrylamide gel. We have used each of these techniques and find them all to be reproducible and consistent with one another. The fraction of the substrate unwound by the helicase can be determined using Eq. (1):

$$\%\ \text{Substrate unwound} = \frac{X_p - B_p}{(X_s - B_s) + (X_p - B_p)} \times 100, \qquad (1)$$

where X_p is the amount of DNA as unwound product; B_p is the background in the unwound product position determined from a no enzyme control; X_s, amount of DNA in the substrate position; and B_s, background in the substrate position determined from a heat denatured control. The fraction of the substrate unwound is then easily converted into base pairs unwound, or femtomoles of DNA fragment unwound, as desired.

In some cases modified partial duplex substrates with ssDNA tails on either the 5′ end (Fig. 3B) or the 3′ end (Fig. 3C) of the DNA fragment to be displaced have been found to be used preferentially by specific DNA helicases. The first example of this was the phage T7 gene 4 protein,[11] which is markedly stimulated by the presence of a ssDNA tail of at least 7 nucleotides in length on the 3′ end of the fragment to be displaced. Since this helicase translocates along ssDNA in the 5′ → 3′ direction,[16] it will first approach the 3′ end of the DNA fragment to be displaced. The results suggest that the T7 gene 4 protein prefers to encounter a "frayed" or partially unwound duplex in order to catalyze an efficient and processive unwinding reaction. Perhaps the enzyme interacts with both DNA strands of the duplex as it translocates through and unwinds the duplex region. Note that the structure presented by such a substrate is the structure expected at the growing point of a replication fork where the single-stranded 3′ tail represents the leading strand template and the circular ssDNA represents the lagging strand template. The gene 4 protein is also a primase that moves along the lagging strand template to synthesize concomitantly primers on the lagging strand and catalyze unwinding of the helix ahead of the advancing T7 DNA polymerase. The phage T4 gene 41 helicase[17] and DnaB

[16] S. Tabor and C. C. Richardson, *Proc. Natl. Acad. Sci. USA* **78,** 205 (1981).

[17] R. W. Richardson and N. G. Nossal, *J. Biol. Chem.* **264,** 4725 (1989).

protein from *E. coli*[18] have also been shown to have preferential activity on partial duplex substrates with single-stranded 3′ tails. It is important to evaluate helicase activity using partial duplex substrates with ssDNA tails in a thorough biochemical characterization of a new helicase.

Partial duplex substrates with ssDNA tails on either end are conveniently constructed using oligonucleotides that contain an appropriate region of DNA that is not complementary to the DNA strand to which the fragment will be annealed. In this case care must be taken to ensure that the duplex region is sufficiently long and to ensure that the ssDNA tail will be of sufficient length. Generally duplex regions of >50 bp in length and ssDNA tails that are about 20 nucleotides long are appropriate, although this may not always be the case and depends on the specific helicase. Alternatively, substrates of this type can be constructed using restriction fragments that contain noncomplementary DNA at one end or the other.[11] This can be achieved by cloning the "noncomplementary DNA" sequence into M13 and then annealing a fragment containing this sequence and additional M13 DNA sequence to phage DNA prepared from M13 lacking the noncomplementary sequence insert. Finally, a 3′ tail can be added to a partial duplex substrate using terminal deoxynucleotidyltransferase.[19] In this case, the length of the noncomplementary ssDNA tail is heterogeneous but adequate for the purpose of measuring helicase activity.

It is also appropriate, in some cases, to use 5′-end-labeled, blunt-ended, fully duplex restriction fragments (Fig. 3D) directly as substrates in helicase reactions. Most helicases do not catalyze the unwinding of this DNA substrate due to a requirement for an initial binding site composed of ssDNA. However, some helicases will unwind fully duplex DNA substrates. *E. coli* DNA helicase II[20] and large T antigen[21] are two cases in point. Helicase II will unwind blunt duplex fragments of any sequence provided the protein is present at sufficiently high concentration. Unwinding in this case has been shown to initiate at the ends of the DNA molecule. Large T antigen will unwind a duplex DNA fragment provided it contains the SV40 origin of replication. In this case unwinding is initiated at the T antigen binding site within the simian virus 40 (SV40) replication origin.

Unwinding DNA–RNA Substrates

The partial duplex substrates described above can be modified to produce a DNA–RNA hybrid by substituting the DNA strand to be displaced

[18] J. H. LeBowitz and R. McMacken, *J. Biol. Chem.* **261,** 4738 (1986).

[19] M. S. Lee and K. J. Marians, *J. Biol. Chem.* **264,** 14,531 (1989).

[20] G. T. Runyon, D. G. Bear, and T. M. Lohman, *Proc. Natl. Acad. Sci. USA* **87,** 6383 (1990).

[21] G. S. Goetz, F. B. Dean, J. Hurwitz, and S. W. Matson, *J. Biol. Chem.* **263,** 383 (1988).

with an RNA transcript.[10,22] The desired region of complementarity is cloned behind an appropriate phage RNA polymerase promoter like T3, T7, or SP6 and transcription reactions are performed in the presence of [α-^{32}P]UTP to generate a labeled RNA. After DNase digestion to destroy the DNA template, the RNA can be isolated on a polyacrylamide gel or on a sizing column. The purified transcript is then annealed onto an appropriate ssDNA to produce the DNA–RNA hybrid substrate. It is important to note that in most cases this substrate will contain a 5′ tail on the RNA that is not hybridized to the DNA since the promoter sequence is often not complementary to any region in the DNA strand. This substrate can be used to measure the ability of a DNA helicase to catalyze the unwinding of a DNA–RNA hybrid.

Polarity of Helicase Reaction

All helicases isolated to date appear to unwind duplex DNA with a specific polarity that is defined with respect to the DNA strand on which the helicase is presumed to be bound. The polarity exhibited by a specific helicase can be determined using the assay shown in Fig. 4.[23] A linear partial duplex substrate is prepared by annealing the complementary strand of an appropriate restriction fragment to the M13 phage chromosome with subsequent digestion by a restriction enzyme that cuts singly within the duplex region. All available 3′ ends are then labeled using an [α-^{32}P]dNTP and the large fragment of DNA polymerase I. This produces a linear molecule with a large internal region of ssDNA to provide an initial binding site for the helicase, and duplex regions of different length at each end. Depending on the polarity of the helicase in question, one fragment or the other will be displaced in an unwinding reaction. This defines the polarity of the unwinding reaction, assuming the helicase bound initially to the ssDNA and translocated in one direction to the junction of ssDNA and duplex DNA. It is important to determine that duplex regions of the lengths represented in the substrate can be unwound by the helicase prior to determining the polarity of the unwinding reactions. It is also important to determine whether the helicase can unwind duplex DNA initiating at a blunt duplex end. In the latter case both of the DNA fragments may be displaced by the protein and care must be taken in assigning a polarity to the unwinding reaction. Finally, the duplex regions at either end of the DNA substrate must be sufficiently long to support a bona fide unwinding reaction catalyzed by the helicase. We routinely construct a 341 bp partial

[22] S. W. Matson, *Proc. Natl. Acad. Sci. USA* **86,** 4430 (1989).
[23] S. W. Matson, *J. Biol. Chem.* **261,** 10,169 (1986).

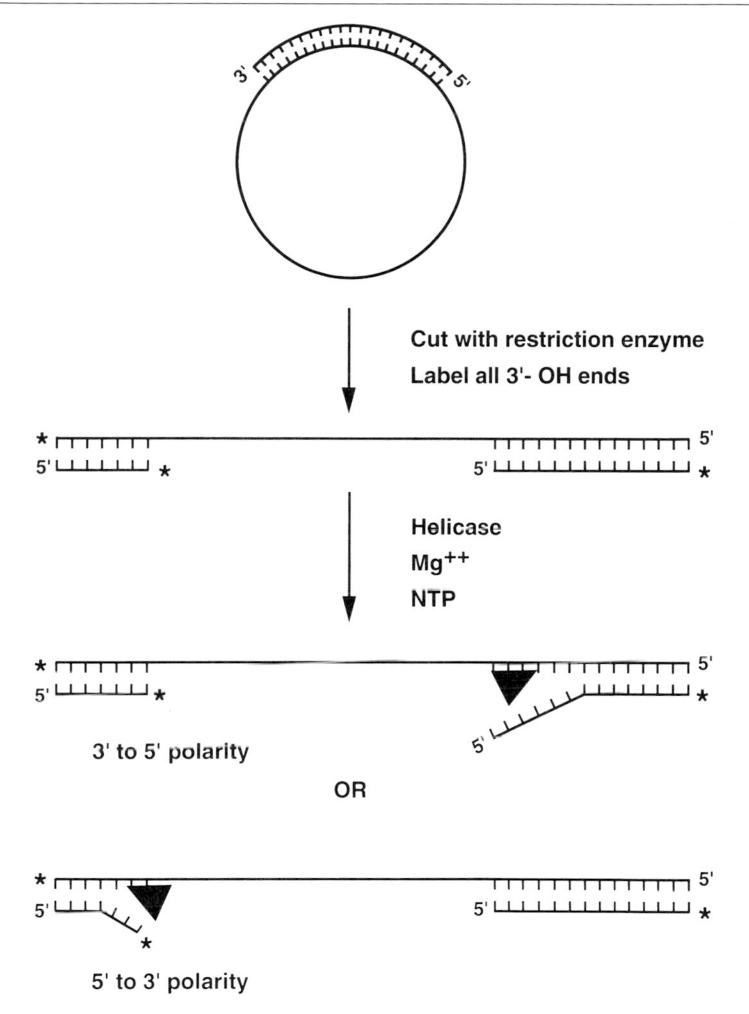

FIG. 4. Scheme for determining the polarity of a helicase unwinding reaction. See text for details.

duplex molecule that, when cleaved with *Cla*I and labeled at all available 3′ ends, yields duplex regions of 141 and 202 bp.

Determining Macroscopic Reaction Mechanism

Most helicases isolated to date exhibit a macroscopic reaction mechanism that can be categorized as processive, distributive (protein concentration dependent), or limited.[24] Assignment of the macroscopic reaction mechanism can be accomplished by characterizing the unwinding reaction catalyzed by a specific helicase using partial duplex DNA substrates with duplex regions of increasing length. We have assigned macroscopic reaction mechanisms to helicases I, II, IV, Rep protein, and SV40 T antigen[9,21,25-27] using partial duplex substrates containing duplex regions of 71, 119, 343, and 851 bp. Initially a simple titration of the helicase in reaction mixtures containing a specific substrate is performed. The data obtained are quantified as the fraction of the substrate unwound at each concentration of protein. These measurements should be made using an excess of substrate and at early times in the reaction. In some cases the use of an ATP regenerating system is important to ensure that a diminishing ATP concentration does not bias the result obtained. Alternatively, high concentrations of ATP and early time points serve to ensure that the ATP concentration has not become limiting.

A processive helicase is expected to catalyze the unwinding of very similar if not identical fractions of each substrate at any given protein concentration with no dependence on the length of the duplex region (Fig. 5A). This indicates that a single active species (monomer, dimer, multimer) can catalyze the unwinding of reasonably long regions of duplex DNA. *Escherichia coli* DNA helicase I[9] and SV40 T antigen[21] both catalyze this type of unwinding reaction and are capable of unwinding duplex regions in excess of 1 kbp in length.

A helicase that catalyzes a protein concentration-dependent (distributive) unwinding reaction unwinds progressively smaller fractions of the substrate as the length of the duplex region increases (Fig. 5B). When these data are expressed as base pairs unwound as a function of protein concentration it becomes apparent that, at a given protein concentration, the same number of base pairs is unwound on each substrate. However, since the length of the duplex region increases from one substrate to the next, the *fraction* of the substrate unwound decreases. *E. coli* DNA helicase II catalyzes this type of unwinding reaction.[26] Electron microscopic studies indicate that the enzyme coats the DNA that has been unwound and thus

[24] S. W. Matson, J. W. George, K. A. Kaiser-Rogers, E. E. Lahue, E. R. Wood, and J. E. Yancey, *in* "Molecular Mechanisms in DNA Replication and Recombination" (C. C. Richardson and I. R. Lehman, eds.), p. 127, Wiley-Liss, New York, 1990.

[25] E. R. Wood and S. W. Matson, *J. Biol. Chem.* **262,** 15,269 (1987).

[26] S. W. Matson and J. W. George, *J. Biol. Chem.* **262,** 2066 (1987).

[27] J. E. Yancey-Wrona, E. R. Wood, J. W. George, K. R. Smith, and S. W. Matson, *Eur. J. Biochem.* **207,** 479 (1992).

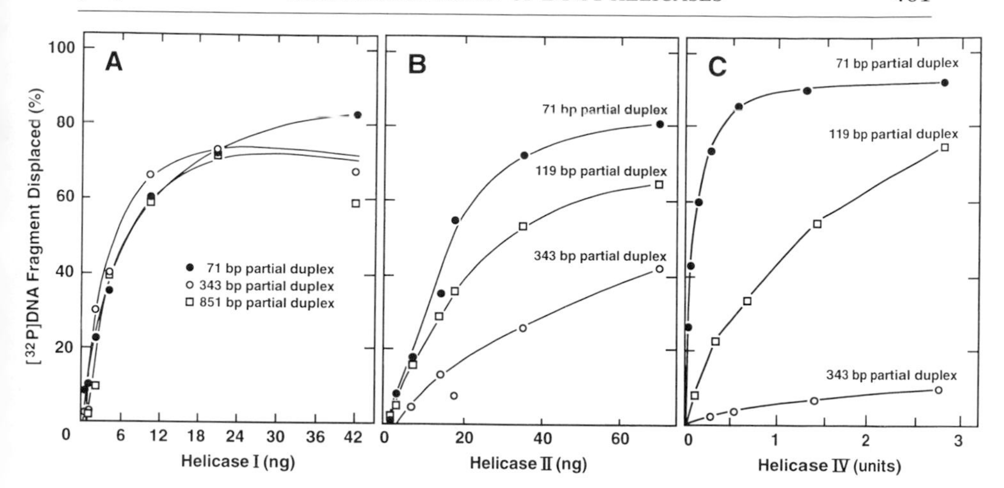

FIG. 5. Determination of the macroscopic unwinding reaction mechanism using partial duplex DNA substrates of increasing length. (A) Processive helicase (*E. coli* DNA helicase I). Reprinted with permission from Ref. 9. Copyright The American Society for Biochemistry & Molecular Biology. (B) Protein concentration-dependent helicase (*E. coli* helicase II). Reprinted with permission from Ref. 26. Copyright The American Society for Biochemistry & Molecular Biology. (C) Limited helicase (*E. coli* helicase IV). Reprinted with permission from Ref. 27.

stoichiometric amounts of protein are required to unwind the duplex region.[28]

Finally, some proteins catalyze an unwinding reaction that we have referred to as a "limited" unwinding reaction.[27] In this case the enzyme unwinds relatively short duplex regions (e.g., 71 bp) efficiently but unwinds longer (e.g., 343 bp) duplex regions poorly (Fig. 5C). Interpretation of this result is difficult, but presumably it is due to the fact that the enzyme encounters some sort of topological or kinetic barrier that prevents unwinding of long duplex regions. Possible explanations for this result include the lack of an additional protein factor(s), which would convert the helicase into a processive enzyme, reannealing of the two strands of DNA behind the advancing helicase, or a substrate that does not accurately mimic the natural *in vivo* substrate. Note that these possibilities are not mutually exclusive. It is useful to determine the macroscopic reaction mechanism exhibited by a purified helicase because it may help to determine the biochemical role of the protein in the cell. It is also useful in the search for additional proteins that may interact with and alter the reaction mechanism

[28] P. Georgi-Geisberger and H. Hoffmann-Berling, *Eur. J. Biochem.* **192,** 689 (1990).

exhibited by a specific helicase. For example, scHelI, a DNA helicase recently isolated from yeast, displays a limited unwinding reaction.[29] The addition of *E. coli* SSB appears to increase the processivity of the enzyme, enabling it to unwind substrates as long as 851 bp while dramatically increasing the efficiency with which it unwinds substrates of 119 and 343 bp in length (D. W. Bean, unpublished results).

Studies of the microscopic reaction mechanism used by helicases to unwind duplex DNA suggest that most, and perhaps all, helicases form dimers or higher order multimers when bound on DNA. Thus, although many helicases behave as monomers in solution, they dimerize (oligomerize) when bound on DNA. Lohman[5] and colleagues have proposed working models of the unwinding reaction catalyzed by a helicase that depend on dimers (oligomers) of the protein as the active species.

Purification of DNA Helicases from Cell Extracts

Assays Used to Monitor Purification

An ATPase assay is commonly used to monitor helicase activity during the course of purification, taking advantage of the fact that helicases catalyze a DNA-dependent NTPase reaction, in addition to the DNA unwinding reaction. The DNA-dependent NTPase assay is easy to perform, highly reproducible, and provides the quantitative data required for effective protein purification. In addition, the NTPase assay is usually not affected by the presence of nucleases. Column fractions may be assayed for DNA-dependent NTPase activity using any one of the several methods discussed above. We prefer to use [^{3}H]ATP as a substrate and resolve reaction products using thin-layer chromatography since this method allows the rapid assay of many fractions and the substrate is stable. In general, helicases are maximally stimulated by ssDNA and therefore phage M13 ssDNA, which is easily purified in large quantities, provides a good DNA effector for use in assays to monitor purification. Since fractions obtained during early purification steps are likely to be heavily contaminated with nuclease activity, a high concentration of ssDNA (>30 μM, nucleotide equivalents) is included in the reaction mixture. This ssDNA concentration is many times greater than that required to stimulate fully the DNA-dependent NTPase activity of the helicases we commonly work with and helps to ensure that the DNA effector is not entirely degraded by nuclease(s) during the incubation period. Fractions may be assayed in the presence and absence of DNA to determine the DNA dependence of the ATPase activity profile.

[29] D. W. Bean, W. E. Kallam, Jr., and S. W. Matson, *J. Biol. Chem.* **268,** 21,783 (1993).

The DNA-dependent NTPase assay can be particularly useful when isolating DNA helicases with unknown nucleic acid substrate requirements. For example, the unwinding activity of some helicases cannot be detected using a simple partial duplex DNA helicase substrate like that shown in Fig. 3A. A DNA helicase like the T7 gene 4 protein[11] requires a forked substrate (Fig. 3B) in order to observe unwinding activity. However, this protein will exhibit NTPase activity in the presence of ssDNA.

The primary disadvantage of the DNA-dependent NTPase assay is the fact that it is neither specific for any one helicase nor for helicases as a class of proteins. Thus, it is possible to monitor an enzyme activity through many purification steps and find that it is not due to a helicase. For this reason assays that directly measure helicase activity should be employed in the purification protocol at the earliest possible step to ensure that the DNA-dependent NTPase activity being monitored is due to a helicase. Before deciding that a DNA-dependent NTPase activity is not likely a DNA helicase, it is important to test it using several substrate types and to include SSBs in the unwinding reaction assays.

The primary reason that a direct helicase assay (described above) is not used initially to monitor enzyme purification is that both the substrate and the product of the helicase reaction are susceptible to nuclease degradation. This makes it very difficult to detect helicase activity during early purification steps, and impossible to quantify that activity in a way that is necessary to optimize purification and protein yields. Thus proteins must usually be taken through several purification steps before a direct demonstration of helicase activity is possible, and often the accurate quantitation of helicase activity requires even further purification. To overcome this problem, we have recently tested a partial duplex substrate constructed using a 52-mer oligonucleotide made with sulfur-substituted nucleotides and annealed to M13 ssDNA (S. W. Matson, unpublished results). The sulfur substitutions in the phosphodiester backbone make this substrate more resistant to the action of nuclease contaminants present during early steps in helicase purification. We have used this substrate to detect individual peaks of helicase activity during the first chromatographic step in the fractionation of yeast whole cell homogenates, a stage at which nucleases obscure helicase activity measured with the usual DNA substrate. The sulfur-substituted DNA substrate thus shows promise in eliminating some of the problems presented by nucleases in helicase purification and detection.

Methods of Fractionation

Most helicases have been purified using standard methods of conventional column chromatography or FPLC. Purification protocols take advantage of the fact that proteins that bind DNA often also bind phosphocellulose and heparin-agarose as well as DNA-cellulose resins. In addition, helicases also necessarily bind an NTP, usually ATP, and ATP-agarose can be exploited as an affinity resin for purification of enzymes with DNA helicase activity.

In our purification protocols designed to obtain pure helicases from crude yeast cell extracts,[29] we have chosen phosphocellulose as the initial chromatography resin because many helicases bind this resin at moderate (0.1 *M*) salt concentrations and because large, high-capacity columns can be poured at relatively little expense. Subsequent columns, which have proven effective for purifying yeast helicases, are single- and double-stranded DNA cellulose, heparin-agarose, and in some cases an anion-exchange resin such as DEAE cellulose. When ATP-agarose is included in the purification procedure, the enzyme can be eluted with a salt gradient, but we have obtained better results using ATP-agarose as a true affinity resin. The following general protocol has been used with good success to isolate yeast DNA helicases:

1. Load fractions from a previous column (e.g., dsDNA cellulose or heparin-agarose) on the ATP agarose column at a flow rate of 5 column volumes per hour and a salt concentration of 100 m*M* NaCl in a buffer containing 20 m*M* KPO_4, pH 7.0, 0.1 m*M* EDTA, 5 m*M* DTT, 0.01% Nonidet P-40 (NP-40), 5 m*M* $MgCl_2$, and 20% glycerol. Load approximately 1 to 2 mg of protein per milliliter of resin.
2. Wash with 5 column volumes of loading buffer + 100 m*M* NaCl.
3. Wash with 10 to 15 column volumes of loading buffer + 150 m*M* NaCl.
4. Elute with 5 m*M* ATP in loading buffer containing 150 m*M* NaCl.

The helicase activity should elute in a sharp peak with the ATP wash. ATP elution is monitored spectrophotometrically at a wavelength of 259 nm.

Following affinity chromatography the helicases are often nearly homogeneous. Because no size-fractionation step has been included to this point, we utilize density gradient ultracentrifugation as the final fractionation step in the purification of yeast helicases. We have chosen ultracentrifugation as a size separation step because size exclusion chromatography, even when performed at various salt concentrations ranging from 0.1 to 0.5 *M*, has proven unsatisfactory. Apparently the combination of low protein concentration, and perhaps a relatively high affinity of the helicases for the carbohydrate matrix of sizing resins, has resulted in large losses of enzymatic activity. In contrast, nearly full recovery of biochemical activity is routinely obtained following ultracentrifugation through glycerol gradients (15 to 35% glycerol, 15 to 30 hr depending on the sedimentation coefficient of the enzyme, 55,000 rpm in a Beckman SWTi55 rotor). Persistent low-molecular-weight impurities are often well resolved from the active helicase during this step. Aliquots of the glycerol gradient fractions can then be precipitated with 10% trichloroacetic acid (TCA) and analyzed on polyacrylamide gels to assess purity and to correlate a specific polypeptide with the peaks of both ATPase and helicase activities. Typically the purified helicase is a single polypeptide, but there are notable exceptions. The *E. coli* RecBCD helicase, for example, is composed of three polypeptides and these are purified as a complex. RuvA and RuvB proteins can be purified separately but exhibit helicase activity only when both proteins are present.[30] This is also true of the UvrAB helicase.[31]

Acknowledgments

We wish to thank Susan Whitfield for the artwork. Work from the author's laboratory has been supported by NIH grants GM33476 and GM50279 and American Cancer Society grant NP833.

[30] I. R. Tsaneva, B. Muller, and S. C. West, *Proc. Natl. Acad. Sci. USA* **90,** 1315 (1993).
[31] E. Y. Oh and L. Grossman, *Proc. Natl. Acad. Sci. USA* **84,** 3638 (1987).

[30] Characterization of DNA Primases

By LYNN V. MENDELMAN

Introduction

DNA primases polymerize ribonucleoside triphosphates into short oligoribonucleotides on single-stranded (ss) DNA templates. The short oligoribonucleotides are used by DNA polymerases as primers to initiate the synthesis of Okazaki fragments during bidirectional DNA replication.[1] The DNA primase from bacteriophage T7, for example, translocates $5' \rightarrow 3'$ along the displaced lagging strand until it recognizes a specific primase recognition site, 3′-CTGG(G/T)-5′ or 3′-CTGTG-5′. The primase stops its $5' \rightarrow 3'$ movement on the single-stranded template and synthesizes an oligoribonucleotide primer of defined sequence:

3′-nnnnnnCTGGGnnnnnn-5′ + ATP + CTP →
pppACCC
3′-nnnnnnCTGGGnnnnnn-5′

[1] A. Kornberg and T. A. Baker, "DNA Replication," Freeman, New York, 1992.

The 3′-cytosine of the primase recognition site is conserved but silent during template-directed primer synthesis by T7 primase. RNA primers are stabilized by T7 DNA primase for the T7 DNA polymerase, which will use deoxynucleoside triphosphates to elongate the primer into nascent DNAs or Okazaki fragments.[1]

In some bacteriophage systems, DNA primase initiated leading strand DNA synthesis. For example, bacteriophage G4 requires the DNA primase of *Escherichia coli,* DnaG, to synthesize a 27–29 base primer to initiate leading strand DNA synthesis.[2] Initiation of leading strand DNA synthesis is template directed and requires the presence of a trinucleotide sequence, 5′-CTG-3′, as a priming recognition signal.[3]

DNA primases are encoded by plasmids, bacteriophages, viruses, and prokaryotic and eukaryotic cells. All DNA primases catalyze the synthesis of oligoribonucleotides. However, the role of DNA template during primer synthesis, the size of RNA products, and the enzymatic activities associated with primase vary drastically from system to system. For example, primases from bacteriophages T7 and T4 initiate the synthesis of primers at conserved sequences on the DNA template, synthesize 4 to 5 base primers, and are stimulated by either intrinsic or extrinsic helicase activities.[1,4] Eukaryotic primases appear to have less stringent requirements for template sequence, synthesize primers between 8 and 10 bases in length, and are tighly associated with DNA polymerase α and a protein of approximately 70 kDa.[5]

The assays described below are guidelines for the biochemical detection and characterization of DNA primases. A more thorough enzymatic characterization of the primase would include descriptions of the influences of other proteins associated with the primase activity, including helicases, DNA polymerases, and single-stranded DNA binding proteins.

Detection of Primases

Coupled Primase Assays

Assays for the detection of primase activity are often coupled to assays that measure DNA polymerase activity in order to provide an assay that has a higher signal-to-noise ratio. Coupled primase assays take advantage

[2] J.-P. Bouche, L. Rowen, and A. Kornberg, *J. Biol. Chem.* **253,** 765 (1978).
[3] H. Hiasa, H. Sakai, K. Tanaka, Y. Honda, T. Komano, and G. N. Godson, *Gene* **84,** 9 (1989).
[4] L. V. Mendelman, S. M. Notarnicola, and C. C. Richardson, *J. Biol. Chem.* **268,** 27,208 (1993).
[5] L. S. Kaguni and I. R. Lehman, *Biochimica Biophysica Acta.* **950,** 87 (1988).

of the fact that DNA polymerases cannot initiate DNA synthesis on ssDNA templates that lack primers. Thus, in a coupled assay, primase uses unlabeled ribonucleoside triphosphates to synthesize a ribonucleotide primer, which is then extended by the DNA polymerase in the presence of a radiolabeled deoxynucleotide substrate.

Coupled Assay for Prokaryotic and Bacteriophage Primase Activity. This assay measures the incorporation of a radioactive dNTP into duplex DNA. In the following example, bacteriophage T7 DNA polymerase specifically extends oligoribonucleotide primers in the presence of the T7 DNA primase.[6] The interactions between the primase and polymerase of bacteriophage T7, like that of bacteriophage T4, are species specific; bacteriophage T7 DNA polymerase specifically extends short RNA primers that are stabilized by the T7 gene 4 protein.[7] Additionally, the bacteriophage T7 gene 4 protein is both a primase and a helicase and does not require the addition of an exogenous helicase, as in the bacteriophage T4 system, to stimulate primase activity.

The coupled primase assay can be modified to reflect assay conditions for other primases. For example, the assay conditions have been described previously for primases from bacteriophage T4, P4 plasmid, and *E. coli* systems.[8–11] Other proteins known to interact with primase, such as helicases, ssDNA binding proteins, or DNA polymerases, can be included in the assay conditions to investigate their effects on primase activity.[8,12–16]

RNA-Primed DNA Synthesis Assay for Bacteriophage T7 Primase. Mix in a total of 50 μl, using water to adjust volume:

10× reaction buffer [400 m*M* Tris–HCl, pH 7.5, 500 m*M* potassium glutamate, 500 m*M* NaCl, 100 m*M* $MgCl_2$, 100 m*M* dithiothreitol (DTT), 500 μg/ml bovine serum albumin (BSA)]	5 μl

[6] S. W. Matson and C. C. Richardson, *J. Biol. Chem.* **258,** 14,009 (1983).
[7] E. Scherzinger, E. Lanka, and G. Hillebrand, *Nucleic Acids Res.* **4,** 4151 (1977).
[8] L. L. Silver and N. G. Nossal, *J. Biol. Chem.* **257,** 11,696 (1982).
[9] E. Lanka, E. Scherzinger, E. Gunther, and H. Schutzer, *Proc. Natl. Acad. Sci. USA* **76,** 3632 (1979).
[10] L. Rowen and A. Kornberg, *J. Biol. Chem.* **253,** 758 (1978).
[11] A. van der Ende, T. A. Baker, T. Ogawa, and A. Kornberg, *Proc. Natl. Acad. Sci. USA* **82,** 3954 (1985).
[12] K.-I. Arai and A. Kornberg, *Proc. Natl. Acad. Sci. USA* **76,** 4308 (1979).
[13] M. Venkatesan, L. L. Silver, and N. G. Nossal, *J. Biol. Chem.* **257,** 12,426 (1982).
[14] T.-A. Cha and B. M. Alberts, *Biochemistry* **29,** 1791 (1990).
[15] L. V. Mendelman and C. C. Richardson, *J. Biol. Chem.* **266,** 23,240 (1991).
[16] E. L. Zechner, C. A. Wu, and K. J. Marians, *J. Biol. Chem.* **267,** 4054 (1992).

3 m*M* ATP, CTP	5 μl
3 m*M* dATP, dCTP, dGTP	5 μl
3 m*M* [^{3}H]dTTP (60 to 80 cpm/pmol)	5 μl
200 μg/ml M13mp18 ssDNA (600 pmol as nucleotide)	1 μl
1.5 μ*M* T7 native gene 5 protein/thioredoxin complex	1 μl
T7 gene 4 primase	0–100 ng

Incubate at 37° for 20 min. Stop the reaction by the addition of 5 μl 200 m*M* EDTA, pH 8.0, to the reaction tube and vortex. Spot 50 μl onto a Whatman DE81 filter paper. Wash in 200 ml per 20 filters 0.3 *M* ammonium formate, pH 8.0, for 5 min, repeat three times, then rinse twice in 1 ml per filter 95% ethanol. Dry the filters, then count in nonaqueous scintillation fluid.

One unit of RNA primed DNA synthesis activity is defined as the activity that catalyzes the incorporation of 1 nmol of total dNTPs under the assay conditions defined.

Coupled Assay for Eukaryotic and Viral DNA Primases. In this assay, the signal from eukaryotic or viral primase activity is stimulated by coupling synthesis of primers to DNA synthesis by the *E. coli* DNA polymerase I.[17] The assay can be performed on natural template such as M13mp18 or ϕX174 ssDNAs or on homopolymers such as poly(dT). In assays using poly(dT) as a template, only ATP and [^{3}H] dATP are included in reaction mixtures. The assay conditions described below for *Drosophila melanogaster* primase[17] can be modified to reflect the origin of the primase. For example, assay conditions have been described for DNA primase from mouse,[18] yeast,[19] human KB cells,[20] and herpes simplex virus.[21]

RNA-Primed DNA Synthesis Assay for Drosophila melanogaster DNA Primase. Mix in a total of 25 μl, using water to adjust volume:

5× reaction buffer (250 m*M* Tris–HCl, pH 8.5, 50 m*M* $MgCl_2$, 20 m*M* DTT, 1 mg/ml BSA)	5 μl
130 μg/ml M13mp18 ssDNA (1000 pmol as nucleotide)	2.5 μl
1 m*M* [^{3}H]dTTP (300 cpm/pmol)	2.5 μl
10 m*M* ATP	2.5 μl
1.25 m*M* GTP, CTP, UTP	2.5 μl
1 m*M* dATP, dCTP, dGTP	2.5 μl
E. coli DNA polymerase I	1 unit
D. melanogaster DNA primase	0–250 ng

[17] R. C. Conaway and I. R. Lehman, *Proc. Natl. Acad. Sci. USA* **79,** 2523 (1982).

[18] M. Suzuki, T. Enomoto, C. Masutani, F. Hanaoka, M. Yamada, and M. Ui, *J. Biol. Chem.* **264,** 10,065 (1989).

[19] P. Plevani, G. Badaracco, C. Augl, and L. M. S. Chang, *J. Biol. Chem.* **7532** (1984).

[20] T. S.-F. Wang, S.-Z. Hu, and D. Korn, *J. Biol. Chem.* **259,** 1854 (1984).

[21] J. J. Crute, T. Tsurumi, L. Zhu, S. K. Weller, P. D. Olivio, M. D. Challberg, E. S. Mocarski, **259,** 7532 and I. R. Lehman, *Proc. Natl. Acad. Sci. USA* **86,** 2186 (1989).

Incubate the reaction for 30 min at 30°. Stop the reaction by the addition of 3 μl 200 mM EDTA, pH 8.0, to the reaction tube and vortex. Spot 25 μl onto a Whatman DE81 filter paper. Wash in 200 ml per 20 filters 0.3 M ammonium formate, pH 8.0, for 5 min, repeat three times, then wash twice in 95% ethanol. Dry the filters, then count in nonaqueous scintillation fluid.

One unit of primase activity is the activity that incorporates 1 nmol of nucleotide in 30 min at 30°.

Direct Primase Assays

Direct assays for primase activity measure the incorporation of a radiolabeled ribonucleotide into oligoribonucleotide products. Oligoribonucleotide synthesis can be assayed directly on a variety of DNA templates, including single-stranded DNA templates from bacteriophages such as M13 or ϕX174, synthetic homopolymers such as poly(dT), and synthetic oligonucleotides of defined sequence and length. Products of the reaction are visualized on denaturing polyacrylamide gels or quantitated directly by adsorption to activated charcoal.

In primase reactions that generate small oligoribonucleotides (less than approximately 8 bases), dephosphorylation of the reaction mixture allows the separation of oligoribonucleotides from unincorporated nucleotide.[7] In reactions using prokaryotic and eukaryotic primases, which synthesize larger primers (8 to 14 nucleotides, or multiples thereof) oligoribonucleotides are visualized directly on denaturing polyacrylamide gels. Alternatively, primase activity can be measured as the total amount of radioactivity incorporated into oligomers by adsorption to activated charcoal.[22]

Oligoribonucleotide Synthesis Assay

Full-length and shorter reaction products catalyzed by DNA primases can be visualized and quantitated directly in the oligoribonucleotide synthesis assay. In this assay, primase incorporates a α-^{32}P-labeled ribonucleotide internally into an oligoribonucleotide of defined or random sequence.[7] Nucleotide and oligoribonucleotide primer 5′ phosphates are removed by alkaline phosphatase and the dephosphorylated reaction products are separated in a high percentage polyacrylamide gel based on their charge-to-base ratios, shorter diribonucleotides migrating more slowly than longer oligoribonucleotides.

The oligoribonucleotide synthesis assay can be easily adapted to examine substrate requirements for primase activity. For example, the role of template sequence in the recognition of primase initiation sites by bacterio-

[22] H. Nakai and C. C. Richardson, *J. Biol. Chem.* **263,** 9818 (1988).

phage T7 and satellite phage P4 primases has been examined by changing the sequence of a synthetic oligonucleotide DNA template.[15,23]

Oligoribonucleotide Synthesis Assay for Bacteriophage T7 Primase. Mix in a total of 10 μl, using water to adjust the volume. dTTP is included in the reaction to stimulate ssDNA binding activities of the T7 gene 4 protein.[24]

10× reaction buffer (400 m*M* Tris–HCl, pH 7.5, 500 m*M* potassium glutamate, 100 m*M* $MgCl_2$, 100 m*M* DTT, 500 μg/ml BSA)	1 μl
ssDNA (250 to 1000 pmol as nucleotide)	1 μl
1 m*M* CTP	1 μl
3 m*M* ATP	1 μl
20 m*M* dTTP	1 μl
40 mCi/ml [α-^{32}P]CTP (800 Ci/mmol)	0.5 μl
T7 primase	0–200 ng

Incubate at 37° for 60 min. Add 10 μl 2× alkaline phosphatase mix (0.1 *M* Tris–HCl, pH 8.5, 0.2 m*M* EDTA, 20 m*M* *N*-ethylmaleimide, 10 units calf intestine alkaline phosphatase): *N*-ethylmaleimide inhibits the T7 gene 4 primase activity but has no effect on alkaline phosphatase activity. Incubate at 50° for 1 hr. Add 40 μl formamide, 0.01% xylene cyanol, 0.01% (w/v) bromphenol blue. Heat to 90° for 3 min immediately prior to loading on a 20% polyacrylamide gel containing 8 *M* urea (0.4 × 35 × 43 cm). Electrophorese the gel at 2300 V for 3.5 hr. Dry the gel, and quantitate with a phosphorimager (Molecular Dynamics) or by densitometry of an autoradiograph.

Oligoribonucleotide Synthesis Assay for Calf Thymus DNA Primase on Poly(dT). This assay is used to visualize oligoribonucleotides synthesized by primase from viral and eukaryotic sources. The assay conditions shown are optimized for the calf thymus DNA primase. Assay conditions for DNA primase from herpes simplex virus,[21] yeast,[25] calf thymus,[26] mouse,[18] human,[20] and multiple vertebrates[27] have been described elsewhere.

Oligoribonucleotide synthesis from eukaryotic sources is affected by the presence of dNTP substrates and DNA polymerase α. The effects of these variables can be examined by including 100 μ*M* dATP and the DNA polymerase in the reaction mixture and examining the products of the reaction in the presence or absence of DNase I digestions.[20]

[23] G. Ziegelin, E. Scherzinger, R. Lurz, and E. Lanka, *EMBO J.* **12,** 3703 (1993).

[24] S. W. Matson and C. C. Richardson, *J. Biol. Chem.* **260,** 2281 (1985).

[25] H. Singh and L. B. Dumas, *J. Biol. Chem.* **259,** 7936 (1984).

[26] L. M. S. Chang, E. Rafter, C. Augl, and F. J. Bollum, *J. Biol. Chem.* **259,** 14,679 (1984).

[27] T. Yagura, T. Kozu, T. Seno, T. Saneyoshi, S. Higara, and H. Nagano, *J. Biol. Chem.* **258,** 13,070 (1983).

Mix in a total of 10 μl, using water to adjust the volume:

10× reaction buffer (500 m*M* Tris–HCl, pH 8.0, 80 m*M* $MgCl_2$, 10 m*M* DTT, 1 mg/ml BSA)	1 μl
0.5 m*M* $poly(dT)_{3000}$	2 μl
0.1 m*M* [α-^{32}P]ATP (10,000 cmp/pmol)	1 μl
Calf thymus DNA primase	0–250 ng

Incubate at 35° for 60 min. Stop the reaction by the addition of 30 μl of formamide, 0.01% xylene cyanol, 0.01% bromphenol blue. Prerun a 20% polyacrylamide gel containing 8 *M* urea (0.4 × 35 × 43 cm) for 90 min at 60 W. Heat sample to 90° for 3 min immediately prior to loading on the gel and run the gel at 60 W for 3 hr. Dry the gel and quantitate with a phosphorimager or by densitometry of an autoradiograph.

Total Primer Synthesis Assay

Norit (activated charcoal) interacts specifically with the bases of nucleosides, nucleotides, and short oligomers. Adsorption of nucleic acids to Norit can be used to precipitate these substrates from solution. The total primer synthesis assay measures the conversion of [α-^{32}P]CTP into Norit absorbable form after treatment with alkaline phosphatase for the T7 gene 4 primase.[22] Reactions are treated with alkaline phosphatase to remove ^{32}P from unincorporated [α-^{32}P]CTP, which is also adsorbed by Norit.

Total Primer Synthesis Assay for Bacteriophage T7 Primase. Mix in a total of 25 μl, using water to adjust the volume. dTTP is included in the reaction as the T7 gene 4 protein uses dTTP to stimulate ssDNA binding activities.[24]

10× reaction buffer (400 m*M* Tris–HCl, pH 7.5, 500 m*M* potassium glutamate, 100 m*M* $MgCl_2$, 100 m*M* DTT, 500 μg/ml BSA)	2.5 μl
200 μg/ml M13mp18 ssDNA (1.5 nmol as nucleotide)	2.5 μl
3 m*M* ATP	2.5 μl
500 μ*M* [α-^{32}P]CTP (1.8 Ci/mmol)	2.5 μl
10 m*M* dTTP	2.5 μl
T7 gene 4 primase	0–100 ng

Incubate at 37° for 20 min. Stop the reaction by adding 2 μl of 200 m*M* EDTA, pH 8.0. Heat to 95° for 2 min. Adjust the volume to 100 μl with 50 m*M* Tris–HCl, pH 8.0, 10 m*M* $MgCl_2$, 0.1 m*M* $ZnSO_4$, 1 m*M* spermidine. Add 20 units calf intestine alkaline phosphatase, incubate at 50° for 30 min, then transfer to ice. Add 90 μl to a 2.0-ml solution of 0.5 *N* HCl, 50 m*M* $NaPP_i$, 100 m*M* phosphoric acid, 2% Norit (packed volume), vortex briefly, and incubate at 0° for 5 to 15 min. Collect the Norit and DNA by filtration through Whatman GF/C glass filter paper (2.4 cm diameter) and wash six times with 2 ml of ice-cold 1 *N* HCl, 100 m*M* $NaPP_i$. Dry the filter and measure radioactivity in nonaqueous scintillation fluid.

One unit of primase activity is defined as the activity required to incorporate 10 pmol CMP under the given assay conditions.

Characterization of Primases

Characterization of Unique Primase Initiation Sites

This assay, first used to identify the bacteriophage T7 primase recognition sites,[28] and subsequently used to identify T4 recognition sites,[29] localizes primase initiation sites on a ssDNA template. In each reaction, primers are labeled at their 5′ termini with [γ-^{32}P]ATP or internally with an [α-^{32}P]NTP, then extended by a processive DNA polymerase. Digestion of the double-stranded product with a restriction endonuclease and subsequent denaturation generates radiolabeled ssRNA–DNA fragments. By separating the reaction products on a denaturing polyacrylamide gel and examining the lengths of labeled DNA fragments, primers that have been extended by the DNA polymerase can be localized to specific primase recognition sequences. The reaction conditions given below are used to determine bacteriophage T7 primase initiation sites on ϕX174 ssDNA.[28]

Characterization of Unique Initiation Sites for Bacteriophage T7 Primase. Preincubate 20 μl of a mixture of primase (e.g., 1 pmol T7 gene 4 protein) and DNA polymerase (e.g., 25 nmol native T7 gene 5 protein/thioredoxin complex) in 50 m*M* Tris–HCl, pH 7.5, 0.1 m*M* EDTA, 1 m*M* DTT, 0.5 mg/ml BSA for 5 min at 30°. Mix the enzymes with the following substrate mix:

10× reaction buffer (400 m*M* Tris–HCl, pH 7.5, 50 m*M* NaCl, 100 m*M* $MgCl_2$, 100 m*M* DTT)	10 μl
3 m*M* dATP, dGTP, dTTP, dCTP	10 μl
1.2 m*M* ATP, CTP	10 μl
1 mg/ml ϕX174 ssDNA	2 μl
[γ-^{32}P]ATP (6000 Ci/mmol)	15 μl
Sterile distilled H_2O	23 μl
	80 μl_{Total}

Incubate the reaction at 30° for 30 min, then terminate the reaction by the

[28] S. Tabor and C. C. Richardson, *Proc. Natl. Acad. Sci. USA* **78,** 205 (1981).
[29] T. Cha and B. M. Alberts, *J. Biol. Chem.* **261,** 7001 (1986).

addition of 20 μl 250 m*M* EDTA, pH 8.0. Extract the reaction with a 1:1 mixture of phenol and chloroform and precipitate DNA in the aqueous phase with ethanol and ammonium acctate as a salt.[30]

Resuspend the DNA in 10 m*M* Tris–HCl, pH 7.5, 1 m*M* EDTA and digest the DNA with 5 units of *Hae*III in a total volume of 50 μl. Incubate at 37° for 1 hr. Add 50 μl 10 m*M* Tris–HCl, pH 7.5, 1 m*M* EDTA, phenol extract, and ethanol precipitate the DNA. Resuspend the DNA in 6 μl formamide containing 0.1% bromphenol blue, 0.1% xylene cyanol, and heat to 95° for 5 min immediately prior to loading on a 6% polyacrylamide gel containing 8 *M* urea. Run the gel for 2 hr at 2000 V. Dry the gel and visualize the reaction products by autoradiography.

DNA markers of known size can be generated by endonucleolytic digestion of a double-stranded template (e.g., φX174 dsDNA) that has been dephosphorylated with calf intestine alkaline phosphatase, and labeled at its 5′ ends with [γ-^{32}P]ATP by T4 polynucleotide kinase.[31]

Characterization of RNA Primers

Identification of Primer 5′-Terminal Ribonucleotide. The ability of primase to initiate RNA primers with a particular ribonuclcotide can be quickly establishcd using a modification of the RNA primed DNA synthesis assay discussed earlier in this chapter. In this assay, a natural DNA template (M13 or φX174 ssDNA) and all four ribo- and deoxyribonucleoside triphosphates are included in the reaction mixture. One of each of the four γ-^{32}P-labeled ribonucleotides is added to four separate reactions. Reaction products are detected by adsorption to Whatman DE81 filter paper, as described earlier in this section for RNA primed DNA synthesis reactions. Incorporation of radioactive label into DNA indicates the use of the particular ribonucleotide at the 5′ terminus of the RNA primer.[23,32]

Sequence of RNA Primer. The sequence of γ-^{32}P-labeled RNA primers can be determined by partial enzymatic digests of the RNA primers with RNases that make endonucleotic incisions at specific ribonucleotide bases.[33,34] The conditions used to sequence RNA primers from bacteriophages T7[32] and T4[35] have been described.

[30] D. M. Wallace, this series, Vol. 152, p. 41.

[31] F. Cobianchi and S. W. Wilson, this series, Vol. 152, p. 94.

[32] L. J. Romano and C. C. Richardson, *J. Biol. Chem.* **254,** 10,476 (1979).

[33] H. Donis-Keller, A. M. Maxam, and W. Gilbert, *Nucleic Acids Res.* **4,** 2527 (1977).

[34] H. Donis-Keller, *Nucleic Acids Res.* **8,** 3133 (1980).

[35] C.-C. Liu and B. M. Alberts, *J. Biol. Chem.* **256,** 2821 (1981).

Characterization of Primase Active Sites

Characterization of the active site of DNA primases can be approached through a mixture of classical genetics, site-directed mutagenesis, and UV photo-cross-linking of radiolabeled NTPs to the active site. Computer-assisted sequence alignments have been used to define conserved residues in prokaryotic DNA primases.[36] Site-directed mutagenesis of conserved residues effectively demonstrates the critical role of the conserved residues for DNA primase activity.[4,37]

In eukaryotic systems, UV photo-cross-linking with [α-^{32}P]ATP of the primase from yeast[38] or [α-^{32}P]GTP of the primase from calf thymus[39] have localized the active site for NTP binding to the smaller, 50-kDa subunit of primase. Future studies of primase active sites will need to combine the techniques of site-directed mutagenesis and affinity labeling to locate specific residues used in enzymatic synthesis of oligoribonucleotides and template recognition.

Characterization of Primase Zinc Motifs

Homology searches of all known primases have shown that DNA primases contain secondary structures similar to the zinc motifs known as zinc fingers or zinc ribbons.[15] Zinc motifs have been postulated to be involved in template recognition because the primase–helicase of bacteriophage T7 will only synthesize primers in a template-directed manner if the zinc motif is present.[40] Because all primases contain a putative zinc motif, it is necessary to determine the metal content of the protein by atomic absorption spectrometry. Techniques for the analysis of metals in biological samples have been described previously.[41]

Further characterization of the zinc motif and its role in template recognition can be examined by site-directed mutagenesis, UV mediated cross-linking to primase recognition sequences, and gel-retardation experiments.

[36] T. V. Ilyina, A. E. Gorbalenya, and E. V. Koonin, *J. Mol. Evol.* **34,** 351 (1992).

[37] B. Strack, M. Lessl, R. Calendar, and E. Lanka, *J. Biol. Chem.* **267,** 13,062 (1992).

[38] M. Fioani, A. J. Lindner, G. R. Hartmann, G. Lucchini, and P. Plevani, *J. Biol. Chem.* **264,** 2189 (1989).

[39] H.-P. Nasheuer and F. Grosse, *J. Biol. Chem.* **263,** 8981 (1988).

[40] J. A. Bernstein and C. C. Richardson, *Proc. Natl. Acad. Sci. USA* **85,** 396 (1988).

[41] K. H. Falchuk, K. L. Hilt, and B. L. Vallee, this series, Vol. 158, p. 422.

[31] Identifying DNA Replication Complex Components Using Protein Affinity Chromatography

By JACQUELINE WITTMEYER and TIM FORMOSA

Introduction

Proteins in cells often act in conjunction with other proteins, in many cases forming complex structures that have been called "protein machines."[1,2] For example, such diverse processes as DNA replication, RNA transcription, transport vesicle budding and fusion, and the incorporation of specific proteins into membranes are all mediated by assemblies containing many proteins, each of which contributes a subactivity that helps to promote the overall catalytic function of the complex. The composition of protein machines can be dynamic, because the activity of the complex is altered by subtle variations in the subunits present. The protein protein interactions that allow the formation of dynamic complexes in the crowded intracellular environment are therefore unlikely to be stable enough to allow purification of the intact assembly.[2]

Determining the composition of a protein machine is then difficult because methods for studying unstable protein interactions *in vitro* are limited. Methods such as cosedimentation and coimmunoprecipitation require high stability (dissociation constants typically smaller than 10^{-9} M), whereas interactions that are biologically significant can be quite weak (dissociation constants as high as 10^{-5} M). One strategy that has been used successfully to study these weaker interactions is protein affinity chromatography, in which one member of a complex is used to identify others by taking advantage of their mutual binding affinity.[1-10]

In this chapter we describe the application of protein affinity chromatography to the analysis of eukaryotic DNA replication. Our studies have been

[1] K. G. Miller, C. M. Field, B. M. Alberts, and D. R. Kellogg, this series, Vol. 196, p. 303.
[2] T. Formosa, J. Barry, B. M. Alberts, and J. Greenblatt, this series, Vol. 208, p. 24.
[3] T. C. Huffaker, M. A. Hoyt, and D. Botstein, *Ann. Rev. Genet.* **21,** 259 (1987).
[4] D. G. Drubin, K. G. Miller, and D. Botstein, *J. Cell. Biol.* **107,** 2551 (1988).
[5] J. Greenblatt and J. Li, *J. Mol. Biol.* **147,** 11 (1981).
[6] M. Sopta, R. W. Carthew, and J. Greenblatt, *J. Biol. Chem.* **260,** 10,353 (1985).
[7] Z. F. Burton, M. Killeen, M. Sopta, L. G. Ortolan, and J. Greenblatt, *Mol. Cell. Biol.* **8,** 1602 (1988).
[8] A. E. Adams, D. Botstein, and D. G. Drubin, *Science* **243,** 231 (1989).
[9] J. Miles and T. Formosa, *Proc. Natl. Acad. Sci. USA* **89,** 1276 (1992).
[10] A. Emili, J. Greenblatt, and C. J. Ingles, *Mol. Cell. Biol.* **14,** 1582 (1994).

performed using the yeast *Saccharomyces cerevisiae* as a model, but the principles should be applicable to other systems in which one member of a suspected complex is available. Similar descriptions of this method using prokaryotic and mammalian cells for studying DNA replication and RNA transcription and cytoskeletal proteins appeared earlier in this series.[1,2] In the current chapter we focus on the potential application of this method for studying DNA replication in yeast. We summarize the techniques used to construct affinity matrices, the preparation of extracts from yeast, and describe methods for analyzing binding fractions. We also suggest strategies for secondary analyses of binding proteins to determine whether the interaction is biologically relevant, and to suggest what roles the proteins might play in DNA replication.

Protein affinity chromatography provides a way to take advantage of the subtle interactions between proteins that play important roles *in vivo* but are difficult to study using other methods. The experiments are easy to perform and can be attempted quickly if purified protein is available. The disadvantage of this method is that relatively large amounts of ligand protein are required to obtain sufficient amounts of binding proteins to allow subsequent analysis. If native ligand can be produced by overexpression or otherwise obtained in milligram quantities, protein affinity chromatography can be performed with no more effort than is required for other types of column chromatography.

Preparing Affinity Columns

Selecting Activated Matrix

Protein affinity chromatography allows sensitive detection of protein–protein interactions because of the high concentration of binding sites presented by the matrix. To achieve optimal sensitivity, it is important to choose a matrix that will couple a maximal amount of protein in a minimal volume under conditions that do not denature the ligand protein molecules. It is also important to minimize interactions between proteins in a cellular extract and the column matrix itself. Agarose is a good neutral chromatographic matrix that is suitable for use with a variety of cross-linking reagents. However, the cross-linking chemistry can produce side reactions that cause the matrix to exhibit unacceptably high levels of nonspecific protein binding. Notably, matrices activated using cyanogen bromide contain high levels of ionic side products that cause high background binding.[11] Cyanogen bro-

[11] M. Wilchek, T. Miron, and J. Kohn, this series, Vol. 104, p. 3.

mide activation can produce acceptable results, but is generally not preferred for detecting weak interactions.

We have had our most consistent successes using an *N*-hydroxysuccinimide-activated agarose matrix sold commercially by Bio-Rad under the trade name Affi-Gel 10 (we have had less favorable experiences with the similarly constructed but cationic matrix Affi-Gel 15). The neutral spacer arm of Affi-Gel 10 is coupled to an agarose base without using cyanogen bromide treatment. The active groups are present at a concentration sufficient to couple proteins to 20 mg/ml, yet produce a neutral matrix with low background binding characteristics. In addition, the commercial availability allows good reproducibility and convenience.

Coupling Proteins to Activated Matrix

To restrict the coupling reaction to the protein of interest, protein samples should be buffered with reagents lacking primary amines or sulfhydryls that react with the activated Affi-Gel 10 matrix. Tris buffers are not suitable, but HEPES, MOPS, and bicarbonate are. The protein sample may contain glycerol, $MgCl_2$, and NaCl. Our standard protocol is to dialyze about 0.5 to 10 mg of protein in a volume of 1 to 20 ml against several changes of a 50-fold excess volume of coupling buffer (25 m*M* HEPES, pH 7.5 to 7.9, 5 to 20% glycerol, 50 to 500 m*M* NaCl), at 4° over the course of 5 to 15 hr. Glycerol is useful for stabilizing nearly all proteins, and is therefore routinely used at a concentration of 10 to 20% in all buffers. High levels of NaCl have been used for coupling proteins that tend to aggregate at low ionic strengths.

The Affi-Gel 10 matrix is supplied as a slurry in 2-propanol and is stored at −70°. Washing and coupling are performed in a single tube to reduce resin losses from sticking to surfaces; polycarbonate or polypropylene centrifuge tubes are suitable (e.g., Falcon 2059 or 2063 tubes or Corning 25319-15 or 25330-50 tubes). The matrix is equilibrated to 4°, shaken to produce a slurry, then aliquots are removed using an Eppendorf pipette with the end of the disposable tip cut off to avoid clogging. (If care is taken to prevent the introduction of water to the stock, the bottle can be reclosed and stored at −70° without noticeable decay of coupling efficiency.) The slurry is about 70% matrix, and a sufficient amount should be washed to produce about 1 ml of washed matrix per milligram of purified ligand protein to be coupled. The slurry is added to about 5 volumes of cold, distilled H_2O, then the matrix is collected by centrifugation at 1000*g* for 2 min. The supernatant is removed, the gel is washed twice with 5 volumes of cold coupling buffer, and finally the matrix is suspended in the dialyzed protein sample. The coupling proceeds with gentle agitation on a rotator at 4° for several hours to overnight. Most of the coupling occurs within a few hours, so the reaction can be terminated after 3 to 4 hr if desired. However, if short coupling times are used, the matrix should be blocked by treatment with neutralized ethanolamine (60 to 200 m*M*) for at least 2 hr. This blocking is not necessary if coupling is allowed to proceed overnight.

We have found that when *S. cerevisiae* Polα or Ctf4 proteins are coupled using this protocol, part of the matrix aggregates to form clumps in which air bubbles are trapped. The amount of material found in the aggregate is roughly proportional to the mass of protein being coupled, suggesting that this represents a by-product of coupling, perhaps a cross-linking of beads through common protein monomers or denatured protein complexes. The aggregation is not observed with albumin or several other proteins, even when similar or greater amounts are coupled. These clumps form even when the level of agitation during coupling is minimized and cannot be removed by degassing the matrix under vacuum. The aggregate can be mechanically disrupted using a rubber policeman but this does not result in a homogenous matrix.

This problem can be avoided by packing the washed matrix into a column, then recirculating the protein ligand over the resin at 2 to 3 column volumes per hour for several hours or overnight. Coupling Polα in this way successfully produces a homogenous matrix without the formation of aggregates. Because the activated matrix decays in aqueous solutions with a half-life of about 0.5 to 1 hr, use of this method necessitates rapid washing and column construction as well as the use of relatively concentrated ligand solution.

Measuring Coupling Efficiency

The efficiency of coupling can be monitored in one of three ways. If the protein is pure, the protein concentration of the initial sample and the supernatant from the coupling reaction can be compared using the Bradford assay.[12] The absorbance at 280 nm of both fractions can also be measured, but in this case the sample and background controls should contain 0.1 *M* HCl to remove absorbance contributed by the unprotonated form of the *N*-hydroxysuccinimide moiety that is released during the coupling reaction. If the protein ligand is significantly heterogenous it is best to compare the aliquots by polyacrylamide gel electrophoresis in the presence of detergent (SDS–PAGE) so that the coupling efficiency of the protein of interest can be determined.

Affi-Gel 10 matrices contain about 15 μmol/ml of active esters, so

[12] M. M. Bradford, *Anal. Biochem.* **72,** 248 (1976).

the active groups are spaced about 5 nm apart. Because globular protein molecules are about this size, it is possible that multiple cross-links to the same protein molecule will form. To prevent this problem, some researchers recommend allowing the activated matrix to hydrolyze in the presence of the aqueous coupling buffer for various amounts of time before adding the ligand protein. Tubulin columns prepared so that only 80% of the ligand is coupled have been shown to have a capacity that is three to four times greater than more heavily coupled columns.[1,5,13]

Constructing an Affinity Column

The matrix is recovered by allowing it to settle for several minutes, removing a sample of the supernatant to check the coupling efficiency as described above, and then suspending the matrix in a suitable column buffer. Affinity matrices can be used to detect binding proteins using batch elution techniques; in this case, matrices are suspended in cell lysates and then collected by centrifugation, washed, and eluted by suspending the matrix in buffers of appropriate composition. However, greater resolution and more efficient washing can be obtained by continuous flow column chromatography. A variety of column sizes, types, and geometries can be used. We describe below methods for preparing standard scale columns. The preparation of smaller and larger columns was described previously.[2]

Columns (0.3 to 10 ml) can be poured in plastic sterile syringes cut to an appropriate length. The bottom of the syringe is plugged with a polyethylene filter disk (Ace glass 5848 filter disk support) or with siliconized glass wool (PhaseSep HGC166) and closed with a stopcock (Bio-Rad 732-8107 3-way stopcocks). The matrix is suspended in about 5 volumes of column buffer, and then placed into the column with a 9-inch Pasteur pipette; the tip of the pipette is inserted all the way to the bottom of the syringe to avoid trapping bubbles in the barrel of the syringe. As the matrix settles and the clear buffer drains from the bottom of the column, more slurry is added so that the packing column bed is always submerged in the slurry. It is important that the packing be continuous so that the column bed is not disrupted by discontinuities that will cause nonuniform flow characteristics. To test a column, a small amount of a solution containing a visible dye can be applied to the column. The dye should migrate through the column as a coherent disk.[1]

Once the bed is settled, it should be topped with a small amount of column buffer, the stopcock closed, and the column cap applied. The matrix bed should never be exposed to air, but rather should always have a small

[13] D. R. Kellogg, C. M. Field, and B. M. Alberts, *J. Cell Biol.* **109,** 2977 (1989).

layer of buffer covering it to avoid improper flow characteristics. The column can be capped either with an appropriately sized silicone stopper or with the tip of the plunger of the syringe after removing the plastic handle. A needle is inserted into the stopper with the sharpened end pointed out of the column and the base of the needle is twisted back and forth until it breaks. This leaves the broken end of the needle inside the column and the tip outside. Tubing from a peristaltic pump can be attached to the tip of the needle and used to deliver solutions at uniform, reproducible flow rates. The optimal flow rate depends on the cross-sectional area rather than the total bed volume; we typically use a rate of about 2 to 3 ml/hr for 1-ml syringes with 0.5- to 1-ml bed volumes, and 10 to 20 ml/hr for 10-ml syringes with 8- to 10-ml bed volumes.

Storing Matrices

Affinity matrices can be equilibrated with buffer containing 50% glycerol, stored at -20° and reused. High concentrations of glycerol are used to prevent the buffer from freezing, which both denatures proteins and destroys agarose matrices. Yeast Pol α protein retains the ability to bind specific proteins for only six to eight cycles of loading and eluting, and long-term storage has not been successful. The bacteriophage T4 gene 32 protein produces affinity columns with much greater stability, and has been found to be reusable indefinitely after this storage regimen.

Control Columns

Because agarose will bind some proteins present in a crude lysate, it is essential to prepare control columns lacking the ligand of interest so that specific interactions can be distinguished from background binding. The nature of an appropriate control is somewhat problematical. The coupling of an unrelated protein can control for nonspecific interactions with the matrix but does not rule out protein-specific artifacts such as binding to a charged or unusually hydrophobic domain. A column prepared from denatured ligand protein could be useful, but it is dificult to know that specific interacting domains are not renatured, or that the denatured protein will not bind a different set of background proteins. Ideally, a protein with a similar isoelectric point should be used, although in practice we have found that serum albumin or a matrix lacking any protein ligand performs as adequate control. If the protein to be used as ligand is significantly contaminated with other proteins, a mock purification from a strain lacking the protein of interest can be coupled to control for interactions not involving the target ligand. As discussed below, it is ultimately necessary to develop independent criteria for the relevance of interactions detected by affinity chromatography, so it is perhaps necessary to control only for the most obvious sources of nonspecific binding.

Preparing Extracts for Chromatography

Choosing a Method of Detecting Proteins

Detecting small amounts of protein normally involves SDS–PAGE followed either by staining with silver or by using fluorography to detect radioactively labeled proteins. In the latter case $^{35}SO_4^{2-}$ or [^{35}S]methionine are usually used to label cells because of the low cost and high specific activity relative to ^{14}C. Radioactive labeling has the advantage of allowing quantitation and detection of fractions containing binding proteins by scintillation counting before polyacrylamide gel electrophoresis. This method also allows a distinction to be made between polypeptides that bind from the extract and proteolytic fragments of the ligand that might be released from the columns, which would otherwise appear to be specific binding proteins. Depending on the ligand protein and the source of the extract, this can be a major consideration that favors the use of radioactive proteins. However, radioactive labeling is relatively expensive and is not more sensitive than silver staining. In addition, the radioactive extracts must be handled more carefully, and fluorography is generally slower than silver staining.

Choosing the Source of Starting Material

In some cases, a total cell extract is not the optimal starting point for affinity chromatography. For instance, replication proteins may be enriched 10-fold by isolation of nuclei. Because replication proteins are also likely to interact with proteins that themselves bind to DNA, it could be beneficial to enrich for such proteins by DNA-cellulose chromatography before attempting affinity chromatography. Such preliminary chromatographic steps can also assist in removing possible interfering substances from an extract.

We have found that preparing extracts from strains lacking certain binding proteins can allow easier access to other binding proteins of similar molecular weight and to alter the relative yield of different binding proteins (Fig. 1). A 120-kDa polymerase α binding protein could not be adequately resolved from a previously identified and more abundant binding protein, Ctf4 (Pob1 protein[9,14]). We therefore prepared an extract from a strain carrying a deletion of the nonessential *CTF4* gene. Surprisingly, the yields of the 120-kDa binding protein (now identified as Cdc68/Spt16) and a 70-

[14] J. Miles and T. Formosa, *Mol. Cell Biol.* **12,** 5724 (1992).

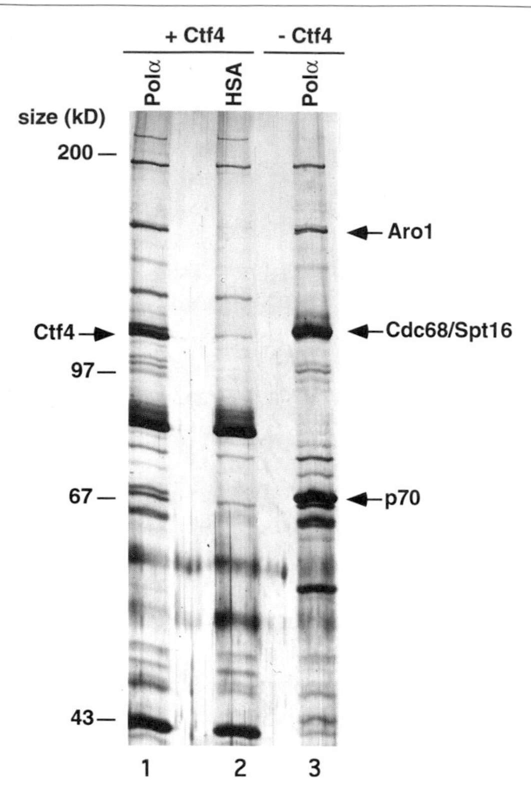

FIG. 1. Removing a major Polα-binding protein from extracts enhances the retention of some other binding proteins. Ten-milliliter columns containing either 15 mg of *S. cerevisiae* Polα catalytic subunit (Pol1 protein, lanes 1, 3) or a similar amount of serum albumin (lane 2) were used to chromatograph extracts prepared from 10-liter cultures of logarithmic-phase yeast cells with either a normal copy of the *CTF4* gene (lanes 1, 2) or containing a deletion of this gene (lane 3). One percent of the eluate was concentrated, subjected to SDS–PAGE on a 7.5% acrylamide gel, and silver stained.

kDa binding protein both increased. This increase suggests that p120 and p70 might compete with Ctf4 for the same or overlapping binding site(s) on Polα. The binding of another polymerase-binding protein, Aro1 (175 kDa), did not change when *CTF4* was deleted. Sequential removal of binding proteins using genetic or immunological methods can therefore provide greater access to remaining binding proteins, and can also be used to demonstrate that interactions are direct (see below).

Preparation of Whole Cell Extracts

For whole cell extracts from yeast, cells are collected by centrifugation and the pellet is suspended in an equal volume of lysis buffer (1 ml/gm of cells; 20 mM Tris–Cl, pH 7.5, 1 mM Na_2EDTA, 1 mM 2-mercaptoethanol, 50 mM NaCl, and protease inhibitors as described below). A volume of glass beads (Sigma G-9268; 500-μm-diameter beads washed with 1 M HCl, rinsed extensively with distilled H_2O until the washes are above pH 5, then dried) equal to the cell slurry is added and the mixture is agitated by vortexing for about 10 min in 1-min bursts with 1-min rests on ice. All subsequent procedures are performed at 0 to 4°.

Extracts are cleared by centrifugation at high speed (1 to 3 hr at greater than 100,000g average centrifugal force). It is crucial to remove particulate matter, which will otherwise cause high levels of apparent nonspecific binding that can obscure specific interactions. We typically recover the high-speed supernatant, add glycerol to 10% (weight/volume), repeat the centrifugation step, then filter the solution through a 0.45-μm cellulose acetate filter. Glycerol stabilizes many proteins and also reduces nonspecific binding, and the second spin and the filtration reduce the level of particulate matter. The lysis buffer is designed to be suitable for direct loading of the supernatant to the affinity column. If this is not feasible (for example, if a high-salt extract of nuclei is used) the extract can be dialyzed into an appropriate buffer, but still must be cleared by centrifugation and filtration prior to loading (at least 30 min at 10,000g; 1 hr at 100,000g is preferred).

Both the lysis buffer and the column buffers should contain protease inhibitors to minimize fragmentation of proteins in the extract and to protect the affinity ligand. While the recipe for an effective protease inhibitor cocktail will vary with the application, a good starting point is 0.2 mM phenylmethylsulfonyl fluoride (PMSF) (from a 100 mM stock in isopropanol), 0.5 mM benzamidine hydrochloride (from a 500 mM stock in H_2O), 0.5 μg/ml leupeptin (from a 0.5 mg/ml stock in H_2O), and 0.7 μg/ml pepstatin (from a 0.7 mg/ml stock in methanol) in a buffer containing 1 mM Na_2EDTA.

In some cases, a crude lysate will contain small molecules that will interfere with chromatography. This can often be remedied by first precipitating proteins with ammonium sulfate (60 to 70% of saturation is usually sufficient), collecting the precipitate by centrifugation, and dissolving the precipitated proteins in affinity column buffer followed by dialysis. Extracts prepared this way should still be centrifuged at high speed (at least 1 hr at greater than 100,000g) and filtered.

Removing polymers such as DNA and RNA that could interfere with affinity chromatography is more problematical. If nucleic acids cause a problem, several approaches for removing them should be attempted, since no one method gives satisfactory results in every case. Passing an extract containing 0.2 M NaCl through a DEAE-cellulose column will remove nucleic acids; this procedure is recommended provided that the proteins of interest do not bind to either nucleic acids or DEAE-cellulose under these conditions. Our experiences with enzymatic removal using nucleases have not been entirely positive; even relatively pure preparations of DNase I tend to show traces of protease contamination. Attempts to remove DNA from extracts with organic polymers such as polyethylene glycol or polyethyleneimine have often had undesirable side effects—such as the inability to resolubilize particular proteins under low-ionic-strength conditions, or a drastic increase in background binding, both of which are assumed to be due to the continued presence of the polymers in the extract even after extensive dialysis. If this approach is to be used, it is necessary to carefully titrate the amount of precipitant needed.[15]

Loading and Eluting Interacting Proteins

Loading the Columns

Once suitable conditions for preparing an extract have been identified, the extract is loaded to the affinity matrix and the control matrix. An extract can be loaded onto a column in a single pass, or can be cycled through the column for several passages. The advantage of the latter strategy is that it allows longer exposure of any binding protein to the ligand. This could be important when attempting to isolate factors that are stably bound to endogenous ligand in the extract.

Our standard affinity chromatography loading buffer contains 10 to 20 m*M* Tris–Cl or HEPES, pH 7.5 to 7.9, 10 to 20% (w/v) glycerol, 1 m*M* Na_2EDTA, 1 to 2 m*M* 2-mercaptoethanol or dithiothreitol, 50 to 100 m*M* NaCl, and protease inhibitors as described above. When the buffer contains 100 m*M* NaCl, there are very few proteins in a whole cell extract that will bind to a control column and be eluted with salt.

Eluting Columns

After washing with at least 5, and preferably 10, column volumes of the low salt loading buffer, bound proteins are eluted with increasing concentrations of salt. Since many interactions have at least a small ionic component, the elution of an interacting protein from a column can usually be accomplished with salt. We have observed many cases in which an interacting protein binds to the ligand in 100 m*M* NaCl and elutes with buffer containing 200 to 300 m*M* NaCl. Increasing the ionic strength in steps is simple and limits the number of fractions that must be checked for

[15] R. R. Burgess, *in* "Protein-DNA Interactions" (R. T. Sauer, ed.), Vol. 208, p. 3. Academic Press, New York, 1991.

binding proteins. However, information about the strength of interactions can be obtained more reliably by eluting with a gradient of increasing salt. It is also easier to distinguish specific from nonspecific binding using a gradient elution. In either case, the conductivity of each fraction is measured to determine the salt concentration as an indication of the fractions that should be assayed for binding proteins.

In pilot experiments, after a sufficiently harsh salt wash (usually 2 *M* NaCl), the matrix should be stripped by eluting at room temperature with 6 *M* urea containing 0.2% SDS to detect any hydrophobic or extremely strong interactions. This, of course, sacrifices the column.

The binding capacity of a column for a particular protein can be determined by saturating the matrix with that protein, assuming that it is available in large quantity. Alternatively, after a column is eluted, the proteins that have flowed through the column are reapplied and the column is reeluted. If more of a given protein is retained in the second pass, then the amount bound in the first pass was probably the maximum that could have bound. If no more binds the second time, no conclusion can be drawn since the column may have been altered by the elution protocol; in this case the experiment can be repeated with two identical columns loaded in series and then eluted separately. If SDS–PAGE reveals that components are completely missing from the flow-through fraction that were present in the load fraction, it can be concluded that the column was not saturated for those components. However, the presence of a given protein in the flow-through fraction does not mean the column was saturated, since some of the protein in question might have been denatured or otherwise unavailable for binding. If binding rate is the problem, the column can be loaded more slowly or by continuous recycling.

Estimates of the total binding capacity of yeast Polα protein columns suggest that about 3% of the protein molecules are available for binding. The bacteriophage T4 gene 32 protein retains about 10% binding (Ref. 2, and J. Hosoda, personal communication, 1984), while other proteins such as the bacteriophage λ N, *Escherichia coli* NusA, and mammalian RNA polymerase II proteins retain 50% or more of their native binding.[5,6,16]

As a rough estimate, we load an amount of extract to a column containing about one-half of a cellular equivalent of the active amount of ligand protein. For example, yeast cells contain about 1000 copies of the Polα protein, and 1 mg of purified Polα protein contains about 1×10^{14} molecules active for binding (since about 3% of the molecules are active). Assuming that five binding proteins are present in yeast cells, and that each

[16] J. Greenblatt and J. Li, *Cell* **24,** 421 (1981).

is found at a concentration similar to Polα, a column with 1 mg of Polα ligand should be loaded with extract prepared from about 1×10^{10} cells.

Analysis of Binding Fractions

Characteristics of Interacting Proteins

Proteins that interact with an affinity matrix can be characterized by SDS–PAGE or by assaying fractions for biochemical activities, assuming the conditions for elution did not alter the protein's activity. Most proteins can be eluted with high concentrations of salt without affecting their activity, but the use of chaotropic agents such as urea or SDS will normally inactivate enzymes.

The presence of proteins can be detected by assaying a sample of each fraction in a scintillation counter if extracts were radioactively labeled. In this case, samples containing proteins are concentrated [for example, by precipitation with acetone or trichloroacetic acid, or by centrifugation against a semipermeable membrane such as Centricon (Amicon) units], separated by SDS–PAGE, and fluorographed to detect proteins specific to the ligand-containing column eluate. For unlabeled extracts, an aliquot from each fraction is separated by SDS–PAGE, and proteins are detected by silver staining. Fractions found to contain proteins can then be concentrated and electrophoresed again to provide greater sensitivity. Weak but specific interactions are often revealed in the initial low salt wash, so this region of the chromatogram must not be overlooked.

An alternative approach is to assay fractions for enzymatic activities. For instance, if a DNA polymerase is used as the ligand, the binding fractions can be added to a replication assay to detect changes in the properties of the polymerase.

Direct versus Indirect Interactions

If cell extracts were used for affinity chromatography, one must eventually determine whether an interacting protein is binding directly to the ligand or indirectly to the ligand via some other molecule. This can be accomplished, once the interacting protein has been purified, by rechromatographing the pure interacting protein on control and ligand columns.[16,17] Alternatively, if one binding protein can be eliminated from the extract by genetic mutation or immunoprecipitation, the dependence of other binding proteins on this protein can be determined (see Fig. 1).

[17] K. F. Stringer, C. J. Ingles, and J. Greenblatt, *Nature* **345,** 783 (1990).

Possible Need for Additional Ligand

If a protein undergoes a change in its conformation when it binds to some ligand inside the cell and only then interacts with another protein, the method described above will fail to detect this interaction unless the needed ligand is present in the extract at a sufficient concentration. It is clear that this situation occurs in biological systems: For example, spectrin, a cytoskeletal protein, binds to actin much more strongly if the spectrin has previously bound to band 4.1 protein.[18] Likewise, *E. coli* RecA protein affinity columns have been found to bind LexA protein only in the presence of DNA.[19] This situation also occurred in the analysis of the primosome of the T4 bacteriophage DNA replication apparatus.[20] Thus, second-generation protein affinity columns should be considered in which a preformed protein–protein or protein–nucleic acid complex is the immobilized affinity ligand, rather than a single protein alone.

Interpreting Binding Interactions

Observation of specific binding between two proteins *in vitro* should be taken as a starting point for studying the relevant process *in vivo*. Independent evidence for an interaction *in vivo* can be obtained using two-hybrid methods,[21–23] or using any of several genetic techniques.[3] Perhaps more informatively, a role for the binding protein in the same cellular process as the ligand protein can be tested directly. Several phenotypes have been associated with defective DNA replication in *S. cerevisiae,* and these phenotypes can be assayed in cells containing mutations in the genes that encode the proteins identified by affinity chromatography. This characterization can produce a picture of the role played by the binding protein *in vivo.*

As a first step, a clone of the gene encoding the binding protein must be identified. Binding fractions can be used to generate antisera or to perform peptide sequencing.[24] Antisera can be used to clone the gene of interest from an expression vector library. Peptide sequence can be used to design oligonucleotide pools for use either directly as hybridization probes to screen libraries or as polymerase chain reaction (PCR) primers

[18] C. S. Potten, W. J. Hume, P. Reid, and J. Cairns, *Cell* **15,** 899 (1978).
[19] N. Freitag and K. McEntee, *Proc. Natl. Acad. Sci. USA* **86,** 8363 (1989).
[20] T.-A. Cha and B. M. Alberts, *in* "Cancer Cells 6; Eukaryotic DNA Replication." Cold Spring Harbor Laboratory Press, New York, 1988.
[21] S. Fields and O. Song, *Nature* **340,** 245 (1989).
[22] J. W. Harper, G. R. Adami, N. Wei, K. Keyomarsi, and S. J. Elledge, *Cell* **75,** 805 (1993).
[23] A. S. Zervos, J. Gyuris, and R. Brent, *Cell* **72,** 223 (1993).
[24] P. Matsudaira, this series, Vol. 182, p. 602.

to amplify a portion of the gene to provide a more specific hybridization probe.[25]

Since many eukaryotic proteins are acetylated at their NH_2 termini and are therefore blocked to NH_2-terminal sequencing, most proteins must be chemically or enzymatically cleaved to generate peptides amenable to the Edman reaction.[24] We have had success transferring proteins from SDS–PAGE separations to PVDF membranes (ProBlott; Applied Biosystems), staining with Coomassie blue or Ponceau S, digesting with trypsin or endoproteinase Lys-C, then recovering the peptides by reversed-phase HPLC.[26] This approach requires approximately 100 pmol of the binding protein (10 μg of a 100-kDa protein), typically necessitating the use of affinity columns containing 1 to 10 mg of ligand protein.

Once a clone is available, mutations can be introduced into the binding protein gene using standard procedures.[27] Yeasts carrying such mutations can be assayed for DNA metabolism defects in several ways. Mutations in DNA polymerases α and δ lead to elevated levels of both recombination and chromosome loss, presumably due to increased DNA repair requirements induced by these mutations.[28] When grown at their maximal permissive temperature, these strains also exhibit a characteristic morphology in which the majority of cells are large-budded with a single nucleus at the neck of the bud and a DNA content near 2C. Since these properties are common to mutants with perturbed DNA metabolism, their observation can be used to indicate loss of normal DNA metabolism. Chromosome loss can be measured using suitably marked disomes ($1n + 1$ cells are assayed for conversion to $1n$[29]), diploids homozygous for the mutation ($2n$ cells become $2n - 1$[28,30]), or by determining the stability of centromere-containing plasmids.[31] The disome and diploid assays can be modified to allow the simultaneous measurement of recombination frequencies. It is important to note that since yields of loss or recombination events can be misleading, frequencies should be measured using fluctuation methods.[32] The morphology of cells is assayed microscopically after staining the DNA,[33] and the DNA content is measured by flow cytometry.[34]

[25] H. A. Erlich, *PCR Technology,* Stockton Press, New York, 1989.

[26] J. Fernandez, M. DeMott, D. Atherton, and S. M. Mische, *Analyt. Biochem.* **201,** 255 (1992).

[27] C. Guthrie and G. R. Fink, this series, Vol. 194, New York, 1991.

[28] L. H. Hartwell and D. Smith, *Genetics* **110,** 381 (1985).

[29] M. A. Hoyt, T. Stearns, and D. Botstein, *Mol. Cell Biol.* **10,** 223 (1990).

[30] S. L. Gerring, F. Spencer, and P. Hieter, *EMBO J.* **9,** 4347 (1990).

[31] J. H. Shero, M. Koval, F. Spencer, R. E. Palmer, P. Hieter, and D. Koshland, this series, Vol. 194, p. 749.

[32] D. E. Lea and C. A. Coulson, *J. Genet.* **49,** 264 (1948).

[33] J. R. Pringle, A. E. M. Adams, D. G. Drubin, and B. K. Haarer, this series, Vol. 194, p. 565.

[34] K. J. Hutter and H. E. Eipel, *Antonie van Leeuwenhoek J. Microbiol. Serol.* **44,** 269 (1978).

Information linking mutations to a role in DNA replication can also be obtained from the phenotype of double mutations with known replication proteins (*cdc2*, *cdc17*, or *cdc9* mutations defective for Polδ, Polα, or DNA ligase, respectively[35]), or DNA-associated checkpoint genes (*rad9*,[36] *mec1*, or *mec2*; Ted Weinert, personal communication, 1994). Mutant cells can also be tested for sensitivity to irradiation or DNA-damaging drugs such as methanesulfonic acid ethyl ester (MMS), or to replication inhibitors such as hydroxyurea.

Conclusion

Protein affinity chromatography has many potential uses for studying complex biochemical systems. Starting with a single element of a complex protein machine, it is possible in principle to reassemble the entire machine. The technique also has some limitations. For example, even though two proteins interact, they will not necessarily have the opportunity to do so on an affinity matrix. The T4 bacteriophage DNA polymerase binds tightly to a gene 32 protein affinity column, consistent with the specific interaction between these two proteins in solution. However, although a T4 DNA polymerase column binds several polymerase accessory proteins, it does not bind gene 32 protein.[37] We assume that this lack of reciprocity reflects a nonrandom orientation of the polymerase molecules on the affinity matrix, with the 32 protein-binding domain in an inaccessible position. Perhaps coupling at a variety of pH values would reveal a condition that provides a more random attachment.

Protein affinity chromatography can detect interactions ranging in strength from 10^{-5} to 10^{-10} M K_d. An interacting protein with a $K_d > 10^{-5}$ M will not remain bound to the column when the column is washed by the large volume of buffer that is necessary to lower the background binding, even if the immobilized protein concentration is 10 mg/ml. Alternatively, an interacting protein with a $K_d < 10^{-10}$ M might not be able to bind to the affinity matrix if it failed to dissociate rapidly enough from the endogenous ligand present in the extract. Such tight interactions might be more reliably detected by coimmunoprecipitation methods. In any case, failure to detect an interaction must be interpreted cautiously.

Detection of binding between two proteins *in vitro* cannot by itself be

[35] J. L. Campbell and C. S. Newlon, *in* "The Molecular and Cellular Biology of the Yeast *Saccharomyces*" (J. R. Broach, J. R. Pringle, and E. W. Jones, eds.), p. 41. Cold Spring Harbor Laboratory Press, New York, 1991.

[36] T. A. Weinert and L. H. Hartwell, *Genetics* **134,** 63 (1993).

[37] B. M. Alberts, J. Barry, P. Bedinger, T. Formosa, C. V. Jongeneel, and K. Kreuzer, *Cold Spring Harbor Symp. Quant. Biol.* **47,** 655 (1983).

interpreted as a demonstration of a biologically significant protein–protein interaction. The two proteins must be shown to be involved in the same process *in vivo;* ideally the interaction should be demonstrated by a functional assay or by any of several genetic techniques that can detect protein–protein interactions. Nevertheless, protein affinity chromatography is a useful technique, both as an initial method for detecting the potential components of protein machines and for the purification of such components. Protein affinity chromatography provides a unique opportunity to exploit the forces that cause protein machines to assemble to allow exploration of the composition and regulation of these complexes that act in so many cellular processes.

[32] Radiolabeling of Proteins for Biochemical Studies

By ZVI KELMAN, VYTAUTAS NAKTINIS, and MIKE O'DONNELL

Introduction

Several processes in nucleic acid metabolism such as replication, transcription, and translation require the coordinated action of numerous proteins. This coordination is manifest through protein–protein and protein–nucleic acid interactions ranging from stabile complexes to transient contacts. Important questions in these multicomponent systems include the following: How much of each protein is present in protein complexes or on the nucleic acid? Is the protein interacting with a specific nucleic acid structure or sequence? How tightly associated are the subunits of a complex, and how are these interactions influenced by other proteins or by ATP? Which surfaces on a protein interact with other proteins? Answers to questions such as these require a convergence of several experimental techniques. Some techniques that address these subjects utilize radioactive proteins.

This chapter describes two radioactive labeling methods and illustrates a few applications of labeled proteins in biochemical studies. One method is reductive methylation,[1–3] which was described in this series using [^{14}C]formaldehyde.[2] We present a modified protocol using NaB^3H_4 that

[1] G. E. Means, this series, Vol. 47, p. 469.
[2] G. E. Means and R. E. Feeney, *Biochemistry* **7,** 2192 (1968).
[3] R. H. Rice and G. E. Means, *J. Biol. Chem.* **246,** 831 (1971).

introduces only 1 or 2 [^{3}H]methyl groups per protein for a specific activity of 20 to 40 Ci/mmol (0.5 to 1.0 $\times$ 10^6 cpm/μg of 50-kDa protein). The second method uses a specific protein kinase to ^{32}P-end-label a protein into which a 5–7 amino acid residue kinase recognition motif has been engineered onto the N or C terminus.[4,5] This procedure results in a specific activity up to 3000 Ci/mmol (>60 $\times$ 10^6 cpm/μg of 50-kDa protein). We have no experience in performing protein labeling using isotopes of iodine and for this the reader is referred to an excellent treatise in this series.[6]

A drawback to chemical labeling is that the modified protein may lose one or more of its activities. Hence, the labeled protein must be tested for activity relative to unlabeled protein. Reductive methylation is the least invasive technique because it introduces only one or two methyl groups. In our experience, 12 of the 13 different proteins that were labeled retained 85 to 100% activity. End-labeling with ^{32}P introduces a larger modification and some proteins may be expected to lose function, but the three proteins we have ^{32}P-end-labeled retain their activity.

Proteins that are radiolabeled metabolically (*in vivo*) with either [^{35}S]methionine or ^{3}H- or ^{14}C-labeled amino acids should retain full activity. However, the drawback to metabolic labeling is the need to purify the radioactive protein, and during the prolonged use of centrifuges and cold room equipment it is difficult to contain the radioactivity. The chemical labeling methods described here also carry a health hazard, but the procedures can be performed in a fume hood in a matter of a few hours. It is important that the operations be performed using all available precautions including gloves, laboratory coat, and film badge, and that proper guidelines are followed to dispose of the radioactive waste.

^{3}H-Labeling by Reductive Methylation

The procedure described here modifies only a few lysine residues per protein molecule and is adapted from methods used in Arthur Kornberg's laboratory.[7] The reaction,[1,2] outlined in Eq. (1), involves the addition of formaldehyde, which forms a Schiff base with lysine followed by reduction using NaB^3H_4 and results in net replacement of a proton for a [^{3}H]methyl group on the primary amine of lysine. Once lysine is methylated, it is more reactive toward a second round of methylation.[1] The charge of the methylated lysine is conserved; the pK of monomethylated lysine is in-

[4] B.-L. Li, J. A. Langer, B. Schwartz, and S. Pestka, *Proc. Natl. Acad. Sci. USA* **86,** 558 (1989).
[5] M. A. Blanar and W. J. Rutter, *Science* **256,** 1014 (1992).
[6] C. W. Parker, this series, Vol. 182, p. 721.
[7] K.-i. Arai, S.-i. Yasuda, and A. Kornberg, *J. Biol. Chem.* **256,** 5247 (1981).

creased by about 0.3 pH unit and the p*K* of dialkylated lysine is lower than lysine by about 0.6 pH units.[2] When only a few methyl groups are incorporated into a protein, the monoalkylated lysine species predominates.[2]

$$\text{Protein-}\ddot{N}H_2 + \overset{\displaystyle H}{\underset{\displaystyle H}{C}}{=}O \xrightarrow[-H_2O]{} \text{Protein-}\ddot{N}{=}\overset{\displaystyle H}{\underset{\displaystyle H}{C}} \xrightarrow{NaB^3H_4} \text{Protein-}\overset{\displaystyle H}{\ddot{N}}\text{-}\overset{\displaystyle H}{\underset{\displaystyle H}{C}}\text{-}^3H \quad (1)$$

The extent and velocity of the reaction increases with pH, very little reaction occurs at pH 7.0 and maximal labeling is achieved at pH 9.5.[2] The elevated pH is needed to form the Schiff base between lysine and formaldehyde (a readily reversible reaction), and to preserve the sodium borohydride, which is rapidly decomposed as the pH is lowered. To prevent extensive labeling the reaction is carried out at pH 8.5. The reaction can be performed at pH 7.0 using cyanoborohydride,[8] although the temperature must be elevated and the reaction time may be longer. Reductive methylation occurs only with lysine and with the amino terminus, no reaction occurs with side chains other than lysine, and disulfide bonds are not reduced.[2] Most of the radioactivity is not incorporated into protein but is consumed by side reactions such as reduction of formaldehyde to methanol and breakdown of borohydride by solvent. Buffers containing primary and secondary amines must be avoided because they form the Schiff base and consume the reactants. We have used sodium borate although buffers with tertiary amines or phosphate buffer can be used. Besides formaldehyde, other aldehydes and ketones can be used (e.g., acetone and acetaldehyde), but the resulting alkyl group on the lysine is then bulkier (isopropyl and ethyl, respectively).[2] Hence we use formaldehyde to produce the smallest modification possible. Note that formaldehyde itself can modify proteins by forming inter- and intraprotein cross-links; however, the destructive reactions of formaldehyde with proteins during reductive alkylation has been examined with none of these products being observed under conditions that are much more extensive in time and temperature than those used here.[8]

Materials for Reductive Methylation

We usually purchase 0.5 Ci of NaB^3H_4 of 50 to 75 Ci/mmol (Du Pont–New England Nuclear, NET-023X). 0.5 Ci of 75 Ci/mmol NaB^3H_4 is 6.7 μmol, which is sufficient to label 3.3 ml of protein. We usually label 1 ml

[8] N. Jentoft and D. G. Dearborn, *J. Biol. Chem.* **254,** 4359 (1979).

of 3 to 5 different proteins on the same occasion. Unfortunately, NaB^3H_4 of this specific activity is not available in smaller quantities. Lower specific activity material can be purchased such as 100 mCi NaB^3H_4 at 5 to 15 Ci/mmol (NET-023H). Because most of the radioactivity is lost in competing side reactions, the higher the concentration of the protein, the more labeled protein is produced per mole of NaB^3H_4 (i.e., the resulting specific activity of the protein is unaffected whether a 1 or a 5 mg/ml solution of protein is used). Protein can also be labeled using [^{14}C]formaldehyde,[3] but the specific activity is 1000-fold lower than NaB^3H_4.

The other necessary materials include a small container for radioactive solid waste, a triangular file, 1 to 200-μl Pipetman with an extended tip (explained below), ice bucket with each dialyzed protein in a separate open tube, 10 m*M* NaOH on ice, 2 *M* formaldehyde on ice, 1 *M* lysine, and an empty Eppendorf tube (for the NaB^3H_4 after it is dissolved in 10 m*M* NaOH). To separate protein from reagents after the reaction, one fraction collector is needed for each protein, 10 ml columns of Sephadex G-25 packed in disposable 10-ml plastic pipettes with a glass wool plug, and gel filtration column buffer. Our typical column buffer is 20 m*M* Tris–HCl, pH 7.5, 0.5 m*M* EDTA, 2 m*M* dithiothreitol (DTT), and 20% glycerol.

Procedure for Reductive Methylation

The night before the labeling reaction, dialyze each protein at a concentration of 1 to 5 mg/ml into 50 m*M* sodium borate, pH 8.5, 0.5 m*M* EDTA, 10% glycerol (DTT can be included). The next day the reagents and hardware must first be arranged in the fume hood. Then the sealed ampule containing the dry NaB^3H_4 is gently tapped to place all the powder at the bottom of the vial and the file is used to score the neck of the ampule. The ampule will have $^3H_2$ gas under pressure and it is important when snapping the neck of the ampule to position it near the hood exhaust port. Dissolve the NaB^3H_4 powder in 10 m*M* NaOH for a final concentration of 100 m*M* NaB^3H_4 (this will be in the range of 67 to 100 μl depending on the specific activity of the NaB^3H_4). Due to the length of the ampule and the narrow opening at the neck, the Eppendorf tip must be modified so that it reaches to the bottom of the ampule. A simple modification is to fasten a 2-in. length of polyethylene tubing onto the tip. The 10 m*M* NaOH is drawn into the Pipetman and then transferred into the ampule to dissolve the NaB^3H_4, then withdrawn and placed into the empty Eppendorf tube on ice. At this point one must work quickly, but steadily, because the reagent is decomposing (as evidenced by the slow appearance of bubbles).

The labeling reaction is initiated upon adding formaldehyde to the protein(s) to a final concentration of 20 m*M* and then NaB^3H_4 to a final concentration of 2 mM. After a further incubation of 15 min on ice, stop the reaction by adding lysine to a final concentration of 50 mM, and separate the protein from the reagents by passing the reaction mixture over a 10-ml column of Sephadex G-25. The column is disposed of as radioactive waste at the end of the procedure. The column is initially prepared in the cold room and the buffer is kept on ice, but just before use it is brought into the fume hood along with a fraction collector. Fractions of 400 μl are collected and placed on ice. A small amount (3 μl) of each fraction is counted to locate the protein peak, these are pooled, the protein concentration is determined, and then the [^{3}H]protein is aliquoted and stored frozen at $-70°$. The labeling procedure should take no more than 15 to 20 min, the gel filtration about 30 min, and the counting and pooling of column fractions, protein determination, and aliquoting may take about 1 hr.

Variables in this reaction include time, pH, temperature, and concentrations of formaldehyde and NaB^3H_4. We find that under the conditions described above, the rate of methylation is nearly linear for 10 min and levels off by 15 min. Also, lowering the formaldehyde from 20 to 10 mM yields half the level of methylation. Finally, use of 2 mM NaB^3H_4 is sufficient for the low level of labeling desired; 4 mM NaB^3H_4 increased the extent of labeling by only 20%, and thus is put to better use at 2 mM to label more protein. We have always performed these procedures on ice and have not experimented with different temperatures. We have examined the effect of increasing the pH to 9.0 and find the extent of methylation increases at least 1.5-fold, but pH 8.5 is less harsh on the protein and is sufficient for incorporation of one to two methyl groups.

^{32}P-End-Labeling of Proteins

An efficient substrate for the cAMP-dependent protein kinase has been identified as a heptapeptide Leu-Arg-Arg-Ala-Ser-Leu (or Val)-Gly (or Ala).[9,10] The K_m of this sequence for phosphate transfer to the Ser by the kinase is 10 to 20 μM.[9,10] Use of only the five inner residues results in an eightfold increase in K_m,[10] and substitution of any of the five inner residues increases the K_m by 10-400 fold, resulting in low to negligible rates of phosphorylation.[9] Radioactive proteins of high specific activity ($\geq$3000 Ci/mmol) have been produced upon cloning five to seven residues of this sequence onto the C terminus of human interferon α^4 and onto the N terminus of segments of c-Fos.[5]

We have engineered the kinase recognition sequence onto either the

[9] B. E. Kemp, D. J. Graves, E. Benjamini, and E. G. Krebs, *J. Biol. Chem.* **252,** 4888 (1977).

[10] Ö. Zetterqvist, U. Ragnarsson, E. Humble, L. Berglund, and L. Engström, *Biochem. Biophys. Res. Commun.* **70,** 696 (1976).

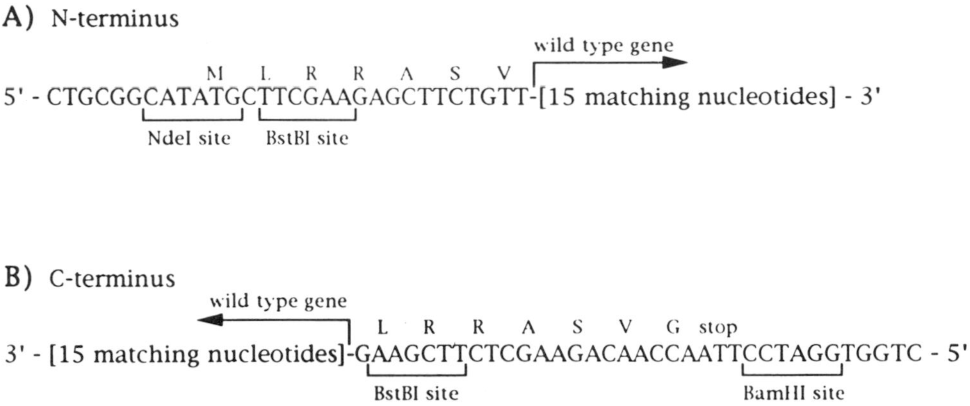

A) N-terminus

wild type gene

M L R R A S V

5' - CTGCGGCATATGCTTCGAAGAGCTTCTGTT-[15 matching nucleotides] - 3'

NdeI site BstBI site

B) C-terminus

wild type gene

L R R A S V G stop

3' - [15 matching nucleotides]-GAAGCTTCTCGAAGACAACCAATTCCTAGGTGGTC - 5'

BstBI site BamHI site

FIG. 1. Oligonucleotide sequences for PCR amplification of either N- or C-terminal PK protein. The oligonucleotides shown are designed for use with a second oligonucleotide that matches the gene of interest to produce a PCR product encoding either (A) an N-terminal PK protein or (B) a C-terminal PK protein. The encoded amino acids of the kinase recognition motif are shown above the nucleotide sequence. In each case, the 15 nucleotides at the 3′ terminus match the gene of interest. The *Nde*I and *Bam*HI cloning sites and the *Bst*BI screening site are marked. The 5–6 nucleotides at the 5′ terminus of these oligonucleotides ensure efficient cleavage of the PCR product.

N or C terminus of three proteins, expressed them in *E. coli*, and purified them to homogeneity. We have mainly studied the β subunit of *E. coli* DNA polymerase III holoenzyme (Pol III) in which a six-residue site, Leu-Arg-Arg-Ala-Ser-Val (followed by Pro), was engineered onto the C terminus. In addition, we have placed the seven-residue site, Leu-Arg-Arg-Ala-Ser-Val-Gly, onto the C terminus of the λ cro repressor and of EBNA1, the latent origin binding protein of the Epstein–Barr virus. We have also placed the seven-residue sequence Met-Leu-Arg-Arg-Ala-Ser-Val onto the N terminus of β and cro. These protein kinase (PK) proteins were engineered by PCR using an oligonucleotide encoding the kinase motif (Fig. 1) and expressed in the pET-3 system.[11] The kinase recognition motif can also be introduced by site-directed mutagenesis.[4] Since the pET-3 system was used for expression, the primer sequences contained an *Nde*I site for N-terminal PK proteins, and a *Bam*HI site for C-terminal PK proteins. The *Bst*BI site is not present in pET-3 and thus aids in screening clones.

Procedure for ^{32}P-End-Labeling

Unlike reductive methylation, this technique can be performed with small amounts of protein and the procedure is compatible with a variety

[11] W. F. Studier, A. H. Rosenberg, J. J. Dunn, and S. W. Dubendorff, this series, Vol. 185, p. 60.

of buffers. We typically use a volume of 30 to 120 μl containing 0.2 to 6 nmol PK protein in 20 m*M* Tris–HCl, pH 7.5, 1 to 60 μM [γ-^{32}P]ATP (specific activity discussed below), 2 m*M* DTT, 100 m*M* NaCl, 12 m*M* $MgCl_2$, and 10 m*M* NaF. Labeling is initiated upon adding cAMP-dependent protein kinase (0.008 U kinase/nmol PK protein) and shifted to 37°. After 10 min at 37°, the reaction is stopped on addition of 30 m*M* EDTA (final concentration). The extent of ^{32}P incorporated into the PK protein can be determined by acid precipitation or by analysis in a SDS–polyacrylamide gel (for the latter, addition of 5 m*M* unlabeled ATP to the sample buffer lowers background radiation). Free [γ-^{32}P]ATP can be removed either by ultrafiltration, spin dialysis, or gel filtration. The catalytic subunit of cAMP-dependent protein kinase from bovine heart can be purchased from Sigma and was used in previous studies,[4,5] however, we have used the murine version expressed and purified from *E. coli*[12] (gift of Dr. Susan S. Taylor, University of California at San Diego).

Under these conditions the stoichiometry of P_i incorporation is typically 0.5 to 0.8 mol P_i/mole protein (except N-terminal labeled β^{PK}, discussed below). Before embarking on these studies, it is important to determine whether the wild-type protein is phosphorylated by protein kinase. Wild-type β was not phosphorylated to a significant extent (<0.9 mmol P_i/mol β^{PK}) and sequence analysis of the radioactive chymotryptic peptide of [^{32}P]β^{PK} confirmed the label was in the kinase motif at the C terminus.

High specific activity protein can be achieved using straight [γ-^{32}P]ATP (3000 Ci/mmol) at approximately 1 μM, or the relative proportion of unlabeled ATP and radioactive ATP can be adjusted to achieve the desired specific activity. This can be very important in experimental designs using both ^{3}H-labeled protein and ^{32}P-labeled protein in the same experiment because the specific activity of ^{3}H-labeled proteins is only 20 to 40 Ci/mmol and therefore one must take care to prepare the ^{32}P-labeled protein with a comparable specific activity so as not to encounter problems of bleed over between windows while counting both isotopes.

Important Considerations

One advantage to the ^{32}P-end-labeling method is that the isotope is available in small quantities and the manipulations are similar to end-labeling DNA, a common laboratory technique. PK proteins can also be labeled using [γ-^{35}S]ATP.[13]

How often do labeled PK proteins retain activity? The modification is

[12] L. W. Slice and S. S. Tylor, *J. Biol. Chem.* **264,** 20,940 (1989).
[13] V. Naktinis and M. O'Donnell, unpublished (1993).

substantial; not only are several amino acids added to the protein, but a large phosphate group is also present. In our experience the activity of both N- and C-labeled $[^{32}P]\beta^{PK}$ is unchanged in several assays including assembly of $[^{32}P]\beta^{PK}$ onto primed DNA by Pol III, and ability of $[^{32}P]\beta^{PK}$ to confer highly processive DNA synthesis onto the polymerase.[14] Likewise, the $[^{32}P]EBNA1^{PK}$ and both N- and C-labeled $[^{32}P]cro^{PK}$ retained their site-specific DNA binding activities. It is important to note that if the amount of ATP used in the labeling reaction is substoichiometric to the protein (often the case in DNA end-labeling reactions), some types of activity assays will not truly evaluate whether the phosphorylated species retains activity.

In theory one should be able to place the kinase motif at any exposed region of a molecule and label it. However, placement of this kinase motif at other positions in the β subunit has revealed limitations to this technique. For example, the N-terminal β^{PK} is labeled approximately 4% at best.[13] The problem is likely due to the inability of the kinase to gain access to the recognition site as addition of 0.5 *M* urea increased the labeling to 16%.[13] Further, we have placed the seven-residue kinase recognition sequence into two different internal positions of β, which are located on the surface according to the crystal structure, but the kinase did not phosphorylate these proteins.[15]

Examples of Applications

Interaction of Proteins with DNA

To identify a particular subunit on DNA in a complex mixture of several proteins, the reaction is analyzed by gel filtration using beads with large pores such that free proteins elute in the included fractions and resolve from proteins bound to the large DNA (e.g., plasmid or M13 ssDNA), which elute in the excluded fractions. For these studies we have used agarose beads Bio-Gel A-5m and A-15m (Bio-Rad) as well as Sepharose 4B (Pharmacia-LKB), which exclude plasmids and ssDNA phage genomes, yet include proteins of up to 1 MDa.

Examples of this technique abound in studies from Arthur Kornberg's laboratory. For example, ^{3}H-labeled proteins were used to identify primosomal proteins that remained on the ϕX174 ssDNA with the finding that the DnaC protein, while essential to assemble the primosome on DNA, does not remain on the DNA.[16] The *E. coli* Pol III contains a five-subunit

[14] P. T. Stukenberg, Ph.D. thesis (1993).
[15] J. Turner and M. O'Donnell, unpublished (1993).
[16] J. A. Kobori and A. Kornberg, *J. Biol. Chem.* **257,** 13,770 (1982).

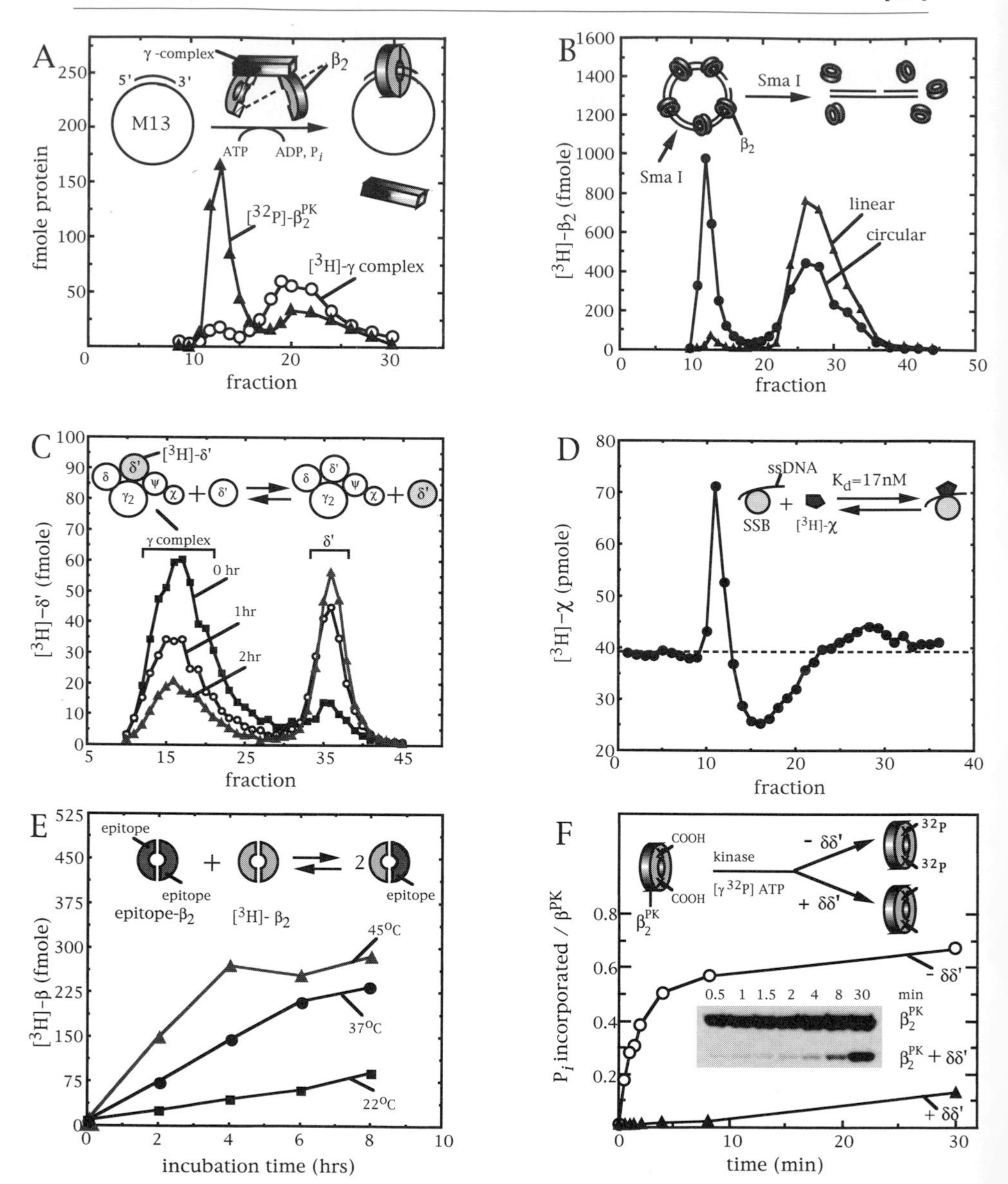

FIG. 2. Applications of radiolabeled proteins to study the molecular dynamics and structural details of multicomponent complexes. (A) The $[^3H]\gamma$ complex matchmaker assembles the $[^{32}P]\beta^{PK}$ dimer onto SSB coated primed M13mp18 ssDNA. Gel filtration on Bio-Gel A-5m resolves the $[^{32}P]\beta^{PK}$ bound to DNA (fractions 10–15) from the $[^3H]\gamma$ complex that dissociates from DNA and elutes in the included fractions. The experiment was performed essentially as described in Fig. 2.3 of Stukenberg.[14] (B) During gel filtration, $[^3H]\beta$ clamps coelute with

subassembly called the γ complex, which acts in a similar fashion as DnaC protein; γ complex places the β subunit of Pol III onto a primed template but then departs from the DNA.[17] Figure 2A shows the use of two labeled subunits in this latter reaction. The ^{3}H-labeled γ complex was used to place [^{32}P]β^{PK} onto a primed template and the reaction was analyzed by gel filtration. Analysis of the column fractions shows a stoichiometric amount of [^{32}P]β^{PK} dimers comigrating with the DNA, but the [^{3}H]γ complex migrates as free protein complex in the included fractions.

Topological Binding of Protein to DNA

The β subunit of Pol III is a ring-shaped dimer that completely encircles DNA.[18] Once the β ring is assembled onto DNA (by the γ complex and ATP) it tethers the rest of the Pol III machinery to the template and slides along with it for highly processive DNA synthesis.[19] This simple solution

[17] P. T. Stukenberg, P. S. Studwell-Vaughan, and M. O'Donnell, *J. Biol. Chem.* **266,** 11,328 (1991).

[18] X.-P. Kong, R. Onrust, M. O'Donnell, and J. Kuriyan, *Cell* **69,** 425 (1992).

[19] M. O'Donnell, J. Kuriyan, X.-P. Kong, P. T. Stukenberg, and R. Onrust, *Mol. Biol. Cell* **3,** 953 (1992).

(bind to) a circular plasmid with a single nick, but upon linearization of DNA the [^{3}H]β clamps fall off and elute in the included fractions. (Adapted from Fig. 3 in Stukenberg *et al.*[17] Copyright The American Society for Biochemistry & Molecular Biology.) (C) Dissociation of [^{3}H]δ' from the γ complex was observed by incubating 14 pmol of γ complex (reconstituted with [^{3}H]δ') with 108 pmol δ' in 60 μl of buffer 20 m*M* Tris–HCl, pH 7.5, 100 m*M* NaCl, and 10% glycerol at 22°. At the indicated times, the samples were analyzed by filtration on a 24-ml Superose-12 column. (D) Equilibrium gel filtration analysis of the interaction between [^{3}H]χ and the SSB-DNA complex was performed on a 5-ml column of P-30 (Bio-Rad) using 500 n*M* [^{3}H]χ in the column buffer and injecting on the column 1.7 nmol SSB in complex with M13mp18 (10 μg ssDNA) in column buffer. Fractions were collected and quantitated by scintillation counting. The K_d of 17 n*M* was calculated from the amount of [^{3}H]χ bound to the SSB–DNA in the peak. (E) Protomer interchange among β dimers using a mixture of 2 pmol [^{3}H]β_2 and 2 pmol hemagglutinin epitope tagged-β_2 in 50 μl of 20 m*M* HEPES, pH 7.5, 150 m*M* NaCl, 0.1% Triton X-100, 10% glycerol followed by immunoprecipitation at the indicated times using monoclonal antibody to the epitope (12CA5, BAbCO) as described by Kolodziej and Young.[23] The [^{3}H]β that coimmunoprecipitated with the epitope tagged-β indicates the extent of subunit exchange to form the heterodimer between the [^{3}H]β and epitope tagged-β. The experiment was repeated at the indicated temperatures. (F) In the kinase protection assay, 180 pmol β^{PK} was preincubated for 30 min at 15° either with or without 1 nmol $\delta\delta'$ complex, followed by treatment with protein kinase in a 30-μl reaction mixture as described in the text. At the indicated times 3-μl aliquots were analyzed by SDS polyacrylamide gel and autoradiography (inset). The kinase motif is located on the C termini, which extrude out from the same face of the β dimer as indicated in the scheme.

to high processivity may generalize to the gene 45 protein processivity factor of the phage T4 replicase, and the PCNA processivity factor of the yeast and human polymerase δ.[19] Protein surrounding DNA seems so basic that it may even generalize to other DNA metabolic machineries. For example, replicative helicases such as those encoded by phage T4 and phage T7, *E. coli* DnaB, and simian virus 40 (SV40) T antigen are all hexamers and may surround duplex DNA for their action (i.e., like the sixfold pseudo-symmetric β dimer).

Initial experiments using [^{3}H]β revealed this "topological binding" mode of β to DNA and predicted its ring shape,[17] thus motivating the crystal structure analysis. A particularly telling experiment is shown in Fig. 2B. Here, [^{3}H]β was placed onto a singly nicked plasmid DNA (by γ complex and ATP) and then gel filtered to reveal that several β dimers coelute with the DNA in the excluded fractions.[17] However, linearization of the DNA results in dissociation of β, implying that it slides along DNA and falls off the end. If β were to bind DNA by direct chemical interaction, as other DNA binding proteins do, then it would have stayed associated with the linear DNA through gel filtration as well as with circular DNA. Hence, β must be bound physically to DNA by virtue of its topology. This type of experiment is rather simple and variations on this theme may help identify the topological binding proteins of other systems.

Subunit Exchange in Multiprotein Complexes

Radiolabeled protein can be used to measure the rate of dissociation of a subunit from within a multiprotein complex. As an example, in Fig. 2C the five-subunit γ complex ($\gamma\delta\delta'\chi\psi$) was labeled by reconstituting it using [^{3}H]δ'. To this was added an eightfold molar excess of unlabeled δ' and the mixture was gel filtered at various times. Whenever the [^{3}H]δ' subunit dissociates from the complex, an unlabeled δ' takes its place. Hence, over time, the column fractions containing γ complex decrease in radioactivity and column fractions containing free δ' increase in radioactivity. This technique can be extended to determine how the rate of subunit exchange is influenced by DNA, nucleotides and other proteins.

Weak Protein–Protein Interactions

Only strong interactions among proteins can be detected by gel filtration because it is not an equilibrium technique. However, weak interactions can be quantitated using the equilibrium gel filtration technique.[20] This technique is normally used to determine the K_d between a protein and a

[20] J. P. Hummell and W. J. Dreyer, *Biochim. Biophys. Acta* **63,** 530 (1962).

small ligand molecule (e.g., a nucleoside triphosphate) where the ligand is radiolabeled and is present throughout the column buffer. The protein binds the radiolabeled ligand and then elutes ahead of the ligand, resulting in a peak of radioactivity that emerges above the baseline level and is followed by a trough where the unbound ligand would have eluted. The K_d can be calculated from this information. This technique has been applied in Arthur Kornberg's laboratory using $[^3H]\beta$ in the column buffer to define the K_d of a weak protein–protein interaction between the β subunit and the Pol III* assembly (Pol III lacking β).[21] Figure 2D shows an analysis of a weak interaction between the χ subunit of Pol III and an SSB–DNA complex using $[^3H]\chi$ in the column buffer.[22]

Exchange Rate Among Subunits of a Dimeric Protein

How rapidly do dimers of identical subunits fall apart and come back together? For example, the protomers of the β dimer form a closed ring, which must open to assemble around DNA. Do the monomer units of a β dimer rapidly come apart and then reassociate, thereby trapping DNA inside? The rate of exchange of monomer units of a dimer cannot be measured by gel filtration as in Fig. 2C since it is always in the dimeric state. An assay to measure stability of the β dimer, shown in Fig. 2E, uses a $[^3H]\beta$ dimer and a β dimer that is genetically tagged with the hemagglutinin epitope (nine amino acids) at its C-terminus.[23] Upon mixing the two β dimers, samples of the reaction were immunoprecipitated at time intervals. As the "heterodimer" of one $[^3H]\beta$ and one epitope tagged-β was formed, the amount of $[^3H]\beta$ in the coimmunoprecipitate increased. The results indicate that monomeric units of the β dimer require a long time to dissociate even at 37°. This assay can be exploited to study the influence of other proteins and of DNA and ATP on the stability of the β dimer.

Kinase Protection Assay

We originally end-labeled proteins with $[^{32}P]$ to develop a "protein footprinting" assay for mapping the interactive surface of one protein with another (or with DNA), but this work is still in progress. However, we have developed a "kinase protection" assay. An example of this assay is shown in Fig. 2F in which β^{PK} is incubated with or without the $\delta\delta'$ complex, a subassembly of the γ complex that binds β.[24] Then kinase is added and

[21] R. S. Lasken and A. Kornberg, *J. Biol. Chem.* **262,** 1720 (1987).
[22] Z. Kelman and M. O'Donnell, unpublished (1993).
[23] P. A. Kolodziej and R. A. Young, this series, Vol. 194, p. 508.
[24] R. Onrust, Ph.D. thesis (1993).

the time course of phosphorylation is monitored by SDS–PAGE followed by autoradiography. The result shows that phosphorylation of β^{PK} is almost completely blocked by the $\delta\delta'$ complex. The two C termini of the β dimer, where the kinase motif is located, extrude from the same face of the β dimer (see scheme in Fig. 2F). Hence, the ability of $\delta\delta'$ complex to block these sites from phosphorylation suggests that $\delta\delta'$ may interact with the "C-terminal" face of the β dimer.

Acknowledgments

We are grateful to Dr. Susan S. Taylor for the catalytic subunit of cAMP-dependent protein kinase. Supported by Public Health Service grant GM38839.

[33] Cycling of *Escherichia coli* DNA Polymerase III from One Sliding Clamp to Another: Model for Lagging Strand

By JENNIFER TURNER and MIKE O'DONNELL

Introduction

The multiprotein replicase of the *Escherichia coli* chromosome, DNA polymerase III holoenzyme (Pol III), achieves a tight ATP-activated grip on DNA through its ring-shaped clamp protein, the β dimer (β_2).[1–4] The β_2 ring encircles DNA, thereby acting as a sliding clamp and continuously holds Pol III to the template for remarkably high speed and processivity.[5] The model of a circular protein clamp riding along in back of the polymerase fits nicely with the continuous synthesis of the leading strand, but conceptually it would hinder the discontinuous mode on the lagging strand where Pol III must rapidly dissociate from the end of one Okazaki fragment to start another fragment at the next RNA primer (i.e., Fig. 1C).

Studies of Pol III showed that it was indeed very slow to replicate more than a stoichiometric number of primed templates.[6] After Pol III completely replicated a circular single-stranded DNA (ssDNA) template, several

[1] A. Kornberg and T. A. Baker, "DNA Replication." Freeman, New York, 1991.
[2] X.-P. Kong, R. Onrust, M. O'Donnell, and J. Kuriyan, *Cell* **69,** 425 (1992).
[3] M. O'Donnell, J. Kuriyan, X.-P. Kong, P. T. Stukenberg, and R. Onrust, *Mol. Biol. Cell* **3,** 953 (1992).
[4] J. Kuriyan and M. O'Donnell, *J. Mol. Biol.* **234,** 915 (1993).
[5] P. T. Stukenberg, P. S. Studwell-Vaughan, and M. O'Donnell, *J. Biol. Chem.* **266,** 11,328 (1991).
[6] P. M. J. Burgers and A. Kornberg, *J. Biol. Chem.* **257,** 11,474 (1982).

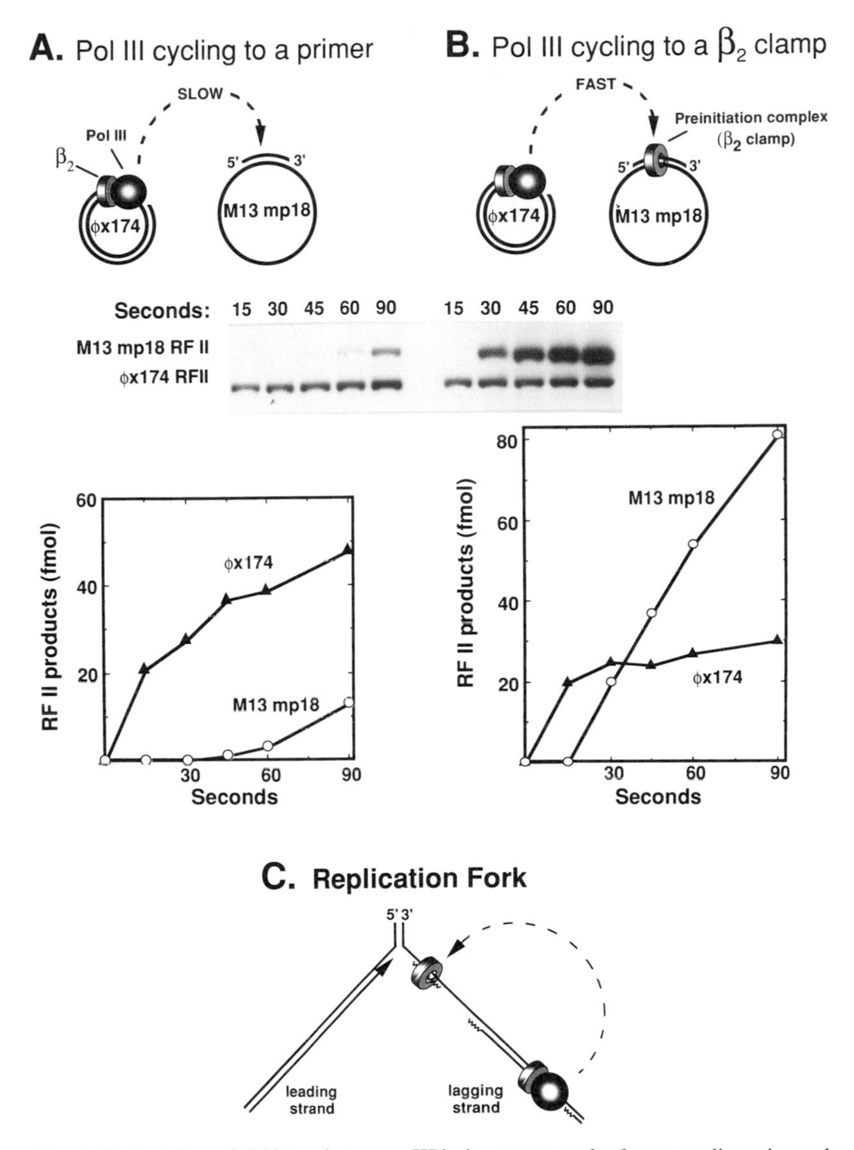

FIG. 1. *Escherichia coli* DNA polymerase III holoenzyme cycles from a replicated template to another DNA containing a β_2 clamp. The donor reaction consists of Pol III bound to its β_2 clamp on a singly primed ϕX174 ssDNA. The acceptor template is a singly primed M13mp18 ssDNA either with a β_2 clamp (B) or without a β_2 clamp (A). Replication is initiated upon mixing the donor and acceptor reactions and products are removed at various times and observed by autoradiography of a native agarose gel. Quantitation of the time course is shown below the autoradiogram. The replication fork diagram in (C) shows the implication of these results for lagging strand replication in which Pol III rapidly cycles to a new RNA primer endowed with a β_2 clamp upon completing an Okazaki fragment. Adapted from Figs. 6 and 8 of O'Donnell.[9]

minutes were required for it to transfer to another primed ssDNA. Pol III was capable of rapid transfer to multiple primers on one circular ssDNA, but it slid over the top of them rather than transferring backward in the antielongation direction as required during lagging-strand synthesis (e.g., Fig. 1C).[7,8] The mechanism that underlies rapid cycling of Pol III from one primed site to another was finally discovered upon asking Pol III to cycle to a primed site that already contained a β_2 clamp[9,10]; further studies showed that the initial β_2 clamp was simply left behind on the original DNA.[11] Hence, Pol III cycling proceeds by a novel pathway in which Pol III dissociates from its β_2 clamp and reassembles with another β_2 clamp (see Fig. 1C). An important characteristic of this reaction is that Pol III only cycles away from its β_2 clamp after it has completely replicated the initial template.[9,11] This property would ensure that each Okazaki fragment is fully extended before Pol III leaves it for the next RNA primer.

How general is this mechanism of protein hopping among other proteins bound to nucleic acid? It seems likely that polymerase cycling from one clamp to another may generalize to other replication systems, especially in light of the similarity of *E. coli* Pol III to the replicase of phage T4, and the polymerase δ holoenzyme of eukaryotes.[3,12] It is tempting to speculate that other nucleic acid metabolic processes may involve a protein hopping from one nucleoprotein complex to another such as in recycling of RNA polymerase on committed promoters, transcriptional activation by protein mediated looping of DNA, recombination acts such as site-specific integration involving different sets of proteins on two DNA molecules, site acceptor choices in RNA splicing, and in recycling of the ribosome on mRNA that contains initiation factors.

Procedures

Two different nucleoprotein complexes are assembled in separate tubes, a donor and an acceptor complex. In the Pol III system, the donor is a complex of Pol III bound to a singly primed ϕX174 ssDNA coated with the ssDNA binding protein (SSB). The acceptor consists of a β_2 clamp that has been assembled by γ complex (a five-protein Pol III subassembly) onto

[7] P. M. J. Burgers and A. Kornberg, *J. Biol. Chem.* **258,** 7669 (1983).

[8] M. E. O'Donnell and A. Kornberg, *J. Biol. Chem.* **260,** 12,875 (1985).

[9] M. E. O'Donnell, *J. Biol. Chem.* **262,** 16,558 (1987).

[10] P. S. Studwell, P. T. Stukenberg, R. Onrust, M. Skangalis, and M. O'Donnell, *UCLA Symp. Mol. Cell. Biol. New Ser.* **127,** 153 (1990).

[11] P. T. Stukenberg, J. Turner, and M. O'Donnell, *Cell* **78,** 877 (1994).

[12] M. O'Donnell, R. Onrust, F. B. Dean, M. Chen, and J. Hurwitz, *Nucleic Acids Res.* **21,** 1 (1993).

a singly primed, SSB-coated M13mp18 ssDNA.[1,5] This β_2 clamp on DNA is also referred to as a "preinitiation complex."[1,9] Replication is initiated upon mixing the two reactions. Aliquots are withdrawn at time intervals and quenched with EDTA and SDS, then one-half is analyzed by electrophoresis in a native agarose gel and the other half is analyzed for total nucleotide incorporation. An example of the results of a "polymerase cycling" reaction are shown in Fig. 1. Pol III replicates the ϕX174 donor within 15 sec, then rapidly cycles to and replicates the acceptor M13mp18 DNA containing the β_2 clamp (Fig. 1B), but not to the M13mp18 acceptor lacking a β_2 clamp (Fig. 1A). This implies that on the lagging strand, Pol III rapidly cycles to a new RNA primer, provided that the primer is already endowed with a β_2 clamp (Fig. 1C). What follows is a description of the design of experiments of this type.

Setting Up Donor

Assembly of the donor nucleoprotein complex requires the enzyme to be bound to DNA such that there is no protein left in solution (e.g., substoichiometric Pol III to ϕX174 DNA). In the Pol III system, ATP is required for Pol III to bind primed ssDNA, and it is bound so tight that there is essentially no Pol III left in solution (provided Pol III is added substoichiometric to the ϕX174 ssDNA template). Alternatively, the donor reaction can be gel filtered to remove any free proteins. The experiment in Fig. 1 utilized 120 fmol Pol III to 243 fmol primed ϕX174 ssDNA (as circles) saturated with SSB (4.2 μg) in a volume of 150 μl of 20 m*M* Tris–HCl, pH 7.5, 8 m*M* $MgCl_2$, 4% (v/v) glycerol, 5 m*M* dithiothreitol (DTT), 40 μg/ml bovine serum albumin, 0.5 m*M* ATP, and 60 μ*M* each dCTP and dGTP. Assembly of the donor complex is complete within 2 min at 30°. Magnesium and ATP are needed for Pol III to bind primed DNA, however magnesium also promotes removal of the primer within a few seconds by the proofreading $3' \rightarrow 5'$-exonuclease of Pol III.[8] This problem is circumvented by including two dNTPs, which allows limited extension by Pol III to compete with the exonuclease activity, thus constraining Pol III to idle at the primed site.[8]

Setting Up Acceptor

We typically assemble a fourfold greater number of acceptor complexes relative to the donor, but in the same volume. Equal volumes of the two reactions facilitate rapid and efficient mixing of acceptor with the donor. The acceptor DNA must be a different size than the donor for product analysis. In the experiment of Fig. 1, the acceptor is a β_2 clamp preinitiation complex and the reaction contained 911 fmol primed M13mp18 ssDNA saturated with SSB (21 μg), 1.6 pmol β_2, and 195 fmol of γ complex in a volume of 150 μl. The same buffer is used for the acceptor as for the donor, except the acceptor reaction contains all 4 dNTPs (60 μM dCTP and dGTP, 120 μM dATP, and 40 μM [α-^{32}P]TTP) to initiate DNA synthesis upon mixing it with the donor. In this reaction the γ complex acts as a molecular matchmaker, hydrolyzing ATP to place β_2 clamps onto DNA and assembly is complete within 2 min at 30°. Assembly of β_2 onto the M13mp18 primed template is dependent on γ complex concentration and can near completion within only 1 sec.[11]

Mixing and Analysis

A running stopwatch is placed by the water bath and the acceptor reaction is withdrawn into a Pipetman tip and positioned over the donor reaction. The reaction is initiated on forcibly expelling the acceptor into the donor reaction, which also effectively mixes the two. Aliquots (45 μl) are then withdrawn and placed into tubes containing an equal volume of 1% SDS and 40 mM EDTA to quench the reaction. In the Pol III system the desired time points are rapid, every 15 sec, necessitating prior set up of the Pipetman and tubes with quench solution. The method of product analysis will depend on the system being studied. In the Pol III system, the acceptor and donor products are large circular duplexes of different sizes and cleanly separate in a 0.8% neutral agarose gel.

Further Considerations

It is important that there not be an excess of the protein under study in the donor reaction or else the extra protein will simply associate with the acceptor complex and the observed products will not reflect cycling of Pol III from one DNA to another. In the Pol III system, a simple test for excess Pol III in the donor reaction is to incubate the donor and acceptor reactions under conditions that do not allow replication (i.e., only dCTP and dGTP present in both reactions). Then at various times an aliquot of the mixture is removed and replication is initiated upon adding the remaining 2 dNTPs, but only 15 sec is allowed for replication before the reaction is quenched. Given this time constraint, only templates on which Pol III has already assembled can be replicated. If Pol III is present in excess, it will associate with the acceptor β_2 clamp in a time-dependent manner, and total synthesis will increase. Experiments of this type in the Pol III system showed no excess Pol III was present and, in fact, agarose gel analysis of this control experiment revealed that Pol III remained with the donor and was very slow to transfer to the acceptor under these idling conditions.[9]

It is important to consider the possibility that the protein reagents used to set up the donor complex may be contaminated with proteins that form the putative acceptor complex. In this event they may assemble onto the acceptor upon addition of the donor, and if this assembly is rapid (i.e., faster than the protein transfer reaction being studied) then it will appear that the acceptor need not require protein assembled on it for cycling to occur. A solution to this problem is to gel filter the donor reaction prior to mixing it with the acceptor. If one is working with large templates, such as the ssDNA phage genomes used here, the large DNA and proteins bound to it elute in the excluded volume of resins such as Bio-Gel agarose A-5m media (5-MDa exclusion limit) and separate from free proteins (i.e., contaminants not bound to the donor DNA) in the included volume.

It is possible that the proteins added to the acceptor DNA may not need to be in the form of a nucleoprotein complex to mediate cycling of protein from the donor template, but may instead act in solution to chaperone the transfer (e.g., of Pol III) from donor to acceptor. The acceptor reaction can be gel filtered to remove free protein as a first step in addressing this issue. Further confirmation that a nucleoprotein complex is required can be made by preparing two acceptors using different sized primed templates (e.g., recombinant M13 phage genomes of different size). One acceptor is treated with proteins and the other is untreated, and both are added to the donor. If cycling requires an acceptor nucleoprotein complex, then only the acceptor template treated with proteins will enter the reaction. If the acceptor proteins act in solution, then both templates will enter the reaction. This "three template" experiment has been used in the Pol III system to show that the β_2 clamp must be on the acceptor DNA for Pol III to cycle to it rather than β_2 and γ complex acting in solution.[11] However, in designing this experiment, the γ complex concentration needed to be kept low enough to require a long time to assemble β_2 onto the acceptor DNA, otherwise β_2 clamps were assembled onto the untreated acceptor during the time required for the donor template to be replicated.

It is important that the ssDNA used in these studies be very pure for complete extension of these primed templates. Hence we doubly band the phage preparations first down and then up in cesium chloride density gradients. Preparation of M13 and ϕX174 phage are slightly different and are briefly described below (adapted from unpublished methods used in Arthur Kornberg's laboratory). M13 phage are first precipitated from the supernatant of 2 liter of infected cells using 5% (w/v) polyethylene glycol

(PEG) 8000 and 0.5 *M* NaCl for 4 hr at 4°. The pellet is then resuspended in 20 ml of 10 m*M* Tris–HCl, pH 7.5, 1 m*M* EDTA (TE buffer) and centrifuged for 3 hr at 153,000*g* at 4° [30,000 rpm in a Sorvall TH-641 rotor using two 12-ml Ultra-Clear transparent thermoplastic tubes (Beckman, #344059)], which pellets the phage and some PEG but leaves most of the PEG in solution. This pellet is resuspended in 4 ml TE buffer and repeatedly spun in a tabletop centrifuge to remove particulate matter, and layered onto two 11-ml 1.3 to 3.3 *M* CsCl gradients in TE and centrifuged 16 hr at 135,000*g* at 4° (27,000 rpm in a Sorvall AH-629 rotor using 17-ml Ultra-Clear Beckman #344061 tubes). The phage band is removed using a syringe with an 18-gauge needle (phage are approximately halfway down the gradient, PEG remains above the phage). If one is unsure which band is phage, the bands can be pulled and analyzed for DNA in a neutral agarose gel (1% SDS in the loading buffer is sufficient to remove the phage coat). Solid CsCl is added to the phage (0.2 g CsCl/ml phage), and the solution is layered under an 11-ml 1.3 to 3.3 *M* CsCl gradient, which is then centrifuged as described for the down banding. The phage are recovered using a syringe and needle and dialyzed against TE. To purify ϕX174, infected cells (30 g wet weight) are suspended in 50 ml of 50 m*M* sodium borate, 50 m*M* EDTA (BE buffer) and lysed using lysozyme treatment (15 mg) for 20 min on ice. Cell debris is removed by low-speed centrifugation and the supernatant (70 ml) is extracted with 50 ml chloroform followed by back extraction with 25 ml BE buffer. The chloroform extraction and back extraction are repeated and the aqueous phases are pooled and layered onto six-step gradients of CsCl in BE buffer in 36 ml Ultra-Clear Beckman #344058 tubes. The composition of the step gradient is density (volume): 1.3 g/ml (5 ml); 1.35 g/ml (5 ml); 1.377 g/ml (5 ml); 1.4 g/ml (4 ml); 1.5 g/ml (4 ml). After 9 hr at 104,000*g* (24,000 rpm in the AH-629 rotor) at 4° the phage are removed from the 1.377 g/ml band and adjusted to 1.5 g/ml with solid CsCl. The phage (4.5 ml per tube) are placed in a 17-ml tube and three CsCl steps are layered on top [density (volume): 1.4 g/ml (3.5 ml); 1.35 g/ml (4.5 ml); 1.3 g/ml (4.5 ml)] followed by 6 hr of centrifugation at 4° at 110,000*g* (24,500 rpm in a AH-629 rotor). The phage are removed from the interface of the 1.4 and 1.3 g/ml steps and dialyzed against BE buffer. The concentration of ssDNA can be estimated from the absorbence at 260 nm (1 OD is approximately 36 μg/ml) since the DNA chromophore outweighs the contribution by the protein coat.

To extract ssDNA, the phage are diluted to an A_{260} of 7 with TE and NaCl is added to 0.1 *M*. An equal volume of phenol is added followed by incubation for 15 min at 37°. The phenol is back extracted with an equal volume of 0.1 *M* NaCl in TE and the aqueous phases are pooled and then extracted with an equal volume of phenol/chloroform (1:1, v:v). Extractions using phenol/chloroform are repeated until the protein at the interface is greatly reduced (usually two extractions are sufficient). This is followed by two chloroform extractions, ethanol precipitation, a 70% ethanol wash, and resuspension of the ssDNA in TE buffer to an A_{260} of 30 to 70.

Acknowledgment

Supported by Public Health Service grant GM38839.

[34] Photochemical Cross-Linking of DNA Replication Proteins at Primer Terminus

By TODD L. CAPSON, STEPHEN J. BENKOVIC, and NANCY G. NOSSAL

Introduction

Photochemical cross-linking can detect weak protein–DNA interactions and is particularly useful in locating components of multisubunit enzymes or multienzyme complexes.[1–15] This chapter describes a primer–template system in which the position of a cross-linkable aryl azide can be moved stepwise through a 30-base region of the primer by selective elongation of the primers

[1] T. L. Capson, S. J. Benkovic, and N. G. Nossal, *Cell* **65,** 249 (1991).
[2] S. Dissinger and M. M. Hanna, *J. Biol. Chem.* **265,** 7662 (1990).
[3] B. Bartholomew, C. F. Mearses, and M. E. Dahmus, *J. Biol. Chem.* **265,** 3731 (1990).
[4] B. Bartholomew, G. A. Kassavetis, B. R. Braun, and E. P. Geiduschek, *EMBO J.* **9,** 2197 (1990).
[5] R. K. Evans and B. E. Haley, *Biochemistry* **26,** 269 (1987).
[6] C. E. Catalano, D. J. Allen, and S. J. Benkovic, *Biochemistry* **29,** 3612 (1990).
[7] J. W. Hockensmith, W. Kubasek, W. R. Vorachek, E. Evertsz, and P. H. von Hippel, *Meth. Enzymol.* **208,** 211 (1986).
[8] J. W. Hockensmith, W. L. Kubasek, W. R. Vorachek, and P. H. von Hippel, *J. Biol. Chem.* **263,** 15,712 (1993).
[9] J. W. Hockensmith, W. L. Kubasek, E. M. Evertsz, L. D. Mesner, and P. H. von Hippel, *J. Biol. Chem.* **268,** 15,721 (1993).
[10] R. L. Tinker, K. P. Williams, G. A. Kassavetis, andd E. P. Geiduschek, *Cell* **77,** 225 (1994).
[11] M. C. Willis, B. J. Hicke, O. C. Uhlenbeck, T. R. Cech, and T. H. Koch, *Science* **262,** 1255 (1993).
[12] E. E. Blatter, Y. W. Ebright, and R. H. Ebright, *Nature* **359,** 650 (1992).
[13] T. D. Allen, K. L. Wick, and K. S. Matthews, *J. Biol. Chem.* **266,** 6113 (1991).
[14] T. Tsurimoto and B. Stillman, *J. Biol. Chem.* **266,** 1950 (1991).
[15] M. Foiani, C. Santocanale, P. Plevani, and G. Lucchini, *Mol. Cell. Biol.* **9,** 3081 (1989).

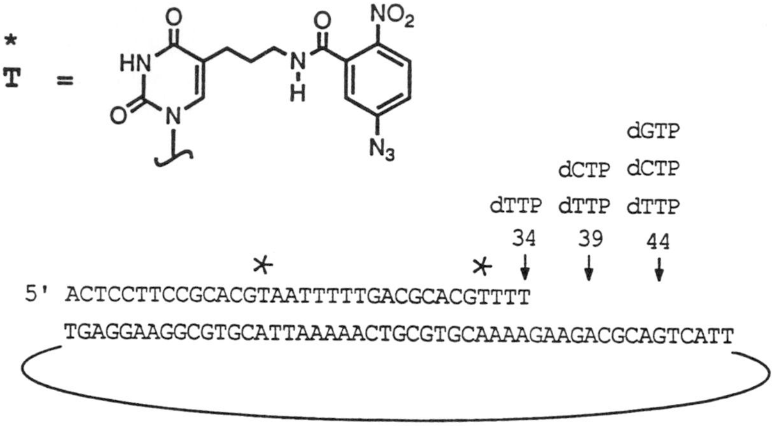

T* =

dGTP
dCTP dCTP
dTTP dTTP dTTP
34 39 44

5' ACTCCTTCCGCACGTAATTTTTGACGCACGTTTT
TGAGGAAGGCGTGCATTAAAAACTGCGTGCAAAAGAAGACGCAGTCATT

Azido 34mer-Φ X174

FIG. 1. Primer–templates with photoactivatable thymidine analogs used for cross-linking experiments. Two 34-mers were used that had identical sequence, complementary to nucleotides 808–841 of the ϕX174 ssDNA template, but differed in the placement of the cross-linkable residue. The 34(4*)-mer and the 34(20*)-mer have the cross-linkable residue −4 and −20 bases, respectively, from the 3′ primer terminus. Adapted from Capson *et al.*[1] with permission of *Cell* press.

with different combinations of dNTPs. Using this system we have shown that T4 DNA polymerase and the T4 genes 44/62 and 45 polymerase accessory proteins assemble into different complexes in the presence of ATP and ATPγS, and have mapped primer–protein contacts in each complex.[1]

Primer–Template Construction

The primer contains a modified thymidine residue to which a 5-azido-2-nitrobenzoyl moiety is attached via a 3-carbon tether, placing the cross-linkable moiety within or close to the major groove of the DNA helix[6,16] (Fig. 1). Irradiation of this cross-linkable residue with ultraviolet light at 302 nm results in formation of a nitrene that can form covalent adducts with adjacent proteins.[17,18]

For the study of T4 polymerase and its accessory proteins, we use two 34-mers of identical sequence complementary to the ϕX174 template, which

[16] K. G. Gibson and S. J. Benkovic, *Nucleic Acids Res.* **15,** 6455 (1987).

[17] H. Bayley, *in* "Laboratory Techniques in Biochemistry and Molecular Biology" (T. S. Work and R. H. Burdon, ed.). Elsevier, New York, 1983.

[18] H. Bayley and J. R. Knowles, *Meth. Enzymol.* **46,** 69 (1977).

differ only in the placement of the cross-linkable residue, which is either −4 (34(4*)) or −20 (34(20*)) residues from the primer terminus (Fig. 1). The template sequence allows the polymerase to extend the primer in steps of 5 bases with different dNTP combinations. Thus, we are able to examine the interaction of the replication proteins at 5 base intervals over a 30-base region of the primer. The T* residue in the 34(4*)-mer is followed by three Ts so that, in the presence of dTTP, it is protected from the $3' \rightarrow 5'$ (editing)-exonuclease activity of the polymerase. A large circular template is used to prevent degradation by the exonuclease and to serve as a trap for proteins bound nonproductively on the single-stranded portion of the template. We chose a sequence where dATP is the last nucleotide to be added, since the ATPase activity of the 44 protein also uses dATP as a substrate[19]; this allowed us to assess the effect of ATP and ATP analogs on the cross-linking reactions.

Chemical Synthesis of Primers

The 34-mers are made using methods developed by Gibson and Benkovic.[16] The photoactivatable analog is introduced at the desired position during automated DNA synthesis with phosphitylated 5-(fluoroenylmethoxycarbonyl13-aminopropyl)-2′-deoxyuridine. The oligonucleotides are purified by reversed-phase HPLC as described,[16] and then further purified on a denaturing (8 *M* urea) 20% polyacrylamide gel and eluted by electrophoresis with an Elutrap (Schleicher and Schuell) in TBE buffer. The detritylated, purified 34-mers are derivatized with *N*-5-azido-2-nitrobenzoyloxysuccinimide and purified by reversed-phase HPLC.[16]

Since the azido moiety is light sensitive, these and all subsequent manipulations are carried out in the dark. We find it convenient to work in a hood closed with an opaque black curtain, illuminated with a 25-W red light. Light-sensitive reagents are stored in amber-colored plastic centrifuge tubes (Eppendorf).

Primer–Template

The 34-mers (25 pmol) are 5′-end-labeled with polynucleotide kinase in a standard reaction except that thiols, such as dithiothreitol (DTT) or mercaptoethanol, which destroy the azide, are not added to the reaction. The kinase reaction mixture is heated to 68° for 20 min to denature the enzyme. The ϕX174 ssDNA template (12 pmol) is heated for 3 min at 100° to reduce intramolecular base pairing, and then added to the kinase reaction at 68°. The final concentration of NaCl in the annealing mixture is 0.2 *M*.

[19] D. C. Mace and B. M. Alberts, *J. Mol. Biol.* **177,** 279 (1984).

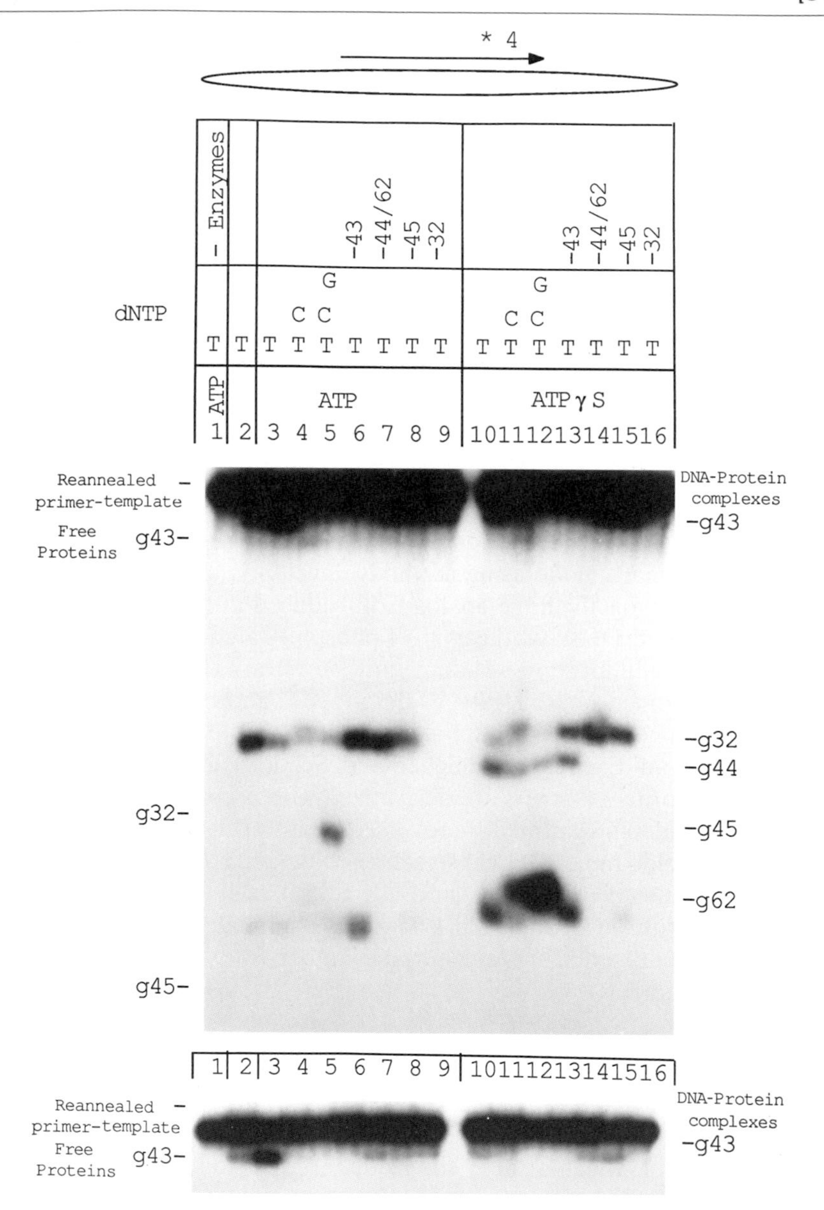

FIG. 2. Cross-linking of the T4 replication proteins to bases −4 to −14 from the 3′ terminus in the presence of ATP or ATPγS. The 34(4*)-mer, present in assays with only dTTP, was elongated to the 39(9*)-mer with dTTP and dCTP or the 44(14*)-mer with dTTP, dCTP, and dGTP. The shorter exposure at the bottom shows that in contrast to relatively strong cross-linking of polymerase (g43) at −4 in the presence of ATP (lane 3), with ATPγS (lane 10), cross-linking of polymerase does not increase above the level seen in the absence of ATP

After 10 min at 68°, the solution is cooled slowly to 25° and then stored briefly on ice. The hybrid is separated from free 34 mer and residual ATP by filtration on a 7-ml column of Sepharose Cl-2B in 20 m*M* ammonium acetate, 50 μM EDTA, and stored at $-20°$.

Replication Reactions and Cross-Linking of Proteins to DNA

In all of the cross-linking experiments with the T4 replication complex the reactions are carried out with an excess of the polymerase and accessory proteins, with 32 protein sufficient to cover about 50% of the template, which is the optimal ratio of 32 protein to DNA. To initiate the reaction, a mixture of the replication proteins is added to a mixture of the primer–template and nucleotides (at room temperature) in a 0.5-ml polypropylene microcentrifuge tube to give final concentrations of 25 μM ATP or ATPγS; 20 μM each dNTP; 25 m*M* Tris acetate, pH 7.5; 60 m*M* potassium acetate; 6 m*M* magnesium acetate; 0.2 mg/ml bovine serum albumin; 2 n*M* (approximately 2×10^6 cpm/pmol) 5′[^{32}P]34(4*) or 34(20*)mer: ϕX174 DNA duplex; 1.5 μg/ml (14.6 n*M*) T4 DNA polymerase; 4 μg/ml, 44/62 protein complex (24.2 n*M*, 44 : 62, 4 : 1[20]); 6 μg/ml (81 n*M* trimer[20]) 45 protein; and 28 μg/ml (814 nM) 32 protein in a total volume of 20 μl. DTT is not added to the reactions, but there is residual DTT ($<$0.05 m*M*) present from the purified replication proteins.[21] The reaction mixtures are incubated for 8 sec at room temperature, a 2-μl aliquot quenched into sequencing gel buffer for analysis of the DNA products on a 16% polyacrylamide, 7 *M* urea gel, and the remainder photolyzed for 1 min at room temperature with a Spectroline EB-280C lamp at 302 nm (860 mW/cm^2 at 15 cm) suspended 3.9 cm above the sample. (The yield of DNA–protein adducts is not increased substantially by increasing the photolysis time to 2 min.) After addition of protein gel sample buffer and boiling for 3 min, the DNA–protein adducts are analyzed on 12% SDS–polyacrylamide gels[22] with no stacking gel. Control lanes con-

[20] T. C. Jarvis, L. S. Paul, and P. H. von Hippel, *J. Biol. Chem.* **264,** 12,709 (1989).
[21] N. G. Nossal, D. M. Hinton, L. J. Hobbs, and P. Spacciapoli, this volume [43].
[22] D. M. Hinton, L. L. Silver, and N. G. Nossal, *J. Biol. Chem.* **260,** 12,851 (1985).

(lane 2) or in the absence of the 44/62 and 45 accessory proteins (lanes 7, 8, 14, 15). The longer exposure at the top demonstrates that the 44 accessory protein is cross-linked in the presence of ATPγS but not ATP, and that ATPγS increases the cross-linking to 62 accessory protein, but abolishes cross-linking of 45 accessory protein to the 44(14*)-mer. Note that cross-linking to the 44 and 62 proteins requires the 45 and 32 proteins (lanes 15 and 16), but occurs in the absence of polymerase (g43) (lane 13). Reprinted from Capson *et al.*[1] with permission of *Cell* press.

taining the free replication proteins are cut off and stained with Coomassie blue. The remaining gel is covered with plastic wrap and exposed to Kodak XAR film with an intensifying screen.

Analysis of the Protein–DNA Adducts

Figure 2 shows the products of cross-linking the T4 replication proteins at positions 4, 9, and 14 nucleotides behind the 3′ primer end, in the presence of ATP or ATPγS. The protein–DNA adducts, labeled on the right side of the gel, are identified by their mobility relative to that of the free proteins, labeled on the left, in complete and partial reactions. For example, the polymerase–DNA adduct is identified in reactions with only polymerase and the single-stranded 34(4*)-mer. 32 Protein is cross-linked to the 34-mer-ϕX174 primer-template in the absence of other proteins. The 62 protein–34-mer adduct runs faster than free 44 protein, and is formed in the presence of only 32 protein and very high concentrations of the 44/62 protein complex. The 44 and 45 protein adducts are identified by their mobilities relative to the free proteins in experiments with ATPγS where all three accessory proteins can be cross-linked.

The major conclusions from this from the gel (see legend to Fig. 2) are that, with ATP, polymerase is most strongly cross-linked −4 nucleotides behind the primer in the reaction with only dTTP (lane 3), whereas 45 protein is cross-linked at −14 in the reaction with dTTP, dCTP, and dGTP (lane 5). Cross-linking of these two proteins is dependent on the gene 32 single-stranded DNA binding protein and the 44/62 polymerase accessory protein. In contrast, with ATPγS, both the 44 and 62 proteins are cross-linked −4, −9, and −14 nucleotides from the 3′ end, but there is no cross-linking of polymerase or 45 protein to this region of the primer (lanes 10–12). Other experiments showed that 45 protein can be cross-linked at −20 with ATPγS. Based on a series of cross-linking experiments with the two 34-mer primers, we have proposed[1] (see Fig. 3) that the initial binding of the three accessory proteins and ATP is followed by ATP hydrolysis, binding of polymerase, and rearrangement of the accessory proteins to form a complex capable of processive DNA synthesis.

Strengths and Weaknesses of this Aryl Azide Cross-Linking System

The major advantages of cross-linking to DNA with the aryl azide are that the photoactivatable residue is located at a specific position in the primer sequence within the major groove of the helix, and is activated at a wavelength (302 nm) above that absorbed by protein and unmodified DNA. Appropriate choice of the template sequence allows the replication

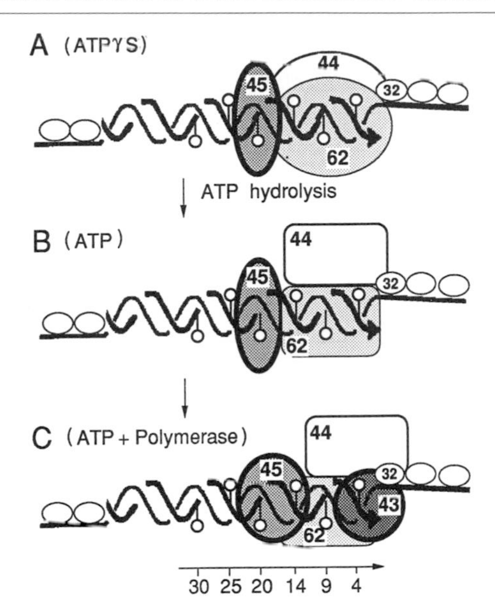

FIG. 3. Model for the assembly of the polymerase accessory proteins and polymerase at a primer terminus based on cross-linking studies.[1] (A) The initial event is binding of the polymerase accessory proteins to a 32 protein-covered primer–template in the presence of ATP, but before ATP hydrolysis. 44 Protein is bound from −4 to −14, most strongly at the −4 position. 62 Protein is bound from −4 to −14, most strongly at the −14 position. 45 Protein is bound at the −20 position. Before ATP hydrolysis there is no accessory protein-dependent binding of the polymerase (43 protein). (B) After ATP hydrolysis, 44 protein moves away from the primer; 62 protein is still present at −4 and weakly bound at −20. 45 Protein is still present at −20. (C) Movement of the accessory proteins as a result of ATP hydrolysis allows polymerase to bind at the primer terminus to generate a complex capable of processive DNA synthesis. Polymerase covers 5–7 bases of the primer. 45 Protein moves forward to the −14 position, and binding of the 62 protein is decreased at −4. The polymerase and 45 proteins are shown closer to the primer than the 62 protein, since cross-linking to 62 protein is weak and is increased at −4 in the absence of polymerase and at −20 in the absence of 45 protein. As discussed in Capson *et al.*,[1] these cross-linking studies do not rule out the possibility that the 44/62 complex is released after binding of polymerase. Reprinted from Capson *et al.*[1] with permission of *Cell* press.

complex to elongate the primer to discrete lengths by varying the available dNTPs. However, because the irradiation time (1 min) is greater than the time needed for synthesis, the protein–DNA complex that is cross-linked is stalled on the template by the absence of the dNTP needed for further synthesis, turning over the last permitted dNTP, rather than moving along the template.

The relative positions of individual members of the replication complex can be identified by selective cross-linking to primers with the activatable moiety in different positions. It is important to appreciate that detection of protein–DNA contacts with this technique requires that the nitrene generated by photolysis be adjacent to an amino acid capable of forming a covalent adduct, and that the resultant adduct be stable under the analysis conditions. Since neither the reactivity of the different amino acids with nitrenes nor the stability of their resultant adducts is well characterized,[17] failure to cross-link a protein at a specific position on the primer does not prove that the protein is absent.

The formation of a covalent cross-link between a replication protein and DNA should allow determination of the amino acid linked to the primer, as has been demonstrated for *Escherichia coli* polymerase I with a primer containing the aryl azide,[6] and for other proteins cross-linked to unmodified DNA,[23] or to DNA with BrU replacing T at specific positions.[12,13] Identification of the cross-linked amino acid(s) should be possible for T4 DNA polymerase, which was cross-linked to 8% of the primers,[1] but not for the 45 and 62 polymerase accessory proteins, which were cross-linked to only 0.6 and 0.2% of the primer, respectively, in the presence of ATP. Furthermore, because the fraction of the T4 accessory proteins cross-linked was so low, it is possible that the protein–DNA complexes detected by this analysis are not the majority species present in each reaction.

DNA containing 5-iodouracil (IU) in place of thymine at a specific position in a telomeric sequence has been shown to be cross-linked to *Oxytricha nova* telomeric proteins after irradiation at 308 or 325 nm.[11] Iodouracil can be inserted during automated DNA synthesis. Cross-linking of replication protein complexes to primer–templates with IU should complement studies with the aryl azide, since the photoactivatable moiety in IU is closer to the DNA helix. Moreover, the higher cross-linking yields with IU–DNA irradiated at 308 nm with a xenon–chlorine (XeCl) excimer laser and at 325 nm with a helium–cadmium (HeCd) laser[11] suggest that the identification of cross-linking amino acids in the replication proteins may be more feasible with IU–DNA.

[23] K. R. Williams and W. H. Konigsberg, *Meth. Enzymol.* **208,** 516 (1991).

[35] DNA Substrates for Studying Replication Mechanisms: Synthetic Replication Forks

By ZEGER DEBYSER

Introduction

The availability of DNA substrates that mimic a DNA replication fork has enabled the study of protein–protein interactions that coordinate DNA replication in reconstituted DNA replication systems. We describe two types of synthetic replication forks. Although the applications listed are based on the study of T7 DNA replication, with modifications these synthetic replication forks can be used in the study of other DNA replication models.

The preformed replication fork (Fig. 1A) is a circular duplex DNA with a protruding 5′-single-stranded tail that supports concurrent leading and lagging strand synthesis in the presence of a processive DNA polymerase, a primase, and a helicase. The DNA polymerase binds to the 3′-hydroxyl terminus, whereas the helicase–primase complex assembles on the 5′ tail. The helicase activity is responsible for the unwinding of the duplex DNA, which enables the DNA polymerase to catalyze rolling circle strand displacement synthesis. This results in the processive synthesis of long leading strands. The primase activity catalyzes the synthesis of ribonucleotide primers at primase recognition sites on the displaced strand. These primers are extended by the lagging strand DNA polymerase, resulting in the formation of short Okazaki fragments that can be separated by alkaline agarose gel electrophoresis. DNA synthesis at the preformed replication fork is stimulated in the presence of single-stranded DNA binding proteins.

The preformed replication fork has been used in the study of DNA replication in the bacteriophage T7,[1–4] T4,[5,6] and *Escherichia coli*[7] systems using the respective replication proteins. At the T7 DNA replication fork four proteins account for the basic reactions: (1) T7 gene 5 protein, the DNA polymerase; (2) thioredoxin, a processivity factor provided by *E. coli*;

[1] R. L. Lechner and C. C. Richardson, *J. Biol. Chem.* **258,** 11,185 (1983).
[2] H. Nakai and C. C. Richardson, *J. Biol. Chem.* **263,** 9818 (1988).
[3] H. Nakai and C. C. Richardson, *J. Biol. Chem.* **263,** 9831 (1988).
[4] Z. Debyser, S. Tabor, and C. C. Richardson, *Cell* **77,** 157 (1994).
[5] T.-A. Cha and B. M. Alberts, *Biochemistry* **29,** 1791 (1990).
[6] N. G. Nossal, *Fasb J.* **6,** 871 (1992).
[7] M. Mok and K. J. Marians, *J. Biol. Chem.* **262,** 16,644 (1987).

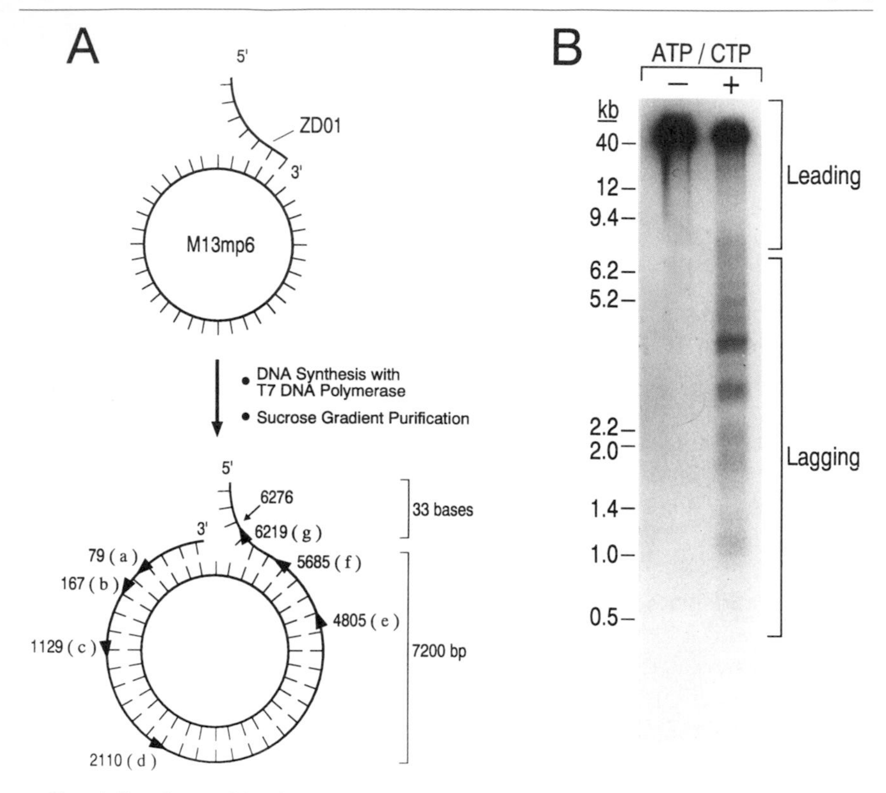

FIG. 1. Leading and lagging strand DNA synthesis at a preformed replication fork. (A) A preformed replication fork. A 52 nucleotide primer is annealed to M13 mp6 single-stranded DNA, in which 33 nucleotides at the 5′ end are noncomplementary. Extension of the primer with T7 DNA polymerase results in a forked circle that mimics a replication fork. The first annealed base of the primer is at position 6276. The major primase recognition sites are indicated with arrowheads: GGGTC (d, f), TGGTC (a, e, g) and GTGTC (b, c) on the M13 mp6 (−) strand (i.e., the lagging strand of the replication fork). Nucleotide positions are those from M13 mp6 plasmid.[7a] (B) DNA synthesis at a preformed replication fork. Reaction mixtures contain 7 n*M* preformed replication fork and 88 n*M* T7 DNA polymerase and 63-kDa gene 4 protein as described. Reactions were carried out at 30° for 10 min and terminated by the addition of EDTA to a final concentration of 20 m*M* and by placing the samples on ice. DNA products were denatured and separated on a 0.8% alkaline agarose gel at 1.5 V/cm for 16 hr. Molecular weight markers were run in parallel. DNA products shorter than 7.2 kb occur only in the presence of ATP, CTP, and primase and are due to the synthesis of lagging strand DNA. DNA products longer than 7.2 kb are predominantly the result of leading strand DNA synthesis. The products of DNA synthesis carried out in the absence (left lane) and presence (right lane) of 300 μM ATP and CTP are shown. Adapted from Z. Debyser, S. Tabor, and C. C. Richardson, *Cell* **77,** 157 (1994).

[7a] J. Messing, B. Gronenborn, B. Muller-Hill, and P. H. Hofschneider, *Proc. Natl. Acad. Sci. U.S.A.* **74,** 3642 (1977).

and (3) the two products of T7 gene 4, a helicase of 56 kDa and a primase of 63 kDa.[1-4] The helicase activity of the gene 4 protein enables the T7 DNA polymerase to catalyze rolling circle strand-displacement synthesis, resulting in the accumulation of DNA products greater than 40 kb in length (Fig. 1B). The displaced strand contains seven primase recognition sites at which the 63-kDa gene 4 protein synthesizes the tetraribonucleotide primers pppACCC, pppACCA, and pppACAC in the presence of ATP and CTP. These primers are extended by the lagging strand DNA polymerase. Lagging strand synthesis results in the formation of Okazaki fragments from 0.5 to 6 kb in length (Fig. 1B).

The minifork (Fig. 2A) is a partially annealed oligonucleotide duplex with 5′- and 3′-single-stranded DNA tails that has been developed for the

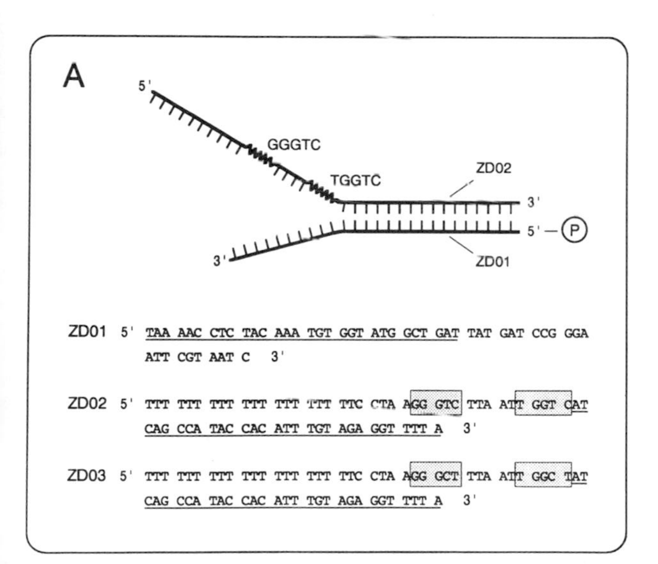

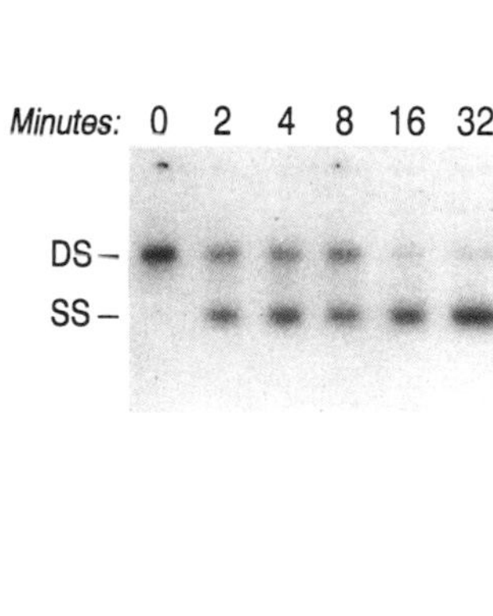

FIG. 2. Helicase activity at the minifork. (A) Scheme of the minifork and the sequences of the oligonucleotides used. Miniforks were constructed by annealing ZD01 with ZD02 (containing two primase recognition sites) and ZD01 with ZD03 (lacking primase recognition sites). The two primase recognition sequences in ZD02 are indicated, as well as the altered sequences in ZD03. The regions of the oligonucleotides that anneal to form duplex DNA are underlined. As indicated, ZD01 is labeled at its 5′ end with ^{32}P. (B) Time course of the strand displacement. Helicase assay was carried out as described in the text in the presence of 8 n*M* of 63-kDa gene 4 protein and 1 n*M* of template (ZD01/ZD02). A time course is shown. After the incubation at 30°, the reactions were stopped by transfer to wet ice and addition of 40 m*M* EDTA and 1% SDS. Aliquots of the samples were subsequently run in high-density polyacrylamide gels in a PhastSystem (see text for detailed running conditions). After drying, the gel was autoradiographed with Kodak X-ray film. Double-stranded DNA (DS) and displaced single-stranded fragments (SS) are indicated. Adapted from Z. Debyser, S. Tabor, and C. C. Richardson, *Cell* **77,** 157 (1994).

study of the functional interactions between helicase and primase activities at a minimal replication fork.[4] The minifork is a helicase substrate that has been optimized for the T7 gene 4 protein. After binding to the 5′-single-stranded DNA tail, T7 gene 4 protein will unwind the duplex DNA. In the presence of primase recognition sequences and ATP/CTP, the primase activity of the T7 gene 4 protein catalyzes the synthesis of tetraribonucleotide primers. This assay enables the study of the effect of primer synthesis on the helicase activity of the gene 4 protein. With a modified DNA substrate depending on the requirements for enzymatic activity of the respective helicase,[8] the minifork assay could be used in other DNA replication systems.

Materials

Chemicals and Buffers

Bovine serum albumin (BSA), 1 mg/ml
Butanol (Fluka Ronkonkoma, NY)
Chloroform (Fluka)
Dithiothreitol (DTT), 10 and 100 m*M*
Ethylenediaminetetraacetic acid (EDTA), 200 m*M*, pH 8
Ethanol, absolute, precooled to −20°
Ethidium bromide, 10 mg/ml
$MgCl_2$, 100 m*M*
NaCl, 1 *M*
Phenol/chloroform: 1 : 1 (v/v) mixture of Tris-saturated phenol (U.S. Biochemicals, Cleveland, OH) with chloroform
3 *M* sodium acetate
Sodium dodecyl sulfate (SDS), 10%
TE buffer: 10 m*M* Tris–HCl, 1 m*M* EDTA, pH 8
Tris–HCl, pH 7.5, 400 m*M*.

DNA and Nucleotides

M13 mp6 DNA, 1 mg/ml in TE
52-mer ZD01 and the 70-mers ZD02 and ZD03 (Fig. 2A) purified by FPLC
dNTPs and rNTPs, 100 m*M* ultrapure solutions (Pharmacia)
[α-^{32}P]dATP (specific activity 800 Ci/mmol) and [γ-^{32}P]ATP (3000 Ci/mmol) (New England Nuclear).

[8] S. W. Matson and K. A. Kaiser-Rogers, *Ann. Rev. Biochem.* **59,** 289 (1990).

Enzymes

T7 DNA polymerase (T7 gene 5 protein in a 1 : 1 mixture with thioredoxin) and T7 gene 4 protein (helicase/primase) were obtained from Stanley Tabor and Benjamin Beauchamp (Harvard Medical School), respectively. The enzymes are stored at −20° in 50% glycerol and diluted in 40 m*M* Tris–HCl, pH 7.5, 1 m*M* DTT immediately prior to addition to the reaction mixture.
Alkaline phosphatase (Boehringer Mannheim)
Polynucleotide kinase (3′-phosphatase free) (Boehringer Mannheim)
Proteinase K (Boehringer Mannheim)
Restriction enzymes *Eco*NI and *Nci*I (New England Biolabs and US Biochemicals, respectively).

Procedures

Preformed Replication Fork

Preparation of Preformed Replication Fork. Construction of a preformed replication fork, a circular duplex DNA with a protruding 5′-single-stranded tail, is a modification of the procedures originally described by Lechner and Richardson[1] and Nakai and Richardson[2] (Fig. 1A). Whereas a linear DNA fragment was annealed to circular M13 DNA in the original procedure,[1] in the present procedure a synthetic 52-mer is annealed to M13 mp6 DNA. Nineteen bases on the 3′ end of this oligonucleotide are complementary to M13 mp6 DNA at the *Bam*HI site. Since the other 33 bases on the 5′ end contain nonhomologous DNA, a 5′-single-stranded tail is created after the primer has been extended the entire length of the M13 molecule by the processive T7 DNA polymerase. The DNA is subsequently deproteinized and the preformed replication forks are purified after sedimentation through a sucrose gradient.

Annealing of the primer to the template is performed in a reaction mixture (300 μl) containing 130 m*M* Tris–HCl, pH 7.5, 33 m*M* $MgCl_2$, 170 m*M* NaCl, 42 pmol M13 mp6 DNA, and 150 pmol oligonucleotide (amount expressed in terms of the number of 3′ ends; primer : template ratio is 3 : 1). The mixture is heated to 90° and cooled over a 1-hr period to 30°. Primer extension is carried out in a 1-ml reaction mixture by the addition of DTT to a final concentration of 10 m*M*, BSA to 50 μg/ml, the 4 dNTPs to a final concentration of 1.5 m*M* each, and 75 μg of T7 DNA polymerase. Incubation is at 30° for 20 min. The reaction is terminated by the addition of EDTA to a final concentration of 20 m*M*. Approximately 90% of the annealed primers is extended the entire length of the M13 molecule, as judged by native agarose gel electrophoresis and ethidium bromide staining. The product of DNA synthesis is deproteinized by addition of 1% SDS and proteinase K to a final concentration of 100 μg/ml and incubation at 50° for 30 min, followed by extraction with phenol : chloroform (1 : 1) and then chloroform. The DNA is precipitated with ethanol and dissolved in TE buffer.

Purification of Forked Circle Template. The preformed replication forks are purified after sedimentation through a high salt sucrose gradient. The DNA is diluted in 400 μl of TE containing 1 *M* NaCl and 20 μg/ml ethidium bromide. Samples (200 μl) are layered onto linear 11-ml sucrose gradients of 5 to 25% sucrose in 50 m*M* Tris–HCl, pH 7.5, 5 m*M* EDTA, 1 *M* NaCl, and 20 μg/ml ethidium bromide. The tubes are centrifuged in a Beckman SW41 rotor at 40,000 rpm for 7 hr at 15°. The DNA is visualized using a long-wavelength ultraviolet lamp. The major band (the top band in the middle of the tube), consists of the preformed replication fork. The DNA in the band is collected by side puncture using a syringe and needle. Ethidium bromide is removed by extraction with butanol (four to six times). The sample is diluted fourfold with water to bring the NaCl concentration below 0.5 *M*, and then the DNA is precipitated with 2 volumes of ethanol. Finally, the preformed replication fork is resuspended in TE, and stored at −20°. The concentration is determined spectrophotometrically (A_{260}) using a conversion factor of 50 μg/ml per *A* unit. Starting from 100 μg single-stranded M13 DNA we usually obtain 40 μg pure preformed replication fork.

DNA Synthesis at Preformed Replication Fork. The reaction mixture (20 μl) contains 40 m*M* reaction buffer (Tris–HCl, pH 7.5, 10 m*M* $MgCl_2$, 50 m*M* NaCl, 10 m*M* DTT, 0.1 m*M* EDTA, 100 μg/ml BSA), 600 μM each of [α-^{32}P]dATP (400 mCi/mmol) or [^{3}H]dATP (50 cpm/pmol), dCTP, dGTP, and dTTP, 300 μM each of ATP and CTP, 350 ng of preformed fork (FC_{33}, 7 n*M* in 3′ ends), 88 n*M* of T7 DNA polymerase, and 88 n*M* of a particular gene 4 preparation. Incubation is at 30° for 10 min. The reaction is terminated by the addition of EDTA to a final concentration of 20 m*M*. Although incorporation of [^{3}H]dAMP can be measured by precipitation with 10% trichloroacetic acid (TCA) and collection of precipitates on Whatman GF/C glass fiber filters and liquid scintillation counting, we carry out routine analysis of ^{32}P-labeled DNA synthesized at the replication fork via alkaline gel electrophoresis (Fig. 1B). This enables the identification of both the products of leading and lagging strand DNA synthesis. DNA products from DNA synthesis reactions are denatured prior to electrophoresis. Reaction mixtures are adjusted to 50 m*M* NaOH, 1 m*M* EDTA, 3% Ficoll (Type 400, Pharmacia), 0.025% (w/v) bromocresol green, and 0.04% xylene cyanol FF. Analysis of leading and lagging strand DNA synthesis is carried out using 0.8% (w/v) agarose gels. The denaturing gels are electrophoresed at 30 V (1.5 V/cm) in the cold room for 16 hr with recirculation of the buffer. Gels are soaked for 30 min with gentle agitation in 7% (v/v) trichloroacetic acid before being dried for autoradiography.

Molecular weight markers are obtained by digesting T7 DNA with *Eco*NI and *Nci*I and radioactively labeling the 5′ ends using [γ-^{32}P]ATP and T4 polynucleotide kinase, following treatment with alkaline phosphatase. Unincorporated ATP is removed by Sephadex G-100 gel filtration.

Minifork Template

Preparation of Minifork. A scheme of the minimal replication fork is shown in Fig. 2A. The minifork contains a duplex of 30 nucleotides, a 3′ tail of 22 nucleotides, and a 5′ tail of 40 nucleotides. In the case of the T7 replication proteins, little activity is observed with a 5′ tail of only 20 nucleotides. We have made miniforks with and without primase recognition sequences located in the 5′ tail. The recognition sequence 5′-GGGTC-3′ starts at position 26 and 5′-TGGTC-3′ at position 36.

The oligonucleotide ZD01 (typically 1 to 2 μM) is incubated with polynucleotide kinase (5 units) for 10 min at 37° in the presence of 50 μCi [γ-^{32}P]ATP (800 ci/mmol) to label the 5′ end. The reaction is stopped by the addition of 20 m*M* EDTA. Unincorporated ATP is removed by gel filtration through a 1-ml Sephadex G-50 column, equilibrated with 20 m*M* Tris–HCl, pH 7.5, 2 m*M* EDTA, 100 m*M* NaCl. Early peak fractions containing the labeled DNA are pooled. Labeled ZD01 is then mixed in a 1 to 1 molar ratio (each at 100 n*M* final concentration) with unlabeled ZD02 or ZD03 in 200 m*M* Tris–HCl, pH 7.5, and the annealing mixture is heated to 90° for 2 min followed by cooling and overnight incubation at 37°.

Purification of Minifork. The template is purified by PAGE. Bromphenol blue (0.04%), xylene cyanol FF (0.04%), and 5% glycerol are added to the annealed oligonucleotide and the sample is loaded onto a nondenaturing 20% polyacrylamide gel. Electrophoresis is carried out at 200 V for 4 to 6 hr at 4°. Labeled DNA is recovered from the gel by the crush and soak method.[3] After cutting out the annealed oligo from the gel guided by an autoradiogram of the gel, the gel is crushed and soaked overnight on a spinning wheel at 37°. The elution buffer contains 0.5 *M* ammonium acetate, 10 m*M* magnesium acetate, 1 m*M* EDTA, and 0.1% sodium dodecyl sulfate (SDS). The sample is then centrifuged at 12,000*g* for 1 min at 4°. The supernatant is recovered and filtered through a Millipore Millex-GV4 filter. DNA is precipitated twice with ethanol and finally resuspended in TE (pH 7.5). From a comparison of the radioactivity in this DNA with the radioactivity prior to purification, the concentration of the annealed oligonucleotide can be estimated. The average recovery of annealed minifork is about 15 to 20% of the total radioactivity. Gel purification reduces the background (i.e., labeled, but not annealed oligonucleotide) in the helicase assay from >30 to 5% of the total amount of labeled oligonucleotide. The minifork is stored at −20°.

Helicase Assay on the Minifork. The standard helicase reaction mixture (10 μl) contains 20 m*M* Tris–HCl, pH 7.5, 10 m*M* $MgCl_2$, 10 m*M* DTT, 50 m*M* NaCl, 50 μg/ml BSA, 1 n*M* of substrate DNA, 2 m*M* TTP, and 8 n*M* of gene 4 protein (helicase). Incubation is for 10 min at 30°. The reaction is stopped by chilling the samples on ice and the addition of 20 m*M* EDTA and 1% SDS. Samples are subsequently loaded on precast nondenaturing high-density gels (7.5% stacking gel, 20% running gel) (Pharmacia). Electrophoresis conditions on PhastSystem (Pharmacia) are prerun at 400 V, 10 mA, 1 W, and 20° for 10 Vhr; loading at 25 V, 5 mA, 1 W, and 20° for 2 Vhr; electrophoresis at 400 V, 10 mA, 1 W, and 20° for 120 Vhr. After electrophoresis the gels are dried for 15 min under a hair dryer and subsequently autoradiographed and/or analyzed by means of a phosphorimager (Molecular Dynamics). Phosphorimaging allowed analysis after a 1- to 2-hr exposure. The whole procedure (helicase assay–PhastSystem electrophoresis–phosphorimaging) takes about 3 to 4 hr.

Quantitation of Helicase Activity. Helicase activity is measured as the enzyme-dependent displacement of a single-stranded 5′-^{32}P-labeled oligonucleotide (SS) from the partially double-stranded oligonucleotide (DS). Background is the amount of labeled single-stranded oligonucleotide present at the outset of the reaction. There is no substantial increase in the background level upon incubation at 30°, implying that DS is stable under reaction conditions. Amounts of SS and DS are determined by volume integration on a phosphorimager. The percentage of displaced oligonucleotide is calculated from the formula:

$$\%\ \text{Displaced oligonucleotide} = 100 \times \frac{\left[\frac{\text{SS}}{\text{SS}+\text{DS}}\right] - \left[\frac{\text{SS}}{\text{SS}+\text{DS}}\right]_0}{1 - \left[\frac{\text{SS}}{\text{SS}+\text{DS}}\right]_0},$$

where the subscript 0 indicates the amounts of SS and DS present at the start of the reaction. The formula accounts for loading errors and also takes into account that background levels of displaced oligonucleotide cannot be enzymatically displaced.

Under our reaction conditions T7 gene 4 protein displaces 60 to 80% of radiolabeled oligonucleotide in 10 to 20 min. The extent of displacement never exceeds 80%. This may be due to the reannealing of the displaced fragments or to the competition of the displaced single-stranded DNA for the binding of gene 4 protein (product inhibition).

Other Enzymatic Activities at Minifork. Primase activity can be measured at the minifork after inclusion of ATP and CTP (300 μM ATP and 300 μM [α-^{32}P]CTP) in the helicase reaction mixture, following the procedure for the measurement of tetraribonucleotide primer synthesis on an oligonucleotide DNA template.[9] Due to the low concentrations of primers synthesized, long exposures are required for visualization of the primers made at the minifork.

It should also be possible to measure primer extension by T7 DNA polymerase at the minifork by including the 4 dNTPs and Sequenase (3′ $\rightarrow$ 5′-exonuclease-deficient T7 DNA polymerase)[10] in the helicase reaction mixture and analyzing the DNA products by electrophoresis in a denaturing 15% polyacrylamide gel.

Discussion

We have described the preparation of two types of synthetic templates that allow the study of DNA replication at the replication fork in reconstituted *in vitro* systems. Each template has its own applications. The preformed replication fork has been used in the study of the protein–protein interactions that coordinate DNA replication at the replication fork of bacteriophages T7 and T4, and *E. coli.*[11] The presence of a single *Bam*HI restriction endonuclease cleavage site in the leading strand has made possible the determination of the microscopic rate of fork movement (300 bases/sec) in the T7 system.[1] Analysis of the processivity of leading and lagging strand synthesis at the T7 replication fork[2–4] and studies on the regulation of Okazaki fragment synthesis in T4 and *E. coli*[11] both have made extensive use of the preformed replication fork.

The minifork has been used in the study of the interaction of the helicase and primase activities of the T7 gene 4 protein at the T7 replication fork.[4] The assay can also be used to investigate the minimal requirements for a helicase substrate (e.g., the minimal length of the 5′ and 3′-single-stranded tails). Moreover, the absence of long stretches of competing single-stranded DNA facilitates kinetic studies. The effect of primer extension by DNA

[9] L. V. Mendelman and C. C. Richardson, *J. Biol. Chem.* **266,** 23,240 (1991).
[10] S. Tabor and C. C. Richardson, *J. Biol. Chem.* **265,** 8322 (1990).
[11] K. J. Marians, *Ann. Rev. Biochem.* **61,** 673 (1992).

polymerase on the helicase activity can be addressed. The ability to measure the interaction of DNA polymerase, helicase, and primase activities at the 5′-single-stranded DNA tail of the minifork makes this a useful template for the study of lagging strand DNA replication.

[36] Using Macromolecular Crowding Agents to Identify Weak Interactions within DNA Replication Complexes

By MICHAEL K. REDDY, STEPHEN E. WEITZEL, SHIRLEY S. DAUBE, THALE C. JARVIS, and PETER H. VON HIPPEL

Introduction

The notion that the cytoplasm is indeed a crowded place has now become part of the collective consciousness of most biochemists and cell biologists. The path to this realization has several key landmarks, including Allan Minton's[1] initial mathematical treatment of "volume-occupied solutions," Alice Fulton's[2] extremely cogent and influential minireview entitled "How Crowded is the Cytoplasm?," and David Goodsell's[3] combined use of art and science to depict the interior of the *Escherichia coli* cell. The central point for biochemists attempting to assemble *in vitro* systems is that there is an approximately two-orders-of-magnitude difference in soluble protein concentration between the intact cytoplasm and the dilute solutions typically used in laboratory reactions. As a consequence, since the volume occupied by macromolecules will affect the activity of other macromolecules in solution, it is to be expected that homo- and heteroassociation reactions will be significantly enhanced in "crowded" environments comparable to the cytoplasm, relative to the situation that prevails in dilute *in vitro* solutions. This fact has been well documented over the years, as highlighted in the selected examples presented in Table I. A more comprehensive list can be found in [4]. In this chapter we describe experiments in which a macromolecular crowding agent was used as a biochemical tool to probe weak, *yet specific,* interactions within the multicomponent bacteriophage T4 DNA replication system (see Ref. 5). We define interactions that are

[1] A. P. Minton, *Biopolymers* **20,** 2093 (1981).

[2] A. B. Fulton, *Cell* **30,** 345 (1982).

[3] D. S. Goodsell, *Trends Biochem. Sci.* **16,** 203 (1991).

[4] S. B. Zimmerman and A. P. Minton, *Ann. Rev. Biophys. Biomol. Struct.* **22,** 27 (1993).

[5] M. C. Young, M. K. Reddy, and P. H. von Hippel, *Biochemistry* **31,** 8675 (1992).

TABLE I
ENHANCEMENT OF PROTEIN–NUCLEIC ACID TRANSACTIONS BY ADDITION OF POLYETHYLENE GLYCOL

In vitro components	Observable
oriC plasmid replication	Essential for cell-free replication[a]
T4 DNA ligase	1000-fold stimulation of cohesive and blunt-end ligation[b]
T4 polynucleotide kinase	Stimulates forward, reverse, and exchange reactions[c]
DNA polymerases	Increases binding to primer–template at high ionic strength[d]
T4 accessory proteins	A factor of 50 increase in the association constant[e]
RecA	Allows strand exchange at physiological concentrations of magnesium[f]
T4 DNA polymerase and "sliding clamp"	Allows assembly of a functional holoenzyme without the hydrolysis of ATP[g]

[a] R. S. Fuller, J. M. Kaguni, and A. Kornberg, *Proc. Natl. Acad. Sci. USA* **78,** 7370 (1981).
[b] B. H. Pheiffer and S. B. Zimmerman, *Nucl. Acids Res.* **11,** 7853 (1983).
[c] B. Harrison and S. B. Zimmerman, *Anal. Biochem.* **158,** 307 (1986).
[d] S. B. Zimmerman and B. Harrison, *Proc. Natl. Acad. Sci. USA* **84,** 1871 (1987).
[e] T. C. Jarvis, D. M. Ring, S. S. Daube, and P. H. von Hippel, *J. Biol. Chem.* **265,** 15,160 (1990).
[f] P. E. Lavery and S. C. Kowalczykowski, *J. Biol. Chem.* **267,** 9307 (1992).
[g] M. K. Reddy, S. E. Weitzel, and P. H. von Hippel, *Proc. Natl. Acad. Sci. USA* **90,** 3211 (1993).

weak, yet specific, as those with K_d values within one order of magnitude of 10^{-6} *M*.

A variety of materials can serve as macromolecular crowding agents (e.g., dextrans and polyvinyl alcohols). We restrict this discussion to experiments with polyethylene glycol (PEG), since PEG was employed as (at least one of) the crowding agents in each of the studies summarized in Table I. In addition, the interactions of PEG with proteins have been studied extensively. It manifests no enthalpic interaction with the reacting components of biological systems (i.e., it is inert), and it is available in a variety of molecular sizes.

Materials and Methods

T4 Proteins

The indicated T4 proteins are dispensed from stocks that had been stored at $-20°$ in buffered solutions containing 50% (v/v) glycerol. All T4 proteins are purified from overproducer strains (gifts of T. C. Lin from the laboratory of W. Konigsberg).[6,7] We report the concentrations of T4 DNA polymerase (gp43) as nanomolar in protein monomers; gp45 in nanomolar trimers; and gp44/62 as nanomolar in 4:1 complexes.

Buffers

Polyethylene glycol (PEG): of molecular weight 12,000 is obtained from Fluka (Ronkonkoma, NY). Stock solutions are dissolved in distilled water and stored at 4° for a period no longer than 2 weeks. All PEG concentrations are given as % w/v (weight of PEG per 100 ml solution). We have found it unnecessary to filter or dialyze these solutions.

T4 replication assay buffer: 25 m*M* *N*-2-hydroxyethylpiperazine-*N'*-2-ethanesulfonic acid (HEPES), pH 7.5, 160 m*M* potassium acetate, 5 m*M* dithiothreitol (DTT), 0.2 m*M* EDTA, 0.2 mg/ml acetylated bovine serum albumin (BSA), and 6 m*M* magnesium acetate.

Equilibration buffer: 25 m*M* HEPES, pH 7.8, 50 m*M* NaCl, 1 m*M* EDTA, 0.5 m*M* DTT, and either 7.5% PEG or 25% glycerol.

Elution buffer: identical to equilibration buffer except the NaCl concentration is 500 m*M*.

DNA–Affinity Chromatography (±PEG)

Sample Preparation. The following components are combined on ice in a 0.5-ml Eppendorf tube. The T4 proteins are mixed first. Then a one-tenth volume (10 μl) of a 10× stock of equilibration buffer (without PEG or glycerol) is added, followed by one-half volume (50 μl) of either a 15% PEG solution or a 50% glycerol solution. Distilled water is added to reach a final volume of solution of 100 μl and therefore a final concentration of 7.5% PEG or 25% glycerol. The sample is then incubated at room temperature for 10 min. One-tenth of the solution is removed for subsequent analysis and the remainder of the sample is loaded directly onto a column that has been equilibrated either in PEG or in glycerol as described below.

Chromatographic Conditions. Single-stranded DNA affinity columns are prepared using P11 cellulose as described[6] or alternatively (as in the experiment shown in Fig. 3) using single-stranded DNA agarose resin (1 to 3 mg/ml coupled DNA) purchased from Pharmacia-LKB Biotechnology (Piscataway, NJ). The overall results are equivalent with either the cellulose or the agarose matrices. Columns with a 0.5- to 1.0-ml packed bed volume are poured into polypropylene columns (0.9 × 10 cm) and

[6] M. K. Reddy, S. E. Weitzel, and P. H. von Hippel, *Proc. Natl. Acad. Sci. USA* **90,** 3211 (1993).
[7] N. G. Nossal, D. M. Hinton, L. J. Hobbs, and P. Spacciapoli, this volume [43].

then washed with several milliliters of elution buffer (i.e., with 500 m*M* NaCl). After the column bed has settled, a small amount of buffer is gently layered onto the bed. The column is then attached to a gradient maker (Hoefer Scientific) via a peristaltic pump and finally to a flow adaptor (Bio-Rad, Richmond, CA) with its outlet positioned just above the top of the column matrix. The column is then equilibrated thoroughly with equilibration buffer.

Although our initial chromatography experiments[6] were run under gravity, we now prefer the use of a pump to maintain a uniform flow rate (30 ml/hr) that is independent of the viscosity of the particular column buffer. In this system the sample is introduced by means of a three-way stopcock connected in-line between the gradient maker and the pump. The column is then washed with 5 to 10 bed volumes of equilibration buffer and developed with a 10 bed volume linear gradient from the equilibration to the elution buffer (i.e., 50 to 500 m*M* NaCl). Chromatography is performed entirely at room temperature (~21°), but each (manually collected) fraction is immediately placed on ice for subsequent processing. The entire process of loading, washing, and elution requires ~60 min at the indicated volumes and flow rates.

Sample Processing. Since each collected fraction contains dilute protein, as well as a high concentration of PEG, whose presence interferes strongly with the detection of proteins by silver staining, we have used the following protocol, which both precipitates the proteins and removes the PEG:

1. An equal volume of cold 5% TCA (freshly made each time from a 100% stock) is added to an aliquot of each eluted fraction, mixed thoroughly, and incubated on ice for 5 min.
2. Protein is collected by centrifugation at 12,000*g* for 10 min.
3. Supernatant is aspirated off using a drawn-out glass Pasteur pipette to facilitate removal of the last small drop of supernatant.
4. Pellets are washed by the addition of 1 ml of chloroform followed by two gentle inversions and a 5-min centrifugation step as above. Supernatant is discarded and samples are dried in a Speed Vac (Savant, Hicksville, NY).
5. The pelleted proteins (barely visible) are resuspended in 10 μl of loading buffer and boiled (3 to 5 min) before electrophoresis on a discontinuous sodium dodecyl sulfate–polyacrylamide gel (SDS–PAGE) system[8] through a 12.5% resolving gel with a 4% stacker. Gels were cast on 10- × 11-cm gel plates with 0.7-mm spacers and subjected to electrophoresis at 10 mA (constant current) for 120 min.

[8] U. K. Laemmli, *Nature (London)* **227,** 680 (1970).

Applications

Accessory Proteins (ATPase) of T4 DNA Replication System

In our initial foray into the use of macromolecular crowding as a biochemical tool to investigate DNA replication complexes we examined the binding equilibria between the gp44/62 complex (containing four gene 44 protein subunits and one gene 62 protein subunit), the trimeric gp45 species, and primer–template (P–T) DNA.[9] This subassembly of T4 polymerase accessory proteins "transforms" the T4 DNA polymerase (gp43) into the high-efficiency—*and physiologically relevant*—holoenzyme form by a mechanism that requires ATP hydrolysis. The gp44/62 complex displays a basal level of ATPase activity that is stimulated approximately 20-fold by the addition of gp45. The ATPase of this three-protein subassembly is additionally stimulated ~100-fold by the addition of P–T DNA. Thus the ATPase activity of the full accessory protein–DNA complex is ~2000-fold greater than that of the gp44/62 species alone.

The fact that gp45 strongly stimulates the ATPase activity of the gp44/62 complex suggests that these species must interact physically. However *in vitro* studies showed the binding of gp45 with gp44/62 to be anomalously weak; in fact, simple calculations show that the measured dissociation constant is too large to support *in vivo* DNA replication. This paradox was resolved by the demonstration that this parameter could be dramatically decreased by the addition of PEG to the reaction solution. Specifically, as illustrated in Fig. 1, the addition of increasing amounts of PEG caused a concomitant decrease in measured K_d values, strengthening the interaction between the accessory proteins approximately 50-fold at the optimal PEG concentration (7.5%). In addition, this approach permitted estimation of the stoichiometry of the entire accessory proteins subassembly as one gp44/62 4:1 complex per gp45 trimer.[9]

T4 DNA Polymerase and the "Sliding Clamp"

We have recently demonstrated[6] the existence of a direct and functionally significant interaction between T4 DNA polymerase (gp43) and gp45. This specific heteroassociation, though weak, can be driven by very high levels of gp45 alone. Alternatively, it can also be driven at reduced—*and physiologically relevant*—protein levels in the presence of PEG. In particular, the functional effect of gp45 was manifested by an increase in the processivity of T4 DNA polymerase in its synthesis ($5' \rightarrow 3'$) mode. In Fig. 2 we extend this initial observation by demonstrating that the addition of

[9] T. C. Jarvis, D. M. Ring, S. S. Daube, and P. H. von Hippel, *J. Biol. Chem.* **265,** 15,160 (1990).

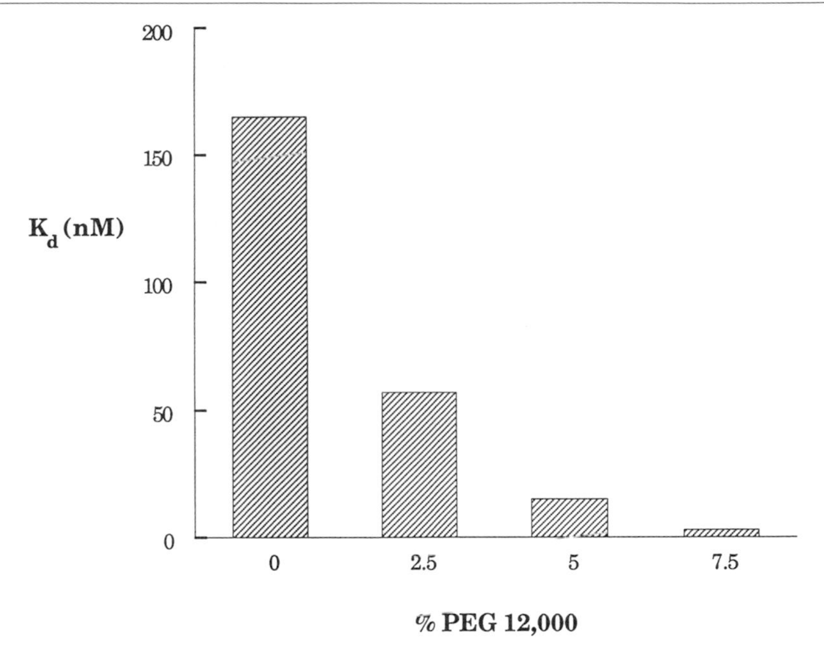

FIG. 1. Addition of PEG increases the binding of gp45 to the gp44/62 ATPase complex. This bar graph (which has been replotted from data originally determined by Jarvis *et al.*[9]) depicts the significant decrease in K_d for the binding of gp45 to gp44/62 that occurs when the amount of PEG added to *in vitro* ATPase reactions is increased as indicated. Since the "window" of stimulation can be quite narrow, it is important to vary, systematically and in small increments, the amount of added PEG when characterizing a new system. Furthermore, one must be aware of the likelihood of protein precipitation above certain levels of PEG. For the conditions depicted here, the T4 accessory proteins begin to precipitate at PEG levels >10%. Reactions were performed in T4 replication buffer using 450 n*M* of primer–template (P–T) DNA (saturating amounts), 95 n*M* gp45, and gp44/62 levels that were varied between 0 to 350 n*M* to determine ATPase rates.

gp45 alone leads to a small, but reproducible, increase in the processivity of gp43 in its exonuclease-only ($3' \rightarrow 5'$) mode as well (compare lanes 2 and 3 of Fig. 2). This observation supports the conclusion that there is a weak, yet specific, association between gp45 and gp43. However, the most significant observation is again (consistent with our initial results[6]) that this increase in processivity is clearly amplified by the addition of PEG (compare lanes 3 and 5 of Fig. 2).

Since gp45 can increase the processivity of the polymerase as an exonuclease, it is relevant to point out that gp45 may affect the ability of the polymerase to proofread errors that occur during replication. This suggests that the phenotype of some *in vivo* mutant bacteriophages (i.e., the so-called

A.

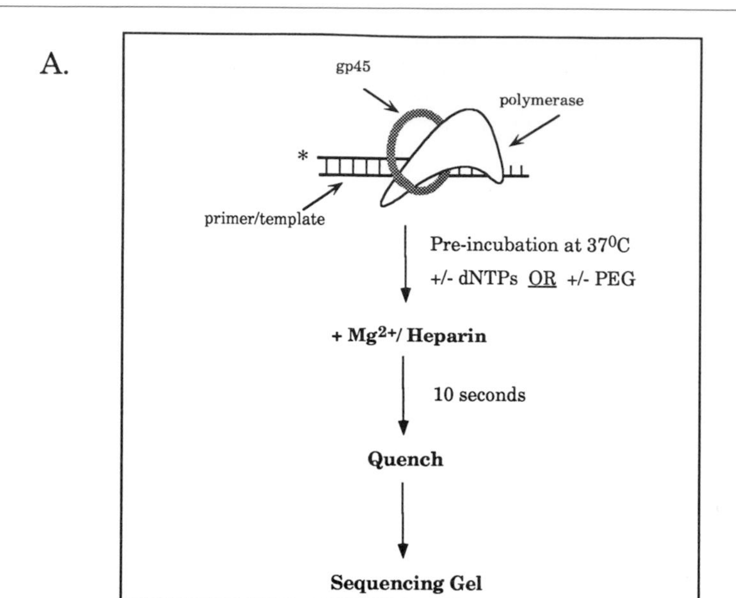

B.

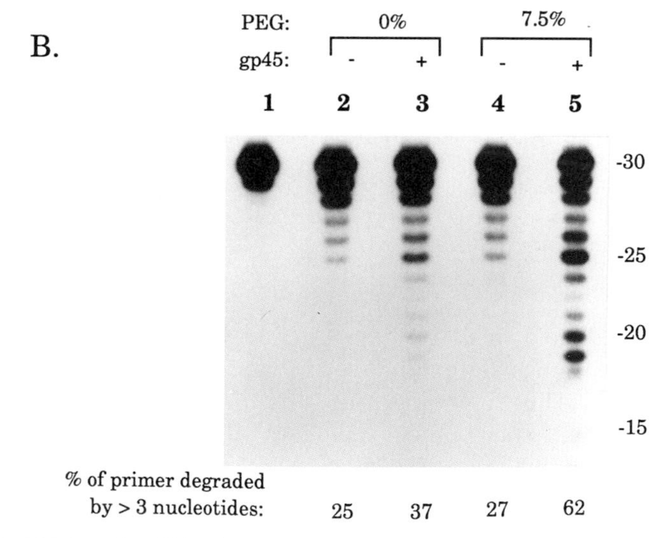

FIG. 2. Addition of gp45 in the presence of PEG increases the exonuclease processivity of gp43 (T4 DNA polymerase). (A) Experimental protocol used to monitor the processivity of gp43 in its synthesis mode (+dNTPs) or in its exonuclease-only mode (−dNTPs). The P–T template substrate was made by annealing a radioactively 5′-end-labeled 30-mer oligonucleotide primer to a 50-mer template oligonucleotide.[6] Preincubation was performed in T4 replica-

mutators and antimutators[10]) may be indicative of an altered interaction of the T4 DNA polymerase with gp45 (for instance, see Ref. 11), rather than reflecting exclusively an intrinsic change in the catalytic activity of the polymerase.

We also note that comparison of lanes 2 and 4 of Fig. 2 demonstrates that the addition of PEG to reactions containing only T4 DNA polymerase does *not* lead to a significant increase in processivity. This result is consistent with previous experimental observations,[6,9,12] as well with theoretical considerations suggesting that macromolecular crowding agents should only increase association rate constants.[1] Processivity is, by definition, independent of the rate of this step in the assembly process.

The results presented in Fig. 2, and our previous observations,[6] establish a functional interaction between the putative sliding clamp of the T4 system (gp45) and its cognate polymerase (gp43). To provide evidence of a direct physical interaction between these two proteins, we studied their adsorption onto an affinity matrix (single-stranded DNA) in the presence or absence of PEG. This protocol produced results like those shown in Fig. 3.

Several conclusions can be drawn from a comparison of the results of the two chromatographic experiments shown in Fig. 3, which were performed in a buffer containing either 25% glycerol (upper gel) or 7.5% PEG (lower gel). One important observation is that the salt concentration required for elution of T4 DNA polymerase is significantly higher in the presence of PEG (peak centered at 0.35 *M* NaCl) than in the presence of glycerol (peak centered at 0.2 *M* NaCl).

However, the central result demonstrated in these experiments is that gp45 was retained on the single-stranded DNA matrix only when it was loaded onto a PEG-equilibrated column in the presence of gp43. The

[10] N. Muzyczka, R. L. Poland, and M. J. Bessman, *J. Biol. Chem.* **247,** 7116 (1972).
[11] M. D. Topal and N. K. Sinha, *J. Biol. Chem.* **258,** 12,274 (1983).
[12] S. B. Zimmerman and B. Harrison, *Proc. Natl. Acad. Sci. USA* **84,** 1871 (1987).

tion buffer containing no Mg^{2+} (which is required for the catalytic function of the polymerase). The reaction is initiated on the simultaneous addition of $Mg(OAc)_2$ and heparin to final levels of 6 m*M* and 0.5 mg/ml, respectively. (B) Products of the exonuclease reactions as fractionated during electrophoresis on a 14% acrylamide (20:1)/8 *M* urea DNA sequencing gel. The reactions were performed in T4 replication buffer using 60 n*M* P–T DNA, 100 n*M* T4 DNA polymerase, and where indicated 250 n*M* gp45. Quantitation of the ^{32}P-labeled products (sizes in nucleotide residues are shown at the right of the gel) was performed using a radioanalytic detector (Ambis, San Diego, CA). Increases in exonuclease processivity were defined by determining the percent of primer degraded by greater than three nucleotides (i.e., products 26 nucleotide residues in length or smaller as a ratio of the total products) and are reported below the gel. Lane 1 contains unreacted primer strand.

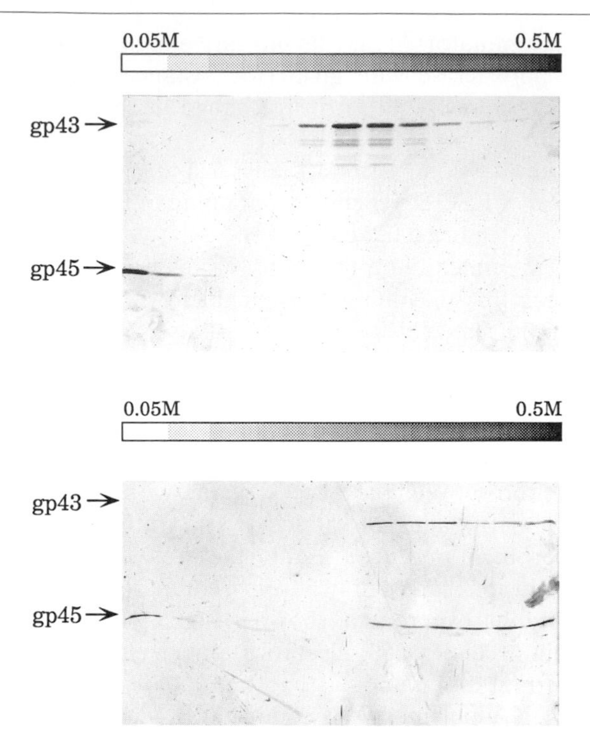

FIG. 3. DNA-affinity chromatography of gp45 and gp43 in the absence and presence of PEG. Two separate chromatographic runs were performed, using a concentration of 2000 n*M* each of gp45 and gp43 in a sample volume of 100 μl (i.e., 0.2 nmol of each protein). In both cases, a 0.5-ml aliquot of each eluted fraction was processed and then electrophoresed on an SDS–polyacrylamide gel (see Materials and Methods section). Proteins were visualized by silver staining. The gel at the top of the figure comprises the chromatographic elution profile in the presence of 25% glycerol (without PEG), whereas the elution profile for chromatography in the presence of 7.5% PEG is shown in the lower gel. Concentrations (molar) of NaCl used during the linear elution of bound proteins are shown across the top bar of each gel.

experiments shown in Figs. 2 and 3, taken together, strongly suggest that the addition of PEG increases the association of gp45 to gp43, which in turn clamps gp43 down onto the DNA, thereby decreasing its dissociation rate. It is important to point out that the use of the present protocol permits detection of complex formation between gp45 and gp43 at gp45 concentrations that are 25-fold lower than those employed in our initial development of the PEG chromatography technique for this purpose.[6]

We also note that we have carried out several control experiments to assure that this gp43–gp45 heteroassociation is specific.[6] In particular, since BSA has a very similar isoelectric point to gp45, we performed identical chromatographic experiments with individual T4 proteins and BSA to monitor the possibility that PEG is simply driving nonspecific associations between proteins and DNA under our conditions. Such control experiments showed that under no conditions could we observe retention of BSA on the single-strand DNA column, nor could we observe association of BSA with any of the T4 proteins.[13]

Finally, it is obvious that the inclusion of PEG can be potentially advantageous in connection with other forms of chromatography. In particular protein–affinity chromatography, as employed previously with the T4 DNA replication system in the Bruce Alberts laboratory,[14] will enable us to determine if the parameters of the gp43–gp45 association are different in the *absence* of DNA. Furthermore, the implementation of PEG chromatography in purification strategies may ultimately allow the direct isolation of multienzyme complexes (see Ref. 15).

Conclusion

We report here the strategic use of a macromolecular crowding agent (PEG) in *in vitro* experiments to uncover important details about the assembly and functioning of the bacteriophage T4 DNA holoenzyme. Although the mechanistic insights we have gained apply directly only to the T4 system, we believe that these approaches will prove successful for the molecular analysis of other multicomponent complexes, as well as in experiments with other *in vitro* replication systems.

For example, the Geiduschek laboratory has independently shown that gp45 alone, in the presence of PEG, can enhance T4 late transcription; thus the addition of a crowding agent has removed the requirement for the ATPase activity of gp44/62 in this process as well.[16] Since the use of PEG has to a large extent allowed us to bypass the ATPase activity of gp44/62, we suggest that during the *in vivo* assembly of the bacteriophage T4 holoenzyme the energy resulting from the hydrolysis of ATP by the gp44/62 complex is used to connect (or to *disconnect*) the sliding clamp (gp45) to (or from) the T4 DNA polymerase. Since the T4 gp45 is considered to be the functional analog of the β subunit of the *E. coli* replication system, and of PCNA in eukaryotic systems (for an overview, see Ref. 17), it will be interesting to determine whether the addition of optimal amounts of

[13] M. K. Reddy, unpublished results.
[14] T. Formosa, J. Barry, B. M. Alberts, and J. Greenblatt, this series, Vol. 208, p. 24.
[15] P. A. Srere and C. K. Mathews, this series, Vol. 182, p. 539.
[16] G. Sanders, G. Kassavetis, and E. P. Geiduschek, *Proc. Natl. Acad. Sci. USA* **91,** 7703 (1994).
[17] M. O'Donnell, *BioEssays* **14,** 105 (1992).

PEG to these other *in vitro* replication complexes will also permit bypass of the ATPase requirement and render unnecessary the subassemblies that carry this activity.

Our studies, together with those of many other investigators, have revealed that specific and functionally important protein–protein and protein–DNA interactions that are masked in dilute *in vitro* conditions can be manifested in volume-occupied (macromolecularly crowded) solutions. We hope that in the future *in vitro* biochemical studies will more often be conducted under crowding conditions (as well as with physiologically relevant anions) in order to mimic more closely the intracellular milieu in which the "protein machines" we study have evolved (see Refs. 18 and 19).

Acknowledgments

These studies were supported in part by USPHS Research grants GM-15792 and GM-29158 (to PHvH) and by a grant to the Institute of Molecular Biology from the Lucille P. Markey Charitable Trust, as well as by Individual USPHS National Research Service award GM-12473 to MKR. SSD was partially supported by the U.S. Department of Education program for Graduate Assistance in Areas of National Need. TCJ was a trainee on USPHS Institutional Service award GM07759. PHvH is an American Cancer Society Research Professor of Chemistry.

[18] P. A. Srere, *Ann. Rev. Biochem.* **56,** 89 (1987).
[19] O. G. Berg, *Biopolymers* **30,** 1027 (1990).

[37] Photochemical Cross-Linking Assay for DNA Tracking by Replication Proteins

By BLAINE BARTHOLOMEW, RACHEL L. TINKER, GEORGE A. KASSAVETIS, and E. PETER GEIDUSCHEK

Introduction

We describe a photochemical method for detecting the tracking of proteins along DNA. A photoactive deoxyribonucleoside 5′-triphosphate is enzymatically incorporated into DNA at a specific site, in close vicinity to a radioactive nucleotide. Protein complexes are assembled on this DNA (the photoprobe) and irradiated to form covalent protein–DNA adducts. The DNA is subsequently extensively digested with nucleases in order to retain only a short length of radioactively labeled oligonucleotide covalently attached to photochemical cross-linked proteins. The latter are then separated and identified by denaturing SDS–PAGE and autoradiography.

In applying this method to the detection of DNA tracking by the processivity-conferring proteins of viral, bacterial, and eukaryotic DNA replication, the photoactive and radioactive nucleotides are incorporated into locations of the DNA probe that are well removed from the DNA entry sites of the tracking proteins. Access of the latter to the vicinity of the photoactive nucleotide is shown to depend on the participation of their specific assembly or loading factors, on hydrolysis of ATP or dATP, and on an unobstructed path along DNA between the entry site for the tracking protein and the photoactive nucleotide.

Synthesis of 5-[*N*′-(*p*-Azidobenzoyl)-3-aminoallyl]dUTP

Materials

100 m*M* sodium borate (Na_3BO_3) adjusted to pH 8.5 with NaOH.

1.5 *M* triethylammonium bicarbonate (TEAB), pH 8.0. In a hood, dilute 104 ml triethylamine (redistilled and stored in a dark bottle under N_2 or Ar at 4°) to 500 ml with deionized water. Adjust pH by bubbling CO_2 through the solution through a fritted glass diffusor, with stirring. (Dry ice in a stoppered side-arm flask serves as a convenient CO_2 source. The two initially immiscible liquids dissolve as the pH approaches 8.)

TE: 10 m*M* Tris–HCl, pH 7.5, 1 m*M* Na_3EDTA.

20 m*M* 5-aminoallyldeoxyuridine triphosphate (Sigma catalog number A5910, ~90% pure, FW = 604.2; abbreviated as 5-aa-dUTP) in 100 m*M* sodium borate, pH 8.5 or 100 m*M* $NaHCO_3$, pH 8.5. The concentration of the solution is verified by absorbance at 289 nm of an aliquot diluted 200-fold into TE. (The extinction coefficient of 5-aa-dUTP at pH 7.5 is 7.3×10^3 M^{-1} cm^{-1}.)

100 m*M* (28 mg/ml) 4-azidobenzoic acid *N*-hydroxysuccinimide ester (ABA; available from Sigma or Pierce) in *N,N*′-dimethylformamide (DMF, HPLC grade, 99.9% pure; Aldrich). This solution is prepared just before use under indirect dim light from small incandescent lamps.

Synthesis of 5-[N-(p-azidobenzoyl)-3-aminoallyl]deoxyuridine triphosphate

This compound (AB-dUTP) was previously also abbreviated as N_3RdUTP.[1] All steps are done under limited, indirect lighting (to facilitate

[1] B. Bartholomew, G. A. Kassavetis, B. R. Braun, and E. P. Geiduschek, *EMBO J.* **9**, 2197 (1990).

work, without danger of photolyzing the aryl azide) from small incandescent lamps (40 to 60 W). The coupling reaction to form AB-dUTP takes place in 100 μl of 100 m*M* ABA mixed with 100 μl of 20 m*M* 5-aa-dUTP at room temperature for 4 hr. (If a precipitate forms, it can be dissolved with additional DMF.) The pH of the reaction, checked with pH paper, should be pH 8.5. Deionized water (200 μl) is added at the end of that time.

The modified nucleotide is purified on a 0.7 × 8-cm (1.6-ml) DEAE A-25 column equilibrated in 100 m*M* TEAB, pH 8.0, washed with the same buffer at a flow rate of 5 ml/hr, and eluted with a 30 ml linear gradient of 0.1 to 1.5 m*M* TEAB, pH 8.0; 500 μl fractions are collected. The desired product elutes at 0.8 to 1.0 *M* TEAB and can be identified by thin-layer chromatography (TLC) on polyethyleneimine (PEI)-cellulose, as follows.

Every second or third fraction is evaporated to dryness with a centrifugal vacuum (CentriVap or Speed-Vac) concentrator, resuspended in 50 μl deionized water, evaporated to dryness, and subjected to two more cycles of resuspension in deionized water and drying. Aliquots (1 to 2 μl) of each concentrated sample (final volume 50 μl) are applied to the PEI cellulose TLC plate with fluorescent indicator, with one lane of 5-aa-dUTP as a marker, and developed with 1 *M* LiCl. The products are visualized with a hand-held UV light source (λ254 nm). If sufficient AB-dUTP is applied to the TLC plate, the aryl azide-containing nucleotide can also be detected by exposing the plate to UV light for several minutes: Spots containing aryl azide turn brown or yellow. The reported R_f for 5-aa-dUTP and AB-dUTP are 0.54 and 0.098, respectively,[1] but large variations depending on the source of the PEI-cellulose TLC plate have been noted. Nevertheless, the R_f value of AB-dUTP is consistently much less than that of 5-aa-dUTP.

All fractions that are indicated to contain AB-dUTP by TLC analysis are combined, and TEAB is removed by repeated drying and resuspension in deionized water, as described above with volumes scaled accordingly. Final resuspension of the sample is in 200 μl of TE. The estimated extinction coefficient of AB-dUTP at 270 nm is $10.3 \times 10^3\ M^{-1}\ \text{cm}^{-1}$ at pH 8.0 (based on summing the extinction coefficients of ABA and 5-aa-dUTP at the indicated pH and wavelength). A stock solution of AB-dUTP in TE is stable for several years, when stored in the dark and at −20 or −80°; we commonly keep a 0.2 m*M* working stock at −20°, while storing the concentrated stock solution at −80°.

Synthesis of DNA Photoaffinity Probe

Summary of Procedure

The photoprobe is synthesized on a single-stranded circular DNA template so as to locate precisely a photoactive nucleotide next to a

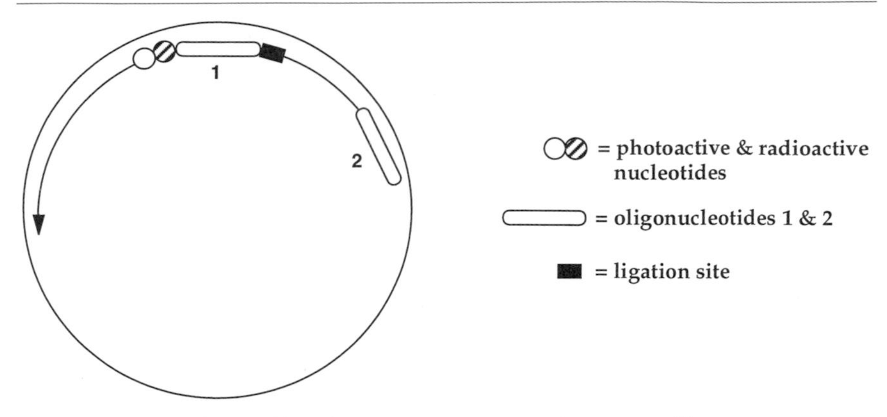

FIG. 1. The scheme of DNA photoprobe synthesis.[2,3]

radioactive nucleotide at a selected site by limited enzymatic extension of an appropriate chemically synthesized oligonucleotide primer[1-3] (Fig. 1; primer 1). DNA synthesis is completed with normal and nonradioactive substrates, DNA upstream of the 5′ end of the primer 1 is filled in by enzymatic extension from a second primer, and the nick remaining at the 5′ end of the first primer is sealed with DNA ligase. The resulting partly double-stranded and partly single-stranded circular DNA contains newly synthesized DNA strands that are heterogeneous at their 3′ ends. An entirely double-stranded and uniquely defined photoprobe can be cut from this product with apporpriate restriction endonucleases.

Materials

Buffer A (made as a 5× concentrated stock, stored at −20°): 30 m*M* Tris–HCl, pH 8.0, 50 m*M* KCl, 7 m*M* $MgCl_2$, 1 m*M* 2-mercaptoethanol, 0.05% Tween 20.

Buffer B: 50 m*M*(P) potassium phosphate, pH 7.0, 10 m*M* 2-mercaptoethanol (freshly added), 50% (v/v) glycerol.

Buffer C: 25 m*M* Tris–HCl, pH 7.5, 100 m*M* NaCl, 10 m*M* 2-mercaptoethanol (freshly added), 0.1 m*M* EDTA, 50% (v/v) glycerol.

10% (w/v) SDS.

[2] R. L. Tinker, K. P. Williams, G. A. Kassavetis, and E. P. Geiduschek, *Cell* **77,** 225 (1994a).

[3] R. L. Tinker, G. A. Kassavetis, and E. P. Geiduschek, *EMBO J.* **13,** 533 (1994).

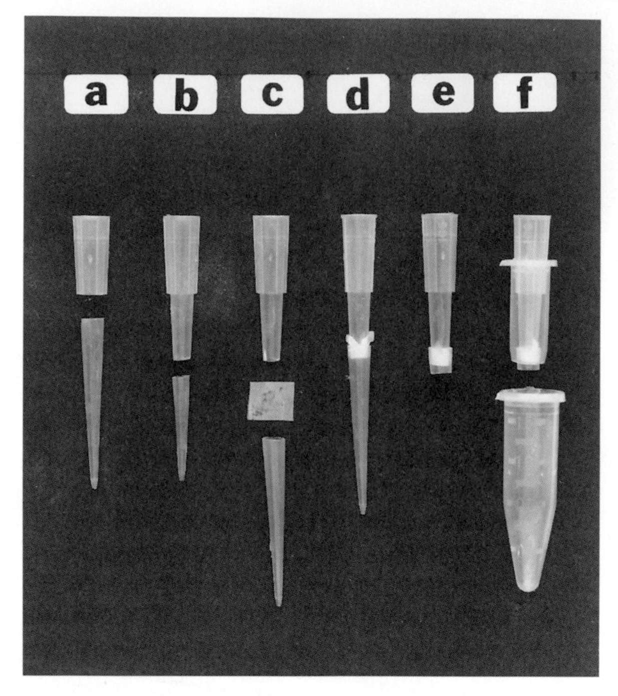

FIG. 2. Assembly of the microspin column. Two plastic pipette tips are sliced with a razor blade as shown in (a) and (b). A 4- to 6-cm square of nylon mesh (Spectra Mesh nylon #146514, Spectrum Laboratory Products, Houston) is placed over the bottom segment of tip (a) and the upper portion of tip (b) is pushed in [see (c) and (d)]. The bottom tip is then trimmed just below the nylon mesh (e). A, 0.5-ml microcentrifuge tube, with the conical portion sliced off, inserted into a 1.5-ml microcentrifuge tube serves as a holder for the column and receptacle for the eluate (f).

Sephacryl S-400 spin columns: Spin columns of various types for larger sample volumes are sold ready-to-use. A simple, readily assembled column that is preferable for the task at hand is shown in Fig. 2. Sephacryl S-400 is equilibrated with Buffer A before loading 200 μl (packed volume) into this column with a Pasteur pipette, making sure to avoid trapping air bubbles. The column is allowed to drip until no head volume remains, and centrifuged at 2000 RCF for 3 min. Adequacy of the column and centrifugation are checked by applying 25 μl of Buffer A, centrifuging as above, and verifying that the volume of liquid yielded up by the tip is approximately that applied. (If it is not, the centrifugation time is adjusted.) Spin columns should be used directly after preparation, with sample volumes that are less than 0.2 column volumes.

Synthesis of DNA Probe: 2 pmol scale

Ten picomoles oligonucleotide No. 1 (5′ phosphorylated) and 2 pmol M13 single-stranded DNA in 10 μl Buffer A are incubated at 90° for 3 min, then transferred to 65° for 10 min and 37° for 20 min. Bovine serum albumin (BSA), AB-dUTP, and the appropriate α-^{32}P-labeled nucleotide (6000 Ci/mmol) are added to a final concentration of 0.1 mg/ml, 10 μ*M*, and 0.3 μ*M* (threefold excess over M13 DNA), maintaining the same buffer, with a total volume of 20 μl. One microliter (1 unit) of exonuclease-free Klenow fragment DNA polymerase I in Buffer B is added and the primer extension proceeds at 37° for 5 min. A 1-μl aliquot is removed at this point and added to 10 μl of 95% formamide, 0.06% bromphenol blue, 0.06% xylene cyanol for later PAGE analysis (10% polyacrylamide gel containing 8.3 *M* urea) to verify proper incorporation of AB-dUMP without read-through. [*Note:* The preceding conditions are for a specific probe in one of the experiments shown below to serve as illustrations of this method (Fig. 3). In general, the optimal conditions for specific and efficient primer extension are best established for each individual primer.]

All four deoxyribonucleoside triphosphates (dNTP; Pharmacia FPLC grade or equivalent; final concentrations 1 m*M* each) and *E. coli* single-stranded DNA binding protein (SSB_E; final concentration 63 μg/ml) (from US Biochemical) are added to the remaining sample (total volume 24 μl) and incubated at 37° for 15 min. Another 1-μl aliquot is removed for PAGE analysis, as just described, to verify efficient primer extension. Klenow enzyme and SSB_E are inactivated in the remaining sample by adding 1 μl of 10% SDS and heating at 65° for 10 min. Inactive Klenow enzyme, SSB_E, and free nucleotides are removed by two passages through Sephacryl spin columns equilibrated in Buffer A.

Oligonucleotide No. 2 (10 pmol) is now added along with BSA (0.14 mg/ml final concentration) and annealed to the M13 DNA template by incubating at 37° for 30 min (in 29 μl total volume). All four deoxyribonucleoside triphosphates (0.7 m*M* each), ATP (0.6 m*M*), 5 units T4 DNA polymerase (the material from GIBCO/BRL is supplied in solution containing 2-mercaptoethanol in place of DDT), and 1 unit T4 DNA ligase are added and incubated at 37° for 60 min (total volume 35 μl). A second addition of 2.5 units T4 DNA polymerase and 0.5 units T4 DNA ligase, with continued incubation at 37° for an additional 60 min can be used to ensure more efficient ligation of DNA (to 85 to 95%). A 1-μl aliquot is removed to analyze the ligation efficiency (by cutting with suitable restriction endonuclease(s), denaturing PAGE, and quantifying radioactivity in unligated and ligated DNA fragments). The reaction is stopped by adding SDS to 0.4%, and heating at 65° for 15 min. The DNA probe is purified by two passages through Sephacryl spin columns equilibrated in the storage buffer of choice (e.g.,

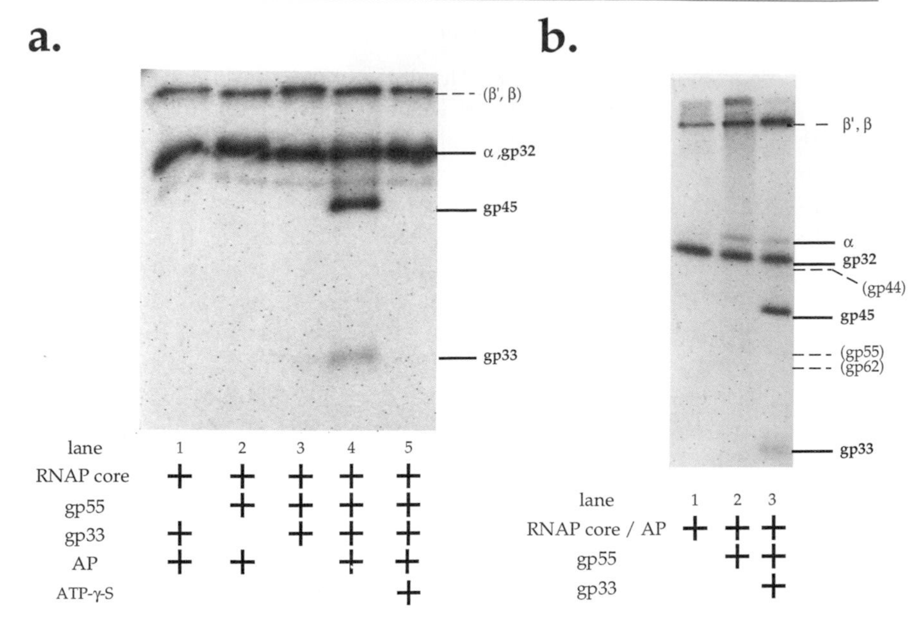

FIG. 3. Demonstration that a T4 DNA polymerase accessory protein (gp45) and RNA polymerase-binding coactivator protein (gp33) are located in the vicinity of the upstream end of an enhanced transcription initiation complex at a bacteriophage T4 late promoter (adapted from Ref. 2, with permission. Copyright Cell Press). The DNA probe has a single AB-dUMP residue incorporated into the nontranscribed strand, 39 bp upstream of the transcriptional start site of a phage T4 late promoter. Transcription complexes assembled as specified below are irradiated, digested, and resolved as described later. The proteins present in each sample are specified below each lane. For the sample analyzed in (a), lane 5, ATPγS replaces dATP. For the analysis in (b), RNA polymerase has been reconstituted from its β', β, and α subunits. Individual proteins are identified at the sides of the panels. AP designates the T4 DNA polymerase accessory proteins gp45, gp44, and gp62. *Method:* DNA (8 fmol), 0.2 pmol *E. coli* RNA polymerase core, 0.5 pmol gp55, 1 pmol gp33, 1.8 pmol (trimer) gp45, 3.3 pmol gp44/62 complex (composition: 4 molecules of gp44 and 1 molecule of gp62[a]); and 20 pmol gp32 (Boehringer Mannheim) in 13 μl of a reaction buffer consisting of 300 m*M* (a) or 250 m*M* (b) potassium acetate, 33 m*M* Tris–acetate, pH 8, 10 m*M* magnesium acetate, 0.8% (v/v) glycerol, 100 μg/ml BSA, 0.5 m*M* PMSF, 1 μg/ml leupeptin, 1 μg/ml pepstatin, and 0.25 m*M* dATP (or ATPγS, as noted) are incubated for 15 min at 37°. Poly(dAT) · poly(dAT) (0.7 μg; Sigma) and shrimp alkaline phosphatase (0.4 units, US Biochemical) in the same buffer are added, bringing the volume to 15 μl. After 15 min at 37°, samples are irradiated for 4 min and digested with DNase I and S1 nuclease, as specified in the text. Cross-linked proteins are resolved by electrophoresis in a standard 12% (w/v) polyacrylamide–SDS gel (37.5 : 1 acrylamide : bisacrylamide) (b) or in a 15.5% (w/v) polyacrylamide-Tricine-SDS gel (15.5 : 1 acrylamide : bisacrylamide) with a 3.8% (w/v) polyacrylamide-Tricine-SDS overlay[b] (A). Gels are silver stained to identify marker proteins (indicated at the sides, but not shown), dried, and subjected to autoradiography with an intensifier screen.

[a] T. C. Jarvis, L. S. Paul, J. W. Hockensmith, and P. H. von Hippel, *J. Biol. Chem.* **264,** 12,717 (1989).

[b] H. Schagger and G. von Jagow, *Anal. Biochem.* **166,** 368 (1987).

Buffer A). DNA probes can be stored at 4 or −20°; their useful lifetime is limited primarily by the decay of the incorporated radioactivity. DNA probes containing multiple radioactive nucleotides have a shorter useful life due to degradation of DNA from radioactivity decay. (Since these probes remain radioactive after they have been chemically altered by a ^{32}P decay event, their potentially compromised protein liaisons are detected.)

To isolate the DNA probe as a restriction fragment cut out of the synthesized construct, a Sephacryl S-400 spin column is used to exchange the above ligation buffer for one that is appropriate to the particular restriction endonuclease(s). The restriction fragment is purified either by native polyacrylamide or agarose gel electrophoresis, located by brief autoradiography of the wet gel, excised from the gel and isolated by passive elution or electroelution. When DNA probes are to be gel purified, it is recommended to include 0.05% Tween 20 in the column and reaction buffers to minimize nonspecific surface adsorption of the DNA. We routinely further purify the DNA by passing over a DNA-selective column such as Qiagen tip 5 or comparable resin (NACS 52).

Several different photoaffinity probes can be generated from a single oligonucleotide primer by appropriately controlling the primer extension. Some situations permit the incorporation of multiple radioactive and AB-dUMP residues. Particular sequences may call for specifically elaborated strategies to achieve specifically designed results. Consider, for example, a primer capable of being extended by adding ATATTAT. The addition of AB-dUTP and [α-^{32}P]dATP together generates a probe containing 4 AB-dUMP and 3 ^{32}P-dAMP residues. However, primer extension can be limited to a single [α-^{32}P]dAMP and a single AB-dUMP residue as follows: After annealing oligonucleotide 1, [α-^{32}P]dATP is added with the Klenow enzyme and BSA, and incubated for 5 min. Unincorporated [α-^{32}P]dAMP is removed on a Sephacryl spin column and the properly limited incorporation is verified on a small aliquot of the reaction mixture by analysis on PAGE. Next, AB-dUTP is added to the flow-through material from the spin column, for incorporation of only a single residue of AB-dUMP (also verified on a small aliquot). Our experience indicates that this kind of stepwise primer extension works best if the oligonucleotide is designed so that the radioactive nucleotide is incorporated first, because this allows each step of the procedure to be followed by PAGE analysis of the radioactively labeled DNA, and because the radioactively labeled nucleotide is routinely used at a much lower concentration than is AB-dUTP and can more effectively be removed by spin column chromatography.

Probes containing multiple AB-dUMP residues may be helpful when trying to survey a longer stretch of DNA for proximity to protein(s). Nevertheless, it is ultimately preferable to restrict the number of AB-dUMP residues in order to optimize the resolution of protein mapping, to take advantage of the uniqueness of photoaffinity labeling afforded by sites separated by as few as 3 to 4 nucleotides, and to probe the polar distribution of protein around DNA. DNA probes containing multiple AB-dUMP residues can also generate cross-linking of two or more proteins to the same strand of DNA as a consequence of multiple, independent photochemical events.[1] The resulting high-molecular-weight products can complicate further analysis because they are resistant to nuclease treatments (described below) that are used to process the irradiated sample.

Special Considerations:

1. Exonuclease-free Klenow fragment DNA polymerase I (US Biochemicals/Amersham) is chosen for extension of oligonucleotide primer 1 with the photoactive and radioactive nucleotide because of its complete lack of $3' \rightarrow 5'$-exonuclease activity, which can potentially remove the incorporated nucleotide, particularly under limiting conditions for chain elongation. The ^{32}P-labeled nucleotide is routinely incorporated at a very low substrate concentration (threefold excess of radioactive nucleotide over template; $<0.5\ \mu M$). To prevent its removal from the 3′ terminus by DNA polymerases with $3' \rightarrow 5'$-exonuclease activity at the next step of probe synthesis, there is a brief chase with all four dNTP at high concentration.
2. Exonuclease-free Klenow, and also Sequenase (exonuclease-deficient T7 DNA polymerase) have the undesirable property of strand displacement, which prevents their use in completing synthesis of the DNA probe. To complete synthesis of upstream and oligonucleotide 1-proximal DNA, a second oligonucleotide (No. 2) is annealed upstream of oligonucleotide 1 and extended with T4 DNA polymerase, which does not do strand displacement, but does possess an active $3' \rightarrow 5'$-exonuclease. Maintaining all four deoxynucleoside triphosphates at a relatively high concentration prevents degradation of the DNA probe at this stage.
3. Sequential changes of enzymes and substrates during multistep DNA synthesis are made on spin columns (Sephacryl S-400), which separate DNA from proteins and free nucleotides rapidly, and without significant sample dilution.
4. Several of the enzymes that are required for probe synthesis are commercially supplied in DTT-containing buffers. DTT rapidly reacts with the aryl azide; 2-mercaptoethanol is $\sim 10^3$ times less reactive

and safe to use as specified earlier.[4] Enzymes in DTT-containing buffers are accordingly diluted from concentrated stocks with 2-mercaptoethanol-containing buffers. We keep the total DTT concentration contributed by all sources below 10 to 12 μM.

Assembly, Irradiation, Nuclease Digestion, and Analysis of Protein–DNA Complexes

Verifying Specificity of Interaction

Since the long and bulky side chain of AB-dUMP might interfere with specific protein–DNA complex formation, or might, conversely, generate nonspecific interactions with proteins, the specificity of protein complex formation with the photoprobe should be verified by (1) gel-retardation analysis of protein-DNA complexes, (2) competition assays with specific and nonspecific normal competitor DNA, analyzed by gel retardation and/or photoaffinity labeling, (3) construction of down-mutant DNA photoaffinity probes, where appropriate, and (4) functional tests such as transcriptional activity of the photoactive DNA. The experience of prior experiments with RNA polymerase III transcription complexes formed on tRNA and 5S rRNA genes,[1,5–7] and of transcription complexes formed at the bacteriophage T4 late promoter[2] has shown that the majority of modified sites appears to have no adverse effects on protein–DNA complex formation. We noted an effect of multiple AB-dUMP-for-TMP substitutions on DNA–protein interaction in one segment of a tRNA gene at which point mutations are known to reduce the affinity of the specific protein for the DNA by factors of 10^2 to 10^3. Modification at this locus reduced the overall stability of the protein–DNA complex, although a considerable measure of specificity was retained.[1]

Materials

Buffer D: 10% SDS, 25% (w/v) 2-mercaptoethanol, 0.44 M Tris–HCl, pH 7.6, 0.044% (w/v) bromphenol blue.
Buffer E: 20 mM sodium acetate, pH 7, 5 mM $CaCl_2$, 50% (v/v) glycerol, 0.1 mM phenylmethylsulfonyl fluoride.

[4] J. V. Staros, H. Bayley, D. N. Standring, and J. R. Knowles, *Biochem. Biophys. Res. Comm.* **80,** 568 (1978).
[5] B. Bartholomew, G. A. Kassavetis, and E. P. Geiduschek, *Mol. Cell. Biol.* **11,** 5181 (1991).
[6] B. Bartholomew, D. Durkovich, G. A. Kassavetis, and E. P. Geiduschek, *Mol. Cell. Biol.* **13,** 942 (1993).
[7] B. R. Braun, B. Bartholomew, G. A. Kassavetis, and E. P. Geiduschek, *J. Mol. Biol.* **228,** 1063 (1992).

Buffer F: 20 m*M* Tris–HCl, pH 7.5, 50 m*M* NaCl, 0.1 m*M* $ZnSO_4$, 50% (v/v) glycerol.

Solution G: 12.5 m*M* $ZnSO_4$, 0.52 *M* acetic acid.

UV Irradiation and Sample Preparation

The UV cross-linking procedure that is described here is technically unsophisticated but has the virtue of simplicity and universal availability. Protein–DNA complexes (in 15-μl volumes) are photochemically cross-linked in microcentrifuge tubes, with tops open, placed in a microcentrifuge tube rack, by irradiating for 1 to 5 min at 380 $\mu W/cm^2$ with short-wavelength UV light. Our most frequently used light source is a bank of three 15-W germicidal lamps (Fotodyne G15T8) mounted in parallel, 7 cm apart (center-to-center) in an enclosure. The principal emission of these lamps is at 254 nm and the appropriate distance of the sample from the light source (~60 cm in our UV box) is determined by calibration of UV intensity with a photometer. Care is taken to position samples reproducibly. Samples can readily be maintained at constant temperature, but that is not part of our routine procedure.

Irradiated samples are extensively treated with nucleases before analysis by SDS–PAGE. Two sets of conditions for nuclease digestion have been used, with one currently preferred.[1–3,5] The objective of the digestion is to retain the radioactive nucleotide and only a very small segment of DNA covalently attached to protein, without degrading the latter. DNase I is used initially because of its relative purity and lack of protease contamination, and is stored at −20° as a 0.5 mg/ml solution in Buffer E. S1 nuclease is chosen for further digestion because it retains activity in 0.5% SDS, allowing one to denature the protein–DNA complex for more extensive access to the nuclease. S1 nuclease is stored at −20° as a 10 unit/μl stock in Buffer E, and optimal amounts for digestion are determined empirically. We have found the following satisfactory even for proteins that are very tightly bound to relatively large segments of DNA: DNase I (GIBCO/BRL) in Buffer E is added to 50 μg/ml and incubated for 10 min at 37°. SDS is added to 0.5%, the sample is held at 90° for 3 min, centrifuged briefly to collect liquid at the bottom of the tube, then placed on ice for 5 min and transferred to room temperature. The pH is adjusted to 4.5 with 0.09 volumes of solution G, 20 units of S1 nuclease in Buffer F are added, and the sample is incubated at 37° for 10 min. Nuclease digestion is stopped by adding 0.26 volumes of Buffer D and heating at 90° for 3 min. Samples are now ready to load for SDS–PAGE.

Other Considerations, Alternative Approaches, and Other Photoactive Nucleotides

Selection of Wavelength and Choice of Aryl Azide

It is tolerable to irradiate photoprobes with 254-nm light, despite general absorption of the radiation by DNA and protein, and the consequent photochemical reactions, because of the much greater reactivity and efficiency of cross-linking by AB-dUMP (for example, 5-azido-2′-deoxyuridine was 7000 and 150 times more efficiently photolyzed at 254 nm than dTTP and 5-BrdUTP, respectively, at 254 nm[8]). Nevertheless, irradiation times should be kept to the minimum that generates the required yield of photoproduct in order to prevent photodamage to the protein or the DNA. A time course to determine the optimal radiation dose is recommended.

Nitro or hydroxyl substitutions on the phenyl ring generate absorption bands at higher wavelengths with generally lower extinction coefficients than the principal band at 260 to 280 nm,[9] permitting photolysis at longer wavelengths so as to minimize radiation damage to protein and unmodified DNA. Photolysis at longer wavelengths characteristically requires a longer irradiation time and may require attention to other problems, such as sample heating.

Photoaffinity labeling with the aryl azide has been presumed to occur through formation of the highly reactive nitrene, and for that reason to be nonselective in regard to amino acid side chains. However, the unsubstituted phenyl azide undergoes rapid ring expansion to the azocycloheptatetraene or benzazirine on photolysis.[10,11] Azocycloheptatetraenes are electrophilic and highly reactive with nucleophiles, but significantly less reactive than the nitrene; they are expected to decrease overall reactivity with hydrophobic groups and potentially yield less photoaffinity labeling. The substitution of electron-withdrawing groups (fluoro- and nitro-) increases the overall reactivity of the aryl azide and helps prevent ring expansion. Changing the phenyl azide moiety to perfluorophenyl azide may be useful for photoaffinity labeling of the highly hydrophobic regions of proteins.[12,13]

[8] R. K. Evans, J. D. Johnson, and B. E. Haley, *Proc. Natl. Acad. Sci. USA* **83,** 5382 (1986).
[9] T. H. Ji, R. Nishimura, and I. Ji, *in* "Methods in Cell Biology" (D. M. Prescott, ed.), Chap. 11, Vol. 32, p. 277. Academic Press, New York, 1989.
[10] H. Bayley and J. V. Staros, *in* "Azides and Nitrenes" (E. F. V. Scriven, ed.), Chap. 9, p. 434. Academic Press, Orlando, FL, 1984.
[11] J. Brunner, *Ann. Rev. Biochem.* **62,** 483 (1993).
[12] J. F. W. Keanna and S. X. Cai, *J. Org. Chem.* **55,** 3640 (1990).
[13] K. A. Schnapp, R. Poe, E. Leyva, N. Soundararajan, and M. S. Platz, *Bioconjungate Chem.* **4,** 172 (1993).

Alternative Approach to Probe Construction

Oligonucleotides containing a modified residue at the desired position can be chemically synthesized.[14] The modified nucleotide typically has a tethered amino or sulfhydryl group that is masked during oligonucleotide synthesis by a protecting group.[15–18] After oligonucleotide synthesis and removal of the protecting group, the aryl azide is coupled to the free amino or sulfhydryl group, and the photoactive oligonucleotide can be subsequently used as a primer for enzymatic incorporation into double-stranded DNA.[15,19] A radioactive label is incorporated at the end of the DNA chain and the cross-linked protein–DNA conjugate is analyzed without digestion of DNA (as described elsewhere [43] in this volume). An alternative approach, which has been presented recently (19a), involves chemically synthesized oligonucleotides in which one or more specific phosphodiester linkages are replaced by phosphorothioate linkages. These can subsequently be coupled to a photoactive function, for example by reaction with *p*-azidophenacyl bromide.

Other Reagents for Site-Specific Placement of Photoactive Nucleotides in DNA

5-Iodo-dUTP and 5-Bromo-dUTP. Absorption spectra that are shifted to longer wavelengths allow more selective photochemical cross-linking than is achieved with unmodified DNA. Both pyrimidines are close analogs of thymine: The van der Waals radii of bromine, iodine, and of the C-5 methyl group of thymine match within 0.15 Å. 5-IUdR has been shown to be more selective for photochemical cross-linking than 5-BrUdR because of its higher absorption at longer UV wavelengths (2.64×10^6 and 3.85×10^5 cm^2/mol, respectively, at 308 nm).[20] Oligonucleotides containing 5-BrdUMP and 5-IdUMP can be chemically synthesized with commercially available phosphoramidites.[20–22] The photosensitivity of DNA containing 5-BrdUMP at 308 nm has been reported to be 10 times that of unmodified

[14] S. L. Beaucage and R. P. Iyer, *Tetrahedron* **49,** 1925 (1993).
[15] K. J. Gibson and S. J. Benkovic, *Nucleic Acids Res.* **15,** 6455 (1987).
[16] D. Bradley and M. M. Hanna, *Tetrahedron Lett.* **33,** 6223 (1992).
[17] J. T. Goodwin and G. D. Glick, *Tetrahedron Lett.* **34,** 5549 (1993).
[18] M. A. Mullen, H. Wang, K. Wilcox, and T. Herman, *DNA Cell Biol.* **13,** 521 (1994).
[19] T. L. Capson, S. J. Benkovic, and N. G. Nossal, *Cell* **65,** 249 (1991).
[19a] S.-W. Yang and H. Nash, *Proc. Natl. Acad. Sci.* **91,** 12183 (1994).
[20] M. C. Willis, B. J. Hicke, O. C. Uhlenbeck, T. R. Cech, and T. H. Koch, *Science* **262,** 1255 (1993).
[21] E. E. Blatter, Y. W. Ebright, and R. H. Ebright, *Nature* **359,** 650 (1992).
[22] B. J. Hicke, M. C. Willis, T. H. Koch, and T. R. Cech, *Biochemistry* **33,** 3364 (1994).

DNA.[22] 5-BrdUMP- and 5-IdUMP-containing DNA can be irradiated at 308 nm with a XeCl excimer laser, or with the 312-nm emission of a UV transilluminator.[22] The greatest photochemical selectivity of 5-IdUMP-containing DNA is achieved by irradiation at 325 nm, using a HeCd laser,[20] since proteins and normal nucleotides are effectively transparent at that wavelength.

4-Thio-dTTP. 4-Thiothymidine triphosphate (4-S-dTTP; available from Amersham/US Biochemicals) is enzymatically incorporated into DNA by the Klenow fragment of DNA polymerase I.[23] Its UV absorption maximum is shifted to ~335 nm[24]; irradiation with a UV transilluminator (312 nm) is effective for cross-linking. Oligonucleotides containing 4-S-dTMP have been chemically synthesized.[25–27]

5-Azido-dUTP and 8-Azido-dATP. Both of these nucleotides can be enzymatically incorporated into DNA and are highly photoreactive.[8,28–30] These placements of the azide favor the probing of close protein contacts with DNA.

Other Reagents. Other reagents have been developed to place the aryl azide at distances ranging from ~9 to 22 Å from the C-5 of deoxyuridine and the N-4 of deoxycytidine using either polar or aliphatic linkers, in order to probe protein-DNA complexes at variable distances from the DNA helix (B. Bartholomew, unpublished results, 1994). The development of new photoactive derivatives of CTP should also be noted,[31,32] since comparable materials can be synthesized as deoxyribonucleotides.

In assessing the comparative merits of the above reagents, it is clear that 5-IdUMP and 5-BrdUMP change DNA structure least. The principal limitation of 5-BrdUMP appears to be its lack of selective photoreactivity relative to dTMP. The selectivity and achievable efficiency of protein cross-linking of 5-IdUMP are very high.[20] The limitations on its use are the requirement for irradiation at the edge of an absorption band, requiring

[23] B. Bartholomew, B. R. Braun, G. A. Kassavetis, and E. P. Geiduschek, *J. Biol. Chem.* **269,** 18,090 (1994).

[24] A. Favre, *in* "Biorganic Photochemistry" (H. Morrison, ed.), p. 379. John Wiley & Sons, New York, 1990.

[25] T. T. Nikiforov and B. A. Connolly, *Tetrahedron Lett.* **32,** 3851 (1991).

[26] S. B. Rajur and L. W. McLaughlin, *Tetrahedron Lett.* **33,** 6081 (1992).

[27] Y-Z. Xu, Q. Zheng, and P. F. Swann, *J. Org. Chem.* **57,** 3839 (1992).

[28] D. K. Lee, R. K. Evans, J. Blanco, J. Gottesfeld, and J. D. Johnson, *J. Biol. Chem.* **266,** 16,478 (1991).

[29] R. Meffert and K. Dose, *FEBS* **239,** 190 (1988).

[30] R. Meffert, G. Rathgeber, H-J. Schafer, and K. Dose, *Nucleic Acids Res.* **18,** 6633 (1990).

[31] M. M. Hanna, Y. Zhang, J. C. Reidling, M. J. Thomas, and J. Jou, *Nucleic Acids Res.* **21,** 2073 (1993).

[32] M. M. Hanna, this series, Vol. 180, p. 383.

prolonged exposure from a laser light source (e.g., ~1 hr from a HeCd laser operating at 37 mW with the entire beam incident on, and reflected back through, the sample for maximum cross-linking[20]). In 4-S-dTMP-containing DNA, the oxygen that is hydrogen bonded to the N-6 amino group of adenine is replaced with the larger sulfur atom. It would not be surprising to see dynamic properties of DNA (breathing and bending) affected by this substitution. Higher levels of misincorporation of 4-S-dTMP in enzymatic synthesis of photoprobes must also be guarded against.[23] 4-S-dTMP offers efficient as well as selective cross-linking with a UV transilluminator.[23,24] A dark reaction that generates cross-links has also been noted. 5-Azido-dUMP and 8-azido-dATP may also be useful in trying to probe for close protein-DNA contacts, although 5-azido-dUMP-containing DNA is moderately unstable upon storage.[8,28] The conformation of 8-azido-dATP has been suggested to be *syn*[8] but the structure of 8-azido-dAMP in DNA has not been determined.

In summary, the different reagents that are now in use for photochemical cross-linking proteins to DNA allow different proximities to be explored, and provide a richly varied set of reagents for probing the space around DNA. Their relative utility for detecting DNA tracking by proteins has not been examined. It would not be surprising to find that zero-distance cross-linking agents are ineffective in this particular application.

Fine Structure Mapping by Identification of Photochemical Cross-Linking Site on a Protein

Cross-linked amino acids have been precisely identified by proteolysis, isolation of the DNA-linked peptide, and microsequencing.[20,28,33,34] The principal limitation of this precise specification method is the requirement for large amounts of protein–DNA complexes and the potential instability of the protein–DNA cross-link under the conditions of analysis. Less precise mapping of the cross-linking site can also be accomplished by partial or single-hit protease digestion and analysis of the sizes of photoaffinity-labeled peptide fragments.[35,36] Placement of epitope tags at the N and C ends of a polypeptide chain make it possible to map cross-links regionally with very small quantities of material.

[33] K. T. Williams and W. H. Konigsberg, this series, Vol. 208, p. 516.

[34] C. E. Catalano, D. J. Allen, and S. J. Benkovic, *Biochemistry* **29,** 3612 (1990).

[35] M. Riva, C. Carles, A. Sentenac, M. A. Grachev, A. A. Mustaev, and E. F. Zaychikov, *J. Biol. Chem.* **265,** 16,498 (1990).

[36] M. A. Grachev, E. A. Lukhtanov, A. A. Mustaev, E. F. Zaychikov, M. N. Abdukayumov, I. V. Rabinov, V. I. Richter, Y. S. Skoblov, and P. G. Chistyakov, *Eur. J. Biochem.* **180,** 577 (1989).

Examples of Use

In conclusion, we show two examples of the use of DNA tagged with AB-dUMP to analyze the properties of a DNA-tracking protein, the bacteriophage T4 gene 45 protein (gp45). T4 gp45 is one of the three accessory proteins that confer processivity on the T4 DNA polymerase, gp43; the other two accessory proteins, gp44 and gp62, are tightly associated in a complex. Gp44 harbors the catalytic site of a DNA-dependent ATPase (dATP is an alternative substrate) and gp45 greatly stimulates the ATPase activity of the gp44/62 complex (reviewed in Ref. 37). The structure of the gp44/62-gp45 complex with primer–template junctions has been analyzed by footprinting[38] and by a related photochemical cross-linking method[19] that is described in this volume [43]. T4 gp43, 44, 62, and 45 are referred to together as the T4 DNA polymerase holoenzyme.

It has been known for a long time that T4 gp45 is directly involved in the activation of the bacteriophage T4 late genes[39] and the outlines of its mechanism of action have recently been elucidated with the help of an *in vitro* system in which the three DNA polymerase accessory proteins greatly stimulate the initiation of transcription at T4 late promoters.[40] This transcriptional activation normally requires a DNA template with a binding site for the gp44/62-45 complex, which serves as a loading site for gp45, and also requires ATP hydrolysis. The gp45 loading site most commonly used for prior experiments is a nick introduced into a specified DNA strand. Because the nick can be located either upstream or downstream of, and at considerable distance from, the promoter, it has the properties of an enhancer. The special feature of this enhancer is that it must be physically connected to the promoter by a continuous and unobstructed path along DNA. That DNA path is used by gp45 in its transit between the enhancer and the promoter, where it ends up bound to the upstream end of the transcriptional initiation complex.[2,3] The ability to recognize the T4 late promoter is conferred on the *E. coli* RNA polymerase core enzyme (E) by the T4-encoded σ-family protein, gp55. The ability of gp45 to interact with and transcriptionally activate the complex of E.gp55 with a T4 late promoter additionally requires another T4-encoded RNA polymerase-

[37] N. G. Nossal, *in* "Molecular Biology of Bacteriophage T4" (J. D. Karam, ed.), p. 43. ASM Press, Washington, DC, 1994.

[38] M. M. Munn and B. M. Alberts, *J. Biol. Chem.* **266,** 20,024 (1991).

[39] R. Wu, E. P. Geiduschek, and A. Cascino, *J. Mol. Biol.* **96,** 539 (1975).

[40] D. R. Herendeen, G. A. Kassavetis, J. Barry, B. M. Alberts, and E. P. Geiduschek, *Science* **245,** 952 (1989).

binding protein, gp33. Thus, this latter protein serves as a prototype transcriptional coactivator.[41]

The first experiment demonstrates the use of this method to analyze a transcription initiation complex, and more specifically to locate gp45 at the upstream end of the replication protein-enhanced T4 late promoter complex. The probe (Fig. 1), prepared as described above, has a single AB-dUMP at bp −39 in the nontranscribed strand of a phage T4 late promoter, and a single [α-^{32}P]dGMP residue at bp −38 (bp +1 designates the transcriptional start site at the T4 late promoter). Once the competence of this DNA for basal and enhanced T4 late transcription is confirmed, transcription initiation complexes are assembled, irradiated, and processed as described earlier and in the legend to Fig. 3. Samples lacking individual components of the system serve as controls, showing that gp55 and gp33 are required for photochemically cross-linking gp45 at bp −39 and that dATP cannot be replaced by ATPγS (Fig. 3A). The gp33 is also specifically cross-linked to DNA at bp −39, but gp62 and gp44 are not (Fig. 3B). Cross-linking of the T4 single-stranded DNA binding protein, gp32, also occurs, perhaps due to template breathing. Gp32 is present in these experiments at a high concentration in order to coat the single-stranded DNA segments of the photoprobe. Background cross-linking of either the β or β' subunits of RNA polymerase is also noted. An appropriate choice of conditions for electrophoresis allows the α subunit of *E. coli* RNA polymerase core and the T4 gp32 to be resolved (Fig. 3B).

In promoter mapping experiments similar to Fig. 3, a puzzling RNA polymerase-independent "background" of gp45 photochemical cross-linking is seen unless residual dATP is degraded with alkaline phosphatase after transcription complex assembly (Fig. 3, legend). The conditions for generating and eliminating this background have led us to analyze whether it might be due to cross-linking of gp45 as it tracks along DNA.[42] The next experiment is taken from work showing that this is the case.[3] The photoprobe has a single AB-dUMP residue and a single [α-^{32}P]dGMP residue placed 120 and 119 nucleotide, respectively, from the 5′ end of the synthesized DNA strand [except in the 5 to 10% of DNA chains in which oligonucleotide 1 (Fig. 1) fails to ligate to the primer extension product of oligonucleotide 2; in this small fraction of molecules, the photoactive nucleotide is located 68 nucleotide from an internal nick]. The synthesized construct with its circular template strand is used intact, and phage T4 DNA polymerase accessory proteins are assembled on it in the presence

[41] D. R. Herendeen, K. P. Williams, G. A. Kassavetis, and E. P. Geiduschek, *Science* **248,** 573 (1990).

[42] D. R. Herendeen, G. A. Kassavetis, and E. P. Geiduschek, *Science* **256,** 1298 (1992).

of phage T4 single-stranded DNA binding protein, which favors the assembly of the DNA polymerase accessory proteins at double-stranded–single-stranded DNA junctions,[19,38] and coats the single-stranded DNA so as to diminish the nonspecific sequestration of other proteins. When all three DNA polymerase accessory proteins are bound to the DNA in the presence of dATP, gp45 is cross-linked (Fig. 4, lanes 1 and 2), whereas gp44 and gp62 are not (gp44 is not resolved from gp32 on the gel that is shown but is on other gels; data not shown). In Fig. 4, photochemical cross-linking of gp45 requires the presence of gp44/62 complex (lane 8) and dATP (lane 6), and is blocked by ATPγS (lane 5). Photochemical cross-linking is retained for 30 min under the conditions of the experiment (lane 1), subse-

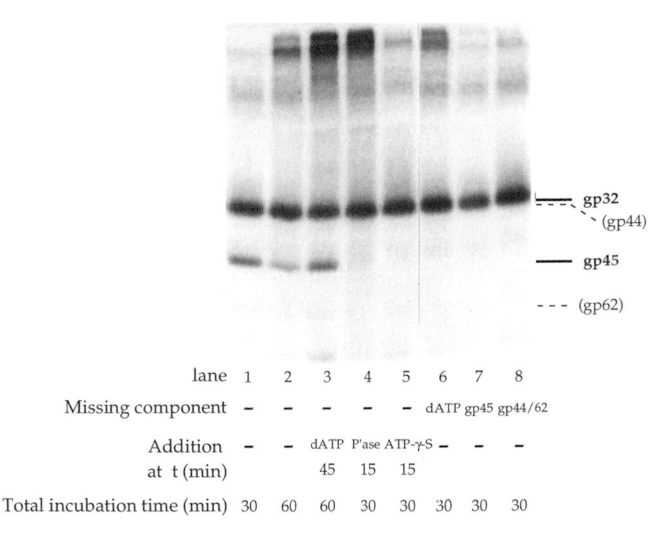

lane	1	2	3	4	5	6	7	8
Missing component	-	-	-	-	-	dATP	gp45	gp44/62
Addition at t (min)	-	-	dATP 45	P'ase 15	ATP-γ-S 15	-	-	-
Total incubation time (min)	30	60	60	30	30	30	30	30

FIG. 4. Demonstrating the ability to photochemically cross-link gp45 under conditions suggestive of tracking along DNA (adapted from Ref. 3). The DNA has a single AB-dUMP residue incorporated 120 nucleotide from the 5′ end of its enzymatically synthesized strand (see text). Transcription complexes are assembled and incubated as specified below each lane. dATP, gp45, and gp44/62 complex are omitted from the samples analyzed in lanes 6, 7, and 8, respectively. For the sample analyzed in lanes 3, 4, and 5, respectively, second additions of dATP, alkaline phosphatase, and ATPγS are made at the time shown. Cross-linked proteins gp32 and gp45 are indicated at the side. GP44 (resolved from gp32 on another gel; results not shown) and gp62 are not detectably cross-linked. *Method:* DNA (8 fmol), 1.8 pmol gp45 trimer, 3.3 pmol gp44/62 complex, and 23 pmol gp32 in 13 μl of the reaction buffer specified in the legend of Fig. 3 (but with 200 m*M* in place of 250 or 300 m*M* potassium acetate) are incubated for the specified time at 37°. Second additions of dATP (to 0.28 m*M* additional), ATPγS (to 0.5 m*M*), or 1 U shrimp alkaline phosphatase are made at the indicated times in 2 μl reaction buffer. Samples are irradiated and prepared as specified in the text and in the legend for Fig. 3. An autoradiogram (with intensifier screen) of a 12% (w/v) polyacrylamide (37.5 : 1, acrylamide : bisacrylamide) SDS gel is shown.

quently declines (lane 2), can be restored by adding more dATP (lane 3), and is quickly abolished if dATP is hydrolyzed with alkaline phosphatase (lane 4). The interpretation, that gp45 (1) is loaded onto DNA by its assembly factor, the gp44/62 complex, at the double-stranded–single-stranded junctions (one of which is a primer–template junction), (2) is released from these assembly sites to track along DNA in an ATP hydrolysis-requiring process, (3) continually falls off the DNA so that continual loading is required to maintain the DNA-tracking traffic, and (4) is cross-linked as it tracks, is supported by additional experiments that are reported elsewhere.[3] Similar experiments with the *E. coli* and human homologs[43,44] of T4 gp45 and gp44/62 have also been reported.[3] The point of the presentation here is to illustrate the use of photochemical cross-linking to probe tracking along DNA by one of these replication proteins.

In the preceding experiments, specificity of photochemical cross-linking was assessed by verifying proper function of the photoprobe as a template for transcription and by dependence of photochemical cross-linking on all the components that are required for transcriptional enhancement. The specificity of gp45 cross-linking during tracking is supported by its dependence on gp44/62 and dATP (Fig. 4), and by the inability of the heterologous bacterial or human gp44/62 analogs, the *E. coli* DNA polymerase III holoenzymes' γ complex, and the replication factor C, respectively,[43] to substitute for gp44/62.[3] It is, however, surprising that the efficiency of photochemical cross-linking during DNA tracking by gp45 is substantial relative to the efficiency of cross-linking in presumably stationary promoter complexes. It is conceivable that the AB-dUMP side chain perturbs DNA-tracking traffic in such a way as to generate more efficient cross-linking than would occur with a less bulky photoactive residue, but this remains to be analyzed.

Acknowledgment

Research on this subject has been supported by grants from the National Institute of General Medical Sciences to BB and EPG. RLT has been a trainee in genetics supported by a training grant from the National Institute of General Medical Sciences.

[43] J. Kuriyan and M. O'Donnell, *J. Mol. Biol.* **234,** 915 (1993).
[44] X. P. Kong, R. Onrust, M. O'Donnell, and J. Kuriyan, *Cell* **69,** 425 (1992).

Section VI

In Vitro Replication Systems: Crude and Reconstituted

[38] Permeabilized Cells

By Robb E. Moses

Introduction

Permeabilized cells allow measurement of authentic replicative DNA synthesis. They are notable for ease of preparation and a minimal disturbance of the replicative apparatus. The limitation is that there is no initiation of new cycles of DNA replication. Such systems serve well to monitor the effects of agents on DNA replication or repair since they allow ready definition of the category of synthesis.

Bacterial Cells

Several methods have been used to measure replicative DNA synthesis in bacteria. With *E. coli* as a model, plasmolyzed cells were studied extensively. This system was not satisfactory for many studies because the method of forming spheroplasts raised background levels of nonspecific DNA repair synthesis to unacceptable levels.[1] This is particularly a problem in cells containing DNA polymerase I.

Several organic solvents have been tried. The best developed system consists of toluene treatment. Replication shows complete ATP dependence and a low background of nonspecific synthesis. The final product is of high molecular weight, made semiconservatively, and discontinuous DNA synthesis can be demonstrated. One advantageous aspect of the toluene-treated cell system is that cells containing DNA polymerase I can be used. This allows testing of strains in which the presence of a *polA* mutation would be lethal. Replicative synthesis is apparently elongation only, with sudden arrest in *dnaB* temperature-sensitive strains. The preparation of cells[2] for assay is by the following steps.

Escherichia coli cells are grown to a concentration of approximately 1×10^9 and harvested by centrifugation at 4°. The cell pellet is suspended in 50 m*M*, pH 7.4, KPO_4 at one-tenth volume of the culture. The cell suspension is made in 1% (v/v) toluene and gently agitated for between 2 and 10 min at room temperature. Individual strains show different tolerance of this procedure and it is important to determine a time course for ATP-dependent synthesis for each.

[1] G. Buttin and M. Wright, *Cold Spring Harbor Symp. Quant. Biol.* **33,** 259 (1968).
[2] R. E. Moses and C. C. Richardson, *Natl. Acad. Sci.* **67,** 674 (1970).

Following exposure, the cells are harvested and the cell pellet is washed thoroughly with the above buffer. Cells are resuspended to a final volume of 0.1 culture volume in 50 m*M* KPO_4, pH 7.4. The cells may be divided into aliquots and quick frozen at −80°. Such cell suspensions are stable for extended periods of time for measuring replicative DNA synthesis.

The assay for DNA replication contains the following in 0.3 ml: 66 m*M* potassium phosphate buffer, pH 7.4, 13 m*M* $MgCl_2$, 1.3 m*M* ATP, 33 μM each of dCTP, dGTP, ATP, dTTP, with a radioactive isotope in one of the triphosphates. Normally, 1 to 5 $\times$ 10^8 toluene-treated cells are used for measurement of replication. Reactions are terminated by the addition of 3 ml of cold 10% trichloroacetic acid–0.1 *M* sodium pyrophosphate. The reaction is allowed to stand for 5 min at 4° and then washed over a Whatman GF/C disk with vacuum aspiration. The filter is washed several times with 10% trichloroacetic acid–0.1 *M* sodium pyrophosphate, followed by 10 ml of cold 0.01 *M* HCl. The disks are dried and radioactivity determined.

Replication is dependent on ATP, requires all 5′-deoxyribonucleoside triphosphates, and occurs at approximately the *in vivo* rate in this system.

The system may be modified by the addition of Triton X-100, a nonionic detergent. The presence of 1% detergent does not alter the rate or extent of synthesis, nor does it change the semiconservative nature of the synthesis. However, the addition of the detergent allows accessibility to macromolecules.[3] With detergent, 10 m*M* 2-mercaptoethanol, final concentration, is added to the reaction.

Mammalian Cells

Several methods have been used to permeabilize mammalian cells to allow assay of DNA replication. One system, which has the advantage of being reversible and allowing cell survival, uses lysolecithin for permeabilization.[4] Fibroblasts are grown on plates, trypsinized, and washed with 0.15 sucrose, 80 m*M* HCl, 5 m*M* $MgCl_2$, and 35 m*M* HEPES buffer, pH 7.2, with protease inhibitor if wished. Cells are resuspended at 8 $\times$ 10^7/ml in the same components and 0.5 mg/ml lysolecithin is added. Synthesis is assayed in the above buffer system (1 to 5 $\times$ 10^7 cells/ml) in 0.4 ml with 1.25 m*M* ATP, 5.0 m*M* phosphoenol pyruvate (PEP), and 10 to 100 μM

[3] R. E. Moses, *J. Biol. Chem.* **247,** 6031 (1972).

[4] M. R. Miller and D. N. Chinault, *J. Biol. Chem.* **257,** 10,204 (1982).

each of dATP, dTTP, dCTP, and dGTP with radioactive isotope label as needed. Reactions are terminated and determined as above.

Applications

Permeable cell systems are useful for defining the biochemical requirements and responses of DNA replication and repair. Access by substrates, DNA-damaging reagents, or inhibitors of synthesis, in combination with the use of mutants defective in known enzymatic function, has made cellular functions apparent. In either the prokaryotic or eukaryotic system, a density analog, BrdUTP, can be substituted for dTTP to differentiate replicative DNA synthesis from nonreplicative. Mutants showing temperature-sensitive DNA replication have also been useful. Each system has been shown to respond to DNA damage with repair synthesis.[5,6] The *E. coli* system has been used to define the role of DNA polymerases in response to ultraviolet radiation or after DNA-damaging drugs, such as bleomycin.[6,7] Permeable cells also have been used to examine the role of DNA polymerases in response to hydrogen peroxide damage, showing that cells have more than one repair pathway in that response.[8,9]

In human fibroblasts the use of inhibitors specific for different polymerases has allowed identification of the roles of these enzymes in response to damage secondary to mutagens.[5] Use of mutants with defined deficiencies has, of course, been much less available in permeable human cells than in bacterial cells.

In summary, permeable cell systems offer the advantage of testing the cellular response to low-molecular-weight agents in a situation representing minimal pertubation of the replisome.

[5] R. A. Hammond, J. K. McClung, and M. R. Miller, *Biochemistry* **29,** 286 (1990).
[6] D. Bowersock and R. E. Moses, *J. Biol. Chem.* **248,** 7449 (1973).
[7] M. R. Miller and C. N. Chinault, *J. Biol. Chem.* **257,** 46 (1982).
[8] M. E. Hagensee and R. E. Moses, *J. Bacteriol.* **168,** 1059 (1986).
[9] M. E. Hagensee and R. E. Moses, *J. Bacteriol.* **171,** 991 (1989).

[39] DNA Synthesis Initiated at *oriC*: *In Vitro* Replication Reactions

By ELLIOTT CROOKE

Introduction

In vitro replication initiated at the *Escherichia coli* chromosomal origin (*oriC*) can occur in a precisely prepared cell extract or be reconstituted with purified proteins. The development of these systems was possible following the identification of *oriC* and the construction of *oriC*-containing plasmids.[1-5] *In vivo* these plasmids act as minichromosomes in that their replication is initiated synchronously with that of the cell chromosome and with similar requirements.[6,7] Behaving as such, these plasmids can serve as templates *in vitro* to study replication at *oriC*.

Another major advance in the development of systems to replicate *oriC* *in vitro* was the preparation of an extract containing all of the required factors and lacking inhibitors.[8] Combined with knowledge gained from studying replication in other systems such as bacteriophage ϕX174, replication in the crude extract permitted the identification, isolation, and characterization of the numerous components necessary to reconstitute *oriC* DNA replication.[9,10]

Replication in cell extracts and in systems of defined components has many physiological features. It requires the addition of template DNA, which contains the *oriC* sequence, and replication is initiated at or near *oriC*. It is dependent on certain replication proteins that have been identified as important for chromosomal replication *in vivo*. Included in these is DnaA,

[1] S. Hiraga, *Proc. Natl. Acad. Sci. USA* **73,** 198 (1976).
[2] S. Yasuda and Y. Hirota, *Proc. Natl. Acad. Sci. USA* **74,** 5458 (1977).
[3] K. von Meyenburg, F. G. Hansen, L. C. Nielsen, and E. Riise, *Mol. Gen. Genet.* **160,** 287 (1978).
[4] M. Meijer, E. Beck, F. G. Hansen, H. F. Bergmans, W. Messer, K. von Meyenburg, and H. Schaller, *Proc. Natl. Acad. Sci. USA* **76,** 580 (1979).
[5] A. Oka, K. Sugimoto, M. Takanami, and Y. Hirota, *Mol. Gen. Genet.* **178,** 9 (1980).
[6] K. von Meyenberg, F. G. Hansen, E. Riise, H. E. Bergmans, M. Meijer, and W. Messer, *Cold Spring Harbor Symp.* **43,** 121 (1979).
[7] A. C. Leonard and C. E. Helmstetter, *Proc. Natl. Acad. Sci. USA* **83,** 5101 (1986).
[8] R. S. Fuller, J. M. Kaguni, and A. Kornberg, *Proc. Natl. Acad. Sci. USA* **78,** 7370 (1981).
[9] J. M. Kaguni and A. Kornberg, *Cell* **38,** 183 (1984).
[10] B. E. Funnell, T. A. Baker, and A. Kornberg, *J. Biol. Chem.* **261,** 5616 (1986).

protein which has been implicated genetically as playing a central role in initiating replication at *oriC*.

Materials and Methods

Replication Templates

Various *oriC* plasmids, or minichromosomes, have been successfully employed as templates for DNA replication *in vitro*. One such, pCM959, is composed solely of *E. coli* DNA (4012 bp) spanning *oriC* (bp −677 to +3335).[4] However, preparations of this plasmid must be done on a large scale inasmuch as yields are minimal due to the low copy number of the plasmid. This limitation is easily avoided by having *oriC* cloned into high copy number vectors, which contain a second origin of replication. A fragment (637 bp) containing *oriC* (bp −189 to +448) has been cloned into the *Pst*I site of M13mp8 to create M13mpRE85.[11] Another minichromosome, pBS*oriC* (also called pTB101),[12] includes the ColE1 origin and was created by cloning an *oriC*-containing fragment (bp −189 to +489; 678 bp) into the pBluescript vector (Stratagene Cloning Systems, La Jolla, CA). The presence of multiple copies of *oriC* has few deleterious effects on host cells, except under certain conditions such as when DnaA protein activity is reduced below normal levels.

Only the covalently closed supercoiled forms of the plasmids (or RF I form of the phages) are suitable as templates. Purification of DNA by two successive bandings in CsCl/ethidium bromide density gradients, followed by ethanol precipitation and gel filtration,[13] in general, produces satisfactory templates.

Preparation of Active Cell Extract

Crude cell extracts (fraction II) that are active for DNA replication can be produced from a variety of strains. However, experience has shown that the reproducibility of obtaining active extracts is strain dependent and needs to be determined empirically. Usually cells can be grown in L-broth[14] + 0.2% glucose to an optical density (A_{600nm}) of 0.8 to 1.5 and then harvested

[11] D. W. Smith, A. M. Garland, G. Herman, R. E. Enns, T. A. Baker, and J. W. Zyskind, *EMBO J.* **4,** 1319 (1985).

[12] D. Bramhill and A. Kornberg, *Cell* **52,** 743 (1988).

[13] J. Sambrook, E. F. Fritsch, and T. Maniatis, *"Molecular Cloning: A Laboratory Manual,"* 2nd ed. Cold Spring Harbor Laboratory Press, New York, 1989.

[14] J. H. Miller, "Experiments in Molecular Genetics." Cold Spring Harbor Laboratory Press, New York, 1972.

by centrifugation, resuspended in a solution of 25 m*M* HEPES–KOH, pH = 7.5, 1 m*M* EDTA, and 2 m*M* dithiothreitol (DTT) to an optical density (A_{600nm}) of 250, frozen in liquid nitrogen, and stored at −80° until used.

To prepare extracts, a modification of the protocol of Fuller *et al.*[8] is followed. Briefly, frozen cells are thawed on ice and the suspension is adjusted to 80 m*M* KCl. To lyse the cells, egg lysozyme is added (300 μg/ml), the mixture is incubated for 30 min on ice, again frozen in liquid nitrogen, and then thawed by swirling in a beaker of room-temperature water. The lysate is clarified by centrifugation at 180,000*g* for 30 min (2°) and the resulting supernatant (fraction I) is collected. Finely ground ammonium sulfate (0.277 g/ml) is added slowly to rapidly stirred (but not foaming) fraction I on ice. The suspension is stirred for an additional 30 min, and the precipitate is collected by centrifugation at 125,000*g* for 30 min (2°). To generate fraction II with a high protein concentration (>100mg/ml), a condition that correlates well with obtaining a replicatively active extract, the ammonium sulfate pellets are transferred directly into a dialysis bag (molecular weight cut-off of 12,000) without resuspension. The bag is sealed with minimal extra space and the precipitate is dissolved by dialysis against a 100-fold volume of 25 m*M* HEPES–KOH, pH 7.6, 0.1 m*M* EDTA, and 2 m*M* DTT for 60 min on ice. The dialysate (fraction II) is frozen in liquid nitrogen as small aliquots and stored at −80°. Samples of fraction II are thawed just prior to use and any unused portions are disposed.

The concentration of ammonium sulfate necessary to prepare active fraction II falls within a very narrow window; too little fails to precipitate all of the necessary proteins, too much and inhibitors of the replication reaction are included.

Replication in Crude Cell Extracts

DNA synthesis is possible in extracts (fraction II) prepared not only from wild-type cells, but also in extracts from cells deficient for an essential replication protein. In such cases, the necessary factor must be supplied in the form of a second extract or as the purified protein. This type of "complementation reaction" has been used to isolate required replication proteins such as DnaA protein.[15]

Besides the importance of having fraction II at a sufficiently high protein concentration, and of it being generated with a precise level of ammonium sulfate (see above), replication in crude cell extracts is dependent on the addition of a large, hydrophilic polymer such as polyvinyl alcohol (PVA) or polyethylene glycol (PEG).[8] These polymers cause macromolecular

[15] R. S. Fuller and A. Kornberg, *Proc. Natl. Acad. Sci. USA* **80,** 5817 (1983).

crowding, thereby creating an environment in which the apparent concentration of macromolecules is increased to approximate those found within the cell.[16]

Reactions (25 μl) for replication of an *oriC* plasmid contain 40 m*M* HEPES–KOH, pH 7.6, 2 m*M* ATP, 0.5 m*M* each of CTP, GTP, and UTP, 100 μ*M* each of dATP, dCTP, dGTP, and [^{32}P-α]dTTP (75 to 150 cpm/pmol dNTP), 11 m*M* magnesium acetate, 40 m*M* creatine phosphate, 200 to 300 μg (as protein) fraction II cell extract; 7% (w/v) PVA, 2.5 μg creatine kinase, and 200 ng (600 pmol as nucleotide) *oriC*-plasmid DNA. The components should be assembled in the order listed; after the addition of PVA, the mixture is centrifuged for 15 sec and the supernatant is transferred to a fresh tube prior to the addition of creatine kinase and template DNA. Typically, samples are incubated for 20 min at 30°; reactions are stopped and replication products are analyzed as described below.

Purified Proteins

References that contain descriptions on how to isolate the proteins required to reconstitute DNA synthesis initiated at *oriC* are listed in Table I. The selected references are ones that have been used successfully by the author and members of the laboratory of Arthur Kornberg; omission of other protocols does not imply that the resulting purified proteins are not functional for *in vitro* DNA replication.

Replication Reconstituted with Purified Components

Solo Primase. Under certain conditions, replication of *oriC* plasmids can be reconstituted with eight purified proteins.[17,18] DnaA (the initiator protein) binds *oriC* and, when optimal low levels of histone-like protein HU are present, it also opens the duplex at an AT-rich region of *oriC*. DnaC protein aids in the delivery of DnaB helicase to the replication fork where DnaB unwinds the helix in both directions. In the presence of SSB, primase (DnaG protein) synthesizes RNA primers, which can be elongated by DNA polymerase III holoenzyme with gyrase serving as a topological swivel.[19] Replication is stimulated by the low level of HU protein, probably by its promoting DNA bending.[20] In the solo primase system the auxiliary

[16] S. B. Zimmerman and S. O. Trach, *BBA* **949,** 297 (1988).

[17] T. Ogawa, T. A. Baker, A. van der Ende, and A. Kornberg, *Proc. Natl. Acad. Sci. USA* **82,** 3562 (1985).

[18] A. van der Ende, T. A. Baker, T. Ogawa, and A. Kornberg, *Proc. Natl. Acad. Sci. USA* **82,** 3954 (1985).

[19] T. A. Baker, K. Sekimizu, B. E. Funnell, and A. Kornberg, *Cell* **45,** 53 (1986).

[20] K. Skarstad, T. A. Baker, and A. Kornberg, *EMBO J.* **9,** 2341 (1990).

TABLE I
PROCEDURES FOR PURIFICATION OF REPLICATION PROTEINS

Protein	Reference
DnaA	K. Selimizu, B. Y. M. Yung, and A. Kornberg, *J. Biol. Chem.* **263,** 7136 (1988).
	D. S. Hwang, E. Crooke, and A. Kornberg, *J. Biol. Chem.* **265,** 19,244 (1990).
DnaB helicase	K. Arai, S. Yasuda, and A. Kornberg, *J. Biol. Chem.* **256,** 5247 (1981).
DnaC	J. Kobori and A. Kornberg, *J. Biol. Chem.* **257,** 13,763 (1982).
Primase (DnaG)	L. Rowen and A. Kornberg, *J. Biol. Chem.* **253,** 758 (1978).
Gyrase A	K. Mizuuchi, M. Mizuuchi, M. H. O'Dea, and M. Gellert, *J. Biol. Chem.* **259,** 9199 (1984).
Gyrase B	K. Mizuuchi, M. Mizuuchi, M. H. O'Dea, and M. Gellert, *J. Biol. Chem.* **259,** 9199 (1984).
SSB	D. A. Soltis and I. R. Lehman, *J. Biol. Chem.* **258,** 6073 (1983).
HU	N. E. Dixon and A. Kornberg, *Proc. Natl. Acad. Sci. USA* **81,** 424 (1984).
DNA Pol III*	H. Maki, S. Maki, and A. Kornberg, *J. Biol. Chem.* **263,** 6570 (1988).
	M. G. Cull and C. S. McHenry, this volume [3].
β Subunit	P. M. J. Burgers and A. Kornberg, *J. Biol. Chem.* **257,** 11,468 (1982).
	K. O. Johanson, T. E. Haynes, and C. S. McHenry, *J. Biol. Chem.* **261,** 11,460 (1986).
RNA Pol	N. Gonzalez, J. Wiggs, and M. J. Chamberlin, *Arch. Biochem. Biophys.* **182,** 404 (1977).
RNase H	J. M. Kaguni and A. Kornberg, *Cell* **38,** 183 (1984).

proteins RNase H and topoisomerase I do not need to be included to maintain dependence on DnaA protein and the *oriC* sequence.[17]

The solo primase system requires that there be sufficient energy, in the form of free superhelicity and adequate temperatures, such that DnaA protein can promote strand opening. Conditions of low temperature, low superhelicity, high levels of HU protein (which restrains the free superhelicity), or the absence of HU protein necessitate the inclusion of RNA polymerase (see below) for efficient DNA replication.[17,20,21]

Conditions for a solo primase reaction (25 μl) are 30 m*M* HEPES–KOH, pH 7.6, 8 m*M* magnesium acetate, 2.0 m*M* ATP, 100 μM each of dATP, dCTP, dGTP, and [^{32}P-α]dTTP (75 to 150 cpm/pmol dNTP), 200 ng (600 pmol as nucleotide) *oriC*-plasmid DNA, 450 ng SSB; 14 ng primase, 400 ng gyrase A subunit, 180 ng gyrase B subunit, 26 ng β subunit, 66 ng DnaB

[21] T. A. Baker and A. Kornberg, *Cell* **55,** 113 (1988).

protein, 28 ng DnaC protein, 60 ng DnaA protein, 120 ng DNA polymerase III*, and 8 ng HU protein. The molar ratios of the different enzymes in the reaction vary from preparation to preparation and need to be optimized for each set of enzymes. Buffers and salts contributed by the enzyme preparations do not need to be considered in the assembly of the *in vitro* synthesis reactions. The components should be assembled on ice; the order of addition is important in that enzymes such as DnaC protein, DnaA protein, DNA polymerase III*, and HU protein best retain their activities if added at the end. The mixture is incubated at 30° for 20 min and the reaction is stopped and products are analyzed as described below.

RNA Polymerase-Dependent. Transcriptional activation by RNA polymerase is necessary under reaction conditions in which the activity of DnaA protein is insufficient to melt the DNA at *oriC*. RNA polymerase forms a RNA–DNA hybrid, or R-loop, which can stimulate the strand separation stage of initiation.[17,20,21] With the inclusion of RNA polymerase, RNase H should be included to maintain the specificity for initiation at *oriC*.[17]

The reaction conditions for RNA polymerase-dependent synthesis are as for the solo primase system except that the level of HU protein is elevated (>150 ng) or omitted, and RNase H and RNA polymerase are included at 0.2 ng and 160 ng, respectively. DNA synthesis is carried out at 30° for 20 min.

Staged Replication Reactions

Initiation of replication at *oriC* can be staged as a succession of three complexes: initial, open, and prepriming.[22] Initial complexes are formed in mixtures (15 μl) containing 50 mM HEPES–KOH, pH 7.6, 2.5 mM magnesium acetate, 0.3 mM EDTA, 20% (v/v) glycerol, 0.007% Triton X-100, 7 mM DTT, 1 μM ATP, 60 ng DnaA protein, and 200 ng (600 pmol as nucleotide) *oriC*-plasmid DNA. The mixtures are incubated at 0° for 10 min and, if desired, the complexes can be isolated by rapid gel filtration through Bio-Gel A5m.[23] The remaining components (including additional magnesium acetate) described in the solo primase system are added for subsequent replication of the initial complexes.

Open complexes are prepared in the conditions described for initial complex formation, except that the concentrations of ATP and magnesium acetate are raised to 5 and 8 mM, respectively, and HU protein (8 ng) is included. Samples are incubated at 38° for 10 min to promote strand opening. Cold-sensitive open complexes can be isolated by sedimentation through a sucrose density gradient at 38°.[22]

[22] K. Sekimizu, D. Bramhill, and A. Kornberg, *J. Biol. Chem.* **263,** 7124 (1988).
[23] B. Y. M. Yung, E. Crooke, and A. Kornberg, *J. Biol. Chem.* **265,** 1282 (1990).

Reactions (20 μl) staged at the formation of prepriming complexes[19] contain 40 m*M* Tricine–KOH, pH 7.6, 5 m*M* ATP, 0.38 m*M* EDTA, 0.3 mg/ml BSA, 8% PVA (20% glycerol if subsequent gel filtration is needed), 0.01% Brij 58, 60 m*M* potassium glutamate, 200 ng (600 pmol as nucleotide) *oriC*-plasmid DNA, 8 ng HU protein, 60 ng DnaA protein, 66 ng DnaB helicase, 28 ng DnaC protein, and 450 ng SSB. Components are assembled at 0° and then incubated at 30° for 10 min. The prepriming complexes can be isolated in sucrose gradients or by gel filtration if desired.[19–21,24] Otherwise, reactions are shifted to 16°, the remaining components are added, and the incubation continued for 20 min.

Assays and Product Analysis

DNA replication in cell-free systems is most often monitored by measuring the incorporation of radiolabeled deoxynucleotides into acid-insoluble material. Synthesis reactions are stopped by the addition of a chilled solution of 10% trichloroacetic acid (TCA) and 100 m*M* pyrophosphate. Acid-insoluble material is then collected on glass filters (GF/C, Millipore), the filters are washed with a solution of 1 *M* hydrochloric acid and 100 m*M* pyrophosphate, followed by ethanol, and dried under an infrared lamp. Radioactivity retained on the filters is measured by liquid scintillation counting. Pyrophosphate is included in the TCA and hydrochloric acid solutions to help diminish the nonspecific binding of unincorporated deoxynucleotides to the filters. In addition, synthesized DNA can be resolved by agarose electrophoresis in either neutral or alkaline conditions and visualized with autoradiography to determine the relative amounts of unit-length product, smaller fragments, and high molecular weight material generated by rolling circle replication.[13]

Conclusion

Inasmuch as the *in vitro* reactions have been valuable in describing the roles the various components play in replication at *oriC*, they will surely play a large part in the characterization of the factors and processes involved in the regulation of replication from the chromosomal origin.

[24] G. C. Allen, Jr., and A. Kornberg, *J. Biol. Chem.* **266,** 22,096 (1991).

[40] ϕX174-Type Primosomal Proteins: Purification and Assay

By KENNETH J. MARIANS

Introduction

Resolution and reconstitution of the enzymatic activities required for conversion *in vitro* of ϕX174 single-stranded (circular) DNA [ss(c)] to the double-stranded replicative form (RF) have resulted in the identification of the two major cellular replication machines, the DNA polymerase III holoenzyme (DNA Pol III HE)[1] and the primosome.[2] The primosome contributes the priming and DNA helicase functions at the replication fork.[3]

Seven proteins are required for primosome assembly: DnaB, DnaC, DnaG, DnaT, PriA, PriB, and PriC.[3,4–7] Assembly occurs at a specific sequence [a primosome assembly site (PAS)[8]] on ϕX174 ss(c) DNA coated with the single-stranded DNA binding protein (SSB).[2] Once assembled, the primosome can both translocate in either direction along the ssDNA and act as a DNA helicase in either direction.[3] The $3' \rightarrow 5'$ movement and helicase activity is fueled by the ATPase (or dATPase) activity of PriA, whereas $5' \rightarrow 3'$ movement and helicase activity is fueled by the NTPase activity of DnaB. Primase (DnaG) does not remain stably associated with the primosome[4,5] (in the absence of DnaG, the mobile protein complex on the DNA, which probably contains PriA, DnaB, and PriB,[3,7] is referred to as the preprimosome[9]). When associated with the preprimosome, forming the primosome, primase will catalyze the synthesis of RNA or DNA primers that can be utilized by the DNA Pol III HE to initiate nascent strand synthesis.[10]

This chapter describes the preparation of crude soluble extracts that are capable of supporting ϕX174 ss(c) $\rightarrow$ RF DNA replication, a comple-

[1] W. Wickner and A. Kornberg, *J. Biol. Chem.* **249,** 6244 (1974).
[2] K.-I. Arai and A. Kornberg, *Proc. Natl. Acad. Sci. USA* **78,** 69 (1981).
[3] M. S. Lee and K. J. Marians, *J. Biol. Chem.* **264,** 14,531 (1989).
[4] J. Weiner, R. McMacken, and A. Kornberg, *Proc. Natl. Acad. Sci. USA* **73,** 752 (1976).
[5] S. Wickner, *in* "The Single-Stranded DNA Phages" (D. T. Denhardt, D. Dressler, and D. S. Ray, eds.), p. 255. Cold Spring Harbor Laboratory Press, New York, 1978.
[6] K.-I. Arai, R. Low, J. Kobori, J. Shlomai, and A. Kornberg, *J. Biol. Chem.* **256,** 5273 (1981).
[7] G. Allen, Jr., and A. Kornberg, *J. Biol. Chem.* **268,** 19,204 (1993).
[8] K. J. Marians, *CRC Crit. Rev. Biochem.* **17,** 153 (1984).
[9] K.-I. Arai, R. Low, and A. Kornberg, *Proc. Natl. Acad. Sci. USA* **78,** 707 (1981).
[10] R. McMacken and A. Kornberg, *J. Biol. Chem.* **253,** 3313 (1978).

mentation assay that can be used to score the activity of some of the replication proteins, the purification of the primosomal proteins from *Escherichia coli* strains engineered to overproduce them, and the reconstitution of ϕX174 ss(c) → RF DNA replication with purified proteins.

Assays with Crude Soluble Extracts

Growth of Cells

Preparation of soluble extracts capable of supporting ϕX174 ss(c) → RF DNA replication requires large amounts of cells that have been resuspended at high densities. It is most convenient to grow these cells in a pilot-plant fermentor.

Our cells are grown in a New Brunswick Scientific model IF400 pilot-plant fermentor. Three hundred liters of deionized H_2O containing 3 kg Bacto-tryptone, 1.5 kg yeast extract, 1.5 kg NaCl, 3 g thiamin, 364 g KH_2PO_4, 254 g K_2HPO_4, and 15 g thymine (when required) is sterilized in place and brought to growth temperature. Then 1.5 kg of glucose (dissolved and sterilized in a minimal volume of H_2O) is added. A 12-liter inoculum of a stationary culture is used to initiate growth. Cells are grown to an OD_{600} of about 4.0 (late log phase) with automatic pH adjustment to pH 7.5. The culture is then chilled using a New Brunswick Scientific model RC-200 rapid chiller to <18°. Cells are harvested using a refrigerated Sharples model AS-26 continuous-flow centrifuge. Harvesting takes about 45 min. The cells are scraped out of the rotor and resuspended using a 1-liter Waring blender at 50% (w/w) in a buffer containing 50 m*M* Tris–HCl, pH 7.5 at 4°, and 10% sucrose (Tris–sucrose). The cell suspension is then frozen using liquid N_2 in 500-ml aliquots in 4-mil-thick plastic bags (Northland Polycleer Bag Corp., New York, NY). The frozen cell suspension is stored at −80°.

Preparation of Crude Soluble Extracts

Frozen cells are broken into small chunks using a mallet and thawed rapidly with continuous stirring in a 30° water bath. The cell suspension should not be allowed to heat up. Twenty milliliters of cell suspension is distributed into each of two tubes on ice that are capable of being centrifuged at 100,000*g* (Sorvall A841 tubes or Oakridge type 30 tubes). The following additions are then made: EDTA to 20 m*M*, Tris–HCl, pH 8.4 at 4°, to 50 m*M*, KCl to 150 m*M*, dithiothreitol (DTT) to 10 m*M*, and lysozyme to 0.2 mg/ml. The suspension is mixed gently by inverting the tubes 10 or 20 times and then incubated on ice for 10 min to allow spheroplast formation. Brij 58 is then added to 0.1% to lyse the spheroplasts. The suspension is mixed gently as before and the incubation continued at 0°.

The time required for lysis after Brij addition varies with the strain used and the length of time the cell suspension has been stored at −80°. For example, PC22 (a *dnaC* strain) is generally completely lysed in <10 min, whereas HMS-83 (a *polA polB* strain) can take 30 to 40 min. A good indicator of lysis is whether large bubbles get trapped when the tube is inverted. Overlysing the cells can result in extract preparations that are inactive.

When sufficient lysis has occurred, the suspension is centrifuged at 100,000*g* at 4° for 1 hr. The supernatant is carefully decanted. Nucleic acid is precipitated by the dropwise addition with rapid stirring of a 20% solution of streptomycin sulfate to a final volume of 4%. The suspension is stirred for an additional 10 min and the precipitated nucleic acid is pelleted by centrifugation in a Sorvall SS-34 rotor at 20,000 rpm for 10 min. Protein is precipitated from the supernatant by the dropwise addition with rapid stirring of a 100% saturated solution of $(NH_4)_2SO_4$ (neutralized to pH 7.5 with NH_4OH) to a final saturation of 45%. The suspension is stirred for an additional 10 min and the precipitate collected as above. The protein is gently resuspended in a volume equal to 1/20 that of the lysate in TEDG buffer (50 m*M* Tris–HCl, pH 7.5 at 4°, 1 m*M* EDTA, 5 m*M* DTT, and 15% glycerol). This fraction is then dialyzed (using narrow dialysis tubing) twice against 500 ml of TEDG for 1.5 hr each time. The crude ammonium sulfate-concentrated fraction is then frozen in 50-μl aliquots using liquid N_2. This preparation of extract was developed from the original protocols of W. Wickner *et al.*[11] and R. Wickner *et al.*[12]

Complementation Assays Using Ammonium Sulfate Receptor Fractions

If the ammonium sulfate fractions are prepared from certain *E. coli* strains, for example, BT1029 (*dnaB polA thyA endoA*), PC22 (*dnaC polA thyA*), or NY73 (*dnaG polA thyA*), they can be used in complementation assays to score the DnaB,[13,14] DnaC,[14,15] and DnaG[16,17] proteins, respec-

[11] W. Wickner, D. Brutlag, R. Schekman, and A. Kornberg, *Proc. Natl. Acad. Sci. USA* **69,** 965 (1972).

[12] R. B. Wickner, M. Wright, S. Wickner, and J. Hurwitz, *Proc. Natl. Acad. Sci. USA* **69,** 3233 (1972).

[13] M. Wright, S. Wickner, and J. Hurwitz, *Proc. Natl. Acad. Sci. USA* **70,** 3120 (1973).

[14] R. Schekman, J. H. Weiner, A. Weiner, and A. Kornberg, *J. Biol. Chem.* **250,** 5859 (1975).

[15] S. Wickner, I. Berkower, M. Wright, and J. Hurwitz, *Proc. Natl. Acad. Sci. USA* **70,** 2369 (1973).

[16] J.-P. Bouché, K. Zechel, and A. Kornberg, *J. Biol. Chem.* **250,** 5995 (1975).

[17] S. Wickner, M. Wright, and J. Hurwitz, *Proc. Natl. Acad. Sci. USA* **70,** 1613 (1973).

tively. In these assays, the temperature-sensitive gene products in the crude fraction are inactivated by heating, and the ability of the crude fraction to support ϕX174 ss(c) → RF DNA replication is complemented by the addition of fractions derived from wild-type strains. In this assay, the crude extract acts as a receptor fraction. If one wishes to prepare a wild-type extract, a *polA* strain should be used.

Assays contain in a 50-μl reaction volume 50 mM Tris–HCl, pH 7.8 at 30°, 10 mM $MgCl_2$, 10 mM DTT, 250 μg/ml bovine serum albumin (BSA), 20 μg/ml rifampicin, 100 μM each GTP, CTP, and UTP, 2 mM ATP, 40 μM dNTPs (one of which should be ^{3}H-labeled at about 100 to 200 cpm/pmol), ϕX174 ss(c) DNA (570 pmol as nucleotide), extract, and fractions for assay. Incubation is at 30°. The amount of extract required to saturate the assay must be determined for each preparation. This can be accomplished by using an excess of purified protein and titrating the extract. Saturation is generally between 4 and 6 μl of extract. Extracts prepared from *E. coli* strains PC22, BT1029, and NY73 must be heat-inactivated before use. The thawed tube of extract is heated at 39°, usually 6 min is more than sufficient—the extracts are often inactive without heating, but this must be determined—and then chilled on ice immediately before use in the assay. The kinetics of ϕX174 ss(c) → RF DNA replication supported by the extracts also varies with the preparation. Thus, the time of assay should be determined by performing a time course for each ammonium sulfate receptor extract.

An example of a complementation assay is shown in Fig. 1. Here, purified DnaC has been added to reaction mixtures containing a receptor extract prepared from the *E. coli* strain PC22. Incubation was at 30° for 10 min.

Overexpression Strains for Primosomal Proteins

Construction of Plasmids

The primosomal proteins are present at very low copy number in the cell. Thus, detailed biochemical analysis required that the genes encoding these proteins be identified and cloned and strains engineered that overproduced them. We have used the bacteriophage T7 transient expression system developed by Studier and colleagues[18] to accomplish this. In the pET series of overexpression plasmids, the gene in question is brought under the control of the bacteriophage T7 ϕ10 promoter and a strong ribosome binding site. T7 RNA polymerase is provided either through infection with

[18] W. F. Studier, A. H. Rosenberg, and J. J. Dunn, *Methods Enzymol.* **185,** 60 (1990).

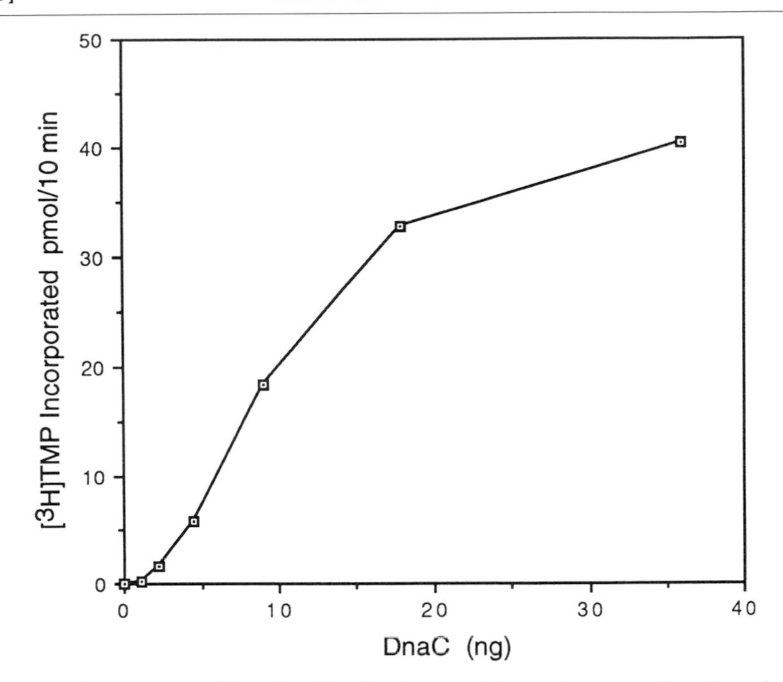

FIG. 1. Measurement of DnaC activity by the complementation assay. Reaction mixtures were as described in the text and contained heat-inactivated (39°, 6 min) ammonium sulfate receptor fraction (~100 μg) prepared from *E. coli* PC22 and the indicated amounts of DnaC (fraction 4, Table II). Incubation was for 10 min at 30°. Acid-precipitable radioactivity was then determined.

a rccombinant bacteriophage λ carrying the T7 RNA polymerase gene or by IPTG induction of the T7 RNA polymerase gene under the control of the *lac* repressor in a defective integrated λ bacteriophage. Proteins that are toxic to the cell are overexpressed using λ infection, whereas those that are tolerated by the cell are overexpressed using IPTG induction.

We have constructed pET3c derivatives carrying all seven primosomal protein genes. These plasmids are referred to as pET3c-*dnaB, -dnaC, -dnaG, -dnaT, -priA, -priB,* and *-priC.* In all cases, the ATG initiation codon of the primosomal protein gene has been spliced directly into the ATG initiator codon of the expression cassette. Of these seven plasmids, only pET3c-*dnaC* is toxic to the cell. Thus, DnaC must be overexpressed using λ infection.

Growth of Overexpression Strains

Either *E. coli* BL21(λDE3)[18] or *E. coli* BL21(λDE3)pLysS[18] carrying the plasmid of interest is grown at 37° in L-broth containing 0.5% glucose,

20 μg/ml thiamin, 500 μg/ml ampicillin, and 25 μg/ml chloramphenicol (pLysS strains only) with vigorous aeration (400 ml in a 2-liter baffled flask shaken at 300 rpm or in the fermentor) to OD_{600} of 0.5. IPTG is then added to 0.4 m*M* and induction is allowed to proceed for 1 to 3 hr. The cells are chilled, harvested, and resuspended in a small volume of Tris–sucrose.

For λ infection, *E. coli* BL21[18] carrying pET3c-*dnaC* is grown in L-broth containing 0.4% maltose, 1 m*M* $MgSO_4$, 20 μg/ml thiamin, and 500 μg/ml ampicillin to OD_{600} = 0.5. Glucose is then added to 0.5% and the $MgSO_4$ concentration is brought to 10 m*M*. The cells are allowed to double once and are then infected with λCE6[18] at multiplicity of infection (MOI) of 10. Two hours after infection the cells are chilled, harvested, and resuspended.

For best results with small quantities of cells, the lysate is prepared immediately without freezing the cells.

Purification of Primosomal Proteins from Overproducer Strains

DnaB

Purification of DnaB exploits the affinity of the protein for ATP-agarose, as discovered by Lanka *et al.*[19] Some steps are also based on the procedure of Arai *et al.*[20] One unit of DnaB activity represents 1 nmol of TMP incorporated into acid-insoluble product in 30 min at 30° using the complementation assay described above.

BL21(DE3)pLyS(pET3c-*dnaB*) was grown in the fermentor as described above. The culture was induced for 2 hr, chilled, harvested, resuspended at 50% w/w with Tris–sucrose, and frozen in liquid N_2. Approximately 500 ml of cell suspension (about 250 g of cells) was thawed overnight at 4°. The suspension was brought to 20 m*M* EDTA, 150 m*M* NaCl, 10 m*M* DTT, and 20 m*M* spermidine. The pH was adjusted to 8.0 by the addition of solid Tris base. The suspension was incubated on ice for 10 min and then centrifuged for 1 hr at 13,000 rpm in a Sorvall GSA rotor. The supernatant (fraction 1) was made 0.075% in Polymin P by the dropwise addition of a 1% solution with stirring. The mixture was stirred an additional 10 min and the precipitate was cleared by centrifugation for 30 min as above. Protein was precipitated by the addition of solid $(NH_4)_2SO_4$ to 50%, the pellet collected by centrifugation for 45 min as above, and resuspended in Buffer A (50 m*M* Tris–HCl, pH 7.5 at 4°, 5 m*M* DTT, 1 m*M* EDTA, and 20% glycerol) + 0.1 *M* NaCl to give fraction 2. Fraction 2 was dialyzed

[19] E. Lanka, C. Edelbluth, M. Schlicht, and H. Schuster, *J. Biol. Chem.* **253,** 5847 (1978).
[20] K.-I. Arai, S.-I. Yasuda, and A. Kornberg, *J. Biol. Chem.* **256,** 5247 (1981).

TABLE I
PURIFICATION OF DnaB FROM BL21(DE3)pLysS(pET3c-*dnaB*) CELLS

Step	Vol (ml)	Protein (mg)	Activity (units)	Yield (%)
1. Lysate	240	5520	ND[a]	
2. $(NH_4)_2SO_4$	60	2352	887,000	100
3. DEAE	24.6	197	311,000	35
4. ATP-agarose[b]	20.5	45	155,000	17.5

[a] Not determined.
[b] Only 50% of fraction 3 was chromatographed on ATP-agarose; however, the values for fraction 4 have been adjusted to reflect the yield expected if all of fraction 3 had been used.

overnight against 4 liters of Buffer A + 30 m*M* NaCl, diluted with Buffer A to a conductivity equal to Buffer A + 30 m*M* NaCl, and applied to a 200-ml DE-52 column that had been equilibrated previously with Buffer A + 30 m*M* NaCl. The column was then washed with 400 ml of equilibration buffer. DnaB was eluted with a 2-liter gradient of 30 to 200 m*M* NaCl in Buffer A. Active fractions (eluting at 125 m*M* NaCl) were pooled and protein was precipitated by the addition of $(NH_4)_2SO_4$ to 50%. The pellet was resuspended in Buffer A (fraction 3) and 50% of it was dialyzed overnight against Buffer A + 10 m*M* $MgCl_2$. The slight precipitate that formed was cleared by centrifugation and fraction 3 was loaded onto a 10-ml ATP-agarose (Sigma, A-6888) column equilibrated previously with Buffer A + 10 m*M* $MgCl_2$. The column was then washed successively with 20 ml of Buffer A + 10 m*M* $MgCl_2$ + 50 m*M* NaCl, followed by 20 ml of Buffer A + 10 m*M* $MgCl_2$ + 50 m*M* NaCl + 10 m*M* AMP. DnaB was eluted with Buffer A + 10 m*M* $MgCl_2$ + 50 m*M* NaCl + 10 m*M* ATP. Active fractions were pooled (fraction 4). This procedure (Table I) yields a preparation of DnaB that is >98% homogeneous for a single 50-kDa polypeptide (Fig. 2).

DnaC

DnaC activity was measured using the complementation assay as described above. Some of the steps in this purification protocol are based on the procedure of Kobori and Kornberg.[21] One unit represents the incorporation of 1 nmol TMP into acid-insoluble product in 30 min at 30°.

One and sixth-tenths liters of BL21(pET3c-*dnaC*) (4 × 400 ml in 2-liter baffled flasks) were grown at 37°. T7 RNA polymerase was delivered

[21] J. Kobori and A. Kornberg, *J. Biol. Chem.* **257,** 13,763 (1982).

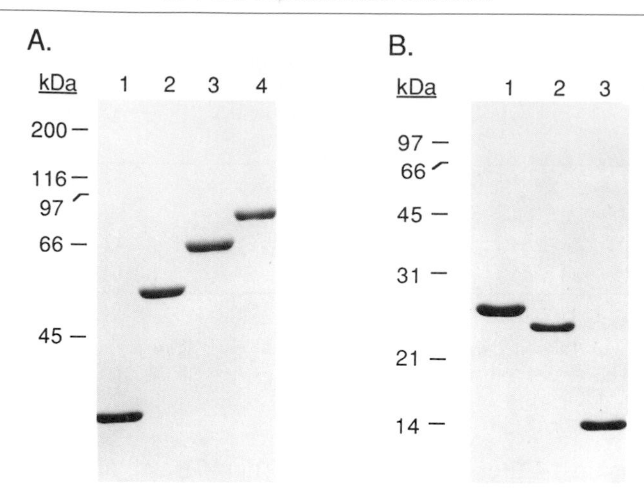

FIG. 2. SDS–PAGE analysis of the purified primosomal proteins. The primosomal proteins (5 μg) were analyzed by electrophoresis through either a 12.5% gel (A) or a 17% gel (B). The gels were stained with Coomassie Brilliant Blue and photographed. A, lane 1, DnaC; lane 2, DnaB; lane 3, DnaG; lane 4, PriA. B, lane 1, DnaT; lane 2, PriC; lane 3, PriB.

by λ infection as described above; induction was for 2 hr. The culture was chilled and the cells were harvested and resuspended in 32 ml of Tris–sucrose. Lysis was as described for the ammonium sulfate receptor fraction. Fraction 1 was treated with Polymin P as for DnaB and protein was precipitated by the addition of $(NH_4)_2SO_4$ to 50%. The pellet was collected by centrifugation and resuspended in Buffer B (50 m*M* Tris–HCl, pH 8.4 at 4°, 1 m*M* EDTA, 5 m*M* DTT, 0.1% Brij 58, and 20% glycerol) + 150 m*M* NaCl to give fraction 2. Fraction 2 was diluted (in 0.5-ml aliquots) with Buffer B to give a conductivity equal to Buffer B + 50 m*M* NaCl and loaded onto a 20-ml phosphocellulose column. The column was washed with 60 ml of equilibration buffer and eluted with a 200-ml linear gradient of 50 to 400 m*M* NaCl in Buffer B. Active fractions (eluting at 125 m*M* NaCl) were pooled (fraction 3), diluted with Buffer B to give a conductivity equal to Buffer B + 100 m*M* NaCl, and loaded onto a 7-ml hydroxylapatite column (containing CF11 cellulose powder at a ratio of HTP : CF11 = 60 : 17) previously equilibrated with Buffer B + 100 m*M* NaCl. The column was washed with 20 ml of equilibration buffer and eluted with a 70-ml linear gradient of 0 to 300 m*M* $(NH_4)_2SO_4$ in Buffer B + 100 m*M* NaCl. Active fractions, eluting at 90 m*M* $(NH_4)_2SO_4$, were pooled and dialyzed against 1 liter of Buffer B (at 30% glycerol) + 150 m*M* NaCl for 3 hr to give fraction 4. This procedure (Table II) yields a preparation of DnaC that is >99% homogeneous for a single 29-kDa polypeptide (Fig. 2).

TABLE II
PURIFICATION OF DnaC FROM BL21(pET3c-*dnaC*) CELLS

Step	Vol (ml)	Protein (mg)	Activity (units)	Yield (%)
1. Lysate	33.5	225	ND[a]	
2. $(NH_4)_2SO_4$	2.5	36	104,000	100
3. Phosphocellulose	32	14.1	90,000	86.5
4. Hydroxylapatite	7	7.9	56,000	54.2

[a] Not determined.

DnaG

DnaG activity can be measured in the complementation assay or because of its requirement for priming at the bacteriophage G4 origin of complementary strand replication.[16] Here, primase synthesizes a 29-nucleotide-long primer at a specific site on SSB-coated G4 ss(c) DNA (or on M13*Gori*1 ss(c) DNA—a recombinant M13 bacteriophage DNA carrying the bacteriophage G4 origin[22]) that is then elongated by the DNA Pol III HE.

Reaction mixtures (25 μl) for the M13*Gori*1 ss(c) $\rightarrow$ RF DNA replication assay containing 50 m*M* HEPES–KOH, pH 8.0, 10 m*M* magnesium acetate, 10 m*M* DTT, 10 μg/ml rifampicin, 200 μg/ml BSA, 1 m*M* ATP, 100 μM each CTP, UTP, and GTP, 80 μM dNTPs (one of which is ^{3}H-labeled at 100 to 300 cpm/pmol), M13*Gori*1 ss(c) DNA (420 pmol as nucleotide), 2.2 μg SSB, 170 ng DNA Pol III HE, and DnaG for assay were incubated at 30° for 5 min. The incorporation of [^{3}H]TTP into acid-insoluble material was then determined. One unit represents the incorporation of 1 nmol dTMP into acid-insoluble product in 30 min.

Twelve liters of BL21(DE3)(pET3c-*dnaG*) were grown at 37° and induced for 2.5 hr as described above. The cultures were chilled and the cells harvested and resuspended in 34 ml of Tris–sucrose. Lysis was as for DnaB except that lysozyme (0.02%) was included and the suspension was centrifuged at 100,000*g* for 1 hr to yield the lysate (fraction 1). Fraction 1 was treated with Polymin P (to 0.04%) and protein was precipitated with $(NH_4)_2SO_4$ (to 50%) as described above. The protein pellet was dissolved in Buffer C (50 m*M* Tris–HCl, pH 7.5 at 4°, 5 m*M* DTT, 1 m*M* EDTA, 10% sucrose) to give fraction 2. Fraction 2 was dialyzed against 4 liters of buffer C + 10 m*M* NaCl for 14 hr and then loaded onto a 60-ml Q-Sepharose Fast Flow column that had been equilibrated previously with Buffer C + 10 m*M* NaCl. The column was washed with 120 ml of equilibration buffer

[22] J. Kaguni and D. S. Ray, *J. Mol. Bio.* **135,** 863 (1979).

TABLE III
PURIFICATION OF DnaG FROM BL21(DE3)(pET3c-*dnaG*) CELLS

Step	Vol (ml)	Protein (mg)	Activity (units)	Yield (%)
1. Lysate	51	1712	ND[a]	
2. $(NH_4)_2SO_4$	15	727	4.8×10^6	100
3. Q-Sepharose	60	205	1.6×10^6	33
4. Hydroxylapatite	50	132	1.5×10^6	31
5. Heparin-agarose	25	63	1.9×10^6	40

[a] Not determined.

and protein was then eluted with a 600-ml linear gradient of 10 to 500 m*M* NaCl in Buffer C. Primase (fraction 3) eluted at 150 m*M* NaCl. Fraction 3 was then loaded onto a 34-ml hydroxylapatite column that had been equilibrated previously with Buffer C. The column was washed with 70 ml of Buffer C, and protein was eluted with a 340-ml linear gradient of 0 to 500 m*M* $(NH_4)_2SO_4$ in Buffer C. Primase (fraction 4) eluted at 125 m*M* $(NH_4)_2SO_4$. Fraction 4 was dialyzed against 6 liters of Buffer C containing 10 m*M* NaCl for 14 hr and loaded onto a 24-ml heparin-agarose column that had been equilibrated previously with Buffer C + 10 m*M* NaCl. The column was washed with 50 ml of equilibration buffer, and protein was eluted with a 240-ml linear gradient of 10 to 500 m*M* NaCl in Buffer C. Primase (fraction 5) eluted at 125 m*M* NaCl. This procedure (Table III), which was described by Tougu *et al.*,[23] yielded a preparation of DnaG that was >95% homogeneous for a single 64-kDa polypeptide (Fig. 2).

DnaT

DnaT activity was measured in the ϕX174 ss(c) → RF DNA replication assay reconstituted with purified proteins as described below. One unit represents the incorporation of 1 nmol dTMP into acid-insoluble product in 30 min at 30°.

Three liters of BL21(DE3)(pET3c-*dnaT*) were grown at 37°, induced with IPTG for 2 hr as described above, harvested, and resuspended in 50 ml of Tris–sucrose. Lysis and the preparation of fraction 2 was as for the ammonium sulfate receptor fraction. Precipitated protein was dissolved in Buffer A and dialyzed overnight against 2 liters of Buffer A + 30 m*M* NaCl. Roughly 50% of the DnaT activity precipitated at this stage, and whereas the precipitated activity could be redissolved in buffers containing

[23] K. Tougu, H. Peng, and K. J. Marians, *J. Biol. Chem.* **269,** 4675 (1994).

TABLE IV
PURIFICATION OF DnaT FROM BL21(DE3)(pET3c-*dnaT*) CELLS

Step	Vol (ml)	Protein (mg)	Activity (units)	Yield (%)
1. Lysate	46	524	ND[a]	
2. $(NH_4)_2SO_4$	16	265	53,000	100
3. DEAE	6.5	9	10,800	20.4
4. Phosphocellulose	6	2.4	4070	7.7
5. KP_i precipitate	0.6	2.0	5000	9.4

[a] Not determined.

300 m*M* NaCl, we did not attempt to recover it. Fraction 2 was applied to a 25-ml DE-52 column equilibrated previously with Buffer A + 30 m*M* NaCl. The column was washed with 75 ml of equilibration buffer and then eluted with a 250-ml linear gradient of 30 to 200 m*M* NaCl in Buffer A. Active fractions (eluting at 100 m*M* NaCl) were pooled, protein was precipitated with $(NH_4)_2SO_4$, resuspended in Buffer D (30 m*M* KP_i, pH 6.5, 1 m*M* DTT, 1 m*M* EDTA, 20% glycerol) and dialyzed against 1 liter of the same buffer overnight. This material (fraction 3) was then applied to a 4-ml phosphocellulose column equilibrated previously with Buffer D. The column was washed with 12 ml of equilibration buffer and eluted with a 40-ml linear gradient of 0 to 0.5 *M* KCl in Buffer D. Active fractions, eluting at 160 m*M* KCl, were pooled (fraction 4) and dialyzed against 500 ml of Buffer E (1 *M* KP_i, pH 7.5, 1 m*M* DTT, 1 m*M* EDTA, 20% glycerol) overnight. Precipitated DnaT was collected by centrifugation and redissolved in Buffer A (30% glycerol) + 150 m*M* NaCl to give fraction 5. This isoelectric precipitation step was originally developed by Roger McMacken (personal communication). This procedure (Table IV) yielded a preparation of DnaT that was >98% homogeneous for a single 22-kDa polypeptide (Fig. 2).

PriA

PriA activity was measured in the ϕX174 ss(c) → RF DNA replication assay reconstituted with purified proteins as described below. One unit represents the incorporation of 1 nmol of dTMP into acid-insoluble product in 30 min at 30°.

One liter of BL21(DE3)pLysS(pET3c-*priA*) was grown at 37°, induced with IPTG for 3 hr as described above, harvested, and resuspended in 30 ml of Tris–sucrose. Lysis was as for the ammonium sulfate receptor fraction. Fraction 1 was treated with 0.29 g/ml $(NH_4)_2SO_4$, and the precipitate was

TABLE V
PURIFICATION OF PriA FROM BL21(DE3)pLysS(pET3c-*priA*) CELLS

Step	Vol (ml)	Protein (mg)	Activity (units)	Yield (%)
1. Lysate	24	86	734,000	100
2. $(NH_4)_2SO_4$	2.8	48	442,000	60.2
3. Bio-Rex 70	4.5	12.6	351,500	47.9
4. Sephacryl S-200	2.6	9.8	313,600	42.7

collected by centrifugation. The protein precipitate was resuspended in 2.5 ml of Buffer F (50 m*M* imidazole hydrochloride, pH 6.7, 5 m*M* DTT, 1 m*M* EDTA, 40 m*M* $(NH_4)_2SO_4$, and 10% sucrose) to give fraction 2. Fraction 2 was diluted with Buffer A to give a conductance equivalent to Buffer F + 25 m*M* NaCl and loaded onto a 4-ml Bio-Rex 70 column equilibrated with Buffer F + 25 m*M* NaCl. This column was washed with 12 ml of Buffer F + 100 m*M* NaCl and developed with a 40-ml linear gradient of 0.1 to 0.8 *M* NaCl in Buffer F. Active fractions (0.3 *M* NaCl) were pooled to give fraction 3. (Separation of PriA, PriB, and PriC by Bio-Rex column chromatography was described originally by Shlomai and Kornberg.[24]) The protein was concentrated by ammonium sulfate precipitation, resuspended in 0.2 ml of Buffer F + 300 m*M* NaCl, and filtered through a Sephacryl S-200 column (1.5 × 83 cm) equilibrated and developed with Buffer F + 300 m*M* NaCl. Active fractions were pooled and dialyzed against Buffer G (50 m*M* imidazole hydrochloride, pH 6.7, 5 m*M* DTT, 0.1 m*M* EDTA, 40 m*M* $(NH_4)_2SO_4$, and 40% glycerol) + 50 m*M* NaCl to give fraction 4. This procedure (Table V), which was described by Zavitz and Marians,[25] yielded a preparation of PriA that was >98% homogeneous for a single 81-kDa polypeptide (Fig. 2).

PriB

PriB activity was measured in the ϕX174 ss(c) → RF DNA replication reaction reconstituted with purified proteins as described below. One unit represents the incorporation of 1 nmol of dTMP into acid-insoluble product in 30 min at 30°.

Three liters of BL21(DE3)(pET3c-*priB*) were grown at 37°, induced with IPTG for 3 hr as described above, harvested, and resuspended in 20 ml of Tris–sucrose. Lysis was as for the ammonium sulfate receptor fraction.

[24] J. Shlomai and A. Kornberg, *J. Biol. Chem.* **255,** 6789 (1980).
[25] K. H. Zavitz and K. J. Marians, *J. Biol. Chem.* **267,** 6933 (1992).

TABLE VI
PURIFICATION OF PriB FROM BL21(DE3)pLysS(pET3c-*priB*) CELLS

Step	Vol (ml)	Protein (mg)	Activity (units)	Yield (%)
1. Lysate	22	103	0.7×10^6	
2. $(NH_4)_2SO_4$	2.5	63	1.3×10^6	100[a]
3. Bio-Rex 70	3.9	17	1.9×10^6	146
4. Sephacryl S-200	16	8	1.1×10^6	84.6
5. Hydroxylapatite	1.7	5.3	1×10^6	77

[a] Because activity determinations with the lysate were unreliable, the activity present in fraction 2 was taken as 100%.

Fraction 1 was treated with 0.29 g/ml solid $(NH_4)_2SO_4$ and the precipitate was collected by centrifugation. The protein precipitate was resuspended in 2.5 ml Buffer F to give fraction 2. Fraction 2 was diluted with Buffer F to give a conductance equivalent to Buffer F + 100 m*M* NaCl and loaded onto a 5-ml Bio-Rex 70 column equilibrated with Buffer F + 100 m*M* NaCl. This column was washed with equilibration buffer and developed with a 50-ml linear gradient of 0.1 to 1.0 *M* NaCl in Buffer F. Active fractions (0.4 *M* NaCl) were pooled to give fraction 3, concentrated by ammonium sulfate precipitation, resuspended in 0.6 ml of Buffer F, and filtered through a Sephacryl S-200 column (1.5 × 83 cm) equilibrated with Buffer F + 300 m*M* NaCl. Active fractions were pooled (fraction 4) and loaded directly onto a 1.5-ml hydroxylapatite column equilibrated with Buffer F + 300 m*M* NaCl. This column was washed with 5 ml of Buffer G (30% glycerol) + 200 m*M* NaCl, and developed with a 15-ml linear gradient of 0 to 300 m*M* KP_i, pH 6.8, in Buffer G (30% glycerol) + 300 m*M* NaCl. Active fractions (100 m*M* KP_i) were pooled and dialyzed against Buffer G (30% glycerol) to give fraction 5. This procedure (Table VI), which was described by Zavitz *et al.*,[26] yielded a preparation of PriB that was >98% homogeneous for a single 12-kDa polypeptide (Fig. 2).

PriC

PriC activity was measured in the ϕX174 ss(c) → RF DNA replication reaction reconstituted with purified proteins as described below. One unit represents the incorporation of 1 nmol of dTMP into acid-insoluble product in 30 min at 30°.

Three liters of BL21(DE3)pLysS(pET3c-*priC*) were grown at 37°, in-

[26] K. H. Zavitz, R. J. DiGate, and K. J. Marians, *J. Biol. Chem.* **266,** 13,988 (1991).

TABLE VII
PURIFICATION OF PriC FROM BL21(DE3)pLysS(pET3c-*priC*) CELLS

Step	Vol (ml)	Protein (mg)	Activity (units)	Yield[a] (%)
1. Lysate	20	44	132,000	
2. $(NH_4)_2SO_4$	3	27	466,500	
3. Bio-Rex 70	4.8	7	645,100	

[a] Yields could not be calculated because activity increased on purification, presumably because of the removal of inhibitors.

duced with IPTG as described above, harvested, and resuspended in 20 ml of Tris–sucrose. Lysis was as for the ammonium sulfate receptor fraction. Fraction 1 was treated with 0.29 g/ml solid $(NH_4)_2SO_4$ and the precipitate collected by centrifugation. The protein precipitate was resuspended in Buffer F to give fraction 2. Fraction 2 was diluted with Buffer F to give a

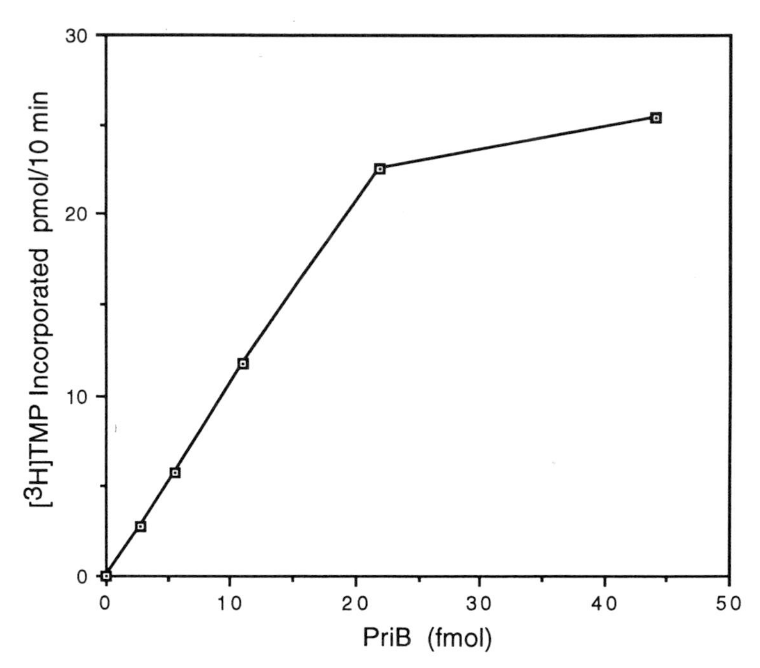

FIG. 3. Measurement of PriB activity by reconstitution of ϕX174 ss(c) → RF DNA replication. Reaction mixtures were as described in the text and contained the indicated amounts of PriB (fraction 5, Table VI). Incubation was for 10 min at 30°. Acid-precipitable radioactivity was then determined.

conductance equivalent to Buffer F + 100 mM NaCl and loaded onto a 3-ml Bio-Rex 70 column that had been equilibrated with Buffer F + 100 mM NaCl. The column was washed with 6 ml of equilibration buffer and eluted with a 30-ml linear gradient of 0.1 to 1.0 M NaCl in Buffer F. Active fractions (0.45 M NaCl) were pooled and dialyzed against Buffer G to give fraction 3. This procedure (Table VII), which was described by Zavitz *et al.*,[26] yielded a preparation of PriC that was >98% homogeneous for a single 21-kDa polypeptide (Fig. 2).

Reconstitution of ϕX174 ss(c) → RF DNA Replication with Purified Proteins

In this reaction, the primosome assembles at the PAS on SSB-coated ϕX174 ss(c) DNA. Primers are synthesized that can then be utilized by the DNA Pol III HE. In the assay, one measures DNA synthesis by determining the incorporation of [^{3}H]TMP into acid-insoluble product. Reaction mixtures (25 μl) contained 50 mM HEPES–KOH, pH 8.0 at 30°, 10 mM magnesium acetate, 10 mM DTT, 100 μg/ml BSA, 10 μg/ml rifampicin, 2 mM ATP, 100 μM each of CTP, GTP, and UTP, 40 μM each of dATP, dGTP, dCTP, and [^{3}H]TTP (200 to 300 cpm/pmol), 220 pmol of ϕX174 ss(c) DNA (as nucleotide), 1.3 μg of SSB, 370 fmol of DnaB (as hexamer), 3.3 pmol of DnaC, 230 fmol of DnaT (as trimer), 80 fmol of PriB (as dimer), 80 fmol of PriC, 1.2 pmol of DnaG, and 200 fmol of DNA Pol III HE (as $(\alpha\varepsilon\theta)_2\tau_2(\beta_4\gamma_2\delta\delta'\psi\chi)$), and were incubated at 30° for 10 min. Acid-insoluble ^{3}H (counts per minute) was then determined. As an example, a titration of PriB in the reaction is shown in Fig. 3.

Acknowledgments

I would like to thank the members of my laboratory who have developed the plasmids and purification procedures described here: Russell DiGate, Donita Dobson, Jenny Ng, Pearl Nurse, Killu Tougu, Carol Wu, and Kenton Zavitz. These studies were supported by NIH grants GM 34557 and GM 34558.

[41] Identification of Eukaryotic DNA Replication Proteins Using Simian Virus 40 *in Vitro* Replication System

By GEORGE S. BRUSH, THOMAS J. KELLY, and BRUCE STILLMAN

Introduction

Mechanistic studies of DNA replication are greatly facilitated by the availability of a cell-free system, which allows for the identification and characterization of the proteins that are involved in the process. At the present time, there is no *in vitro* DNA replication assay suitable for the direct analysis of chromosomal DNA replication in eukaryotes. Except in simple organisms such as yeast, the identification of replication origins and the proteins that bind to these sites has been extremely difficult. Furthermore, eukaryotic DNA replication is tightly controlled during the cell cycle, and clearly depends on the action of many regulatory proteins in addition to those directly involved in synthesizing DNA. One approach to circumvent these difficulties has been to investigate the replication of viral DNA within eukaryotic cells. Viruses have small, well-defined genomes and the replication of viral DNA is free of some of the normal cellular constraints. In spite of this simplicity, viral DNA replication serves as a good model for chromosomal DNA replication because it depends on the host replicative apparatus. Simian virus 40 (SV40) DNA replication requires only one viral protein, the large T antigen, and has proven to be an extremely useful tool for studying cellular DNA replication. The development of an *in vitro* SV40 DNA replication system[1] has led to the identification and characterization of many human replication proteins and has allowed for the isolation of homologous proteins from other eukaryotic species. This work has led to the realization that the basic replication machinery is conserved in all eukaryotes from yeast to man.

SV40 DNA replication takes place within the nucleus of a permissive primate cell, where T antigen binds to and locally unwinds the circular SV40 minichromosome at the origin. This allows the host replication proteins to initiate bidirectional DNA synthesis and complete the duplication of the entire DNA molecule in a semidiscontinuous manner.[2-4] SV40 DNA repli-

[1] J. J. Li and T. J. Kelly, *Proc. Natl. Acad. Sci. USA* **81,** 6973 (1984).

[2] M. D. Challberg and T. J. Kelly, *Ann. Rev. Biochem.* **58,** 671 (1989).

[3] B. Stillman, *Ann. Rev. Cell Biol.* **5,** 197 (1989).

[4] J. A. Borowiec, F. B. Dean, P. A. Bullock, and J. Hurwitz, *Cell* **60,** 181 (1990).

cation *in vitro* closely resembles the *in vivo* reaction and is dependent on the viral origin of replication, T antigen, and primate cell cytoplasmic extract. The cellular components necessary and sufficient for the replication of plasmid DNA containing the SV40 origin have been identified through fractionation of the cytoplasmic extract and reconstitution of activity. In purified form, these proteins are capable of catalyzing the entire replication reaction, as outlined in Fig. 1.[5-8] This has allowed for a detailed dissection of the various stages of replication, and a greater understanding of the process. Nonetheless, many questions remain unanswered regarding the mechanism of eukaryotic DNA replication. The cell-free system described here serves as a useful tool for investigating these problems.

Preparation of Replication Extracts from Human Cells

Although SV40 DNA replication takes place within the nucleus of the host cell, crude nuclear extract is ineffective in catalyzing *in vitro* SV40 DNA replication due to the presence of inhibitory factors. However, cytoplasmic extract prepared by hypotonic lysis contains a significant quantity of proteins that have leaked from the nucleus, and supports efficient cell-free SV40 DNA replication. The cytoplasmic extract has served as the basic source of the replication proteins that have been identified to date, and it is likely to contain other factors that affect replication. This is supported by the fact that replication with purified proteins is less efficient than replication with crude cytoplasmic extract.

The crude and partially fractionated systems have been used extensively to study the replication mechanism, and some aspects of the reaction are better studied with all components present rather than with the purified proteins alone. For example, the crude system has been particularly useful for monitoring the effects of a wide variety of factors on replication activity. In this manner, it has been shown that wild-type human p53 inhibits SV40 DNA replication whereas a mutant form does not.[9] Similarly, studies have shown that an inhibitor of cyclin-dependent kinases (p21) inhibits SV40

[5] D. H. Weinberg, K. L. Collins, P. Simancek, A. Russo, M. S. Wold, D. M. Virshup, and T. J. Kelly, *Proc. Natl. Acad. Sci. USA* **87,** 8692 (1990).
[6] T. Tsurimoto, T. Melendy, and B. Stillman, *Nature* **346,** 534 (1990).
[7] Y. Ishimi, A. Claude, P. Bullock, and J. Hurwitz, *J. Biol. Chem.* **263,** 19,723 (1988).
[8] S. Waga, G. Bauer, and B. Stillman, *J. Biol. Chem.* **269,** 10,923 (1994).
[9] P. N. Friedman, S. E. Kern, B. Vogelstein, and C. Prives, *Proc. Natl. Acad. Sci. USA* **87,** 9275 (1990).

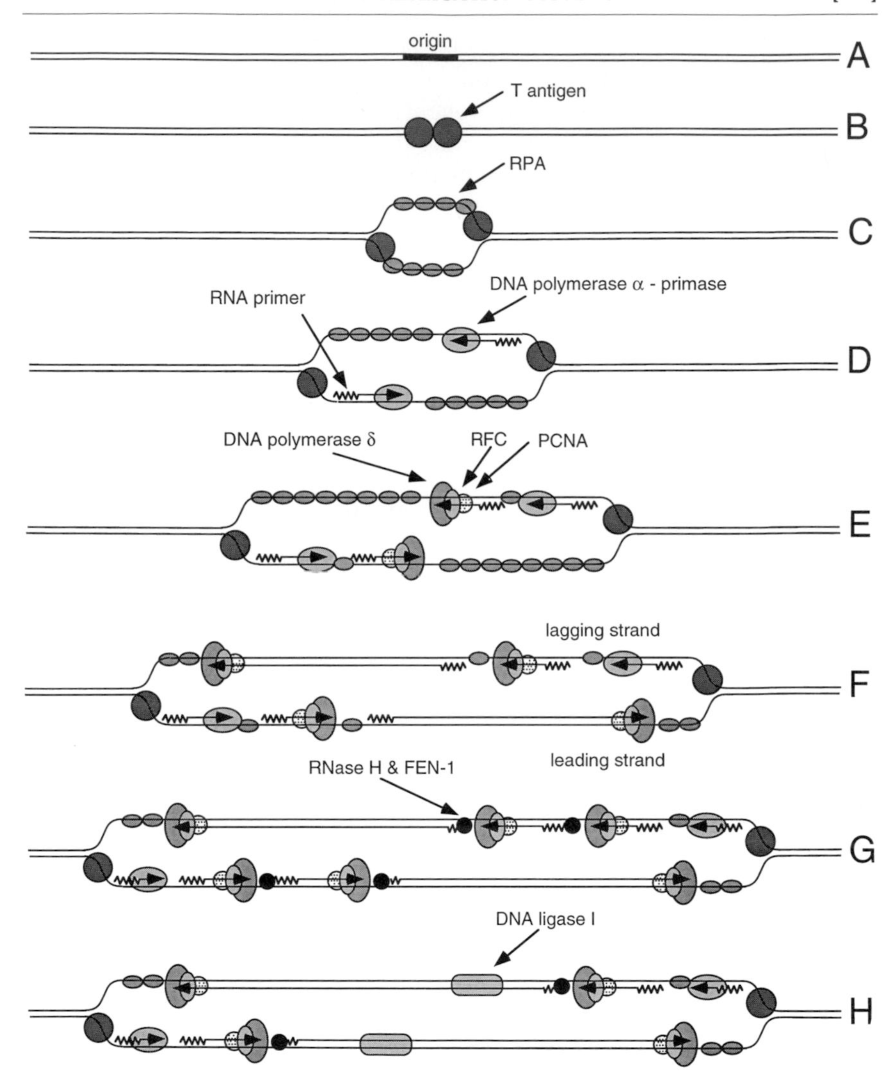

FIG. 1. Pathway for the replication of SV40 DNA (see text for details).

DNA replication in the crude system by interacting with proliferating cell nuclear antigen (PCNA).[10,11]

The crude system has also been used to explore cell-cycle control of DNA replication. For instance, whereas extracts from cells at S phase

[10] S. Waga, G. J. Hannon, D. Beach, and B. Stillman, *Nature* **369,** 574 (1994).

[11] H. Flores-Rozas, Z. Kelman, F. B. Dean, Z.-Q. Pan, J. W. Harper, S. J. Elledge, M. O'Donnell, and J. Hurwitz, *Proc. Natl. Acad. Sci. USA* **91,** 8655 (1994).

are replication competent, G_1 extracts do not support replication unless supplemented with protein phosphatase 2Ac (PP2Ac) or cdc2 kinase.[12,13] In addition, the S-phase phosphorylation of replication protein A (RPA)[14] has been investigated. Phosphorylation of RPA occurs in an SV40 DNA replication-dependent manner[15] consistent with the timing during the cell cycle. Recent studies employing immunodepletion of the cytoplasmic extract have shown that DNA-activated protein kinase is required for this phosphorylation event.[16]

The utility of the unpurified system is not limited to the measurement of effects on overall DNA replication. For example, several studies have focused on initiation, which involves only a subset of the identified replication proteins. Such investigation has demonstrated that DNA replication in the cell-free system initiates at the SV40 origin,[17,18] and has defined the regions within the origin that are important for activity.[19] In addition, pulse–chase experiments with a modified cell-free system[20] and inhibition studies with aphidicolin[21] have been employed to monitor the behavior of the initiation complex during the early stages of replication. Other work has centered on the activities of individual components. In particular, studies using the unfractionated cell-free system have demonstrated that the phosphorylation state of T antigen has a profound effect on its ability to initiate replication.[12,22]

In addition to processes directly involved in DNA replication, related functions have been studied with the crude and partially fractionated assays. For example, *in vitro* replication has been demonstrated with SV40 chromatin isolated from infected cells,[23] and the effects of chromatin assembly on *in vitro* replication have been investigated.[24,25] Further studies using a partially fractionated system have demonstrated that parental nucleosomes

[12] D. M. Virshup, M. G. Kauffman, and T. J. Kelly, *EMBO J.* **8,** 3891 (1989).
[13] G. D'Urso, R. L. Marraccino, D. R. Marshak, and J. M. Roberts, *Science* **250,** 786 (1990).
[14] S. Din, S. J. Brill, M. P. Fairman, and B. Stillman, *Genes Dev.* **4,** 968 (1990).
[15] R. Fotedar and J. M. Roberts, *EMBO J.* **11,** 2177 (1992).
[16] G. S. Brush, C. W. Anderson, and T. J. Kelly, *Proc. Natl. Acad. Sci. USA* **91,** 12,520 (1994).
[17] J. J. Li and T. J. Kelly, *Mol. Cell. Biol.* **5,** 1238 (1985).
[18] B. W. Stillman and Y. Gluzman, *Mol. Cell. Biol.* **5,** 2051 (1985).
[19] J. J. Li, K. W. C. Peden, R. A. F. Dixon, and T. Kelly, *Mol. Cell. Biol.* **6,** 1117 (1986).
[20] P. A. Bullock, Y. S. Seo, and J. Hurwitz, *Mol. Cell. Biol.* **11,** 2350 (1991).
[21] R. S. Decker, M. Yamaguchi, R. Possenti, and M. L. DePamphilis, *Mol. Cell. Biol.* **6,** 3815 (1986).
[22] D. McVey, L. Brizuela, I. Mohr, D. R. Marshak, Y. Gluzman, and D. Beach, *Nature* **341,** 503 (1989).
[23] R. S. Decker, M. Yamaguchi, R. Possenti, M. K. Bradley, and M. L. DePamphilis, *J. Biol. Chem.* **262,** 10,863 (1987).
[24] B. Stillman, *Cell* **45,** 555 (1986).
[25] Y. Ishimi, *J. Biol. Chem.* **267,** 10,910 (1992).

stay associated with replicating DNA.[26] Other areas of research have included the effects of transcription factors[27,28] and ultraviolet radiation[29] on replication activity.

Methods

A description of crude cytoplasmic preparation from human cells is provided below, along with the conditions for assaying DNA replication. Extracts prepared in this manner contain all the cellular components necessary for the complete replication of SV40 DNA, and can be fractionated or used directly to investigate the mechanism of DNA replication.

Preparation of Cytoplasmic Extract[17,30]

HeLa cells are grown in suspension at 37° in Eagle's minimal essential medium supplemented with 5% horse serum. At mid-log phase (4 to 5 $\times$ 10^5 cells/ml), the cells are harvested by centrifugation at 500 to 1000*g* for 3 to 5 min at 4°. The supernatant is discarded and the pellet is washed first with ice-cold isotonic buffer [20 m*M* HEPES, pH 7.8, 1.5 m*M* $MgCl_2$, 5 m*M* KCl, 250 m*M* sucrose, 1 m*M* dithiothreitol (DTT), 0.1 m*M* phenylmethylsulfonyl fluoride (PMSF)] and then with ice-cold hypotonic buffer (isotonic buffer without sucrose). These washes are routinely carried out with 1/10 culture volume, and the resuspended cells are centrifuged at 500 to 1000*g* for 3 to 5 min at 4°. The final pellet is resuspended in ice-cold hypotonic buffer to give a cell density of 7 $\times$ 10^7 to 1 $\times$ 10^8 cells/ml, and the suspension is kept on ice for 15 min. The swollen cells are lysed with 3 to 7 strokes of a tightly fitting Dounce homogenizer, and the lysate is kept on ice for another 30 to 60 min. Following this incubation, the nuclei are pelleted by centrifugation at 1700*g* for 10 min at 4°. The supernatant is clarified by centrifugation at 12,000*g* for 10 min at 4°, and the resulting cytoplasmic extract is dripped directly into liquid nitrogen and stored in frozen bead form at −70°. The pelleted nuclei, which are also frozen with liquid nitrogen and stored at −70°, can be used as a source of various replication factors (see Purified Proteins section). The protein concentration of the crude cytoplasmic extract prepared in this manner will be ~10 mg/

[26] S. K. Randall and T. J. Kelly, *J. Biol. Chem.* **267,** 14,259 (1992).
[27] L. Cheng and T. J. Kelly, *Cell* **59,** 541 (1989).
[28] L. Cheng, J. L. Workman, R. E. Kingston, and T. J. Kelly, *Proc. Natl. Acad. Sci. USA* **89,** 589 (1992).
[29] M. P. Carty, M. Zernik-Kobak, S. McGrath, and K. Dixon, *EMBO J.* **13,** 2114 (1994).
[30] M. S. Wold, D. H. Weinberg, D. M. Virshup, J. J. Li, and T. J. Kelly, *J. Biol. Chem.* **264,** 2801 (1989).

ml. Note that extracts accumulate insoluble material inhibitory to replication after freeze–thawing, and should be clarified with a 5- to 10-min full-speed spin in an Eppendorf centrifuge at 4° prior to use.

SV40 DNA Replication Assay[1,18,31]

In vitro SV40 DNA replication depends on the viral T antigen, SV40 origin-containing plasmid DNA, and cellular proteins (provided by the cytoplasmic extract). Standard 25-μl reactions contain the following components: 30 m*M* HEPES, pH 7.5, 7 m*M* $MgCl_2$, 50 μM [α-^{32}P]dCTP (2.5 μCi), 100 μM each dATP, dGTP, and dTTP, 200 μM each CTP, GTP, and UTP, 4 m*M* ATP, 40 m*M* creatine phosphate, 2.5 μg creatine kinase, 15 m*M* potassium phosphate, pH 7.5 (optional), 50 to 100 ng supercoiled plasmid containing the SV40 origin (e.g., pUC · HSO[32] or pSV011[33]), ~100 μg cytoplasmic extract, and 0.15 to 1.0 μg T antigen (depending on its source—see Purified Proteins, T Antigen). In measuring SV40 DNA replication *in vitro,* it is important to test for T antigen and origin dependence by including one reaction lacking T antigen and another containing a plasmid with a mutant origin (e.g., pUC · HSOd4[32,34]). Reaction mixtures are assembled on ice, incubated at 37° for 2 hr, and terminated with 25 μl of 2% SDS/50 m*M* EDTA/0.2% proteinase K. Each sample is then incubated for another 30 min at 37° to effect proteolysis. At this stage, aliquots (usually 1/5 of the sample) are analyzed for trichloroacetic acid insoluble radioactivity, which gives a measure of total DNA synthesis. The remainder of each sample is then used to determine the nature of the replication products. Samples are routinely processed by ethanol precipitation, RNAse treatment, and electrophoresis through a 1.0 to 1.2% neutral agarose gel. Total DNA is visualized under UV light after ethidium bromide staining. The gel is then dried and the reaction products are detected by autoradiography.

In some cases, it may be necessary to determine if the radioactive products have been formed from *bona fide* DNA replication rather than a DNA repair process. To test this, the samples are treated with *Dpn*I prior to gel electrophoresis.[35] This enzyme will digest unreplicated and repaired plasmids, since these DNA molecules still contain fully methylated *Dpn*I recognition sites. Replicated plasmids will be hemimethylated at all the *Dpn*I sites, rendering them resistant to cleavage. To digest effectively the

[31] M. S. Wold and T. Kelly, *Proc. Natl. Acad. Sci. USA* **85,** 2523 (1988).

[32] M. S. Wold, J. J. Li, and T. J. Kelly, *Proc. Natl. Acad. Sci. USA* **84,** 3643 (1987).

[33] G. Prelich and B. Stillman, *Cell* **53,** 117 (1988).

[34] Y. Gluzman, R. J. Frisque, and J. Sambrook, *Cold Spring Harbor Symp. Quant. Biol.* **44,** 293 (1980).

[35] K. W. C. Peden, J. M. Pipas, S. Pearson-White, and D. Nathans, *Science* **209,** 1392 (1980).

DNA with *Dpn*I (or with a linearizing restriction enzyme), the samples should be phenol/chloroform extracted prior to ethanol precipitation. Furthermore, an excess of restriction enzyme should be used (5 units/reaction).

In addition to the above methods, the size of nascent strands can be determined by employing alkaline rather than neutral agarose gel electrophoresis. The denatured replication products will separate according to size and can be detected by autoradiography.

General Approach to Subcellular Fractionation

Complementation Assay

The strategy for purifying the cellular proteins required for DNA replication is based on the reconstitution of T antigen- and origin-dependent *in vitro* DNA replication following fractionation. The crude extract is initially resolved into two fractions [termed cellular fraction I (CF I or CF I′) and cellular fraction II (CF II)] by chromatography over phosphocellulose.[32,36] CF I contains all of the cellular proteins needed for the presynthesis reaction, whereas CF II contains the polymerases and replication factor C (RFC). Neither fraction supports SV40 DNA replication individually; however, when the two fractions are mixed, activity is restored. CF I can be further fractionated by this method, with CF II required for the reconstitution of replication activity, and vice versa. This method allows for detection of both essential and stimulatory factors.

Complementation assays are identical to crude cell-free replication assays (see Preparation of Replication Extracts from Human Cells section), except that the crude extract is replaced by cellular protein fractions, and all reactions are supplemented with topoisomerase I.[31] Therefore, a standard complementation reaction contains 30 m*M* HEPES, pH 7.5, 7 m*M* $MgCl_2$, 50 μM [α-^{32}P]dCTP (2.5 μCi), 100 μM each dATP, dGTP, and dTTP, 200 μM each CTP, GTP, and UTP, 4 m*M* ATP, 40 m*M* creatine phosphate, 2.5 μg creatine kinase, 15 m*M* potassium phosphate, pH 7.5 (optional), 50 to 100 ng pUC·HSO, 100 ng topoisomerase I (Promega), the indicated cellular protein fractions, and 0.15 to 1.0 μg T antigen. A typical reconstitution reaction with CF I and CF II contains 10 to 20 μg of each fraction. Because the cellular protein components added to a complementation reaction will depend on the replication protein that is being purified, further

[36] G. Prelich, M. Kostura, D. R. Marshak, M. B. Mathews, and B. Stillman, *Nature* **326,** 471 (1987).

details regarding the cellular fractions used in these assays are provided below (see Purified Proteins section).

Purification Tree

The general purification scheme as originally carried out by reconstitution of replication activity is shown in Fig. 2. Although the cytoplasmic extract is capable of supporting cell-free SV40 DNA replication, some proteins are more efficiently prepared from nuclear extract or heterologous expression systems. Nonetheless, the cytoplasmic extract serves as a major source of replication proteins, and a description of the early stages in the purification of the cytoplasmic extract is provided below. Further details on the purification of individual proteins are provided in the Purified Proteins section, including procedures for preparation from nuclear extract and heterologous systems. Similar techniques can be used to identify other proteins that are involved in replication.

Note that complete fractionation of the cytoplasmic extract is not necessary for the study of individual components. This is exemplified by the complementation assays themselves, which make use of crude cellular fractions in combination with purified proteins to reconstitute replication activity. Because these systems contain factors other than those that are abso-

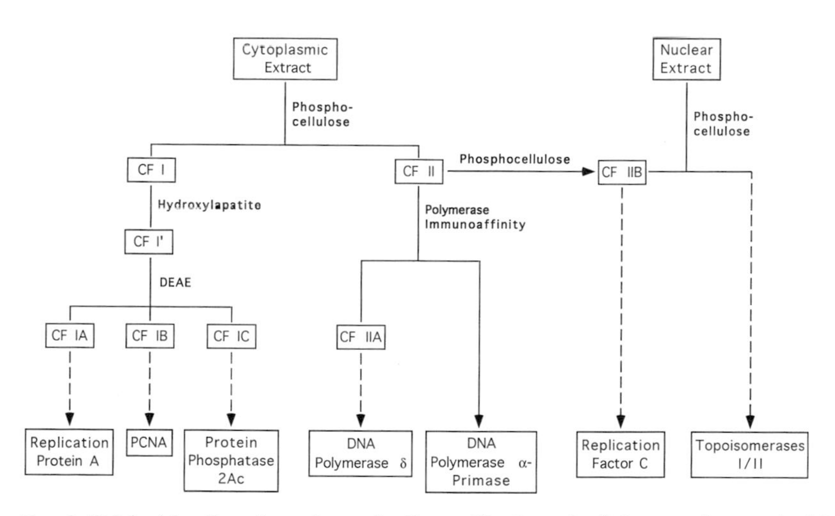

FIG. 2. Original fractionation scheme for the purification of cellular proteins required for *in vitro* SV40 DNA replication. Three additional proteins (RNase H, FEN-1, and DNA ligase I) are required for the processing of Okazaki fragments (see text for details). Dashed lines represent multiple fractionation steps.

lutely essential for replication, they support more efficient replication *in vitro* and may be preferable for certain studies.

Initial Fractionation[30,37]

Buffers

Buffer F: 30 m*M* HEPES, pH 7.8, 0.25% *myo*-inositol, 1 m*M* DTT
Buffer P: 80 m*M* potassium phosphate, pH 7.2, 0.25 m*M* EDTA, 0.25% *myo*-inositol, 1 m*M* DTT
Buffer H: 30 m*M* HEPES, pH 7.8, 0.25 m*M* EDTA, 0.50% *myo*-inositol, 0.01% Nonidet P-40 (NP-40), 1 m*M* DTT.

CF I (CF I′) and CF II Preparation. The first step in the fractionation of the cytoplasmic extract results in the preparation of CF I (CF I′) and CF II. Frozen cytoplasmic extract is thawed and PMSF is added to a final concentration of 0.2 mg/ml (all steps in this and subsequent fractionations are carried out at 4°). After clarification of the extract at 10,000*g* for 30 min, KCl is added to a final concentration of 100 m*M* from a 3 *M* stock. The extract is then passed over a phosphocellulose column (5 mg protein/ml matrix) equilibrated in Buffer F containing 100 m*M* KCl. To avoid flow problems due to aggregation, it is advisable to prepare an S100 fraction prior to chromatography by centrifuging the extract at 100,000*g* for 1 hr. Otherwise, a small precolumn of Sepharose CL-4B should be placed upstream of the phosphocellulose column. After the sample has been loaded (0.2 to 0.4 column volumes/hour), the column is washed with 3 column volumes of Buffer F containing 100 m*M* KCl. Bound protein is then eluted with Buffer P containing 500 m*M* KCl. The flow-through fraction is designated CF I and the eluted fraction CF II.

Following phosphocellulose chromatography, CF I is concentrated by hydroxylapatite chromatography, yielding CF I′. The above procedure can be modified so that the concentration step is included in the initial fractionation. In this setup, the phosphocellulose column is upstream of a hydroxylapatite column of similar dimensions. After the loading and washing steps, the columns are separated. Fraction CF I′ is eluted from the hydroxylapatite column with Buffer F containing 70 m*M* potassium phosphate and 100 m*M* KCl. CF II is eluted from the phosphocellulose column as described previously. Following chromatography, CF I′ and an aliquot of CF II are dialyzed versus Buffer H containing 15 m*M* KCl. These dialyzed samples can be used in complementation assays. Dialyzed CF I′ and undialyzed CF II are used for further purification of the replication proteins (see Purified Proteins section).

[37] M. P. Fairman and B. Stillman, *EMBO J.* **7,** 1211 (1988).

CF IA, CF IB, and CF IC Preparation. CF I′ is further fractionated by DEAE chromatography, resulting in the preparation of fractions CF IA, CF IB, and CF IC. Dialyzed CF I′ is loaded onto a DEAE-Sephacel column (5 to 10 mg protein/ml matrix) equilibrated in Buffer H containing 15 m*M* KCl, and the column is then washed with 1/3 column volume of equilibration buffer followed by 2 column volumes of Buffer H containing 100 m*M* KCl. The fractions of interest are then eluted with a linear gradient of 100 to 350 m*M* KCl in Buffer H (~10 column volumes). A flow rate of ~1/3 column volume/hour is routinely used for loading and elution. CF IA, CF IB, and CF IC elute at approximately 150, 250, and 200 m*M* KCl, respectively. These fractions are used for further purification as well as for complementation assays.

Purified Proteins

T Antigen

The phosphorylation state and resulting specific activity of T antigen depends on the source from which it is purified.[38,39] The most commonly used preparations are derived from recombinant adenovirus-infected human cells and from recombinant baculovirus-infected insect cells. While the enzyme expressed in insect cells is more active, HeLa T antigen can be activated by the catalytic subunit of protein phosphatase 2A (see Purified Proteins, PP2Ac section). The preparation of T antigen from these two sources is reviewed below.

Assays

DNA replication. The cell-free SV40 replication system serves as an assay for T antigen function within the context of DNA replication. This includes both the crude and fractionated systems, as previously outlined.

Initiation. The proteins involved in the initiation of DNA synthesis in the SV40 system are T antigen, RPA, DNA polymerase α–primase, and topoisomerase. A modified complementation assay is used to monitor this reaction. The following protein components are added to a standard complementation reaction mix: ~400 ng RPA (or 25 to 30 μg CF I′); 0.1 to 2.5 units DNA polymerase α–primase; and 0.35 to 1.0 μg T antigen (plus 100 to 200 ng PP2Ac, if necessary). Reactions are carried out under the normal conditions, and reaction

[38] C. Prives, *Cell* **61,** 735 (1990).
[39] E. Fanning, *J. Virol.* **66,** 1289 (1992).

products are analyzed by agarose gel electrophoresis and autoradiography. The products are either heat denatured prior to electrophoresis or subjected to alkaline gel electrophoresis so that the size of the products can be determined.

Presynthesis (unwinding). Two methods are used to monitor the unwinding of SV40 origin-containing DNA by T antigen:

1. *Supercoiled DNA.*[32,40] Reactions (25 μl) contain the following components: 30 m*M* HEPES, pH 7.5, 7 m*M* $MgCl_2$, 15 m*M* potassium phosphate, pH 7.5, 0.5 m*M* DTT, 0.05% NP-40, 4 m*M* ATP, 40 m*M* creatine phosphate, 2.5 μg creatine kinase, and 80 to 100 ng supercoiled DNA containing the SV40 origin. A single-stranded DNA-binding protein must be provided: RPA (~250 ng), CF I (90 μg), or *Escherichia coli* SSB (~600 ng) can be used. Reactions are incubated for 10 to 20 min at 37° to allow for relaxation of plasmid DNA by topoisomerase I. T antigen is then added and incubation at 37° is continued for another 30 min. If the T antigen is derived from human cells, up to 2 μg may be required for significant unwinding to occur. PP2Ac, which is present in CF I or can be added separately (100 to 200 ng), will significantly stimulate this reaction, thereby reducing the amount of T antigen required. It may be useful to preincubate the T antigen with phosphatase prior to the reaction. Because the T antigen purified from insect cells is highly active, 0.5 to 1 μg of this protein should be sufficient. Reactions are terminated with 25 μl of 2% SDS/50 m*M* EDTA/0.2% proteinase K and incubated for another 30 min at 37°. The samples are then ethanol precipitated and electrophoresed through a 1.0 to 1.2% neutral agarose gel. The DNA is visualized under UV light after ethidium bromide staining. For more quantitative measurement, the DNA can be detected by Southern blot hybridization. Unwound product is detected as a fast-migrating form.
2. *Linear DNA.*[41,42] Reactions (10 μl) contain the following components: 30 m*M* HEPES, pH 7.5, 7 m*M* $MgCl_2$, 15 m*M* potassium phosphate, pH 7.5, 0.5 m*M* DTT, 0.05% NP-40, 4 m*M* ATP, 40 m*M* creatine phosphate, 1 μg creatine kinase, 0.4 ng pUC.HSO (origin-containing) DNA, which has been linearized with *Hin*dIII, ^{32}P end labeled with radioactive deoxynucleoside triphosphates and DNA polymerase, and digested with *Bam*HI, and 200 ng RPA (or 300 ng

[40] F. B. Dean, P. Bullock, Y. Murakami, C. R. Wobbe, L. Weissbach, and J. Hurwitz, *Proc. Natl. Acad. Sci. USA* **84,** 16 (1987).

[41] D. M. Virshup and T. J. Kelly, *Proc. Natl. Acad. Sci. USA* **86,** 3584 (1989).

[42] G. S. Goetz, F. B. Dean, J. Hurwitz, and S. W. Matson, *J. Biol. Chem.* **263,** 383 (1988).

E. coli SSB). As before, T antigen from either source can be used: 200 ng T antigen from human cells plus 100 to 200 ng PP2Ac or 50 ng T antigen from insect cells. Reactions are incubated for 30 min at 37° and terminated with 10 μl of 2% SDS/50 m*M* EDTA/0.2% proteinase K. After further incubations at 37° for 30 min and 65° for 5 min, the samples are adjusted to 10% glycerol and 0.02% bromphenol blue, and electrophoresed through an 8% polyacrylamide gel in 89 m*M* Tris–base/89 m*M* boric acid/2 m*M* EDTA. The gel is dried on DEAE-paper and the DNA is detected by autoradiography. In this system, single-stranded DNA migrates slower than duplex DNA of the same length.

Note that the assays employing purified proteins (e.g., initiation and unwinding) can be modified to assay the other components of the reaction. For example, samples can be assayed for RPA by omitting this protein from the reaction mixture while including T antigen.

Purification

BUFFERS

PBS+Ca^{2+}+Mg^{2+}: phosphate-buffered saline (137 m*M* NaCl, 3 m*M* KCl, 6.4 m*M* Na_2HPO_4, 1.5 m*M* KH_2PO_4, 0.5 m*M* EDTA, pH 7.0) containing 0.9 m*M* $CaCl_2$ and 0.5 m*M* $MgCl_2$

Lysis buffer: 20 m*M* Tris–HCl, pH 8.0, 200 m*M* LiCl, 1 m*M* EDTA, 0.5% NP-40, 0.2 m*M* DTT, 0.1 m*M* PMSF

Wash I: 20 m*M* Tris–HCl, pH 8.0, 500 m*M* NaCl, 1 m*M* EDTA, 10% glycerol, 1% NP-40, 1 m*M* DTT, 0.1 m*M* PMSF

Wash II: 20 m*M* Tris–HCl, pH 8.0, 500 m*M* NaCl, 1 m*M* EDTA, 10% glycerol, 1 m*M* DTT, 0.1 m*M* PMSF

Wash III: 20 m*M* Tris–HCl, pH 9.0, 500 m*M* NaCl, 1 m*M* EDTA, 10% glycerol, 1 m*M* DTT, 0.1 m*M* PMSF

Elution buffer: 20 m*M* Tris–HCl, pH 11, 500 m*M* NaCl, 1 m*M* EDTA, 10% glycerol, 1 m*M* DTT, 0.1 m*M* PMSF

Adjustment buffer: 500 m*M* Tris–HCl, pH 7.0, 1 m*M* EDTA, 10% glycerol, 1 m*M* DTT, 0.1 m*M* PMSF

Rinse buffer: 100 m*M* Tris–HCl, pH 7.4, 150 m*M* NaCl, 1 m*M* EDTA, 0.02% NaN_3

Column storage buffer: 10 m*M* Tris–HCl, pH 8.0, 150 m*M* NaCl, 1 m*M* EDTA, 0.02% NaN_3

T antigen storage buffer: 10 m*M* Tris–HCl, pH 8.0, 50 m*M* NaCl, 1 m*M* EDTA, 50% glycerol, 1 m*M* DTT.

ADENOVIRUS RECOMBINANT INFECTED-CELL LYSATE.[30] Mid-log HeLa cells (4 to 5 × 10^5 cells/ml), grown in suspension as described for cytoplasmic extract preparation, are infected with an adenovirus recombinant[43] at a multiplicity of 10 to 20. The infection is allowed to proceed for 48 to 50 hr at 37°, and the cells are then harvested by centrifugation at 500*g* for 5 min at 4°. The cell pellet is washed twice with 1/25 culture volume of PBS+Ca^{2+}+Mg^{2+}, and then resuspended with vigorous agitation in 10 pellet volumes of lysis buffer. The lysate is incubated on ice for 20 min, and then clarified by centrifugation at 33,000*g* for 25 min at 4°.

BACULOVIRUS RECOMBINANT INFECTED-CELL LYSATE. Sf9 cells are grown at room temperature in Grace's medium supplemented with yeastolate and lactalbumin hydrolyzate (GIBCO, Grand Island, NY) and 10% fetal bovine serum. At 80 to 90% confluency, the medium is removed and the cells are infected with a baculovirus recombinant[44] at a multiplicity of 10 (~2 ml/150 cm^2 flask). After an adsorption period of 1 hr at room temperature, during which time the cells are briefly rocked at 15-min intervals, 20 ml of media is added to each flask, and the infection is allowed to proceed for 30 to 72 hr. The cells are scraped from the flasks, and lysate is prepared by the same method described for infected HeLa cells.

IMMUNOAFFINITY CHROMATOGRAPHY.[45] T antigen is purified from infected-cell lysates by a one-step immunoaffinity procedure. Purified monoclonal antibody PAb419 is linked to Affi-Gel 10 (Bio-Rad) according to manufacturer's instructions (resulting in ~3 to 4 mg antibody/ml gel). An antibody column and an equivolume Sepharose CL-4B precolumn are equilibrated with lysis buffer, and cell lysate (~200 mg/ml antibody matrix) is loaded at 3 antibody column volumes per hour. The columns are then washed with 10 antibody column volumes of Wash I over a 1-hr period and the precolumn is disconnected. The antibody column is then washed with 10 column volumes of Wash II followed by 10 column volumes of Wash III (1 hr each). Buffer is removed from the top of the gel bed and replaced with 6 column volumes of elution buffer. T antigen is then eluted at 60 column volumes/hour. One column volume (1-min) fractions are collected in tubes containing 1/4 column volume adjustment buffer, and each fraction is gently mixed immediately after collection. Five fractions are collected, and the column is then immediately washed with 10 column volumes of rinse buffer followed by 3 column volumes of column storage buffer (10 column volumes/hour). Peak fractions containing T antigen are pooled, dialyzed extensively versus T antigen storage buffer, and stored at −20°. SDS–PAGE analysis of the pooled fractions should reveal a major polypeptide of ~90 kDa.

[43] C. Thummel, R. Tjian, S.-L. Hu, and T. Grodzicker, *Cell* **33,** 455 (1983).
[44] D. R. O'Reilly and L. K. Miller, *J. Virol.* **62,** 3109 (1988).
[45] R. A. F. Dixon and D. Nathans, *J. Virol.* **53,** 1001 (1985).

Replication Protein A (RPA)

Early studies using the partially fractionated system indicated that CF I contained proteins involved in the origin- and ATP-dependent unwinding of DNA by T antigen.[32,37,46] Since *E. coli* SSB could substitute for CF I in this presynthetic reaction, there was strong evidence that CF I contained the eukaryotic homolog to the prokaryotic helix-destabilizing protein. On further fractionation using the complementation of DNA replication as an assay, RPA was purified and characterized as a heterotrimeric single-stranded DNA-binding protein with subunit molecular weights of 70, 32, and 14 kDa.[31,37] In contrast to the presynthetic reaction, the initiation of DNA synthesis cannot be supported by *E. coli* SSB, but relies on RPA.[31,47–49]

Assays

Fractions containing RPA will reconstitute DNA replication when added to a standard reaction mixture containing 20 μg CF II and 5 μg CF IB.[31] Fraction CF IB contains enough PP2Ac to stimulate HeLa-cell derived T antigen[31] (see Purified Proteins, T Antigen section). Therefore, CF IC is not required for this complementation reaction. For more purified fractions, the single-stranded DNA binding activity of RPA can be detected.[30,37,50]

Purification

BUFFERS

Buffer H: 30 m*M* HEPES, pH 7.8, 0.25 m*M* EDTA, 0.50% *myo*-inositol, 0.01% NP-40, 1 m*M* DTT

Buffer A: 25 m*M* Tris–HCl, pH 7.5, 1 m*M* EDTA, 10% glycerol, 0.01% NP-40, 1 m*M* DTT, 0.1% PMSF.

RPA can be purified from CF IA (see General Approach to Subcellular Fractionation section) by affinity chromatography with Affi-Gel Blue (Bio-Rad)[31]: CF IA is loaded at 1 column volume per hour onto an Affi-Gel Blue column (6 mg protein/ml matrix) equilibrated in Buffer H containing 15 m*M* KCl. The column is then washed with 4 column volumes of each of the following (in Buffer H): (a) 1 *M* KCl, (b) 0.5 *M* KSCN, (c) 0.75 *M* KSCN, (d) 1.3 *M* KSCN. Typically, RPA elutes with 1.3 *M* KSCN, but

[46] C. R. Wobbe, L. Weissbach, J. A. Borowiec, F. B. Dean, Y. Murakami, P. Bullock, and J. Hurwitz, *Proc. Natl. Acad. Sci. USA* **84,** 1834 (1987).

[47] M. K. Kenny, S.-H. Lee, and J. Hurwitz, *Proc. Natl. Acad. Sci. USA* **86,** 9757 (1989).

[48] T. Matsumoto, T. Eki, and J. Hurwitz, *Proc. Natl. Acad. Sci. USA* **87,** 9712 (1990).

[49] T. Melendy and B. Stillman, *J. Biol. Chem.* **268,** 3389 (1993).

[50] C. Kim, R. O. Snyder, and M. S. Wold, *Mol. Cell Biol.* **12,** 3050 (1992).

exact elution conditions will depend on the batch of Affi-Gel Blue. Because the RPA is relatively dilute at this stage, concentration by hydroxylapatite chromatography is often necessary. The Affi-Gel pool is loaded directly onto a small hydroxylapatite column (5 mg protein/ml matrix) equilibrated in Buffer H containing 15 m*M* KCl. The bound RPA is then eluted in a small volume with Buffer F containing 70 m*M* potassium phosphate and 100 m*M* KCl, and dialyzed versus Buffer H containing 15 m*M* KCl. The purified RPA is divided into small aliquots, frozen with liquid nitrogen, and stored at −70°.

Alternatively, RPA can be purified from CF IA by single-stranded DNA-cellulose chromatography[37]: CF IA is initially fractionated by ammonium sulfate precipitation (40% saturation). The precipitate is pelleted by centrifugation, redissolved in Buffer A containing 25 m*M* NaCl, and dialyzed extensively versus the same buffer. The sample is then loaded onto a denatured DNA-cellulose column (8 mg protein/ml matrix) equilibrated in the same buffer. After the sample is loaded, the column is washed with 2 column volumes of each of the following (in Buffer A): (a) 25 m*M* NaCl, (b) 750 m*M* NaCl, (c) 1.5 *M* NaCl/50% ethylene glycol. Fractions containing RPA (wash c) are dialyzed versus Buffer A containing 25 m*M* NaCl and 20% sucrose. As in the method described above, it may be necessary to concentrate the purified RPA prior to storage at −70°.

Although HeLa cells contain a significant quantity of RPA, recent studies indicate that functional RPA can be produced in *E. coli.*[51] Purification of this recombinant protein is achieved by Affi-Gel Blue affinity chromatography (see above) followed by fractionation over Mono Q.

DNA Polymerase α–Primase Complex

As described previously, initial fractionation of the cytoplasmic extract separated the proteins involved in presynthesis from those involved directly in DNA polymerization.[30,37,52] The polymerase activities were found in CF II, and immunoaffinity chromatography of this fraction resulted in the identification and purification of the well-studied DNA polymerase α–primase complex. Activity that did not bind to the antipolymerase α column was later purified and shown to be the proliferating cell nuclear antigen (PCNA)-dependent DNA polymerase δ.[53]

[51] L. A. Henricksen, C. B. Umbricht, and M. S. Wold, *J. Biol. Chem.* **269,** 11,121 (1994).

[52] J. J. Li, M. S. Wold, and T. J. Kelly, *in* "DNA Replication and Recombination" (R. McMacken and T. J. Kelly, eds.), p. 289. A. R. Liss, Inc., New York, 1987.

[53] D. H. Weinberg and T. J. Kelly, *Proc. Natl. Acad. Sci. USA* **86,** 9742 (1989).

Assays

DNA polymerase α–primase activity can be measured using the standard procedures developed previously (see [8] in this volume). However, activity can also be measured in the context of the cell-free SV40 DNA replication system using the standard complementation reaction mixture containing 25 to 30 μg CF I′ and 10 μg of CF IIA (described below).[30]

Purification

BUFFERS

Buffer W: 30 m*M* HEPES, pH 7.8, 365 m*M* KCl, 0.25 m*M* EDTA, 0.25% myoinositol, 10% glycerol, 1 m*M* DTT

Buffer E: 20 m*M* Tris–HCl, pH 8.5, 500 m*M* KCl, 0.25 m*M* EDTA, 50% ethylene glycol, 10% glycerol, 1 m*M* DTT

Buffer D: 20 m*M* Tris–HCl, pH 8.5, 50 m*M* KCl, 0.25 m*M* EDTA, 50% glycerol, 1 m*M* DTT.

DNA polymerase α–primase is purified from CF II (see General Approach to Subcellular Fractionation section) by a one-column immunoaffinity method[30,54]: Antibody-linked agarose (10 mg antibody/ml gel) is prepared using Affi-Gel 10 (Bio-Rad) and monoclonal antibody SJK-237 according to Bio-Rad instructions. A column is then prepared (1 ml of antibody matrix per 25 mg CF II should be sufficient), which is equilibrated with Buffer W (10 column volumes). Prior to equilibration, it is advisable to wash the column thoroughly with Buffer E to reduce the eventual level of antibody contamination in the purified polymerase fractions. Undialyzed CF II is brought to 20 m*M* Tris–HCl, pH 8.5/10% glycerol/0.2 m*M* PMSF, and then loaded onto the column from the bottom at less than 1 column volume per hour. After the sample has been loaded, the column is washed with at least 20 column volumes of Buffer W at 3 column volumes/hour. Polymerase α–primase is then eluted from the top with Buffer E at 0.5 column volume/hour. Active fractions are dialyzed versus Buffer D and stored at −20°. The flow-through containing DNA polymerase δ is dialyzed versus Buffer F (see p. 530) containing 15 m*M* KCl and 10% glycerol, frozen with liquid nitrogen in aliquots, and stored at −70°. Dialysis of this fraction for longer than 3 to 4 hr can lead to a significant loss in activity. The dialyzed fraction is called CF IIA and can be fractionated further or used in complementation assays as described above.

Proliferating Cell Nuclear Antigen (PCNA)

PCNA was first identified as a human autoantigen using sera from patients who had systemic lupus erythematosus and separately as a protein

[54] Y. Murakami, C. R. Wobbe, L. Weissbach, F. B. Dean, and J. Hurwitz, *Proc. Natl. Acad. Sci. USA* **83,** 2869 (1986).

that had a variable nuclear fluorescence pattern throughout the cell cycle.[55] The protein was purified from human tissue culture cells using an assay based on its ability to complement a fractionated SV40 DNA replication system[56] and was quickly found to be identical to an accessory protein for DNA polymerase δ.[56,57] Subsequently, PCNA was shown to stimulate the DNA-dependent ATPase activity of RFC and to bind to DNA in an RFC-dependent manner.[6,58] PCNA is also required for nucleotide excision repair of UV-induced lesions in DNA.[59]

Assay

PCNA can be assayed using a reconstituted SV40 DNA replication reaction. However, the protein (denatured M_r of 37,000) is routinely detected by Western blotting using the anti-PCNA monoclonal antibody PC10,[60] which is available from a number of commercial sources.

Purification

BUFFERS. The following protease inhibitors are added to the buffers listed below at each step in the purification: 2 m*M* benzamidine, 2 μM pepstatin A, 10 m*M* sodium bisulfite, and 1 m*M* PMSF.

- Buffer A: 25 m*M* Tris–HCl, pH 7.5, 1 m*M* EDTA, 10% glycerol, 0.01% NP-40, 1 m*M* DTT
- Buffer L: 25 m*M* Tris–HCl, pH 7.5, 25 m*M* NaCl, 1 m*M* EDTA, 0.01% NP-40, 1 m*M* DTT
- HAP buffer: potassium phosphate, pH 7.0 (variable concentrations as indicated below), 10% glycerol, 0.01% NP-40, 1 m*M* DTT.

A cDNA encoding human PCNA[61] was cloned into a bacterial expression vector containing a bacteriophage T7 promoter, and the protein was expressed in *E. coli* BL21(DE3), which contains an inducible T7 RNA polymerase.[62] This strategy was similar to that used for the expression

[55] M. B. Mathews, *in* "Growth Control During Cell Aging" (H. R. Warner and E. Wang, eds.), p. 89. CRC Press, Boca Raton, FL, 1989.

[56] G. Prelich, C.-K. Tan, M. Kostura, M. B. Mathews, A. G. So, K. M. Downey, and B. Stillman, *Nature* **326,** 517 (1987).

[57] C.-K. Tan, C. Castillo, A. G. So, and K. M. Downey, *J. Biol. Chem.* **261,** 12,310 (1986).

[58] T. Tsurimoto and B. Stillman, *J. Biol. Chem.* **266,** 1950 (1991).

[59] M. K. K. Shivji, M. K. Kenny, and R. D. Wood, *Cell* **69,** 367 (1992).

[60] N. H. Waseem and D. P. Lane, *J. Cell Sci.* **96,** 121 (1990).

[61] J. M. Almendral, D. Huebsch, P. A. Blundell, H. Macdonald-Bravo, and R. Bravo, *Proc. Natl. Acad. Sci. USA* **84,** 1575 (1987).

[62] F. W. Studier, A. H. Rosenberg, J. J. Dunn, and J. W. Dubendorff, *Methods Enzymol.* **185,** 60 (1990).

of the *Saccharomyces cerevisiae* PCNA protein.[63,64] Both proteins can be purified with a yield of approximately 11 mg of PCNA from 2 liters of *E. coli* culture. The yeast PCNA, however, cannot substitute for the human PCNA in the reconstituted SV40 DNA replication system. The purification procedure for the human PCNA protein expressed in *E. coli* is as follows:

1. *Growth and expression.* Grow a 15-ml culture of BL21(DE3) harboring the pT7-hPCNA plasmid in LB medium containing ampicillin (LB/Amp) overnight. Inoculate 2 liters of LB/Amp with the overnight culture and grow for 3 hr at 37° (to an OD at 600 nm of 0.34). Induce the T7 promoter by addition of IPTG to 1 m*M* and grow for another 3 hr. The cells are then harvested by centrifugation and washed once with either distilled H_2O or Buffer L (the washed cell pellet can be frozen at −70° at this stage).
2. *Extract preparation.* The cells are resuspended in 40 ml of Buffer L and lysed with a French press. The lysate is adjusted to 0.1 *M* NaCl (final concentration) and centrifuged at 100,000*g* for 1 hr at 4°. After this high-speed spin, the supernatant is adjusted to 0.2 *M* NaCl (final concentration).
3. *Q-Sepharose chromatography.* The protein sample is loaded onto a Q-Sepharose column (Pharmacia, 1.2 × 16 cm) equilibrated in Buffer A containing 0.2 *M* NaCl. The column is washed with 2 column volumes of equilibration buffer, and PCNA is eluted with a 200-ml linear gradient of 200 to 700 m*M* NaCl in Buffer A (PCNA elutes at 350 to 450 m*M* NaCl). The PCNA peak is pooled and dialyzed versus HAP buffer containing 25 m*M* potassium phosphate.
4. *S-Sepharose chromatography.* The Q-Sepharose pool is loaded onto an S-Sepharose column (Pharmacia, 1.2 × 16 cm) equilibrated in HAP buffer containing 25 m*M* potassium phosphate. The column is washed with equilibration buffer, and fractions equivalent to 1/5th column volume are collected. The PCNA will flow through this column and the protein peak can be collected and pooled. This step removes an *E. coli* DNA endonuclease, which, if not removed, causes an accumulation of abnormal replication products in the reconstituted SV40 system.
5. *Hydroxylapatite chromatography.* The pooled flow-through fraction from the S-Sepharose column is loaded onto a hydroxylapatite column (Bio-Rad, 2.5 × 11 cm) equilibrated in HAP buffer containing 25 m*M* potassium phosphate. The column is washed with 2 column

[63] K. Fien and B. Stillman, *Mol. Cell. Biol.* **12,** 155 (1992).
[64] K. Fien and B. Stillman, unpublished results (1994).

volumes of equilibration buffer, and PCNA is then eluted with a 500-ml linear gradient of 25 to 500 m*M* potassium phosphate HAP buffer (PCNA begins to elute at approximately 150 m*M* phosphate). The PCNA peak fractions are pooled and dialyzed versus Buffer A (without glycerol) containing 1.2 *M* NaCl.

6. *Phenyl-Sepharose chromatography.* The dialyzed pool is loaded onto a phenyl-Sepharose column (Pharmacia, 1.2 × 11 cm) equilibrated in Buffer A (without glycerol) containing 1.2 *M* NaCl, and the column is washed with 2 column volumes of equilibration buffer. PCNA is then eluted with a linear gradient of 1.2 to 0.0 *M* NaCl in Buffer A (without glycerol). The purified PCNA is pooled and dialyzed versus Buffer A containing 25 m*M* NaCl and 20% sucrose. The protein can be stored in small aliquots at −70° for at least one year.

Replication Factor C

Replication factor C (RFC) was identified as an essential DNA replication factor required for reconstitution of SV40 DNA replication *in vitro*.[65] The protein was also identified by its ability to activate the SV40 DNA replication system under specific conditions and has been alternatively called activator 1 (A1).[66] The protein contains polypeptide subunits with apparent molecular masses of 140, 41, 40, 38, and 37 kDa based on SDS–polyacrylamide gel electrophoresis (PAGE). The cDNAs encoding all five human subunits have been cloned and, although they are related to each other, they are encoded by separate genes.[67–70]

RFC has multiple functions in DNA replication that are all related to its function as a DNA polymerase accessory factor.[6,58,71,72] The five-subunit protein has a DNA-dependent ATPase activity that is stimulated by PCNA, and binds to DNA in a structure-specific manner. Some of the functions have been investigated by expressing the individual subunits in *E. coli*,[68,73,74]

[65] T. Tsurimoto and B. Stillman, *Mol. Cell. Biol.* **9,** 609 (1989).
[66] S.-H. Lee, A. D. Kwong, Z.-Q. Pan, and J. Hurwitz, *J. Biol. Chem.* **266,** 594 (1991).
[67] M. Chen. Z.-Q. Pan, and J. Hurwitz, *Proc. Natl. Acad. Sci. USA* **89,** 5211 (1992).
[68] M. Chen, Z.-Q. Pan, and J. Hurwitz, *Proc. Natl. Acad. Sci. USA* **89,** 2516 (1992).
[69] F. Bunz, R. Kobayashi, and B. Stillman, *Proc. Natl. Acad. Sci. USA* **90,** 11,014 (1993).
[70] M. O'Donnell, R. Onrust, F. B. Dean, M. Chen, and J. Hurwitz, *Nucleic Acids Res.* **21,** 1 (1993).
[71] S.-H. Lee and J. Hurwitz, *Proc. Natl. Acad. Sci. USA* **87,** 5672 (1990).
[72] T. Tsurimoto and B. Stillman, *Proc. Natl. Acad. Sci. USA* **87,** 1023 (1990).
[73] Z.-Q. Pan, M. Chen, and J. Hurwitz, *Proc. Natl. Acad. Sci. USA* **90,** 6 (1993).
[74] P. D. Burbelo, A. Utani, Z.-Q. Pan, and Y. Yamada, *Proc. Natl. Acad. Sci. USA* **90,** 11,543 (1993).

but these studies are just beginning. Thus, for SV40 DNA replication studies, the protein is still purified from human cells in culture.

Assays

RFC is assayed using a standard complementation reaction with fraction I* and topoisomerase II (in addition to the standard topoisomerase I).[65] Fraction I* is the fraction of cytoplasmic S100 that elutes from a phosphocellulose column with 0 to 600 m*M* NaCl. Alternatively, a standard reaction can be carried out in the presence of cellular protein components CF I (19 μg) and CF II′ (8.7 μg).[5] CF II′ is the fraction of CF II that elutes from a phosphocellulose column with 0.4 *M* KCl, and is devoid of cytoplasmic RFC activity.

Purification

BUFFERS

Extraction buffer: 25 m*M* Tris–HCl, pH 7.5, 200 m*M* NaCl, 1 m*M* EDTA, 1 m*M* DTT, 10 m*M* sodium bisulfite, 2 μg/ml leupeptin, 10 m*M* benzamidine, 0.1 m*M* PMSF

Buffer A: 25 m*M* Tris–HCl, pH 7.5, 1 m*M* EDTA, 10% glycerol, 0.01% NP-40, 1 m*M* DTT, 10 m*M* sodium bisulfite, 0.5 μg/ml leupeptin, 11 m*M* benzamidine, 0.1 m*M* PMSF

HAP buffer: potassium phosphate, pH 7.5 (variable concentrations as indicated below), 10% glycerol, 0.01% NP-40, 1 m*M* DTT, 10 m*M* sodium bisulfite, 0.5 μg/ml leupeptin, 11 m*M* benzamidine, 0.1 m*M* PMSF

TE: 10 m*M* Tris–HCl, pH 7.5, 1 m*M* EDTA.

The procedure described below is modified from the original purification protocol.[65]

1. *Preparation of nuclear extract.* Frozen unwashed nuclear pellets prepared from 64 liters of suspension cells by Dounce homogenization (see Preparation of Replication Extracts from Human Cells section) are mixed with 20 ml of extraction buffer and the suspension is stirred at room temperature until the pellets are thawed (~30 min). Another 20 ml of ice-cold extraction buffer is added to the suspension, and after 10 min of stirring, an additional 160 ml is added. Stirring is continued for 30 min, whereupon the extract is centrifuged at 10,000*g* for 10 min at 4°. The supernatant (~250 ml, 4 mg/ml protein) is adjusted to a conductivity of 440 μS/cm with 5 *M* NaCl (~1/100 dilution).
2. *Phosphocellulose chromatography.* The nuclear extract is loaded onto a phosphocellulose column (4 × 13 cm, 160 ml) equilibrated in Buffer

A containing 330 m*M* NaCl. The column is washed with 2 column volumes of this same buffer followed by 2 column volumes of Buffer A containing 660 m*M* NaCl. The peak protein fractions from the 660 m*M* NaCl wash are pooled (~130 ml, 1.5 mg/ml protein).

3. *Hydroxylapatite chromatography.* The pooled protein is then loaded directly onto a hydroxylapatite column (30-ml packed volume in a 50-ml syringe) equilibrated in Buffer A containing 660 m*M* NaCl. The column is washed with 3 column volumes of HAP buffer containing 200 m*M* potassium phosphate and RFC is eluted with a linear gradient of 200 to 500 m*M* potassium phosphate HAP buffer. Three-milliliter fractions are collected in tubes that contain 30 μl of 100 m*M* EDTA/100 m*M* EGTA. The peak RFC activity, eluting at approximately 300 m*M* potassium phosphate, is pooled conservatively (36 ml, 1.1 mg/ml protein) and dialyzed versus Buffer A containing 100 m*M* NaCl.
4. *Single-stranded DNA cellulose chromatography.* The resin is prewashed as follows: 4 ml of ssDNA cellulose (United States Biochemical Corp.) is packed into a 10-ml syringe and washed successively with 30 ml of 0.1 mg/ml BSA (crystallized) in TE, 60 ml of 5 *M* NaCl, 60 ml of TE, and 15 ml of Buffer A containing 100 m*M* NaCl. The dialyzed pool from the hydroxylapatite column is then loaded and the column is washed with 10 ml of Buffer A containing 100 m*M* NaCl. RFC is then eluted with a 100-ml linear gradient of 100 m*M* to 1.0 *M* NaCl in Buffer A. Activity should elute in two peaks (~250 and ~450 m*M* NaCl). The peak eluting at higher salt is pooled and used in the next step.
5. *Mono Q chromatography.* The pooled protein is diluted into three volumes of Buffer A (pH 8.0 rather than pH 7.5), and loaded onto a Mono Q 5/5 (Pharmacia) column equilibrated in Buffer A (pH 8.0) containing 100 m*M* NaCl. The column is washed with 3 ml of equilibration buffer and the RFC is then eluted with a 10-ml linear gradient of 0.1 to 0.5 *M* NaCl in Buffer A (pH 8.0). RFC should elute at approximately 180 m*M* NaCl. If further purification is necessary, glycerol gradient sedimentation can be performed as described.[65] The Mono Q peak fractions are divided into small aliquots, frozen with liquid nitrogen, and stored at −70°. The protein will lose activity if dialyzed to low NaCl concentrations.

DNA Polymerase δ

The discovery that PCNA was required for the reconstitution of SV40 DNA replication immediately suggested that DNA polymerase δ was also required.[36,56] This DNA polymerase was first purified from calf thymus[75] and was subsequently shown to be required for the reconstituted SV40 replication system.[5,8,48,72,76–79] Analysis of the DNA replication products in the presence and absence of Pol δ have demonstrated that it is required for replication of the continuously synthesized leading strand and for the completion of the Okazaki fragments during replication of the lagging strand.[5,6,8,53,71,76,78–82] The properties of the two-subunit polymerase δ enzyme have been reviewed.[83]

For reconstitution of SV40 DNA replication, the Pol δ enzyme is usually purified from fetal calf thymus following the procedure originally described.[75] A more complete description of this protocol is reported in this volume [9]. Briefly, 300 g of frozen fetal calf thymus tissue (Pel-Freez Biologicals, Rogers, AR) are disrupted with a hammer and the tissue thawed in buffer. Extract is then prepared in a blender and by Dounce homogenization. Polymerase δ is then purified from the resulting extract by chromatography on DEAE-cellulose, phenyl-Sepharose, phosphocellulose, hydroxylapatite, and single-stranded DNA cellulose. The polymerase activity is followed by a PCNA-stimulated DNA synthesis assay using poly(dA) · oligo(dT) as a template–primer (see [9], this volume).

Protein Phosphatase 2Ac (PP2Ac)

CF I was found to contain factors involved in the presynthetic stage of replication, as previously discussed (see General Approach to Subcellular Fractionation section). Purification using reconstitution assays revealed that one protein involved in this reaction was the single-stranded DNA-binding protein, RPA. In addition, a 34 kDa polypeptide was isolated that was highly stimulatory to T antigen- and origin-dependent unwinding in the presence of RPA and ATP. Preincubation of T antigen with ATP, origin-containing DNA, and this protein [originally named replication protein C (RPC)], reduced the lag time associated with the presynthetic reaction.[41] Further characterization revealed that RPC was identical to the catalytic subunit of protein phosphatase 2A (PP2Ac) and indicated that this enzyme activated T antigen by dephosphorylating key serine residues involved in

[75] M. Y. W. T. Lee, C.-K. Tan, K. M. Downey, and A. G. So, *Biochemistry* **23,** 1906 (1984).
[76] S.-H. Lee, T. Eki, and J. Hurwitz, *Proc. Natl. Acad. Sci. USA* **86,** 7361 (1989).
[77] T. Melendy and B. Stillman, *J. Biol. Chem.* **266,** 1942 (1991).
[78] T. Tsurimoto and B. Stillman, *J. Biol. Chem.* **266,** 1961 (1991).
[79] S. Waga and B. Stillman, *Nature* **369,** 207 (1994).
[80] T. Tsurimoto and B. Stillman, *EMBO J.* **8,** 3883 (1989).
[81] T. Eki, T. Matsumoto, Y. Murakami, and J. Hurwitz, *J. Biol. Chem.* **267,** 7284 (1992).
[82] V. N. Podust and U. Hubscher, *Nucleic Acids Res.* **21,** 841 (1993).
[83] A. G. So and K. M. Downey, *Crit. Rev. Biochem. Mol. Biol.* **27,** 129 (1992).

DNA binding.[12,84,85] Because the phosphorylation state of purified T antigen varies depending on its source, the degree of activation by PP2Ac is also variable.

Assay

Active fractions will stimulate DNA replication when added to a standard complementation reaction mix containing the following cellular protein components: 200 ng RPA, 30 ng PCNA, and 15 μg CF II.[41] T antigen expressed in HeLa cells from a recombinant adenovirus vector (see Purified Proteins, T Antigen section) is used in this complementation assay.

Purification

The original procedure for the purification of RPC (PP2Ac)[41] from CF IC (see General Approach to Subcellular Fractionation section) is presented below. However, it should be noted that another protocol has been developed for the purification of PP2Ac,[86] and a modification of this method is now routinely used to purify the phosphatase from bovine heart.

1. *Ammonium sulfate precipitation.* CF IC is adjusted to 0.2 mg/ml PMSF, and ammonium sulfate is added to 25% saturation. The precipitated protein is removed by centrifugation, and the ammonium sulfate saturation of the supernatant is adjusted to 65%. The resulting precipitate is pelleted by centrifugation and dissolved in Buffer HN (30 m*M* HEPES, pH 7.8, 0.50 m*M* EDTA, 0.50% *myo*-inositol, 0.01% NP-40, 1 m*M* NaN_3, and 1 m*M* DTT) containing 100 m*M* KCl to give a final protein concentration of ~10 mg/ml (the pellet should contain 70 to 75% of the CF IC protein).
2. *Gel filtration.* The ammonium sulfate fraction is passed over a Sephacryl S-300 HR column equilibrated in Buffer HN containing 100 m*M* KCl, resulting in a broad peak of activity.
3. *Hydroxylapatite chromatography.* The pooled Sephacryl fractions are loaded onto a hydroxylapatite column (5 mg protein/g matrix) equilibrated in Buffer HN (without EDTA) containing 100 m*M* KCl. The column is then washed with 3 column volumes equilibration buffer followed by 6 column volumes of equilibration buffer containing 2 *M* $MgCl_2$. The $MgCl_2$ eluate contains the majority of the activity.
4. *FPLC gel filtration.* The hydroxylapatite pool ($MgCl_2$ eluate) is concentrated by ammonium sulfate precipitation (60% saturation). The precipitated protein is dissolved in a small volume of Buffer HN

[84] K. H. Scheidtmann, D. M. Virshup, and T. J. Kelly, *J. Virol.* **65,** 2098 (1991).
[85] D. M. Virshup, A. A. R. Russo, and T. J. Kelly, *Mol. Cell. Biol.* **12,** 4883 (1992).
[86] A. DeGuzman and E. Y. C. Lee, *Methods Enzymol.* **159,** 356 (1988).

containing 100 mM KCl and 2 M $MgCl_2$ and is then chromatographed on a Superose 12 FPLC column equilibrated in the same buffer. Fractions are dialyzed versus Buffer HN containing 15 mM KCl and 50% glycerol and stored at −20°.

DNA Ligase

DNA ligase is required for the joining of nascent Okazaki fragments during replication of the lagging strand template.[7,8] It appears that of all the eukaryotic DNA ligase enzymes, only DNA ligase I can function efficiently in the SV40 DNA replication system.[8]

Two different DNA ligase assays are employed for ligase I purification. One assay measures the phosphatase-resistant radioactivity remaining after reaction with ^{32}P-labeled oligo(dT) · poly(dA),[87,88] while the other measures the formation of an AMP-ligase complex.[88] The enzyme is purified from fetal calf thymus tissue following published methods[89] with minor modifications. Although the expression of an active form of the full-length human DNA ligase I in *E. coli* has been reported,[90] this protein has not yet been tested for its activity in the reconstituted SV40 DNA replication reactions.

FEN-1 Endonuclease

Removal of the RNA primers during maturation and ligation of Okazaki fragments to form the complete lagging strand requires a nuclease that functions in different contexts as an endonuclease or an exonuclease. This enzyme was identified as an exonuclease with a M_r of 45,000 using a variety of replication assays that detect ligation of Okazaki fragments.[7,8,91–93] The same protein was also identified as a structure-specific endonuclease that may function during recombination[94] and recently cDNAs encoding the human, murine, and *S. cerevisiae* proteins have been cloned.[95,96] The protein, which was previously termed MF-1 and will be henceforth called the

[87] B. M. Olivera and I. R. Lehman, *J. Mol. Biol.* **36,** 261 (1968).

[88] A. E. Tomkinson, D. D. Lasko, G. Daly, and T. Lindahl, *J. Biol. Chem.* **265,** 12,611 (1990).

[89] A. E. Tomkinson, E. Roberts, G. Daly, N. F. Totty, and T. Lindahl, *J. Biol. Chem.* **266,** 21,728 (1991).

[90] H. Teraoka, H. Minami, S. Iijima, K. Tsukada, O. Koiwai, and T. Date, *J. Biol. Chem.* **268,** 24,156 (1993).

[91] M. Goulian, S. H. Richards, C. J. Heard, and B. M. Bigsby, *J. Biol. Chem.* **265,** 18,461 (1990).

[92] G. Siegal, J. J. Turchi, T. W. Myers, and R. A. Bambara, *Proc. Natl. Acad. Sci. USA* **89,** 9377 (1992).

[93] J. J. Turchi and R. A. Bambara, *J. Biol. Chem.* **268,** 15,136 (1993).

[94] J. J. Harrington and M. R. Lieber, *EMBO J.* **13,** 1235 (1994).

[95] J. J. Harrington and M. R. Lieber, *Genes Dev.* **8,** 1344 (1994).

[96] S. Waga, R. Kobayashi, and B. Stillman, unpublished results (1994).

FEN-1 endonuclease, has been expressed in *E. coli.* The recombinant protein has endonuclease activity but has not yet been tested to determine whether it will support Okazaki fragment maturation during SV40 DNA replication. Thus the purification of the enzyme from human cell extract is described.

Assay

FEN-1 nuclease activity is followed using an exonuclease assay with ^{3}H-labeled poly(dA-dT). Preparation of the substrate and reactions conditions are as described previously.[91]

Purification

BUFFERS

The following protease inhibitors are added to the buffers listed below at each step in the purification: 0.5 m*M* benzamidine, 2 μM pepstatin A, and 0.1 m*M* PMSF.

HAP buffer: potassium phosphate, pH 7.2 (variable concentrations as indicated below), 10% glycerol, 0.01% NP-40, 1 m*M* DTT

Buffer S: 25 m*M* potassium phosphate, pH 7.2, 1 m*M* EDTA, 10% glycerol, 0.01% NP-40, 1 m*M* DTT

Buffer Q: 20 m*M* triethanolamine, pH 7.4, 1 m*M* EDTA, 10% glycerol, 0.01% NP-40, 1 m*M* DTT.

1. *Hydroxylapatite chromatography.* Nuclear extract is passed over a phosphocellulose column as described for the purification of chromatin assembly factor 1 (CAF1).[97] The flow-through fraction (450 ml, 225 mg) is loaded onto a hydroxylapatite column (2.8 × 10 cm) equilibrated in HAP buffer containing 25 m*M* potassium phosphate. The column is developed with a 60-ml gradient of 25 to 500 m*M* potassium phosphate HAP buffer and active fractions (eluting at 200 m*M* potassium phosphate) are pooled and dialyzed versus Buffer S containing 25 m*M* NaCl (two 1-liter changes).
2. *S-Sepharose chromatography.* The dialyzed fraction (45 ml, 1.3 mg/ml) is loaded onto an S-Sepharose column (Pharmacia, 1.8 × 4 cm) equilibrated in Buffer S containing 25 m*M* NaCl. FEN-1 is eluted with a 100-ml linear 25 to 500 m*M* NaCl gradient in Buffer S. Active fractions (eluting at 270 m*M* NaCl) are pooled and dialyzed against 1 liter of Buffer Q.
3. *Q-Sepharose chromatography.* This dialyzed fraction (20 ml, 0.36 mg/ml) is loaded onto a Q-Sepharose column (Pharmacia, 1.1 × 3.2 cm)

[97] S. Smith and B. Stillman, *Cell* **58,** 15 (1989).

equilibrated in Buffer Q, and the flow-through fraction containing the activity (28 ml, 0.04 mg/ml) is collected. The purified protein is dilute at this stage, and should be concentrated prior to storage. The pooled flow-through is passed over a small hydroxylapatite column (0.5 ml), and the bound protein is eluted with a small volume of HAP buffer containing 500 m*M* potassium phosphate. The concentrated nuclease is stored in aliquots at −70°.

Topoisomerases I and II

DNA topoisomerases I and II are required for the reconstitution of SV40 DNA replication with partially purified protein fractions[31,40,65] and in the reaction reconstituted with purified proteins.[5,7,8] These enzymes function to relieve the torsional strain generated by the action of the T antigen DNA helicase activity and to segregate the two daughter DNA molecules. The topoisomerases can be purified from fetal calf thymus tissue or human cell nuclei according to standard procedures.[98,99]

Reconstitution of SV40 DNA Replication with Purified Proteins

Initial fractionation of human cytoplasmic extract resulted in the identification of seven cellular proteins necessary and sufficient for the replication of SV40 origin-containing DNA (Fig. 1, A–F; Fig. 2).[5–7] Together, these purified proteins can reconstitute the initiation of DNA synthesis and elongation of DNA chains in a T antigen- and SV40 origin-dependent manner. Specifically, T antigen that has been dephosphorylated by PP2Ac unwinds the origin in concert with RPA and topoisomerase, allowing DNA polymerase α–primase to initiate DNA synthesis *de novo*. Once DNA synthesis has started, DNA polymerase α–primase and DNA polymerase δ, with its accessory proteins PCNA and RFC, elongate the nascent DNA chains. For the joining of Okazaki fragments and completion of the replication reaction, additional enzymes are required. FEN-1 nuclease (MF1) hydrolyzes the RNA primers that have been synthesized by primase, and DNA ligase I catalyzes the joining of Okazaki fragments (Fig. 1, G and H).[7,8] Turchi *et al.*[100] have studied the enzymatic completion of lagging strand DNA replication using a synthetic template. They report that the removal of the RNA from an Okazaki fragment requires ribonuclease H1 (RNase H1) in addition to the FEN-1 endo-/exonuclease. RNase H1 re-

[98] L. F. Liu and K. G. Miller, *Proc. Natl. Acad. Sci. USA* **78,** 3487 (1981).

[99] U. Schomburg and F. Grosse, *Eur. J. Biochem.* **160,** 451 (1986).

[100] J. J. Turchi, L. Huang, R. S. Murante, Y. Kim, and R. A. Bambara, *Proc. Natl. Acad. Sci. USA* **91,** 9803 (1994).

moves most of the RNA and the 5′→3′ nuclease activity of FEN-1 removes the last ribonucleotide prior to ligation of the Okazaki fragment to the adjacent fragment. Only trace amounts of the RNase activity are required for the maturation of Okazaki fragments.[79]

To reconstitute the complete replication reaction resulting in the production of covalently closed daughter molecules, reactions are usually performed with 6 μg/ml origin-containing plasmid DNA under the standard reaction conditions described previously. Typical amounts of each protein used in a 50-μl replication reaction are 4 μg of T antigen, 2.5 μg of RPA, 0.1 μg of polymerase α–primase, 0.4 μg of PCNA, 60 ng of RFC (glycerol gradient purified), 0.15 μg of polymerase δ, 0.5 μg of DNA ligase I, 40 ng of FEN-1 (MF1), 0.2 μg of topoisomerase I, and 90 ng of topoisomerase II.[8] The amounts given serve only as a guide, and titrations of each protein should be performed. For example, the ratio of polymerase α to polymerase δ greatly influences the coordination of the leading and lagging strand replication.[78,81] The native and denatured replication products can be best analyzed by agarose gel electrophoresis using neutral and alkaline buffer conditions, respectively.

Although much more needs to be accomplished before the replication of DNA from the SV40 origin is understood, it is also clear that much progress has been made since the initial reports describing the development of a cell-free system for SV40 DNA replication. The ability to reconstitute this complex process has enabled a detailed understanding of the functions of these proteins and how they combine into an efficient protein machine.

[42] Adenovirus DNA Replication in a Reconstituted System

By FRANK E. J. COENJAERTS and PETER C. VAN DER VLIET

Introduction

Employing crude nuclear extracts of HeLa cells infected with adenovirus type 5 (Ad5) Challberg and Kelly[1] demonstrated *in vitro* initiation and replication of exogenous adenovirus DNA. This was the first system in higher eukaryotes for which such an assay became possible, presumably due to the high level of replication proteins present in these nuclei. Subsequent studies led to the characterization of the mechanism of DNA replication,

[1] M. D. Challberg and T. J. Kelly, *Proc. Natl. Acad. Sci. USA* **76,** 655 (1979).

which occurs by protein priming, and the purification of the various replication proteins. Based on, among others, mutant complementation, three viral proteins are required, all encoded by one early transcription unit (E2). These are the viral DNA-binding protein (DBP), the precursor terminal protein (pTP), and the DNA polymerase (Pol). The latter two form a strong heterodimer in infected cells (pTP–Pol). In addition, essential origin sequences have been defined (Fig. 1).

The Ad2/Ad5 genome consists of a 36 kbp double-stranded linear genome with 102 bp long inverted terminal repeats (ITRs). Moreover, the viral terminal protein (TP) is covalently attached to both 5′ ends and enhances template activity. The two origins are located in the ITRs and encompass approximately 50 bp. A core origin can be distinguished located in a region (9–18) conserved in the various adenovirus serotypes. A second region, partially conserved, functions as an auxiliary region to enhance replication up to 200-fold. Interestingly, this region in Ad2/Ad5 is recognized by two cellular transcription factors, NFI and NFIII/Oct-1 (Fig. 1). Finally, a third nuclear protein is required for replication of full-length DNA. This protein, NFII, has topoisomerase I activity. Thus, optimal replication requires the interaction of three viral and three cellular proteins. These proteins, with the exception of NFII, have now all been overexpressed and purified to homogeneity. The 3-D structures of essential regions of two of these (DBP and the NFIII/Oct-1 POU domain) are known. The proteins have been used to reconstitute initiation of DNA replication effectively and to ensure a rapid elongation rate, comparable to the rates measured *in vivo*. Below, we describe both the purification of these proteins and the optimal replication conditions. Based on these reconstitution experiments and previous *in vivo* data, the events taking place during replication can be summarized as follows (Fig. 2).

The TP–DNA template forms a multiprotein–DNA complex with DBP. This enables NFI to bind more efficiently to its recognition sequence in the auxiliary origin. Also NFIII/Oct-1 binds, displacing DBP. NFI attracts

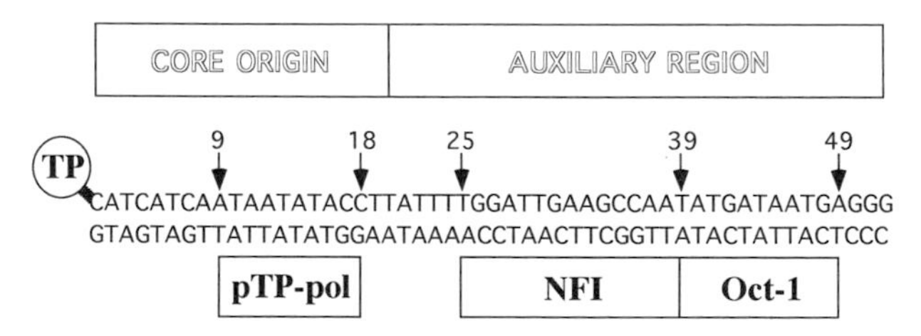

FIG. 1. The adenovirus type 2/5 origin of replication. The terminal protein (TP) is covalently attached to the 5′-dCMP residue through Ser^{-580}. Binding sites for pTP–Pol in the core origin and for NFI and NFIII/Oct-1 in the auxiliary region are indicated.

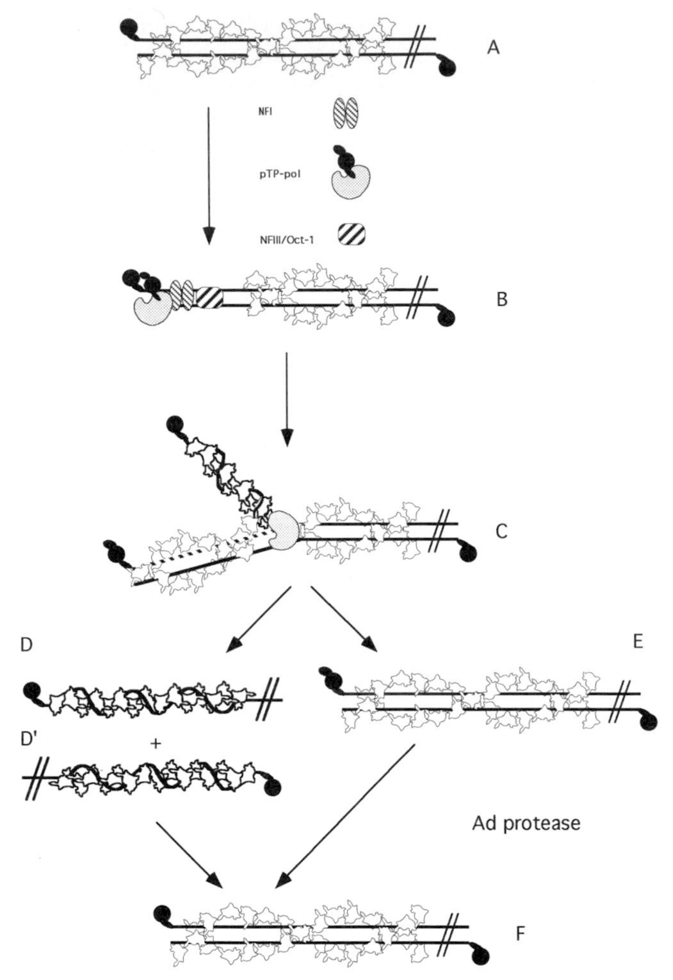

FIG. 2. General outline of the first round of Ad DNA replication. The parental TP–DNA forms a multiprotein complex with DBP (A). A preinitiation complex is assembled with the various replication proteins (B). Initiation occurs by covalent coupling of a dCMP residue to pTP. In the presence of dNTPs elongation starts by a displacement mechanism, requiring Pol and DBP (C). Replication can proceed from both origins. Here, we have drawn a molecule that started replication at the left origin only. Elongation gives rise to partially single-stranded intermediates and finally to a duplex daughter molecule containing TP at one end and pTP at the other end (E). Later in infection these daughter molecules are effectively used as templates for second and further rounds. Displaced single-stranded DNA wound around a DBP polymer (D) can renature with a displaced strand of the opposite polarity (D′) to generate a double-stranded molecule (F). Late in infection the pTP moiety of newly formed pTP-containing templates is processed to TP by a virally encoded protease, also generating (F).

the pTP–Pol complex by a direct interaction. The pTP–Pol complex is bound and correctly positioned assisted by recognition of the core origin, by the parental TP, and possibly by NFIII/Oct-1. Presumably the core origin becomes partially unwound. This preinitiation complex can now, in the presence of dNTPs, initiate by forming a covalent complex between pTP and the first nucleotide, a dCMP residue. This reaction is catalyzed by the polymerase and enhanced by DBP through a reduction of the K_m. Elongation occurs by displacement of the parental strand and is presumably accompanied by dissociation of the initiation complex. Both TP- and pTP-containing origins are competent for reinitiation. On completion of replication, the double-stranded daughter molecules contain pTP and TP. pTP is finally processed to TP late in infection by a viral protease. The displaced strand can either renature with a partner of the opposite polarity originating from starts at the other origin, or form a panhandle through intramolecular renaturation. This panhandle regenerates an origin that can be used in a similar way to synthesize a second double-stranded daughter molecule. Details can be found in several reviews.[2–4]

Purification of Different Replication Components

Preparation of Terminal Protein-Containing DNA Template (TP–DNA)

Plasmid DNA, linearized at or near the origin, can serve as a template, but this process is rather inefficient and requires high DNA concentrations. Therefore, the preferred template is isolated from virion particles and contains the covalently bound terminal protein. TP–DNA is up to 100-fold more efficient than protein-free DNA. Preparation and handling of this protein–DNA complex is complicated by technical problems since common nucleic acid handling techniques such as phenol extraction and ethanol precipitation cannot be employed. Nevertheless it has been possible to isolate short TP-containing origin fragments by Mono Q column chromatography.[5]

The preparation of TP–DNA starts with purified virion particles and includes dissociation of the virion with 4 *M* guanidine hydrochloride followed by sucrose gradient centrifugation to remove noncovalently bound proteins. We routinely employ the following procedure. To isolate virions from 2 liters of HeLa cells, harvested 48 hr after infection with Ad5, cells are sonicated and cellular debris is removed by centrifugation. The superna-

[2] P. C. Van der Vliet, *Sem. Virol.* **2,** 271 (1991).
[3] M. Salas, *Ann. Rev. Biochem.* **60,** 39 (1991).
[4] M. L. DePamphilis, *Ann. Rev. Biochem.* **62,** 29 (1993).
[5] R. Pronk and P. C. Van der Vliet, *Nucleic Acids Res.* **21,** 2293 (1993).

tant is layered on a discontinuous CsCl gradient in 10 m*M* Tris–HCl, pH 8.1, consisting, from bottom to top, of 4 ml CsCl with an η_D of 1.375, 3 ml η_D = 1.3657, and 3 ml η_D = 1.3572. After centrifugation for 2 hr at 18,000 rpm in an SW25.1 tube, the virus band (η_D = 1.3657) is collected and applied to a continuous CsCl gradient with a starting η_D of 1.3657 followed by centrifugation for 18 hr at 35,000 rpm in SW41 tubes. The pooled, whitish virus bands are collected and a second discontinuous CsCl gradient is used to remove remaining contaminants. This gradient, run in a buffer containing 50 m*M* Tris–HCl, pH 7.2, 1 m*M* EDTA, and 100 m*M* NaCl, consists of 4 ml CsCl (η_D = 1.375) and 4 ml CsCl (η_D = 1.3572) in an SW41 tube to which a virus suspension is added that has been diluted 2:3 by slowly adding 0.5 volume of Tris buffer. Centrifugation is as described above. The pooled virus bands contain approximately 2×10^{13} virus particles per milliliter as measured spectroscopically after disruption of the particle with 0.5% SDS. One A_{260} unit corresponds to 1.1×10^{12} particles per milliliter and 38 μg/ml DNA. To prepare TP–DNA, the concentrated virus is dialyzed against 100 m*M* Tris–HCl, pH 7.6, 10 m*M* EDTA, and 1 m*M* phenylmethylsulfonyl fluoride (TE/PMSF), subsequently mixed with an equal volume of 8 *M* guanidine hydrochloride (p.a. quality), and 500 μl is layered on a 12.5-ml 5 to 20% sucrose gradient containing 4 *M* guanidine chloride in an SW41 tube. After centrifugation for 18 hr at 29,000 rpm at 4°, fractions are collected and screened for the presence of TP–DNA by agarose gel electrophoresis. TP–DNA is present in a broad band in the bottom half of the gradient whereas mainly hexon and other virion proteins stay on top. DNA-containing fractions are dialyzed extensively against TE/PMSF. Finally we add 20% (w/v) glycerol and 500 μg/ml BSA and store TP–DNA in aliquots at −80°. This template is stable for at least 2 yr under these storage conditions.

Purification of Viral Proteins

Adenovirus DNA-Binding Protein (DBP). The Ad DBP is expressed at a high level (approximately 2×10^7 molecules per infected HeLa cell) and accumulates late in infection in the form of a nucleoprotein complex in nuclei. The purification of DBP is based on isolation of this nucleoprotein complex followed by ammonium sulfate fractionation and column chromatography using phosphocellulose and single-stranded DNA cellulose columns.[6,7] During the latter step, the small DNA fragments bound to DBP are competed away by the ssDNA in the matrix, which is bound more strongly due to the cooperative ssDNA binding mode.

[6] N. M. Schechter, W. Davies, and C. W. Anderson, *Biochemistry* **19,** 2802 (1980).
[7] D. Tsernoglou, A. D. Tucker, and P. C. Van der Vliet, *J. Mol. Biol.* **172,** 237 (1984).

Typically, we purify DBP from 100 liters of HeLa cells 48 hr postinfection (5×10^{10} cells; approximately 100 g wet weight). Cells are washed in PBS and stored at $-80°$ in 10-g aliquots. Pellets are slowly thawed on ice, suspended in 900 ml 10 m*M* $NaHCO_3$, pH 8.0, 150 m*M* NaCl, 1 m*M* 2-mercaptoethanol, 1 m*M* PMSF, and 10 μg/ml L-1-chlor-3-(4-tosylamido)-4-phenyl-2-butanon (TPCK) (Buffer A) containing 0.25% Nonidet P-40 (NP-40) and stirred for 10 min at 4°. Nuclei are collected by centrifugation for 5 min at 2000 rpm and washed once with Buffer A without NP-40. Pellets are resuspended in 600 ml 25 m*M* NaCl, 8 m*M* EDTA, pH 8.0, 1 m*M* 2-mercaptoethanol, 0.5 m*M* PMSF, and 5 μg/ml TPCK (Buffer B), centrifuged, and washed once with Buffer B. Finally, the pellets are suspended in 500 ml 10 m*M* Tris–HCl, pH 8.0, 1 m*M* 2-mercaptoethanol, 0.2 m*M* PMSF, and 2 μg/ml TPCK (low salt Buffer C) and washed three times, during which procedure a thick gel is formed (approximately 400 ml). Buffer C is added to a total volume of 500 ml and the suspension is divided into 50-ml aliquots and sonicated 10 times, each for 30 sec at 70 W with a rod sonicator using a 4-mm tip. Care is taken that the temperature does not exceed 6°, a procedure essential to obtain high yields. The suspension is adjusted to 150 m*M* NaCl by slow addition of 5 *M* NaCl and clarified by centrifugation for 15 min at 20,000 rpm in an SW27 rotor. To the supernatant 1% (v/v) 1 *M* Tris–HCl, pH 8.0, is added followed by a two-step ammonium sulfate precipitation. Proteins precipitating between 25 and 60% saturation (approximately 1100 mg) are dissolved in 100 ml 20 m*M* Tris–HCl, pH 8.0, 50 m*M* NaCl, 1 m*M* 2-mercaptoethanol, 0.2 m*M* PMSF, and 2 μg/ml TPCK (Buffer D) and dialyzed overnight against 2 liters of this buffer. After centrifugation for 30 min at 20,000 rpm (SW27) the supernatant is applied to a freshly prepared 100-ml Whatman P11 phosphocellulose column (i.d. 2.2 cm), equilibrated with Buffer D. Sample is applied at 25 ml/hr $\cdot$ cm^2, washed with 200 ml Buffer D/150 m*M* NaCl and eluted using a 250-ml linear gradient of 150 to 800 m*M* NaCl in Buffer D. Forty fractions are collected and 10-μl aliquots are analyzed by SDS–PAGE and Coomassie Brilliant Blue staining. Peak fractions, eluting between 320 and 460 m*M* NaCl and containing 60% DBP (65 ml, 110 mg protein) are diluted 1:4 with a buffer containing 10 m*M* Tris–HCl, pH 8.8, 1 m*M* 2-mercaptoethanol, 0.2 m*M* PMSF, and 2 μg/ml TPCK (Buffer E) and applied slowly (15 ml/hr) to a 15-ml ssDNA cellulose column (i.d. 1.4 cm). The column is washed with 30 ml Buffer C containing 200 m*M* NaCl followed by three successive wash steps with 15 ml of this buffer containing 300 m*M* NaCl + 200 μg/ml potassium dextran sulfate, 300 m*M* NaCl, and 500 m*M* NaCl, respectively. The latter contains DBP but this is contaminated with nucleic acids. Finally, elution with 15 ml of this buffer containing 2 *M* NaCl yields 30 mg of 95% pure DBP. To obtain 99% pure DBP free of contaminating enzymatic activities, the DBP is diluted 1:8 with a buffer containing 25 m*M* HEPES–KOH, pH 8.0, 1 m*M* DTT, 0.1 m*M* PMSF, 0.02% NP-40, and 20% (w/v) glycerol (Buffer F) and applied to an 8-ml heparin-Sepharose column (i.d. 1.4 cm) equilibrated with Buffer F/250 m*M* NaCl. The column is washed at 20 ml/hr with Buffer F/370 m*M* NaCl and step-eluted with Buffer F/440 m*M* NaCl. DBP-containing fractions are diluted to 100 m*M* NaCl and applied to an 8-ml FPLC Mono S column at 40 ml/hr. DBP is eluted in one step with Buffer F/300 m*M* NaCl. Top fractions contain 2 to 3 mg/ml of DBP.

Adenovirus DNA Polymerase–Precursor Terminal Protein Complex (pTP–Pol)

These two essential viral replication proteins are present as a heterodimer in infected cells. Due to their low abundance the complex can only be purified in reasonable amounts by overexpression. Their large size (140 kDa for the polymerase and 80 kDa for the precursor terminal protein) has prevented functional expression in bacterial systems and therefore the complex is overexpressed by recombinant vaccinia virus[8,9] or baculovirus.[9,10] We routinely use vaccinia virus-expressed pTP and Pol that are fully functional in DNA replication[8] and can be produced in high amounts, up to 8 mg/liter infected HeLa cells, employing the T7-EMCV system. The construction of the viruses has been described.[8,9] To isolate the Ad5 pTP–Pol complex, HeLa cells are coinfected with recombinant viruses [5 plaque-forming units (pfu) of each construct per cell] expressing the pTP and Pol open reading frames as well as a recombinant virus expressing T7 polymerase (vv/T7[11]) and harvested 22 hr postinfection. Purification takes place in a four-step procedure modified from Ref. 12. Cells (5×10^9) from a 10-liter culture are washed in 500 ml PBS and in 200 ml hypotonic Buffer G (20 m*M* HEPES–KOH, pH 7.5, 5 m*M* KCl, 0.5 m*M* $MgCl_2$, 0.1% NP-40, 1 m*M* PMSF, 10 μg/ml TPCK, and 5 m*M* $Na_2S_2O_5$). Subsequently, cells are swollen for 20 min in 150 ml of Buffer G and Dounce homogenized. KCl is added to 300 m*M* and proteins are extracted from the nuclei for 60

[8] H. G. Stunnenberg, H. Lange, L. Philipson, R. T. Van Miltenburg, and P. C. Van der Vliet, *Nucleic Acids Res.* **16,** 2431 (1988).

[9] R. Nakano, L. J. Zaho, and R. Padmanabhan, *Gene* **105,** 173 (1991).

[10] C. J. Watson and R. T. Hay, *Nucleic Acids Res.* **18,** 1167 (1990).

[11] B. Moss, T. Elroy-Stein, T. Mizukami, W. A. Alexander, and T. R. Fuerst, *Nature* **348,** 91 (1990).

[12] Y. M. Mul, R. T. Van Miltenburg, E. De Clercq, and P. C. Van der Vliet, *Nucleic Acids Res.* **17,** 8917 (1989).

min on ice, followed by centrifugation for 10 min at 3500 rpm and subsequently for 40 min at 100,000*g* (29,000 rpm) in the SW41 rotor. The supernatant (whole cell extract) is first adjusted to 250 m*M* KCl by adding glycerol and DTT to final concentrations of 15% and 1 m*M*, respectively. Then, the extract (200 ml; 900 mg protein; approximately 80 mg pTP–Pol) is adjusted to 100 m*M* KCl by adding 300 ml of a buffer containing 25 m*M* MES, pH 6.2, 1 m*M* DTT, 1 m*M* EDTA, 0.02% NP-40, 0.1 m*M* PMSF, 5% (w/v) sucrose, 20% (w/v) glycerol (Buffer H) and applied to a 50-ml DEAE column at 100 ml/hr. The column is washed with 100 ml Buffer H/100 m*M* KCl and developed with a 250-ml linear gradient of 100 to 500 m*M* KCl in Buffer H. The column fractions (5 ml) are screened for pTP–Pol activity using the *in vitro* DNA replication assay. Peak activity elutes around 200 m*M* KCl. Fractions eluting between 120 and 300 m*M* KCl are pooled, diluted to 100 m*M* KCl with Buffer H, and applied to an 8-ml Mono S FPLC column run at 0.5 ml/min. The column is washed with 16 ml Buffer H/100 m*M* KCl and developed with an 80-ml linear gradient of 100 to 600 m*M* KCl in Buffer H. Three-milliliter fractions are collected and screened for pTP–Pol activity. Peak activity elutes at 360 m*M* KCl. Peak fractions are combined, diluted to 140 m*M* KCl with Buffer H, and applied to a 16-ml ssDNA cellulose column at 3 ml/hr. This column is washed with 80 ml Buffer H/150 m*M* KCl at 0.5 ml/min, and eluted with a 120-ml linear gradient of 150 to 600 m*M* KCl in Buffer H. Three-milliliter fractions are collected and screened for pTP–Pol activity. Peak activity eluted at 280 m*M* KCl. Peak fractions are combined, diluted to 100 m*M* KCl with Buffer H, and applied to a 1-ml Mono S (FPLC column) at 0.5 ml/min. This column is washed with 5 ml Buffer H/100 m*M* KCl and eluted with a 20-ml linear gradient of 100 to 600 m*M* KCl in Buffer H. The total yield is approximately 4 mg of 95% pure pTP–Pol; main losses occur during the ssDNA column. One unit of pTP–Pol, defined according to Ref. 13 equals approximately 800 ng protein.

Transcription Factors Involved in Adenovirus DNA Replication

Minimal initiation requires the core origin (see Fig. 1) and the viral replication proteins. Initiation of Ad5 DNA replication is 100- to 200-fold enhanced, at least at low pTP–Pol concentrations[13] by binding of NFI and NFIII/Oct-1 to the auxiliary region. These proteins function by improving the positioning of pTP–Pol to the core origin and by stabilizing a preinitia-

[13] Y. M. Mul, C. P. Verrijzer, and P. C. Van der Vliet, *J. Virol.* **64,** 5510 (1990).

tion complex. For stimulation, the DNA-binding domains are sufficient[14-16] and therefore we routinely use these domains, called NFIBD and the Oct-1 POU domain. NFIBD contains amino acids 4–240 from rat NFI[17] and the POU domain contains amino acids 1 to 23 and 269 to 440 from human Oct-1.[16] We routinely purify these polypeptides from recombinant vaccinia virus-infected cells although isolation of replication competent POU domain is also possible from bacterial expression systems.[18] Here we describe the purification from HeLa cells infected with the appropriate recombinant vaccinia viruses.

Purification of DNA-Binding Domain of NFI (NFIBD)

Recombinant vaccinia viruses containing the rat NFIBD cDNA have been constructed as described[15] and are used to infect HeLa cells (5 pfu per cell). A whole cell extract is prepared from 10 liters of HeLa cells (containing approximately 20 mg NFIBD) 22 hr postinfection as described for pTP–Pol and applied to a 40-ml DEAE column connected to a 20-ml heparin-Sepharose column equilibrated in 25 m*M* HEPES–KOH, pH 7.5, 1 m*M* DTT, 15% (w/v) glycerol, 1 m*M* PMSF, 10 μg/ml TPCK, 5 m*M* $Na_2S_2O_5$, and 0.02% NP-40 (Buffer I) containing 250 m*M* KCl. Both columns are washed with 60 ml of Buffer I/250 m*M* KCl and disconnected. After an additional 20 ml wash, the heparin-Sepharose column is developed with a 100-ml linear gradient of 0.25 to 1 *M* KCl in Buffer I. NFIBD containing fractions elute around 525 m*M* KCl as monitored by bandshift analysis. Bandshift analysis is performed by incubating 3 fmol of the Klenow end-labeled 114 bp *Eco*RI–*Xba*I fragment from plasmid pHRI, containing the NFI recognition sequence,[19] with 0.1-μl aliquots of each fraction for 30 min at 20°. One binding unit (b.u.) is defined as the amount of protein that can bind 50% of the probe added in this assay, corresponding to 1.5 ng NFIBD. Samples are analyzed by 6% PAGE. NFIBD-containing fractions are pooled, diluted with Buffer I to 150 m*M* KCl, and applied onto a 1-ml Mono S column run at 0.25 ml/min. After washing with 10 ml Buffer I/150 m*M* KCl, NFIBD is eluted with a 20-ml linear gradient of 150 to 600 m*M* KCl in Buffer I. Peak activity elutes around 320 m*M* KCl. NFIBD-

[14] N. Mermod, E. A. O'Neill, T. J. Kelly, and R. Tjian, *Cell* **58,** 741 (1989).

[15] F. Gounari, R. De Francesco, J. Schmidt, P. C. Van der Vliet, R. Cortese, and H. G. Stunnenberg, *EMBO J.* **9,** 559 (1990).

[16] C. P. Verrijzer, A. J. Kal, and P. C. Van der Vliet, *EMBO J.* **9,** 1883 (1990).

[17] G. Paonessa, F. Gounari, R. Frank, and R. Cortese, *EMBO J.* **7,** 3115 (1988).

[18] C. P. Verrijzer, M. Strating, Y. M. Mul, and P. C. Van der Vliet, *Nucleic Acids Res.* **20,** 6369 (1992).

[19] E. De Vries, W. Van Driel, S. J. Van den Heuvel, and P. C. Van der Vliet, *EMBO J.* **6,** 161 (1987).

containing fractions are adjusted to 100 m*M* KCl with 20 m*M* Tris–HCl, pH 8.0, 1 m*M* DTT, 15% (w/v) glycerol, 0.02% NP-40, 1 m*M* PMSF, 10 μg/ml TPCK, and 5 m*M* $Na_2S_2O_5$ (Buffer J) and applied to a 1-ml Mono Q column. This column is washed with 5 ml Buffer J/100 m*M* KCl and developed with a 20-ml linear gradient of 100 to 500 m*M* KCl in Buffer J. NFIBD is present mainly in the flow-through and wash fractions. These are pooled, an equal volume of 50 m*M* KPO_4, pH 6.5, is added, and the extract is applied to a hydroxylapatite column. This column is washed with 25 m*M* KPO_4, pH 6.5, 1 m*M* DTT, 5 m*M* $Na_2S_2O_5$, 15% (w/v) glycerol (Buffer K) and developed with a linear gradient of 25 to 500 m*M* KPO_4 in Buffer K. NFIBD elutes around 180 m*M* phosphate. This schedule yields 3 mg NFIBD purified to homogeneity.

Purification of NFIII/Oct-1 POU Domain

Several members of the POU domain family of transcription factors can stimulate adenovirus DNA replication.[18] We routinely use the Oct-1 POU domain and purify this from vaccinia virus-infected cells using a protocol slightly modified from Mul *et al.*[13] Ten liters of HeLa cell culture are infected with the appropriate vaccinia virus at 5 pfu per cell, and a nuclear extract, rather than a whole cell extract, is prepared 24 hr postinfection. Cells are treated as described above for the purification of pTP–Pol but after Dounce homogenization nuclei are first spun down for 5 min at 3500 rpm. The resulting pellet is resuspended in Buffer G adjusted to 300 m*M* KCl and proteins are extracted for 60 min on ice, followed by centrifugation for 10 min at 3500 rpm and 40 min at 100,000*g*. The nuclear extract (containing approximately 15 mg POU domain) is adjusted to 100 m*M* KCl with Buffer I (see above) and applied to a 40-ml DEAE column connected to a 20-ml Fast Flow S column equilibrated in Buffer I/100 m*M* NaCl. Both columns are washed with 80 ml of Buffer I/100 m*M* NaCl and disconnected. After an additional 20 ml wash, the Fast Flow S column is developed with a 120-ml linear gradient ranging from 100 to 400 m*M* NaCl in Buffer I. Fractions eluting between 200 and 240 m*M* NaCl, containing the POU domain, are pooled, diluted to 30 m*M* NaCl, and loaded onto a 20-ml Fast Flow Q column equilibrated with Buffer J (see above) containing 30 m*M* NaCl. The flow-through plus 20 ml wash are loaded onto a 10-ml heparin-Sepharose column equilibrated with Buffer I/30 m*M* NaCl. The column is washed with 50 ml of Buffer I/100 m*M* NaCl and eluted with a 60-ml linear gradient of 100 to 600 m*M* NaCl in Buffer I. POU-domain-containing fractions—eluting around 300 m*M* NaCl—are pooled, diluted to 100 m*M* NaCl with Buffer I, and applied to a 5-ml ssDNA column equilibrated with Buffer I/100 m*M* NaCl. The column is washed with 100 ml of Buffer I/100 m*M* NaCl followed by a stepwise elution with I/400 m*M* NaCl, on which the POU domain elutes. Total yield of this purification is approximately 4 mg POU domain purified to homogeneity; one b.u. corresponds to 1.7 ng POU domain.

Assay of Various Steps in DNA Replication

Several assays exist to measure the level of initiation, partial elongation, or complete elongation *in vitro*. These assays take advantage of the covalent attachment of pTP to the newly synthesized strand. This changes the electrophoretic mobility of the DNA slightly, thus enabling us to distinguish between authentic replication and random incorporation. For initiation, ^{32}P-labeled dCTP is added as the only substrate and ddATP to block elongation. This leads to the formation of an 80-kDa pTP–dCMP complex, which can be separated by SDS–PAGE and visualized by autoradiography. Partial elongation up to nucleotide 26 can be measured by adding dCTP, dATP, dTTP, and ddGTP. This leads to an elongation block after nucleotide 26, the first G-residue in the new strand (Fig. 1). Finally, an easy and sensitive assay that is routinely used is the synthesis of origin-containing restriction fragments. For that purpose, an *Xho*I digest of Ad5 TP–DNA is added. This contains two origin fragments of approximately 5700 and 6200 bp in addition to five internal fragments varying in length between approximately 580 and 15,100 bp. After incubation in the presence of dNTPs, the products are separated on an SDS-containing agarose gel, followed by autoradiography. Under these conditions the pTP–DNA interaction remains intact, leading to a slightly reduced mobility. Labeling of internal fragments serves as a control for a specific DNA synthesis. Moreover, the second round of replication can be detected by the appearance of labeled single strands containing pTP. These have a mobility that differs from pTP–dsDNA and their presence indicates a high level of replication.

Initiation Assay and Partial Elongation

Reaction mixtures (25 μl) contain 50 ng of pTP–Pol, 50 ng of Ad5 TP–DNA, 1 μg of DBP, and 3 b.u. of NFIBD or 5 b.u. of the POU domain, in a buffer containing 25 m*M* HEPES–KOH, pH 7.5, 1 m*M* DTT, 1.5 m*M* $MgCl_2$, and NaCl to a final concentration of 55 m*M*. The initiation reaction is allowed to proceed for 1 hr at 37° in the presence of 40 μM ddATP, and 0.5 μM [α-^{32}P]dCTP (600 Ci/mmol). For partial elongation we add 40 μM dATP, dTTP, and ddGTP in addition to 0.5 μM [α-^{32}P]dCTP. The reaction is stopped by the addition of 15 μl stopmix giving final concentrations of 75 m*M* sodium pyrophosphate, 10 m*M* EDTA, and 100 ng/μl bovine serum albumin (BSA). Protein is precipitated with 7.5 μl trichloroacetic acid (TCA; 15% final concentration) for 30 min at 4°, centrifuged for 15 min at 12,000 rpm, and washed with 15% TCA. Products are analyzed on a 7.5% SDS–polyacrylamide gel followed by autoradiography. A band at 80 kDa represents the pTP–dCMP initiation product and a band around 90 kDa represents the pTP–26 nucleotides partial elongation product. Note that, in the partial elongation assay performed under these conditions, always an intermediate in initiation formed as well, consisting of pTP–CAT.[22] This can be prevented by using higher dCTP concentrations (e.g., 5 μ*M*), which favor elongation.[20,22]

Combined Initiation and Elongation Assay Using TP–DNA Restriction Fragments

Reaction mixtures (15 μl) contain 30 ng of *Xho*I digested TP–DNA, 15 ng pTP–Pol, 1 μg DBP, 2 b.u. NFIBD, and 3 b.u. POU domain in a buffer containing 25 m*M* HEPES–KOH, pH 7.5, 1 m*M* $MgCl_2$, 40 μ*M* of dATP, dGTP, and dTTP, 220 n*M* dCTP (300 Ci/mmol), and 0.4 m*M* DTT. Reactions are allowed to proceed for 1 hr at 37° and the reaction is stopped by addition of 2 μl of stopmix (30% sucrose, 1% SDS, 0.1% bromphenol blue, and 0.05% xylene cyanol). The products are analyzed by 1% agarose gel electrophoresis in the presence of 0.1% SDS. Gels are partially dehydrated and subjected to autoradiography. Incorporation is measured as described above. The two origin-containing bands are indicative of replication. Under these conditions approximately 10 nCi is incorporated corresponding to approximately 40 pg new DNA. Higher levels can be obtained by increasing the dCTP concentration, which is below the K_m for elongation (1.4 μ*M*[20]).

Conclusion

Reconstitution of Ad DNA replication using purified proteins leads to efficient initiation followed by rapid elongation at rates comparable to those measured in infected cells. *In vitro* initiation requires a short lag period[21] and is linear for at least 30 min, indicating that possibly other factors are involved in this rate-limiting process. In this respect it is noteworthy that *in vivo* replication occurs in localized foci containing presumably high concentrations of replication proteins and that DNA is attached to the nuclear matrix through the terminal protein. Nevertheless, the reconstituted replication system described here has enabled a much better understanding

[20] Y. M. Mul and P. C. Van der Vliet, *Nucleic Acids Res.* **21,** 641 (1993).
[21] Y. M. Mul and P. C. Van der Vliet, *EMBO J.* **11,** 751 (1992).
[22] A. J. King and P. C. van der Vliet, *EMBO J.* **13,** 5706 (1994).

of protein primed initiation of DNA replication. Initiation requires a well-balanced formation of a multiprotein initiation complex stabilized by multiple protein–protein and protein–DNA interactions. The mapping of the various interacting domains is still in its infancy, as are the conformational changes that will undoubtedly accompany the various interaction steps. The solving of the three-dimensional structures of two proteins (DBP and the POU domain) will enable a detailed description of this process in the near future.

Acknowledgments

We wish to thank Wim van Driel, Audrey King, Wieke Teertstra, Diederik Zijderveld, Job Dekker, Roel Schiphof, and Hans van Leeuwen for useful comments. This work was supported in part by the Netherlands Foundation for Chemical Research (SON) and the Medical Research Council (G-MW) with financial support from the Netherlands Organization for Scientific Research (NWO).

[43] Purification of Bacteriophage T4 DNA Replication Proteins

By NANCY G. NOSSAL, DEBORAH M. HINTON, LISA J. HOBBS, and PETER SPACCIAPOLI

Introduction

The bacteriophage T4 DNA replication system is a relatively simple system of ten T4 encoded proteins that together catalyze rapid and highly accurate copying of the two strands of a replication fork *in vitro*. The genes for most of the T4 replication proteins were first identified in studies of conditionally lethal phage mutants.[1] These proteins were initially purified from T4 infected *Escherichia coli* using either complementation assays, which measured their ability to stimulate DNA synthesis by a crude

[1] R. H. Epstein, A. Bolle, C. M. Steinberg, E. Kellenberger, E. Boy de la Tour, R. Chevalley, R. S. Edgar, M. Susman, G. H. Denhardt, and A. Lielausis, *Cold Spring Harbor Symp. Quant. Biol.* **28,** 375 (1963).

extract of cells infected with a replication-defective T4 mutant, or functional assays of their ability to catalyze or stimulate specific replication reactions.[2-5]

The T4 proteins required for leading and lagging strand synthesis *in vitro* have now been cloned, sequenced, and highly purified (Table I). Beginning from a preformed forked primer–template, this DNA synthesis requires T4 DNA polymerase, three polymerase accessory proteins (the 44/62 complex and 45 protein) that increase the time polymerase remains bound to the nascent strand, the gene 32 single-stranded DNA-binding protein, and a primase–helicase composed of the gene 41 and 61 proteins. The gene 41 protein by itself is a $5' \rightarrow 3'$ DNA helicase. Both the 61 and 41 proteins are required to make the pentamer RNA primers that initiate T4 lagging strand fragments (reviewed in Refs. 6–9). The gene 59 helicase assembly factor increases the binding of 41 protein to the template.[10,11] The dda helicase can substitute for the gene 41 helicase in unwinding the duplex ahead of the leading strand polymerase, but does not function as a primase.[12,13] T4 RNase H removes RNA primers from lagging strand fragments *in vitro*.[14] The resulting gaps are filled in by polymerase, and the adjacent lagging strand fragments joined by DNA ligase. The process by which T4 phage selects from multiple origins to initiate DNA synthesis *in vivo* is not yet clear (reviewed in Ref. 15) and has not been reconstituted *in vitro*.

Methods to assay and purify the T4 replication proteins from T4 infected *E. coli* and from *E. coli* with expression plasmids have been developed in several laboratories. This chapter describes procedures for most of the T4

[2] J. Barry and B. Alberts, *Proc. Natl. Acad. Sci. U.S.A.* **69,** 2717 (1972).
[3] N. G. Nossal, *J. Biol. Chem.* **254,** 6026 (1979).
[4] C. F. Morris, H. Hama-Inaba, D. Mace, N. K. Sinha, and B. M. Alberts, *J. Biol. Chem.* **254,** 6787 (1979).
[5] C. F. Morris, L. A. Moran, and B. M. Alberts, *J. Biol. Chem.* **254,** 6797 (1979).
[6] N. Nossal and B. M. Alberts, *in* "Bacteriophage T4" (C. K. Matthews, E. M. Kutter, G. Mosig, and P. B. Berget, eds.), p. 71. American Society for Microbiology, Washington, DC, 1983.
[7] B. M. Alberts, *Phil. Trans. R. Soc. Lond. B.* **317,** 395 (1987).
[8] N. G. Nossal, *FASEB J.* **6,** 871 (1992).
[9] N. G. Nossal, *in* "Molecular Biology of Bacteriophage T4" (J. Karem, ed.), p. 43. American Society for Microbiology, Washington, DC, 1994.
[10] J. Barry and B. Alberts, *J. Biol. Chem.* **269,** 33,049 (1994).
[11] P. Spacciapoli and N. G. Nossal, *J. Biol. Chem.* **269,** 477 (1994).
[12] P. Bedinger, M. Hochstrasser, C. V. Jongeneel, and B. M. Alberts, *Cell* **34,** 115 (1983).
[13] K. Hacker and B. Alberts, *J. Biol. Chem.* **267,** 20,674 (1992).
[14] H. C. Hollingsworth and N. G. Nossal, *J. Biol. Chem.* **266,** 1888 (1991).
[15] K. N. Kreuzer and S. W. Morrical, *in* "Molecular Biology of Bacteriophage T4" (Jim Karem, ed.), p. 28. American Society for Microbiology, Washington, DC, 1994.

TABLE I
BACTERIOPHAGE T4 DNA REPLICATION PROTEINS

Protein	Gene	Function	Monomer size (kDa) (subunit composition)	Refs. for *S*(equence), *E*(xpression plasmids), *P*(urification)
DNA polymerase	43	Synthesis of leading and lagging strands	103.6	*S*[a]; *E*[b,c]; *P*[c,d]
Polymerase accessory proteins		Retain polymerase on nascent chain		
44/62 complex	44		35.8	*S*[e,f]; *E*[f,g]; *P*[f,h,i]
	62		21.4	
			(44:62, 4:1)[j]	
45 protein	45		24.8	*S*[k]; *E*[f,g]; *P*[f,h,i]
Single-stranded DNA binding protein	32	Binds ssDNA Increases rate of polymerization Increases DNA binding of polymerase and accessory proteins Modulates primase	33.5	*S*[l]; *E*[m]; *P*[m,n,o]
Helicase–primase	41	Unwinds duplex DNA Essential for pentamer primer synthesis	53.8	*S*[p]; *E*[q]; *P*[h,q,r,s]
Primase	61	Primer synthesis	39.8	*S*[p]; *E*[t]; *P*[t,u,v]
Helicase assembly protein	59	Facilitates binding of 41 protein helicase	26.0	*S*[w]; *E*[x,y]; *P*[x,y]
DNA helicase	*dda*	Unwinds duplex DNA	49.9	*S*[z]; *E*[z]; *P*[z]
RNase H	*rnh*	Removes primers	35.6	*S*[w]; *E*[aa]; *P*[aa,bb]
DNA ligase	30	Joins lagging strand fragments	55.3	*S*[cc]; *E*[dd]; *P*[ee]

[a] E. K. Spicer, J. Rush, C. Fung, L. J. Reha-Krantz, J. D. Karam, and W. H. Konigsberg, *J. Biol. Chem.* **263,** 7478 (1988).
[b] T.-C. Lin, J. Rush, E. K. Spicer, and W. H. Konigsberg, *Proc. Natl. Acad. Sci. U.S.A.* **84,** 7000 (1987).
[c] P. Spacciapoli and N. G. Nossal, *J. Biol. Chem.* **269,** 448 (1994).
[d] W. H. Konigsberg, this volume [26].
[e] E. K. Spicer, N. G. Nossal, and K. R. Williams, *J. Biol. Chem.* **259,** 15,425 (1984).
[f] J. Rush, T. C. Lin, M. Quinones, E. K. Spicer, I. Douglas, K. R. Williams, and W. H. Konigsberg, *J. Biol. Chem.* **264,** 10,943 (1989).
[g] T. Hsu, R. Wei, M. Dawson, and J. D. Karam, *J. Virol.* **61,** 366 (1987).
[h] N. G. Nossal, *J. Biol. Chem.* **254,** 6026 (1979).

replication proteins. Alternative methods can be found in the references listed in Table I. Methods to purify T4 dda helicase and DNA ligase are not described here, and can be found in Hacker and Alberts[13] and Weiss *et al.*,[16] respectively.

Assays for T4 Replication Proteins

Unless otherwise indicated the enzymes are diluted, if necessary, just prior to the assay in a buffer containing 50 m*M* Tris–acetate, pH 7.5, 100 m*M* KCl, 5 m*M* MgCl, 1 m*M* Na_2EDTA, and 25% glycerol, 10 m*M* dithiothreitol (DTT), and 200 μg/ml bovine serum albumin (BSA).

[16] B. Weiss, A. Jacquemin-Sablon, T. R. Live, G. C. Fareed, and C. C. Richardson, *J. Biol. Chem.* **243,** 4543 (1968).

---

[i] C. F. Morris, H. Hama-Inaba, D. Mace, N. K. Sinha, and B. M. Alberts, *J. Biol. Chem.* **254,** 6787 (1979).
[j] T. C. Jarvis, L. S. Paul, and P. H. von Hippel, *J. Biol. Chem.* **264,** 12,709 (1989).
[k] E. K. Spicer, J. A. Noble, N. G. Nossal, W. H. Konigsberg, and K. R. Williams, *J. Biol. Chem.* **257,** 8972 (1982).
[l] K. R. Williams, M. LoPresti, M. Setoguchi, and W. H. Konigsberg, *Proc. Natl. Acad. Sci. U.S.A.* **77,** 4614 (1980).
[m] Y. Shamoo, H. Adari, W. H. Konigsberg, K. R. Williams, and J. W. Chase, *Proc. Natl. Acad. Sci. U.S.A.* **83,** 8844 (1986).
[n] M. Bittner, R. Burke, and B. M. Alberts, *J. Biol. Chem.* **254,** 9565 (1979).
[o] M. Venkatesan and N. G. Nossal, *J. Biol. Chem.* **257,** 12,435 (1982).
[p] M. Nakinishi, B. Alberts, H. E. Selik, and B. M. Abremski, GenBank volume 72, accession no. K03113. Geneworks Data Banks, IntelliGenetics, Inc., Mountain View, CA (1991).
[q] D. M. Hinton, L. L. Silver, and N. G. Nossal, *J. Biol. Chem.* **260,** 12,851 (1985).
[r] R. W. Richardson and N. G. Nossal, *J. Biol. Chem.* **264,** 4732 (1989b).
[s] C. F. Morris, L. A. Moran, and B. M. Alberts, *J. Biol. Chem.* **254,** 6797 (1979).
[t] D. M. Hinton and N. G. Nossal, *J. Biol. Chem.* **260,** 12,858 (1985).
[u] L. L. Silver and N. G. Nossal, *J. Biol. Chem.* **257,** 11,696 (1982).
[v] R. L. Burke, M. Munn, J. Barry, and B. M. Alberts, *J. Biol. Chem.* **260,** 1711 (1985).
[w] S. Hahn and W. Ruger, *Nucleic Acids Res.* **17,** 6729 (1989).
[x] P. Spacciapoli and N. G. Nossal, *J. Biol. Chem.* **269,** 447 (1994).
[y] J. Barry and B. Alberts, *J. Biol. Chem.* **269,** 33,049 (1994).
[z] K. Hacker and B. Alberts, *J. Biol. Chem.* **267,** 20,674 (1992).
[aa] H. C. Hollingsworth and N. G. Nossal, *J. Biol. Chem.* **266,** 1888 (1991).
[bb] L. J. Hobbs and N. G. Nossal, unpublished experiments (1992).
[cc] J. Armstrong, R. S. Brown, and A. Tsugita, *Nucleic Acids Res.* **11,** 7145 (1983).
[dd] G. G. Wilson and N. E. Murray, *J. Mol. Biol.* **132,** 471 (1979).
[ee] B. Weiss *et al., J. Biol. Chem.* **243,** 4543 (1968).

Complementation Assay

This complementation assay measures the ability of a crude extract or a purified protein to stimulate DNA synthesis from an undefined DNA template present in crude extracts of *E. coli* infected with a T4 amber mutant in gene 41, 44, 62 or 45.[3] This assay, and similar procedures developed by Alberts and coworkers,[2,4,5] allowed the T4 41, 45, and 44/62 proteins to be purified at a time when genetic studies had demonstrated that these gene products were essential for DNA replication, but the functions of the proteins had not yet been defined, and all of the other proteins required to reconstitute fork synthesis were not available. The complementation assay is also useful for assaying these replication proteins in crude extracts of T4 infected cells, because high levels of DNA and RNA nucleases in these extracts interfere with measuring synthesis from defined templates.

Extracts. The assay requires a crude extract (receptor) made from cells infected with phage defective in the protein to be assayed. The receptor extract provides both the DNA template and all other replication proteins. The enzyme of interest is provided by a DNA-depleted extract from cells infected with phage producing this enzyme (donor extract) or by more purified enzyme fractions. To prepare receptor and donor extracts, *E. coli* D110 (DNA *endoI*$^-$, *polA1*)[17] is grown to 5×10^8/ml at 37° in M9 medium supplemented with 0.1 μg/ml of thiamine, 20 μg/ml of thymidine, and 0.3% casamino acids. It is convenient to use 400 ml of cells to prepare both the receptor and donor fractions from each T4 mutant, and to process several mutants simultaneously. Bacteria are infected at a multiplicity of seven with T4 bacteriophage containing a mutation in *regA* (SP62) as well as an amber mutation in genes 41, 44, 62, 45, or 42 (dCMP hydroxymethylase). As noted below, the *regA* (SP62) mutation results in overproduction of the 44/62 and 45 proteins.[18] Fifteen minutes after infection, the cultures are poured on top of 400 ml of cold (4°) complementation lysis buffer (50 m*M* Tris–HCl, pH 7.5, 1 m*M* Na_2EDTA, pH 7.0, 20% sucrose) in centrifuge bottles and centrifuged at 4° for 5 min at 6000*g*. The cells are gently resuspended in 8 ml of lysis buffer, transferred to glass tubes, centrifuged for 10 min at 8000*g*, and resuspended in 0.4 ml of cold 0.1 *M* NaCl, 10 m*M* Tris–HCl, pH 7.8, 20% sucrose. After the addition of 0.1 ml of a solution of 4 mg/ml of lysozyme in 0.1 *M* Tris–HCl, pH 8.0, 50 m*M* Na_2EDTA, the cells are left on ice with occasional gentle stirring for 20 min; 0.5 ml of a solution of 0.2 *M* KCl and 0.01 *M* Na_2EDTA is then added, and the cells left on ice 20 min longer. The cells are then lysed by freezing and thawing

[17] R. E. Moses and C. C. Richardson, *Proc. Natl. Acad. Sci. USA* **67,** 674 (1970).

[18] J. S. Wiberg, S. Mendelsohn, V. Warner, K. Hercules, C. Aldrich, and J. L. Munro, *J. Virol.* **12,** 775 (1973).

three times, alternating between dry ice–ethanol and water at room temperature.

The portion of the extract to be used as a receptor is immediately divided into small aliquots, frozen in dry ice, and stored at −80°. The receptor extracts are extremely viscous and must be transferred with wide-bore pipettes. Donor preparations are made by adding $MgCl_2$ to the remaining extract to a final concentration 0.02 *M*, and centrifuging for 90 min at 100,000*g* to remove DNA. The clear supernatant is divided into small aliquots and stored at −80°.

Complementation Assay. The receptor extract is thawed just prior to the assay and diluted with an equal volume of 0.06 *M* $MgCl_2$ in lysis buffer. A homogeneous suspension is formed by repeatedly pipetting with a 100-μl disposable glass pipette. A 10-μl aliquot of the diluted receptor is mixed at 4° with 10 μl of donor extract (or purified protein) and 20 μl of a solution of 80 m*M* Tris–HCl, pH 7.4, 80 m*M* KCl, 0.33 m*M* deoxyadenosine, 2.5 m*M* dCTP, 30 μM dTTP and dGTP, 10 μM [^{3}H]dATP (400 cpm/pmol), 3.3 m*M* rATP, and 30 μM each rCTP, rUTP, and rGTP. (A higher concentration of dCTP is required due to the T4 dCTPase in the extracts.) After 10 min of incubation at 37°, acid precipitable DNA is determined by the following procedure. A grid of about 25 squares is drawn with a number 1 pencil on a 12.5-cm-diameter glass fiber filter (Whatman GF/C), which is then saturated with a solution of 10% trichloroacetic acid (TCA) and 0.01 *M* sodium pyrophosphate, and placed on an open petri dish. Each complementation assay mixture is transferred to a square on the filter. The filter is washed in a funnel on a vacuum flask, first with 50 ml 10% TCA, then three times with 50 ml ice-cold H_2O, and finally with 25 ml ethanol. The filter is cut into squares, which are counted in a scintillation solution. A unit of complementing activity is defined as the incorporation of 1 nmol of dAMP in 10 min at 37°, after subtracting that incorporated by the receptor alone, assuming no dilution by dATP in the extract.

T4 DNA Polymerase Activity with Denatured Sperm DNA

During enzyme purification, polymerase activity is determined using alkali denatured salmon sperm DNA (0.6 μmol nucleotide/ml), 100 μM of each dNTP including [^{3}H]dTTP (300 cpm/pmol), 25 m*M* Tris–acetate, pH 7.5, 60 m*M* potassium acetate, 6 m*M* magnesium acetate, 10 m*M* DTT, and 200 μg/ml BSA in a final volume of 10 μl.[19] After 15 min at 30°, the DNA product is precipitated as described for the complementation assay. A unit is defined as the incorporation of 1 nmol of dTMP under these conditions.

[19] P. Spacciapoli and N. G. Nossal, *J. Biol. Chem.* **269,** 438 (1994).

Primer-Dependent DNA Synthesis on Circular Single-Stranded DNA

In this reaction pentamer RNA primers, whose synthesis requires both the T4 61 and 41 proteins,[20,21] are elongated by T4 DNA polymerase in conjunction with the T4 genes 44/62 and 45 DNA polymerase accessory proteins, and the gene 32 DNA-binding protein. This is a convenient assay for the genes 61 and 41 primase components, and for each of the polymerase accessory proteins, when all of the other required proteins are present at nonlimiting concentration.[22,23] Reaction mixtures (10 μl) contain 25 mM Tris–acetate, pH 7.5, 60 mM potassium acetate, 6 mM $MgCl_2$, 10 mM DTT, 100 μg/ml BSA, 230 pmol (as nucleotide) of ϕX174 or M13 viral DNA, 0.5 mM rATP, 0.1 mM each rCTP, rGTP, rUTP, dATP, dCTP, and dGTP, and 0.1 mM [^{3}H]dTTP at 300 cpm/pmol. The nonlimiting concentrations (μg/ml) of the required T4 replication proteins used in this assay are T4 DNA polymerase (2), 32 protein (90), 44/62 and 45 polymerase accessory proteins (40 and 6, respectively), 41 protein helicase–primase (20), and 61 protein primase (1.2). Reactions are incubated at 30° for 30 min, and the DNA precipitated as described for the complementation assay. Rifampicin at 40 μg/ml is added to partially purified extracts to inhibit any RNA polymerase activity.

DNA Synthesis on DNA Oligonucleotide-Primed Single-Stranded DNA Templates

Primer Elongation. Polymerase and the polymerase-accessory proteins can be assayed in the absence of the primase–helicase by using an oligonucleotide primer (25- to 35-mer) annealed to ϕX174 or M13 DNA under the reaction conditions above. The 4dNTP and rATP are present at the concentrations above, but rUTP, rCTP, and rGTP are omitted.[11,19,24]

Coupled Leading and Lagging Strand Synthesis. This reaction is measured with the same buffer used for primer-dependent DNA synthesis, except that a preformed fork is provided by annealing the circular DNA (16 fmol) to an oligonucleotide with 34 complementary bases at the 3′ end and 50 noncomplementary bases at the 5′ end.[11,14,25] Reaction mixtures (10 μl) contain 16 fmol (circular molecules) primer–template, 1 mM rATP, 250 μM rCTP, rUTP, and rGTP, and 250 μM of each dNTP including a [^{32}P]dNTP (400 cpm/pmol). When nonlimiting, the concentrations of the

[20] D. M. Hinton and N. G. Nossal, *J. Biol. Chem.* **262,** 10,873 (1987).
[21] T.-A. Cha and B. M. Alberts, *Biochemistry* **29,** 1791 (1990).
[22] D. M. Hinton and N. G. Nossal, *J. Biol. Chem.* **260,** 12,858 (1985).
[23] D. M. Hinton, L. L. Silver, and N. G. Nossal, *J. Biol. Chem.* **260,** 12,851 (1985).
[24] T. L. Capson, S. J. Benkovic, and N. G. Nossal, *Cell* **65,** 249 (1991).
[25] R. W. Richardson, R. L. Ellis, and N. G. Nossal, *UCLA Symp. Mol. Cell Biol.* **127,** 247 (1990).

T4 proteins (μg/ml) used in this assay are polymerase (2), 32 protein (35), 44/62 and 45 polymerase accessory proteins (40 and 12, respectively), and the primase–helicase proteins 61 and 41 (12 and 85, respectively). The gene 59 helicase assembly factor, which stimulates but is not required for this synthesis, is present at 1.2 μg/ml.[11] Reaction mixtures with all proteins except polymerase and 61 protein are assembled on ice, incubated for 2 min at 30 or 37°, and synthesis is begun by adding a mixture of polymerase and 61 protein. Products are analyzed by acid precipitation as described above, or by electrophoresis on alkaline agarose gels.[26] In the absence of the 61 primase component, this forked primer–template can be used to measure helicase-dependent leading strand synthesis.

Purification Procedures

The procedures that follow use autoclaved buffers (with 2-mercaptoethanol, DTT, and $MgSO_4$ added, if indicated, after sterilization), and sterile columns, plastic, and glassware. Unless otherwise indicated the purification steps are carried out at 4°. During sonication the extract is kept in a salt–ice water bath, and the sonication interrupted periodically to maintain the temperature below 8°. Extracts and intermediate fractions are frozen in dry ice and stored at −80° if there will be a delay in going to the next step. The final purified proteins are stored in small aliquots at −80°.

Purification of T4 DNA Polymerase[19]

T4 DNA polymerase by itself copies single-stranded DNA templates, but requires the three polymerase accessory proteins and 32 protein for strand-displacement synthesis on duplex DNA. The polymerase has a potent proofreading 3′ → 5′-exonuclease (see chapters by Konigsberg[27] and Reha-Krantz[28]). This polymerase is a useful reagent for filling in 5′ overhangs, removing 3′ overhangs, and labeling duplex DNA by exonucleolytic hydrolysis followed by replacement synthesis.

Plasmids Encoding T4 DNA Polymerase. T4 gene 43, encoding T4 DNA polymerase, was first cloned by Lin *et al.*[29] in the expression plasmid pTL43W, in which polymerase expression is controlled by the heat inducible λ P_L promoter. pPST4Pol[19] contains T4 DNA polymerase controlled by

[26] M. Venkatesan, L. L. Silver, and N. G. Nossal, *J. Biol. Chem.* **257,** 12,426 (1982).
[27] W. H. Konigsberg, this volume [26].
[28] L. J. Reha-Krantz, this volume [25].
[29] T.-C. Lin, J. Rush, E. K. Spicer, and W. H. Konigsberg, *Proc. Natl. Acad. Sci. U.S.A.* **84,** 7000 (1987).

TABLE II
PURIFICATION OF WILD-TYPE T4 DNA POLYMERASE[a]

Fraction	Total protein[b] (mg)	Total activity[c] ($U \times 10^{-5}$)	Specific activity (U/mg)
I. High-speed supernatant	820	3.36	410
II. DEAE	159	2.87	1800
III. Phosphocellulose	77	2.34	3040

[a] Adapted from P. Spacciapoli and N. G. Nossal, *J. Biol. Chem.* **269,** 438 (1994).
[b] From 3 liters *E. coli* MV1190/pPST4Pol.
[c] A unit is 1 nmol TMP incorporated in 15 min at 30° using alkali denatured salmon sperm DNA.

the isopropyl-β-D-thiogalactopyranoside (IPTG) inducible *lac* promoter, in a vector with an f1 origin to facilitate site-directed mutagenesis.

The following method has been used to prepare wild-type T4 DNA polymerase from pTL43W and pPST4Pol (Table II), and mutant T4 DNA polymerases from plasmids derived from pPST4Pol.[19] An alternative method for T4 DNA polymerase purification is described in this volume by Konigsberg.[27]

Cell Growth. E. coli 71-18 containing pTL43W[29] is grown in L-broth with 35 μg/ml ampicillin in a fermentor at 29° to mid-log phase, then shifted to 40° to inactivate the heat-sensitive cI857 λ repressor. The cells are harvested after 2 hr and stored at −80°.

To produce polymerase from the pPST4Pol plasmids,[19] *E. coli* MV1190[30] containing a plasmid encoding wild-type or a mutant polymerase is streaked on minimal agar plus glucose and ampicillin (35 μg/ml), and grown overnight at 37° (30° for temperature-sensitive polymerases) until small colonies are visible. We have found that this selection by growth on minimal medium, which ensures that the MV1190 cells retain the f1 episome encoding both the *lac* repressor and the *proAB* genes needed for growth in the absence of proline, is essential for consistent induction of high levels of polymerase from plasmids with the lac promoter. A single fresh colony is used to inoculate 3 ml of minimal media containing glucose and ampicillin. After growing overnight at 30 or 37°, the entire 3 ml is used to inoculate 60 ml of L-broth containing ampicillin. This culture is grown at 30 or 37° to 3×10^8 cells/ml; 8 ml is then used to inoculate each of six flasks containing 500 ml of L-broth plus ampicillin. When the cells reach 3×10^8/ml, IPTG is added to a final concentration of 1 m*M*, the culture harvested after 3 hr

[30] J. Geisselsoder, F. Witney, and P. Yuckenberg, *Biotechniques* **5,** 786 (1987).

at 30° for the temperature-sensitive polymerases, or 2 hr at 37° for the other polymerases, and the cells stored at −80°.

Extract. The packed cells (about 12 g) are suspended in 70 ml 43-sonication buffer [50 m*M* Tris–HCl, pH 7.5, 25% glycerol, 1 m*M* Na_2EDTA, 10 m*M* 2-mercaptoethanol, 1 mg/ml phenylmethylsulfonyl fluoride (PMSF)], and then sonicated until the absorbance at 560 nm has decreased to about 40% of its initial value. The extract is centrifuged for 20 min at 10,000*g*, and then centrifuged for 2 hr at 100,000*g* (high-speed supernatant) (Table II).

DEAE Chromatography. A 100-ml column of DEAE-Sephacel (Pharmacia) is equilibrated with 43-DE buffer (43-sonication buffer without PMSF). The high-speed supernatant is diluted with an equal volume of a solution containing 25% glycerol, 1 m*M* Na_2EDTA, and 10 m*M* 2-mercaptoethanol to lower the conductivity to that of 43-DE buffer with 50 m*M* KCl, and then loaded on the column. The column is washed with 150 ml 43-DE buffer and the polymerase is eluted in the middle of a linear gradient (1 liter) of 0 to 0.15 *M* KCl in 43-DE buffer. Polymerase is identified by SDS–gel electrophoresis, and its activity in the pooled fractions assayed with the sperm DNA template as described above.

Phosphocellulose Chromatography. A 25-ml column of phosphocellulose P11 (Whatman) is equilibrated with 43-PC buffer (50 m*M* Tris–HCl, pH 7.5, 25% glycerol, 10 m*M* KCl, 0.1 m*M* Na_2EDTA, 10 m*M* 2-mercaptoethanol). The pooled DEAE fraction is diluted with 1.5 volumes of a solution containing 25% glycerol, 0.1 m*M* Na_2EDTA, and 10 m*M* 2-mercaptoethanol to reduce the conductivity to that equivalent to 43-PC buffer with 50 m*M* KCl. The diluted fraction is loaded on the column, which is then washed with 43-PC buffer until no further protein is eluted (about 50 ml). Polymerase is eluted in the middle of a linear gradient (250 ml) of 0 to 0.4 *M* KCl in 43-PC buffer.

The purification of homogeneous wild-type polymerase from pPST4Pol by this method[19] is summarized in Table II.

Rapid Small-Scale Purification of T4 DNA Polymerases for Screening Mutants

In this procedure[31] mutant or wild-type T4 DNA polymerase is adsorbed on phosphocellulose by adding the resin directly to the crude extract, and the enzyme purified by batchwise elution. It is convenient for quickly preparing almost homogeneous polymerase from multiple mutants simultaneously for initial screening.

[31] N. G. Nossal and L. J. Hobbs, unpublished experiments (1991).

A single colony of *E. coli* MV1190 containing pPST4Pol or a mutant derivative is isolated and grown overnight as described above. The overnight culture is diluted 1:50 into 70 ml L-broth plus ampicilin, and then grown at 30 or 37° to 3×10^8 cells/ml. IPTG is added to a final concentration of 1 m*M*, the cells harvested after 2 hr, and frozen at −80°. The cells are suspended in 5 ml 43-sonication buffer, sonicated with a microprobe, centrifuged for 25 min at 10,000*g* (low-speed supernatant) and then 2 hr at 100,000*g* (high-speed supernatant).

The salt concentration of the high-speed supernatant (approximately 5 ml) is diluted by adding 10 ml of 43R-PC buffer (50 m*M* Tris–HCl, pH 7.5, 10% glycerol, 1 *M* Na_2EDTA, 1 m*M* DTT) containing 1 m*M* PMSF. One milliliter of a 1:1 suspension of phosphocellulose P11 equilibrated in 43R-PC buffer is added to the diluted extract. After mixing for 1 hr on a rotary shaker in the cold room, the suspension is centrifuged for 20 min at 10,000*g*. The supernatant is discarded, and the resin washed with 20 ml 43R-PC buffer containing 0.1 *M* KCl but no PMSF by rotating for 30 min, and then centrifuged for 20 min. After discarding the supernatant, the P11 resin is transferred to a 1.5-ml microcentrifuge tube using 1 ml 43R-PC buffer with 0.1 *M* KCl, and pelleted by centrifuging for 5 min at 10,000*g* in a microcentrifuge in the cold room. After discarding the supernatant, the polymerase is eluted by adding 0.5 ml 43R-PC buffer with 0.3 *M* KCl, mixing the P11 suspension on a rotary shaker for 30 min in the cold room, and then centrifuging for 5 min at 10,000*g*. Figure 1 shows a gel of the purification of the wild-type and exonuclease-defective (Asp219A1a)[32] mutant T4 DNA polymerases by this procedure.

Other Applications of This Rapid Phosphocellulose Purification Procedure

A large-scale version (3-liter culture) of this rapid procedure has been used to purify the exonuclease-defective (Asp219Ala) mutant[32] of T4 DNA polymerase from the high-speed supernatant in a single step.[31] This polymerase is essentially homogeneous, and is suitable for most applications such as end-labeling DNA. However, T4 DNA polymerase isolated by this procedure has low levels of contaminating RNase H activity not present in polymerase made by the two-column method described above.

The ability to adsorb proteins on phosphocellulose resin added directly to a crude extract has proved to be a powerful tool for the rapid purification of several T4 proteins. Appropriate modifications of the procedure de-

[32] M. W. Frey, N. G. Nossal, T. L. Capson, and S. J. Benkovic, *Proc. Natl. Acad. Sci. USA* **90,** 2579 (1993).

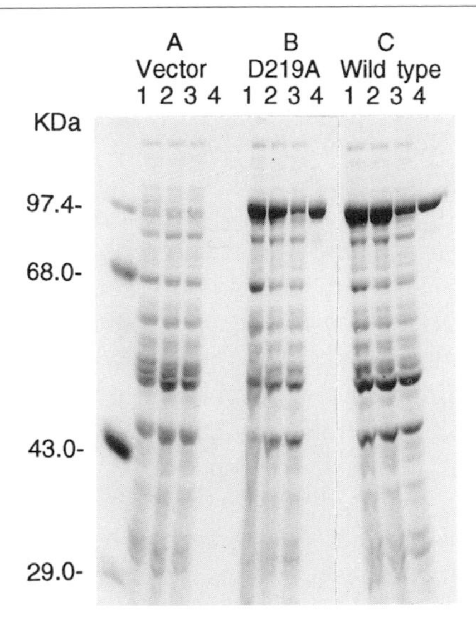

FIG. 1. Rapid purification of wild-type and mutant T4 DNA polymerase expressed from pPST4Pol and its derivatives. (A) pTZ18U vector. (B) Exonuclease-defective (Asp219Ala) mutant. (C) Wild type. Within each set: lane 1, 10,000*g* supernatant; lane 2, 100,000*g* supernatant; lane 3, protein not bound to phosphocellulose; lane 4, protein eluted from phosphocellulose by buffer with 0.3 *M* KCl. See text for details of the purification procedure. Proteins were separated on a 12% SDS–polyacrylamide gel.

scribed here for T4 DNA polymerase have been used to purify the T4 gene 61 primase,[22] RNase H,[33] and the gene 59 helicase assembly protein[11] (each described below), as well as the T4 SegA endonuclease,[34] and the T4 MotA transcription factor[35] from *E. coli* with expression plasmids. The best purification of the last four of these proteins was achieved by adsorbing the protein from the crude extract on to the PC resin, washing the resin extensively by the batchwise procedure, packing the washed resin into a column, and eluting the protein using a linear salt gradient.

Purification of T4 Gene 44/62 and 45 Polymerase Accessory Proteins[3,44]

The T4 gene 44/62 and 45 polymerase accessory proteins increase the time that T4 DNA polymerase remains bound on the primer–template, and thus increase the processivity of both leading and lagging strand synthesis *in vitro*.[2–5,11,36] The 44/62 complex is a DNA-dependent ATPase that is

[33] L. J. Hobbs and N. G. Nossal, unpublished experiments (1992).

[34] M. Sharma, R. L. Ellis, and D. M. Hinton, *Proc. Natl. Acad. Sci. U.S.A.* **89,** 6658 (1992).

[35] D. M. Hinton, *J. Biol. Chem.* **266,** 18,034 (1991).

[36] T. C. Jarvis, J. W. Newport, and P. H. von Hippel, *J. Biol. Chem.* **266,** 1830 (1991).

strongly stimulated by the 45 protein.[2–4,37–39] Structurally and functionally the 44/62 complex resembles the *E. coli* γ complex and eukaryotic RF-C (activator 1), whereas 45 protein resembles *E. coli* β protein and eukaryotic PCNA (reviewed in Refs. 9 and 40–42).

The complex of the gene 44 and 62 proteins remains tightly bound during purification, and complements extracts defective in either gene 44 or gene 62.[2–4] The ratio of the 44:62 subunits in the complex has been estimated to be 4:1.[43] The 44/62 and 45 polymerase accessory proteins are conveniently prepared from the same extract, since they are easily separated from each other on DEAE cellulose. The following method was developed for extracts of *E. coli* infected with mutants in the T4 *regA* gene.[3,44] The T4 RegA protein is a translational regulator of several T4 genes. The *regA* mutants produce high levels of both the 44/62 and 45 proteins.[18] Similar procedures have been reported by Morris *et al.*,[4] and an alternative procedure for purifying 45 protein separately has been described by Nossal.[3] The replication activity of these proteins is assayed during purification with the complementation assay described above. These purification procedures have been used to prepare 44/62 and 45 proteins from extracts of *E. coli* 71-18 with the expression plasmid pTL151WX, which contains T4 genes 44, 62, and 45 under the control of the heat-inducible λ P_L promoter.[39] In this case, the purification of the proteins can be followed simply by SDS–gel electrophoresis.

Extract. E. coli ER21,[45] defective in DNA endonuclease I, is grown to a concentration of 1.6×10^9/ml in a glycerol–salts medium[46] at 37° in a 300-liter fermentor and infected at a multiplicity of 8 with T4 *amN55*(gene 42), SP62 (*regA*).[18] The mutation in gene 42, encoding dCMP hydroxymethylase, prevents T4 DNA replication, and thus extends the time of synthesis of the polymerase accessory proteins, and the time before cell lysis. Sixty minutes after infection, the culture is cooled by circulating

[37] J. R. Piperno and B. M. Alberts, *J. Biol. Chem.* **253,** 5174 (1978).

[38] T. C. Jarvis, L. S. Paul, J. W. Hockensmith, and P. H. von Hippel, *J. Biol. Chem.* **264,** 12,717 (1989).

[39] J. Rush, T. C. Lin, M. Quinones, E. K. Spicer, I. Douglas, K. R. Williams, and W. H. Konigsberg, *J. Biol. Chem.* **264,** 10,943 (1989).

[40] X.-P. Kong, R. Onrust, M. O'Donnell, and J. Kuriyan, *Cell* **69,** 425 (1992).

[41] M. O'Donnell, R. Onrust, F. B. Dean, M. Chen, and J. Hurwitz, *Nucleic Acids Res.* **21,** 1 (1993).

[42] T. Tsurimoto, and B. Stillman, *J. Biol. Chem.* **266,** 1950 (1991).

[43] T. C. Jarvis, L. S. Paul, and P. H. von Hippel, *J. Biol. Chem.* **264,** 12,709 (1989).

[44] E. K. Spicer, J. A. Noble, N. G. Nossal, W. H. Konigsberg, and K. R. Williams, *J. Biol. Chem.* **257,** 8972 (1982).

[45] J. Eigner and S. Block, *J. Virol.* **2,** 320 (1968).

[46] D. Fraser and E. A. Jerrel, *J. Biol. Chem.* **205,** 291 (1953).

cold water in the jacket of the fermentor, and the cells are harvested by centrifugation, and the cell paste stored at −80°.

The cell paste (200 g) is thawed by suspension in 360 ml of complementation lysis buffer (see Complementation Assay section above) at 4° and lysed by the addition of 4 ml of 1 *M* Tris–HCl, pH 7.9, and 40 ml of lysozyme (4 mg/ml in 10 m*M* Tris–HCl, pH 8.0, 50 m*M* Na_2EDTA). After 45 min at 4° with occasional stirring, the extract is frozen and thawed three times as described for the complementation assay above, $MgCl_2$ is added to a final concentration of 20 m*M*, and the extract centrifuged for 2 hr at 100,000*g*.

DEAE-Cellulose Chromatography. The supernatant is loaded onto a column of DE23 (Whatman) (740 ml) that has been equilibrated with 44-DE buffer (10 m*M* Tris–HCl, pH 7.4, 1 m*M* Na_2EDTA, 10 m*M* 2-mercaptoethanol, 5 m*M* $MgSO_4$, 25% glycerol). The column is washed first with 200 ml 44-DE buffer and then developed with a linear gradient (8 liters) of 0 to 0.5 *M* KCl in 44-DE buffer. 44/62 Protein does not bind to DEAE under these conditions, whereas 45 protein is eluted at 0.25 *M* KCl (Table III).

44/62 Protein Purification[3]

Hydroxylapatite Chromatography. The DEAE fraction containing the 44 protein complementing activity is applied to a hydroxylapatite column (550 ml) equilibrated with 44-HA buffer (20 m*M* potassium phosphate, pH 7.0, 10 m*M* 2-mercaptoethanol, 1 m*M* Na_2 EDTA, 5 m*M* $MgSO_4$, and 10% glycerol). The column is washed with 200 ml of this buffer, and the 44/62 protein eluted at 0.15 *M* potassium phosphate using a linear gradient (4 liters) of 0.02 to 0.2 *M* potassium phosphate in 44-HA buffer.

TABLE III
PURIFICATION OF T4 GENE 44/62 DNA POLYMERASE ACCESSORY PROTEIN COMPLEX FROM T4 INFECTED CELLS[a]

Fraction	Protein (mg)	44 Complementing activity	
		($U \times 10^{-5}$)	(U/mg)
I. Crude extract	10,033	0.27	2.7
II. DEAE	2,772	1.29	46.4
III. Hydroxylapatite	472	3.08	654.
IV. Phosphocellulose	312	1.83	588.

[a] Adapted from N. G. Nossal, *J. Biol. Chem.* **254,** 6026 (1979). From 205 g *E. coli* ER21 infected with T4 *amN55*(gene 42), SP62(*reg A*).

Phosphocellulose. The hydroxylapatite fraction is dialyzed against three changes each of 3 liters of 44-PC buffer (20 m*M* potassium phosphate, pH 6.5, 1 m*M* Na_2EDTA, 1 m*M* 2-mercaptoethanol, and 10% glycerol) (14 hr total) and then applied to a phosphocellulose P11 column (180 ml) equilibrated in the same buffer. The column is washed with 300 ml of the 44-PC buffer and 44/62 protein eluted at 0.15 *M* KCl using a linear gradient (3.2 liters) of 0 to 0.5 *M* KCl in 44-PC buffer.

The 44/62 protein appears to be homogeneous after the phosphocellulose column. There is cochromatography of the 35.8-kDa gene 44 and 21.4-kDa gene 62 proteins, the activity complementing the gene 44 and gene 62 mutant extracts, and the activity necessary for synthesis on single-stranded ϕX174 DNA (as described above) in the presence of T4 DNA polymerase, the gene 45 polymerase accessory protein, gene 32 protein, and the genes 41 and 61 primase-helicases. Table III summarizes a preparation of 44/62 protein using this procedure.[3]

DNA Cellulose. This step is used to remove traces of single-stranded DNA endonuclease found in the PC fraction. The protein (7 mg) is dialyzed against 44-DC buffer (50 m*M* NaCl, 20 m*M* Tris–HCl, pH 8.1, 1 m*M* Na_2EDTA, 10% glycerol, 1 m*M* DTT) and applied to a 2.5 ml column of single-stranded calf thymus DNA cellulose[47] equilibrated in the same buffer. 44/62 Protein is not bound under these conditions, and is separated from the retained nuclease.

45 Protein Purification[3,44]

Hydroxylapatite. The DEAE fractions with 45 protein are applied to a hydroxylapatite column (500 ml) that has been equilibrated with 45-HA buffer (20 m*M* potassium phosphate, pH 7.0, 10% glycerol, 5 m*M* $MgSO_4$, and 1 m*M* 2-mercaptoethanol). The column is washed with 700 ml of this buffer and the 45 protein eluted at 0.14 *M* potassium phosphate using a linear gradient (4 liters) of 0.02 to 0.25 *M* potassium phosphate in 45-HA buffer.

DNA-Cellulose Chromatography. The 45 protein from the HA column is dialyzed against 45-DC buffer (50 m*M* NaCl, 1 m*M* Na_2EDTA, 1 m*M* 2-mercaptoethanol, 20 m*M* Tris–HCl, pH 8.1, and 10% glycerol), and then applied to a 20-ml single-stranded sperm DNA-cellulose[47] column equilibrated in the same buffer. 45 Protein does not bind under these conditions and is thus separated from a trace of *E. coli* RNA polymerase present in the HA fractions from infected cells. 45 Protein is stable when stored in this buffer at −80°.

[47] B. M. Alberts and G. Herrick, *Methods Enzymol.* **21D,** 198 (1971).

The homogeneous 27-kDa gene 45 protein cochromatographs on the last two columns with the activity complementing the gene 45 mutant extract, and with the activity necessary for synthesis on single-stranded ϕX174 DNA, as described above, in the presence of T4 DNA polymerase, the gene 44/62 polymerase accessory protein complex, gene 32 protein, and the genes 41 and 61 primase–helicase. Using this procedure 45 mg of 45 protein was obtained from 200 g of cells infected with a T4 *regA* mutant.

Purification of Gene 32 Single-Stranded DNA-Binding Protein

The T4 gene 32 single-stranded DNA binding protein was discovered by Alberts[48] who found that it was the T4 protein with the strongest affinity for single-stranded DNA cellulose. Bittner *et al.*[49] have reported the most comprehensive study of methods for purifying this protein. The following procedure, modified from Bittner *et al.*,[49] employs chromatography on DEAE cellulose to remove DNA from the extract, prior to affinity chromatography on single-stranded DNA cellulose. This procedure has been used to purify the 32 protein in infected cells and that encoded by a plasmid.[50]

Cell Growth. The plasmid pYS6[51] contains T4 gene 32 under the control of the λ P_L promoter. *E. coli* N4830[52] contains the heat-sensitive cI857 λ repressor. *E. coli* N4830/pYS6 is grown in L-broth with 35 μg/ml ampicillin in a fermentor at 29° to mid-log phase, then shifted to 42°. The cells are harvested after 2 hr, and stored at −80°.

Extract. The cell paste (25 g) is suspended in 100 ml 32-sonication buffer (10 m*M* Tris–HCl, pH 8.0, 50 m*M* NaCl, 1 m*M* Na_2EDTA, 5 m*M* 2-mercaptoethanol, 10% glycerol, and 5.6 m*M* PMSF). The cells are broken by sonication in pulses until the absorbance at 580 nm has decreased to 40% of its initial value. The extract is centrifuged for 2 hr at 100,000*g* (high-speed supernatant).

DEAE Chromatography. A 120-ml column of DEAE (Whatman DE23) is equilibrated with 32-DE buffer (20 m*M* Tris–HCl, pH 8.0, 1 m*M* Na_2EDTA, 5 m*M* 2-mercaptoethanol, 10% glycerol) containing 50 m*M* NaCl. The extract (110 ml) is diluted with 32-DE buffer to reduce its conductivity to that of 32-DE buffer with 0.1 *M* NaCl, before loading it on the column. The column is washed with 225 ml 32-DE buffer containing 50 m*M* NaCl, and the 32 protein eluted beginning at 0.25 *M* NaCl using a

[48] B. M. Alberts and L. Frey, *Nature* **227,** 1313 (1970).

[49] M. Bittner, R. L. Burke, and B. M. Alberts, *J. Biol. Chem.* **254,** 9565 (1979).

[50] N. G. Nossal, unpublished experiments (1988).

[51] Y. Shamoo, H. Adari, W. H. Konigsberg, K. R. Williams, and J. W. Chase, *Proc. Natl. Acad. Sci. U.S.A.* **83,** 8844 (1986).

[52] S. Adhya and M. Gottesman, *Cell* **29,** 939 (1982).

2-liter linear gradient of 0.05 to 0.5 *M* NaCl in 32-DE buffer. 32 Protein is identified by SDS–gel electrophoresis.

DNA-Cellulose Chromatography. A 20-ml column of single-stranded calf thymus DNA cellulose[47] is washed with 32-DC buffer (20 m*M* Tris–HCl, pH 8.0, 1 m*M* Na_2EDTA, 10 m*M* 2-mercaptoethanol, 10% glycerol) until no further material absorbing at 260 nm (free DNA) is eluted, and then washed with 20 ml of the same buffer with 5 m*M* Na_2EDTA. The DEAE fraction of 32 protein is dialyzed against three changes of 2 liters each of 32-DC buffer with 50 m*M* NaCl and 5 m*M* Na_2EDTA, and then loaded on the column. The high concentration of Na_2EDTA is used to decrease hydrolysis of the bound DNA by nucleases in the DEAE fraction. The column is washed with 40 ml of the dialysis buffer, and then with 40 ml of the same buffer containing 0.1 *M* NaCl. At this point the absorbance at 280 nm of the eluant should be equivalent to that of the buffer. The column is then developed successively with a 120-ml linear gradient of 0.1 to 0.6 *M* NaCl in 32-DC buffer, 40 ml of 32-DC buffer with 0.6 *M* NaCl, and finally 50 ml of 32-DC buffer with 2.0 *M* NaCl. 32 Protein binds cooperatively to the single-stranded DNA on the column. With high concentrations of 32 protein, such as that in extracts of cells with the pYS6 plasmid, most of the 32 protein is not eluted until 2.0 *M* NaCl. With the lower concentrations present in some extracts of infected cells, there is significant 32 protein eluted beginning at about 0.6 *M* NaCl. Using this procedure, 43 mg of apparently homogeneous 32 protein were purified from 25 g of cells with the pYS6 plasmid.

Phenyl-Sepharose Chromatography. Traces of nucleases often present in the DNA cellulose fraction of 32 protein can be removed by phenyl-Sepharose chromatography as described by Bittner *et al.*[49] The phenyl-Sepharose resin (Pharmacia LKB) is washed extensively with deionized H_2O, and then equilibrated with 32-PS buffer (30 m*M* Tris–HCl, pH 8.0, 1 m*M* DTT, 0.1 *M* NaCl, 10 m*M* $MgCl_2$, and 10% glycerol) containing 1.05 *M* (14%) $(NH_4)_2SO_4$. The 32 protein is dialyzed against 32-PS buffer without $MgCl_2$, and then diluted with an equal volume of 32-PS buffer containing 2.1 *M* (28%) $(NH_4)_2SO_4$ and 20 m*M* $MgCl_2$, before loading it on the column. The column is developed with 4 column volumes of 32-PS buffer with 14% $(NH_4)_2SO_4$, then with a linear gradient of 20 column volumes of 14 to 0% $(NH_4)_2SO_4$ in 32-PS buffer, and finally with 2 volumes of 32-PS buffer. 32 Protein is eluted toward the end of the gradient or in the 32-PS buffer without $(NH_4)_2SO_4$. We have found that the yield of 32 protein varies greatly with the batch of phenyl-Sepharose, and it is important to use the smallest column that will adsorb the protein. For example, in one preparation, only 20% of 40 mg of 32 protein was recovered from a 5-ml column.

TABLE IV
PURIFICATION OF T4 GENE 41 HELICASE FROM pDH518 EXPRESSION PLASMID[a]

Fraction	Volume (ml)	Protein (mg)	DNA synthesis[b] (U/mg)	Recovery (%)
I. 130,000*g* supernatant	1280	4990	64	100
II. DEAE pool	360	970	180	53
III. DNA cellulose eluate	760	1140	180	66
IV. Hydroxylapatite pool	250	450	310	44
V. Precipitated 41 protein	78	133	650	27

[a] Adapted from D. M. Hinton, L. L. Silver, and N. G. Nossal, *J. Biol. Chem.* **260,** 12,841 (1985).
[b] Primer-dependent DNA synthesis in the presence of T4 DNA polymerase, the gene 44/62 and 45 polymerase accessory protein, gene 32 single-stranded DNA-binding protein, and the gene 61 primase component.

Purification of T4 Gene 41 DNA Helicase[23,53]

The T4 gene 41 protein is both a DNA helicase and an essential component of the primase. The 5′ → 3′-helicase of 41 protein opens the duplex ahead of polymerase, increasing the rate of leading strand synthesis about tenfold *in vitro.*[26] This helicase activity requires a nucleoside triphosphate (ATP, GTP, dATP, or dGTP) that is hydrolyzed by 41 protein in the presence of single-stranded DNA.[26,54] Both the 41 and 61 proteins are needed to make the pentamer RNA primers that initiate lagging strand synthesis.[21,22] In the absence of 41 protein, the major products of synthesis by 61 protein are dimers that cannot be elongated by T4 DNA polymerase.[20,55]

Two different procedures have been developed to purify the T4 gene 41 helicase from *E. coli* OR1265 containing pDH518, a plasmid with gene 41 controlled by the heat-inducible λ P_L promoter.[23] Both procedures use the same method to grow the cells and prepare the high-speed supernatant fraction. In the first, 3 columns are used to prepare homogeneous 41 protein[23] (Table IV). In the second rapid procedure, essentially homogeneous 41 protein is precipitated from the supernatant with a low concentration of $(NH_4)_2SO_4$.[53] The rapid procedure is much simpler, but as noted below, we have had variable success in redissolving the 41 protein.

Extract. A 50-liter fermentor culture of pDH518/OR1265[23] is grown at 29° in L-broth plus 25 μg/ml ampicillin to 1×10^9/ml, the temperature

[53] R. W. Richardson and N. G. Nossal, *J. Biol. Chem.* **264,** 4732 (1989).
[54] C. Liu and B. M. Alberts, *J. Biol. Chem.* **256,** 2813 (1981).
[55] N. G. Nossal and D. M. Hinton, *J. Biol. Chem.* **254,** 10,879 (1987).

raised to 42° to inactivate the cI857 repressor, and the cells harvested after 2.5 hr and stored at −80°. Twenty-seven grams of cell paste is suspended in 500 ml of complementation lysis buffer (see above). A solution of 100 mg lysozyme in 25 ml of 100 m*M* Tris–HCl, pH 8.0, 50 m*M* Na_2EDTA is added, the mixture kept on ice for 45 min, and finally frozen and thawed twice by transferring between a dry ice–ethanol bath and H_2O at room temperature. After the addition of 12 ml of 1 *M* $MgCl_2$, the extract is centrifuged for 90 min at 100,000*g*. About 50% of the 41 protein remains soluble after this centrifugation.

DEAE-Cellulose Chromatography. The supernatant (520 ml) is diluted with 1280 ml of 41-DE buffer (20 m*M* Tris–HCl, pH 7.4, 1 m*M* Na_2EDTA, 10 m*M* 2-mercaptoethanol, 5 m*M* $MgSO_4$, 25% glycerol), and then loaded on a 430-ml column of DE23 (Whatman) equilibrated with the same buffer. The column is washed with 350 ml of 41-DE buffer, and 41 protein eluted at 0.12 *M* KCl using a linear gradient (4 liters) of 0 to 0.35 *M* KCl in 41-DE buffer (Table IV).

Single-Stranded DNA-Cellulose Chromatography. The DEAE fraction of 41 protein (360 ml) is diluted twofold with 41-DC buffer (20 m*M* Tris–HCl, pH 7.5, 1 m*M* Na_2EDTA, 10 m*M* 2-mercaptoethanol, 25% glycerol) and then loaded on a 17-ml column of single-stranded calf thymus DNA cellulose[47] equilibrated with 41-HA buffer (41-DC buffer plus 50 m*M* KCl). 41 Protein does not bind to the column and is found in the first fractions eluted.

Hydroxylapatite Chromatography. 41 Protein from the DNA cellulose column (760 ml) is loaded on a 240-ml column of Bio-Gel HT (Bio-Rad) equilibrated with 41-HA buffer. The column is washed with 250 ml of this buffer, and 41 protein eluted at 0.18 *M* $(NH_4)_2SO_4$ using a linear gradient (4 liters) of 0 to 0.76 *M* $(NH_4)_2SO_4$ in 41-HA buffer.

Precipitation of 41 Protein. The hydroxylapatite fraction of 41 protein is diluted with an equal volume of 2.3 *M* (30%) $(NH_4)_2SO_4$ in 41 HA buffer. 41 Protein precipitates overnight when stored in the cold room at about 4°. (Note that it will not precipitate under these conditions if stored on ice.) The protein is collected by centrifugation at 20,000*g* for 30 min, resuspended in a third of its original volume of 41-HA buffer, and dialyzed against 41 dialysis buffer (10 m*M* Tris–acetate, pH 7.5, 25 m*M* potassium acetate, 5 m*M* magnesium acetate, 1 m*M* DTT, and 25% glycerol). Table IV summarizes a purification of the T4 41 protein helicase by this procedure.[23]

Rapid Purification of T4 Gene 41 Protein.[31,53] *E. coli* OR1265 (pDH518) is grown as described above. Sixteen grams of cell paste is suspended in 350 ml of complementation lysis buffer, to which 3.5 ml of 0.1 *M* PMSF and 80 mg of lysozyme dissolved in 20 ml 10 m*M* Tris–HCl, pH 8.0, and 1 m*M* Na_2EDTA have been added. The suspension is frozen and thawed three times as above, $MgCl_2$ is added to 20 m*M*, and the extract centrifuged for 2 hr at 100,000*g*.

The supernatant is diluted with 0.5 volumes of lysis buffer containing 20 m*M* $MgCl_2$. PMSF and 2-mercaptoethanol are then added to final concentrations of 1 and 10 m*M*, respectively. An equal volume of 2.7 *M* (36%) $(NH_4)_2SO_4$ dissolved in 41-HA buffer is added. After 24 hr at 4° (in refrigerator or cold room, but not on ice) the precipitate is collected by centrifugation for 30 min at 20,000*g* at 4°. The supernatant is carefully removed and stored separately at 4°. If more 41 protein precipitates, it can be harvested like the first batch.

The precipitate is washed with 16% $(NH_4)_2SO_4$ in 41-HA buffer, centrifuged as above, and the pellet is dissolved in 40 ml 41-HA buffer. (If necessary more 41-HA buffer is added to try to dissolve the protein. Protein that does not dissolve at this stage is discarded. The amount of this protein has varied in different preparations, for reasons that are not yet clear).[31] The precipitation of 41 protein is repeated by adding an equal volume of 32% $(NH_4)_2SO_4$ in 41-HA buffer to give a final concentration of 16%. After 24 hr at 4°, the suspension is centrifuged 15 min at 10,000*g*. The pellet is dissolved in 30 ml 41-dialysis buffer (see above), and dialyzed against three changes of 2 liters of 41-dialysis buffer over a total of 4 hr. The yield of 41 protein from 16 g of cells is usually about 70 mg.

Purification of T4 Gene 61 Primase Component[22]

The T4 gene 61 protein by itself makes the dimers pppApC and pppGpC in a template directed reaction.[20,21] In conjunction with 41 protein it makes the pentamer primers (pppApCpNpNpN and pppGpCpNpNpN) that initiate T4 lagging strand fragments *in vitro.* Only pentamers starting with A are made using T4 DNA, since the hydroxymethylated and glucosylated cytosines in T4 DNA do not act as a template for the T4 primase.[20] pppApCpNpNpN is the predominant primer found on lagging strand fragments *in vivo.*[56]

Extract. T4 gene 61 has been cloned under the control of the heat-inducible λ P_L promoter in the expression plasmid pDH911.[22] *E. coli* N4830[52] containing this plasmid is grown at 29° in L-broth plus 25 μg/ml ampicillin to mid-log phase, the temperature shifted to 42°, and the cells harvested after 2.5 hr and stored at −80°. Frozen cells (178 g) are suspended in 61-sonication buffer (20 m*M* Tris–HCl, pH 8.0, 50 m*M* NaCl, 5 m*M* Na_2EDTA, 1 m*M* 2-mercaptoethanol, 10% glycerol, and 1 m*M* PMSF), and then sonicated in pulses until the absorbance at 560 nm has decreased to

[56] Y. Kurosawa and T. Okazaki, *J. Mol. Biol.* **135,** 841 (1979).

about 50% of its initial value. The extract is centrifuged for 20 min at 11,000*g* (low-speed supernatant). (Table V).

Phosphocellulose Batch Absorption and Chromatography. Phosphocellulose resin (120-ml packed volume, Whatman P11), equilibrated with 61-PC buffer (20 m*M* Tris–HCl, pH 8.0, 1 m*M* Na_2EDTA, 10 m*M* 2-mercaptoethanol, and 10% glycerol) containing 0.2 *M* NaCl, is added to the supernatant (650 ml) and the suspension stirred for 40 min. The resin is collected by centrifugation for 20 min at 13,000*g*, washed by resuspension and stirring for 20 min in 240 ml more 61-PC buffer with 0.2 *M* NaCl, and collected by centrifugation. 61 Protein and other proteins binding under these conditions are eluted by stirring the resin for 30 min with 240 ml 61-PC buffer with 0.8 *M* NaCl, followed by centrifugation. The batch-eluted PC fraction is dialyzed against 61-PC buffer with 0.1 *M* NaCl until its conductivity is reduced to that of 61-PC buffer with 0.2 *M* NaCl. Precipitated protein is removed by centrifugation and discarded. The dialyzed fraction is then loaded on a 100-ml P11 column previously equilibrated with 61-PC buffer containing 0.2 *M* NaCl. The column is washed with 950 ml of the equilibrating buffer, and 61 protein eluted at 0.4 *M* NaCl using a linear gradient (2 liters) of 0.2 to 0.6 *M* NaCl in 61-PC buffer. 61 Protein activity is measured by primer-dependent DNA synthesis on single-stranded ϕX174 DNA, as described above, in the presence of the other replication proteins.

TABLE V

PURIFICATION OF T4 GENE 61 PRIMASE COMPONENT FROM pDH911 EXPRESSION PLASMID[a]

Fraction	Volume (ml)	Protein (mg)	DNA synthesis activity[b] ($U \times 10^{-7}$)	DNA synthesis activity[b] ($U \times 10^{-4}$/mg)	Recovery (%)
I. Sonicated extract[c]	720	40,000			
II. 11,000*g* supernatant[c]	650	32,000			
III. Batch phosphocellulose	270	1,500	5.7	3.8	100
IV. Phosphocellulose column	126	240	1.9	7.9	33
V. Hydroxylapatite column[d]	85	100	0.8	8.3	15

[a] Adapted from D. M. Hinton and N. G. Nossal, *J. Biol. Chem.* **260,** 12,858 (1985). From 178 g *E. coli* N4830/pDH911.

[b] Primer-dependent DNA synthesis on ϕX174DNA in the presence of T4 DNA polymerase, the genes 44/62 and 45 polymerase accessory proteins, 32 single-stranded DNA binding protein, and the gene 41 helicase–primase.

[c] 61 protein cannot be assayed quantitatively in the crude fractions.

[d] The hydroxylapatite pool represents approximately two-thirds of the total 61 protein eluted from the column.

Hydroxylapatite Chromatography. The phosphocellulose fraction of 61 protein is loaded on a 90-ml column of Bio-Rad Bio-Gel HT, equilibrated with 61-HA buffer (20 m*M* Tris–HCl, pH 7.5, 1 m*M* Na_2EDTA, 10 m*M* 2-mercaptoethanol, 25% glycerol, and 100 m*M* KCl). The column is washed with 270 ml of this buffer. 61 Protein is eluted at 0.16 *M* $(NH_4)_2SO_4$ using a linear gradient (2 liters) of 0 to 0.4 *M* $(NH_4)_2SO_4$ in 61-HA buffer. A purification of 61 protein using this procedure is summarized in Table V.[22]

Purification of T4 Gene 59 Helicase Assembly Factor[11]

59 Protein Helicase Stimulating Activity. In the *in vitro* T4 DNA replication system there is a significant lag before the gene 41 DNA helicase adds to all the forked molecules in the reaction.[11,26,57] The T4 gene 59 protein has been called a helicase assembly factor because it increases the rate at which 41 protein adds to 32 protein-covered DNA, but does not increase the rate of leading strand synthesis beyond that achieved by replication forks to which 41 protein has added in the absence of 59 protein.[10,11] The activity of 59 protein is conveniently assayed with a preformed fork primer–template, as described above. Under these conditions all of the molecules are replicating at the faster rate dependent on the 41 protein helicase within 2 min at 30° in the presence of 59 protein, whereas only about 50% are replicating at the faster rate by 10 min in its absence.[11,58] 41 Protein-dependent primer synthesis is also evident at earlier times with the 59 protein.

Cell Growth. The expression plasmid pNN2859[11] has T4 gene 59 cloned under control of the phage T7 promoter in the vector pVEX11.[59] *E. coli* BL21 (DE3)plysS has T7 RNA polymerase under the control of the IPTG inducible *lac* promoter.[60] *E. coli* BL21 (DE3)pLysS/pNN2859 is grown overnight at 37° in 50 ml L-broth with 35 μg/ml ampicillin and 30 μg/ml chloramphenicol. The overnight culture (5 ml) is used to inoculate each of six 2-liter flasks containing 500 ml L-broth with ampicillin and chloramphenicol. The cultures are grown at 37° to 5×10^8 cells/ml. IPTG is added to a final concentration of 1 m*M*. After 4 more hours, the cells are harvested by centrifugation and frozen at −80°.

Extract. The cell paste from 3 liters is suspended in 100 ml 59-sonication buffer (50 m*M* Tris–HCl, pH 7.5, 0.2 *M* KCl, 5 m*M* $MgCl_2$, 1 m*M* DTT, 10% glycerol, and 1 m*M* PMSF), broken by sonication and centrifuged for

[57] T.-A. Cha and B. M. Alberts, *J. Biol. Chem.* **264,** 12,220 (1989).

[58] N. G. Nossal, unpublished experiments (1990).

[59] V. K. Chandhardy, I. Pastan, and S. Adhya, personal communication (1990).

[60] F. W. Studier, A. H. Rosenberg, J. J. Dunn, and J. W. Dubendorff, *Methods Enzymol.* **185,** 60 (1990).

20 min at 8000*g* (low-speed supernatant), and then for 2 hr at 100,000*g* (high-speed supernatant).

Phosphocellulose Chromatography. Phosphocellulose (Whatman P11) is equilibrated with 59-PC buffer (50 m*M* Tris–HCl, pH 7.5, 1 m*M* Na_2EDTA, 1 m*M* DTT, and 10% glycerol). The high-speed supernatant (100 ml) is diluted with 100 ml 59-PC buffer in order to reduce its conductivity to that of 59-PC buffer with 0.1 *M* KCl. PC resin (25 ml) is added to the diluted supernatant. The suspension is stirred slowly (magnetic stirrer) for 1 hr in the cold room, and then centrifuged for 20 min at 8000*g*. The supernatant is decanted and discarded. Proteins binding weakly to the resin are removed by washing it first with 200 ml 59-PC buffer, and then with 100 ml 59-PC buffer with 0.2 *M* KCl, stirring for 30 min, and centrifuging for 20 min at 8000*g* after each addition. The resin is then transferred into a column using 10 ml 59-PC buffer with 0.2 *M* KCl, and 59 protein eluted beginning at 0.3 *M* KCl using a linear gradient (200 ml) of 0.2 to 1.0 *M* KCl in 59-PC buffer. Fractions with 59 protein are identified by SDS–gel electrophoresis.

Single-Stranded DNA-Cellulose Chromatography. A 5-ml column of denatured calf thymus DNA cellulose[47] is equilibrated with 59-PC buffer containing 0.05 *M* KCl. The phosphocellulose fractions from the peak of 59 protein are pooled (20 ml at about 0.48 *M* KCl) and diluted with 80 ml 59-PC buffer. The diluted protein is loaded on the column, and the column washed with 50 ml 59-PC buffer with 0.1 *M* KCl. The 59 protein is eluted beginning at 0.3 *M* KCl using a linear gradient (150 ml) of 0.1 to 1.0 *M* KCl in 59-PC buffer. Using this procedure, 23 mg of homogeneous 59 protein were obtained from 3 liters of cells.[11]

59 Protein activity in stimulating both DNA and primer synthesis on the forked DNA templates described above is highest in the phosphocellulose and DNA cellulose fractions with the highest concentrations of 59 protein (measured by SDS–gel electrophoresis). However, it is not possible to determine from this assay whether the activity coincides exactly with the elution of the 59 protein since the stimulatory activity does not increase linearly with 59 protein concentration under these conditions.[11,58]

Purification of T4 RNase H[14,33]

The T4 *rnh* gene, located just upstream of gene 33, encodes an enzyme that degrades the RNA strand of RNA–DNA hybrids, but not single-stranded RNA.[14] It also has $5' \rightarrow 3'$-exonuclease activity on double-stranded DNA. T4 RNase H is able to remove the pentamer RNA primers from DNA chains initiated by the T4 primase–helicase *in vitro.*[14,33] T4 RNase

TABLE VI
PURIFICATION OF T4 RNase H FROM pNN2202 EXPRESSION PLASMID

Fraction	Volume (ml)	Protein (mg)	RNase H activity	
			($U \times 10^{-3}$/mg)	($U \times 10^{-5}$)
I. Sonicated extract	200	1400[a]	0.24	3.3
II. 100,000*g* supernatant	200	660	0.26	1.7
III. Phosphocellulose column pool[b]	28	59	1.7	1.0
IV. Hydroxylapatite column pool[c]	34	19	2.1	0.40[c]

[a] From 6 liters *E. coli* BL21(DE3)pLysS/pNN2202.
[b] The phosphocellulose pool represents approximately three-fourths of the total RNase H protein eluted from the column.
[c] Only two-thirds of the phosphocellulose pool was applied to the hydroxylapatite column.

H is essential for T4 DNA replication in *E. coli* hosts that are defective in the 5′ → 3′-exonuclease activity of DNA polymerase I.[33]

RNase H Activity. During purification RNase H activity is assayed by the degradation of the poly(rA) strand of poly(rA) · poly(dT).[14,33] Reaction mixtures contain 40 m*M* Tris–HCl, pH 7.8, 10 m*M* DTT, 20 m*M* NaCl, 10 m*M* $MgCl_2$, 0.5 mg/ml BSA, and 200 pmol nucleotide equivalents of poly(dT), 0.6 pmol ^{3}H-labeled poly(rA) (500 cpm/pmol) annealed to poly(dT), and enzyme in a total volume of 10 μl. After 20 min at 30°, reactions are terminated by the addition of 40 μl of salmon sperm DNA (2.5 mg/ml) and 50 μl of 0.5 *M* perchloric acid. Tubes are chilled on ice for 10 min and centrifuged for 5 min in a microcentrifuge. The 80-μl aliquots of the supernatant are counted in an appropriate scintillation solution. A unit of activity is defined as 1 nmol of ribonucleotide hydrolyzed in 20 min at 30°.

Cell Growth. The plasmid pNN2202[14] contains the T4 *rnh* gene under control of the phage T7 promoter in a vector derived from pT7-7.[61] *E. coli* BL21 (DE3)plysS[60] containing pNN2202 is grown to mid-log phase (5×10^8/ml) at 37° in L-broth with 35 μg/ml ampicillin and 30 μg/ml chloramphenicol. IPTG is added to a final concentration of 0.4 m*M* to induce the expression of T7 RNA polymerase. The cells are harvested after 2 hr and stored at −80°.

[61] S. Tabor, *in* "Current Protocols in Molecular Biology" (F. M. Ausubel, R. Brent, R. E. Kingston, D. D. Moore, J. G. Seidman, J. A. Smith, and K. Struhl, eds.), p. 16. John Wiley & Sons, New York, 1990.

Extract. The cell paste from 6 liters of culture is resuspended in 200 ml of RNH-sonication buffer (50 m*M* Tris–HCl, pH 7.5, 0.2 *M* KCl, 5 m*M* $MgCl_2$, 1 m*M* DTT, 10% (v/v) glycerol, and 1 m*M* PMSF). The cells are broken by sonication in pulses until the absorbance at 580 nm has decreased to 40% of its initial value. The extract is centrifuged for 20 min at 8600*g*, and then for 2 hr at 100000*g* (high-speed supernatant). (Table VI).

Phosphocellulose Chromatography. Phosphocellulose (Whatman P11) is equilibrated with RNH-PC buffer (50 m*M* Tris–HCl, pH 7.5, 1 m*M* Na_2EDTA, 1 m*M* DTT, and 10% glycerol). The high-speed supernatant (200 ml) is diluted with RNH-PC buffer containing 1 m*M* PMSF (approximately 550 ml) in order to reduce its conductivity to that of RNH-PC buffer with 0.1 *M* KCl. Then 50 ml of the packed PC resin is added to the diluted supernatant. The suspension is stirred slowly (magnetic stirrer) for 1 hr in the cold room, and then centrifuged for 20 min at 10,000*g*. The supernatant is decanted and discarded. Proteins binding weakly to the resin are removed by washing it first with 200 ml RNH-PC buffer with 1 m*M* PMSF, and then with 100 ml RNH-PC buffer with 50 m*M* KCl and 1 m*M* PMSF, stirring for 30 min, and centrifuging for 20 min at 10,000*g* after each addition. The resin is then transferred into a column using 10 ml RNH-PC buffer with 50 m*M* KCl. T4 RNase H is eluted between 0.17 and 0.32 *M* KCl using a linear gradient (400 ml) of RNH-PC buffer with 0.05 to 0.8 *M* KCl. The enzyme is monitored by SDS–gel electrophoresis, and by the degradation of the poly(rA) chain in poly(rA) · poly(dT) as described above.

Hydroxylapatite Chromatography. The phosphocellulose fractions from the first half of the T4 RNase H peak, which contain the fewest contaminating proteins, are pooled and loaded on a 30-ml column of Bio-Rad Bio-Gel HT, equilibrated with RNH-HA buffer (20 m*M* Tris–HCl, pH 8.0, 0.1 m*M* Na_2EDTA, 1 m*M* DTT, 10% glycerol, and 100 m*M* NaCl). The column is washed with 30 ml of this buffer. T4 RNase H is eluted between 0.21 and 0.31 *M* $(NH_4)_2SO_4$ using a 600-ml linear gradient of 0 to 0.6 *M* $(NH_4)_2SO_4$ in RNH-HA buffer. In the preparation of T4 RNase H using this procedure that is summarized in Table VI,[33] two-thirds of the pooled phosphocellulose fraction was chromatographed on a 20-ml column of Bio-Gel HT, and the buffers were reduced proportionately.

Section VII

DNA Synthesis *in Vivo*

[44] DNA Replication of Bacteriophage T4 *in Vivo*

By GISELA MOSIG and NANCY COLOWICK

Introduction

Various methods used to investigate bacteriophage T4 DNA replication *in vivo* and interpretations of the results have to take into account the fact that T4 uses multiple modes to initiate replication forks. In wild-type T4, each of the different modes occurs at different stages of the developmental program, but in some mutants in which any one of these modes is selectively inactivated, another mode may substitute as a bypass mechanism under certain conditions.[1–4]

The T4 genome is large for a bacterial virus. It comprises about 168,800 base pairs. It encodes most of the T4 DNA replication and recombination proteins (except for host RNA polymerase) and enzymes that degrade the host DNA[5] (Fig. 1). Because the host DNA is degraded, phage DNA is selectively labeled after infection. The breakdown products of the host DNA and RNA are rapidly converted into precursors for T4 DNA replication by an enzyme complex that is associated with the replication apparatus.[6,7] This recycling process greatly reduces incorporation of exogenously added precursors into the earliest replicating phage DNA.

Two properties of the T4 chromosomes are important to recall when considering DNA replication. First, the ends of individual chromosomes are nearly, but not completely, randomly permuted over the circular map (Fig. 1). Mature T4 virions contain linear, double-stranded DNA molecules of about 173,000 base pairs. The difference between this size and the genome size is due to a so-called terminal redundancy, a repeat of 3000 to 5000

[1] A. Luder and G. Mosig, *Proc. Natl. Acad. Sci. U.S.A.* **79,** 1101 (1982).

[2] G. Mosig, A. Luder, A. Ernst, and N. Canan, *The New Biologist* **3,** 1195 (1991).

[3] G. Mosig, N. Colowick, M. E. Gruidl, and A. J. Harvey, *FEMS Microbiology Reviews*, in press (1995).

[4] K. N. Kreuzer and S. W. Morrical, *in* "The Molecular Biology of Bacteriophage T4" (J. Karam, ed.), p. 28. American Society for Microbiology, Washington, DC, 1994.

[5] E. Kutter, T. Stidham, B. Guttman, E. Kutter, D. Batts, S. Peterson, T. Djavakhishvili, F. Arisaka, V. Mesyanzhinov, W. Rüger, and G. Mosig, *in* "The Molecular Biology of Bacteriophage T4" (J. Karam, ed.), p. 491. American Society for Microbiology, Washington, DC, 1994.

[6] C. K. Mathews, *Prog. Nucleic Acid Res. Mol. Biol.* **44,** 167 (1993).

[7] G. R. Greenberg, P. He, J. Hilfinger, and M.-J. Tseng, *in* "The Molecular Biology of Bacteriophage T4" (J. Karam, ed.), p. 14. American Society for Microbiology, Washington, DC, 1994.

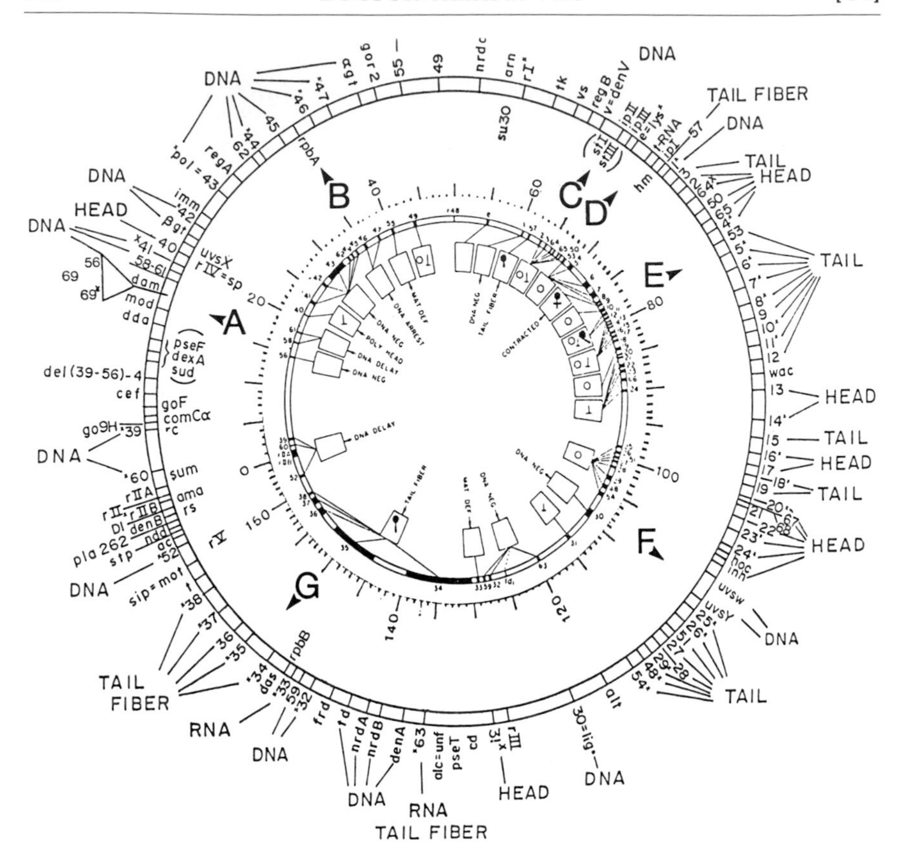

FIG. 1. A map of the T4 genome with the positions of replication origins (labeled A through G) relative to known genes. Modified from G. Mosig and F. Eiserling, *in* "The Bacteriophages" (R. Calendar, ed.), Vol. 2, p. 521. Plenum, New York, 1988.

base pairs at both ends of most T4 chromosomes.[8] The terminal redundancy is sufficiently large to allow efficient homologous recombination between the two copies of a single chromosome. Because recombination-dependent DNA replication is essential, this aspect is important for successful infection of a host cell by a single T4 particle.

Second, the glucosylation and hydroxymethylation of the cytosine residues protect T4 DNA from the T4 enzymes that degrade the host DNA, as well as from most of the known restriction enzymes. This property impedes direct analysis of nascent T4 DNA with most restriction enzymes.

[8] G. Streisinger, *in* "Phage and the Origins of Molecular Biology" (J. Cairns, G. S. Stent, and J. D. Watson, eds.), p. 335. Cold Spring Harbor Laboratory Press, NY, 1966.

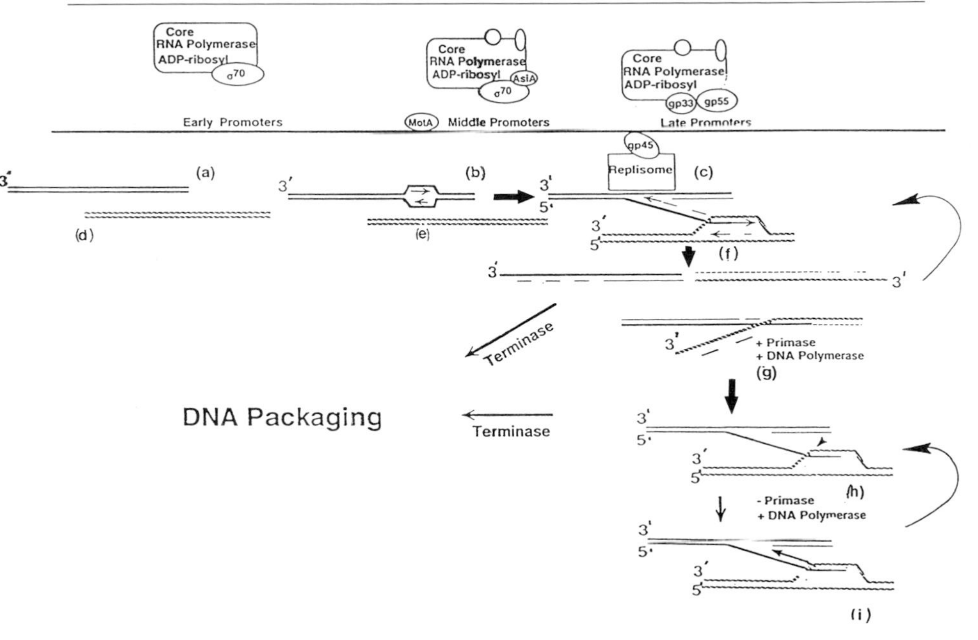

FIG. 2. A diagram to illustrate the relationship of different transcription modes and replication modes, discussed in the text. (a–c) represent the different modes of transcription. (d) Two unreplicated T4 chromosomes. (e) One of the chromosomes has initiated bidirectional replication at an origin. (f) Once the rightward fork has reached an end, the 3′ terminus of the template for lagging strand synthesis invades a homologous region in the other chromosome and initiates join–copy replication/recombination. The products of this replication (g) can reiterate the process. If primase is defective, no Okazaki pieces are synthesized. The displaced strand can invade as in (f), but no join–copy synthesis occurs (h). After a cut by endonuclease VII, the strand that was initially displaced can be copied in the retrograde direction, using the 3′ end at the cut as primer (i). This process can also be reiterated. At late times, packaging proteins compete with the replication proteins for binding to partially single-stranded DNA.

On the other hand, these modifications facilitate the separation of phage and host DNA in Cs_2SO_4 density gradients, because they increase the buoyant density of the phage DNA.

The first mechanism to initiate T4 DNA replication after infection resembles initiation at origins of certain other replicons (Fig. 2e). In any individual chromosome it is initiated from one[9,10] of several origins (labeled A through G in Fig. 1). Because individual T4 chromosomes are circular permutations of the map, origins are located at variable distances from the ends. Transcription by RNA polymerase generates primers for leading

[9] R. Dannenberg and G. Mosig, *J. Virol.* **40,** 890 (1981).
[10] R. Dannenberg and G. Mosig, *J. Virol.* **45,** 813 (1983).

strand DNA synthesis (Fig. 2). Okazaki pieces for lagging strand synthesis are then primed by primase. σ^{70}-containing host RNA polymerase, the only known host protein directly required for T4 DNA replication, synthesizes primers for leading strand DNA synthesis at origins.[1,3]

Origin initiation in most wild-type T4 chromosomes is bidirectional. In primase- or topoisomerase-deficient mutants, initiation from *oriA* or *oriF* is unidirectional in the direction of transcription, as expected if transcripts prime leading strand synthesis in one direction, but primase is required to prime Okazaki pieces so that the first Okazaki piece becomes the leading strand in the direction opposite to that of transcription. Probably, topoisomerase has to resolve the ensuing entanglement of DNA strands.[2,11,12] Bidirectional initiation from *oriE*, being investigated in our laboratory, probably depends on bidirectional transcription.

The T4 replication origins have been mainly characterized by hybridizations of *in vivo* labeled nascent T4 DNA, isolated early after infection, to Southern blots of cloned T4 fragments or restriction digests of unmodified T4 DNA[4,12–18] and by electron microscopy.[19] In addition, transformation potential of nascent T4 DNA has been used for origin mapping.[20] The results, taken together, led to the origin assignments indicated in Fig. 1. However, because of the limited origin use during infection and because the earliest T4 replication preferentially incorporates breakdown products of bacterial DNA, quantitative measurements of relative usage of different origins or of different initiation modes are difficult, if not impossible.

Kreuzer and co-workers[4,21] have used an alternative approach to characterizing T4 origins: They cloned two T4 DNA segments that confer the potential for autonomous replication to the vector plasmids. These two cloned origins correspond to segments of two origins, *oriF* and *oriG* (see Fig. 1). It now appears[3] that a larger T4 sequence is required for *oriF* function in the T4 genome, than in cloning vectors, suggesting that T4

[11] G. Mosig, G. Lin, and A. Chang, *J. Cell. Biochem.*, **Supplement 16B,** 77 (1992).

[12] B. L. Stitt and G. Mosig, *J. Bacteriol.* **171,** 3872 (1989).

[13] M. B. Halpern, T. Mattson, and A. W. Kozinski, *Proc. Natl. Acad. Sci. U.S.A.* **76,** 6137 (1979).

[14] A. W. Kozinski, *in* "Bacteriophage T4" (C. K. Mathews, E. M. Kutter, G. Mosig, and P. B. Berget, eds.), p. 111. American Society for Microbiology, Washington, DC, 1983.

[15] G. J. King and W. M. Huang, *Proc. Natl. Acad. Sci. U.S.A.* **79,** 7248 (1982).

[16] G. Mosig, *in* "Bacteriophage T4" (C. K. Mathews, E. M. Kutter, G. Mosig, and P. B. Berget, eds.), p. 120. American Society for Microbiology, Washington, DC, 1983.

[17] G. Mosig, P. Macdonald, G. Lin, M. Levin, and R. Seaby, *in* "Mechanism of DNA Replication and Recombination" (N. R. Cozzarelli, ed.), p. 173. A. R. Liss, NY, 1983.

[18] P. M. Macdonald, Ph.D. Thesis, Vanderbilt University, Nashville, TN (1983).

[19] J. K. Yee and R. C. Marsh, *J. Virol.* **54,** 271 (1985).

[20] R. C. Marsh, A. M. Breschkin, and G. Mosig, *J. Mol. Biol.* **60,** 213 (1971).

[21] K. N. Kreuzer, H. W. Engman, and W. Y. Yap, *J. Biol. Chem.* **263,** 11,366 (1988).

origins are multipartite and that surrogate sequences may substitute for some components of the T4 origins.

Except for promoter motifs, no obvious sequence similarity is apparent among the different origins. However, a sequence near the 5′ start of each origin (primer) transcript is complementary to sequences near several transcription termination regions, where DNA polymerase can be recruited to initiate leading strand synthesis.[3]

Three of the four sequenced origins (*A*, *F*, and *G*) require a *motA*-dependent middle promoter[3,4]; primers for *oriE* are apparently initiated from early promoters.[3] The transition from prereplicative (early or middle) to late transcription also leads to inactivation of origin initiation, either by default or because a late protein actively inhibits origin activity or for both reasons.

Subsequently, at least two different nonorigin initiation modes that depend on recombination use 3′ DNA ends as primers for nascent DNA chains.[1–4,14,16,22] To appreciate the importance of the subsequent recombination-dependent initiation modes for T4 DNA replication, one has to recall that the host RNA polymerase is used throughout T4 infection, but that it is sequentially modified during T4's development.[23] The substitution of the host's σ^{70} by T4-encoded σ^{gp55} and other accessory proteins leads (directly or indirectly) to cessation of transcription from early and middle promoters, including the primer promoters. Therefore, other priming mechanisms become essential. As outlined in Fig. 2, we propose that usages of different recombination-dependent initiation modes at different developmental stages are in large part responses to changing patterns of gene expression.

Recombinational intermediates resembling D-loops (Figs. 2f and h) are formed when partially unreplicated 3′ termini of replicative intermediates (catalyzed by UvsX, UvsY, and gene 32 proteins) invade homologous regions of another chromosome or in the terminal redundancy at the other end of the same chromosome. There are at least two ways in which such recombinational junctions can prime DNA synthesis. An early mode, which requires only proteins synthesized prior to the onset of replication, can start as soon as a replication fork initiated at an origin has reached a chromosomal end (Fig. 2f). In this pathway, from the invading 3′ DNA end, the invaded chromosome can be copied by a process that we have called join–copy recombination/replication.[1]

[22] G. Mosig, *in* "The Molecular Biology of Bacteriophage T4" (J. Karam, ed.), p. 54. American Society for Microbiology, Washington, DC, 1994.

[23] K. P. Williams, G. A. Kassavetis, D. R. Herendeen, and E. P. Geiduschek, *in* "The Molecular Biology of Bacteriophage T4" (J. Karam, ed.), p. 161. American Society for Microbiology, Washington, DC, 1994.

An additional late recombination-dependent pathway also relies on homologous invasions. But in contrast to join–copy recombination, the join–cut–copy pathway[2] requires, among others, T4 endonuclease VII, gp49, a late protein that cuts recombinational branches.[24] From the 3′ ends of endonuclease VII-catalyzed cuts, DNA synthesis can be initiated to copy single-stranded regions in the invading DNA strand. In primase- and topoisomerase-deficient mutants, both synthesis of Okazaki pieces and initiation by the join–copy pathway are defective, but eventually, with some delay, double-stranded DNA synthesis occurs. The requirement for endonuclease VII-induced nicks and the predominantly late synthesis of this enzyme[25] explains why the bypass DNA replication in primase- and topoisomerase-deficient mutants is delayed.

Of course, recombinational junctions can be resolved and joined by ligase, without initiating DNA synthesis.

Alternatively, and probably competitively, single-stranded regions, left uncopied at that time, are recognized by the gp17 subunit of the T4 terminase to initiate packaging into preformed heads that accommodate chromosomes with terminal redundancies (Fig. 2). Because T4 has no site-specific packaging mechanism, this process restores the circular permutation of packaged chromosomes.

Considerations for Choice of Methods

For reasons related to the different initiation modes, we usually compare DNA replication of wild-type T4 and different mutants and we apply several different methods. We *estimate* total T4 DNA replication by incorporation of ^{3}H-labeled thymidine.[1,2] We assay the first round of replication of infecting T4 chromosomes by labeling the parental chromosomes with density and radioactive isotopes and analyzing the shift in density (due to replication in light medium) of replicative intermediates in Cs_2SO_4 density gradients,[2,9,10,20,26,27] and we measure priming of leading strand synthesis at origins by a repetitive primer extension assay on these leading strand DNA templates,[3] described below. The hybridization of nascent T4 DNA labeled with radioisotopes to T4 restriction fragments was mentioned above.

Because the life cycle of T4 is short (30 min after infection at 37°), and each of the replication pathways is active only within a narrow time span,

[24] B. Kemper, F. Jensch, M. U. Depka-Prondzynski, H. J. Fritz, R. U. Borgmeyer, and M. Mizuuchi, *Cold Spring Harbor Symp. Quant. Biol.* **49,** 815 (1984).

[25] K. A. Barth, D. Powell, M. Trupin, and G. Mosig, *Genetics* **120,** 329 (1988).

[26] A. Breschkin and G. Mosig, *J. Mol. Biol.* **112,** 279 (1977).

[27] A. Breschkin and G. Mosig, *J. Mol. Biol.* **112,** 295 (1977).

the distinction between different pathways depends less on timing than on the use of specific mutants. Nevertheless, it is desirable to synchronize infections, because a bacterium first infected will eventually exclude super-infecting T4 phage and degrade their DNA.

The synchronization attempt requires some compromises. Adsorption is faster at high bacterial concentrations. However, bacteria grown to a concentration of more than 3×10^8/ml approach the stationary phase and poorly support T4 growth, probably because σ^{70} becomes limiting for phage transcription. As a compromise, the bacteria are grown to 1×10^8/ml and concentrated by rapid centrifugation and resuspension in 1/10 of the original volume, or they are grown to 3×10^8/ml and less synchrony and lower phage production are tolerated. (Monitor growth by counting in a Petroff–Hauser chamber under a microscope; OD measurements are too inaccurate.)

It is important to monitor the number of phage particles per bacterium, i.e., the multiplicity of infection (MOI), in each experiment. If it is too high, bacteria lyse without producing progeny. If it is too low, bacterial DNA synthesis may contribute to the incorporation of [^{3}H]thymidine. In many experiments it is desirable to infect at an MOI of approximately five. Predicted by the Poisson distribution, and experimentally found, at that MOI, 0.67% (e^{-5}) of the bacteria remain uninfected.

Low-multiplicity experiments require that the products of phage and bacterial DNA synthesis be distinguished, for example, by differential hybridization of the nascent DNA to immobilized phage or bacterial DNA,[28] or to T4 sequence-specific oligonucleotides followed by primer extensions (described below), or by equilibrium density gradient centrifugation[9,10,20,26,27] taking advantage of the fact that T4 and host DNA have different buoyant densities in Cs_2SO_4. "Density shift" experiments, which measure the decrease in density of heavy parental DNA during replication in light medium (or vice versa), are specifically useful to monitor the first round of DNA replication, because the results are not influenced by the complications due to the preferential incorporation of host-derived precursors or the coupling of the precursor-synthesizing complex to the replication machine, just mentioned.

Monitoring Total T4 DNA Replication by Incorporation of [^{3}H]Thymidine

Bacteria are grown in a synthetic medium, that is, phosphate-buffered M9 or Tris-buffered low-phosphate medium,[29] to a density of 1×10^8/ml

[28] G. Mosig and R. Ehring, *Virology* **35,** 171 (1968).
[29] A. D. Hershey and M. Chase, *J. Gen. Physiol.* **36,** 39 (1952).

at a desired temperature (between 25 and 42°). Usually, we supplement the basic medium with 50 μg/ml thiamine chloride and 1 μg/ml biotin. The bacteria are briefly pelleted by centrifugation at room temperature and resuspended to a titer of 1×10^9/ml in prewarmed fresh medium containing 50 μg/ml tryptophan, to facilitate adsorption. It is important to work fast.

The bacteria are infected with wild-type or desired T4 mutants. After 2.5 min to allow adsorption and cessation of host DNA synthesis, an equal volume of prewarmed medium, containing, in addition, 250 μg/ml of unlabeled deoxyadenosine and 20 μCi/ml of ^{3}H-labeled thymidine (~82Ci/m*M*, ICN), is added to the infected bacteria (this manipulation dilutes the infected bacteria as well as the isotope twofold). The infected bacteria are incubated in a shaking water bath.

At various times (usually between 7 and 120 min after infection) 50-μl aliquots are pipetted onto 25-mm-diameter glass fiber filters (e.g., Gelman A/E), which are immediately immersed into ice-cold 10% trichloroacetic acid (TCA) solution in a Teflon device (Waschkörbchen, Schleicher and Schuell, Dassel, Germany) that holds 24 filters in precise order. The filters are left in 10% TCA for at least 10 min and are subsequently rinsed (10 min each) in 5% TCA, 1% TCA, and methanol. They are dried and isotopes are counted in a scintillation counter.

Infected and uninfected bacteria and unadsorbed phage and progeny phage are routinely assayed from aliquots of the cultures.

Comments

This method is fast and simple and although it underestimates the earliest T4 DNA synthesis, it can be used to estimate relative DNA synthesis of different mutants (Fig. 3).

T4 mutants that are defective in DNA polymerase synthesize no DNA; they are called DO. Any host DNA that might be labeled is rapidly degraded. T4 mutants that are defective in the join–copy recombination-dependent initiation of DNA replication shown in Fig. 2f arrest DNA replication prematurely when origin initiation ceases (due to the switch in transcription). They have a so-called DNA arrest (DA) phenotype. The DA phenotype of a temperature-sensitive recombination-deficient mutant can be reversed late after infection by a shift to the permissive temperature.[1] This shows that the DNA arrest is not caused by limiting availability of isotopes. It is noteworthy that there are additional reasons for cessation of origin initiation, which are not necessarily related to the cessation of origin transcription. The corresponding mutants (e.g., defective in the gene 41 helicase or its gene 59 helper protein) also have a DA phenotype, but they have less residual DNA synthesis than the recombination-deficient mutants (Fig. 3).

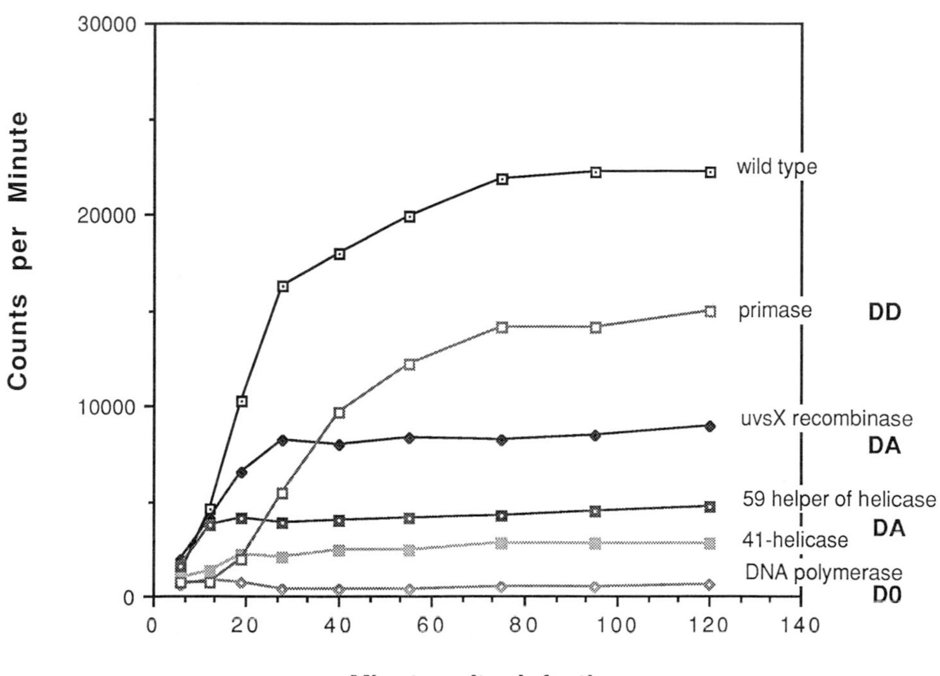

FIG. 3. An example of different patterns of thymidine incorporation into the DNA of different T4 mutants after infection of *E. coli* B at 37°. The DO DNA polymerase mutant was *amB22*, the DA gene *41* helicase mutant was *amN81*, the gene 59 mutant was *amC5* (*amHL628* gave identical incorporation), the *uvsX* mutant was *amX-1*, kindly provided by K. Ebisuzaki, University of Ontario. The DD primase mutant was *amE219*. All mutants, except *amX-1*, were initially obtained from R. Edgar and backcrossed with wild type by the authors.

T4 mutants that cannot reinitiate from origins nor use the join–copy pathway (e.g., primase- and topoisomerase-deficient mutants) have to wait until the T4 endonuclease VII (which is a late protein) has been synthesized, to use the alternative join–cut–copy recombination-dependent initiation shown in Fig. 2H.[2] Therefore, they have a so-called DNA delay (DD) phenotype.

Density shift experiments,[2,30] electron micrographs,[2] and primer extensions,[3] taken together, have shown that the DD mutants initiate leading strand synthesis from origins at the same time as wild-type T4; however, they incorporate relatively little [^{3}H]thymidine into the origin-initiated rep-

[30] C. D. Yegian, M. Mueller, G. Selzer, V. Russo, and F. W. Stahl, *Virology* **46,** 900 (1971).

licated DNA. Similar experiments with gene *32* mutants, which fail to reinitiate at origins and do not enter the recombination-dependent mode of DNA replication,[27] also led to the conclusion that exogenous thymidine is poorly incorporated during the earliest T4 replication.

One might expect that M9 or Tris medium could be used interchangeably. However, the apparent synthesis delay of DD mutants as compared with wild-type T4 is far less severe in Tris than in M9 medium. The reasons are unknown. Possibly, the different ratios of monovalent ions and Mg^{2+} in the different media affect, among others, the structure of recombinational intermediates and thereby cutting by endonuclease VII[31] and functioning of recombination enzymes, as has been shown for *E. coli*'s RecA protein.[32]

If one suspects excessive DNA degradation, for example, in certain mutants,[33] the DNA of the infecting phage is labeled with a different isotope (e.g., ^{32}P). Degradation of the parental DNA can then be measured simultaneously with the [^3H]thymidine incorporation. Since the nascent DNA is also degraded in most of these mutants, [^3H]thymidine incorporation during short pulses can be estimated, but even those experiments give only estimates of the extent of DNA synthesis.

Gentle Isolation of Replicating T4 DNA

In many types of experiments (e.g., repetitive primer extensions, sequencing and electron microscopy) it is desirable to isolate the complex replicating T4 DNA with little or no shearing damages during extraction.

Inoculate 100 to 150 ml medium (per desired time point) with 1 ml overnight bacteria. [The medium can be LB (Luria broth) or HB (Hershey broth), when no labeling of the DNA with density or radioisotopes is necessary, or M9 or Tris medium, when the nascent DNA is to be labeled.]

Grow the bacteria at the desired temperature to a titer of 2 to 3 $\times$ 10^8/ml. Assay colony-forming bacteria before infection.

Infect with an average MOI of 5.

Assay unadsorbed phage, infected bacteria, and surviving bacteria (to double-check the MOI) within 5 to 10 min after infection by plaque assays or colony assays, respectively.

Harvest the infected bacteria at the desired times (e.g., 10 min after

[31] J. B. Welch, D. R. Duckett, and D. M. Lilley, *Nucleic Acids Res.* **21,** 4548 (1993).

[32] J. P. Menetski, A. Varghese, and S. C. Kowalczykowski, *J. Biol. Chem.* **267,** 10,400 (1992).

[33] G. Mosig and S. Bock, *J. Virol.* **17,** 756 (1976).

infection at 30°, or later) by pouring them into centrifuge bottles containing 50 to 100 ml frozen 10 mM KCN.

Spin 15 min at 7000 rpm in a refrigerated Sorvall GSA rotor or equivalent. (During this time the ice should melt.)

Decant supernatant (not into the sink!) and resuspend the pellet in 5 ml ice-cold Tris medium without glucose.

Add 0.27 ml freshly made egg-white lysozyme solution (15 mg/ml) and incubate 30 min on ice with occasional gentle shaking to prevent clumping of the cells.

Transfer the samples to a *slowly* shaking waterbath at 37°, incubate for an additional 5 min, then add 25 μl Triton-X100, 0.27 ml (w/v) (10%) Sarcosyl NL97 (Geigy), 0.4 ml (w/v) (0.1 M) EDTA, 0.66 ml (w/v) (5 M) NaCl, and 0.27 ml (w/v) (15 mg/ml) Proteinase K.

Incubate 5 min at 37°, or until the lysate clears, then heat gradually to 62°, while shaking slowly. Leave 5 min at 62°, then chill.

Extract three times with equal volumes of cold phenol buffered with 10 mM Tris, pH 7.3, then two times with chloroform–isoamyl alcohol (24 : 1), to remove residual phenol. (Alternatively, the phenol can be removed with ethyl ether.) Note: *Never pipette* the DNA in the aqueous phase. Pour the sample into the extraction tube, add the phenol or chloroform, and rotate the tube each time 15 to 30 min at 15 to 30 rpm in the cold; then remove the organic phase below the aqueous phase with a Pasteur pipette.

Do the following steps at room temperature (to reduce precipitation of RNA):

Measure the final volume (pour!).

Add 1/10 volume (3 M) potassium acetate.

Add three volumes (95%) ethanol.

Spool out the DNA with a glass rod and transfer it to an Eppendorf tube.

Rinse the DNA with 70% ethanol; pellet the DNA in an Eppendorf centrifuge and decant the ethanol.

Allow the residual ethanol to evaporate (without vacuum to avoid DNA denaturation).

Resuspend in 0.5 to 1 ml TE buffer (10 mM Tris, pH 8.0, and 1 mM EDTA). The rehydration of the DNA may require 24 to 48 hr. Because the DNA is large and complex and sensitive to shearing, it must not be vortexed.

(Digest with DNase-free RNase [bovine pancreas, Boehringer Mannheim, Cat. No. 1119915] and, when T4 mutants were used that do not degrade host DNA, with *Eco*RI, which digests host but not T4 DNA. Reextract with phenol and reprecipitate with ethanol.)

We found that the steps in parentheses have to be included with some nuclease-deficient T4 mutants that fail to degrade host DNA, but they can be omitted in wild-type T4 infections, if the protocol is followed exactly.

After the DNA is completely rehydrated, measure DNA concentrations by ethidium bromide spot tests or with DNA Quick Strips (Kodak-IBI IB 73000). Other dip sticks are not sufficiently sensitive and spectrophotometers damage too much of the precious DNA.

The host DNA and most of the RNA is apparently degraded by endogenous nucleases. Any undigested T4 RNA remains in the ethanol phase and can be recovered by chilling at −20° overnight and spinning down at 20,000 rpm for 30 min in a Sorvall SS34 rotor. As an alternative to phenol and chloroform extractions, followed by ethanol precipitation, the DNA solution can be dialysed at 4° against 0.05 *M* Tris buffer, pH 7.4, 0.02 *M* EDTA.

We found that omission of treatments with lysozyme or proteinase K leads to considerable loss of T4 DNA during isolation (monitored by the loss of labeled parental DNA). The loss is more serious, the earlier the DNA is isolated after infection, i.e., when T4-encoded lysozyme has not yet been synthesized. In these cases, precipitates are formed in the interphase between the phenol and aqueous phases during extraction. It is difficult, if not impossible, to recover the DNA from these precipitates, once they have been formed.

When replicating T4 DNA, isolated without prior lysozyme or proteinase K treatment and without phenol extraction, is subjected to equilibrium density gradient centrifugation, much of the DNA floats to the top of the gradient. Apparently it remains associated with proteins and membrane components[20] (and unpublished observations).

Equilibrium Density Gradient Centrifugation

The isolated replicative DNA (and associated RNA) can be subjected to equilibrium density gradient centrifugation. Cs_2SO_4 gradients are particularly useful for separating host and T4 DNA, because, as mentioned before, in such gradients T4 DNA bands at a heavier position than *E. coli* DNA, and RNA–DNA copolymers of even higher buoyant density can be separated from DNA and RNA.[34]

After dialysis, or after complete rehydration of ethanol-precipitated DNA, DNA solutions are mixed with a saturated Cs_2SO_4 solution (1.1 ml per 1.4 ml of DNA solution) and spun in a Beckman type 40 rotor (or equivalent) at 35,000 rpm for approximately 60 hr at 10°. Gradients are

[34] P. M. Macdonald, R. M. Seaby, W. Brown, and G. Mosig, *in* "Microbiology 1983" (D. Schlessinger, ed.), p. 111. American Society for Microbiology, Washington, DC, 1983.

fractionated by collecting drops from the bottom of the tube. Aliquots are assayed for radioactivity, viewed in the electron microscope,[9,10] or assayed for gene content by transformation.[20]

Labeling of DNA in Parental T4 Particles

To label parental DNA with ^{3}H, phage are grown in minimal medium containing [^{3}H]thymidine and deoxyadenosine for a single infection cycle as described for thymidine incorporation.

To obtain parental phage whose DNA is labeled with heavy isotopes, phage are grown in medium containing ^{15}N and ^{13}C, for example, in Tris minimal medium[29] containing 0.021 *M* $^{15}NH_4Cl$ (99.5 at.%; Bio-Rad) and 0.2% ^{13}C glucose (56 at.%; Merck, Sharp and Dohme of Canada, Ltd). Parental phage can also be labeled with ^{32}P in this Tris medium (with or without density label). It is important to use low specific activities (0.2 to 0.5 μCi/mg P in the medium) to avoid apparent anomalies in subsequent DNA and RNA synthesis.[35] In retrospect, double-strand breaks caused directly and indirectly by ^{32}P decay probably lead to premature arrest of replication forks and enhanced recombination-repair. Addition of trace amounts of ethanol minimizes damages due to free radicals.

The labeled progeny phage particles are purified by pelleting bacterial debris by centrifugation in a Sorvall SS34 rotor for 10 min at 10,000 rpm and subsequent filtering of the supernatant through nitrocellulose filters of 0.6-μm pore size (filters with smaller pores retain T4 particles). Then the phage particles are pelleted in a Beckman type 30 rotor or equivalent for 90 min at 20,000 rpm. They are gently resuspended overnight (without vortexing!) in, for example, Tris medium without glucose, containing 0.001% gelatin, and stored in Eppendorf tubes in the dark.

Analysis of Initiation Sites in Origin Regions by Repetitive Primer Extensions

We are mapping 5′ ends of nascent T4 DNA corresponding to transition points from RNA primers to leading strand DNA in RNA–DNA copolymers (Fig. 4) of three origin regions (A, E, and F in Fig. 1).[3] We gently isolate early replicating T4 DNA from bacteria infected with either wild-type or mutant T4 phage. The nucleic acids, containing parental and newly synthesized DNA, RNA–DNA copolymers, and RNA–DNA hybrids, are denatured and annealed to short end-labeled synthetic oligonucleotides corresponding to different sequences in the origin regions. The oligonucleo-

[35] A. D. Hershey, personal communication (1962).

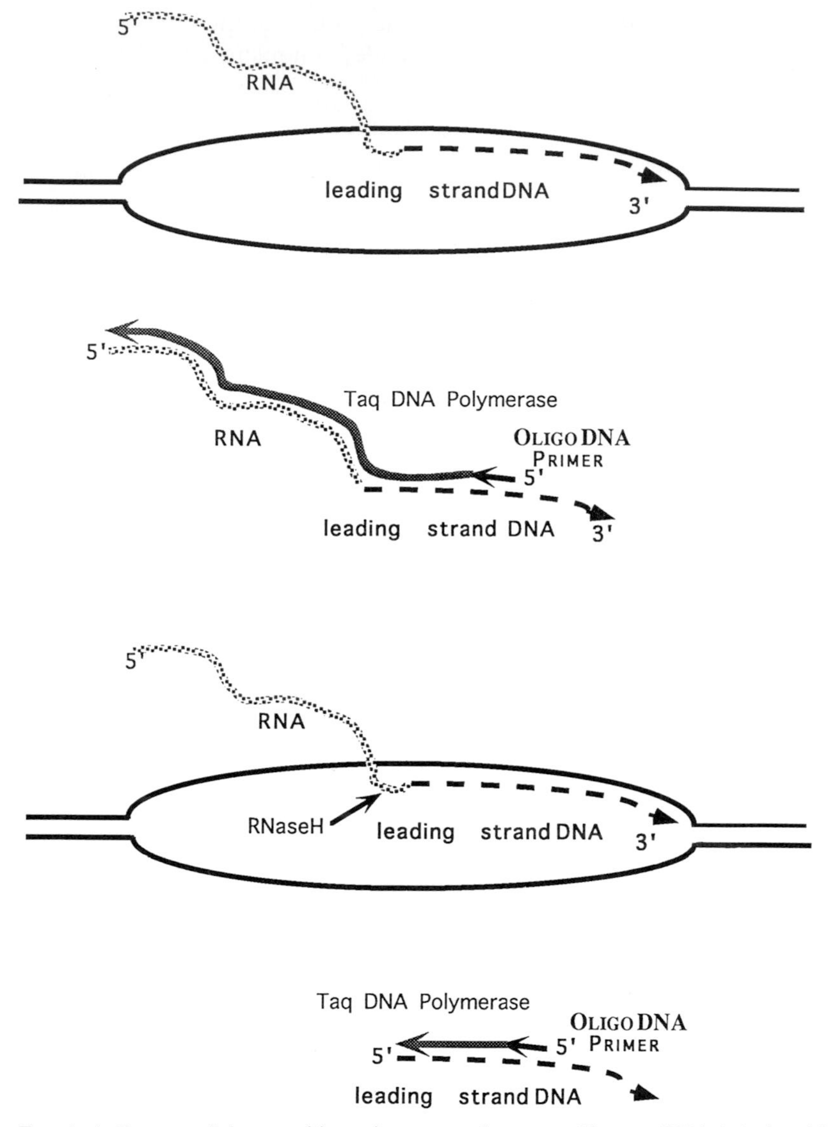

FIG. 4. A diagram of the repetitive primer extension assay. Nascent DNA is isolated by the gentle method. The nucleic acids are heat denatured, and end-labeled short oligonucleotides are annealed that are complementary to the nascent DNA downstream of the suspected initiation sites. *Taq* DNA polymerase extends the primers in a thermocycler. If the nucleic acids have not been treated with RNase H prior to denaturation, some extensions proceed into the RNA template. If the nucleic acids have been cleaved by endogenous or added RNase H, the primer extensions will stop at the 5′ ends of the nascent DNA.

tides are then repeatedly extended with *Taq* DNA polymerase in a Perkin–Elmer thermocycler. This procedure amplifies the signal *linearly*. The extension products are separated on polyacrylamide sequencing gels together with products of sequencing reactions obtained with the same primers on homologous cloned T4 DNA (Figs. 4 and 5).

In preliminary experiments, we found unexpectedly that *Taq* DNA polymerase can efficiently copy RNA templates when, and only when, it

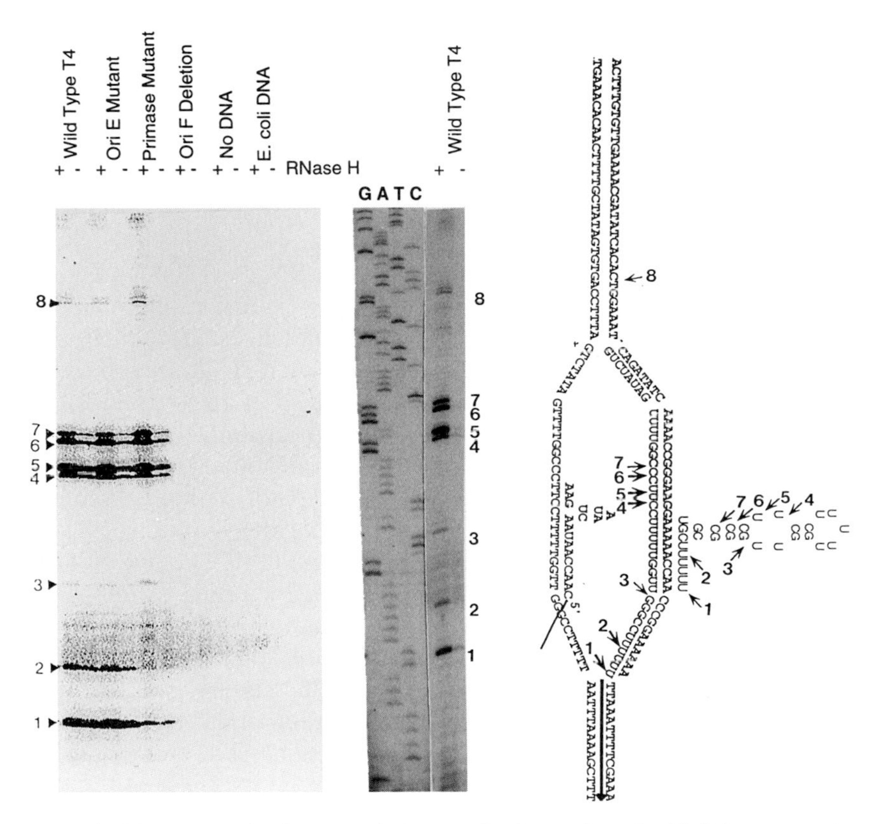

FIG. 5. 5′ DNA ends in the transcription termination region of *oriF*. Primer extensions and sequencing ladders from an oligonucleotide (underlined in the sequence) annealed to denatured, nascent DNA from the mutants indicated above the lanes, with (+) or without (−) prior RNase H treatment. The numbered arrows in the sequence correspond to the numbered bands (stops of the primer extension reactions). The sequence of the transcript ends is drawn as two alternative structures: RNA hybridized to the DNA or as a stem-loop, thought to be a termination signal. [See M. E. Gruidl and G. Mosig, *Genetics* **114,** 1061 (1986).] Note that base-pairing between the "mismatched" bases is possible. [See W. B. Cruse, T. P. Saludjian, E. Biala, P. Strazewski, T. Prange, and O. Kinnard, *Proc. Natl. Acad. Sci. U.S.A.* **91,** 4160 (1994).]

gets a jump start on DNA covalently linked to the RNA (e.g., in the postulated RNA–DNA copolymers, Fig. 4B).[3] Of several RNases tested, only RNase H digested the RNA segments of the suspected copolymers in our nascent DNA preparations. T4-encoded RNase H,[36] kindly provided by Drs. Hollingsworth and Nossal, or *E. coli* RNase H, purchased from USB, is effective. These results suggest that in our DNA preparations some nascent DNA strands are still attached to their RNA primers, which in turn are base-paired to their DNA templates. Therefore, we perform all primer extensions on nascent T4 DNA with and without prior RNase H treatments. Because the synthetic oligonucleotides also anneal to the parental T4 chromosomes, breakage points in that DNA would appear as false stops to the primer extensions. Therefore, it is critically important that the DNA is not sheared during isolation. Most of the extension products from unbroken parental DNA remain close to the well of the separating gel. (The rare exceptions are the random ends of T4 chromosomes that terminate in the region under study.)

The gently isolated T4 DNA is treated with RNase H (USB), 1.5 units per μg of DNA, for 30 min at 37°. Primer extensions are then done on such treated DNA as well as on DNA not digested with RNase H.

In a 0.5-ml microfuge tube, 1 or 2 μg of isolated DNA are mixed with 1.5 pmol of an oligonucleotide (synthesized in an Applied Biosystems 281A DNA synthesizer) that hybridizes with DNA downstream of the RNA–DNA transition sites. The oligonucleotide has been end-labeled with 50 μCi γ^{32}P-ATP per 10 pmol, 5 μl of 5× sequencing buffer (Promega fmol DNA sequencing kit), 4 μl of a dNTP mix that is 186 μM in each dNTP (dGTP, dATP, dTTP, and dCTP), 1 μl of *Taq* DNA polymerase (5 units per μl; Promega fmol DNA sequencing kit), and H_2O to bring the final volume to 25 μl.

This reaction mix is then overlaid with mineral oil, spun briefly in a microcentrifuge, and placed in a thermocycler (Perkin–Elmer). The thermocycling program is as follows: An initial denaturation at 95° for 5 min, then 20 cycles of denaturation at 95° for 30 sec, annealing at 42° for 30 sec, and extension at 70° for 1 min.

The reaction tube is then chilled on ice, and 12 μl of loading dye (Promega fmol DNA sequencing kit) is added. The tube is spun briefly in a microcentrifuge. The reaction product is then boiled for 2 min, and 4 μl is loaded on an 8% polyacrylamide sequencing gel. The sample is electrophoresed until the labeled oligonucleotide has run off the gel. Size standards, or sequencing reactions done with the same primer on cloned T4 DNA, using termination mixes from the Promega fmol DNA sequencing

[36] H. C. Hollingsworth and N. G. Nossal, *J. Biol. Chem.* **266,** 1888 (1991).

kit, are loaded into adjacent wells of the same gel. (Of course, reagents for sequencing or primer extensions can be made up individually; however, reagents in the kit are less expensive. We use non-kit dNTP solutions that are adjusted to the high AT ratio of T4 DNA.)

The gel is fixed in 5% methanol–5% acetic acid, dried and autoradiographed (either at room temperature or at −70° with a Dupont Cronex intensifying screen).

Figure 5 shows a representative result of such a repetitive primer extension from the *oriF* region. Clearly, there are several transition sites within one origin region, and the same transition sites occur with similar frequencies in wild-type and primase-deficient T4, as expected if primers for T4 origin leading strands are synthesized by RNA polymerase.

Comments

This assay depends on prior identification of origin regions by other means, but it avoids the problems of labeling biases mentioned before.

Primase-dependent 5′ DNA ends would be detected by comparing DNA from wild-type and primase-deficient mutants. Control primer extensions using cloned T4 DNA as a template are used to reveal artifactual *in vitro* stops due to certain template sequences. Misinterpretations due to possible cross-hybridization to contaminating bacterial DNA can be avoided by primer extensions from each oligonucleotide on DNA of uninfected bacteria.

The repetitive primer extension assay can be used to compare relative initiation frequencies from different origins or from different sites within one origin. Different oligonucleotides, complementary to different regions but labeled with the same specific activity, and at the same molar concentrations, can be used on the same DNA preparation.

Results obtained thus far,[3] taken together, show that in each of the three analyzed origin regions there are several RNA–DNA transition points, most of which are more than 1000 nucleotides downstream of the promoters from which the primer transcripts were initiated. These sites are at or close to transcription termination sites, suggesting that little or no processing of the transcripts is required before they can become primers. However, only a small minority of transcripts become primers; the vast majority serve as mRNAs for certain proteins. This is in contrast to the priming of leading DNA strands by processed transcripts in ColE1 plasmids.[37–39] Possible reasons for the difference are discussed elsewhere.[3] In

[37] B. Polisky, X.-Y. Zhang, and T. Fitzwater, *EMBO J.* **9,** 295 (1990).
[38] Y. Eguchi and J. Tomizawa, *J. Mol. Biol.* **220,** 831 (1991).
[39] G. Cesareni, M. Helmer-Citterich, and L. Castagnoli, *Trends Genet.* **7,** 230 (1991).

E. coli, multiple transcript termination sites within the *oriC* region have been proposed to act directly as primers for leading strand DNA synthesis as an alternative to priming by primase that requires binding of DnaA protein to the iterons.[40]

Redundancies in initiation mechanisms of DNA replication are becoming apparent in other systems.[41,42] We believe that the redundancies of the T4 initiation mechanisms and their interrelationship with other DNA transactions can serve as an example of integrated regulatory circuits that connect different physiological processes during development.

[40] J. W. Zyskind, *in* "The Bacterial Chromosome" (K. Drlica and M. Riley, eds.), p. 269. American Society for Microbiology, Washington, DC, 1990.
[41] T. Asai and T. Kogoma, *J. Bacteriol.* **176,** 1807 (1994).
[42] M. Filutowicz, S. Dellis, I. Levchenko, M. Urh, F. Wu, and D. York, *in* "Progress in Nucleic Acid Research and Molecular Biology" (W. E. Cohn and K. Moldave, eds.), p. 239. Academic Press, San Diego, 1994.

[45] Analysis of DNA Replication *in Vivo* by Flow Cytometry

By KIRSTEN SKARSTAD, ROLF BERNANDER, and ERIK BOYE

Introduction

The cellular DNA content and the kinetics of *in vivo* DNA replication can be analyzed by flow cytometry. The DNA content of several thousand individual cells is rapidly measured, which results in a DNA histogram in which the DNA content of individual cells is displayed. Other methods of measuring DNA content give values that represent an average for the population; the advantage of flow cytometry is that it yields information about the distribution within a population. The measurements can give information about (1) cell cycle parameters, (2) timing of initiation of replication, (3) the speed of replication fork movement, and (4) DNA content and cell size at the time of initiation. In this chapter we focus on DNA replication in microorganisms, the examples given being mainly from work with *E. coli.* The ease with which mutants with biochemically well-characterized defects are obtained in microorganisms makes it possible to address specific questions about the regulation and kinetics of DNA replication *in vivo.*

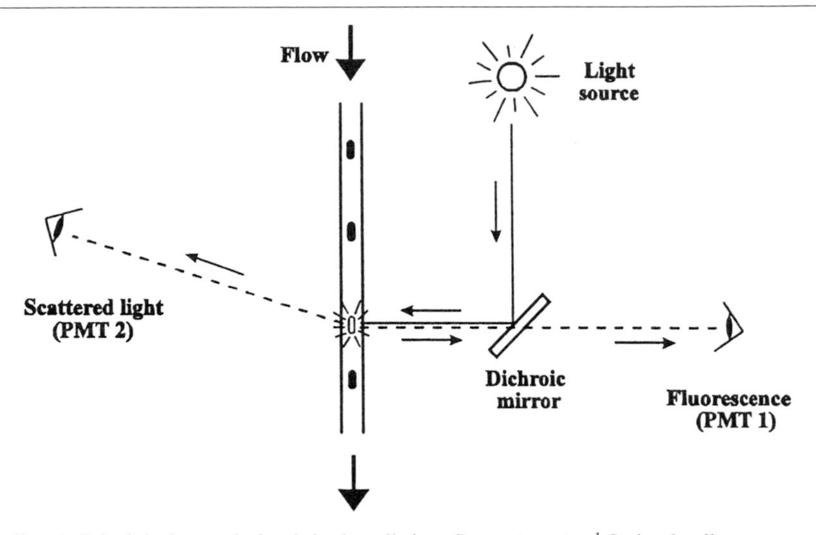

FIG. 1. Principle for analysis of single cells in a flow cytometer.[1] Stained cells pass one by one, in a flow of water, through the focus of the excitation light beam. The cells give rise to a pulse of fluorescence light, the size of which is proportional to the amount of dye bound. The fluorescence is registered in a photomultiplier tube (PMT 1). The amount of light scattered by each cell is registered simultaneously in a second photomultiplier tube (PMT 2).

Cells are fixed and stained with a DNA-specific stain. In the flow cytometer, each cell passes through the focus of an excitation light beam and gives rise to a pulse of fluorescence light, the size of which is proportional to the cellular DNA content (Fig. 1). A sensitive photomultiplier tube (PMT) transforms the fluorescence pulses into electrical pulses, which are sized and stored as counts in a multichannel analyzer. In addition, the light scattered by each cell is proportional to the mass and protein content of the cell, and is simultaneously detected and stored in the same multichannel analyzer. Thus, a three-dimensional histogram is accumulated for DNA content and cell size of individual cells in the population (Fig. 2).

DNA distributions with adequate resolution for cell cycle analysis of bacteria were first obtained with a microscope-based flow cytometer,[1a] now commercially available (Bryte HS from Bio-Rad). This instrument is equipped with a mercury arc lamp, which provides excitation light suitable for several DNA-specific dyes. The procedures given here were developed for this particular flow cytometer, but are also applicable to other flow cytometers with appropriate excitation light source and sensitivity.

[1] H. M. Shapiro, "Practical Flow Cytometry." Alan R. Liss, NY, 1988.
[1a] H. B. Steen and T. Lindmo, *Science* **204,** 403 (1979).

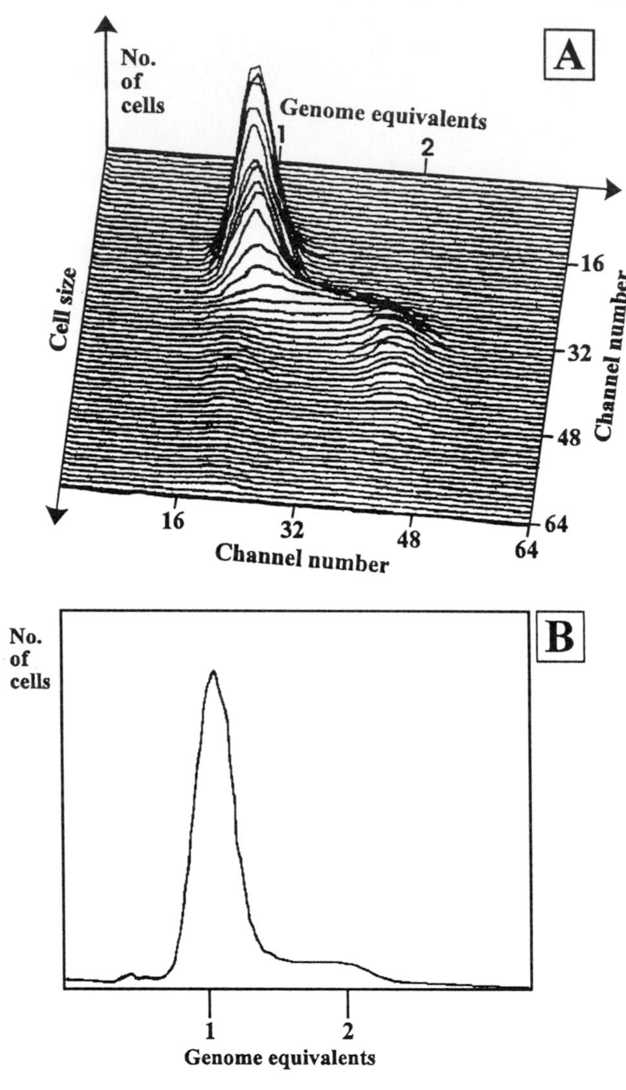

FIG. 2. Dual-parameter DNA and size histogram (A) and single-parameter DNA histogram (B) of slowly growing *E. coli* B/r A cells stained with MI and EB.

Materials and Methods

For DNA staining, a combination of mithramycin (MI) and ethidium bromide (EB) is used in our laboratory. MI is DNA specific but the fluorescence yield is low (about 0.05%[2]). It is therefore used in combination with EB to obtain a higher fluorescence yield. Ethidium bromide has negligible absorption at the excitation wavelength used (mainly the strong 436-nm line of the mercury arc lamp), and is therefore excited primarily by energy transfer from MI molecules bound nearby (<5 nm away[2]). Because of the higher fluorescence yield of EB, energy transferred from MI to EB produces more fluorescence than MI alone. Fluorescence from RNA-bound EB is negligible at this excitation wavelength, for measurements of bacteria. For DNA measurements of yeast, the RNA must be removed by treatment with RNase (see below).

The DNA-specific dyes DAPI (4′,6-diamidino-2-phenylindole) and Hoechst 33258 also give good results, and require staining protocols similar to the one given below for MI and EB. Note, however, that it is more difficult to obtain good light scatter histograms with the excitation light used for these dyes.

Solutions

I. Fixing solution: 77% (v/v) ethanol in water
II. Wash buffer (TE): 10 m*M* Tris, pH 8.0, 1 m*M* EDTA
III. Staining buffer: 10 m*M* Tris, pH 7.4, 10 m*M* $MgCl_2$
IV. EB stock solution: 1 mg/ml EB in water
V. Staining solution: 180 μg/ml MI and 40 μg/ml EB in staining buffer. Prepare by dissolving the contents of one 2.5-mg ampule of Mithracin (Pfizer) into 13.3 ml of the staining buffer and add 0.56 ml of the EB stock solution.

The EB stock solution is stable for several months when stored in the dark at 4°. The staining solution should also be stored in the dark at 4° and be used within a month.

All solutions should be filtered through 0.22-μm filters. Likewise, the sheath water of the flow cytometer must be filtered to avoid particles that contribute to the light scattering background.

Fixation

A sample (1 ml) of bacteria grown to an optical density of $OD_{600} = 0.1$ to 0.2 is harvested in a microcentrifuge, resuspended in 1 ml of wash buffer (II), reharvested, and resuspended in 0.1 ml wash buffer. One milliliter of fixing solution (I) is added to the suspension and mixed by vortexing. Fixation is complete within a few minutes. The cells may be stored at 4° in the fixing solution for a few months and at −20° for much longer.

[2] R. G. Langlois and R. H. Jensen, *J. Histochem. Cytochem.* **27,** 72 (1979).

Staining

Fixed cells are washed once in staining buffer (III). Centrifugation is carried out for 3 min at 10,000*g* in a standard refrigerated microcentrifuge. Resuspend the cell pellet in 300 μl of ice-cold staining buffer. The cell density should be around 10^8 cells/ml. Equal volumes of the cell suspension and the staining solution are mixed to ensure a reproducible dye concentration in the samples. The cells should be protected from light and kept on ice in staining solution for 30 min before measurement. Stained samples can be stored on ice for several hours before any deterioration (in terms of reduced quality of the measuring data, probably due to DNA degradation) can be observed.

Flow Cytometric Measurement

For measurement of cells stained with MI/EB in the microscope-based flow cytometer, a 100-W high-pressure mercury arc lamp is used as the excitation light source. A combination of filters is used that permits transmission of excitation light of wavelengths from 390 to 440 nm. Fluorescent light passes through a long-pass filter such that only light with wavelengths above 470 nm is detected by the photomultiplier tube collecting fluorescence (PMT 1 in Fig. 1). Scattered light collected by the second photomultiplier tube is proportional to cell mass and protein content.[3]

Interpretation of Results

A cell of a certain size and DNA content gives rise to a count in the multichannel analyzer (MCA); for instance, in DNA-channel No. 20 and light scatter channel No. 16 (Fig. 2A). A cell of twice the size and with twice as much DNA will give rise to a count in DNA channel 40 and light scatter channel 32. Measurement of several hundred cells per second rapidly yields a three-dimensional DNA and light scatter histogram.

Note that in all histograms, the height of the peaks is merely a function of how many cells are measured, the important features being the heights of the peaks relative to one another and their placement on the axes.

The histogram of Fig. 2A resulted when measuring cells from a slowly growing chemostat culture of *Escherichia coli* B/r A with doubling time of 17 hr.[4] Most of the cells (counts) were found in a peak with a DNA amount equivalent to one single chromosome. These cells were in the prereplication (B) period of growth. Cells in the postreplication (D) period are seen as

[3] E. Boye, H. B. Steen, and K. Skarstad, *J. Gen. Microbiol.* **129,** 973 (1983).

[4] K. Skarstad, H. B. Steen, and E. Boye, *J. Bacteriol.* **154,** 656 (1983).

a smaller peak at a DNA content corresponding to two chromosomes and with a larger size or mass than the B period cells. The ridge connecting the one- and two-chromosome peaks represents cells in the C period, that is, the period of DNA synthesis. From the histogram it is seen that cells increase in mass throughout the C period. The average cell mass and the average DNA content per cell are easily found from the histogram, and thus also the DNA per mass ratio. Also, the mass at which the cells initiate DNA replication (start of C) can be found from the histogram.

Duration of Cell Cycle Periods

From the DNA histogram of the slowly growing cells, it is possible to determine the duration of the cell cycle periods from the fraction of cells in each peak and in the ridge. By also taking into consideration the exponential age distribution of cells growing in steady-state cultures (i.e., that there are twice as many newborn as dividing cells), B, C, and D periods of 13.6, 1.7, and 1.7 hr, respectively, were calculated for this culture. The dual-parameter histogram of Fig. 2A was integrated with regard to light scatter, yielding a single-parameter DNA histogram (Fig. 2B).

Most often *E. coli* cells are grown in batch cultures with doubling times of less than an hour. Such rapid growth results in a DNA replication pattern in which rounds of replication are initiated before the previous ones are completed.[5] DNA histograms from rapidly growing cells do not look like the ones from slowly growing cells, where peaks representing the different cell cycle periods are easily distinguished. Rather, DNA contents may range from, for instance, 2 to 4 chromosome equivalents or from 2.5 to 5 chromosome equivalents, depending on the growth rate and the duration of the cell cycle periods. Information about cell cycle periods is contained also in these DNA histograms, but is more difficult to extract. One method of determining the cell cycle periods of rapidly growing cells is to employ a computer simulation routine, which compares the experimental DNA distributions with theoretical ones in which the durations of periods C and D are varied until the best fit is found.[6] Another method involves first determining the age at which initiation occurs, and how many generations one round of replication spans (by rifampicin run-out; see below) and then calculating the duration of C + D, the time from initiation to the corresponding cell division.[6a]

[5] S. Cooper and C. E. Helmstetter, *J. Mol. Biol.* **31,** 519 (1968).

[6] K. Skarstad, H. B. Steen, and E. Boye, *J. Bacteriol.* **163,** 661 (1985).

[6a] K. Skarstad, R. Bernander, S. Wold, H. B. Steen, and E. Boye, *in* "Flow Cytometry Applications in Cell Culture" (M. Al-Rubeai and A. N. Emery, eds.). Marcel Dekker, New York (1995).

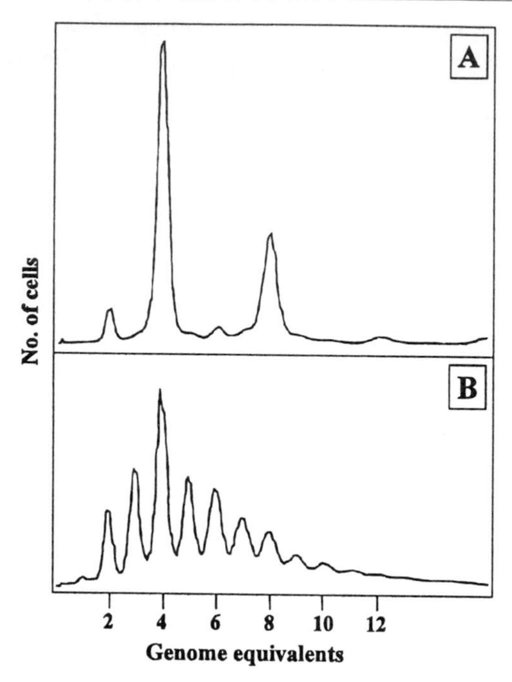

FIG. 3. DNA histograms of *E. coli* K12 wild-type (A) and *dnaA46* (B) cells grown at 30° to an OD_{600} of 0.2, treated with rifampicin and cephalexin, fixed after a 4-hr incubation with the drugs, and then stained with MI and EB.

Number of Replication Origins per Cell

To determine the number of origins per cell, a specific inhibitor of initiation of replication is added to the culture, for example, rifampicin or chloramphenicol, and the cells are allowed to complete ongoing rounds of replication. All cells will then contain integral numbers of chromosomes, which correspond to the number of origins present at the time of drug addition. Cell division must not occur after drug addition because this would reduce the number of chromosomes per cell. To prevent cell division, cephalexin (1 to 10 μg/ml) is added at the same time as the initiation inhibitor. Rapidly growing cells normally contain either two or four origins at the time of initiation, and these origins are initiated simultaneously.[7] The DNA histogram of a culture treated with rifampicin and cephalexin (Fig. 3A) shows that the cells contain four origins when newborn and eight origins after initiation of replication. The 4-chromosome peak represents

[7] K. Skarstad, E. Boye, and H. B. Steen, *EMBO J.* **5,** 1711 (1986).

cells that had not yet initiated replication at the time of drug addition, whereas the 8-chromosome peak represents cells that had initiated. From the fraction of cells in the 4-chromosome peak, the uninitiated cells, the age at which initiation occurs can be calculated (also taking the exponential age distribution into consideration). From the DNA and cell size histogram of the exponentially growing culture (i.e., before drug addition), the DNA content and cell size at the time of initiation of replication can be found by applying the information about the fraction of uninitiated cells.

Timing of Initiation of Replication

If initiation at the four origins within each rapidly growing cell does not occur simultaneously, drug addition may trap some cells when only one, two, or three of the four origins have been initiated. This results in cells containing not only four or eight fully replicated chromosomes after replication run-out, but also cells with five, six, or seven chromosomes. Some *dnaA* mutants are unable to coordinate initiations[8] and yield such DNA histograms with significant peaks also at three, five, six, and seven fully replicated chromosomes (Fig. 3B).

Speed of Replication Fork

The speed of fork movement can be monitored by first adding rifampicin and cephalexin and then taking samples for flow cytometry at regular intervals until replication is completed, thus measuring how long it takes for all replication forks to finish.

Flow Cytometry of Organisms Other than *Escherichia coli*

The protocols described above can also be used for other bacteria and for yeast cells. For some species, for example, *Bacillus subtilis*, sonication may be required to obtain single-cell suspensions. To obtain satisfactory DNA histograms of MI and EB stained yeast cells, treatment with RNase A is necessary. Fixed cells are washed once in TE (with centrifugation at 3000*g* for 5 min) and incubated with 400 μg/ml RNase A in TE for 60 min. Cells are then resuspended in staining buffer and stained as described above. Because yeast cells are larger, a larger sample volume should be collected at the above-mentioned OD, to obtain the same number of cells. In a histogram of exponentially growing *Schizosaccharomyces pombe*

[8] K. Skarstad, K. von Meyenburg, F. G. Hansen, and E. Boye, *J. Bacteriol.* **170,** 852 (1988).

(Fig. 4), the large peak represents cells with two genome equivalents, the ridge represents S-phase cells, and the small peak represents cells with four genome equivalents. From such a histogram, the duration of S and the cell mass at the start of S can easily be determined.

Standards and Controls

Monodisperse fluorescent beads are run when the instrument is set up to check the nozzle and water jet and to optimize the instrument with regard to sensitivity and resolution. The coefficient of variation of the fluorescence peak should be around 1%. A standard cell sample is run to monitor staining and to calibrate the DNA axis. The sample is usually rifampicin-treated bacteria (Fig. 3), yielding sharp peaks at integral numbers of chromosomes, thus giving information about which channels correspond to the different amounts of DNA. The standard is run at regular intervals between other samples to monitor the stability of the instrument and to recalibrate if necessary. If calibration of the DNA axis is especially critical, for instance, if absolute DNA contents per cell are required rather than relative values, it is recommended to run more than one type of calibration sample [e.g., chloramphenicol-treated cells or a *dnaA* (*t*s, temperature-sensitive) mutant incubated at nonpermissive temperature, in addition to

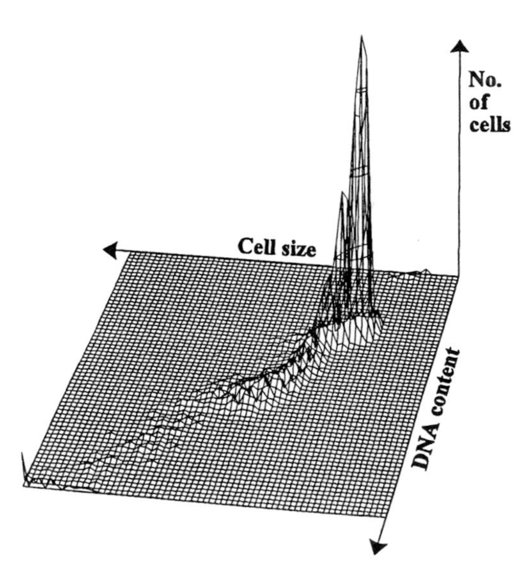

FIG. 4. DNA and cell size histogram of *Schizosaccharomyces pombe* grown to an OD_{600} of 0.2, fixed, treated with RNase A, and stained with MI and EB.

rifampicin-treated cells]. All the calibration samples should give the same amount of fluorescence per chromosome. For repeated monitoring of instrument stability, the inclusion of beads in all samples is recommended.

Acknowledgments

We are grateful to Harald B. Steen for providing flow cytometry expertise. This work was supported by grants from the Norwegian Research Council (to KS), the Swedish Natural Science Research Council (to RB) and the Norwegian Cancer Society (to EB).

[46] Analysis of Replication Intermediates by Two-Dimensional Agarose Gel Electrophoresis

By KATHERINE L. FRIEDMAN and BONITA J. BREWER

Introduction

The shape of a DNA molecule influences its electrophoretic mobility in an agarose gel. This influence is most often seen when analyzing circular DNA on agarose gels: neither the supercoiled nor the nicked circles necessarily have the mobility of linear molecules of the same mass. Branched DNA molecules also run anomalously in agarose gels, and these aberrant migration properties are strongly influenced by the agarose concentration and the strength of the electric field. Based on this property of agarose gels, Bell and Byers[1] developed a two-dimensional (2-D) agarose gel electrophoresis method to separate linear molecules from the branched intermediates that are generated during recombination. The first dimension gel is run under conditions that minimize the effect of molecular shape, whereas the second dimension gel is run under conditions that emphasize the effect of shape. As a result, branched recombination structures run more slowly than linear molecules of the same mass during the second dimension of electrophoresis, effectively separating the two forms. Our laboratory has modified this technique for the analysis of branched replication intermediates. An alternative 2-D gel technique that involves an alkaline second dimension has also been developed. The denaturing conditions in the second dimension release the nascent strands from replication intermediates. Analysis of these nascent strands with different hybidization probes permits

[1] L. Bell and B. Byers, *Anal. Biochem.* **130,** 527 (1983).

the mechanism of replication to be deduced. For a discussion of this technique, see Huberman *et al.*[2]

Restriction fragments derived from replicating DNA fall into three basic shapes, depending on the sites of initiation and termination of replication. If a restriction fragment lies between an origin and terminus, a series of Y-shaped molecules is formed as a single replication fork moves through the restriction fragment. Initiation of replication within the fragment results in a series of bubble-shaped molecules, whereas termination within the restriction fragment leads to double Y molecules with two approaching forks. These three types of molecules have different shapes, and thus can be distinguished from each other and from nonreplicationg molecules by their migration behavior in 2-D agarose gels[3] (Fig. 1). In this chapter, we first discuss the technique and second the interpretation of 2-D gels, as well as a variation of this technique that allows the direction of replication fork movement through a DNA fragment to be determined.

2-D Agarose Gel Electrophoresis

The first step in the analysis of replication intermediates on 2-D gels involves the isolation of branched molecules in a way that preserves their structure. Branched molecules are delicate, and shearing must be minimized by use of large-bore pipette tips and gentle hand mixing. Shearing can lead to loss of replication intermediates, as well as generation of novel branched artifacts.[4,5] Exposure of the DNA to excessive heat must also be avoided in order to preserve the integrity of DNA at the fork. Therefore, heat inactivation of restriction enzymes and the use of thermophilic restriction enzymes is discouraged. Because replication intermediates are generally rare, some method of enrichment may be helpful. In our studies with the yeast *Saccharomyces cerevisiae,* we find that synchronizing the cells in G_1 phase and harvesting DNA during the S period greatly aids in visualizing replication intermediates of single-copy sequences. When synchronization is not possible and amounts of replicating intermediates are low, several techniques can be used to enrich for replication intermediates. Both the use of BND-cellulose and the isolation of nuclear matrix-associated DNA

[2] J. A. Huberman, L. D. Spotila, K. A. Nawotka, S. M. El-Assouli, and L. R. Davis, *Cell* **51,** 473 (1987).

[3] B. J. Brewer and W. L. Fangman, *Cell* **51,** 463 (1987).

[4] B. J. Brewer, in "DNA Replication: The Regulatory Mechanisms" (P. Hughes, E. Fanning, and M. Kohiyama, eds.), p. 97. Springer-Verlag, Berlin and New York, 1992.

[5] L. Martín-Parras, P. Hernández, M. L. Martínez-Robles, and J. B. Schvartzman, *J. Biol. Chem.* **267,** 22,496 (1992).

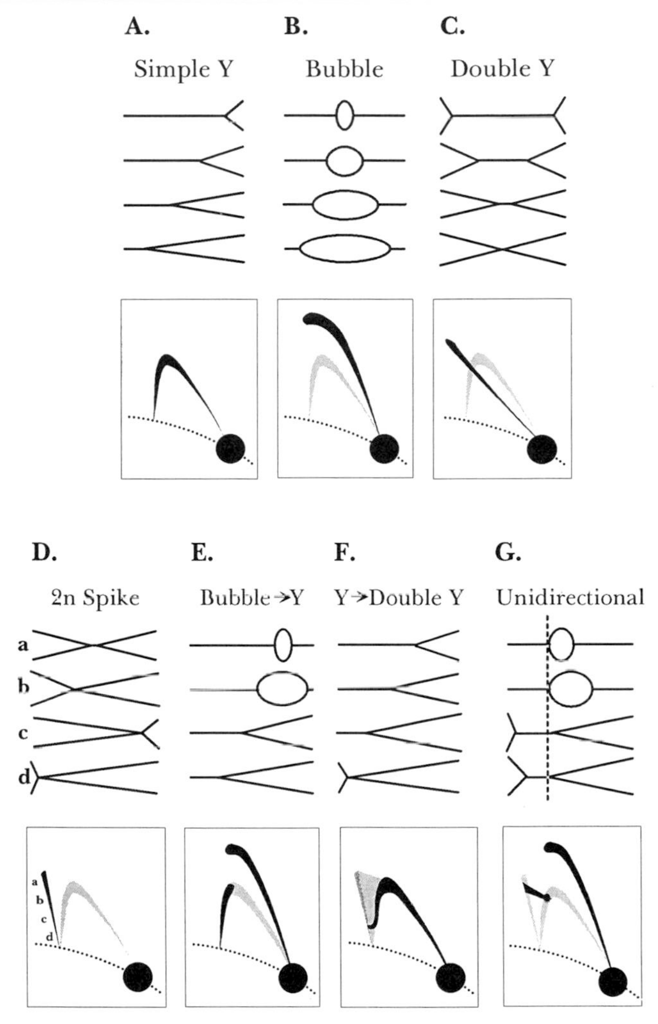

FIG. 1. Schematic representation of 2-D gel patterns produced by a restriction fragment undergoing the indicated modes of replication. Replication intermediates of the restriction fragment are shown above each 2-D gel. The relevant 2-D gel pattern is indicated in black in each drawing; the lightly shaded arcs are included for reference. The vertical dotted line in part G indicates the position of the unidirectional origin of replication. See the text for a discussion of each type of pattern.

have been shown to result in preferential recovery of replication intermediates.[6]

The choice of the precise gel conditions to be utilized is often crucial to the success of an analysis. Only certain conditions of electrophoresis accentuate the differences between simple Y and bubble-shaped molecules. Furthermore, the optimal electrophoresis conditions differ according to fragment size. For fragments in which the nonreplicating molecules are ~2.0 to 5.5 kb, the first dimension is run in a 0.4% gel without ethidium bromide, in 1× TBE buffer (85 m*M* Tris base, 89 m*M* borate, and 2.5 m*M* EDTA) at 1 V/cm for approximately 20 to 24 hr. Our gels are 13.5 cm wide by 20 cm long and 0.7 cm deep; the wells are 0.4 cm wide and 0.125 cm thick. This combination results in a long, narrow gel lane with very tight bands. Volts per centimeter are measured with a hand-held voltmeter. Alternatively, volts per centimeter can be calculated by dividing the total volts read from the power supply by the total path length from electrode to electrode. An empty well is left between samples in order to avoid cross-contamination. After electrophoresis, the gel is stained in 1× TBE with 0.3 μg/ml ethidium bromide and photographed. To limit nicking of the DNA, excessive exposure to UV light should be avoided and only long-wavelength UV light is recommended. The position of the unreplicated fragment (ideally 9 cm or greater from the well) is calculated from the position of molecular weight markers and the lane is carefully excised with a ruler and a clean razor blade beginning 1 cm below the desired fragment and extending 10 cm up the lane. This portion of the gel lane will include linears of more than twice the size of the fragment of interest, as well as all of the fragment's replication intermediates.

The excised lanes are rotated 90 degrees from the original direction of electrophoresis and placed in a new gel form. We rotate the slabs so that the top of the first dimension is toward the left in the second dimension. The fragile gel slabs can be transferred on a thin, flexible ruler or piece of plastic. The orientation of the original top and bottom surfaces should be maintained so that the DNA will exit the first dimension gel slab through a cut gel surface. The second dimension gel forms are 20 cm wide by 25 cm long. This large size allows four samples to be run at once, two across the top, and two in the middle (Fig. 2). If smaller gel platforms are used for the second dimension, a minimum gel length of 15 cm is recommended for good separation. After alignment of the first dimension gel lanes, the slabs are sealed in place by pipetting a small amount of the agarose to be used in the second dimension (1.1% agarose in 1× TBE containing 0.3 μg/ml ethidium bromide, equilibrated to 55°) behind the first dimension gel

[6] P. A. Dijkwel, J. P. Vaughn, and J. L. Hamlin, *Mol. Cell. Biol.* **11,** 3850 (1991).

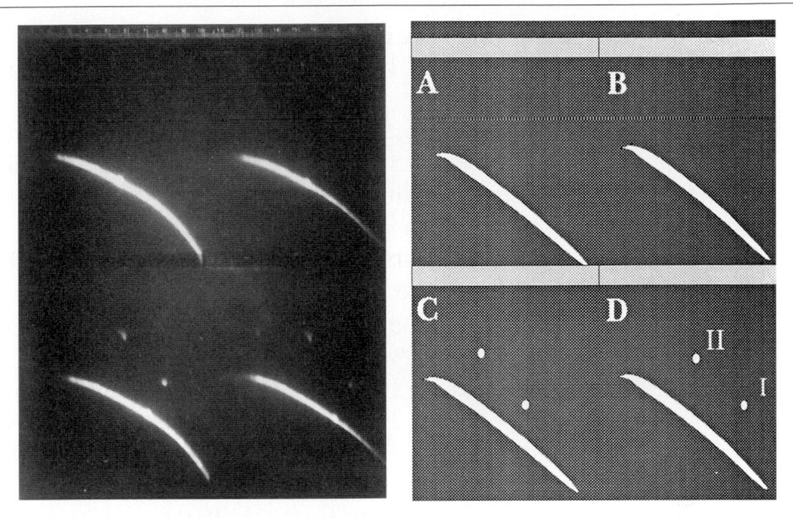

FIG. 2. Photograph of an ethidium bromide stained 2-D gel. Linear restriction fragments of yeast chromosomal DNA (2.5 μg/sample) are seen as diagonal arcs. Letters A–D on the right-hand side indicate each of the four samples. The positions of the strips of low-percentage agarose from the first dimension gel are also indicated on the right-hand side. Slowly migrating circular forms of the yeast 2-μm plasmid (form I and form II) appear as spots of fluorescence above arcs C and D. Samples A and B were digested with *Bgl*I and *Sna*BI; samples C and D were digested with *Bgl*II.

slabs. When hardened, this agarose holds the gel slabs in place as the remaining agarose is slowly poured around the first dimension slices. This second dimension gel should be poured until it just covers the first dimension gel slabs (a depth of about 0.8 cm). Submerging the first dimension gel slabs in the new agarose helps stabilize the joint between the first and second dimension gels for later handling. The second dimension gel is run in the cold room (4°) at 5 to 6 V/cm in an electrophoresis box with large buffer reservoirs (≥2 liters). The 1× TBE buffer containing 0.3 μg/ml ethidium bromide is circulated from anode to cathode at a rate of approximately 50 to 100 ml/min to maintain a constant ethidium bromide concentration along the gel length. The gel is covered by no more than ~2 mm of TBE/ethidium bromide buffer. The progress of the second dimension electrophoresis can be monitored with a hand-held, long-wavelength UV lamp to ensure that the visible diagonal arc of linear molecules does not run off the gel. Electrophoresis is continued until the smallest linears to be analyzed have run 10 to 12 cm (about 3.5 to 6 hr). Following electrophoresis, the gels are photographed (Fig. 2) and treated for Southern blotting—first with two 15-min washes in 0.25 *N* HCl, followed by two 15-min washes in 0.5 *N* NaOH, 1 *M* NaCl, and finally for 30 min in 0.5 *M* Tris,

3 *M* NaCl. We have found that it is beneficial to invert the gels for blotting, because the upper surface of high concentration gels tends to be irregular.

Restriction fragments of larger or smaller size can be analyzed on 2-D gels if some modifications are made to the electrophoresis conditions. Finding the correct electrophoresis conditions for larger or smaller fragments may require some experimentation using fragments of known size and replication pattern. We have successfully analyzed fragments as small as 1 kb. For these small fragments, the first dimension is typically run at a slightly higher agarose concentration (0.6 or 0.7%), whereas the second dimension is run in a 2% gel. The 2% agarose is not easily poured at 55°. Since it is not advisable to expose the DNA in the first dimension gel slabs to high temperatures, we pour the 2% agarose of the second dimension before inserting the first dimension slabs. When the 2% gel has hardened, the top 2 cm of the gel are removed with a clean razor blade, the first dimension slices are sealed against the cut surface with 0.6 or 0.7% agarose containing 0.3 μg/ml ethidium bromide equilibrated to 55°, and the remaining portion of the gel is filled with the 0.6 or 0.7% agarose containing ethidium bromide. The second dimension is run under conditions similar to those outlined above.

Fragments that are larger than 5.5 to 6 kb must be run under conditions of lower agarose concentration and lower voltage in both dimensions in order to separate successfully the bubble and simple Y arcs.[7–9] If gels are run under the conditions intended for smaller fragments (2 to 5.5 kb), the bubble and simple Y patterns are not clearly separated, and the signals for each are distorted. Again, finding the correct electrophoresis conditions for large fragments may require some experimentation using fragments of known size and replication pattern. It is crucial that conclusions about the presence or absence of initiation within a fragment not be drawn unless it is demonstrated that the electrophoresis conditions utilized for those particular fragment sizes are able to separate bubble and simple Y forms.

Interpretation of 2-D Gels

After the second dimension of electrophoresis, nonreplicating molecules are seen as a shallow arc (arc of linears; Fig. 2), whereas the rare replicating molecules run above this arc. Probing for a specific DNA fragment reveals

[7] O. Hyrien and M. Méchali, *Nucleic Acids Res.* **20,** 1463 (1992).

[8] P. J. Krysan, "Using Autonomous Replication to Physically and Genetically Define Human Origins of Replication," p. 45. Ph.D. Thesis, Stanford University (1993).

[9] P. J. Krysan and M. P. Calos, *Mol. Cell. Biol.* **11,** 1464 (1991).

a spot of nonreplicating molecules on the arc of linears at the size predicted for the fragment (called the monomer or 1*n* spot). Y-shaped intermediates (Fig. 1A) rise from the 1*n* spot as the newly replicating arms increase the mass and structural complexity of the fragment. When the fragment is half replicated, it reaches a state of maximal structural complexity in which all of the arms are of equal length. As replication of the fragment continues, the replicated arms can align to form a long linear with a centrally located side branch of unreplicated DNA that decreases in length as the linear portion increases in length. As this process proceeds to the end of the fragment, the shape becomes less extensively branched, and therefore the replication intermediates migrate more quickly in the second dimension. The replicating molecules rejoin the arc of linears when the fragment is nearly fully replicated, or 2*n* in size, completing the characteristic simple Y arc.

Bubble and double-Y patterns differ from the simple Y arc as shown in Figs. 1B and C, respectively. Unlike the simple Y arc, bubble and double-Y patterns do not intersect with the linear arc at the 2*n* position. As a bubble completes replication, its shape resembles a nicked circle of 2*n* mass. As the forks in a double-Y intermediate meet, the shape of the replication intermediate resembles the shape of a recombination intermediate of 2*n* mass. The position of meeting forks can vary in a double-Y replication intermediate, just as the position of crossed strands in a recombination intermediate can occur at any position within the fragment. Both types of intermediates form a nearly vertical streak arising from the 2*n* position on the arc of linears[10] (2*n* spike; Fig. 1D). Molecules are arrayed along the 2*n* spike according to the position of their crossed strands or approaching forks. Molecules consisting of four arms of equal length are maximally retarded in the second dimension and thus are located at the apex of the 2*n* spike.

Combinations of these patterns may exist for a single restriction fragment. For example, if the fragment is not replicated in the same way by all cells in the population, then a mixture of overlapping patterns is observed. An origin that is not used efficiently will produce bubbles in some cells, but simple Y's or double Y's in other cells of the population. In this case, all three complete 2-D gel patterns will exist simultaneously (Fig. 3A). A second type of complex pattern occurs when an origin (or terminus) does not lie in the center of the restriction fragment. For example, if replication initiates asymmetrically within a fragment or if the two forks diverging from a centrally located origin move at different rates, only the initial replication intermediates are bubbles. After one fork reaches the

[10] B. J. Brewer, E. P. Sena, and W. L. Fangman, *Cancer Cells* **6,** 229 (1988).

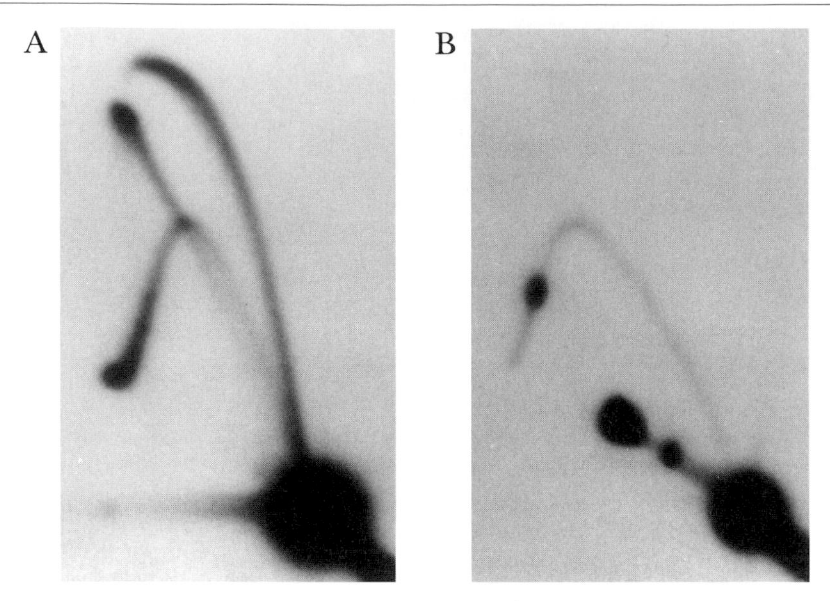

FIG. 3. 2-D gels of yeast replication intermediates. (A) Autoradiogram of a 2-D gel in which the restriction fragment of interest is replicated as a bubble, simple Y, or double Y in different cells of the population. See text for discussion. (B) Autoradiogram of a 2-D gel in which the probed restriction fragment contains a region in which DNA polymerase complexes stall during replication. This replication fork arrest results in a large spot of hybridization in the middle of the descending arm of the simple Y arc, indicating that the replication fork pause occurs when the fragment is approximately three-quarters replicated. The prominent spots along the linear arc result from partial digestion of the DNA.

restriction site forming the end of the fragment, the intermediates are converted into simple Y's. This mode of replication produces a discontinuous 2-D gel pattern in which the signal begins as a bubble and ends on the simple Y arc (Fig. 1E). Similarly, the 2-D gel pattern produced by a replicating fragment can begin as a simple Y arc and be converted to a double-Y arc as a second fork enters the unreplicated portion of the fragment (Fig. 1F). The shape of the double-Y pattern, particularly the point at which it terminates along the $2n$ spike, depends on the position within the fragment where forks meet. If the second fork enters the fragment very late in its replication, then the intermediates will produce a nearly complete simple Y arc, but will terminate as an asymmetric double Y ending at a position along the $2n$ spike very near the $2n$ linears (as shown by the bold line in Fig. 1F). The double-Y intermediates occupy the region of the gel between these two points. The earlier the second fork enters a restriction fragment, the earlier that fragment deviates from the simple Y arc and the higher up the $2n$ spike it terminates.

In some ways simple Y intermediates are the least informative since they indicate the absence of both an origin and a terminus. However, they provide an opportunity to detect variations in the rate of fork movement. Since simple Y molecules have a single fork, the relative intensity of hybridization along the Y arc provides a history of that fork's movement. For example, if the replication fork stalls at a specific site in the fragment, then the relative abundance of replication intermediates of that particular shape will be increased. These stalled intermediates will be detected as an increased darkening at a specific point along the simple Y arc (Fig. 3B). More subtle changes in the intensity of hybridization along the arc are also detectable and can be used to infer the positions of changes in the rate at which the DNA polymerase complex moves along the template. It is important to note that all of the replication arcs show characteristic variations in hybridization intensity that are unrelated to replication fork pause sites. For example, the apex of the Y arc often appears darker than the ascending and descending arms. Similarly, the ascending arm is almost always less intense than the descending arm. These variations most likely result from compression of larger fragment sizes in the gel. To verify that variation along the arc is biologically significant, it is imperative to examine the region of interest in overlapping fragments and to confirm that the pause site moves to predicted positions along the arc. Note also that pauses and barriers can only be detected in simple Y arcs, since the second moving fork in bubble and double-Y fragments disguises the fact that one fork has paused and thus prevents the accumulation of intermediates of a specific size.

The bubble and double-Y 2-D gel patterns that have been described to this point are generated during bidirectional replication. Bubble and double-Y intermediates produced by unidirectional replication differ from those generated by bidirectional replication in that one fork moves, while the other remains fixed. The populations of molecules that contribute to unidirectional bubble and double-Y arcs are therefore different from those generated by bidirectional replication. During unidirectional replication, a fragment containing an origin has a bubble that expands in only one direction. As one fork moves off the end of the DNA fragment, this bubble is converted to a simple Y form. Because the remaining fork in the simple Y is stationary, no larger simple Y forms are produced. This intermediate is then eventually converted to a double Y as a fork enters the opposite end of the fragment and termination occurs at the original site of initiation (Fig. 1G). Interestingly, Schvartzman *et al.*[11] have shown that the location

[11] J. B. Schvartzman, M. L. Martínez-Robles, and P. Hernández, *Nucleic Acids Res.* **21,** 5474 (1993).

of a unidirectional origin in the restriction fragment affects the shape of the bubble arc. The shape of the bubble arc is subtly different if the origin is located in the center of the fragment than if it is located toward one end of the fragment. For an excellent 2-D gel study of unidirectional replication of the *Escherichia coli* plasmid pBR322 also see Martín-Parras *et al.*[12]

Limitations of 2-D Gels

2-D gels are extremely useful for preliminary characterization of the replication pattern of a restriction fragment. However, several cautions must be observed in the interpretation of these gels. First, replication bubbles arising from initiation may not be visible if the origin is located very near one end of the restriction fragment.[3,13] In this case intermediates are quickly converted into simple Y's. Because there is very little difference in the migration of the bubble and simple Y arcs as they emerge from the 1*n* spot, bubble forms can be easily missed. Therefore, in looking for the presence of initiation, it is critical to examine overlapping restriction fragments. Second, breakage of DNA at replication forks which occurs during the isolation and handling of the DNA can result in the loss of bubbles. Although this breakage can be reduced by following the precautions suggested above, some extent of shearing is probably unavoidable. When a branch is broken from a simple Y, the replication intermediate is converted into two linear molecules. However, when one of the forks of a bubble is broken, the molecule is converted into a shape similar to a simple Y. Since broken bubble forms are similar, but not identical to simple Y's, they may form aberrant signals near or on the simple Y arc (for a discussion of this problem, see Martín-Parras *et al.*[5]). Therefore, these artificially produced Y's may limit the reliability of origin efficiency estimates based on quantitation of simple Y and bubble arcs. Third, even under conditions that eliminate fork breakage, quantitation of bubble and simple Y arcs is problematic. It is often impossible to determine which parts of the two arcs should be compared since some of the larger simple Y's may reflect the discontinuous pattern caused by an asymmetrically located origin or unequal fork rates from a centrally located origin. These discontinuities do not occur at a unique mass in the replicating fragment due to cell-to-cell variations in fork rate. Further complicating the comparison of bubbles and simple Y's is the fact that bubbles contain two replication forks, and therefore complete replication at twice the rate of simple Y's. Therefore, if a fragment is

[12] L. Martín-Parras, P. Hernández, M. L. Martínez-Robles, and J. B. Schvartzman, *J. Mol. Biol.* **220,** 843 (1991).

[13] M. H. K. Linskens and J. A. Huberman, *Nucleic Acids Res.* **18,** 647 (1990).

replicated in one-half of cells as a bubble, and in the other half of cells as a simple Y, the simple Y arc will appear twice as strong as the bubble arc.

2-D gels can be used to map discrete termination sites, since double-Y forms can be seen extending to the nearly vertical $2n$ spike. However, if termination is not discrete, but rather occurs over a large interval, it is often only detectable as a faint smearing between the descending arc of the simple Y and the $2n$ spike (gray shading in Fig. 1F). This smearing corresponds to a second fork entering the restriction fragment when the first fork is at many different positions in the fragment. As a result, many different double-Y arcs diverge from the descending side of the simple Y arc and join the $2n$ spike at many corresponding points. This broad termination zone can be difficult to distinguish from background hybridization on 2-D gel blots.

Determining Direction of Fork Movement in Simple Y Intermediates

If a fragment is replicated by a single fork, identical simple Y arcs are produced regardless of the direction in which that fork is moving. The ability to determine the direction of fork movement can provide information on the location of the nearest origin and can also overcome several of the limitations of 2-D gels that were discussed above. For example, determination of the direction of fork movement in restriction fragments on either side of an origin of replication can provide estimates of origin efficiency that are not affected by the random breakage of bubbles. If all replication forks are moving out from the origin, then the origin must be very efficient, whereas the presence of replication forks moving into the origin region, from one side or the other, reveals that the efficiency of initiation is lower. The direction of fork movement can also be used to confirm broad zones of replication termination, since the ratio of forks moving in the two directions in different fragments from the termination zone can be used to map the area over which termination occurs.[14]

Directional information is derived from 2-D gels by an "in-gel" digestion with a second restriction enzyme between the first and second dimensions of electrophoresis.[15] This second enzyme site should be approximately a quarter to a third of the distance from one end of the original fragment. After the second dimension is run, the gel is blotted and probed for the

[14] B. J. Brewer, J. D. Diller, K. L. Friedman, K. M. Kolor, M. K. Raghuraman, and W. L. Fangman, *Cold Spring Harbor Symp. Quant. Biol.,* **58,** 425 (1993).

[15] B. J. Brewer, D. Lockshon, and W. L. Fangman, *Cell* **71,** 267 (1992).

larger fragment. As shown in Figs. 4A and C, this type of gel produces two different simple Y arcs, depending on the direction in which replication is proceeding relative to the internal restriction site. Both of these arcs originate below the initial arc of linears, since the fragments from the first dimension gel have been shortened. If replication forks first enter the original fragment in the portion that is being probed, then an arc of replication intermediates rises immediately from the new linear spot. This simple Y arc terminates when the replication forks pass the site of the second restriction digest and enter the unprobed portion of the original fragment. If, on the other hand, replication forks first enter the small, unprobed segment and only later enter the probed region, the detected intermediates are first linear, and only begin to form a Y arc after the replication forks enter the probed fragment. The intensity of parallel, nonoverlapping portions of these two arcs can be quantitated in order to estimate roughly the percentage of forks moving in either direction. In this case, two simple Y arcs of identical size are being compared, eliminating the problems inherent in determining bubble-to-simple Y ratios. Analysis by phosphor imaging yields more accurate results than scanning of films. Since the phosphor imager directly measures the amount of radioactive decay, it achieves a linearity of signal that is difficult or impossible to obtai ı on standard film, especially in light of the long exposure times required to detect replication intermediates.

Restriction enzyme sites must be carefully chosen for fork direction analysis. Ideally, the initial digest should yield a fragment of ~3.5 to 5 kb. The second restriction digest should remove a quarter to a third of this fragment. The larger of the resulting fragments should be ≥2 kb in size. The type of agarose used for the first dimension gel influences the efficiency of restriction enzyme cleavage *in situ.* We have found that using Beckman LE agarose for the first dimension gel permits efficient restriction enzyme cleavage, although restriction enzyme grade agarose products from other sources are also likely to work well. Suppliers of restriction enzymes, such as New England Biolabs, provide information on "in-gel" cleavage by their enzymes. When in doubt, it is advisable to assay either the agarose or the enzyme, using embedded plasmid DNA.

After running the first dimension in Beckman LE agarose under the normal first dimension gel conditions, the lane of interest is excised as outlined above. This gel slab is placed in a disposable pipette reagent reservoir from Costar (No. 4870) with the top and bottom of the slab clearly marked on the reservoir. The reservoir is then filled with TE 10/0.1 (10 m*M* Tris, pH 8, 0.1 m*M* EDTA), placed in a glass baking dish, and gently agitated at room temperature on a shaker for 0.5 hr. This wash is repeated once. After the second TE wash, the dish is again emptied and any excess

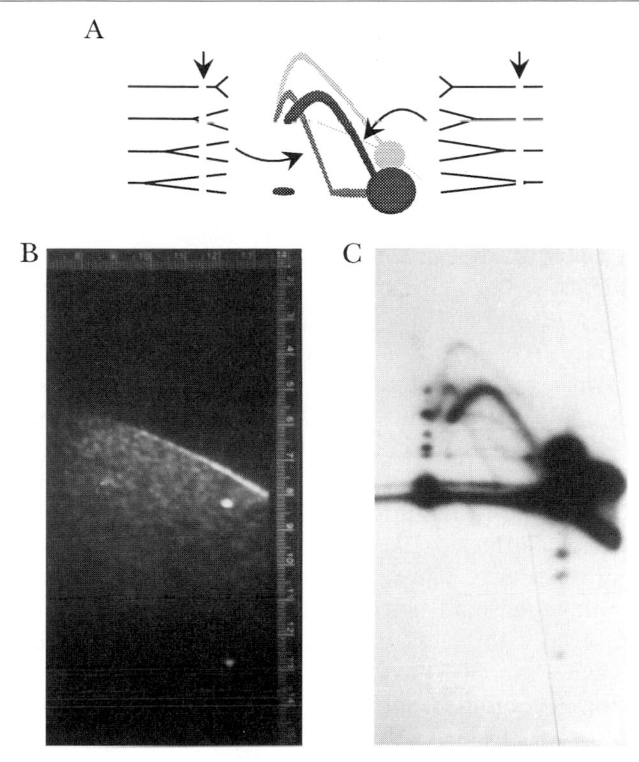

FIG. 4. Determining the direction of replication fork movement. (A) Schematic diagram of a 2-D gel used to determine the direction of replication fork movement. Simple Y replication intermediates containing replication forks moving in both directions were digested in the gel at the indicated site (↓), and the larger remaining fragment was probed. If the replication fork was moving to the left, then the arc of intermediates is displaced from the 1*n* spot; if the fork was moving to the right, then the arc of intermediates emerges from the 1*n* spot. The upper, lightly shaded arc consists of molecules that were not cut by the second restriction enzyme. See text for further explanation. (B) Ethidium bromide-stained 2-D gel following "in-gel" digestion and electrophoresis in the second dimension. Yeast DNA was digested with *Xab*I prior to the first dimension and was subsequently digested in the gel with *Cla*I before electrophoresis in the second dimension. *Xba*I fragments that do not contain a *Cla*I site can be seen as an arc of linear molecules, whereas those *Xba*I fragments that do contain *Cla*I sites migrate at a lower mass after the second digestion, forming a haze below the arc. The haze is derived from single-copy sequences; the prominent spots are derived from repeated sequences in the genome. (C) Autoradiogram of 2-D gel used to demonstrate the direction of replication fork movement. As indicated in the schematic drawing and by the different intensities of hybridization of the two arcs, this fragment is replicated by a fork moving from left to right more frequently than by a fork moving in the opposite direction.

TE is removed with a Pasteur pipette. The dish is filled with restriction enzyme buffer made to the manufacturer's specifications for the enzyme being used [bovine serum albumin (BSA) is included as required, but spermidine is not included in the buffer because it might alter the mobility of the DNA in the second dimension]. [We have found that older catalogs (such as the 1988–1989 New England Biolabs catalog) are a good source of restriction enzyme buffer recipes.] Recipes from universal buffer systems are not recommended because they may not be optimized for the specific enzyme being used. The gel slabs are incubated for 1 hr at room temperature, with gentle agitation. This incubation in fresh restriction enzyme buffer is repeated once. After the washes are completed, the restriction enzyme buffer is drawn off, and excess buffer is removed with a Pasteur pipette. Because it is important to achieve the highest enzyme concentration possible in the gel slab, excess buffer should be removed from the surfaces of the gel and reservoir by blotting briefly with a Kimwipe. Twenty microliters of restriction enzyme (a minimum of 5 units per microliter) are pipetted directly onto a cut surface of the gel slab, distributing the aliquot of enzyme along the entire surface. The reservoirs are then placed in a glass baking dish that contains a small amount of water in the bottom to maintain humidity, covered with plastic wrap, and placed at 37° for 3 hr. After this incubation, an additional 10 to 20 μl of enzyme is added to each slab, and the incubation is continued another 2 to 4 hr. Incubation overnight may be acceptable as well; however, long incubations can lead to diffusion of the fragments within the gel slab and result in horizontal streaks in the second dimension gel. At the completion of the digest, the slab is rinsed by incubation in TE 10/1 (10 m*M* Tris, pH 8, 1 m*M* EDTA) for 15 min and the slabs are placed in the second dimension gel as described previously. Because the fragments of interest have been truncated by the "in-gel" digestion, the original arc of linears is no longer a reliable marker for the extent of electrophoresis. The restriction fragment of interest now lies in the haze of DNA fragments extending below the arc of linears (Fig. 4B). For this reason we allow the full length of 25 cm for a single fork direction gel and include a well for molecular weight markers in the second dimension gel.

Several factors must be considered in the interpretation of fork direction gels. Complete digestion in the gel is often not achieved. After probing for the subfragment of interest, it is frequently possible to see the newly formed fork direction arcs, the original, larger simple Y arc, and several other arcs of intermediate size that result from cleavage of only one of the two arms of a replicating molecule. It is important not to misinterpret these arcs of partial digestion as evidence of forks moving in opposite directions. The interpretation of these partials has been discussed elsewhere.[14]

Conclusion

In this chapter, we have discussed only the common types of replication intermediates found among DNA restriction fragments. Bubbles, simple Y's, and double Y's are certainly not the only possible replication intermediates. For example, branched intermediates with large stretches of single-stranded DNA are produced by displacement synthesis, which occurs in the replication of mitochondrial DNA of mammals. These intermediates might be expected to display distinct migration properties in 2-D gels. Similarly, the multiple microbubbles that are thought to be intermediates in the replication of chromosomal DNA in early cleavage divisions of *Drosophila* embryos might be expected to produce novel 2-D gel patterns. Neither of these types of intermediates has yet been analyzed on 2-D gels. Finally, replication intermediates of circular molecules, both theta[10] and rolling circle forms, would be expected to produce novel hybridization patterns on 2-D gels.

2-D gels can provide useful qualitative information about the mode of replication. As with any new technique, one must be cautious when interpreting these 2-D images. It is prudent to analyze the intermediates from duplicate samples and to examine the region of interest in overlapping restriction fragments. Deriving quantitative information from standard 2-D gels is not advised. Rather, results obtained from standard 2-D gels can be confirmed and quantitated by analyzing the direction of fork movement. However, since quality of the DNA, variations in hybridization along an arc, and differences in the rate of replication can all influence quantitation, values obtained through this analysis must be considered estimates.

Acknowledgments

We would like to thank Walt Fangman, John Diller, Katherine Koler, and M. K. Raghuraman for comments on the manuscript, and David Friedman for assistance with the figures. Work in the Fangman/Brewer laboratory is supported by NIGMS grant 18926 to WLF and BJB. KLF is a Howard Hughes Medical Institute Predoctoral Fellow.

[47] Specific Labeling of Newly Replicated DNA

By MELVIN L. DEPAMPHILIS

Introduction

Methods for specific labeling of newly replicated DNA are based on characteristics that distinguish replication of DNA from repair of damaged DNA. DNA replication is a semiconservative process in which one DNA strand acts as the template for the newly synthesized DNA strand. Most genomes are replicated by the replication fork mechanism (Fig. 1) in which DNA synthesis occurs concomitantly on both templates. DNA synthesis in the direction of fork movement (forward arm of fork or leading strand template) occurs by continuous incorporation of deoxyribonucleotide precursors to form long nascent DNA strands, whereas synthesis in the direction opposite fork movement (retrograde arm of fork or lagging strand template) is carried out discontinuously by the repeated synthesis of short nascent DNA chains referred to as Okazaki fragments.[1-5] Okazaki fragments in eukaryotic cells and their viruses may be assembled from a cluster of even shorter pppRNA-p-DNA chains of $\leq$40 nucleotides that are referred to as "DNA primers" (Fig. 2).[6,7] One consequence of the asymmetrical distribution of Okazaki fragments at replication forks is that an origin of bidirectional replication (OBR) is revealed by the transition from discontinuous to continuous DNA synthesis that must occur on each template (Fig. 1), a fact that has been used to identify sites where replication begins in both viral and cellular chromosomes.[8-11] Examples are found in prokaryotic and eukaryotic cell chromosomes, plasmids that can replicate in pro-

[1] S. Anderson and M. L. DePamphilis, *J. Biol. Chem.* **254,** 11,495 (1979).
[2] M. L. DePamphilis and P. M. Wassarman, *Ann. Rev. Biochem.* **49,** 627 (1980).
[3] R. T. Hay, E. A. Hendrickson, and M. L. DePamphilis, *J. Mol. Biol.* **175,** 131 (1984).
[4] E. A. Hendrickson, C. E. Fritze, W. R. Folk, and M. L. DePamphilis, *EMBO J.* **6,** 2011 (1987).
[5] W. C. Burhans, L. T. Vassilev, M. S. Caddle, N. H. Heintz, and M. L. DePamphilis, *Cell* **62,** 955 (1990).
[6] T. Nethanel, T. Zlotkin, and G. Kaufmann, *J. Virol.* **66,** 6634 (1992).
[7] D. Denis and P. A. Bullock, *Mol. Cell. Biol.* **13,** 2882 (1993).
[8] M. L. DePamphilis, E. Martínez-Salas, D. Y. Cupo, E. A. Hendrickson, C. E. Fritze, W. R. Folk, and U. Heine, *in* "Eukaryotic DNA Replication" (B. Stillman and T. Kelly, eds.), pp. 164–175, Vol. 6, *Cancer Cells.* Cold Spring Harbor Laboratory Press, New York, 1988.
[9] M. L. DePamphilis, *Ann. Rev. Biochem.* **62,** 9 (1993).
[10] M. L. DePamphilis, *Curr. Opin. Cell Biol.* **5,** 434 (1983).
[11] M. L. DePamphilis, *J. Biol. Chem.* **268,** 1 (1993).

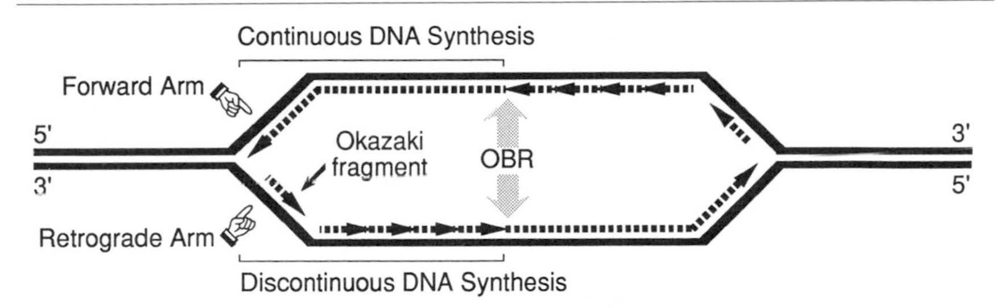

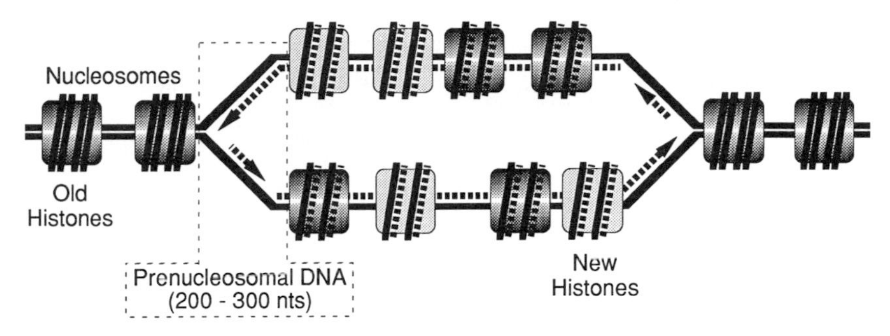

FIG. 1. Structure of replication forks.

karyotic cells or in the nuclei of eukaryotic cells, and the double-strand DNA forms of most bacteriophage and animal virus genomes.[12]

In eukaryotic cells and their viruses (Fig. 1), the old histone octamers in front of the fork are distributed randomly to both arms,[13–15] and new histone octamers are rapidly assembled on both arms by a two-step reaction in which histones H3/H4 are deposited first followed by histones H2A/H2B and later histone H1.[16] Newly replicated DNA is rapidly organized into nucleosomes, although a region of 200 to 300 nucleotides on each arm of the fork that includes Okazaki fragments is nucleosome free.[17]

Okazaki fragments provide a useful handle for distinguishing newly replicated DNA from newly repaired DNA.

[12] A. Kornberg and T. Baker, "DNA Replication." W. H. Freeman & Co., New York, 1992.

[13] W. C. Burhans, L. T. Vassilev, J. Wu, J. M. Sogo, F. Nallaseth, and M. L. DePamphilis, *EMBO J.* **10,** 4351 (1991).

[14] T. Krude and R. Knippers, *Mol. Cell. Biol.* **11,** 6257 (1991).

[15] K. Sugasawa, Y. Ishimi, T. Eki, J. Hurwitz, A. Kikuchi, and F. Hanaoka, *Proc. Natl. Acad. Sci. U.S.A.* **89,** 1055 (1992).

[16] C. Gruss and J. M. Sogo, *BioEssays* **14,** 1 (1992).

[17] M. F. Cusick, P. M. Wassarman, and M. L. DePamphilis, *in* "Methods in Enzymology—Nucleosomes" (P. M. Wassarman and R. D. Kornberg, eds.), Vol. 170, pp. 290–316. Academic Press, New York, 1989.

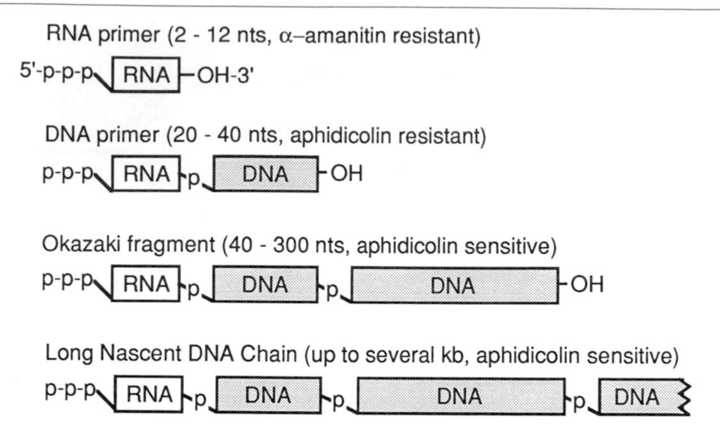

FIG. 2. Structure of newly replicated nascent DNA chains from replication forks.

Characteristics of Okazaki Fragment Synthesis in Eukaryotic Cells and Their Viruses

Eukaryotic Okazaki Fragments

1. Are transient intermediates in replication ($t_{1/2} \sim 1$ min, 30°).
2. Are short nascent DNA chains (30 to 300 nucleotides).
3. Contain an oligoribonucleotide covalently attached to their 5′ ends ("RNA primer"). RNA primers begin with rATP (~80%) or rGTP (~20%) at 3′-PuT and 3′-PuC sites, respectively, in the DNA template. All 16 possible rN-p-dN junctions are found at frequencies reflecting the dN-p-dN linkages in the template, revealing that the transition from RNA to DNA synthesis is sequence independent. RNA primers are not uniform in length, but vary from 2 to 12 nucleotides with a mean length of 10 ± 2 nucleotides. RNA primer synthesis is insensitive to α amanitin.
4. Originate predominantly (exclusively?) from the retrograde arm of replication forks (80% to 95%).
5. Appear to be synthesized from short RNA-p-DNA chains ("DNA primers"). DNA primers are up to 40 nucleotides long (mean length ~35 nucleotides). Their synthesis is insensitive to aphidicolin[6] and so is the first 30 to 40 nucleotides carried out by DNA primase : DNA polymerase α,[18] the enzyme complex responsible for synthesis of both DNA primers and Okazaki fragments. So far, DNA primers

[18] R. S. Decker, M. Yamaguchi, R. Possenti, and M. L. DePamphilis, *Mol. Cell Biol.* **6,** 3615 (1986).

have been reported only in SV40 replicating intermediates. However, since SV40 DNA replication depends almost entirely on cellular proteins, cellular DNA replication is also likely to use DNA primers.

However, not all DNA replication systems involve Okazaki fragments or RNA primers.[12] Adenovirus, ϕ29 phage, and mitochondria genomes replicate by a strand-displacement mechanism in which DNA synthesis proceeds on the forward arm in the absence of DNA synthesis on the retrograde arm. DNA synthesis in both directions is a continuous process. Nascent DNA strands in adenovirus, ϕ29 phage and synthesis of the "−" strand in hepatitis virus utilize a specific protein primer that remains covalently attached to their 5′ ends. Mitochondrial and hepatitis virus "+" strand nascent DNA chains contain an RNA transcript covalently attached to their 5′ ends. Parvoviruses represent another unique replication system in which a hairpin structure at the 3′ end of this template is used as a primer to initiate DNA synthesis. Thus, during the initial stage of replication, nascent DNA is covalently attached to the 3′ end of the parental DNA template.

The characteristics of newly replicated DNA molecules that utilize the replication fork mechanism provide the basis for four experimental strategies to label newly replicated DNA in whole cells or cell extracts.

Strategies for Labeling Newly Replicated DNA

1. Incorporation of radiolabeled deoxyribonucleotides into nascent DNA followed by characterization of labeled DNA replication intermediates.
2. Incorporation of density-labeled or affinity labeled deoxyribonucleotides into nascent DNA following by analysis of changes in DNA density, purification of DNA by affinity chromatography, or staining of nascent DNA *in situ*.
3. Identification of replication fork polarity.
4. Detection of changes in methylation during DNA replication.
5. Identification of RNA primers covalently attached to the 5′ ends of DNA.

The experimental conditions described below are for studies on mammalian cells, simian virus 40 (SV40), and polyomaviruses, but are readily adapted to other situations. Refer to Sambrook *et al.*[19] for additional experimental details of procedures generally used in analyses of DNA and RNA,

[19] J. Sambrook, E. F. Fritsch, and T. Maniatis, "Molecular Cloning: A Laboratory Manual." Cold Spring Harbor Laboratory Press, New York, 1989.

and to the Merck Index for an awareness of biohazards associated with the reagents involved.

Incorporation of Radiolabeled Deoxyribonucleotides into Nascent DNA Chains

The simplest approach to labeling newly replicated DNA is to label newly synthesized DNA by incorporation of labeled dNTPs at the 3′ end of growing DNA chains, and then to characterize the DNA products in order to distinguish DNA replication from DNA repair. Radioactively labeled substrates provide the most sensitive means for detecting nascent DNA. In subcellular systems, any one of the four dNTP substrates for DNA synthesis can be used to label nascent DNA, although addition of two noncomplementary [α-^{32}P]dNTPs produces uniformly labeled DNA, regardless of sequence composition. Metabolic redistribution of these radioisotopes to other molecules is not a problem, because DNA polymerases incorporate dNTPs directly and rapidly into DNA. Nevertheless, consideration of endogenous nucleotide pool sizes can be useful, because they affect the specific radioactivity of the labeled nucleotide and therefore the amount of label that appears in nascent DNA chains. Pool sizes can vary considerably with the phase of the cell division cycle, and with the method used to synchronize cells.[20,21]

Average Nucleotide Concentrations in Mammalian Cells. The following tabulation indicating nucleotide concentrations is adapted from Kornberg and Baker.[12]

[dTTP] $\approx$ 23 μM	[rATP] $\approx$ 2.8 mM
[dCTP] $\approx$ 22 μM	[rUTP] $\approx$ 0.48 mM
[dATP] $\approx$ 13 μM	[rGTP] $\approx$ 0.48 mM
[dGTP] $\approx$ 5 μM	[rCTP] $\approx$ 0.21 mM

[^{3}H]Deoxythymidine (Thd) is used routinely to provide rapid and specific labeling of DNA in eukaryotic cells, although this approach cannot be used with fungi since they lack thymidine kinase. High concentrations of thymidine (>1 mM) inhibit dTTP synthesis, and a high specific radioactivity of [^{3}H]Thd can perturb DNA replication and cell cycle progression.[22] Deoxyribonucleosides other than Thd can radiolabel other molecules in addition to DNA. For example, [^{3}H]deoxycytidine rapidly labels dCDP-

[20] B. Nicander and P. Reichard, *J. Biol. Chem.* **260,** 9216 (1985).

[21] J. M. Leeds and C. K. Mathews, *Mol. Cell. Biol.* **7,** 532 (1987).

[22] C. A. Hoy, E. D. Lewis, and R. T. Schimke, *Mol. Cell. Biol.* **10,** 1584 (1990).

choline and dCDP-ethanolamine as well as DNA,[23] both of which can appear as acid precipitable radioactivity. Caution should be taken when attempting to increase the specific radioactivity of intracellular [^{3}H]TdR by reducing the dTTP pool through inhibition of dTTP synthesis, because this will increase the intracellular dUTP : dTTP ratio. When dUTP is incorporated into DNA in place of dTTP, dUMP is subsequently glycosylated and excised. This results in the appearance of short nascent DNA chains that can be mistaken for Okazaki fragments.[2,24]

Labeling Okazaki Fragments. Nascent DNA at replication forks can be distinguished from nascent DNA produced at repair sites by the immediate appearance of label in Okazaki fragments. The fraction of label found in Okazaki fragments depends on the length of time DNA is exposed to the labeled nucleotide, the temperature at which labeling is carried out, the presence or absence of cytosol, the use of nonionic detergents, the presence or absence of chloride anion, and the biological system studied. The chloride anion is a common inhibitor of enzyme activities (e.g., DNA polymerase α). Therefore, substitution of acetate or glutamate salts for chloride salts permits characterization of DNA replication at higher ionic strengths that are more characteristic of "physiological" salt levels.[25] The amount of DNA synthesis observed in nuclei prepared from cells lysed with nonionic detergents such as Nonidet P-40 (NP-40) or Triton X-100 that disrupt the outer nuclear membrane is often about sixfold less than in nuclei that have not seen detergents or in cells that have been lysed with streptolysin O,[26] although this effect does not appear to be true for plasmid DNA replication in isolated nuclei.[27] The rate of joining of Okazaki fragments to long nascent DNA chains is about 10-fold slower in isolated nuclei washed free of cytosol than in a complete cell lysate,[1,2] resulting in a greater fraction (~50%) of the pulse-label appearing in Okazaki fragments. For example, Okazaki fragments are labeled for 90 sec at 26° in mammalian nuclei isolated in the absence of nonionic detergents and free of cytoplasm, but 90 sec at 34° in cell lysates prepared with NP-40. *Xenopus* prefers to live at ~19° and *Xenopus* eggs have an S phase of ~30 min. Accordingly, Okazaki fragments are labeled in extracts of *Xenopus* eggs for 20 sec at 12°.[27a] In general, the time required to observe a particular DNA substrate replicate ranges from

[23] G. Spyrou and P. Reichard, *J. Biol. Chem.* **264,** 960 (1989).

[24] M. L. DePamphilis and P. M. Wassarman, *in* "Organization and Replication of Viral DNA" (A. S. Kaplan, ed.), pp. 37–114. CRC Press, Boca Raton, FL, 1982.

[25] Z.-S. Guo, C. Gutierrez, U. Heine, J. M. Sogo, and M. L. DePamphilis, *Mol. Cell. Biol.* **9,** 3593 (1989).

[26] P. Hozák, A. B. Hassan, D. A. Jackson, and P. R. Cook, *Cell* **73,** 361 (1993).

[27] M. L. DePamphilis and P. Berg, *J. Biol. Chem.* **250,** 4348 (1975).

[27a] D. M. Gilbert, H. Miyazawa, and M. L. DePamphlis, *Mol. Cell. Biol.* **15,** 2942 (1995).

minutes to days, depending on the efficiency of the initiation process (e.g., SV40 > polyomavirus > papillomavirus > cellular chromosomes) and the ability of the host cell to replicate DNA (e.g., amphibian or fly eggs > synchronized somatic cells in S phase > exponentially proliferating somatic cells and transformed cells such as human HeLa and 293 cells > well-behaved primary and secondary cultured cell lines).

Labeling DNA Primers. DNA primers can be observed under the following conditions: (1) Nascent DNA is radiolabeled in a subcellular system with [α-^{32}P]rNTPs for 30 sec at 30° to label their 5′-RNA moiety.[28,29] Transcription by RNA polymerases II and III is suppressed by addition of 200 μg/ml of α amanitin; DNA primase is insensitive to this inhibitor. (2) Cellular DNA polymerases α, δ, and ε are specifically inhibited by addition of 10 μg/ml of aphidicolin to a subcellular DNA replication system.[28,29] (3) A subcellular DNA replication system is depleted of ATP by preincubating the complete reaction mixture with 100 μg of hexokinase and 1 m*M* glucose for 90 sec at 25°.[6] (4) Antibodies that neutralize proliferating cell nuclear antigen are added to a subcellular DNA replication system.[30]

Labeling Nascent DNA in Small Circular Genomes Replicating in Mammalian Nuclei

DNA Replication in Cultured Cells. To label Okazaki fragments during their synthesis in cultured mammalian cells, advantage is taken of the fact that DNA synthesis is strongly temperature-dependent relative to equilibration of cellular nucleotide pools with nucleosides present in the culture medium.[31] The culture medium is aspirated from plastic 100-mm-diameter tissue culture dishes of mammalian cells or mammalian cells infected with a virus or transfected with a plasmid, and the dishes are then floated for about 10 sec on ice water to arrest DNA synthesis.[32] Floating plastic dishes containing monolayers of cells on a water bath allow the temperature of the cells to be changed "instantaneously," thus ensuring that DNA synthesis is arrested as rapidly as possible. Each dish then receives 100 μCi of [^{3}H]Thd in 0.5 ml of ice-cold 20 m*M* Tris–HCl, pH 7.4, 1 m*M* Na_2HPO_4, 5 m*M* KCl, 137 m*M* NaCl, and the incubation continued at 0° for 10 min to allow the cellular deoxythymidine pools to equilibrate with the added radiolabeled DNA precursor in the absence of DNA synthesis. DNA synthesis is resumed by floating the dishes on a 20° water bath for 30 sec. The specific

[28] T. Nethanel, S. Reisfeld, G. Dinter-Gottlieb, and G. Kaufmann, *J. Virol.* **62,** 2867 (1988).
[29] T. Nethanel and G. Kaufmann, *J. Virol.* **64,** 5912 (1990).
[30] P. A. Bullock, Y. S. Seo, and J. Hurwitz, *Mol. Cell Biol.* **11,** 2350 (1991).
[31] D. Perlman and J. A. Huberman, *Cell* **12,** 1029 (1977).
[32] T. M. Herman, M. L. DePamphilis, and P. M. Wassarman, *Biochemistry* **18,** 4563 (1979).

radioactivity of replicating SV40 DNA labeled in this way is two- to threefold greater than that obtained when cells are not preequilibrated with [^{3}H]Thd.

To separate plasmid DNA from cellular DNA, mammalian cells are frequently lysed according to the procedure of Hirt.[33] Dishes of cells are washed free of culture medium and decanted. To each dish of cells is added 2 ml of 0.6% sodium dodecylsulfate (SDS), 20 m*M* EDTA, and 10 m*M* Tris–HCl, pH 7.6. Lysate is collected in 10-ml polypropylene centrifuge tubes, brought to a concentration of 1.5 *M* NaCl, and allowed to form a precipitate overnight at 4°. If the DNA is to be run directly in a CsCl density equilibrium gradient (Method A, below), then CsCl can be substituted for NaCl to allow the density to be adjusted later using a refractometer. Low-molecular-weight DNA in the supernatant is separated from cellular genomic DNA in the precipitate by centrifugation at 12,000*g* for 1 hr at 4°. The supernatant is poured into 15-ml polypropylene tubes and digested with proteinase K (100 μg/ml) at 37° for 6 to 12 hr. DNA is then extracted vigorously with an equal volume of phenol for 1 min before adding an equal volume of chloroform : isoamyl alcohol (24 : 1), extracting and then separating phases by centrifugation. This procedure is repeated followed by a final extraction with chloroform : isoamyl alcohol to remove residual phenol. Phenol should be preequilibrated with 10 m*M* Tris, pH 8.0, 0.1% 8-hydroxyquinoline, 0.2% 2-mercaptoethanol, and 5% m-cresol to saturate it with water and to eliminate any free radicals or oxidants.

A useful improvement of the Hirt method allows DNA synthesis to be stopped abruptly by adding to each dish a lysis buffer containing 0.7 ml of 20 m*M* Tris–HCl, pH 7.6, 1.2% lithium dodecyl sulfate, and 40 m*M* EDTA[1] and then returning the dishes to 0°. Lithium dodecyl sulfate remains soluble at 0°. When cell lysis is complete, 0.7 ml of 2.5 *M* NaCl is added to each dish and the lysates are gently scraped into centrifuge tubes, stored at 4° for 12 hr, and then sedimented at 23,000*g* for 30 min at 2° to remove cellular DNA.

If chromatin is to be analyzed instead of DNA, then strong detergents cannot be used and DNA synthesis is arrested simply by returning the dishes to ice–water and adding 10 ml of ice-cold 15 m*M* Tris–HCl, pH 7.6, 2 m*M* $MgCl_2$, and 25 m*M* NaCl.[32]

DNA Replication in Cell Extracts. Circular DNA molecules that depend on a viral origin of replication (e.g., SV40) and viral origin recognition proteins [e.g., SV40 large tumor antigen (T-Ag)] and that have initiated replication in whole cells will continue replication in isolated nuclei supplemented with a cytoplasmic fraction. SV40 DNA replication can continue

[33] B. Hirt, *J. Mol. Biol.* **26,** 365 (1967).

in nuclei isolated from infected cells when supplemented with cytosol,[27,34] and replicating SV40 chromosomes extracted from these nuclei will also continue replication when supplemented with cytosol,[35] but initiation of new rounds of replication is not detected under these conditions.[36,37] These systems faithfully reproduce *in vitro* the process of SV40 DNA replication observed *in vivo*. SV40 Okazaki fragments in cell lysates are rapidly labeled by incubating for 90 sec at 30° (25 to 35% of the label appears in Okazaki fragments). Addition of a 100-fold excess of the unlabeled dNTP at this time allows one to follow the fate of Okazaki fragments, as further incorporation of radiolabel is limited to <1% of the total.

SV40 *ori*-dependent DNA replication can be observed when the low salt cytosol fraction from virus-infected cells used in these systems is combined with a high salt nuclear extract from virus-infected cells and supplemented with polyethylene glycol to concentrate the reaction components that include endogenous T-Ag.[18,25,36,37] If the nuclear extract is prepared from uninfected cells, it must be supplemented with SV40 T-Ag. PEG is not necessary when T-Ag is added to extracts, or when extracts from SV40 cs1085-infected CV-1 cells are used, because they contain 5- to 20-fold higher levels of endogenous T-Ag than extracts from SV40 wt-infected cells. This system initiates replication in bare DNA or in SV40 chromosomes, replicates bidirectionally from the SV40 origin, begins DNA synthesis on the template encoding early mRNA, produces normal replicating intermediates (rolling circles were observed only with extracts from uninfected cells), reinitiates replication within the same DNA molecule, assembles nucleosomes (20 to 60% of maximum number allowed depending on the type of extract used), and is α amanitin resistant. Most importantly, this system allows transcription factor DNA binding sites that function as *ori*-auxiliary sequences to stimulate replication *in vitro* to the same extent they do *in vivo*.[25,38]

The ratio of replication rates for the minimal origin (*ori*-core) relative to the complete origin *(ori)* varies from 10- to 30-fold, depending on the conditions used. The maximum ratio is observed with saturating levels of T-Ag, the optimum ratio of DNA substrate to cell extract, and with extract prepared from either SV40-infected CV-1 cells or uninfected HeLa cells. Replication rates vary from 10 to 200 pmol dNMP incorporated into full-length DNA product in 1.5 hr, depending on the T-Ag concentration and

[34] S. Anderson, G. Kaufmann, and M. L. DePamphilis, *Biochemistry* **16,** 4990 (1977).

[35] R. T. Su and M. L. DePamphilis, *J. Virol.* **28,** 53 (1978).

[36] R. S. Decker, M. Yamaguchi, R. Possenti, M. Bradley, and M. L. DePamphilis, *J. Biol. Chem.* **262,** 10,863 (1987).

[37] M. Yamaguchi and M. L. DePamphilis, *Proc. Natl. Acad. Sci. U.S.A.* **83,** 1646 (1986).

[38] C. Gutierrez, Z.-S. Guo, J. Roberts, and M. L. DePamphilis, *Mol. Cell. Biol.* **10,** 1719 (1990).

the cell type used. Using the endogenous T-Ag concentration in extracts from SV40-infected CV-1 cells, the rate of *ori*-core replication is stimulated from 10- to 15-fold by the presence of *ori*-auxiliary sequences. Supplementing endogenous T-Ag with enough purified T-Ag to give the maximum rate of DNA replication increases the rate of plasmid replication three- to five-fold, but the ratio of *ori* to *ori*-core replication remains unchanged.

Preparation of Cell Extracts. Efficient virus infection yields the most effective cytosol component.[18,36] CV-1H African Green monkey kidney cells are propagated in Dulbecco's modified Eagle's medium (DMEM) plus 5% fetal calf serum in 5% CO_2 at 37° and infected at 80% confluency ($\sim 1 \times 10^7$ cells/15-cm dish) with SV40 wt800 (wt, wild-type) or SV40 cs1085 (a cold-sensitive mutant that overproduces T-Ag). Extracts from CV-1H and CV-1P cell lines gave excellent results, whereas extracts from CV-1 cells obtained from the American Type Culture Collection gave poor results. This presumably reflects the "transformed state" of various cell lines. The amount of virus used to infect a dish of cells is predetermined as the amount of virus required to produce the maximum rate of viral DNA synthesis at 36 to 38 hr postinfection as measured by [^{3}H]Thd incorporation in a 30-min pulse. Swirl the virus innoculum (~0.5 ml) over the cells every 15 min while incubating dishes at 37° in 5% CO_2 for 1 hr. Then add 20 ml/dish of DMEM with 5% FCS and penicillin/streptomycin and continue incubation. Significant changes in cell morphology should be evident by ~30 hr postinfection if the infection is efficient.

Ten 15-cm dishes of CV-1H cells are harvested 36 to 38 hr after infection with SV40 by transferring them to a cold room (4°), washing them twice with 3 ml of cold TS Buffer (20 m*M* Tris–HCl, pH 7.4, 137 m*M* NaCl, 5 m*M* KCl, 1 m*M* $CaCl_2$), and then incubating them for 5 min in 2 ml of cold hypotonic buffer [20 m*M* HEPES, pH 7.8, 5 m*M* potassium acetate, 0.5 m*M* $MgCl_2$, 0.5 m*M* dithiothreitol (DTT)]. Excess buffer is removed by stringent aspiration to reduce lysate volume to a minimum. Set dishes at an angle to drain excess liquid to one edge. Excess buffer must be removed several times so that the cytosol volume will be no more than 2.5 ml. It is important to keep cells cold and work fast (~20 min).

Cells are scraped free with a rubber policeman, collected in a 7-ml Dounce homogenizer resting in ice, and lysis completed with 10 strokes of pestle B. The lysate is incubated on ice for 1 hr with gentle agitation every 15 min before centrifuging at 10,000*g* (Dupont-Sorval SM24 rotor, 9000 rpm) for 10 min in a polypropylene tube. The supernatant ("cytosol," ~2 ml, 4 to 7 mg protein/ml) is immediately stored in convenient aliquots at −80°. The pellet is resuspended in 0.5 ml of high salt buffer (20 m*M* HEPES, pH 7.8, 500 m*M* potassium acetate, 0.5 m*M* $MgCl_2$, 0.5 m*M* DTT), incubated on ice for 1 hr, and then centrifuged at 10,000*g* for 10 min. The supernatant ("high salt nuclear extract," ~1 ml, 3 to 6 mg protein/ml) is aliquoted and stored at −80°.

DNA Replication Conditions. Nucleotide stocks (20 mM) are prepared in 100 mM HEPES, pH 7.8, and 0.1 mM EDTA. ATP stock (500 mM) is adjusted to pH 7 to 8 with NaOH. Store aliquots at −20° to prevent slow hydrolysis. Buffers are stored at −20° to prevent growth of microorganisms. Preequilibrate water bath at 37°, and dry down [α-^{32}P]dCTP and [α-^{32}P]dTTP (3000 Ci/mmol) in a Savant Speed-Vac. Radiolabeled nucleotides are redissolved in 5× (30 mM HEPES, pH 7.8, 0.5 mM DTT, 7 mM magnesium acetate, 1 mM EGTA, 10 mM phosphoenolpyruvate, 3 μg pyruvate kinase, 4 mM ATP, 200 μM each of CTP, GTP, and UTP, 100 mM each of dATP and dGTP, 40 μM each of dCTP and dTTP). This 5× reaction buffer should contain 5 μCi of each radiolabeled nucleotide per 10-μl aliquot. Transfer aliquots of cytosol and nuclear extract from −70° to ice and allow them to thaw. Mix assay components quickly in the following order:

1. When cytosol and nuclear extract aliquots are almost completely thawed in ice, combine them in the ratio 20 μl cytosol to 10 μl HS nuclear extract for each 50 μl assay by transferring them to 0.5-ml siliconized Eppendorf tubes that were prechilled on ice.
2. Add 10 μl of 5× reaction buffer containing radiolabeled nucleotides and mix contents by flicking the tube with your fingers; do not vortex. Cell extract contributes an additional 12 mM HEPES, pH 7.8, 102 mM potassium acetate, 0.3 mM $MgCl_2$, and 0.3 mM DTT to the final reaction.
3. Add 2 μl containing 100 ng DNA substrate in 10 mM Tris–HCl, pH 7.6, 1 mM EDTA.
4. Add 3 μl double-distilled, deionized water.
5. Add 5 μl of 30% polyethylene glycol [PEG 12,000 (Fluka) in water] to give a 50 μl reaction volume.
6. Mix well, preincubate on ice for 15 min to allow assembly of replication complexes, then incubate at 37° for up to 1.5 hr.
7. Reactions are terminated by placing tubes on ice and adding one volume of 1% SDS and 30 mM EDTA. Yeast tRNA (10 μg) is added as a carrier, and the mixture digested with 200 U of proteinase K at 37° for 1 hr. Samples are extracted once with an equal volume of phenol–chloroform–isoamyl alcohol (25:24:1), once with chloroform–isoamyl alcohol (24:1), and twice with ether. Nucleic acids are then precipitated at −70° in 70% ethanol once in the presence of 300 mM sodium acetate and twice in the presence of 2 M ammonium

acetate to concentrate the DNA and remove free ^{32}P-labeled nucleotides before subjecting the DNA to further analysis.

Purification of Nascent DNA from Circular Molecules

Purification of Replicating Viral or Plasmid DNA. Two methods are commonly used to isolate circular DNA replicating intermediates (RI).[39] Method A first separates SV40(I) SV40(RI) DNA in a CsCl/ethidium bromide density gradient at equilibrium, and then separates SV40(II) and SV40(III) DNA from SV40(RI) DNA by sedimentation in a neutral sucrose gradient. Method B isolates SV40(RI) DNA by chromatography on benzoylated naphthylated DEAE-cellulose (BND-cellulose), a resin whose preferential affinity for single-stranded DNA and RNA selectively binds DNA containing replication forks. While gel electrophoresis provides greater resolving power than sucrose gradients, sucrose gradients provide greater capacity and ease of recovery when purifying large amounts of DNA.

METHOD A. Viral DNA extracted from eight dishes (10 cm) of infected cells is adjusted to a final concentration of ethidium bromide of 600 μg/ml, and a final density of 1.564 g/cm^3 with CsCl. Density gradients are formed in a Beckman 50Ti rotor at 38,000 rpm for 48 hr at 18°. Ethidium bromide is removed from the [^{3}H]DNA by passing the sample through Bio-Rad AG50 resin in the Na^+ form. CsCl is removed by dialysis against 10 m*M* Tris–HCl, 1 m*M* EDTA, and 0.1 *M* NaCl or by extraction with 1-butanol equilibrated against 10 m*M* Tris–HCl, 1 m*M* EDTA, and 0.1 *M* NaCl in order to saturate it with water. DNA is precipitated at −35° in 75% ethanol and then resuspended in 0.5 ml of the dialysis buffer and treated with 40 μg of RNase per milliliter for 2 hr at 35°. The sample is layered on a 5 to 20% linear sucrose gradient containing 10 m*M* Tris–HCl, pH 7.8, 1 *M* NaCl, and 2 m*M* EDTA, and centrifuged at 35,000 rpm for 12 hr at 4° in a Beckman SW41 rotor. SV40(RI) [^{3}H]DNA is precipitated in 67% ethanol in the presence of 0.1 *M* sodium acetate (acetate is more soluble in ethanol than chloride, and the pellet is washed with 70% ethanol, resuspended in 0.05 ml of 0.1× electrophoresis buffer, 5% glycerol, and 0.025% bromphenol blue in preparation for electrophoresis in agarose gels.

METHOD B. Viral DNA extracted from eight dishes of cells was adjusted with solid CsCl to a final density of 1.700 g/cm^3. A density gradient was

[39] D. P. Tapper and M. L. DePamphilis, *J. Mol. Biol.* **120,** 401 (1978).

formed in a Beckman 50Ti rotor at 38,000 rpm for 48 hr at 18°. [^{3}H]DNA is pooled, dialyzed against 10 m*M* Tris–HCl, pH 7.6, 1 m*M* EDTA, and 0.3 *M* NaCl for 24 hr at 4°, and then adsorbed to a 1-ml BND-cellulose column previously washed with the same buffer. The column is first eluted with 20 ml of 10 m*M* Tris–HCl, pH 7.6, 1 m*M* EDTA, 0.65 *M* NaCl, and then with 20 ml of 10 m*M* Tris, pH 7.6, 1 m*M* EDTA, 1 *M* NaCl, and 2% caffeine. SV40(I) [^{3}H]DNA and SV40(II) [^{3}H]DNA are recovered in the initial eluent while 90 to 99% of the [^{3}H]DNA in the 2% caffeine eluent is SV40(RI) DNA. SV40(RI) [^{3}H]DNA is dialyzed against 1 ml Tris–HCl, pH 7.6, 0.25 m*M* EDTA, lyophilized, resolubilized in 10 m*M* Tris–HCl, pH 7.6, 1 m*M* EDTA, and 0.1 *M* NaCl, incubated for 2 hr at 35° with 40 μg of RNase A/ml and finally precipitated in 75% ethanol. The pellet is treated as in Method A. Note that plasmid DNA can be desalted whenever necessary simply by exclusion chromatography on Sepharose CL-4B (29 × 3.2 cm) in 10 m*M* Tris–HCl, pH 7.6, and 0.1 m*M* EDTA.

Labeling Nascent DNA in Mammalian Chromosomes

Exponentially proliferating cells can be used to label nascent DNA chains, but only 50% or less will be in S phase. Cells synchronized at their G_1/S boundary not only allow more efficient labeling of newly replicated DNA, but also allow mapping of chromosomal origins of replication by virtue of the asymmetric distribution of Okazaki fragments between the two arms of a replication fork. The mapping of origins by this method requires well-synchronized cells if Okazaki fragment asymmetry is to be observed in single-copy sequences. The following methods are modified from Burhans *et al.*[5]

Cell Synchronization. Monolayers of Chinese hamster ovary (CHO) cells are grown in Dulbecco's modified Eagle medium supplemented with 5% fetal bovine serum and nonessential amino acids using 150-mm tissue culture dishes. When cell monolayers reach 80 to 90% confluency, approximately 90 to 95% of the cells can be synchronized at the G_1/S phase boundary by first collecting them in their G_0 phase by isoleucine deprivation as described by Heintz and Hamlin,[40] followed by release into complete medium containing 10 μg/ml aphidicolin for 12 hr. The optimum time allowed for isoleucine deprivation has been found to vary between 30 and 45 hr in different laboratories. Therefore, the effectiveness of this or any other synchronization procedure must be checked by fluorescent-activated

[40] N. H. Heintz and J. L. Hamlin, *Proc. Natl. Acad. Sci. U.S.A.* **79,** 4083 (1982).

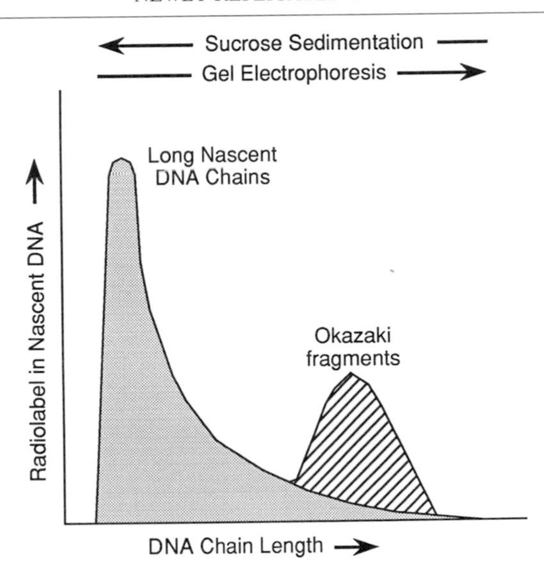

FIG. 3. Isolation of Okazaki fragments.

cell sorting.[41,42] Cells collected at the G_1/S boundary are washed three times with 20 ml/dish of ice-cold medium without serum to remove the aphidicolin. Cells are then allowed to transit into S phase by adding 40 ml/dish of prewarmed complete medium and reincubating under CO_2 at 37° for a brief period of time. The amount of time is determined by how long it takes to observe a typical distribution of nascent DNA between long DNA strands and Okazaki fragments (Fig. 3). Usually it is 3 to 5 min, but longer times may be required depending on the cell line and effectiveness of the temperature shifts. The need for this procedure may reflect the time required to reverse inhibition by aphidicolin as well as to avoid aberrant synthesis of short fragments of labeled DNA that result from DNA damage due to prolonged arrest of DNA synthesis.[43–45] Cells are washed once again with ice-cold medium without serum. Excess medium is aspirated and the cells are scraped into the residual medium (~1 ml/150-mm dish/~2×10^7 cells). The cell suspension from each dish (maintained at

[41] D. M. Gilbert and S. N. Cohen, *Cell* **50,** 59 (1987).

[42] S. M. Carroll, M. L. DeRose, J. L. Kolman, G. H. Nonet, R. E. Kelly, and G. M. Wahl, *Mol. Cell. Biol.* **13,** 2971 (1993).

[43] G. Dinter-Gottlieb and G. Kauffman, *J. Biol. Chem.* **258,** 3809 (1983).

[44] R. M. Snapka, C. G. Shin, P. A. Permana, and J. Strayer, *Nucl. Acids Res.* **19,** 5065 (1991).

[45] D. M. Gilbert, A. Neilson, H. Miyazawa, M. L. DePamphilis, and W. C. Burhans, *J. Biol. Chem.* **270,** 9597 (1995).

0 to 4° until the labeling period begins) is transferred to a 1.5-ml microcentrifuge tube, and the cells pelleted by centrifugation for 3 min in a Beckman microfuge (setting 3). Supernatants are removed by aspiration, leaving 50 to 100 μl of packed cells per tube.

DNA Replication Conditions. Each 100 μl of packed cells is resuspended in 120 μl of ice-cold 2× replication cocktail [60 m*M* potassium HEPES, pH 7.8, 0.2 m*M* each dGTP, dCTP, and BrdUTP, 0.4 m*M* each GTP, CTP and UTP, 8 m*M* ATP, 20 m*M* $MgCl_2$, 0.2 mg/ml nuclease free bovine serum albumin (BSA), 2 m*M* DTT, and 30% (v/v) glycerol]. Nucleotide stocks are as described above for plasmid replication. The volume of each suspension is adjusted to 200 μl/tube in order for the temperature of each reaction to be equilibrated rapidly at the same rate, and 20 μl [α-^{32}P]dATP (10 mCi/ml, 3000 Ci/mmol, Amersham) is added to each tube. The nonionic detergent NP-40 is then added to a final concentration of 0.4%. Labeling reactions are started by transferring the tubes to a 34° water bath. Okazaki fragment synthesis is too rapid at higher temperatures to allow sufficient labeling. Reactions are stopped after precisely 1.5 min of incubation by adding 800 μl of 50 m*M* Tris–HCl, pH 7.8, 10 m*M* EDTA, 0.4 *M* NaCl, 0.6% SDS, 0.2 mg/ml proteinase K (Boehringer Mannheim). One dish of cells (~1 to 2 × 10^7 cells) generally incorporates from 1 to 10 × 10^5 Cherenkov cpm. About 40% of the label appears in Okazaki fragments. To demonstrate that labeled Okazaki fragments join to long nascent DNA chains (pulse–chase experiment), 0.1 m*M* dATP can be added at the end of the labeling and the incubation continued for 10 min, at which point reactions are stopped.

Purification of Long Nascent DNA Strands from Mammalian Cells. DNA is purified by treating terminated labeling reactions with 50 μg/ml RNAse A for 3 hr at 37°. Proteins are then digested by adding 0.2 mg/ml proteinase K and incubating at 37° overnight. DNA is extracted with phenol and chloroform–isoamyl alcohol (24:1 v/v), and then precipitated with ethanol. DNA precipitates are collected by centrifuging for 30 min at 10,000 rpm in a Sorvall HB4 rotor at 4°, and redissolved in 3 ml of 10 m*M* Tris–HCl, pH 7.5, and 1 m*M* EDTA per 10^8 cells. To ensure that DNA preparations are free of residual RNA that may interfere with hybridization, DNA is treated with 0.3 *N* NaOH for 24 hr at 37°, neutralized and then precipitated with ethanol.

Purification of Okazaki Fragments from Mammalian Cells. Branch migration and DNA extrusion can occur during the process of DNA purification. This results in dissociation of Okazaki fragments from their templates and failure to recover these short pieces of single-stranded DNA during subsequent ethanol precipitation steps. To minimize this problem, the salt concentration is increased and exposure to elevated temperatures during enzyme digestion is minimized. A second problem is contamination of the Okazaki fragment population by small pieces of unreplicated DNA that result from fragmentation of high-molecular-weight DNA during purification, some of which will contain sequences complementary to those of the labeled Okazaki fragments. This will reduce the bias Okazaki fragments exhibit toward the retrograde arm template in subsequent hybridization studies. Therefore, minimize shearing forces on the DNA during purification.

Terminated DNA labeling reactions are incubated for 1 to 2 hr at 37° to allow protein degradation. Residual peptides are precipitated in a polypropylene microfuge tube by adding 330 μl of 6 *M* NaCl (saturated NaCl solution) to each 1 ml of lysate to give a final salt concentration of 1.7 *M* Na^+. Closed tubes are mixed by gentle shaking until a uniform precipitate is formed. Avoid shaking the DNA vigorously, since this will cause shearing of high-molecular-weight DNA. Precipitates are removed by centrifugation at 2000*g* for 15 min at room temperature, and supernatants are combined. Precipitate the DNA from the supernatant with 2 volumes of absolute ethanol at room temperature. If samples are too viscous to allow soluble DNA to be separated from insoluble cell debris, then treat the entire sample with ethanol. The visible precipitate is collected by centrifuging for 30 min at 10,000 rpm in a Sorvall HB4 rotor at 4°. The DNA pellet is rinsed twice with 5 to 10 ml of 70% ethanol, redissolved in 10 m*M* Tris–HCl, pH 7.8, and 1 m*M* EDTA using 50 μl/150-mm dish of cells, and then denatured in boiling water for 3 min to release nascent DNA chains from their templates.

Isolation of Okazaki Fragments and DNA Primers

Okazaki fragments that have been released from their templates can be separated from long nascent DNA chains either by sedimentation in sucrose gradients or by gel electrophoresis (Fig. 3). Sucrose gradients provide a greater capacity, broader size range distribution, and easier recovery of the separated DNA products than gel electrophoresis, but gel electrophoresis offers a simpler procedure with higher resolution. DNA primers are isolated by gel electrophoresis.

Alkaline Sucrose Gradients.[1,46] DNA was first dialyzed against 10 m*M* Tris–HCl, pH 7.8, 100 m*M* NaCl, and 1 m*M* EDTA before layering 100-μl samples over 5 to 20% alkaline sucrose gradients. Sucrose gradients were formed in Beckman SW60 tubes from 5% sucrose containing 0.2 *M* NaOH and 0.8 *M* NaCl to 20% sucrose with 0.8 *M* NaOH and 0.2 *M* NaCl. In

[46] M. L. DePamphilis, P. Beard, and P. Berg, *J. Biol. Chem.* **250,** 4340 (1975).

addition, the gradients contained 2 m*M* EDTA and 0.015% sodium dodecyl sarcosinate. Gradients are centrifuged for 6.5 hr at 55,000 rpm (4°) and then collected dropwise from the bottoms of the tubes onto Whatman 3 MM paper disks. The disks are dried and radioactivity measured in a toluene-based scintillation fluid.

Alkaline Gel Electrophoresis.[5,34] To separate Okazaki fragments from long nascent DNA strands while displaying simultaneously the size range and relative amounts of both replication intermediates, purified DNA samples are fractionated by electrophoresis through 1.8 to 2% agarose alkaline gels[19] at 1 V/cm for 16 hr at room temperature. Okazaki fragments comigrate with bromphenol blue dye; unincorporated labeled nucleotides migrate faster than this dye marker. To confirm the size of Okazaki fragments, DNA size standards should be run in a parallel lane. A useful 123 bp ladder of DNA size standards can be purchased from Gibco Life Sciences. These DNA standards are visualized by excising the lane and staining it for 30 min in 300 ml 3 μg/ml ethidium bromide in 0.5 *M* Tris–HCl, pH 7.8. [^{32}P]DNA is localized by covering the wet gels with plastic wrap and exposing them to individually wrapped Kodak X-Omat AR films for 1 hr through the film wrapping. Appropriate regions of the gel can be excised and the labeled DNA electroeluted into 0.5 ml 0.5× TBE buffer[19] using an electroelution trap (Schleicher and Schuell) according to the manufacturer's instructions. Greater than 99% of the DNA is recovered at this step. Contaminating RNA is then hydrolyzed by adjusting the solutions to 0.2 *N* NaOH and incubating at 37° for 24 hr. Solutions are neutralized by adding Tris–HCl, pH 7.8, to a final concentration of 100 m*M* and adjusting the pH to 7.2 with dilute HCl.

Glyoxal Gel Electrophoresis.[47] Alkali degrades RNA and therefore its use is not appropriate if the object is to analyze RNA-primed Okazaki fragments. One solution is to treat DNA with glyoxal. Glyoxal forms a stable adduct with guanosine at or below pH 7 and prevents renaturation of RNA or DNA by hindering formation of G : C base pairs. This eliminates formation of secondary structure at neutral pH that can dramatically affect the rate of migration of a polynucleotide chain during gel electrophoresis. DNA samples are denatured in a final volume of 5 μl of 10 m*M* $NaPO_4$, pH 7.0, 0.1 m*M* EDTA, 1.0 *M* freshly deionized glyoxal (to eliminate free radicals that induce breaks in polynucleotides), and 50% methyl sulfoxide for 1 hr at 50°, and then fractionate the DNA in agarose gels made up in 10 m*M* $NaPO_4$, pH 7.0, with 0.1 m*M* EDTA. Each cooled sample receives 2.5 μl of 40% sucrose containing bromphenol blue as a tracking dye before it is loaded onto the gel. Neutral gels are run at 3 V/cm at room temperature

[47] G. G. Carmichael and G. K. McMaster, *Methods Enzymol.* **65,** 380 (1980).

until the bromphenol blue dye that is used to monitor gel electrophoresis migrates approximately 8 cm. Both buffer reservoirs contained 10 m*M* $NaPO_4$, pH 7.0, and 0.1 m*M* EDTA, which was continuously recirculated during electrophoresis. Denaturation of DNA and RNA in glyoxal is a hazardous protocol, and must be carried out under protective conditions (see Merck Index).

To determine accurately the size range of glyoxal-treated Okazaki fragments, gels of 5% polyacrylamide were made up in 10 m*M* $NaPO_4$, pH 7.0, containing 0.1 m*M* EDTA and allowed to harden for 12 to 16 hr. Preelectrophoresis was carried out for 1 hr at 10 V/cm before loading the samples. Electrophoresis is carried out at room temperature at 10 V/cm until the bromphenol blue migrates ~13 cm.

Urea Gel Electrophoresis.[28,30] An alternative to glyoxal treatment is to denature purified DNA in formamide, and then fractionate it by electrophoresis in 12% polyacrylamide gels containing 7 *M* urea in 25 m*M* Tris–borate buffer, pH 8.3, 0.6 m*M* EDTA, and 7 *M* urea. Size analysis of DNA primers is also achieved by electrophoresis in 10% polyacrylamide containing 8 *M* urea in a Tris–borate–EDTA buffer.[19]

Incorporation of Density-Labeled or Affinity-Labeled Deoxyribonucleotides into Nascent DNA

5-Bromo-2'-deoxyribouridine (BUdR)

Nascent DNA can be labeled with BUdR in whole cells or with BrdUTP in a subcellular system. BrdUTP is incorporated into DNA in place of dTTP. This allows newly synthesized DNA to be separated from unreplicated DNA either by the difference in their densities, or by the affinity of BrdU-DNA for antibodies directed against BUdR. The same antibodies can be used to detect BrdU-DNA *in situ*.[48,49]

The density (ρ) of DNA in CsCl is a function of its GC content and is described by the equation $\rho = (0.098)(GC) + 1.660$ g/ml, in which (GC) is the mole fraction of G + C in native DNA.[50] Thus, DNA that is 50% GC has a density in CsCl of 1.71 g/cm^3. In CsCl solutions, single-stranded DNA has a density that is from 0.015 to 0.020 g/cm^3 greater than that of double-stranded DNA. In Cs_2SO_4 solution, the density of DNA is insensitive to base composition, and therefore only isotope differences are detected. Another advantage of Cs_2SO_4 solution is that RNA has a much

[48] J. Ellwart and P. Dormer, *Cytometry* **6,** 513 (1985).

[49] M. Vanderlann and C. B. Thomas, *Cytometry* **6,** 501 (1985).

[50] C. L. Schildkraut, J. Marmur, and P. Doty, *J. Mol. Biol.* **4,** 430 (1962).

higher density than in CsCl and therefore is more easily separated from DNA.

Incorporation of BUdR or BrdUTP into nascent DNA increases the density of DNA as much as 0.119 g/cm^3 in a CsCl density equilibrium gradient. One round of semiconservative DNA replication in the presence of BrdU produces plasmid DNA containing one light strand with Thd (parental DNA template, Fig. 1) and one heavy strand substituted with BrdU (nascent DNA strand, Fig. 1). More than one round of replication produces DNA molecules with two heavy strands.

Labeling Nascent DNA with BUdR. BrdU substituted nascent DNA is sensitive to light-induced damaged. Therefore, if maintaining the original lengths of nascent DNA is critical, culture dishes and reaction vessels must be kept in the dark (wrapping with aluminum foil works well). When this is not possible, work under an orange safety light. Close to 100% incorporation is generally observed with 100 μM BrdU (Boehringer Mannheim) in the cell culture medium. Including 1 μM [^{3}H]dCyt is an inexpensive way to radiolabel the nascent BrdU–DNA as well. The appropriate incubation time depends on the time required for the DNA substrate to complete one round of replication.

Purification of BrdU-DNA on the Basis of Density. Purified cellular DNA from ~10^7 cells is fragmented either by sonication or by digestion with an appropriate restriction enzyme such as *Sal*I or *Eco*RV, and then a 100-μl sample is added to 4 ml Cs_2SO_4 [0.671 g/ml; refractive index 1.3713 (25°)] and centrifuged in a Beckman VTi65 rotor at 27,000 rpm for 48 to 72 hr at 25°, or in a Beckman VTi80 rotor (5.2-ml tubes) at 30,000 rpm.[41,51] The relationship between density and refractive index will vary with different grades of Cs_2SO_4, because refractive index is influenced by all the salts present, not just Cs^+. Optimum conditions are usually determined empirically: Faster speeds produce steeper gradients, which produce narrower peaks that are separated by fewer fractions. Sal1 digested [^{14}C]DNA is included to locate the position of light/light (LL) DNA. Gradients are collected from the bottom of the tube into 30 fractions. Dripping gradients into green Eppendorf microfuge tubes helps to protect DNA from UV light. Aliquots of each fraction are diluted with Aquasol Universal Liquid Scintillation Cocktail (Dupont NEF-934) and [^{14}C]DNA detected in a Beckman scintillation counter. Specific DNA sequences are detected by blotting-hybridization using vector-specific ^{32}P-labeled probes.

Purification of BrdU–DNA on Basis of Affinity for Anti-BUdR Antibodies. Nascent DNA that has been labeled with BrdU as described above

[51] F. S. Nallaseth and M. L. DePamphilis, *J. Virol.* **68,** 3051 (1994).

can be purified by immunoprecipitation with anti-BrdU antibodies[52] in order to eliminate unlabeled parental DNA strands that can interfere with subsequent experiments in which nascent DNA is to be hybridized to specific genomic sequences. High-molecular-weight nascent BrdU–DNA is sonicated to an average length of 300 bp. It can then be combined with the Okazaki fragment fraction. BrdU–DNA is heat denatured in boiling water for 5 min and then rapidly cooled to 0° in an ice bath. Solutions are adjusted to 0.5% Triton X-100, and incubated with 10 μl of mouse anti-BrdU monoclonal antibody (Becton-Dickenson, 5 μg/ml) per 150-mm dish of cells for 45 min at 4° with constant agitation. Rabbit IgG directed against mouse IgG (2 μl/dish of 2.3 mg/ml, Sigma) is then added to precipitate BrdU–DNA–Ab complexes under the same conditions. Immunoprecipitates are collected by centrifugation at 4° for 5 min at 12,000 rpm in a microcentrifuge and then resuspended in 0.2 ml 10 m*M* Tris–HCl, pH 7.8, and 1 m*M* EDTA. Generally, about 50% of the [^{32}P]DNA is precipitated, and its specific activity increased by about 100-fold. Typical preparations of labeled, nascent DNA for hybridization analysis are performed with cells from ten 150-mm dishes. Yields of labeled Okazaki fragments or high-molecular-weight DNA are at least 1×10^6 Cherenkov cpm/ten 150-mm dishes of cells or $\sim 2 \times 10^8$ cells.

Contreas *et al.*[53] find that immunoaffinity chromatography after size fractionation in neutral sucrose gradients is more efficient for recovery of BrdU–DNA. Neutral gradients were used to avoid alkaline degradation of BrdU–DNA. Goat anti-mouse IgG (Zymed Labs, San Francisco, CA) is coupled to CNBr-activated Sepharose 4B (Pharmacia) according to the manufacturer's protocol (2.5 mg protein/ml gel). The coupled Sepharose (0.5 ml) is poured into a 2-ml disposable chromatography column (Bio-Rad, Richmond, CA), washed with >30 ml of TSE (10 m*M* Tris–HCl, pH 7.5, 150 m*M* NaCl, and 0.1 m*M* EDTA), left with 0.5 ml of TSE above the surface, mixed with 3 μg/ml of mouse anti-BrdU monoclonal antibody for 2 hr at room temperature with slow agitation, and then washed with 10 ml TSE. From 0.1 to 10 μg DNA in 1.5 to 4 ml TSE is denatured in boiling water, chilled on ice, and added to the column. After 2 hr incubation at room temperature with slow agitation, unbound DNA was allowed to drain out. The Sepharose is washed with 4 ml TSE, and then bound DNA is eluted with 2 ml 150 m*M* NaCl (adjusted to pH 11.5 with NH_4OH). Greater than 90% of the [^{3}H]BrdU–DNA is recovered with virtually no contamination by a [^{14}C]DNA control. The column can be stored at 4° in the presence of 0.05% thimerosal (Sigma) and reused many times.

[52] L. Vassilev and G. Russev, *Nucl. Acids Res.* **16,** 10,397 (1988).

[53] G. Contreas, M. Giacca, and A. Falaschi, *BioTechniques* **12,** 824 (1992).

5-Mercury Deoxyribocytosine Triphosphate (Hg-dCTP)

Nascent DNA can be labeled with 0.05 m*M* Hg-dCTP in addition to [α-^{32}P]dNTPs in a subcellular system and then isolated by its affinity to the -SH groups in thiol-agarose.[54,55] Labeled DNA is first purified and then sheared by sonication to 400 to 800 bp fragments. The DNA fragments are denatured at 100° for 5 min and then chromatographed on Affi-Gel 401 (Bio-Rad). After washing the column to elute nonbound DNA, at least 85% of the bound nascent Hg-DNA is eluted with 2-mercaptoethanol. Hg-DNA binds efficiently to thio-agarose when the DNA contains at least two Hg-dCMP/400 bp.

Biotinylated-dUTP

Nascent DNA can be labeled by incorporation of biotinylated dUTP {5-[(*N*-biotinyl)-(3 to 10)-aminoallyl]-2′-deoxyuridine 5′-triphosphate}. The resulting biotinylated DNA can be bound to streptavidin-coated beads or a column matrix, but in contrast to BUdR and Hg-dCTP, it is very difficult to recover the purified DNA. Therefore, the most common application of biotinylated DNA is to label it with fluorescein conjugated streptavidin or antibiotin antibody in order to visualize the location of nascent DNA *in situ*.[26,56]

It is possible to isolate replicating SV40 DNA or chromosomes using the chemically cleavable biotinylated nucleotide analog, Bio-19-SS-dUTP.[57] This biotinylated nucleotide is first incorporated into replicating DNA during a brief pulse–label in a subcellular system. Replicating chromosomes are separated from mature chromosomes by affinity chromatography using streptavidin to cross-link the biotinylated DNA to biotin-cellulose. Replicating chromosomes are then eluted by cleaving the -S-S- linkage in Bio-19-SS-dUMP-labeled DNA with DTT.

Identification of Replication Fork Polarity

Replication forks have a definite polarity (i.e., direction) that can be determined either from hybridization of Okazaki fragments to the lagging strand DNA template (template on the retrograde arm of replication forks) or hybridization of long nascent DNA strands to the leading strand DNA template (template on the forward arm of replication forks) (Fig. 1). The

[54] G. Banfalvi, S. Bhattacharya, and N. Sarkar, *Anal. Biochem.* **146,** 64 (1985).

[55] J. Taljanidisz, J. Popowski, and N. Sarkar, *Mol. Cell. Biol.* **9,** 2881 (1989).

[56] H. Nakayasu and R. Berezney, *J. Cell. Biol.* **108,** 1 (1989).

[57] T. M. Herman, *Methods Enzymol.* **170,** 41 (1989).

results of these assays have been used to confirm that nascent DNA chains are Okazaki fragments, to identify the direction of fork movement, and to locate an origin of bidirectional replication in virus[3,4,8,58,59] and cellular[5,42,60] chromosomes.

Okazaki Fragment Distribution

Okazaki fragments can be radiolabeled and isolated from cells as described above. The templates to which labeled Okazaki fragments are annealed consist of the two separated strands of DNA restriction fragments representing unique DNA sequences. The simplest approach is to denature DNA restriction fragments in 0.2 *M* NaOH for 10 min at room temperature, and then to separate individual single-strand DNA fragments by electrophoresis in 2% agarose gels in 0.1 *M* Tris–borate, pH 8.3, and 2.5 m*M* EDTA.[61–63] The gel pattern is then transferred to a suitable membrane, hybridized with ^{32}P-labeled Okazaki fragments, and the results recorded and quantified. The two earlier studies used nitrocellulose membranes on which to blot the DNA, but part or all of the faster migrating DNA band can occasionally peel off the paper during the hybridization reaction, leading to artifactual results, a problem that was solved by chemically cross-linking the DNA to the membrane.[63] This problem can now be more easily solved using commercially available nylon transfer membranes such as Zeta-Probe (Bio-Rad) or Hybond-N+ (Amersham). DNA is transferred using a Positive Pressure Blotter (Stratagene) and cross-linked to the membrane by irradiation at 0.12 J in a UV Stratalinker Model 1800 (Stratagene). Nevertheless, two problems that remain are first that not all DNA fragments are readily separated into two strands by gel electrophoresis and, second, that each of the two separated strands remains contaminated to some degree with its complementary strand.

A strategy with greater application involves cloning each strand of a unique DNA restriction fragment into bacteriophage M13[3,4,59] blotting the M13/DNA template onto an appropriate membrane and hybridizing this DNA to radiolabeled Okazaki fragments isolated by either gel electrophoresis or sucrose sedimentation gradients. Improved methods for cloning DNA into bacteriophage M13 through the use of a plasmid intermediary

[58] E. A. Hendrickson, C. E. Fritze, W. R. Folk, and M. L. DePamphilis, *Nucl. Acids. Res.* **15,** 6369 (1987).

[59] R. T. Hay and M. L. DePamphilis, *Cell* **28,** 767 (1982).

[60] E. S. Tasheva and D. J. Roufa, *Mol. Cell. Biol.* **14,** 5628 (1994).

[61] G. Kaufmann, R. Bar-Shavit, and M. L. DePamphilis, *Nucl. Acids Res.* **5,** 2535 (1978).

[62] M. M. Seidman, A. J. Levine, and H. Weintraub, *Cell* **18,** 439 (1979).

[63] M. E. Cusick, M. L. DePamphilis, and P. M. Wassarman, *J. Mol. Biol.* **178,** 49 (1984).

are described in Sambrook *et al.*[19] Note that any degradation of the nascent DNA that occurs during radiolabeling or purification will artifactually increase the fraction of short [^{32}P]DNA chains that anneal to the forward arm template. With viruses, this problem can be avoided by first isolating Okazaki fragments from replicating chromosomes and then specifically radiolabeling their RNA-p-DNA junctions using the unmasking assay. With cellular DNA, contamination by broken DNA chains can be detected from densitometric tracings of film exposed to the preparative gel. In general, the Okazaki fragment peak should be fivefold or greater than the level of DNA between 0.25 kb and the high-molecular-weight peak.[5]

Burhans *et al.*[5] optimized the method for blotting-hybridization so that it was sensitive enough to analyze Okazaki fragments synthesized at a unique single-copy locus in mammalian cells. Blots are prepared using M13 cloned DNA dissolved in 0.4 *N* NaOH at 6 μg/ml. Samples containing no less than 3 μg of DNA are applied in duplicate to prewetted Zeta-Probe membranes (Bio-Rad) using a Schleicher and Scheull dot blot manifold. A negative control should be included consisting of prokaryotic sequences (e.g., M13 DNA) that do not hybridize to mammalian Okazaki fragments. The membranes are rinsed briefly in 2× SSC,[19] once in 100 ml of water to remove salt that interferes with subsequent hybridization. Membranes are air-dried overnight, and then baked for 0.5 hr at 80° under vacuum before using them for hybridization. Baked membranes are incubated in no more than 5 ml of prehybridization buffer.[64] The large excess of inorganic phosphate in this buffer improves hybridization efficiency.[65] Membranes must be completely neutralized to avoid high background signals during hybridization. Labeled Okazaki fragments (30 to 300 nucleotides; smaller DNA lengths will give less strand specificity) and high-molecular-weight DNA (0.2-ml samples) are denatured in boiling water for 3 min and then added directly to the prehybridization buffer. For M13 [^{32}P]DNA probes, M13 vector DNA is labeled with [α-^{32}P]dCTP by random primer extension,[66] heat denatured and then added to the prehybridization buffer (final concentration = 10^6 cpm/ml). Hybridization is carried out for 16 hr at 65° with agitation. Membranes are then washed in 2× SSC and 2% SDS once at room temperature for 5 min and once at 68° for 30 min.[19] This was followed by washing them in 0.2× SSC and 0.2% SDS once at 68° for 30 min and five times at room temperature for 1 min. Hybridization can also be done with single-stranded complementary RNA on the filter.[42,60]

[^{32}P]DNA is detected by exposing membranes to Kodak X-Omat AR

[64] G. M. Church and W. Gilbert, *Proc. Natl. Acad. Sci. U.S.A.* **81,** 1991 (1984).
[65] M. Mahmoudi and V. K. Lin, *Biotechniques* **7,** 331 (1989).
[66] A. P. Feinberg and B. Vogelstein, *Anal. Biochem.* **137,** 266 (1984).

film with a Dupont Cronex Lightning Plus intensifying screen for up to 2 weeks at −80°, and the amount of radioactivity per spot can be quantified using a scanning densitometer. Alternatively, the blot can be analyzed directly using a phosphorimager.

Long Nascent DNA Strand Distribution

Nascent DNA that originates exclusively from the forward arm of only one of the two forks in a single replication bubble is difficult to radiolabel, because radiolabeled Okazaki fragments are rapidly joined to the 5′ ends of long nascent DNA strands during the radiolabeling process. To circumvent this problem, cells are treated with emetine, a general inhibitor of protein synthesis that does not kill the cells during prolonged treatment. Inhibition of protein synthesis *in vivo* results in preferential inhibition of Okazaki fragment synthesis.[13,67] Under these conditions, synthesis of Okazaki fragments is inhibited 90% in the first 1 hr of incubation, resulting in preferential labeling of long nascent DNA strands on forward arms of replication forks (Fig. 4).

Chinese hamster ovary cells are seeded at 50% confluency ($\sim 5 \times 10^6$ cells in 150-mm tissue culture dishes) and cultured in Dulbecco's modified Eagle's medium supplemented with 5% fetal bovine serum and nonessential amino acids. Approximately 12 hr after seeding, exponentially proliferating cultures containing a total of $\sim 10^8$ cells are incubated in the same medium supplemented with freshly prepared 2 μM emetine (the purest grade is provided by Fluka) and 10 μM each of fluorodeoxyuridine (FUdR), bromodeoxyuridine (BrdU), and 2 μCi/ml [^{3}H]dC. FUdR is included in the incubation medium to repress synthesis of thymidine, in order to allow maximum substitution of nascent DNA with bromodeoxyuridine. The greatest imbalance between synthesis on forward and retrograde arms of forks occurs between 1 and 2 μM emetine. After 24 hr, monolayers are washed twice with phosphate buffered saline, and cells are lysed in 3 ml/dish of 100 mM Tris–HCl, pH 7.5, 10 mM EDTA, and 0.6% SDS. DNA is purified from this cell lysate as described above. Since total cellular DNA synthesis is inhibited 90% by 10 hr of incubation, one should be able to harvest DNA at earlier times, although this supposition has not been tested.

To separate labeled nascent DNA from its template and from unreplicated DNA, the purified DNA (treated again with RNase) is sonicated to an average size of 300 to 500 bp, adjusted to 0.1 M NaOH and 10 mM EDTA in a final volume of 4.5 ml, and then solid Cs_2SO_4 added to give a final density of 1.749 gm/ml. [^{14}C]TdR-labeled "light" DNA is used as an

[67] D. Kitsberg, S. Sellg, I. Keshet, and H. Cedar, *Nature* **366,** 588 (1993).

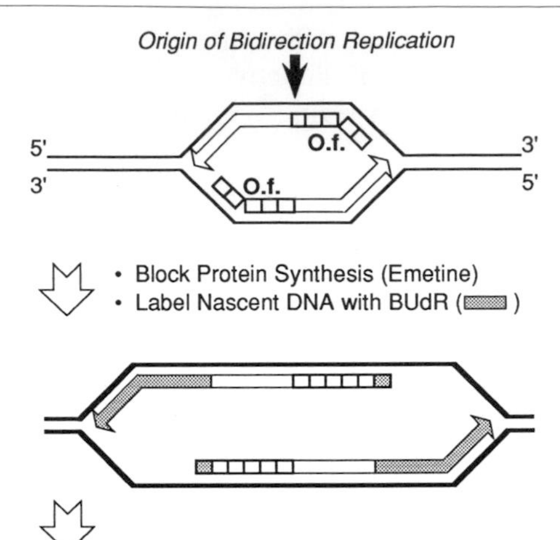

• Purify DNA and Sonicate
• Isolate Nascent DNA in Alkaline Cs_2SO_4 Gradients

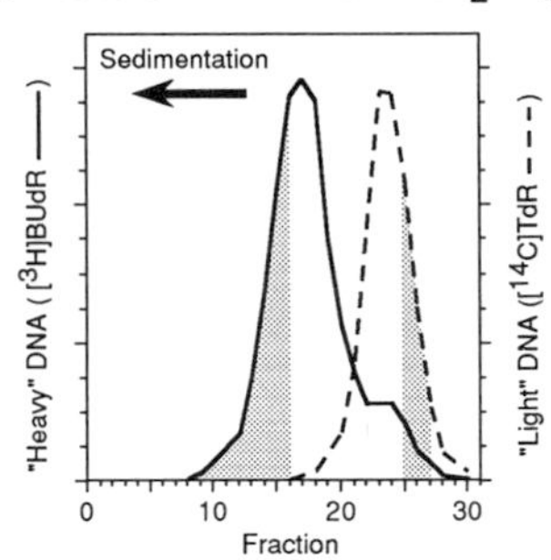

• Hybridize DNA in Shaded Areas to Strand-Specific ^{32}P-Probes

FIG. 4. Preferential labeling of nascent DNA on forward arms of replication forks.

internal standard during isopycnic gradient centrifugation. DNA is centrifuged to equilibrium at 54,000 rpm for 24 hr at 20° in a Beckman VTi 65 rotor. Fractions of 200 μl are collected from the bottom of each gradient, and radioactivity is measured in 5-μl aliquots using a liquid scintillation counter. Fractions of "heavy" and "light" DNA that are cleanly separated from one another are pooled separately (use the leading side of the heavy peak and the lagging side of the light peak). To avoid any contamination of the small amount of nascent BrdU-DNA with unlabeled DNA, repeat the isopycnic gradient step on the "heavy" DNA fraction.

These fractions can be used to determine replication fork polarity by hybridization with strand-specific probes as described above. Samples con-

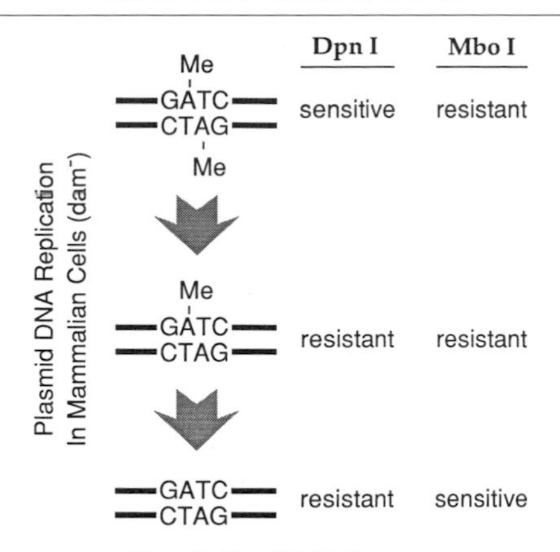

FIG. 5. *Dpn*I/*Mbo*I assay.

taining 1 to 5 μg of DNA are applied to Zeta-Probe membranes (Bio-Rad) prewetted with water using a Schleicher and Schuell slot-blot manifold as described above for Okazaki fragments. Preparation of [^{32}P]RNA probes has been described in detail.[13,67]

Detection of Changes in Methylation During DNA Replication

One of the most widely used and simplest assays for detecting and quantifying newly replicated DNA in mammalian cells or cell extracts is based on changes in the sensitivity of DNA to *Dpn*I and *Mbo*I,[68] two restriction endonucleases that recognize the same DNA sequence but cut different methylated forms of the recognition sequence (Fig. 5). Under appropriate assay conditions, *Dpn*I cleaves only fully methylated DNA, whereas *Mbo*I cleaves only unmethylated DNA, and neither enzyme cleaves hemimethylated DNA. In general, the *Dpn*I/*Mbo*I assay is limited to DNA substrates such as plasmids that can be propagated in *E. coli,* because the DNA substrate must be methylated at its *Dpn*I/*Mbo*I restriction sites by the enzyme deoxyadenosine methylase (dam). Since most metazoa lack this enzyme, these plasmids become hemimethylated if they undergo one round of DNA replication in a metazoan cell or cell extract and then unmethylated if they undergo two or more rounds of replication. Thus, the fraction of *Dpn*I-resistant DNA mesaures the amount of DNA that underwent at least one round of replication, whereas the fraction of *Mbo*I-resistant DNA measures the amount of DNA that underwent at least two rounds of replication.

[68] K. W. Peden, J. M. Pipas, S. Pearson-White, and D. Nathans, *Science* **209,** 1392 (1980).

Like the density shift assay described above, the *Dpn*I/*Mbo*I assay also can be used to distinguish between origins of replication that can initiate replication more than once during a single S phase (e.g., viral origins) and those that are restricted to one initiation event per S phase (e.g., eukaryotic cellular origins). Exponentially proliferating cells are cultured in the presence of nocodazole in order to restrict them to a single S phase. Nocodazole blocks microtubule assembly and thereby prevents cells from progressing through mitosis.[69] The optimum concentration of nocodazole for arresting mouse fibroblasts is ~0.1 μg/ml.[51] This concentration produced ~15% mitotic cells after 3.5 hr of incubation, 50% after 13.5 hr, and >90% after 18 hr, at which time most of the cells were in suspension. Higher concentrations of nocodazole (1 mg/ml) appeared toxic as judged by cell morphology. In the absence of nocodazole, ~5% of C127 cells were in mitosis.

Assay Conditions

Three features of the plasmid substrate are critical: (1) Plasmid substrates should lack "poison sequences" that may inhibit their replication in mammalian cells.[51] (2) Plasmid substrates should contain multiple (≥10) *Dpn*I/*Mbo*I restriction sites in order that DNA damage and repair at one or two sites does not produce *Dpn*I-resistant plasmids. (3) Plasmid substrates should be accompanied by another plasmid that is resistant to the indicated restriction enzyme in order to provide an internal standard for comparison. A plasmid that does not replicate in the eukaryotic host because it lacks a functional eukaryotic origin of replication will remain *Dpn*I sensitive and *Mbo*I resistant. If the same plasmid is propagated in a *dam*$^-$ strain of *E. coli,* it will remain resistant to *Dpn*I and sensitive to *Mbo*I. This is particularly useful in systems such as *Xenopus* eggs where virtually any circular or linear plasmid replicates. These standards provide a way to correct the final data for variations in plasmid substrate that do not result from DNA replication.

To minimize rearrangements during cloning, plasmid substrates are propagated in *E. coli* SURE (Stratagene), a *dam*$^+$ bacterial host that has been inactivated in three different recombination/repair pathways. A *Dpn*I-resistant plasmid is produced in dam$^-$ *E. coli* (SCS110, Stratagene). At various times posttransfection, plasmid DNA is recovered, purified, and then cut at a single restriction site in order to convert all replication products into linear DNA molecules of unit length. This serves to collect all *Dpn*I-resistant plasmid molecules into a single DNA band that is more easily detected following gel electrophoresis. One aliquot of DNA is then digested

[69] G. W. Zieve, D. Turnbull, J. M. Mullins, and J. R. McIntosh, *Exp. Cell Res.* **126,** 397 (1980).

with either *Dpn*I or *Mbo*I, and one aliquot is left undigested to reveal the total amount of plasmid present at each time point. Although it has not been our experience, others have reported that the amount of *Dpn*I resistant superhelical plasmid DNA provides a more accurate assessment of replication than the amount of *Dpn*I resistant linear DNA.[70] This observation may reflect the ionic strength of the reaction conditions employed (see below) as well as a sensitivity of *Dpn*I to the conformational state of its substrate.

Digestion conditions for each restriction endonuclease are those recommended by the supplier with the following suggestions. If digestion conditions are compatible, *Dpn*I can be included in the reaction mix used to linearize the plasmid (e.g., *Sal*I and *Dpn*I), allowing both reactions to be carried out concurrently. Addition of the polycation spermidine frequently facilitates complete digestion of plasmids recovered from mammalian cells and embryos.[71–73] The optimal conditions for digestion of replicated plasmid DNA with *Dpn*I include 100 m*M* NaCl.[73] *Dpn*I digests fully methylated DNA most efficiently at low salt concentrations (50 m*M* KCl + <50 m*M* NaCl), but under these conditions, hemimethylated DNA also is partially digested.[73,74] In the presence of 50 m*M* KCl + 100 m*M* NaCl, nicked circular and linear hemimethylated DNA molecules remain intact, but covalently closed, superhelical DNA molecules are converted to either nicked circular or linear DNA molecules. In the presence of 50 m*M* KCl + >100 m*M* NaCl, the ability of the *Dpn*I to cut superhelical molecules decreases. *Dpn*I is unable to cleave unmethylated DNA under any condition. Thus, in the presence of ~150 m*M* K^+ and Na^+, one can distinguish hemimethylated DNA (one round of replication) from unmethylated DNA (two or more rounds of replication) because all hemimethylated plasmid DNA is converted into nicked circular and linear DNA molecules, while all unmethylated plasmid DNA remains as undigested superhelical DNA.[73] In the presence of ~250 m*M* K^+ and Na^+, all forms of hemimethylated and unmethylated plasmid DNA are resistant to cleavage by *Dpn*I.[73] Under these conditions, cellular DNA that has undergone one or more rounds of replication in the presence of BUdR to form heavy/light and heavy/heavy BrdU-DNA is completely resistant to digestion with *Dpn*I.[27a]

DpnI Digestion Conditions. In a 25-μl reaction volume, 0.5 to 10 ng of

[70] B. S. Rao and R. G. Martin, *Nucl. Acids Res.* **16,** 4171 (1988).

[71] J. P. Bouche, *Anal. Biochem.* **115,** 42 (1981).

[72] E. Martínez-Salas, D. Y. Cupo, and M. L. DePamphilis, *Genes & Devel.* **2,** 1115 (1988).

[73] J. A. Sanchez, D. Marek, and L. J. Wangh, *J. Cell Sci.* **103,** 907 (1992).

[74] C. R. Wobbe, F. Dean, L. Weissbach, and J. Hurwitz, *Proc. Natl. Acad. Sci. U.S.A.* **82,** 5710 (1985).

plasmid DNA is digested with 2 to 4 units *Dpn*I/ng plasmid DNA in either Bethesda Research Labs buffer 4 (20 m*M* Tris–HCl, pH 7.4, 50 m*M* KCl, and 5 m*M* $MgCl_2$) or New England Biolabs buffer 4 (20 m*M* Tris–acetate, pH 7.9, 10 m*M* magnesium acetate, 50 m*M* potassium acetate, and 1 m*M* DTT) supplemented with 100 μg/ml BSA or gelatin (prevents loss of enzyme to reaction surfaces), 0.01% Triton X-100, and 3 m*M* spermidine (promotes accessibility of enzyme to substrate[62,71,73]), 300 ng bacteriophage λ DNA (*Dpn*I digestion appears to require a high enzyme to substrate ratio[73]), and either 100 m*M* NaCl if one wishes to distinguish between hemimethylated and unmethylated superhelical DNA or 200 m*M* NaCl if one simply wishes to prevent digestion of all hemimethylated forms of DNA (see above).

*MboI Digestion Conditions. Mbo*I digestion was carried out as described above except that BRL buffer 2 was used instead of buffer 4. *Mbo*I also works well in New England Biolabs buffer 3 (50 Tris–HCl, pH 7.9, 10 m*M* $MgCl_2$, 100 m*M* NaCl, and 1 m*M* DTT).

Identifying DpnI-Resistant and MboI-Sensitive Plasmid DNA by Gel Electrophoresis

The amounts of *Dpn*I- and *Mbo*I-resistant DNA are generally measured by fractionating the DNA samples by electrophoresis through agarose gels. Plasmid DNA digested with *Dpn*I frequently is fractionated by electrophoresis through 0.8% agarose in 135 m*M* Tris–HCl, pH 8.0, 45 m*M* boric acid, 2.5 m*M* EDTA. For each *Dpn*I or *Mbo*I digested sample, there should be a control lane showing DNA that was not digested by either enzyme but was cut at a single restriction site. This allows one to calculate the total amount of plasmid DNA that remains at that time point relative to the starting time point ($t = 0$). If the internal standard was propagated in *dam*$^-$ bacteria, then the additional lane is not necessary. Since a *dam*$^-$ plasmid such as pML-1 that lacks a eukaryotic origin of replication does not replicate in mammalian cells, the amount of pML-1(*dam*$^-$) in the +*Sal*I alone sample should be the same as the amount in the +*Sal*I, +*Mbo*I sample for each time point. This provides a simple way in which to normalize the fraction of *Mbo*I-resistant DNA present at each time point (e.g., cpm in +*Sal*I, +*Mbo*I band/cpm in +*Sal*I alone band) to the total amount of plasmid substrate present at the beginning of the transfection by multiplying the fraction of *Mbo*I resistant DNA at each time point by ratio of pML-1(*dam*$^-$) in the +*Sal*I alone sample to pML-1(*dam*$^-$) in the +*Sal*I, +*Mbo*I digested sample for each time point.

Plasmids are generally identified by blotting-hybridization using sequence specific [^{32}P]DNA probes or using autoradiography to detect DNA containing [α-^{32}P]dNMPs that were incorporated during DNA synthesis in a subcellular system containing one or more [α-^{32}P]dNTPs. In the second case, the plasmid standard is detected by staining the gel with ethidium bromide.[19] Complete digestion can be checked by ethidium bromide staining of the agarose gel prior to processing it for blotting-hybridization.

Blotting-hybridization can be carried out by transferring DNA bands to a nylon membrane (Hybond-N$^+$, Amersham) in a Positive Pressure Blotter (Stratagene), and then cross-linking it to the membrane by irradiation at 0.12 J in a UV Stratalinker Model 1800 (Stratagene). Membranes are prehybridized at 42° for 2 hr in 0.15 ml/cm^2, 50% formamide, 5× SSPE (175.3 g/liter NaCl, 27.6 g/liter $NaH_2PO_4.H_2O$, 7.4 g/liter EDTA, pH 7.4 with NaOH), 1% SDS, 0.25% glycine, 0.1% polyvinylpyrrolidone (PVP, molecular weight 360,000), 0.1% Ficoll (MW 400,000), 400 μg/ml total yeast RNA, 20 μg/ml sonicated salmon sperm or calf thymus DNA, 50 μg/ml BSA (fraction V) and 0.1% sodium pyrophosphate.

Hybridization probes are made by digesting plasmids with an appropriate restriction enzyme and then radiolabeling the DNA with [α-^{32}P]dATP (6000 Ci/mmol, Amersham) by replacement synthesis using T4 DNA polymerase optimized for labeling ~600 bp fragments.[75] Labeled plasmids are separated from unincorporated nucleotides by sequential centrifugation through two Sephadex G-50 Quick-Spin columns (Boehringer Mannheim), adjusting the samples to 2 *M* ammonium acetate, precipitating the DNA with two volumes of ethanol, and then resuspending the DNA in water. Labeled DNA is denatured at 100° for 10 min before adding hybridization buffer [50% formamide, 5× SSPE, 0.2% SDS, 10% dextran sulfate (molecular weight 500,000) 50 μg/ml BSA, and 100 μg/ml total yeast RNA] to give final concentration of 10 ng DNA/ml.

Membranes are hybridized at 42° for 12 to 20 hr in hybridization buffer (0.15 ml/cm^2 membrane). Each membrane is washed at 65° on a rotating platform (~40 rpm) as follows: (1) 2.5 ml/cm^2 of 6× SSPE, 1% SDS, 0.1% PVP, 0.1% Ficoll, 0.25% glycine for 4 hr, (2) twice in 1× SSPE, 1% SDS for 30 min each, and (3) twice in 0.2× SSPE, 1% SDS for 30 min each. Membranes are dried, checked for background with a Geiger counter, and exposed to Kodak XAR-5 film for up to 1 hr at room temperature. If necessary, the membrane is then exposed at −70° with an intensifying screen.

The presence of newly replicated DNA is generally evident by simple inspection of the resulting autoradiograms; the amount of *Dpn*I-resistant

[75] P. O'Farrell, *Focus* **3,** 1 (1981).

DNA increases with time, whereas the amount of *Mbo*I-resistant DNA decreases with time, concomitant with the appearance of rapidly migrating *Mbo*I DNA digestion products. However, these data can be quantified simply by measuring the amount of radiolabel associated with each full-length *Dpn*I or *Mbo*I plasmid DNA band in the gel and comparing it to the amount of radiolabel associated with the nondigested, full-length internal standard. Radiolabel is quantified by one of three methods: (1) Several exposures of the autoradiogram are subjected to densitometry to ensure that the intensity of the exposure is proportional to the amount of radiolabel.[5] (2) Individual bands in the gel can be excised and the radioactivity measured in a scintillation counter.[25] (3) The original gel can be scanned directly by an appropriate Geiger detection device such as the Betagen Model 603 Betascope or by a phosphorimager.

Identifying DpnI-Resistant Plasmid DNA by Propagation in E. coli

An alternative method for detecting *Dpn*I-resistant plasmid DNA is to measure its ability to reestablish itself as a plasmid in bacterial cells.[76] If the putative ARS element is cloned into an ampicillin-resistant pBR vector, it can be cotransfected into competent bacterial cells together with a control vector that is both ampicillin and tetracycline resistant. Following transformation, bacteria are plated onto agar containing both antibiotics. The number of bacterial colonies that form are proportional to the number of plasmid molecules recovered from mammalian cells that survived *Dpn*I digestion. A modification of this approach uses pUC vectors.[77] These simplify colony counting by producing blue (vector alone) and white (vector plus recombinant DNA) colonies on the same plate. Using highly transformation-competent cells, one can detect levels of *Dpn*I-resistant plasmids that are not easily detectable by blotting-hybridization. Retransformation of bacteria provides an easy method for quantitation as well as an internal control for completion of *Dpn*I digestion because retransformation measures primarily form I plasmid DNA.

Identification of RNA Primers Covalently Attached to the 5′ Ends of DNA

Radiolabeling the Internal Nucleotides of RNA Primers

RNA primers in SV40,[29,78] polyomavirus,[75] and mammalian cell nuclei[79] have been radiolabeled internally by incorporation of [α-^{32}P]rNTPs (Fig.

[76] L. Vassilev and E. M. Johnson, *Nucl. Acids Res.* **16,** 7742 (1988).
[77] S. Rusconi, Y. Severne, O. Georgiev, I. Galli, and S. Wieland, *Gene* **89,** 211 (1990).
[78] G. Kaufmann, *J. Mol. Biol.* **147,** 25 (1981).
[79] B. Y. Tseng and M. Goulian, *J. Mol. Biol.* **99,** 317 (1975).

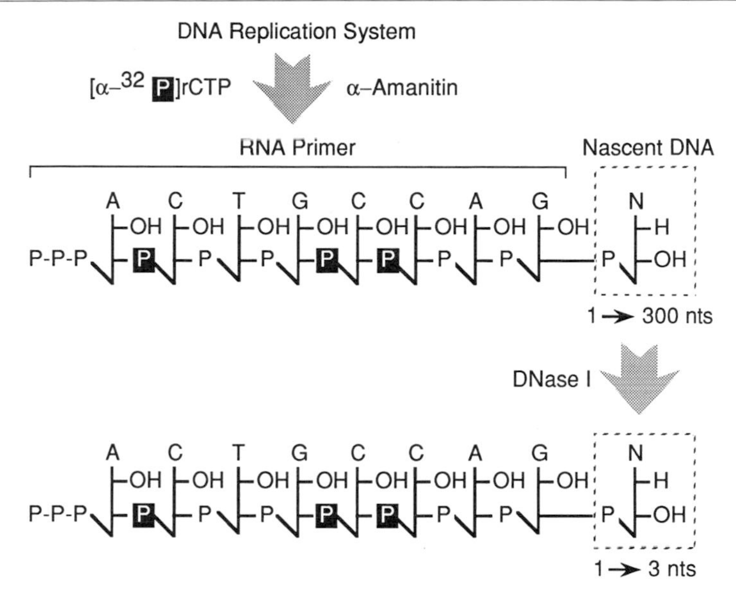

FIG. 6. Labeling all RNA primers.

6). Although this approach to identifying RNA primer DNA appears the most direct, it suffers from the fact that the relative concentrations of rNTPs and dNTPs in the reaction mixture can strongly affect the composition of RNA primers. The ratio of ATP to GTP influences the selection of RNA primer initiation sites, and therefore affects the nucleotide composition of RNA primers and the frequency with which they initiate with either A or G.[80] At low rNTP concentrations, the corresponding dNTP can be incorporated in place of the correct rNTP.[75] Transcription by RNA polymerases II and III is suppressed by addition of 200 μg/ml of α amanitin. The radionucleotide is commonly 1 to 10 μM UTP with 2 μM of each of the dNTPs. The fraction of radiolabel incorporated selectively into RNA primers is determined by purifying replicating DNA intermediates and then digesting them with 10 U/ml of pancreatic DNase I for up to 4 hr at 25° (lower temperatures helps to control digestion) in 50 mM Tris–HCl, pH 7.6, 5 mM $MgCl_2$, and 500 μg carrier RNA/ml.[75,79,81] Digestion is monitored by subjecting the sample to gel electrophoresis. RNA primers retain 2 or 3 deoxyribonucleotides covalently linked to their 3′ ends and therefore migrate as short oligonucleotides (10 ± 2 residues).

[80] M. Yamaguchi, E. A. Hendrickson, and M. L. DePamphilis, *J. Biol. Chem.* **260,** 6254 (1985).
[81] B. Y. Tseng, J. M. Erickson, and M. Goulian, *J. Mol. Biol.* **129,** 531 (1979).

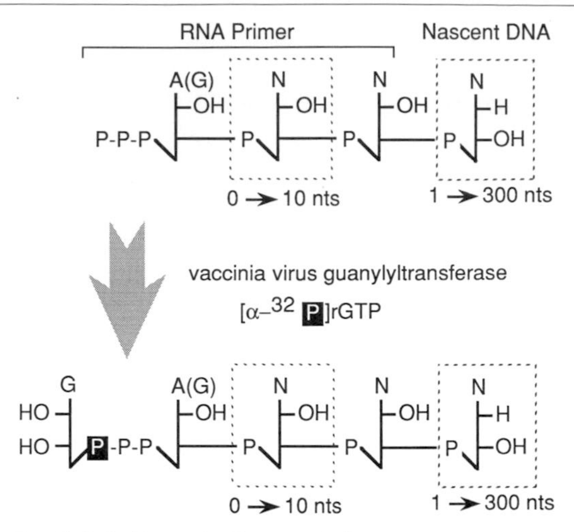

FIG. 7. Labeling only 5′-p-p and 5′-p-p-p-RNA primers.

Radiolabeling the 5′ Ends of RNA Primers

The 5′ ends of RNA primers can be radiolabeled in two ways (Fig. 7). The first is to incorporate [β-^{32}P]rATP or [β-^{32}P]rGTP into RNA primers during DNA replication in a cell lysate in order to produce 5′-p-32p-p-RNA-p-DNA and 5′-32p-p-RNA-p-DNA chains.[76,78]

The second method isolates newly replicated DNA and then radiolabels the 5′ ends of RNA that contain either a di- or triphosphate using the vaccinia virus capping enzyme. This method is preferred for its simplicity, specificity, and efficiency of labeling. RNA-p-DNA chains containing a 5′-terminal di- or triphosphate can be radiolabeled using vaccinia virus guanylyltransferase, the enzyme responsible for capping mRNA with GTP.[82] The advantage of using the capping reaction to label RNA primed DNA chains is that only those chains retaining their original 5′ nucleotide are radiolabeled, and the amount of radioactivity per chain is independent of chain length or composition. Moreover, this method avoids labeling nascent DNA chains that were initiated at the 3′ ends of DNA fragments present as a result of DNA damage. Its disadvantage is that it also labels mRNA, a problem that is largely eliminated by addition to a subcellular DNA replication system of 200 μg/ml of α amanitin, a specific inhibitor of RNA polymerases II and III. Therefore, its application is recommended for characterization of purified proteins and DNA templates, or when

[82] S. Venkatesan, A. Gershowitz, and B. Moss, *J. Biol. Chem.* **255,** 903 (1980).

nascent DNA is affinitied labeled and then purified from the reaction mixture.

Purified DNA samples are denatured for 2 min at 100° in 10 m*M* Tris–HCl, pH 7.6, and 1 m*M* EDTA before rapidly cooling them in ice water.[83] The capping reaction is carried out in a 20-μl volume containing 50 m*M* Tris–HCl, pH 7.9, 1.25 m*M* $MgCl_2$, 6 m*M* KCl, 2.5 m*M* DTT, 10 μM [α-^{32}P]dGTP (2 to 3 Ci/μmol), and 6 U of vaccinia virus guanylyltransferase (Bethesda Research Laboratories). After incubation at 37° for 45 min, the reaction was terminated by the addition of 1 μl 0.3 *M* EDTA. Labeled products are purified by chromatography on Bio-Gel P-60 (Bio-Rad Labs, 0.7 by 18 cm) with 10 m*M* Tris–HCl, pH 8.0, 1 m*M* EDTA, and 20 m*M* NaCl. DNA in the pass-through was adjusted to 0.3 *M* sodium acetate, pH 7.0, and concentrated by precipitation with 75% ethanol for 16 hr at −20°.

The 5′-terminal GpppN residue can be identified by digesting purified capped products with 2 μg nuclease P1 in 10 μl 30 m*M* sodium acetate, pH 5.3, containing 2 μg of tRNA for 2 hr at 37°. The products are chromatographed on polyethyleneimine (PEI)-cellulose (Brinkmann Instruments) in 1.6 *M* LiCl together with unlabeled cap standards. The location of each GpppN spot is identified by autoradiography and the GpppN standards are visualized under short-wavelength UV light. Each spot is excised and the amount of radioactivity present is measured by scintillation counting.

Exposing 5′ Ends on Nascent DNA: Unmasking Assay

DNA chains that contain RNA covalently attached to their 5′ ends (Fig. 8) can be radiolabeled specifically by digesting away their RNA to "unmask" the 5′-terminal hydroxyl group that can then be selectively radiolabeled.[3,61,84] This method has two important advantages. First, RNA-p-DNA chains can be selectively radiolabeled after purifying DNA from cells. Second, since each nascent DNA chain carries a single radiolabeled nucleotide, the amount of radioactivity is proportional to the number of DNA chains, not to their length. However, this can also be a disadvantage because one must isolate more nascent DNA chains in order to have sufficient labeled DNA for analysis.

When low-molecular-weight viral or plasmid DNA is being studied, advantage should be taken of the ability to first isolate intact replicating intermediates[85] (see above) before labeling nascent DNA chains. This helps to reduce the background of nonspecific labeling. Otherwise, Okazaki frag-

[83] M. Yamaguchi, E. A. Hendrickson, and M. L. DePamphilis, *Mol. Cell. Biol.* **5,** 1170 (1985a).
[84] R. Okazaki, S. Hirose, T. Okazaki, T. Ogawa, and Y. Kurosawa, *Biochem. Biophys. Res. Commun.* **62,** 1018 (1975).
[85] D. P. Tapper and M. L. DePamphilis, *Cell* **22,** 97 (1980).

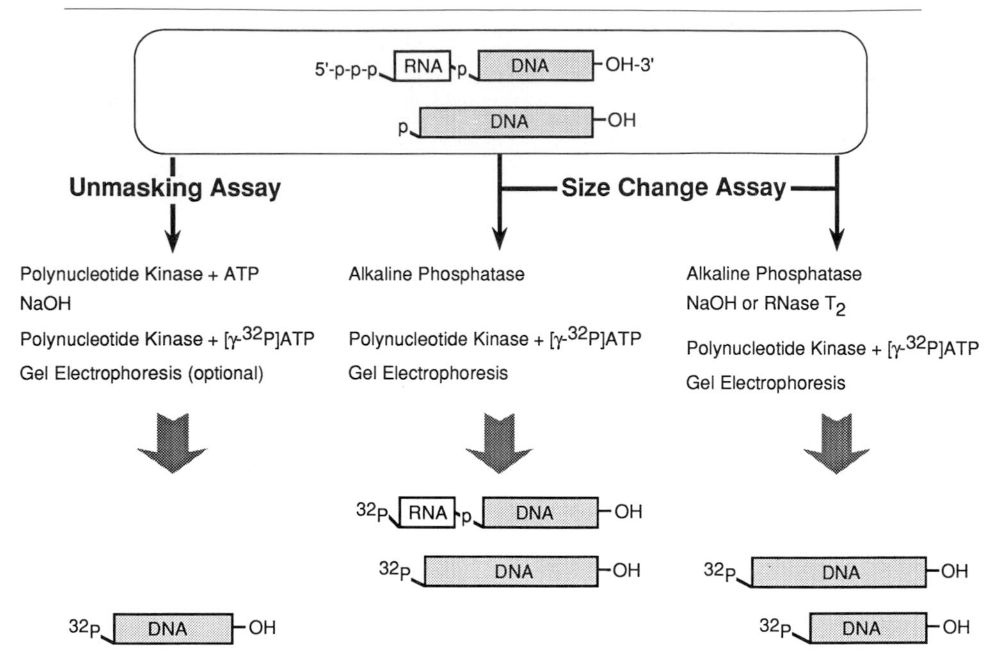

FIG. 8. Identifying RNA primed nascent DNA by unmasking the 5′-OH termini of nascent DNA chains and by measuring the change in size of nascent DNA chains after degrading RNA.

ments can be isolated from nuclei[13] (see above). To prevent degradation of RNA primers, omit RNase treatment of DNA, use RNase-free reagents (e.g., RNase-free sucrose in sucrose gradients), and, if desired, one can include the general RNase inhibitor, 0.1% diethyl pyrocarbonate, in the purification procedure. However, as long as the RNA primers remain annealed to their DNA template, they are resistant to most RNases. Thus, low temperature and high salt conditions during DNA purification help prevent loss of RNA primers from RNase activity.

Purified DNA is resuspended in 10 m*M* Tris–HCl, pH 8.3, 0.1 m*M* spermidine, and 0.1 m*M* EDTA, denatured at 100° for 2 min, and then cooled to 0° in ice water. All 5′-OH termini that might be present in the sample are phosphorylated with unlabeled phosphate in a 200-μl reaction containing 60 m*M* Tris–HCl, pH 8.0, 10 m*M* $MgCl_2$, 5 m*M* DTT, 0.1 m*M* EDTA, 0.1 m*M* spermidine, 40 μ*M* ATP, and 8 U of bacteriophage T4 polynucleotide kinase at 37° for 1 hr. This reaction is terminated in 0.2% sarcosyl, 20 m*M* EDTA, and 50 μg/ml proteinase K for 1 hr at 37°. Nucleic acids are extracted once with phenol : chloroform : isoamyl alcohol, once with chloroform : isoamyl alcohol and then precipitated with ethanol as described above. The precipitate is resuspended in 100 μl 0.15 *M* NaOH,

1 m*M* EDTA, and incubated at 37° for 16 hr to hydrolyze completely any RNA, thereby exposing 5′-OH ends on any DNA chains that were previously covalently linked to RNA. The reaction is cooled to 0°, neutralized with 1 *M* HCl and then immediately applied to a 6-ml column of Sepharose CL-4B (use a disposable 10-ml pipette that has been siliconized to prevent adsorption of ssDNA) in 100 m*M* NaCl, 10 m*M* Tris–HCl, pH 8.3, and 1 m*M* EDTA. DNA in the *included* volume is pooled, concentrated by sedimentation in a Beckman SW41 rotor at 40,000 rpm for 20 hr (4°), resuspended in 50 μl, and precipitated with 67% ethanol in the presence of 0.1 *M* sodium acetate. The exposed 5′-OH termini are then selectively radiolabeled in a 100-μl reaction containing 50 m*M* glycine, pH 9.0, 10 m*M* $MgCl_2$, 5 m*M* DTT, 0.1 m*M* EDTA, 0.1 m*M* spermidine, 4 μ*M* [γ-^{32}P]ATP (7000 Ci/mmol), and 8 U bacteriophage T4 polynucleotide kinase at 0° for 1 hr. The low temperature inhibits the exchange reaction between [γ-^{32}P]ATP and 5′-P-DNA termini to less than 2%. The reaction was terminated and the DNA extracted and precipitated as described above.

To determine the fraction of DNA chains that contained RNA primers, the 5′ ends of all DNA chains, regardless of the presence or absence of RNA primers, can be radiolabeled by removing all terminal RNA and phosphates prior to ^{32}P-phosphorylation. DNA (1 μg) is resuspended in 10 m*M* Tris–HCl, pH 8.3, and 1 m*M* EDTA, denatured at 100° for 2 min, chilled in ice water, and then incubated at 37° for 30 min in 50 m*M* sodium acetate, pH 6.0, containing 1 U RNase T2. The reaction is terminated and the DNA extracted as described above. Alternatively, RNA is hydrolyzed by incubating DNA in 0.15 *M* NaOH and 1 m*M* EDTA at 37° for 20 hr. This reaction is neutralized as described above and DNA precipitated with ethanol. DNA is resuspended in 100 μl of 10 m*M* Tris–HCl, pH 7.6, and 20 m*M* NaCl containing 0.1 U bacterial alkaline phosphatase to remove all 5′-terminal phosphates, and then radiolabeled with ^{32}P as described above in the procedure for RNA-p-DNA covalent linkages.

Removing RNA Primers from Nascent DNA: Size Change Assay

Removal of a 5′-terminal oligoribonucleotide from nascent DNA chains increases their electrophoretic mobility in polyacrylamide gels (Fig. 8). This change in DNA mobility reveals what fraction of radiolabeled nascent DNA chains contain an RNA primer as well as the size of RNA primers associated with nascent DNA chains of a particular length.[86] This method has revealed that ~40% of SV40 Okazaki fragments contain an oligoribo-

[86] G. Kaufmann, S. Anderson, and M. L. DePamphilis, *J. Mol. Biol.* **116,** 549 (1977).

nucleotide of 6 to 8 residues,[86] ~50% of nascent DNA chains synthesized in *Xenopus* oocyte extract on a single-stranded DNA template contained an oligoribonucleotide of 9 to 10 residues,[87] and ~90% of Okazaki fragments in mammalian chromosomes cotain an oligoribonucleotide of 8 to 12 residues.[13]

Okazaki fragments can be radiolabeled with [α-^{32}P]dNTPs during their synthesis in a cell lysate. With small genomes such as plasmid DNA and viral DNA, replicating intermediates are then purified. Purified DNA is dissolved in 98% (v/v) formamide and heated for 2 min at 50° to release Okazaki fragments. Portions of 20 μl are fractionated by electrophoresis in 12% (w/v) polyacrylamide, 7 *M* urea slab gels (15 × 15 × 0.15 cm) in 0.1 *M* Tris–borate buffer, pH 8.3, and 2.5 m*M* EDTA at 10 V/cm, 25° until the bromphenol blue tracking dye has migrated 90% of the gel's length. Gels can be calibrated with markers such as *Msp*I or *Hpa*II [5′-^{32}P]DNA restriction fragments from pBR322, and a homologous series of oligo(Ap)$_n$. The gel is sliced into 1-mm fractions, and the DNA is electroeluted into 0.4 ml of 20 m*M* Tris–HCl, pH 7.5, 40 m*M* sodium acetate, and 1 m*M* EDTA. The solutions are adjusted to 0.5 *M* NaCl and 30 to 50 mg sonicated DNA/ml (for size change assay) or yeast tRNA (for RNA-p-DNA linkage assay) and the DNA precipitated with 3 volumes of ethanol.

Selected sections containing at least 2000 cpm are divided into two portions. One portion is left untreated, and one portion is dissolved in 10 μl of 0.01 mg RNase T2/ml in 0.1 *M* sodium acetate and 0.01 *M* EDTA before incubating the sample for 4 hr at 37° to digest RNA. Alternatively, one portion is dissolved in 10 μl of 0.1 *M* NaOH and incubated for 4 hr at 60° to hydrolyze RNA and then neutralized with 1 μl of 1 *M* acetic acid. Both the treated and untreated samples were mixed with 10 μl of 50% (v/v) glycerol containing 0.01% each of bromphenol blue and xylene cyanol FF in 0.1 *M* Tris–borate, pH 8.3, 2.5 m*M* EDTA and analyzed by electrophoresis in polyacrylamide gels containing 7 *M* urea. DNA chains longer than 70 nucleotides are fractionated in 12% polyacrylamide while DNA chains smaller than 70 nucleotides are fractionated in 15% polyacrylamide.

An alternative protocol is to radiolabel the 5′ ends of one fraction of a unique size class of DNA chains with [γ-^{32}P]ATP and polynucleotide kinase before hydrolysis of RNA and then radiolabel another fraction of the same size class after hydrolysis of RNA. The methods are the same ones used in the unmasking assay. In this case,[3,61] RNA-p-DNA chains in purified replicating intermediate DNA are first dephosphorylated at 65° for 1 hr in 10 μl of 10 m*M* Tris–HCl, pH 7.6, 20 m*M* NaCl containing 50 U bacterial alkaline phosphatase. The reaction is terminated by addition

[87] M. Méchali and R. M. Harland, *Cell* **30,** 93 (1982).

of 1 μl of 10 mM EGTA. The DNA is denatured in 40 μl of 10 mM glycine, pH 9.0, 0.1 mM EDTA, 1 mM spermidine for 2 min at 100°, then rapidly cooled to 0° and phosphorylated with ^{32}P.

Identifying RNA-p-DNA Covalent Linkages

The identities and relative amounts of RNA-p-DNA covalent linkages can be determined by alkaline degradation of the RNA component and characterization of the nucleotide products (Fig. 9). DNA is stable in alkali, but RNA is hydrolyzed into a mixture of 2′- and 3′-rNMPs that result from

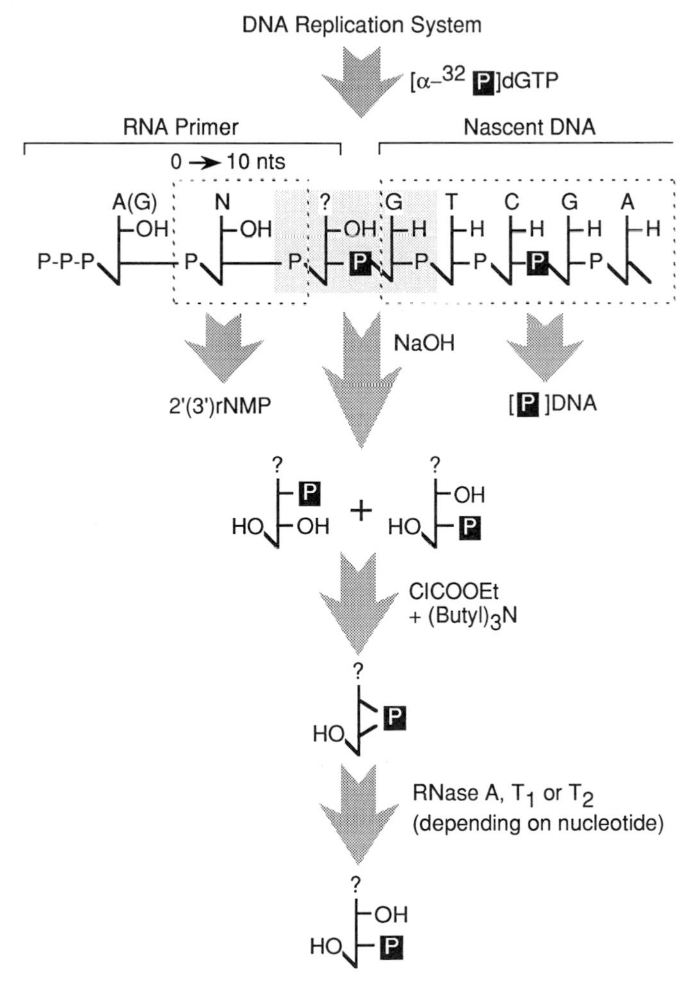

FIG. 9. Identifying RNA-p-DNA covalent linkages.

random cleavage of a cyclic 2′:3′-rNMP intermediate. Thus, when RNA primed DNA chains are radiolabeled with a [α-^{32}P]dNTP substrate and then incubated in alkali, the 5′-terminal ^{32}P in the DNA moiety is transferred to a mixture of 2′- and 3′-[^{32}P]rNMPs. These products can be identified and quantified by various chromatography methods. The amount of ^{32}P "transferred" from nascent DNA to 2′(3′)rNMPs depends on the amount of label incorporated per DNA chain and the fraction of DNA chains that contain RNA. The identity and amount of each of the four [2′(3′)-^{32}P]rNMPs reveals the frequency of each of the 16 possible rN-p-dN linkages ("nearest neighbor analysis"). This assay is easiest when carried out with purified replicating intermediates of animal viruses such as SV40[34,80] and polyomavirus[88,89] and of mitochondrial genomes,[90] although it has been used successfully with cellular chromosomes.[78,91]

In view of the relatively small amount of 32P label that appears in RNA-p-DNA linkages (10^{-5} to 10^{-6} of the total [α-^{32}P]dNTP substrate, or 0.1 to 1% of the [^{32}P]DNA), one must take special precautions to avoid experimental artifacts. As with the unmasking and size change assays described above, purification of replicating DNA intermediates in the absence of RNase is recommended. As controls, purified [^{32}P]DNA that has completed replication and therefore should not contain any RNA-p-DNA covalent linkages (e.g., SV40 covalently closed, superhelical [^{32}P]DNA) can be subjected to the same treatments. In addition, newly replicated [^{32}P]DNA that is not incubated in KOH should not yield radiolabeled nucleotides, demonstrating that none of the original ^{32}P-labeled deoxyribonucleotide substrates contaminated the purified DNA.

Transferring ^{32}P Label from DNA to RNA. Nascent DNA is first radiolabeled with *one* of the four [α-^{32}P]dNTP substrates using a cell lysate system. To exclude the possibility that [α-^{32}P]dNTPs are contaminated with [α-^{32}P]rNTPs (which would label RNA internally), each [α-^{32}P]dNTP can be treated with periodate and then repurified prior to use.[92] Purified [^{32}P]DNA is dialyzed against 10 m*M* Tris–HCl, pH 8.0, 0.4 *M* K_2HPO_4, 0.125 *M* potassium acetate, and 2 m*M* EDTA. Approximately 1-ml portions are incubated with 3 ml of 2 mg/ml salmon sperm DNA (previously extracted with $CHCl_3$-isoamyl alcohol), 6 μl of 10 mg/ml yeast soluble RNA to provide internal markers, and 30 μl of 1% hexadecyltrimethylammonium chloride for 30 min at 4°. The nucleic acid precipitate is collected in a 1.5-ml polypropylene microcentrifuge tube by centrifugation, first in 1 ml of

[88] G. Magnusson, V. Pigiet, E. L. Winnacker, R. Abrams, and P. Reichard, *Proc. Natl. Acad. Sci. U.S.A.* **70,** 412 (1973).
[89] T. Hunter and B. Francke, *J. Mol. Biol.* **83,** 123 (1974).
[90] D. D. Chang, W. W. Hauswirth, and D. A. Clayton, *EMBO J.* **4,** 1559 (1985).
[91] M. A. Waqar and J. A. Huberman, *Cell* **6,** 551 (1975).
[92] R. Wu, *J. Mol. Biol.* **51,** 501 (1970).

cold 70% ethanol containing 0.1 *M* potassium acetate, then in 1 ml of cold absolute ethanol, and finally dried under vacuum. Alkaline hydrolysis of RNA is accomplished by dissolving the dried pellet in 60 ml of 0.3 *N* KOH and incubating for 20 hr at 37° in a sealed microcentrifuge tube. Total ^{32}P in the sealed tube can be determined from Cerenkov radiation. KOH is then neutralized with 0.8 ml of 30 m*M* acetic acid.

The sample is then diluted with 5 ml of water and adsorbed onto a 21 × 0.8 cm DEAE-Sephadex A-25 column previously equilibrated with 1 *M* sodium acetate, pH 5.0, and washed thoroughly with water. The 2′(3′)rNMPs are separated from DNA with a 160-ml linear gradient of 0 to 1 *M* triethylammonium acetate, pH 5.0, at a flow rate of 8 ml/hr. Cerenkov radiation and A_{260} are measured in each 5-ml fraction. The internal RNA standard should be quantitatively converted into 2′(3′)rNMPs; all four 2′(3′)rNMPs are present in the major A_{260} peak. When a large number of ^{32}P-transfer analyses are done, the DEAE-Sephadex columns (2 to 2.5 ml bed volume in disposable 5-ml pipettes) are eluted stepwise with triethylammonium acetate, pH 5.0, consisting of twenty 1.5-ml increments of 0.025 *M* triethylammonium acetate (0.025 to 0.5 *M*). The results are comparable to those obtained with a continuous linear gradient.

Identifying [^{32}P]rNMP Products. Two simple methods for fractionation of ribonucleotides have been used, although more sophisticated methods are now available. Method 1[34] is to lyophilize the 2′(3′)rNMP pool from the DEAE-Sephadex column in order to remove the volatile salt, dissolve it in 70 ml of water, and adsorb it onto a Bio-Rad AG1-X8 column (27 × 0.8 cm) previously equilibrated with 1 *M* formic acid and thoroughly washed with water. Optimum separation of rNMPs is achieved by eluting with two consecutive 200-ml linear gradients; the first is 0 to 1 *M* formic acid and the second is 1 to 10 *M* formic acid. Nucleotides are eluted in the following order: 2′(3′)Cp, 2′Ap, 3′Ap, 2′(3′)Gp, and 2′(3′)Up. Fractions of 6 ml were collected at a flow rate of 13 ml/hr at room temperature. No detectable radioactivity appears upon further elution with 1 *N* NaOH. The distribution of radioactivity in the four rNMPs is calculated from the radioactivity associated with each A_{260} peak. Greater than 96% of the ^{32}P-label should chromatograph with the individual 2′(3′)rNMPs.

Method 2[85] is to dissolve [^{32}P]DNA samples containing 20,000 to 200,000 cpm and 10 mg yeast tRNA in 10 ml of 0.3 *M* NaOH and incubate it for 16 hr at 37° to hydrolyze RNA. This solution is neutralized with 10 volumes of 0.03 *M* acetic acid containing 0.005% (w/v) of the tracking dye, xylene cyanol FF, and then chromatographed on polyethyleneimine (PEI)-cellulose (Brinkmann Instruments) thin-layer sheets. Hydrolysates are applied to thin layers (10 × 20 cm) that are then rinsed with water and dried. A triple-thick pad of Whatman's 3MM paper (1 × 10 cm) is attached to the upper end of the thin-layer sheet to absorb excess effluent. The thin layers are then developed with 5 ml of "monomixture" until the tracking dye reaches 0.6 to 0.7 of the layer's height. Monomixture is prepared by dissolving 5 g yeast RNA (Sigma type VI) in 100 ml of 0.27 *M* KOH and incubating at 50° for 3 days before titrating to pH 7.7 with glacial acetic acid. NH_4HCO_3 (0.25 *M*) can also be used as the solvent, but it does not separate Cp from Up. The layers are dried, coated with plastic wrap, and autoradiographed. The position of each rNMP is determined with appropriate ^{32}P-labeled standards. The fraction of label transferred from DNA to each rNMP is determined by excising each rNMP and DNA spot from the chromatogram and then measuring the ^{32}P present in a toluene-based scintillation fluid. DNA should never be exposed to acidic conditions in order to avoid depurination and subsequent breakdown of DNA[93] into low-molecular-weight compounds that chromatograph like 2′(3′)rNMPs.

To confirm the identity of the radioactive material associated with each of the 2′(3′)rNMPs, each peak of A_{260} material was chemically and enzymatically modified (Fig. 9). The 2′(3′)rNMPs are converted first into cyclic 2′:3′-rNMPs by a chemical treatment,[94] and the resulting products purified by chromatography on DEAE-Sephadex. Individual rNMPs, isolated from a Bio-Rad AG1-X8 column, are lyophilized to remove formic acid and then converted to the sodium salt by lyophilization with equimolar amounts of $NaHCO_3$. The 2′(3′)rNMPs are converted into the 2′:3′-cNMPs by adding 5 ml of ethyl chloroformate and 15 ml of tri-n-butylamine to 50-ml samples and vortexing for 10 min in a sealed 1.5-ml Eppendorf tube. The reaction mixture is then neutralized with 1 ml of 70 m*M* KH_2PO_4, pH 6.0, diluted with 10 ml of water, and adsorbed onto a DEAE-Sephadex A-25 column (9 × 0.6 cm) previously equilibrated with 1 *M* $NaHCO_3$ and washed extensively with water. The nucleotides are eluted in 20 steps consisting of 1.5-ml increments of 0.015 *M* NH_4HCO_3 from 0.115 to 0.4 *M*. Each fraction is monitored for ^{32}P Cerenkov radiation and A_{260}, and the fractions containing 2′:3′-cNMPs are lyophilized to remove salt. At neutral pH, cyclic nucleotides, with a charge of −1, are cleanly resolved from 2′(3′)rNMPs with a charge of −2. The ^{32}P-labeled material chromatographed with the 2:3′-cNMPs.

To verify further the identity of the 2′:3′-cNMPs, each cyclic nucleotide can be converted into a 3′-rNMP by treatment with an appropriate RNase. Cyclic nucleotides are dissolved in 1.2 ml of 0.2 *M* KH_2PO_4, pH 6.1, and 5 m*M* EDTA. RNase A (15 ml of 4 mg/ml) is added to cyclic UMP and cyclic CMP, RNase T1 (50 ml of 0.4 mg/ml) is added to cyclic GMP, and

[93] H. Türler, *in* "Procedures in Nucleic Acid Research" (Canntoni & Davis, eds.), Vol. 2, pp. 680–702. Harper & Row, New York, 1971.

[94] A. M. Michelson, *J. Chem. Soc.* (*London*) **3–4,** 3655 (1959).

RNase T2 (50 ml of 0.4 mg/ml) is added to cyclic AMP. These solutions are incubated for 2 hr at 30° and chromatographed on DEAE-Sephadex as described above for cyclic nucleotides.^{32}P-labeled material should chromatograph with the A_{260} material. No hydrolysis of cyclic nucleotides should be observed in the absence of added RNase.

Confirming Source of ^{32}P-Labeled rNMPs. To demonstrate directly that [2′(3′)-^{32}P]rNMPs received their ^{32}P label from [α-^{32}P]dNTPs incorporated into nascent DNA, rather than from contaminating ^{32}P-labeled rNTPs, the amount of ^{32}P appearing in rNMPs can be compared with the specific radioactivity of the [α-^{32}P]dNTP.[34] When the concentration of [α-^{32}P]dNTPs is saturating ($\geq$10 m*M*), the amount of ^{32}P label incorporated into DNA is proportional to the specific radioactivity of the substrate. Under these conditions the amount of ^{32}P label transferred to 2′-(3′)rNMPs also should be proportional to the specific radioactivity of the [α-^{32}P]dNTP substrate. Hence, the fraction of ^{32}P transferred to 2′(3′)rNMPs should be independent of the specific radioactivity of the [α-^{32}P]dNTP substrate. If, however, labeled RNA–RNA junctions, in addition to RNA–DNA junctions, had contributed [2′(3′)-^{32}P]rNMPs, then the percentage of ^{32}P label transferred will increase as the specific radioactivity of the [α-^{32}P]dNTP substrate is decreased by addition of unlabeled dNTP. Moreover, no change in the amount of [2′(3′)-^{32}P]rNMPs should be observed when the concentration of unlabeled rNTPs is varied in the DNA synthesis reaction mixture.

Applications

The ability to label specifically newly synthesized DNA chains that are products of DNA replication has several general applications. The most frequent applications involve characterization of DNA replication in the genomes of bacteria, bacteriophage, plasmids, eukaryotic cells, mitochondria, and animal viruses. In addition, these techniques can be used to characterize proteins and enzymes involved in DNA replication. Examples of these applications along with literature references that give experimental details can be found in Kornberg and Baker.[12] Other applications include analysis of chromatin structure at the sites of DNA synthesis[17] and mapping origins of DNA replication to specific genomic sites both in small chromosomes[12] as well as in the large chromosomes of eukaryotic cells.[95] One can even identify locations of initiation sites[3,59,83] and termination sites[85,96,97] for DNA synthesis at specific genomic sites in viral and plasmid genomes at the resolution of single nucleotides.

[95] L. T. Vassilev and M. L. DePamphilis, *Crit. Rev. Biochem. Mol. Biol.* **27,** 445 (1992).
[96] D. T. Weaver and M. L. DePamphilis, *J. Biol. Chem.* **257,** 2075 (1982).
[97] D. T. Weaver and M. L. DePamphilis, *J. Mol. Biol.* **180,** 961 (1984).

Author Index

Numbers in parentheses are footnote reference numbers and indicate that an author's work is referred to although the name is not cited in the text.

A

B

D

E

F

G

H

I

J

K

M

N

O

P

V

W

Subject Index

A

B

C

D

O

P

R

S

T

U

V